PRENTICE-HALL INFORMATION AND SYSTEMS SCIENCES SERIES

Thomas Kailath, editor

BARTON
Radar System Analysis

DI FRANCO AND RUBIN
Radar Detection

DOWNING
Modulation Systems and Noise

DUBES
The Theory of Applied Probability

FRANKS
Signal Theory

GOLOMB, BAUMERT, EASTERLING, STIFFLER, AND VITERBI
Digital Communications with Space Applications

RAEMER
Statistical Communication Theory and Applications

STIFFLER
Theory of Synchronous Communications

VAN DER ZIEL
Noise: Sources, Characterization, and Measurement

PRENTICE-HALL INTERNATIONAL, INC., *London*
PRENTICE-HALL OF AUSTRALIA, PTY. LTD., *Sydney*
PRENTICE-HALL OF CANADA, LTD., *Toronto*
PRENTICE-HALL OF INDIA PRIVATE LIMITED, *New Delhi*
PRENTICE-HALL OF JAPAN, INC., *Tokyo*

Theory of Synchronous Communications

J. J. Stiffler

Prentice-Hall, Inc., Englewood Cliffs, New Jersey

13-914739-x

Current printing (last digit):
10 9 8 7 6 5 4 3 2 1

Library of Congress Catalog Card Number:
78-101202

Printed in the United States of America

Preface

This book is concerned with synchronization in communication systems. In an earlier stage of its evolution, in fact, these last four words were to be the title of the book. The present title came to be preferred for two reasons: First, although the subject matter is devoted exclusively to those communication systems which require synchronism between the transmitter and receiver to operate effectively, the discussion is not limited to means for achieving this synchronism; a sizable portion of the book concerns the description and analysis of the systems themselves. Second, some topics which might be inferred from the earlier title, notably the *network* synchronization problem when a number of users require simultaneous access to a limited-capacity system, are not explicitly investigated.

The aim of the book is to present the basic theory of the operation and performance of synchronous communication systems and, at the same time, to exhibit the conclusions of this theory in a form useful to the communication system designer. In accordance with the first of these goals, the theoretical aspects of the subject are developed at length and all of the important results are derived explicitly. In deference to the second goal, simplifying approximations are introduced into the theoretical development whenever tractable results (i.e., results which can be interpreted without recourse to a large-scale computer) cannot otherwise be obtained.

The book is divided into three parts. In the first part, Chapters One through Five, the fundamentals of statistical detection and estimation theory are developed and applied to the analysis of representative synchronous communication systems. A number of additional concepts are also introduced, including search theory and the theory of phase-locked loops, which are basic to the second part of the book. The second part, consisting of Chapters Six through Nine, concerns methods for providing the synchronization required for the communication systems in Part I to operate efficiently. In this part, at least partial answers are provided to such questions as "What is the best method for providing the needed synchronization?", "How quickly can synchronization be acquired initially?", and "How accurately can it be maintained?". The third part of the book is devoted to coding theory, including both source encoding and channel encoding. In compliance with the general theme of this book, considerable attention is given to methods for constraining codes so as to facilitate the acquisition and maintenance of synchronization between the encoder and decoder.

Problems have been included following each chapter (excluding Chapter One). These problems are intended primarily to extend the material of their associated chapters and in many cases develop new material which, for reasons of space, could not be included in the body of the text.

The reader is assumed to have some background in the theory of random processes and noise. The material presented in the first nine chapters of *Random Signals and Noise* by Davenport and Root (McGraw-Hill Book Co., N.Y., 1958), for example, or in Chapters Two and Three of *Principles of Communication Engineering* by Wozencraft and Jacobs (John Wiley & Sons, Inc., N.Y., 1965), is entirely adequate for this purpose. For a reader with this background, the book is intended to be self-contained. Two appendices are included, one on some special topics in statistics and a second on the algebraic background necessary for the study of coding theory, since this material, while essential to the development in the body of the text, has not been assumed necessarily to be in the reader's background. This of course is not to suggest that supplemental reading is not recommended; it is, however, intended that a reader with the stipulated background would find it unnecessary to consult other books in order to read this one.

There are many to whom I would like to express my sincere gratitude for their help in this endeavor. First and foremost among these is my wife who not only tolerated my preoccupation with this project for these past few years, but, to compound the felony, was burdened with virtually all of the typing and re-typing of the several versions of the manuscript. I would also especially like to thank Professors R. Scholtz and I. Blake, both of whom read the entire manuscript and offered many useful suggestions for its improvement. Professor Blake's thorough proofreading of the page proofs was additionally helpful and greatly appreciated. I am grateful, too, to my colleagues who volunteered, or allowed themselves to be persuaded, to read and comment upon various portions of the manuscript. These include Doctors E. Brookner, S. Butman, R. Esposito, S. Farber, R. King, J. Layland, J. Salz, G. Solomon, and A. Viterbi. I am particularly grateful to Dr. R. Turyn, who was especially helpful during the final revision, and to Mr. P. Schottler, who not only read and commented upon several chapters of the manuscript but in addition carefully proofread the entire set of galley proofs. Thanks also are due to Mr. R. Hall and to Mr. R. Jacobson and his colleagues for their cooperation in providing all the line drawings. Finally, I wish to thank Professor T. Kailath who first encouraged me to write this book and who offered a number of helpful suggestions and criticisms throughout its inception, and Professor S. Golomb, who not only stimulated my original interest in this area, but who contributed much to the subject matter of this book. (He also, during one of my periodic moods of despair concerning the progress of, in particular, the synchronization portions of the manuscript, called my attention to the appropriateness of the words of Hamlet with which I conclude this preface.)

"The time is out of joint; O cursed spite,
That ever I was born to set it right!"

[*Hamlet*, Act I, Scene 5.]

J. J. STIFFLER
Concord, Mass.

Contents

chapter eleven

Synchronizability 319

chapter twelve

Synchronizable Block Codes 351

chapter thirteen

Coding to Prevent Errors 383

PART ONE

FUNDAMENTALS

chapter one

Synchronous Communication Systems

1.1 Introduction

A communication system can be classified as synchronous if the existence of a time reference common to both the transmitter and the receiver is a necessary condition for its satisfactory operation. The purpose of this introductory chapter is to emphasize the distinguishing features of such systems and to indicate the different functions this common time reference may be expected to fulfill. In the process, the relevance of the various topics to be discussed in subsequent chapters to the general subject of synchronous communications will be indicated.

1.2 A Communication System Model

The block diagram in Figure 1.1 depicts a generalized communication system model. This particular configuration was chosen not only because it is sufficiently general to encompass a wide variety of systems, but also because each block shown represents a specific synchronization constraint, i.e., a specific requirement that the common time reference must satisfy. In many systems, one or more of the blocks indicated in Figure 1.1 will be omitted.

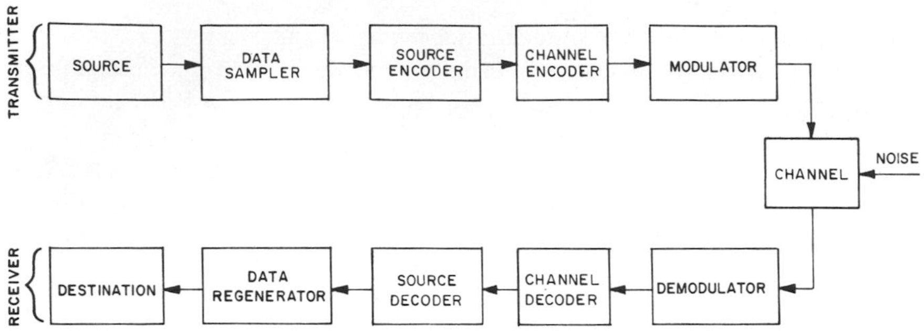

Figure 1.1 Model of a communication system.

Since the elimination of any of these blocks simplifies the resulting system, as far as synchronization is concerned, we shall first discuss the system in which all of the blocks are present.

The *source* simply represents the source of information to be transmitted. Its output is assumed to be a random process, although it may be either amplitude-continuous or amplitude-discrete and either time-continuous or time-discrete. The *data sampler* converts this random process to a random sequence by a sampling operation.

The *source encoder* serves several functions. In its most elemental form it is simply an analog-to-digital converter, representing the data samples at its input as a sequence of digits in some convenient base, usually, but not always, the binary base. More generally, it serves as a device for mapping from data samples onto *data words*, that is, onto sequences of digits or *data symbols*, a mapping which may be quite different from the straightforward one usually associated with analog-to-digital converters. It may be designed, for example, to reduce the average number of symbols needed to represent a typical sequence of samples by decreasing the redundancy almost inevitably present in the unprocessed data, or it may be governed by any number of other considerations.

Ironically, the *channel encoder* is designed to *add* redundancy to the digital sequence presented at its input. The purpose of this, of course, is to enable the message, now represented by the *channel code words*, to be correctly identified at the receiver even though some errors may occur in transmission. The two encoders do not nullify each other although they may appear to do so at first glance. In general, the natural redundancy removed by the source encoder will be much less effective in mitigating transmission errors than the controlled redundancy added by the channel encoder. Even if this were not the case, the mere fact that the redundancy is controlled allows it to be exploited more systematically.

The function of the *modulator*, of course, is to convert the sequence of symbols at the encoder output into a sequence of waveforms suitable for transmission over the communication *channel*. Each waveform generally comprises a *carrier* (or *subcarrier*), usually a sinusoid, the amplitude, phase, or frequency of which is altered (*modulated*) in accordance with the information to be transmitted.

Physical channels can often be characterized by a transfer function $H_c(\omega, t)$ followed by an adder through which the signal is perturbed by noise. Whenever, in the chapters to follow, it is necessary to be more explicit concerning the nature of the channel, it will generally be assumed to be both wideband and time-invariant; i.e., $H_c(\omega, t)$ will be assumed independent both of ω (over the relevant frequency band) and of t. Moreover, the noise will generally be assumed to be a sample function of a white (i.e., having a spectral density uniform over all frequencies of interest) Gaussian process. The resulting channel is a convenient, but seldom precise, model of a real channel. (The deep-space channel is possibly the only physical channel for which all of these conditions actually obtain.) Its importance, however, is two-fold: (1) It is amenable to relatively straightforward analysis, and (2) the results of such analyses are at once indicative of the results to be expected with more complex channels, and in many cases, a necessary prerequisite for obtaining these more general results.

The procedure is reversed at the receiver; each block on the receiver side of Figure 1.1 performs the inverse of the operation of the corresponding block on the transmitter side. The *demodulator* interprets the received waveform as a sequence of symbols which the *channel decoder* converts, hopefully, into the form impressed at the input to the channel encoder. The *source decoder* then transforms its input into a sequence of pulses of various amplitudes. These are in turn used in the *data regenerator*, to construct an estimate of the original source output which is transferred to the *destination*.

This communication system as just described will be referred to as a *coded digital communication system*. If the channel coder-decoder blocks are omitted, the system is still digital in nature (i.e., only a finite set of waveforms is used) and will be designated as a *digital* or *discrete-amplitude pulse communication system*. If the source encoder and decoder blocks are also bypassed, the system will be called a *continuous-amplitude pulse communication system*. Although no longer in digital form (unless the source is already digital) the information is still sampled prior to transmission. Finally, if the data processor is also eliminated, the resulting system is an *analog communication system*.

It is instructive, in discussing synchronous communication systems, to assume initially that a common time reference satisfying all of the necessary constraints is indeed available. Accordingly, after an introductory chapter (Chapter Two) in which are presented some of the fundamental concepts needed in the study of communication systems in general, the next two chapters (Chapters Three and Four) are devoted to the description and analysis of pulse communication systems, and in particular, the form of the optimum demodulator for such systems. Coding techniques are discussed in Chapters Ten and Thirteen. Chapter Ten covers the formatting and redundancy removal aspects of coding (source encoding) and Chapter Thirteen the error-control aspects (channel encoding). Analog systems, however, will not be the subject of further investigation here because they are generally not synchronous and hence fall outside the scope of this book. If synchronization is required in an analog system it is invariably a requirement imposed by the source (as in television transmission, for example), not by the communication system itself.† Thus, the method used to satisfy this requirement is also, in effect, independent of the particular modulation scheme employed, and its discussion does not entail a correspondingly detailed examination of the remainder of the system.

1.3 The Synchronization Problem

Two sequences of events are said to be *synchronous* if corresponding events in the two sequences occur simultaneously. *Synchronization* is defined simply as the process of bringing about, and retaining, a synchronous situation.

To relate these terms more explicitly to a discussion of communication systems, it is only necessary to identify one of the two sequences of events to be synchronized with that taking place at the transmitter, and the other with that taking place at the receiver. Although in some instances it is rather artificial to do so, it is generally convenient, both conceptually and operationally, to separate the synchronization process into two distinct modes. In the first mode, the *clock synchronization* mode, the clocks which regulate the two sequences being synchronized (i.e., the transmitter and receiver clocks) are forced to run at the same rate. In the second mode, the *higher-order synchronization* mode, a corresponding pair of events in the two sequences are identified and made to occur simultaneously. Clearly, if the same event occurs in two identical sequences simultaneously, and if the sequences are progressing at the same rate, the sequences are, and will remain, synchronized.

If the transmitter and receiver clocks are both sufficiently stable relative to the required synchronization accuracy, the clock synchronization mode may be bypassed. Most generally, however, this will not be the case and

†An exception to this statement could be made with reference to phase-coherent amplitude modulation systems. The carrier does represent a time reference common to both transmitter and receiver, so the system is, at least in a rudimentary sense, synchronous. In any event, such systems are closely related to the coherent pulse-amplitude modulation systems which are examined in detail in the ensuing chapters, and hence will not be discussed explicitly.

some technique must be devised to provide the needed clock synchronization. The most straightforward method is to represent the transmitter clock as a periodic signal and to transmit this signal along with the information. A properly filtered version of this received signal can then be used as the receiver clock. This approach gives rise to two rather basic questions. The first, concerning the nature of the optimum receiver filter for this purpose, is the subject of Chapter Five. The second, having to do with the relationship between the signal used and the quality of the recovered clock, is one of the topics explored in Chapter Six.

The carrier itself, of course, could be used as a clock. In a *phase-coherent* system,† in which knowledge of the carrier phase is assumed at the demodulator, the received carrier and the local carrier reference must also be synchronized. If the carrier phase bears some fixed and known relationship to the transmitter clock, the local carrier reference can be used as the local clock as well. If the transmitted signal contains a spectral component at the carrier frequency (i.e., if the carrier is not 100% modulated), the situation is as described in the previous paragraph. Even if no such carrier component is transmitted, however, it is often possible to recover the carrier from the received modulated signal. Techniques for accomplishing this are presented in Chapter Eight. Note, incidentally, that carrier synchronization is just a special case of clock synchronization, and is, a fortiori, also discussed in the aforementioned chapters. When no distinction is required, the terms *carrier* (or *subcarrier*) *synchronization* and *clock synchronization* will be used synonymously.

Once the transmitter and receiver clocks have been synchronized, the second mode of the synchronization process begins. Some events taking place in each of the blocks in the receiver portion of Figure 1.1 must be synchronized with the corresponding event taking place in the analogous block in the transmitter portion. Efficient demodulation requires the demodulator to be synchronized with the modulator so as to know when the waveform representing one channel symbol ceases and the next one begins (*symbol synchronization*). The channel decoder cannot decode correctly unless it can identify the beginning of each code word (*code word synchronization*). Similarly, the source decoder is useless unless the digits appearing at its input can be separated into groups, one group corresponding to each data sample (*data word synchronization*). Finally, since the significance of a particular data sample may be defined only in terms of its position in a sequence of samples, sometimes called a *frame*, the data regenerator must

†Systems in which the demodulator does not rely upon knowledge of the carrier phase will be called *phase-incoherent*. Some authors use the term non-coherent rather than incoherent in this context, apparently in deference to the more conventional meaning of the latter term. We shall stick to the term incoherent, however, with much the same bravado mathematicians display in referring to certain numbers as irrational.

frequently be synchronized with the data samples (*frame synchronization*).

If it is impossible to specify whether an event takes place at time t or at time $t + nT_0$ for any integer n, a T_0 second *time ambiguity* is said to exist. The minimum acceptable time ambiguity in a communication system is the periodicity of the event corresponding to the highest-order synchronization required (e.g., the frame period). If the clock period were equal to, or some multiple of, the minimum acceptable time ambiguity, the phase of the recovered clock signal would itself provide all of the needed synchronization information and the second synchronization mode could be superfluous. Unfortunately, it is rarely practical to use a clock having such a long period. Thus, if the clock period is ΔT seconds and the minimum acceptable time ambiguity T_0 seconds, with $T_0/\Delta T \triangleq N > 1$,† there remains an N-fold ambiguity which must be resolved before the synchronization process is completed. The purpose of the higher-order synchronization mode is to effect this resolution.

As with clock synchronization, one obvious method for providing the information needed in the higher-order synchronization mode is to use a separate channel solely for this purpose. A less obvious but potentially more efficient approach is to attempt to recover the required synchronization directly from the modulated signal. Both of these methods will be investigated in detail in succeeding chapters. Problems to be considered include the signal design problem when a separate synchronization channel is to be used (Chapter Six), and the question of the optimum synchronization detection or estimation procedure both when this information is to be transmitted over a separate channel (Chapter Six) and when it is to be obtained directly from the modulation (Chapter Seven). The cost of imperfect synchronization and the question of the optimum allocation of power between the synchronization channel and the information channel are considered in Chapter Nine. Finally, Chapters Eleven, Twelve, and Fourteen are devoted to analyses of methods for imposing additional constraints on the way in which symbols representing the data are to be grouped into words so as to facilitate word synchronization.

Before we conclude these introductory remarks, it might be well to comment on the method whereby timing ambiguities are most frequently resolved, and why this is the case. If each frame contains, say, n_1 words, each word contains n_2 symbols, and each symbol has exactly n_3 clock cycles, and if these numbers n_1, n_2, and n_3 are known integers, all the required synchronization is in principle available once frame synchronization has been established. This suggests concentrating exclusively on frame synchronization and obtaining all of the other synchronization information as a by-product. While this would be a reasonable approach if the equipment available at the receiver

†The symbol $\triangleq$ will be used to denote equal by definition.

were unlimited, it is most impractical when it is necessary to operate under more usual constraints. As will be shown, the time needed to resolve an N-fold synchronization ambiguity under typical equipment constraints is directly proportional to N and may even be proportional to some integer power of N depending upon the circumstances. Thus, if $N = n_1 n_2 n_3$, and if all N ambiguities are to be resolved in one operation, the synchronization time would be of the order of $n_1 n_2 n_3$. In contrast, if symbol synchronization were first established, followed by word and frame synchronization in that order, the time needed for synchronization would be proportional to $n_1 + n_2 + n_3$. That is, the total synchronization delay is just the delay in resolving the n_3-fold symbol synchronization ambiguity, plus the delay in resolving the n_2-fold word synchronization ambiguity, plus the delay in resolving the n_1-fold frame synchronization ambiguity. If the inherent substructure which is necessarily a part of each frame is exploited, the total synchronization time can be reduced by the sizable factor of $n_1 n_2 n_3/(n_1 + n_2 + n_3)$. For this reason, the most expeditious synchronization strategy is to begin generally with the lowest-order synchronization (i.e., synchronization involving the shortest ambiguity) and to proceed to the successively higher orders.

chapter two

Hypothesis Testing, Decision Theory, and Search

2.1 Introduction

Much of the subject matter of this and subsequent chapters will be concerned with the following generic problem. A random process $y(t)$ is observed. As a result of this observation a set of observable coordinates or, more briefly, *observables*† $Y = \{y_i\}$ is abstracted. These observables emanate from a source known to be in one of several possible *states* and their statistical properties are known as a function of the source state. It is desired to determine from these observables the present state of the source. The ground rules by which this is to be accomplished will vary. Sometimes, the number of observables in the set Y will be fixed. In other cases, the number of observables will be essentially unlimited, but a decision is to be made by using as few of them as possible, subject to a constraint on the reliability of this decision. Sometimes, equipment limitations will require the decision to be made in a number of stages; perhaps only one state can be tested at a time. On other occasions, all the states may be tested simultaneously.

†The set of observables Y is also sometimes called a *statistic*, the term statistic used to refer to any mapping from one space of observables [here the space of the original process $y(t)$] to another (in this case, the space of observable coordinates Y). A statistic is called *sufficient* if no information of significance to the user is lost in the mapping (cf. Reference 2.1, for example, for a more precise definition of this term).

In any event, the following questions are of significance: (1) Subject to the ground rules of the problem, how can the observables at hand be best used to ascertain the state of the source? and (2) How reliable is the final decision?

In this chapter we will attempt to answer these questions for a variety of situations which will be encountered in later chapters. We begin with perhaps the simplest case of all: the source has only two possible states, and the number of observables is fixed.

2.2 Deciding between Two Simple Hypotheses

Suppose there is available a set of observables $Y = \{y_1, y_2, \ldots, y_M\}$. The problem is to determine from these observables which of the following two hypotheses is in effect: (1) the *null hypothesis* H_0 that the source is in the *null state* and the observables are characterized by the conditional density function $p(Y \mid H_0)$ or (2) the *alternative hypothesis* H_1 that the source is in the *alternative state* and the density function is $p(Y \mid H_1)$. Since each hypothesis refers to only one state of the source, each is called a *simple* hypothesis. The decision between them is sometimes referred to as a *test* of hypothesis H_0 against the alternative hypothesis H_1, but whether one decides between two hypotheses and accepts one of them, or, instead, tests one of them and accepts or rejects it, is clearly a question of terminology. There are two kinds of errors inherent in this hypothesis testing procedure. An *error of the first kind* is said to occur when hypothesis H_0 is true but is rejected by the test. Similarly, an *error of the second kind* results when H_1 is the true hypothesis but H_0 is accepted. The probability of an error of the first kind is also known as the *level* of the test and will be denoted by α. The probability of an error of the second kind will be designated β. Finally, to complete the terminology, we mention that the probability $1 - \beta$ is called the *power* of the test of hypothesis H_0 against H_1.

Let S denote the space of all possible values of the observables $Y = \{y_1, y_2, \ldots, y_M\}$. The problem in a hypothesis test is to divide the space S into two non-overlapping regions R_0 and R_1 ($R_0 + R_1 = S$) so that if Y is in R_0, H_0 is accepted, and if Y is in R_1, H_1 is accepted (H_0 is rejected). The error probabilities α and β can then be expressed in terms of integrals† over these M-dimensional regions R_0 and R_1:

$$1 - \int_{R_1} p(Y \mid H_0)\, dY = \int_{R_0} p(Y \mid H_0)\, dY = 1 - \alpha$$

†If the observables y_i are discrete random variables, the integrals can be replaced by sums and the density functions by probabilities. Both situations, of course, could be accommodated by using Stieltjes integrals.

and

$$1 - \int_{R_1} p(Y\,|\,H_1)\,dY = \int_{R_0} p(Y\,|\,H_1)\,dY = \beta \tag{2.2.1}$$

In order to determine the optimum choice of the region R_0, it is necessary to specify a criterion for optimality. To this end we assign a numerical value or *cost* to each of the possible decisions. This cost is to represent the loss (or gain, if the cost is negative) resulting from a particular decision. While the cost of a specific decision is often obscure, it does seem reasonable to assume that the cost of an erroneous decision is greater than that of a correct decision. This assumption is actually sufficient to determine the general form of the optimum decision rule, here defined as the rule minimizing the expected cost of a decision.

Let C_{ij} denote the cost of accepting hypothesis H_i when the true hypothesis is H_j, and let $P(H_i)$ be the a priori probability that H_i is the correct hypothesis. The expected cost of a particular decision is, then,

$$\begin{aligned} E(\text{cost}) = C_{00} \int_{R_0} p(Y\,|\,H_0)P(H_0)\,dY + C_{01} \int_{R_0} p(Y\,|\,H_1)P(H_1)\,dY \\ + C_{10} \int_{R_1} p(Y\,|\,H_0)P(H_0)\,dY + C_{11} \int_{R_1} p(Y\,|\,H_1)P(H_1)\,dY \end{aligned} \tag{2.2.2}$$

Since some decision must be made,

$$\int_{R_0} p(Y\,|\,H_i)\,dY + \int_{R_1} p(Y\,|\,H_i)\,dY = 1$$

so

$$\begin{aligned} E(\text{cost}) = \int_{R_0} [(C_{01} - C_{11})P(H_1)p(Y\,|\,H_1) - (C_{10} - C_{00})P(H_0)p(Y\,|\,H_0)]\,dY \\ + C_{11}P(H_1) + C_{10}P(H_0) \end{aligned} \tag{2.2.3}$$

The expected cost of a decision is evidently minimized by defining R_0 as the region containing all those points Y for which the integrand in Equation (2.2.3) is negative. As the terms $C_{01} - C_{11}$ and $C_{10} - C_{00}$ are both assumed positive, the optimum region R_0 contains all points Y satisfying the inequality:

$$\frac{p(Y\,|\,H_0)}{p(Y\,|\,H_1)} > \frac{(C_{01} - C_{11})P(H_1)}{(C_{10} - C_{00})P(H_0)} \tag{2.2.4}$$

If the probability ratio

$$\frac{p(Y\,|\,H_0)}{p(Y\,|\,H_1)}$$

exceeds the constant $(C_{01} - C_{11})P(H_1)/(C_{10} - C_{00})P(H_0)$, the hypothesis H_0 is accepted according to this test, and is rejected otherwise. Such a test is known as a *Bayes test*.

It may happen that $C_{01} - C_{11} = C_{10} - C_{00}$, or that this equality is assumed in lieu of a better estimate of the true costs. The optimum decision then is to accept the hypothesis H_i corresponding to the larger of the quantities $p(Y|H_i)P(H_i)$. A decision of this kind is, for obvious reasons, referred to as a *maximum a posteriori probability* decision. [The term *ideal observer* is also sometimes used in reference to this decision rule. Notice that when $C_{01} - C_{11} = C_{10} - C_{00}$, minimizing the expected cost is equivalent to minimizing the unconditional probability $P_e = \alpha P(H_0) + \beta P(H_1)$ of an erroneous decision.] If, in addition, the a priori probabilities $P(H_i)$ are (assumed) equal, $P(H_0) = P(H_1)$, the decision is based solely on the functions $p(Y|H_i)$, sometimes called the *likelihood functions*. A decision to accept the hypothesis H_i for which the likelihood function $p(Y|H_i)$ is a maximum is commonly called a *maximum-likelihood* decision.

If it is not possible to assign values to the costs of the various decisions, another criterion, the *Neyman–Pearson* criterion, is sometimes more meaningful. According to this criterion, the probability of a false rejection α is stipulated in advance and, of all tests having the level α, the test with the maximum power $1 - \beta$ is to be used. Again, regardless of the value of α, the optimum test under this criterion turns out to be a probability ratio test. For in order to minimize β subject to a fixed α we introduce the Lagrange multiplier λ and seek that acceptance region R_0 for which the quantity

$$\lambda\alpha + \beta = \lambda\left[1 - \int_{R_0} p(Y|H_0)\,dY\right] + \int_{R_0} p(Y|H_1)\,dY$$

is minimized. But this expression is identical to Equation (2.2.3) with $P(H_0) = P(H_1) = \frac{1}{2}$, $C_{10} = 2\lambda$, $C_{01} = 2$, and $C_{11} = C_{00} = 0$. Since $C_{01} - C_{11}$ and $C_{10} - C_{00}$ are both positive for any positive λ, the conditions necessary for the optimality of the probability ratio test Equation (2.2.4) hold. The optimum test, according to the Neyman–Pearson criterion, is therefore to accept hypothesis H_0 if

$$\frac{p(Y|H_0)}{p(Y|H_1)} > \frac{1}{\lambda} \tag{2.2.5}$$

and to reject it otherwise. The parameter λ, of course, is determined by the condition:

$$\alpha = 1 - \int_R p(Y|H_0)\,dY$$

where R contains all those sets Y which satisfy Equation (2.2.5).

Finally, if the cost functions are known but there is no meaningful way to determine the a priori probabilities, one may prefer a decision rule which minimizes the cost of a decision under the most adverse choice of these probabilities. Since the expected cost of a decision is a linear function of $P(H_0) = 1 - P(H_1)$, the maximum cost must occur either when $P(H_0) = 1$ or when $P(H_1) = 1$. Thus, an equivalent rule is one for which the maximum of the expected decision costs when $P(H_0) = 1$ and when $P(H_1) = 1$ is minimized. The decision made according to this rule is called a *minimax decision*. From Equation (2.2.3):

$$E(\text{cost} \mid P(H_0) = 1) = C_{10} - \int_{R_0} (C_{10} - C_{00})p(Y \mid H_0)\, dY$$

and

$$E(\text{cost} \mid P(H_1) = 1) = C_{11} + \int_{R_0} (C_{01} - C_{11})p(Y \mid H_1)\, dY$$

If $E(\text{cost} \mid P(H_0) = 1)$ were the larger of the two terms, the maximum expected cost could be decreased by increasing the region R_0; if $E(\text{cost} \mid P(H_1) = 1)$ were the larger it could be decreased by decreasing R_0. The optimum region R_0, therefore, must be one of the regions R for which

$$\begin{aligned} E(\text{cost} \mid P(H_0) = 1) &= E(\text{cost} \mid P(H_1) = 1) \\ &= \gamma E(\text{cost} \mid P(H_0) = 1) + (1 - \gamma)E(\text{cost} \mid P(H_1) = 1) \\ &= \int_{R_0} [(C_{01} - C_{11})p(Y \mid H_1)(1 - \gamma) \\ &\quad - (C_{10} - C_{00})p(Y \mid H_0)\gamma]\, dY + \gamma C_{10} + (1 - \gamma)C_{11} \end{aligned} \tag{2.2.6}$$

with γ an arbitrary constant. Since this last expression is of the same form as Equation (2.2.3), the maximum expected cost is minimized by defining R_0 as the region for which

$$\frac{p(Y \mid H_0)}{p(Y \mid H_1)} > \frac{(C_{01} - C_{11})(1 - \gamma)}{(C_{10} - C_{00})\gamma} \tag{2.2.7}$$

where γ is so chosen as to satisfy the first equality in Equation (2.2.6).

The important point is that all of these decisions, regardless of the particular rule invoked, rely on the comparison of the probability ratio $p(Y \mid H_0)/p(Y \mid H_1)$ with some predetermined constant. This ratio is commonly called the *likelihood ratio* and the test based on this ratio a *probability ratio test*. The optimum hypothesis test under any of the above definitions of optimality is a probability ratio test.

In the next section, we shall reexamine the test between two simple hypotheses when the number M of observables is not fixed in advance.

2.3 The Sequential Probability Ratio Test

The test evolved in the previous section relied upon comparison of the ratio $p(Y \mid H_0)/p(Y \mid H_1)$ with some predetermined constant C. All M observables in the set Y are obtained before a decision is made. But if M is sufficiently large to insure a reliable decision in nearly every case, it would seem that in some of these instances, the correct decision will become apparent long before all M observables are available. That is, let $Y(n)$ be the set composed of the first n observables and suppose that the ratio $p(Y(n)\,|\,H_0)/p(Y(n)\,|\,H_1)$ is considerably greater (or considerably less) than the decision constant C for some $n < M$. It would seem reasonably safe to accept (or reject) the null hypothesis even though fewer than M observables are available for the decision. In allowing this flexibility, we will find that the expected number of observations needed for a decision can be decreased significantly without increasing the probability of making an incorrect decision.

The *sequential probability ratio test* is designed so that a decision will be made as soon as it can be done with sufficient assurance of being correct. The procedure is as follows: Let $Y(m)$ denote the set of observables $\{y_1, y_2, \ldots, y_m\}$. For each m the ratio,

$$\lambda(m) = \frac{p(Y(m)\,|\,H_0)}{p(Y(m)\,|\,H_1)} \tag{2.3.1}$$

is determined. If $\lambda(m)$ is greater than some threshold value A, the test is completed with the acceptance of hypothesis H_0. If $\lambda(m)$ is less than a second threshold value B, H_0 is rejected. Finally, if $B < \lambda(m) < A$, additional observations are made, the ratio $\lambda(m+1)$ is formed and again compared to A and B. Thus the test ends with the smallest value of $m = M$ for which $\lambda(M) \geq A$ or $\lambda(M) \leq B$. The sets of observables $Y(m)$ can presumably be smaller on the average than those necessary for the fixed-sample-size test. Now, of course, the number of observables used in a given test is a random variable.

Clearly, the probabilities α and β will be functions of the threshold values A and B. To determine the form of this dependence, assume, first, that the test has concluded with the acceptance of H_0. Let $Y(M) = \{y_1, y_2, \ldots, y_M\}$ be a specific set satisfying all of the inequalities $B < \lambda(m) < A$, $m = 1, 2, \ldots, M-1$, and $\lambda(M) \geq A$, and let R_M be the space of such sets $Y(M)$. Then†

$$\begin{aligned} 1 - \alpha &= \sum_{M=1}^{\infty} \int_{R_M} p(Y(M)\,|\,H_0)\, dY(M) \\ &\geq A \sum_{M=1}^{\infty} \int_{R_M} p(Y(M)\,|\,H_1)\, dY(M) = A\beta \end{aligned} \tag{2.3.2}$$

Similarly, letting S_M represent the space of the sets $Y(M)$ leading, after M observations, to the rejection of hypothesis H_0, we have:

$$\begin{aligned} \alpha &= \sum_{M=1}^{\infty} \int_{S_M} p(Y(M)\,|\,H_0)\, dY(M) \leq B \sum_{M=1}^{\infty} \int_{S_M} p(Y(M)\,|\,H_1)\, dY(M) \\ &= B(1-\beta) \end{aligned} \tag{2.3.3}$$

†It is implicitly assumed here that the test does end with probability one for some value of M. This is shown to be true in Appendix A, Lemma A.1, when the observables are statistically independent. It can actually be proved under more general conditions (Reference 2.2).

Thus, $A \leq (1 - \alpha)/\beta$ and $B \geq \alpha/(1 - \beta)$

These relationships are inequalities due to the fact that the ratio $\lambda(m)$ may exceed A or be smaller than B at the time of acceptance or rejection. But if the observables are such that $\lambda(m)$ is not rapidly changing with m, then since

$$B < \lambda(M - 1) < A$$

the ratio $\lambda(M)$ cannot differ greatly from A or B. Clearly, replacing B by $\alpha/(1 - \beta)$ and A by $(1 - \alpha)/\beta$ may increase the length of the test. Nevertheless, under the conditions just described this increase will be slight. Replacing B by $\alpha/(1 - \beta)$ and A by $(1 - \alpha)/\beta$ will also alter the error probabilities somewhat. However, letting α' and β' be the actual probabilities of error of the first and second kinds, respectively, when A and B are approximated in this way, we find from Equations (2.3.2) and (2.3.3) that

$$\frac{1 - \alpha'}{\beta'} \geq A = \frac{1 - \alpha}{\beta}$$

and

$$\frac{\alpha'}{1 - \beta'} \leq B = \frac{\alpha}{1 - \beta}$$

Consequently,

$$\beta' \leq \frac{\beta'}{1 - \alpha'} \leq \frac{\beta}{1 - \alpha}, \qquad \alpha' \leq \frac{\alpha'}{1 - \beta'} \leq \frac{\alpha}{1 - \beta}$$

and

$$\alpha' + \beta' \leq \alpha + \beta$$

Because α and β will both generally be quite small, the possible increase in error probabilities due to these approximations for A and B is insignificant. In some cases, we will be interested in the sequential test when $\beta \longrightarrow 0$. Then $\beta' \longrightarrow 0$ and $\alpha' \leq \alpha$ regardless of the value of α. A similar statement holds true when $\alpha \longrightarrow 0$.

The optimum fixed-sample-size test involved the comparison of a probability ratio to a threshold determined by the various cost functions and a priori probabilities involved. The error probabilities α and β were in turn determined by this threshold. This section started with the observation that if the probability ratio in question either sufficiently exceeds or is significantly smaller than this threshold value before all the observables have been collected, it would seem reasonably safe to truncate the test at that point. Consequently, *two* thresholds A and B were defined, and the test allowed to terminate when the ratio reached either of them. Again, the error probabilities α and β could be expressed in terms of these thresholds. The discussion thus far, however, has ignored the question of the costs involved. In a sequential test, there is another cost which did not enter into the analysis of fixed-sample-size tests, viz., the cost of continued observation. In the fixed-sample-size test, the observation cost was independent of the decision made. Here, since the length of the test will depend upon the probabilities of the various decisions, the cost of the observation is also a function of the particular test. The expected cost of a sequential test can be written [cf. Equation (2.2.3)]

$$E(\text{cost}) = [\beta C_{01} + (1-\beta)C_{11}]P(H_1) + [\alpha C_{10} + (1-\alpha)C_{00}]P(H_0) + C_0 E(M \mid H_0)P(H_0) + C_0 E(M \mid H_1)P(H_1) \tag{2.3.4}$$

where C_0 is the cost of an observation and $E(M \mid H_0)$ and $E(M \mid H_1)$ are the conditional expectations of the length of the test when H_0 and H_1, respectively, are the true hypotheses. It can be shown (cf. Reference 2.3) that regardless of the specific values of these cost coefficients (assuming of course, that C_0, $C_{01} - C_{11}$, and $C_{10} - C_{00}$ are all positive) a sequential probability ratio test does exist which minimizes the expected cost of the decision. Furthermore, using this fact, it can be proved that of all tests yielding a particular set of probabilities α and β, the expected number of observables M needed for a decision, under either hypothesis, is minimized by the sequential probability ratio test. These proofs are rather lengthy, however, and the reader is referred to the references for details.

The efficacy of the sequential probability ratio test vis-à-vis other tests having the same decision error probabilities can be measured in terms of the expected number M of observations required to complete the test. While the derivation of a general expression for this parameter seems an insuperable task, it is possible in the important special case in which the observables y_i are identically distributed, statistically independent random variables, and when "the excesses at the boundaries" [$\lambda(M) - A$ when H_0 is accepted and $B - \lambda(M)$ when it is rejected] can be ignored. Specifically, it is shown in Appendix A.1 that, if† $z_i \triangleq \log [p_0(y_i)/p_1(y_i)]$, with $p_j(y_i) \triangleq p(y_i \mid H_j)$, and if $E(z \mid H_0) \triangleq E(z_i \mid H_0) \neq 0$,

$$E(M \mid H_0) = \frac{(1-\alpha)\log A + \alpha \log B}{E(z \mid H_0)} = \frac{(1-\alpha)\log\left(\frac{1-\alpha}{\beta}\right) + \alpha \log\left(\frac{\alpha}{1-\beta}\right)}{E(z \mid H_0)} \tag{2.3.5a}$$

Similarly, if $E(z \mid H_1) \neq 0$,

$$E(M \mid H_1) = \frac{\beta \log \frac{1-\alpha}{\beta} + (1-\beta)\log \frac{\alpha}{1-\beta}}{E(z \mid H_1)} \tag{2.3.5b}$$

It is often helpful to be able to characterize a test more completely than

†Since the results of interest here will involve ratios of logarithms, the logarithms can be taken with respect to any convenient base without affecting the conclusions. In general, however, when no other base is indicated, the natural base e will always be assumed.

by the expected number of observables needed for a decision. One can show, for example, that under certain conditions the distribution of M is approximately Gaussian. In any event, the distribution of M can be at least partially characterized by two parameters: its expected value and its variance. An expression for the variance of M is also derived in Appendix A.1. Specifically,

$$\begin{aligned}\text{Var}\,(M\,|\,H_0) &= \frac{E(M\,|\,H_0)\,\text{Var}\,(z\,|\,H_0)}{E^2(z\,|\,H_0)} - \frac{\alpha(1-\alpha)(\log A - \log B)^2}{E^2(z\,|\,H_0)} \\ &\quad + \frac{2\alpha(1-\alpha)}{1-\alpha-\beta}\frac{\log A - \log B}{E(z\,|\,H_0)}[E(M\,|\,H_0) - E(M\,|\,H_1)]\end{aligned} \tag{2.3.6a}$$

and

$$\begin{aligned}\text{Var}\,(M\,|\,H_1) &= \frac{E(M\,|\,H_1)\,\text{Var}\,(z\,|\,H_1)}{E^2(z\,|\,H_1)} - \frac{\beta(1-\beta)(\log A - \log B)^2}{E^2(z\,|\,H_1)} \\ &\quad + \frac{2\beta(1-\beta)}{1-\alpha-\beta}\frac{\log A - \log B}{E(z\,|\,H_1)}[E(M\,|\,H_0) - E(M\,|\,H_1)]\end{aligned} \tag{2.3.6b}$$

Finally, we state without proof the following result (cf. Reference 2.2). When the lower threshold B is set equal to zero (and the upper threshold A is finite) and when the independent random variables z_i are normally distributed with mean $E(z) = \mu$ and variance $\text{Var}(z) = \sigma^2$, the distribution of the number of observables needed to complete the test, given it is completed, is approximately Gaussian. The mean and variance of this distribution are given by Equations (2.3.5) and (2.3.6), respectively. This approximation becomes increasingly valid as $(\mu/\sigma^2)\log A$ becomes large as compared to one. A similar statement holds when B is finite and $A = \infty$. Note that when $B = 0$, $\alpha = 0$, and only the first term remains in the expression for $\text{Var}(M\,|\,H_0)$ in Equations (2.3.6). Likewise, when $A = \infty$, $\beta = 0$, the second and third terms in the expression for $\text{Var}(M\,|\,H_1)$ vanish.

2.4 Composite Hypothesis Tests

In the previous discussion, the observables being investigated were assumed to satisfy one of two conditions: either H_0 was true and the observables were characterized by the density function $p(Y\,|\,H_0)$ or H_1 was true and $p(Y\,|\,H_1)$ was the governing density function. Each of the two hypotheses corresponded to a single state of the source of the observables; both were *simple* hypotheses. Frequently, however, the source of the observed phenomena may be in any one of several states, the observables pertaining to the state θ distributed according to the density function $p(Y\,|\,\theta)$. In such an event, it is sometimes satisfactory to divide the set of possible states into two disjoint subsets and to seek to identify only the subset to which the state of the source presently belongs. The null hypothesis H_0 might be that the source state θ is an element in the subset Ω_0, while the alternative hypothesis H_1 is that θ is in the subset Ω_1 with $\Omega_0 + \Omega_1$ containing the complete set of possible states. Each of these hypotheses is said to be a *composite* hypothesis if the corresponding subset contains more than one value of the state parameter θ.

Composite two-hypothesis tests arise in numerous contexts. One might, for example, be interested only in whether or not θ falls within some subset, not in its specific value. Or θ may be a vector $\boldsymbol{\theta} = (\phi_1, \phi_2, \ldots)$ with ϕ_1 the only parameter of interest. If ϕ_1 can assume only two values, the test is clearly a test between two hypotheses, but because the other parameters are unknown, both hypotheses are composite.

In any event, we must re-examine both the fixed-sample-size test and the sequential test to see how they are altered when the hypotheses are composite.

2.4.1 *Fixed Sample-Size Composite Hypothesis Tests*

Let $C_0(\theta)$ and $C_1(\theta)$ be the costs of accepting the hypotheses H_0 and H_1, respectively, given that the source is in state θ, and let $\pi(\theta)$ be the a priori probability density function of the state parameter θ. Further, let R_0 denote the acceptance region of hypothesis H_0, and $P(\theta)$ the probability of accepting H_0 given the state θ; i.e.,

$$P(\theta) = \int_{R_0} p(Y\,|\,\theta)\,dY \triangleq \begin{cases} 1 - \alpha(\theta) & \theta \in \Omega_0 \\ \beta(\theta) & \theta \in \Omega_1 \end{cases}$$

Then the expected cost of a decision is

$$\begin{aligned} E(\text{cost}) &= \int_{\theta\in\Omega_0} \pi(\theta)[C_0(\theta)(1 - \alpha(\theta)) + C_1(\theta)\alpha(\theta)]\,d\theta \\ &\quad + \int_{\theta\in\Omega_1} \pi(\theta)[C_0(\theta)\beta(\theta) + C_1(\theta)(1 - \beta(\theta))]\,d\theta \\ &= K + \int_{\theta\in\Omega_0} \pi(\theta)(C_1(\theta) - C_0(\theta))\alpha(\theta)\,d\theta \\ &\quad + \int_{\theta\in\Omega_1} \pi(\theta)(C_0(\theta) - C_1(\theta))\beta(\theta)\,d\theta \end{aligned} \tag{2.4.1}$$

where K is a constant independent of the decision rule. Define $\lambda_0 w_0(\theta) \triangleq (C_1(\theta) - C_0(\theta))\pi(\theta)$ and $\lambda_1 w_1(\theta) \triangleq (C_0(\theta) - C_1(\theta))\pi(\theta)$ with

$$\int_{\theta\in\Omega_0} w_0(\theta)\,d\theta = \int_{\theta\in\Omega_1} w_1(\theta)\,d\theta = 1$$

Then, if λ_0 and λ_1 are both positive (this is analogous to the condition that $C_{01} > C_{11}$ and $C_{10} > C_{00}$ in Section 2.2), the optimum decision rule involves choosing an acceptance region so as to minimize the quantity

$$\int_{\theta \in \Omega_1} w_1(\theta)\beta(\theta)\, d\theta + \lambda \int_{\theta \in \Omega_0} w_0(\theta)\alpha(\theta)\, d\theta \tag{2.4.2}$$

with $\lambda = \lambda_0/\lambda_1$. Now let H_0' and H_1' be defined in such a way that

$$p(Y\,|\,H_0') = \int_{\theta \in \Omega_0} p(Y\,|\,\theta) w_0(\theta)\, d\theta \tag{2.4.3a}$$

and

$$p(Y\,|\,H_1') = \int_{\theta \in \Omega_1} p(Y\,|\,\theta) w_1(\theta)\, d\theta \tag{2.4.3b}$$

If R_0 were the region of acceptance (and $\bar{R}_0$ the region of rejection) of both the simple hypothesis H_0' and the composite hypothesis H_0, we would have

$$\alpha \triangleq \int_{\bar{R}_0} p(Y\,|\,H_0')\, dY = \int_{\bar{R}_0} \int_{\theta \in \Omega_0} p(Y\,|\,\theta) w_0(\theta)\, d\theta\, dY = \int_{\theta \in \Omega_0} w_0(\theta)\alpha(\theta)\, d\theta \tag{2.4.4a}$$

and

$$\beta \triangleq \int_{R_0} p(Y\,|\,H_1')\, dY = \int_{\theta \in \Omega_1} w_1(\theta)\beta(\theta)\, d\theta \tag{2.4.4b}$$

The region R_0 which minimizes the quantity $\beta + \lambda\alpha$ is defined, referring back to the discussion of the Neyman-Pearson criterion, by the inequality

$$\frac{p(Y\,|\,H_0')}{p(Y\,|\,H_1')} < \frac{1}{\lambda} \tag{2.4.5}$$

Accordingly, letting the acceptance region of H_0 be defined by Equation (2.4.5) also minimizes the expected cost of the composite hypothesis decision [see Equation (2.4.2)]. If

$$C_1(\theta) - C_0(\theta) = \begin{cases} C_0 & \theta \in \Omega_0 \\ -C_1 & \theta \in \Omega_1 \end{cases}$$

with C_0 and C_1 positive constants, this test leads to the acceptance of H_0 only if

$$\frac{\int_{\theta \in \Omega_0} p(Y\,|\,\theta)\pi(\theta)\, d\theta}{\int_{\theta \in \Omega_1} p(Y\,|\,\theta)\pi(\theta)\, d\theta} > \frac{C_1}{C_0} \tag{2.4.6}$$

This is sometimes called the *average-likelihood* test.

If the functions $C_i(\theta)$ or the a priori probability density $\pi(\theta)$ or both are unknown, the functions $w_i(\theta)$ are undefined. Nevertheless, they can often be defined in a meaningful way by some other criterion. The weighted average error probability $\int_{\theta \in \Omega_0} \alpha(\theta) w_0(\theta)\, d\theta + \int_{\theta \in \Omega_1} \beta(\theta) w_1(\theta)\, d\theta$ is still minimized by the test (2.4.5) regardless of the criterion by which the weight functions are determined. One possible choice for the weight functions, in the absence of sufficient a priori or cost information, is $w_0(\theta) = \delta(\theta - \theta')$ and $w_1(\theta) = \delta(\theta - \theta'')$, where $\delta(x)$ is the Dirac delta function, θ' is in Ω_0, and θ'' in Ω_1. In some important situations, the acceptance region defined by Equation (2.4.5) with these definitions of $w_0(\theta)$ and $w_1(\theta)$ can be made independent of both θ' and θ'' for some $\lambda = \lambda(\theta', \theta'')$. The resulting test, called a *uniformly most powerful test* at the level $\alpha(\theta)$, is particularly attractive in this case, at least if $\alpha_0 = \max_{\theta \in \Omega_0} \alpha(\theta)$ and $\beta_0 = \max_{\theta \in \Omega_1} \beta(\theta)$ are both acceptably small. The term uniformly most powerful test is used because, for any θ_0 in Ω_0 and θ_1 in Ω_1, it is of all the tests at the level $\alpha(\theta_0)$ the one having the maximum power, $1 - \beta(\theta_1)$.

Other meaningful tests can be defined which are not of the form given in Equation (2.4.5). One such test, the *maximum-likelihood* test, involves the determination of the maximum of the probabilities $p(Y|\theta)$. If this maximum is attained for some $\theta \in \Omega_0$, H_0 is accepted; if not H_0 is rejected.† A slightly more general form of this same test might involve the comparison of the ratio

$$\frac{\max_{\theta \in \Omega_0} p(Y|\theta)}{\max_{\theta \in \Omega_1} p(Y|\theta)} \tag{2.4.7}$$

to some threshold C, accepting H_0 if and only if the threshold is exceeded.

2.4.2 *Sequential Composite Hypothesis Tests*

As just discussed, the acceptance regions of the composite hypotheses H_0 and H_1 can often be defined in terms of a probability ratio test. If a meaningful test of this sort can be prescribed, the next obvious step is to investigate a sequential decision procedure. However, if the probability ratio is determined by taking a weighted average of the functions $p(Y|\theta)$, Equations (2.4.3), the successive observables y_i will often not be independent under either hypothesis. That is, even if $p(Y|\theta) = \prod_i p(y_i|\theta)$,

$$\int_{\theta \in \Omega_\nu} p(Y|\theta) w_\nu(\theta)\, d\theta = \int_{\theta \in \Omega_\nu} \prod_i p(y_i|\theta) w_\nu(\theta)\, d\theta$$

is in general not equal to

$$\prod_i \int_{\theta \in \Omega_\nu} p(y_i|\theta) w_\nu(\theta)\, d\theta$$

for either $\nu = 0$ or $\nu = 1$. Nevertheless, the argument leading to Equations (2.3.2) and (2.3.3) used the assumed independence of the observables y_i only

†Note that this definition of a maximum-likelihood test is equivalent to that in Section 2.2 when both H_0 and H_1 are simple hypotheses.

in the proof of the eventual termination of the test. This property can actually be proved under considerably more general conditions (cf. Reference 2.2). Provided it does hold, then, Equations (2.3.2) and (2.3.3) are valid for composite hypotheses tests as well. If the upper threshold A is equated to $(1-\alpha)/\beta$ and the lower threshold B to $\alpha/(1-\beta)$ Equations (2.4.4) will also be satisfied for the sequential test. The same comments concerning the choice of the weight functions apply here as in the case of the fixed-sample-size test.

The derivation of the expressions (see Section 2.3) for the mean and variance of the number of observables needed for a decision in a sequential test, however, did use the assumed independence of the observables in a rather fundamental way. It is generally difficult to obtain comparable results when they are not independent.

Still, the observables y_i can be independent for some special classes of weight functions, as when $w_v(\theta) = \delta(\theta - \theta_v)$. More generally, if θ is a vector, $\boldsymbol{\theta} = (\theta, \phi_1, \phi_2, \ldots, \phi_k)$, if

$$w_v(\boldsymbol{\theta}) = \delta(\theta - \theta_v) f_1(\phi_1) f_2(\phi_2) \ldots f_k(\phi_k),$$

and if

$$p(Y \mid \boldsymbol{\theta}) = \prod_i p(y_i \mid \theta, \phi_i),$$

then

$$\int p(Y \mid \boldsymbol{\theta}) w_v(\boldsymbol{\theta})\, d\boldsymbol{\theta} = \int \ldots \int \prod_i p(y_i \mid \theta_v, \phi_i) f_i(\phi_i)\, d\phi_1 d\phi_2 \ldots d\phi_k$$

$$= \prod_i \int p(y_i \mid \theta_v, \phi_i) f_i(\phi_i)\, d\phi_i = \prod_i p(y_i \mid \theta_v)$$

and the observables are statistically independent. The test is equivalent to a test between two simple hypotheses.

When the observables are independent, and identically distributed, the probability $P(\theta)$ of a sequential probability ratio test ending in the acceptance of the null hypothesis H_0 when the source is in the state θ is shown in Appendix A.1 to be

$$P(\theta) = \frac{1 - B^{h(\theta)}}{A^{h(\theta)} - B^{h(\theta)}} \tag{2.4.8}$$

where $h(\theta)$ is the unique non-zero solution to the equation

$$\phi(h) = E(e^{hz} \mid \theta) = 1 \qquad \left(z = \log_e \frac{p_0(y)}{p_1(y)}\right)$$

(see Theorem A.1). The expected number of observables needed for a decision under these same conditions is

$$E(M \mid \theta) = \frac{P(\theta) \log A + (1 - P(\theta)) \log B}{E(z \mid \theta)} \tag{2.4.9}$$

An expression for the variance of M when the source is in state θ is also derived in Appendix A.1.

Note that if θ_0 is the assumed null hypothesis state, $h(\theta_0) = -1$ is a solution to the equation $\phi(h) = 1$ and because this equation has only one non-zero solution

$$P(\theta_0) = \frac{A(1 - B)}{A - B} = 1 - \alpha$$

When θ_1 is the alternative hypothesis state, $h(\theta_1) = 1$ and

$$P(\theta_1) = \frac{1 - B}{A - B} = \beta$$

Since $A > 1$ and $B < 1$, if α and β are to be positive, $P(\theta)$ is a monotonically decreasing function of $h(\theta)$. Thus, if the inequality $h(\theta) \geq h(\theta_1) = 1$ is satisfied for all θ in Ω_1, the maximum probability of an acceptance of H_0 will be β and will occur when θ_1 is indeed the observed state. Similarly, if $h(\theta) < -1$, the null hypothesis will be accepted with a probability greater than $1 - \alpha$.

2.5 Search

In the preceding sections, the problem has been to identify to which of two possible subsets, Ω_0 or Ω_1, the state θ actually belongs. But suppose more information concerning θ is desired; in particular, suppose the set of states Ω is to be divided into N subsets $\omega_1, \omega_2, \ldots, \omega_N$, and that the problem is to determine the subset ω_i best characterizing the source state θ. If $N > 2$, the resulting test is called a *multiple-hypothesis* test. If ω_i represents only one of the possible states θ, the corresponding hypothesis H_i is simple; if not, it is composite. In either event, the multiple-hypothesis test may be considerably more complex, computationally, than a two-hypothesis test. Indeed, practical consideration may rule out the possibility of conducting such a test. If this is the case, it may be necessary first to group the hypotheses ω_i into two larger subsets, say Ω_1 and $\bar{\Omega}_1$, with $\Omega_1 + \bar{\Omega}_1$ containing the totality of subsets ω_i. The test would then involve selecting a set of observables Y_1 and deciding, on the basis of these observables, whether to accept the null hypothesis H_1 (that θ is in the subset Ω_1) or the alternative hypothesis $\bar{H}_1$ (that θ is in $\bar{\Omega}_1$). Such a decision reduces the contending hypotheses to those contained either in subset Ω_1 or in subset $\bar{\Omega}_1$. If one then divides these remaining hypotheses into two new subsets Ω_2 and $\bar{\Omega}_2$, representing the two hypotheses H_2 and $\bar{H}_2$, a second two-hypothesis test, based on the set of observables Y_2, further reduces the number of hypotheses to be considered. If this procedure is continued, and the successive hypotheses H_1, $H_2, H_3, \ldots$, are properly chosen, eventually only one hypothesis will remain in contention. We shall refer to this process of reducing an inherently multiple-hypothesis test to a series of two-hypothesis tests as a *search*.

The purpose of a search, then, is to reduce the complexity of a multiple-hypothesis test, at the possible cost of increasing the amount of time needed for a decision. Although the ultimate goal is to determine the specific subset ω_i of the source state, this goal is to be achieved by answering a series of hopefully simpler questions. This approach is typical, for example, when the state of the source refers to its location in space or time. Often, the most efficacious procedure is to ask not "Where is it?" but rather" Is it here?".

If the computational complexity is to be reduced significantly through a search routine, the hypotheses H_i and $\bar{H}_i$ must be such that the resulting tests are much simpler than the original multiple-hypothesis test. Since both H_i and $\bar{H}_i$ are likely to be composite if the procedure just outlined is used, further simplifications are often necessary in practice. For this reason, the discussion in this and the next section will be restricted to the situation in which each subset ω_i represents a single source state θ_i. At the jth stage of the search, the hypothesis H_j, that $\theta = \theta_i$ with $i = j$, modulo N, will be tested against some alternative (presumably simple) hypothesis $\bar{H}_j$. (We shall refer to this as a test of the state θ_i.) The test terminates as soon as some state is accepted as the correct one. A variation on this strategy, in which the search is truncated after all N states have been tested once, will also be discussed. If no state has been accepted at this point some other decision criterion could be used; the state corresponding to the largest likelihood ratio, for example, might be chosen. Other variations will be encountered in the problems at the end of this chapter and in later chapters.

We first determine the probability that the search is successful when it is allowed to continue indefinitely and when, at each stage of the search, the probability α of a false dismissal is independent of the source state, and the probability β of a false acceptance is independent of the state being tested. (These restrictions will be relaxed subsequently.) Let $P(\mu) = 1/N$ denote the a priori probability that the source is in the μth state. Then the probability that the source state is correctly identified is

$$\begin{aligned} P_c &= \sum_{\mu=0}^{N-1} \sum_{j=0}^{\infty} P(\mu)(1 - \beta)^{j(N-1)+\mu} \alpha^j (1 - \alpha) \\ &= \frac{(1 - \alpha)[1 - (1 - \beta)^N]}{N\beta[1 - \alpha(1 - \beta)^{N-1}]} \end{aligned} \tag{2.5.1}$$

This result leads to an approximation which will prove useful in the sequel. Specifically, if the error probability is to be small, the inequality

$$N\beta \ll 1 - \alpha \qquad (P_e \ll 1) \tag{2.5.2}$$

must hold. Because P_c is clearly a decreasing function of α, and since, when $\alpha = 0$,

$$P_c = \frac{1 - (1 - \beta)^N}{N\beta} = 1 - \frac{N - 1}{2}\beta + \frac{(N - 1)(N - 2)}{6}\beta^2 + \ldots$$

the error probability $P_e = 1 - P_c$ is small compared to one only if $N\beta \ll 1$. But if $N\beta \ll 1$, we can simplify Equation (2.5.1) by retaining only the most significant terms in the power series expansion of the functions $(1 - \beta)^N$ and $(1 - \beta)^{N-1}$. Doing so yields the relationship

$$\frac{N-1}{2}\beta \approx \frac{(1-\alpha)P_e}{1+\alpha-2\alpha P_e} \qquad (\text{for } P_e \ll 1) \tag{2.5.3}$$

and the stated inequality follows for all $P_e \ll 1$. For future reference, we also note that approximation (2.5.3) remains valid for any value of $P_e < 1$ when $\alpha \to 1$. This follows because, from Equation (2.5.1), $(1 - \beta)^{N-1}$ must approach unity as $\alpha \to 1$ if the probability P_c is not to approach zero. Thus, again $N\beta \ll 1$, and this is the only assumption needed to justify the approximation (2.5.3).

The probability of testing $\nu = jN + i + 1$ states before a final decision is

$$\begin{aligned}\Pr(\nu = jN + i + 1) &= \sum_{\mu=0}^{N-1} \Pr(\nu = jN + i + 1 \mid \mu) P(\mu) \\ &= \frac{1}{N}(1-\beta)^{j(N-1)}\alpha^j[(1-\beta)^i(1-\alpha) \\ &\quad + i\alpha\beta(1-\beta)^{i-1} + (N-i-1)\beta(1-\beta)^i]\end{aligned} \tag{2.5.4}$$

where

$$\Pr(\nu = jN + i + 1 \mid \mu) = \begin{cases}(1-\beta)^{j(N-1)+i}\alpha^j(1-\alpha) & i = \mu \\ (1-\beta)^{j(N-1)+i-1}\alpha^{j+1}\beta & i > \mu \\ (1-\beta)^{j(N-1)+i}\alpha^j\beta & i < \mu\end{cases}$$

Consequently, the expected number of states tested is

$$E(\nu) = \sum_{\nu=0}^{\infty} \nu \Pr(\nu) = \frac{1}{N}\sum_{i=0}^{N-1}\sum_{j=0}^{\infty}(jN + i + 1)\Pr(\nu = jN + i + 1) \tag{2.5.5}$$

While this summation is easily evaluated, the resulting expression is rather inconvenient for use in further manipulations. However, if $P_e \ll 1$, then from Equation (2.5.3), $N\beta \ll 1 - \alpha$ and we obtain the useful approximation

$$E(\nu) \approx \frac{N\alpha}{1-\alpha} + \frac{N+1}{2} \tag{2.5.6}$$

Under this same condition ($P_e \ll 1$), we find that

$$\operatorname{Var}(\nu) \approx N^2\frac{\alpha}{(1-\alpha)^2} + \frac{N^2-1}{12} \tag{2.5.7}$$

The preceding expressions have been derived under the assumption that the search would be allowed to continue indefinitely. If the search is terminated after the Nth state has been investigated, as discussed earlier, the expected number of states which must be tested is

$$E(\nu) = \frac{1}{N}\sum_{i=0}^{N-1}(i+1)\Pr(\nu = i+1) + N\alpha(1-\beta)^{N-1} \tag{2.5.8}$$

[cf. Equation (2.5.5)]. When $N\beta$ is small, this reduces approximately to

$$E(\nu) \approx \frac{N+1}{2} + \frac{N-1}{2}\alpha \tag{2.5.9}$$

The probability P_e of an error is

$$\begin{aligned} P_e &= 1 - \frac{1}{N}\sum_{i=0}^{N-1}(1-\beta)^i(1-\alpha) \\ &= 1 - (1-\alpha)\frac{1-(1-\beta)^N}{N\beta} \approx \alpha + \frac{N-1}{2}\beta \end{aligned} \tag{2.5.10}$$

where the absence of an acceptance after all N states have been tested is also identified as an error.

If α and β are not independent of the states involved, the above results are still useful in bounding the search parameters. The approximate mean and variance of the number of states which must be tested [Equations (2.5.6) and (2.5.7)] are monotonically increasing functions of α, and the error probability [Equation (2.5.1)] is an increasing function of both α and β. Thus, upper and lower bounds on these quantities are easily established in terms of the upper and lower bounds on α and β.

If fixed-sample-size tests are conducted at each stage of the search, with M observables needed to complete each test, the total number M_T of observables required for the search is just M times the expected number $E(\nu)$ of tests. Both M_T and P_c, the probability of a successful conclusion to the search, are functions of α and β. Presumably these two parameters can be adjusted, subject to a constraint on P_c, so as to minimize M_T. This can be done, however, only if the test statistics are specified, for only then is the relationship between M and α and β known. Further discussion of this aspect of the problem must therefore be deferred until we encounter more specific problems (cf. Chapter Six). This conclusion does not apply when sequential tests are made at each stage. This situation is investigated in some detail in the following section.

Before concluding this section, however, we mention two variations on the preceding search strategy which are not only of interest in themselves but which also point out some of the deficiencies inherent in the fixed-sample-size approach. The first variation involves subdividing the search into two (or more) modes. In the first mode, a relatively rapid search is made, the speed achieved at the cost of a rather larger value of the erroneous acceptance probability β than one is willing to accept in the long run. If one or more states are provisionally accepted, the second mode of operation is begun in which each accepted state is monitored for a somewhat longer period than that needed to accept it in the first place. At the end of this probationary period the state is finally accepted or rejected. The search ends either when one state has been accepted or when all have been rejected. Extension of this procedure to more than two modes is obvious.

The advantage of this approach is readily apparent. Suppose the first of the tests involve M_1 observables and that the probabilities of an erroneous rejection and an erroneous acceptance using this test are α_1 and β_1, respectively. Let M_2, α_2, and β_2 be the corresponding parameters for the second test. The probability of an erroneous rejection using the combined test is

$$\alpha = \alpha_1 + \alpha_2(1 - \alpha_1)$$

while the probability of an erroneous acceptance is

$$\beta = \beta_1\beta_2$$

The expected number of observables needed for a decision when the correct state is being tested is

$$M_c = M_1 + (1 - \alpha_1)M_2$$

while that needed in testing any other state is

$$M_e = M_1 + \beta_1 M_2$$

Since presumably β_1, β_2, α_1 and α_2 are all small, the erroneous states can be rejected much more reliably than they were with the first test alone, while the correct state is rejected with only slightly greater probability than before. Furthermore, if $M_2 \gg M_1$, the time spent testing erroneous states has decreased considerably relative to the time spent testing the correct state. When the total number of states is large (and there is only one correct state) these are significant advantages.

The advantage inherent in increasing the relative time spent in testing the correct state is also realized in another modification of the fixed-sample-size search (Reference 2.4). This approach also involves two modes of operation. In the first mode, each state is again tested for a relatively short period of time, but no decision is made at this point. Rather, the likelihood ratio itself is recorded. In the second mode, the states are investigated in order of decreasing values of these likelihood ratios, and a decision is made to accept or reject each state as in a single mode search. Thus, the most probable states are tested first with the net effect that the time spent testing erroneous states is likely to be decreased.

The most serious deficiency of fixed-sample-size searches is that they tend to use inefficiently the information which is available. In the first place, as the search progresses and a number of states have already been rejected, the effective a priori probabilities of the remaining states increase. The acceptance threshold should presumably decrease correspondingly. Taking this factor into account, however, would tend to nullify the primary advantage of this search procedure, namely, its simplicity. An even grosser inefficiency occurs when some of the states are tested more than once before a decision is made. For to ascertain for the ith time whether to accept or reject a state involves a probability ratio of the form $p_0(Y_{\nu i})/p_1(Y_{\nu i})$, $Y_{\nu i}$ denoting the ith set of observables pertaining to the νth stage. This is all the information generally used at each stage, even though after i independent observations, the probability ratio $\prod_{j=1}^{i}[p_0(Y_{\nu j})]/[p_1(Y_{\nu j})]$ is actually a much better measure of whether or not the νth state is the one being observed. Since each of the terms in this product must have been determined earlier, it seems singularly inefficient to use only one of them to make the decision. This, indeed, suggests the use of some sort of sequential test.

2.6 Sequential Search

Many of the objections and difficulties associated with the fixed-sample-size search can be eliminated if a sequential test can be made for each state investigated. First, the inefficiencies inherent in the fixed-sample-size test when multiple observations of each state are necessary are no longer a factor. This is because (to the extent that the excesses at the boundaries can be neglected) the probability ratio $p_0[Y_{\nu i}(M_i)]/p_1[Y_{\nu i}(M_i)]$ at the end of the ith test is equal to a constant B for all but the final test. Thus,

$$\prod_{j=1}^{i-1} \frac{p_0[Y_{\nu j}(M_j)]}{p_1[Y_{\nu j}(M_j)]} = B^{i-1}$$

assuming B is kept constant. Since this factor is the same for each rejected state, the decisions during the ith series of tests would not be altered by taking it into account.

Another advantage of the sequential search, which can be significant, rests in its potential elimination of the verify mode of operation often required with fixed-sample-size searches. The capability of recognizing an erroneous decision in order to reinstigate the search mode, and of simultaneously minimizing the possibility of doing so when the decision is in fact a correct one, can be an automatic part of the sequential search.

Before elaborating on this, we will first determine the expected number of observables M_T needed to complete a sequential search. The probability $\Pr(\nu)$ of exactly ν states being tested before a decision is made is as given in Equation (2.5.4) regardless of whether a sequential or a fixed-sample-size decision is made. If the distribution of the test observables $Y_j = \{y_i^{(j)}\}$ depends not on the specific state θ_j being tested, but only on whether or not it is the true source state,† the conditional distributions $P(M \mid A)$ and $P(M \mid R)$ of the numbers of observables needed to accept and to reject a particular state are both independent of the state actually being tested (see Appendix A.1). The expected value of M_T can therefore be expressed in the form

$$E(M_T) = \sum_{\nu=1}^{\infty} \Pr(\nu) E(M_T \mid \nu) = \sum_{\nu=1}^{\infty} [E(M \mid R)(\nu - 1) + E(M \mid A)] \Pr(\nu)$$
$$= E(\nu - 1) E(M \mid R) + E(M \mid A) \tag{2.6.1}$$

with $E(M \mid R)$ and $E(M \mid A)$, the expected number of observations needed to reject and to accept, respectively, the state being tested, as determined in Appendix A.1 [Equations (A.1.18)]. Similarly, the variance of the number of observables needed for a decision is

$$\mathrm{Var}(M_T) = E(M_T^2) - E^2(M_T) = \sum_{\nu=0}^{\infty} E(M_T^2 \mid \nu) \Pr(\nu) - E^2(M_T)$$

But

$$E(M_T^2 \mid \nu) = E(\sum_{i=1}^{\nu} M_i)^2 = (\nu - 1) E(M^2 \mid R) + E(M^2 \mid A)$$
$$+ (\nu - 1)(\nu - 2) E^2(M \mid R) + 2(\nu - 1) E(M \mid R) E(M \mid A)$$

so that

$$\mathrm{Var}(M_T) = E(\nu - 1) \mathrm{Var}(M \mid R) + E^2(M \mid R) \mathrm{Var}(\nu - 1) + \mathrm{Var}(M \mid A) \tag{2.6.2}$$

Both $E(\nu)$ and $\mathrm{Var}(\nu)$ are evaluated in the preceding section and the quantities $E(M \mid A)$, $E(M \mid R)$, $\mathrm{Var}(M \mid A)$, and $\mathrm{Var}(M \mid R)$ are evaluated in Appendix A.1. The mean and variance of the number of observables M_T required for a decision are functions of both α and β. These two probabilities, however, are not independent parameters; for any given error probability P_e they are constrained by Equation (2.5.1). Using this equation and Equations (2.6.1) and (2.6.2), we can obtain expressions for the mean and variance of M_T in terms of α(or β), the error probability P_e, the number of states N, and the conditional moments of the (assumed independent) variables $z_i = \log [p_0(y_i)/p_1(y_i)]$.

Since the resulting general expressions for $E(M_T)$ and $\mathrm{Var}(M_T)$ are rather cumbersome, some of the more important special cases are summarized in Table 2.1. The simplifications result upon application of the relevant approximations developed in the preceding section, viz., Equations (2.5.3), (2.5.6), and (2.5.7). The situations when α and β are both small are of interest both for purposes of contrast with the other extreme (when $\alpha \to 1$) and for later comparison with the fixed-sample-size search which is generally practical only under this condition. Since $E(M_T)$ becomes infinite as $\alpha \to 0$, a meaningful comparison is possible only for $\alpha > 0$. For this reason, we have somewhat

---

†A more general case will be considered presently.

arbitrarily equated α to $P_e/2$ and β to $P_e/(N - 1)$ to represent this extreme. Note that when α and β are so defined, Equation (2.5.10) is satisfied. The probability P_e therefore includes here the event in which no acceptance has been made by the time each state has been investigated exactly once.

Referring to Table 2.1, we see that in the usual situation in which $E(z \mid H_0)$ and $E(z \mid H_1)$ are of the same order of magnitude, and $N \gg \log [(N - 1)/P_e]$, the search time when α is kept small is approximately $\frac{1}{2} \log (2/P_e)$ times longer than that attainable when $\alpha \to 1$. Similarly, the search time variance is decreased by a factor of roughly $\frac{1}{12} \log^2 (2/P_e)$ by allowing α to approach unity.

The possibility of an error in accepting a state as the true source state can be rather awkward in a search procedure in that it suggests the use of a verify mode which can erroneously start the whole search over again. In a sequential probability ratio test, this inconvenience can sometimes be avoided by setting β equal to zero. This is equivalent to letting A be infinite and consequently never accepting any state as the correct one. The true state, on the other hand, will be rejected with probability α while all the others will be rejected with probability 1. The result is that with probability $1 - \alpha$ the search apparatus will be unable to reject the true state and hence will observe it indefinitely. In many applications, the true state is sought only for purposes of continued observation. In such instances, this approach may be entirely satisfactory. This notion of never completing the search is tantamount to recognizing that new information could be received which would result in the rejection of the present tentatively accepted state. Because of this feature, and in order to distinguish it from the more conventional method just investigated, we title this search procedure the *continual* search (cf. Reference 2.5). The probability of a correct decision will be identified with the probability that at some instant of time, the true state is encountered and is subsequently never rejected.

When the search is never formally truncated, the definition of search time must be altered somewhat to have significance. Since the state of the source is effectively determined as soon as the search apparatus begins to test a state which it cannot subsequently reject, the search time can be defined as the time needed to reach this point. This is equivalent to setting $E(M \mid A)$ and $\text{Var}(M \mid A)$ equal to zero in the expression for the mean and variance of the total search time. The resulting expressions for $E(M_T)$ and $\text{Var}(M_T)$ (with $\beta = 0$) are also included in Table 2.1.

As in the case of the fixed-sample-size search, the above results can be used to bound the search parameters even when the false acceptance probability is a function of the specific state being tested. This approach, however, while generally quite satisfactory for fixed-sample-size searches, can result in rather poor sequential search bounds. Better bounds can be obtained by taking into account the fact that, when the alternative hypothesis is actually

Table 2.1

Conditions	$E(M_T)$	$\text{Var}(M_T)$
General	$E(\nu - 1)E(M \mid R) + E(M \mid A)$	$E(\nu - 1)\,\text{Var}\,(M \mid R) + \text{Var}\,(\nu - 1)E^2(M \mid R) + \text{Var}\,(M \mid A)$
$P_e \ll 1$	$\frac{(N-1)(1+\alpha)}{2(1-\alpha)}\frac{\log\frac{1}{\alpha}}{\mu_1} - \frac{\alpha}{1-\alpha}\frac{\log\frac{1}{\alpha}}{\mu_0} + \frac{\log\left(\frac{N-1}{P_e}\frac{1+\alpha}{2}\right)}{\mu_0}$	$\left(\frac{N^2-1}{12} + \frac{N^2\alpha}{(1-\alpha)^2}\right)\frac{\log^2\frac{1}{\alpha}}{\mu_1^2} + \frac{(N-1)(1+\alpha)}{2(1-\alpha)}\frac{\sigma_1^2}{\mu_1^3}$ $\log\frac{1}{\alpha}\frac{\sigma_0^2}{\mu_0^3}\left[\log\left(\frac{N-1}{P_e}\frac{1+\alpha}{2}\right) - \frac{\alpha}{1-\alpha}\log\frac{1}{\alpha}\right]$ $-\frac{\alpha\log^2\frac{1}{\alpha}}{(1-\alpha)^2}\left(\frac{1}{\mu_0}+\frac{1}{\mu_1}\right)^2 \quad \left(P_e \ll \frac{N\mu_0}{\mu_1}\right)$
$P_e \ll 1$, $\alpha = \frac{P_e}{2}$, $\beta = \frac{P_e}{N-1}$	$\frac{N-1}{2}\frac{\log\frac{2}{P_e}}{\mu_1} + \frac{\log\frac{N-1}{P_e}}{\mu_0}$	$\frac{N^2-1}{12}\left(\frac{\log\frac{2}{P_e}}{\mu_1}\right)^2 + \frac{N-1}{2}\frac{\sigma_1^2}{\mu_1^3}\log\frac{2}{P_e} + \frac{\sigma_0^2}{\mu_0^3}\log\frac{N-1}{P_e}$
$\alpha \to 1, \beta \to 0$, $P_e \leqslant (N-1)/N$	$\frac{P_e}{\mu_1}(A^* - 1 - \log A^*) + \frac{1}{\mu_0}(1-P_e)\left(\log A^* - \frac{A^*-1}{A^*}\right)$	
$\alpha \to 1, \beta \to 0$, $P_e \ll 1$	$\frac{N-1}{\mu_1} + \frac{\left(\log\frac{N-1}{P_e} - 1\right)}{\mu_0}$	$\frac{N^2}{\mu_1^2} + (N-1)\frac{\sigma_1^2}{\mu_1^3} + \frac{\sigma_0^2}{\mu_0^3}\left(\log\frac{N-1}{P_e} - 1\right) - \left(\frac{1}{\mu_0}+\frac{1}{\mu_1}\right)^2 \left(P_e \ll \frac{N\mu_0}{\mu_1}\right)$
$P_e \ll 1, \beta = 0$ (continual search)	$\left\{\frac{\alpha N}{1-\alpha} + \frac{N-1}{2}\right\}\frac{\log\frac{1}{\alpha}}{\mu_1}$	$\left(\frac{N^2-1}{12} + \frac{N^2\alpha}{(1-\alpha)^2}\right)\frac{\log^2\frac{1}{\alpha}}{\mu_1^2} + \left(\frac{N-1}{2} + \frac{\alpha N}{1-\alpha}\right)\frac{\sigma_1^2}{\mu_1^3}\log\frac{1}{\alpha}$
$\alpha = P_e \ll 1, \beta = 0$ (continual search)	$\frac{N-1}{2}\frac{\log\frac{1}{P_e}}{\mu_1}$	$\frac{N^2-1}{12}\frac{\log^2\frac{1}{P_e}}{\mu_1^2} + \frac{N-1}{2}\frac{\sigma_1^2}{\mu_1^3}\log\frac{1}{P_e}$
$\alpha \to 1, \beta = 0$, $P_e \ll 1$ (continual search)	$\frac{N}{\mu_1}$	$\frac{N^2}{\mu_1^2} + \frac{N\sigma_1^2}{\mu_1^3}$

Legend: $\mu_i = \lvert E(z \mid H_i) \rvert$, $\sigma_1^2 = \text{Var}\,(z \mid H_i)$ $\qquad A^* = \lim_{\substack{\alpha\to 1\\ \beta\to 0}} \left(\frac{1-\alpha}{\beta}\right) = \frac{(N-1)(1-P_e)}{P_e}$

composite (even though for testing purposes it is replaced by a simple hypothesis), some erroneous states may be much more quickly dismissed, on the average, than others.

In the two situations of greatest interest, when $\alpha \to 1$ and $\beta \to 0$, and when $\alpha = P_e/2$ and $\beta = P_e/(N-1)$, in both cases with $P_e \ll 1$, more precise estimates of the number M_T of observables required to complete the test are not difficult to derive, even under the more general conditions just described. To this end, let $y^{(i)}$ be an observable pertaining to the ith state to be tested, and let $z^{(i)} = \log [p_0(y^{(i)})/p_1(y^{(i)})]$ be the corresponding test statistic. If the source is in the jth state, let the statistic $z^{(i)}$ be characterized by the density function $p(z^{(i)} \mid j)$, with $p(z^{(i)} \mid i) = p_0(z^{(i)})$ for all i. The density function $p(z^{(i)} \mid j)$ can be a function of both i and j, but the *set* of functions $\{p(z^{(i)} \mid j), j = 0, 1, \ldots, N-1\}$ will be assumed independent of i. [This last condition will always be satisfied in later applications of the sequential search algorithm. In fact, $p(z^{(i)} \mid j)$ will be in every case a function only of the difference $i - j$, modulo N.] The alternative hypothesis density function $p_1(y^{(i)})$ is assumed to be so defined that a test of the ith state will end in its acceptance with the probability $1 - \alpha$ when the source is actually in this state and with a probability of *at most* β when it is in the jth state for any $j \neq i$, modulo N.

Let ν denote the number of states tested before a decision is made, and M_i the number of observables needed for the ith test. Then

$$E(M_T) = E\left(\sum_{i=1}^{\nu} M_i\right) = \sum_{n=1}^{\infty}\left[\Pr(\nu = n)\sum_{i=1}^{n} E(M_i \mid \nu = n)\right]$$

$$= \sum_{i=1}^{\infty}\sum_{n=i}^{\infty}\Pr(\nu = n)E(M_i \mid \nu = n) \tag{2.6.3}$$

Noting that $\sum_{n=i}^{\infty}\Pr(\nu = n \mid \nu \geq i)E(M_i \mid \nu = n) = E(M_i \mid \nu \geq i)$ and that $\Pr(\nu = n) = \Pr(\nu = n \mid \nu \geq i)\Pr(\nu \geq i)$ for all $n \geq i$, we have

$$E(M_T) = \sum_{i=1}^{\infty}\Pr(\nu \geq i)E(M_i \mid \nu \geq i) \tag{2.6.4}$$

In general, the random variable M_i is not independent of the event $\{\nu \geq i\}$ even when the erroneous state statistics are all identically distributed. If $i = N$, for example, the fact that $N - 1$ states have already been rejected will tend to increase the probability of the Nth state being the correct one. And since the number of observables needed for a decision is a function of the state being tested, this fact will affect the quantity $E(M_i \mid \nu \geq i)$. When $\alpha \to 1$, however, this dependence no longer exists. Since all states, the correct one as well as the erroneous ones, are rejected with probability one, the probability that the ith state to be tested is either the source state, or any particular one of the erroneous states, is unaltered by the previous $i - 1$ observations. In this case, then

$$E(M_i \mid \nu \geq i) = E(M_i) = \frac{1}{N}\sum_{j=0}^{N-1}E(M_i \mid j) \tag{2.6.5}$$

where j is the index of the correct state. Since the set of density functions $\{p(z^{(i)} | j), j = 0, 1, \ldots, N-1\}$ is independent of i so also is $E(M_i)$. Consequently, from Equation (2.4.9),

$$E(M_i) = E(M_0) \triangleq \frac{1}{N} \sum_{j=0}^{N-1} \frac{P_j \log A + (1 - P_j) \log B}{E(z^{(0)} | j)} \tag{2.6.6}$$

with P_j the probability of accepting the zeroth state when the source is in the jth state. Combining Equations (2.6.4), (2.6.5), and (2.6.6) yields, when $\alpha \to 1$,

$$E(M_T) = \frac{1}{N} \sum_{i=1}^{\infty} P(\nu \geq i) \sum_{j=0}^{N-1} E(M_0 | j) = \frac{E(\nu)}{N} \sum_{j=0}^{N-1} \frac{P_j \log A + (1 - P_j) \log B}{E(z^{(0)} | j)} \tag{2.6.7}$$

But by hypothesis, $P_0 = (1 - \alpha)$ and $P_j \leq \beta$ for all $j \neq 0$. Moreover, from Equation (2.5.6), when $P_e \ll 1$, $E(\nu) \approx N\alpha/(1 - \alpha)$ and is virtually independent of the probabilities P_j, $j \neq 0$. Thus, if $A = (1 - \alpha)/\beta$ and $B = \alpha/(1 - \beta)$, and if $\alpha \to 1$, $\beta \to 0$ in such a way that Equation (2.5.3) is satisfied (with $P_e \ll 1$) we find

$$E(M_T) = \sum_{j=1}^{N-1} \frac{1}{|E(z^{(0)} | j)|} + \frac{\left(\log \dfrac{N-1}{P_e} - 1\right)}{E(z^{(0)} | 0)} \tag{2.6.8}$$

(Compare this result with that in Table 2.1 under the same condition.) The term P_e, of course, is an upper bound on the actual error probability since Equation (2.5.3) was derived for the case in which $P_j = \beta$ for all $j \neq 0$.

At the other extreme, when both α and β are small, ($\alpha = P_e/2$, $\beta = P_e/(N - 1)$) an approximate expression for $E(M_T)$ follows from the observation that

$$E(M_i | \nu \geq i) \approx \sum_{j=i}^{N} E(M_i | j) P(j | \nu \geq i) \approx \sum_{j=i}^{N} \frac{1}{N + 1 - i} E(M_i | j) \tag{2.6.9}$$

j again denoting the index of the correct state, and $P(j | \nu \geq i)$ the probability of the jth state being the correct one conditioned by the event $\{\nu \geq i\}$. This approximation is valid because, with high probability, a state is accepted if, and only if, it is the correct one. For the same reason, $P(\nu \geq i) \approx (N + 1 - i)/N$, $i = 1, 2, \ldots, N$. Combining Equations (2.6.4) and (2.6.9), and using this last approximation, we find

$$E(M_T) \approx \frac{1}{N} \sum_{i=1}^{N} \sum_{j=i}^{N} E(M_i | j) \tag{2.6.10}$$

This expression can be further simplified when the distribution $p\,(z^{(i)}| j)$ is a function only of the magnitude of $i - j$, modulo N, for then

$$E(M_T) \approx \frac{1}{N} \sum_{k=0}^{N-1} \sum_{i=1}^{N-k} E(M_0 | k) = \sum_{k=0}^{N-1} \frac{N-k}{N} E(M_0 | k) \tag{2.6.11}$$

Finally, since $P_0 = 1 - \alpha$ and $P_j \leq \beta$ for all $j \neq 0$, and since α and β are both small, ($\alpha = P_e/2$, $\beta = P_e/(N-1)$, $P_e \ll 1$).

$$E(M_0 \mid j) = \frac{P_j \log A + (1 - P_j) \log B}{E(z^{(0)} \mid j)} \approx \begin{cases} \dfrac{\log \dfrac{2}{P_e}}{|E(z^{(0)} \mid j)|} & j \neq 0 \\[2ex] \dfrac{\log \dfrac{N-1}{P_e}}{E(z^{(0)} \mid 0)} & j = 0 \end{cases}$$

and

$$E(M_T) \approx \log \frac{2}{P_e} \sum_{k=1}^{N-1} \frac{N-k}{N} \frac{1}{|E(z^{(0)} \mid k)|} + \frac{\log \dfrac{N-1}{P_e}}{E(z^{(0)} \mid 0)}$$

$$= \frac{1}{2} \log \frac{2}{P_e} \sum_{k=1}^{N-1} \frac{1}{|E(z^{(0)} \mid k)|} + \frac{\log \dfrac{N-1}{P_e}}{E(z^{(0)} \mid 0)} \qquad (2.6.12)$$

the last expression following because, under the assumptions here, $E(z^{(0)} \mid k) = E(z^{(0)} \mid N - k)$. Again, of course, P_e is an upper bound on the actual error probability.

2.7 On the Optimum Sequential Search

It might be supposed that to minimize the search time the probability α of an erroneous dismissal should be kept small in order to accept the correct state with high probability once it is encountered (or in the case of the continual search in order not to reject it). Actually, Table 2.1 suggests that quite the opposite is true. Upon differentiating the expected search time with respect to α, in fact, one discovers it to be a monotonically decreasing function of α for all values of $N (N \geq 2)$ for both the conventional and the continual sequential searches. The expected value of M_T is plotted in Figure 2.1 as a function of α when $N\mu_0/\mu_1$ is large and P_e is small as compared to unity (see Table 2.1). Note that under this same condition the expected search time is virtually identical using either the conventional or the continual search algorithm.

It must be emphasized that as $\alpha \to 1$, the number of times each state is tested in the course of the search approaches infinity. The search time remains finite because the amount of time spent investigating each state simultaneously approaches zero. Nevertheless, this fact imposes several obvious limitations on this procedure. First of all, the observables were tacitly assumed to be continuous in nature; each state could be tested until the moment its probability ratio dropped below α and the observation instantly stopped at that point. In practice, the observables are generally discrete (M is an integer) and

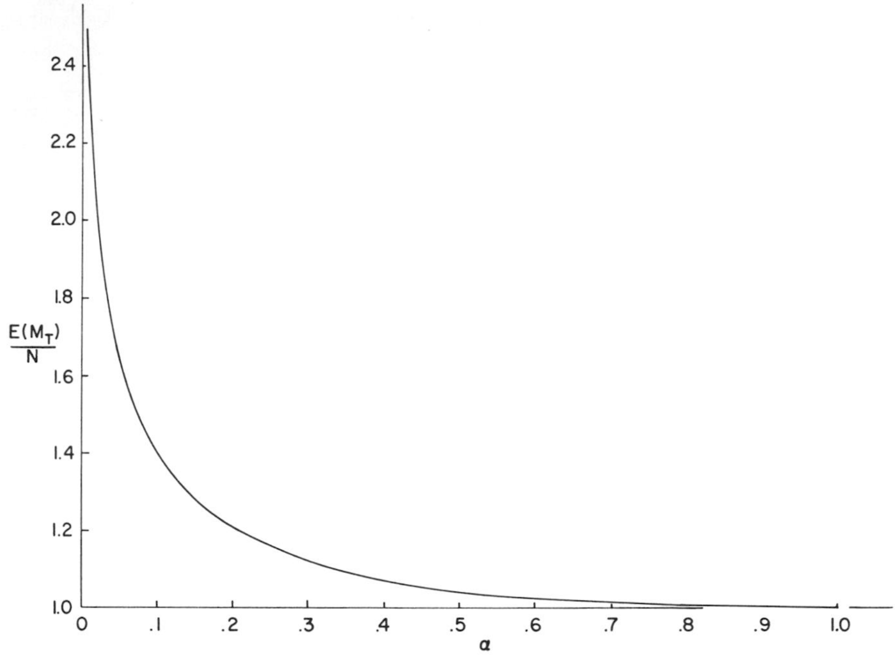

Figure 2.1 Expected sequential search time as a function of α.

hence at least one observable must be available before a decision can be made. This quantization effect will prevent the number of observations for each state from approaching zero as $\alpha \to 1$. Further, it was assumed that no time was consumed in passing from the investigation of one state to the next. Suppose t_0 seconds were spent in making each observation and t_1 seconds were actually necessary to shift from one state to the next. Then the expected search time would be

$$E(T_s) = E(\nu - 1)t_1 + E(M_T)t_0 \tag{2.7.1}$$

As $\alpha \to 1$, the first term becomes infinite regardless of how small t_1 is as long as it is not zero. Nevertheless, the results when $\alpha \to 1$ are still useful in providing bounds on how well an ideal scheme might be expected to work. Moreover, a useful test still results even when α is somewhat less than one; when, as a consequence, the amount of time spent testing each state does not approach zero, and the number of such tests does not approach infinity (cf. Equation 2.5.5). Both of the above mentioned objections are overcome, at least to some extent, when α is somewhat less than one. Note if the amount of time t_1 necessary to shift the investigation from one state to another is known, then it would in fact be possible to determine an optimum value for α. Often, however, the value of t_1 will be sufficiently small so that the optimum value of α satisfying this equation will still be too close to unity to justify neglecting the quantization effect mentioned earlier. In any event, the increased time necessitated by decreasing α below unity is not excessive as can be seen from Figure 2.1.

The average number of observables needed for each test can be determined by using the results of Section 2.3. As this number approaches, say, ten, it can reasonably be concluded that the quantization effect is no longer significant.

Subject to these reservations, the expected time needed to complete a sequential search is minimized by letting the probability of a false dismissal approach one. In this sense, therefore, the sequential search with $\alpha = 1$ is the optimum of this class of search procedures. But can this method be said to be optimum in any larger sense? More generally, if only one state can be tested at a time, what search algorithm minimizes the expected search time?

Although a completely general answer to this question seems elusive, a promising, but not always successful, contender for the optimum strategy is the following (cf. Problem 2.7): always test the state having the *currently* greatest probability of being the correct one. Suppose the probabilities of the μth state θ_μ and of the νth state θ_ν are at some point in the search P_μ and P_ν, respectively, with $P_\mu = P_\nu \geq P_i$ for all $i \neq \mu, \nu$. We then make the observation $y_j^{(\nu)}$ pertaining to the νth state. Then, if

$$p(y_j^{(\nu)} \mid \theta_l) = \begin{cases} p_0(y_j^{(\nu)}) & l = \nu \\ p_1(y_j^{(\nu)}) & l \neq \nu \end{cases}$$

the new probability of the νth state is

$$\begin{aligned} P'_\nu &= \frac{p_0(y_j^{(\nu)})P_\nu}{p_0(y_j^{(\nu)})P_\nu + p_1(y_j^{(\nu)})(1 - P_\nu)} \\ &= \frac{P_\nu}{P_\nu + \dfrac{1}{\lambda_{\nu j}}(1 - P_\nu)} \end{aligned} \tag{2.7.2}$$

where $\lambda_{\nu j} = p_0(y_j^{(\nu)})/p_1(y_j^{(\nu)})$ is the likelihood ratio. If P'_ν is less than P_ν, then the μth state becomes the most probable and if this rule of always testing the currently most probable state is to be implemented the investigation of the νth state will cease for the present. But P'_ν is less than P_ν if, and only if, $\lambda_{\nu j}$ is less than one. If $\lambda_{\nu j}$ is greater than one, the νth state retains its position as the most probable state and its investigation continues. It continues, in fact, until the product $\lambda_{\nu j}\ \lambda_{\nu, j+1}\ \lambda_{\nu, j+2} \ldots \lambda_{\nu, j+l}$ drops below unity. The condition that the search always investigates the most probable state is obtained, therefore, in a sequential search with the lower threshold $B = \alpha/(1 - \beta)$ equal to one. Accordingly, the sequential search with $\alpha = 1 - \beta$ [and, consequently, from Equation (2.5.3) with $\alpha \approx 1$] can be regarded as an algorithm for implementing this intuitively appealing strategy.

2.8 Multiple-Hypothesis Tests

If computational or other limitations do not preclude the possibility, it is often advantageous to make a multiple-hypothesis test, simultaneously testing all of the contending hypotheses, rather than engaging in a generally more time consuming search. In this section, we shall discuss briefly some of the aspects of multiple-hypothesis tests of both the fixed-sample-size and the sequential type. Again, the discussion will be limited to simple hypotheses only. The generalization to composite hypotheses is in most cases straightforward.

2.8.1 Fixed-Sample-Size Tests

If the test is to last for a predetermined length of time (i.e., if a fixed number of observations are to be made), the optimum test is easily ascertained by generalizing the results of Section 2.2. Let C_{ij} again represent the cost of accepting the hypothesis $H_i(\theta = \theta_i)$ when the true hypothesis is $H_j(\theta = \theta_j)$. Then the expected cost of a decision is

$$E(\text{cost}) = \sum_j \sum_i C_{ij} P(H_j) \int_{R_i} p(Y \mid H_j)\, dY \tag{2.8.1}$$

where R_i is the region of acceptance of the hypothesis H_i. By using Bayes' rule, and interchanging the order of integration and summation, we obtain

$$E(\text{cost}) = \sum_i \int_{R_i} \sum_j C_{ij} P(H_j \mid Y) p(Y)\, dY \tag{2.8.2}$$

The expected cost is therefore minimized by assigning Y to the region R_ν if, and only if,

$$\sum_j C_{ij} P(H_j \mid Y) \tag{2.8.3}$$

attains its minimum value when $i = \nu$. Thus, the νth hypothesis H_ν is accepted if, and only if,

$$\frac{\sum_j C_{\nu j} P(H_j \mid Y)}{\sum_j C_{ij} P(H_j \mid Y)} = \frac{\sum_j C_{\nu j} p(Y \mid H_j) P(H_j)}{\sum_j C_{ij} p(Y \mid H_j) P(H_j)} < 1 \tag{2.8.4}$$

for all $i \neq \nu$. This is the *Bayes test* in the case of multiple hypotheses [cf. Equation (2.2.4)].

One again encounters difficulties in determining the optimum test when either the costs or the a priori probabilities $P(H_i)$ or both are unknown. We shall restrict our attention to the case in which the a priori probabilities are all known, and in which the costs of all erroneous decisions are equal:

$$C_{ij} = \begin{cases} C_0 & i = j \\ C_1 & i \neq j \end{cases}$$

Then we wish to minimize the expression

$$\sum_j C_{ij} P(H_j \mid Y) = C_1 \sum_j P(H_j \mid Y) - (C_1 - C_0) P(H_i \mid Y)$$

$$= C_1 \sum_j P(H_j \mid Y) - (C_1 - C_0) \frac{p(Y \mid H_i) P(H_i)}{P(Y)} \tag{2.8.5}$$

The cost is minimized (if $C_1 > C_0$) by choosing that hypothesis H_i for which the probability

$$p(Y \mid H_i) P(H_i) \tag{2.8.6}$$

attains its maximum. This, of course, is the *maximum a posteriori probability* test for multiple hypotheses.

Denoting by ξ_ν the random variable $p(Y \mid H_\nu) P(H_\nu)$, we can express the probability of an error in making a decision of this kind as

$$P_e = \sum_{\mu=1}^{N} P(H_\mu) P_e(\mu) \tag{2.8.7}$$

where

$$P_e(\mu) = 1 - \Pr\left\{\xi_\mu > \max_{\nu \neq \mu} \xi_\nu \mid H_\mu\right\}$$

$$= 1 - \int_0^\infty \int_0^{\xi_\mu} \cdots \int_0^{\xi_\mu} p(\xi_1, \xi_2, \ldots, \xi_N \mid H_\mu)\, d\xi_1\, d\xi_2 \ldots d\xi_N\, d\xi_\mu$$

The integrand is the conditional joint probability density function of the random variables ξ_ν. Since these variables are generally not independent, an exact evaluation of the error probability is often difficult. Nevertheless, it is possible to bound this probability. One such bound is afforded by noting that

$$\max_{i \neq \mu} \Pr\{\xi_i - \xi_\mu \geq 0 \mid H_\mu\} \leq P_e(\mu) \leq \sum_{\substack{i=1 \\ i \neq \mu}}^{N} \Pr\{\xi_i - \xi_\mu \geq 0 \mid H_\mu\}$$

And, since $\Pr\{\xi_i - \xi_\mu \geq 0 \mid H_\mu\} = \Pr\{\xi_\mu / \xi_i \leq 1 \mid H_\mu\}$ (the terms ξ_i are probability densities and hence cannot be negative), we have

$$\max_{i \neq \mu} \int_{\lambda_{\mu i} \leq r_{\mu i}} p(Y \mid H_\mu)\, dY \leq P_e(\mu)$$

$$\leq \sum_{\substack{i=1 \\ i \neq \mu}}^{N} \int_{\lambda_{\mu i} \leq r_{\mu i}} p(Y \mid H_\mu)\, dY \leq N \max_{i \neq \mu} \int_{\lambda_{\mu i} \leq r_{\mu i}} p(Y \mid H_\mu)\, dY \tag{2.8.8}$$

where

$$\lambda_{\mu i} = \frac{p(Y \mid H_\mu)}{p(Y \mid H_i)}$$

and

$$r_{\mu i} = \frac{P(H_i)}{P(H_\mu)}$$

2.8.2 Sequential Tests

In order to be able to make a sequential multiple hypothesis decision, we must somehow define a probability ratio of the form

$$\frac{p(Y(m)\,|\,H_j)}{p(Y(m)\,|\,H_j^*)} \tag{2.8.9}$$

for each hypothesis H_j. The alternative hypothesis H_j^* may or may not be one which could actually occur. The probability density $p(Y\,|\,H_j^*)$ might, for example, be some weighted average of the density functions corresponding to all the alternative hypotheses:

$$p(Y(m)\,|\,H_j^*) = \sum_{i \neq j} w_i(m) p(Y(m)\,|\,H_i) \tag{2.8.10}$$

where $\sum_{i \neq j} w_j(m) = 1$. We have indicated a dependence of the weights w_i on the number of observables in the set $Y(m)$. A test to be examined momentarily involves simultaneously testing each hypothesis H_j until all but one are rejected; hence one form this dependence might take is in the requirement that $w_i(m) = 0$ if H_i has already been rejected. In the multiple-hypothesis sequential test just mentioned, all the ratios (2.8.9) are observed simultaneously, and the hypothesis H_j is rejected as soon as the corresponding ratio drops below some threshold B_j. It follows that

$$\alpha_\mu = \sum_{m=1}^{\infty} \int_{\bar{R}_\mu(m)} p(Y(m)\,|\,H_\mu)\,dY(m) = B_\mu \sum_{m=1}^{\infty} \int_{\bar{R}_\mu(m)} p(Y(m)\,|\,H_\mu^*)\,dY(m)$$

where $\bar{R}_\mu(m)$ is the rejection region of the μth hypothesis, as defined by the ratios (2.8.9). Regardless of the alternative hypothesis H_μ^*, the events indicated in the summation on the right side of this expression are mutually exclusive; if $Y(m)$ is in $\bar{R}_\mu(m)$, then $Y(l)$ could not have been in $\bar{R}_\mu(l)$ for any $l < m$ or the test of the μth state would have ceased before the mth observation was made. Therefore this sum cannot exceed unity,

$$\alpha_\mu \leq B_\mu \tag{2.8.11}$$

and the probability of erroneously rejecting the hypothesis H_μ is bounded by the value of the threshold B_μ. Unfortunately, it is generally difficult to determine the expected number of observables needed to complete a multiple-hypothesis sequential test. Most results in this area are obtained empirically.

2.9 Estimation of Statistical Parameters

The essential difference between the estimation of statistical parameters and the testing of statistical hypotheses lies in the fact that there are presumably only a discrete set of hypotheses, while the parameters can generally take on a continuum of values. Consider the problem of estimating the single parameter x on the basis of the observables Y. To extend the remarks concerning the expected cost of a multiple-hypothesis decision to the estimation problem, we need only define a cost *function* $C(\hat{x}, x)$, the cost of accepting the estimate $\hat{x} = \hat{x}(Y)$ when the true parameter is actually x. Then by adapting equation (2.8.2) to the present context, we have

$$E(\text{cost}) = \int E(\text{cost} \mid Y)p(Y)\, dY \tag{2.9.1}$$

where

$$E(\text{cost} \mid Y) = \int C(\hat{x}, x)p(x \mid Y)\, dx$$

Thus, for every Y for which $p(Y) \neq 0$ we wish to define an estimate $\hat{x}(Y)$ such that $E(\text{cost} \mid Y)$ is minimized. Such estimates are referred to as *Bayes estimates.*

One of the most commonly used cost functions is the square of the error of the estimate:

$$C(\hat{x} - x) = (\hat{x} - x)^2 \tag{2.9.2}$$

It is easy to see that when this is the cost function, the optimum estimate $\hat{x}$ of x is its conditional expectation

$$\hat{x} = E(x \mid Y) \tag{2.9.3}$$

For consider any other estimate $\tilde{x}$; then

$$\begin{aligned} E(\text{cost} \mid Y) &= E((\tilde{x} - x)^2 \mid Y) = E[((\tilde{x} - \hat{x}) - (x - \hat{x}))^2 \mid Y] \\ &= (\tilde{x} - \hat{x})^2 + E[(x - \hat{x})^2 \mid Y] \geq E[(x - \hat{x})^2 \mid Y] \end{aligned}$$

the equality holding only when $\tilde{x} = \hat{x}$.

Actually, it can be shown that subject to some rather general conditions on the density function $p(x \mid Y)$, the conditional expectation $E(x \mid Y)$ is the optimum estimator† for a fairly large class of cost functions (Reference 2.6). In particular, this statement holds if the cost function is a symmetric function of the difference $\hat{x} - x$, and a non-decreasing function of its magnitude, and if $p(x \mid Y)$ is unimodal and symmetric about its mean. But if $p(x \mid Y)$ is unimodal and symmetric about its mean, then clearly,

$$\max_x p(x \mid Y) = p(\hat{x} \mid Y) \tag{2.9.4}$$

where $\hat{x} = E(x \mid Y)$; i.e., the conditional mean of x is equal to its conditional mode. Accordingly, the optimum estimator, under these conditions, is the *maximum a posteriori probability estimator.*

As in the testing of hypotheses, the a priori probabilities may not be known, and, as there, the choice of a criterion for an estimator in the absence of this information is less obvious. Two measures commonly used in the

†The term *estimator* will be used to denote the function $\hat{x} = \hat{x}(Y)$. An *estimate* of x is the value of $\hat{x}$ for any specific set Y.

evaluation of estimators are: (1) The *bias* of the estimate. If $\hat{x}$ is an estimate of x, and $E(\hat{x}\,|\,x) = x + b(x)$, $b(x)$ is called the bias of the estimate $\hat{x}$. (2) The conditional variance $E[(\hat{x} - x - b(x))^2\,|\,x]$ of the estimate. It is shown in Appendix A.2 that any estimate $\hat{x}$ of x based on the set of independent, identically distributed observables $Y = \{y_i\}$, $i = 1, 2, \ldots, N$, has a variance underbounded by

$$E[(\hat{x} - x - b(x))^2\,|\,x] \geq \frac{\left(1 + \dfrac{db(x)}{dx}\right)^2}{NE\left[\left(\dfrac{\partial \log p(y\,|\,x)}{\partial x}\right)^2 \,\middle|\, x\right]} \tag{2.9.5}$$

where $p(y\,|\,x) = p(y_i\,|\,x)$ is the conditional density function of the observables y_i. This result is known as the Cramér-Rao inequality.

Most generally, one would prefer an estimate which is unbiased, and of all unbiased estimates, has a variance attaining the lower bound of Equation (2.9.5). Such estimates are called *efficient* estimates. The *maximum-likelihood estimate*, the value (or values)† of x satisfying the *likelihood equation*

$$\frac{\partial}{\partial x} \log p(Y\,|\,x) = 0 \tag{2.9.6}$$

has two rather attractive properties in this regard. First, if an efficient estimate exists, it is the maximum-likelihood estimate. Second, even if an efficient estimate does not exist, a maximum-likelihood estimate does exist which is *asymptotically efficient* as the number of observables becomes large. Both of these statements are also proved in Appendix A.2, subject to some quite general conditions.

The *maximum-likelihood principle* used throughout the earlier sections of this chapter would, in the present context, suggest selecting as the estimate of x that value which maximizes, over the allowable range of x, the function $p(Y\,|\,x)$ (or, equivalently, $\log p(Y\,|\,x)$). While this estimate will often be a solution to the likelihood equation, it of course need not be (e.g., the likelihood equation may not have a solution over the prescribed range of x). Moreover, the likelihood equation may have solutions which do not maximize $p(Y\,|\,x)$. In order to distinguish between these two estimators in the sequel, we shall refer to any estimator based on the maximum of the function $p(Y\,|\,x)$ as a *maximum-likelihood-principle* estimator and reserve the term maximum-likelihood estimator for those estimators indicated by Equation (2.9.6).

Extension to the more general case in which several parameters are to be estimated simultaneously is, for the most part, straightforward. The unknown parameter x now becomes a vector $\mathbf{x} = (x_1, x_2, \ldots, x_m)$. The maximum a posteriori probability estimates of the parameters $x_1, x_2, \ldots, x_m$,

†It sometimes happens that $p(Y\,|\,x)$ is maximized by some x which is independent of Y. Such estimates are usually discarded in favor of some other estimate $\hat{x} = \hat{x}(Y)$ which is a function of Y and for which $p(Y\,|\,x)$ attains, at least, a relative maximum.

for example, given the observables $Y = \{y_i\}$ are those values simultaneously maximizing the expression

$$p(\mathbf{x} \mid Y) = p(x_1, x_2, \ldots, x_m \mid Y) \tag{2.9.7}$$

while the maximum-likelihood estimates are those values of x_i satisfying†

$$\frac{\partial}{\partial x_i} p(Y \mid x_1, x_2, \ldots, x_m) = 0 \qquad i = 1, 2, \ldots, m \tag{2.9.8}$$

Notes to Chapter Two

The foundations of statistical decision theory can be traced to the work of Neyman and Pearson (References 2.7 and 2.8) and much of the subsequent development to Wald (References 2.9 and 2.10). Many of the concepts, and much of the terminology of estimation theory originated with the even earlier work of Fisher (Reference 2.11). The material on sequential detection can be attributed almost entirely to Wald (Reference 2.2). For more comprehensive treatments of these and related topics, as well as extensive bibliographies, the reader is referred to the books by Cramèr (Reference 2.1), Wilks (Reference 2.12), and Lehman (Reference 2.13). In contrast to most of the material in this chapter, the search problem, at least as presented here, has received attention only recently. The "most probable state" search algorithm first was investigated (although from a rather different approach than that used here) by Posner and Rumsey (Reference 2.14) and the "continual" search algorithm by Gumacos (Reference 2.5).

Problems

2.1 Let S be a binary source, its output consisting of a sequence of statistically independent *zeros* and *ones* with $\Pr(0) = 1 - \Pr(1) = p$. It is to be determined, on the basis of six consecutive outputs from S, whether it is in state $\theta_0 : p = \frac{2}{3}$ or in state $\theta_1 : p = \frac{1}{2}$. Show that, in each of the following situations, the appropriate test is to accept the hypothesis H_0, that $\theta = \theta_0$, if and only if the number

---

†The minimum possible variance of the estimate of a particular parameter ϕ_i may be larger when several parameters are to be estimated simultaneously than it would be were it the sole unknown parameter. In particular, if two parameters ϕ_1 and ϕ_2 are to be estimated by the maximum-likelihood method, the asymptotic variances of the two estimates will both be larger than the value given in Equation (2.9.5) unless the two parameters are inpendent in the sense that (cf. reference 2.1)

$$E\left(\frac{\partial \log p(y \mid \phi_1, \phi_2)}{\partial \phi_1} \frac{\partial \log p(y \mid \phi_1, \phi_2)}{\partial \phi_2}\right) = 0$$

of observed *zeros* exceeds some integer ν, and determine ν in each case. (The notation is as defined in Section 2.2.)

(a) $P(H_0) = P(H_1) = \frac{1}{2}$; $C_{01} - C_{11} = C_{10} - C_{00}$.
(b) $P(H_0) = P(H_1) = \frac{1}{2}$; $C_{01} - C_{11} = 1$, $C_{10} - C_{00} = 10$.
(c) $P(H_0) = P(H_1) = \frac{1}{2}$; $C_{01} - C_{11} = 10$, $C_{10} - C_{00} = 1$.
(d) α is not to exceed $\frac{1}{9}$.
(e) $C_{11} = C_{00} = 0$, $C_{01} = 1$, $C_{10} = 10$; the a priori probabilities are unknown.
Determine α and β in each case.

2.2 (a) Describe a sequential probability ratio test for determining the state of the source defined in the previous problem. Find the mean and variance of the number of source outputs needed for a sequential decision with $\alpha = \beta = \frac{1}{3}$. (Note that under these circumstances, the excesses at the boundaries are not insignificant so these answers will be only approximate.) (b) Suppose the source is actually emitting *zeros* with the probability $\frac{5}{12}$. Find the probability of accepting H_0 under these conditions and determine the expected number of observations needed to complete the test.

2.3 What conditions does the minimax criterion place on the decision thresholds in a sequential test?

2.4 Using Wald's identity (cf. Appendix A.1), show that, when $E(z\,|\,\theta) = 0$, $\operatorname{Var}(z\,|\,\theta) \neq 0$, Equations 2.4.8 and 2.4.9 become, respectively,

$$P(\theta) = \frac{\log B}{\log B - \log A}$$

and

$$E(M\,|\,\theta) = \left|\frac{\log A \log B}{\operatorname{Var}(z\,|\,\theta)}\right|$$

2.5 The state θ of a source is to be determined by a sequential search. The distribution of the observables z is known to be $p_0(z)$ if the source state is the state actually being tested, and $p_1(z)$ otherwise. The situation differs from that discussed in Section 2.6 in that the a priori probability that the source is in state θ_i is P_0 for all i, with P_0 an arbitrary probability not exceeding $1/N$. The search is designed to determine the correct source state, when it is actually one of those being tested, with an error probability P_e. Show that, as $\alpha \to 1$, $\beta \to 0$, subject to the condition imposed by Equation (2.5.1), the expected number of observations needed to complete the search is (cf. Table 2.1)

$$E(M_T; P_0) = E\left(M_T; \frac{1}{N}\right)$$
$$+ (1 - P_e)(1 - NP_0)\left\{\frac{A^* - 1 - \log A^*}{\mu_1} - \frac{\log A^* - \dfrac{A^* - 1}{A^*}}{\mu_0}\right\}$$

Show that the term on the right is positive provided that the "quantization effect" mentioned on page 36 is negligible.

2.6 (a) The search described in the preceding problem is to be modified by concluding it either when $P_i(M) \geq P_c$, some i, or when $P_i(M) \leq P_r$, all i, $i = 1, 2, \ldots, N$, where $P_i(M)$ is the a posteriori probability after M observations that the source is in state i. Show that, if the search is designed to test the most probable state at every instant, the expected number of observations required is

$$E(M_T; P_0, P_r, P_c) = \int_{P_r}^{P_0} E(M_T; x)p(x)\,dx$$

where $E(M_T; x)$ is as defined in Problem 2.5, but with $A^* \to A^*(x) = (1 - x)P_c/x(1 - P_c)$, $P_e = 1 - P_c$, and where $p(x) = dP(x)/dx$, with

$$P(x) = \exp\left\{-\int_x^{P_0} \frac{(1-\xi)}{P_c - \xi}\frac{N}{1 - N\xi}\,d\xi\right\}$$

[Hint: Consider a series of truncated sequential searches, the ith search conducted with upper and lower thresholds A_i and B_i, respectively, and truncated not only when one test statistic reaches A_i but also when all N test statistics reach B_i. Show that the expected number of observations needed to complete such a search is just $1 - \Pr(B_i \mid A_i, B_i)$ times the expected number needed to complete the corresponding untruncated search, where $\Pr(B_i \mid A_i, B_i)$ is the probability that all N test statistics reach the lower threshold. Further, show that if P_i denotes the a posteriori probability of each of the tested states at the end of the ith truncated search, if $A_i \triangleq (1 - P_{i-1})P_c/P_{i-1}(1 - P_c)$, $B_i \triangleq (1 - P_{i-1})(P_{i-1} - \Delta_{i-1})/P_{i-1}(1 - P_{i-1} + \Delta_{i-1})$, and if $\Delta_i = (1 - P_i)\Delta/(1 - NP_i)$, then, as $\Delta \to 0$, $P_i \to P_0 - i\Delta$. The stated result then follows upon taking the limit as $\Delta \to 0$ of a weighted sum of the expected number of observations needed to complete each successive truncated search, the weighting determined by the probability that that search is need to reach a decision.]

(b) Using the result of part (a), derive an expression for the expected number of observables needed to conclude a sequential search when the a priori probabilities of the different states are not equal.

2.7 One method for determining whether a source is in state θ_0 or in state θ_1, when the a priori probability of each state being correct is $\frac{1}{2}$, is simply to test one of the states until its a posteriori probability reaches either P_c or $1 - P_c$. Compare this strategy, in terms of the expected number of observations needed to conclude the test, with the sequential search strategy discussed in Section 2.7. Under what conditions is the latter superior? Is it ever more efficient to adapt a strategy of always testing the *less* probable of the two states?

2.8 Let $Y = \{y_i\}$ be a set of N independent samples of a random variable y, with $p(y) = (1/\sqrt{2\pi}\,\sigma)\exp\{-(y - \mu)^2/2\sigma^2\}$, but with both μ and σ unknown. Find the maximum-likelihood estimates of μ, σ, and σ^2. Are these estimates biased? Are they efficient?

References

2.1 H. Cramér, *Mathematical Methods of Statistics*, Princeton University Press, Princeton, N. J., 1946.

2.2 A. Wald, *Sequential Analysis*, John Wiley & Sons, Inc., New York, N. Y., 1947.

2.3 A. Wald, and J. Wolfowitz, "Optimum Character of the Sequential Probability Ratio Tests," *Ann. Math. Stat.*, **19**, p. 326, 1948.

2.4 E. C. Posner, "Optimal Search Procedures," *IRE Trans. on Inf. Theory*, IT-11, p. 157, 1963.

2.5 C. Gumacos, "Analysis of an Optimum Sync Search Procedure," *IEEE Trans. on Comm. Syst.*, CS-11, p. 89, 1963.

2.6 A. J. Viterbi, *Principles of Coherent Communication*, McGraw-Hill Book Co., New York, N. Y., 1967.

2.7 J. Neyman, and E. S. Pearson, "On the Use and Interpretation of Certain Test Criteria for Purposes of Statistical Inference," *Biometrika*, **20A,** Part I, p. 175; Part II, p. 268, 1928.

2.8 ———, "On the Problem of the Most Efficient Tests of Statistical Hypotheses," *Phil. Trans. Royal Soc.*, London, Ser. A., 231, p. 289, 1933.

2.9 A. Wald, "Contributions to the Theory of Statistical Estimation and Testing of Hypotheses," *Ann. Math. Stat.*, **10,** p. 299, 1939.

2.10 ———, *Statistical Decision Functions*, John Wiley & Sons, Inc., New York, N. Y., 1950.

2.11 R. A. Fisher, "On the Mathematical Foundations of Theoretical Statistics," *Phil. Trans. Royal Soc.*, London, Ser. A., 222, p. 309, 1922.

2.12 S. S. Wilks, *Mathematical Statistics*, John Wiley & Sons, Inc., New York, N. Y., 1962.

2.13 E. L. Lehmann, *Testing Statistical Hypotheses*, John Wiley & Sons, Inc., New York, N. Y., 1959.

2.14 E. C. Posner, and H. Rumsey, Jr., "Continuous Sequential Decision in the Presence of a Finite Number of Hypotheses," *IEEE Trans. on Inf. Theory*, IT-12, p. 248, 1966.

Supplementary Bibliography

Chu, W. W., "Optimal Adaptive Search," Tech. Rept. No. 6252-1, Stanford Electronics Labs., Stanford, Calif., 1966.

Davenport, W. B., Jr., and W. L. Root, *An Introduction to the Theory of Random Signals and Noise*, McGraw-Hill Book Co., New York, N. Y., 1958.

Deutsch, R., *Estimation Theory*, Prentice-Hall, Inc., Englewood Cliffs, N. J., 1965.

Gilbert, E. N., "Optimal Search Strategies," *J. Soc. Indus. Appl. Math.*, **7**, p. 413, 1959.

Koopman, B. O., "The Optimum Distribution of Searching Effort," *Operations Res. Soc. America*, **5,** p. 613, 1957.

Selin, I., *Detection Theory*, Princeton University Press, Princeton, N. J., 1965.

Sherman, S., "Non-Mean-Square Error Criteria," *IRE Trans. Inf. Theory* IT-4, p. 125, 1958.

Van Trees, H. L., *Detection, Estimation and Modulation Theory, Part I*, John Wiley & Sons, Inc., New York, N. Y., 1968.

chapter three

Amplitude-Continuous, Time-Discrete Communication Systems

3.1 Introduction

In this and the following chapter are discussed some of the communication systems of greatest interest from the synchronization point of view, systems which will be collectively referred to as *pulse* or *time-discrete communication systems*. In such systems, one of a set of waveforms is transmitted over each time interval $iT_s < t < (i + 1)T_s$, the particular transmitted waveform being determined by the information source.

Pulse communication systems can be subdivided into two categories, depending upon whether a finite (or countably infinite) set of waveforms is to be used or whether the set is uncountably infinite (e.g., the set of waveforms may be a family of time functions dependent upon some continuous parameter). This second category of pulse communication systems, the *amplitude-continuous* systems, is the subject of this chapter. The first category, commonly called digital communication systems, will be deferred to Chapter Four.

The communication systems to be considered in the present chapter include coherent pulse-amplitude-modulation, incoherent pulse-amplitude-modulation, and coherent phase-shift-keyed modulation. In all of these systems, the received signal will be assumed to be of the form

$$y(t) = Ax(t - iT_s, v_i) + n(t) \qquad iT_s < t < (i + 1)T_s \qquad (3.1.1)$$

where v_i is a parameter representing the information to be transmitted and $n(t)$ denotes white, Gaussian noise with the single-sided noise spectral density N_0. The source is assumed stationary, so the statistics of the signal $x(t - iT_s, v_i)$ are independent of i. We will proceed to determine the form of the optimum demodulator and to analyze the resultant performance of these systems. In order to do this, we obviously need some criterion for optimality, and some basis on which different systems can be compared. This is discussed in Section 3.3. Even more basic to the discussion of pulse modulation systems is the presupposed capability of satisfactorily representing the information by a sequence of numerical values at the average rate of one number every T_s seconds. If the information is already of this time-discrete form it presents no fundamental problem. But if it is inherently a continuous function of time, it must somehow be converted to such a sequence, hopefully with a minimal loss of information in the process. One method for accomplishing this is discussed in the following section.

3.2 The Sampling Theorem

Let $f(t)$ represent a sample function of a stationary random process and denote by $b_\tau(t - \eta)$ the periodic function depicted in Figure 3.1. The pulses, each τ seconds wide, are centered about the time instants $t = \eta + iT$, $i = \ldots, -1, 0, 1, 2, \ldots$, where η is a random variable with the density function $p(\eta)$:

$$p(\eta) = \begin{cases} \dfrac{1}{T} & 0 \leq \eta \leq T \\ 0 & \text{otherwise} \end{cases} \tag{3.2.1}$$

Since $b_\tau(t - \eta)$ is periodic, it can be expanded in a Fourier series yielding

$$b_\tau(t - \eta) = a_0 + 2 \sum_{n=1}^{\infty} a_n \cos \omega_n(t - \eta) \tag{3.2.2}$$

where

$$a_0 = 1$$

$$a_n = \frac{\sin \omega_n \tau}{\omega_n \tau}$$

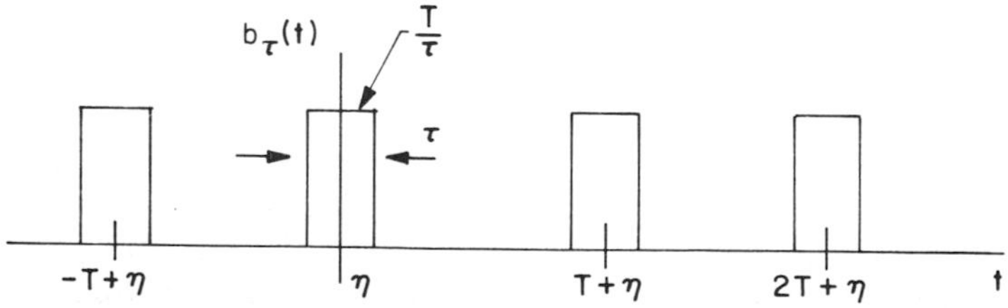

Figure 3.1 Periodic sampling function.

and

$$\omega_n = \frac{2\pi n}{T}$$

Now consider the product $g(t) = f(t)b_\tau(t - \eta)$. If the process $f(t)$ is independent of η, the autocorrelation function of the product is just the product of the autocorrelation functions

$$\begin{aligned} R_g(\gamma) &= E\{f(t)f(t+\gamma)\, b_\tau(t-\eta) b_\tau(t-\eta+\gamma)\} \\ &= R_f(\gamma) R_b(\gamma) \end{aligned} \tag{3.2.3}$$

But

$$R_b(\gamma) = a_0^2 + 2\sum_{n=1}^{\infty} a_n^2 \cos \omega_n \gamma \tag{3.2.4}$$

Consequently, the spectral density of the product is

$$\begin{aligned} S_g(\omega) &= \int_{-\infty}^{\infty} R_g(\gamma)\, e^{-j\omega\gamma}\, d\gamma \\ &= \sum_{n=-\infty}^{\infty} a_n^2 S_f(\omega - \omega_n) \end{aligned} \tag{3.2.5}$$

where we have defined $a_{-n} = a_n$. If the power spectral density $S_f(\omega)$ of the process $f(t)$ is as illustrated in Figure 3.2(a) then $g(t)$ has the spectral density shown in Figure 3.2(b).

The important conclusion from this observation is that if the spectral density $S_f(\omega)$ vanishes for all frequencies $|\omega/2\pi| \geq W$ as implied in Figure 3.2, then none of the terms on the right side of Equation (3.2.5) overlap in frequency so long as $W \leq 1/2T$. It is therefore possible, in principle, to pass the process $g(t)$ through a filter having the transfer function $H(j\omega) = \begin{cases} 1, & |\omega| < 2\pi W \\ 0, & |\omega| > 2\pi W \end{cases}$ leaving only the low-frequency term $a_0^2 S_f(\omega) = S_f(\omega)$. But this is just the spectral density of the process $f(t)$. It would seem that no information has

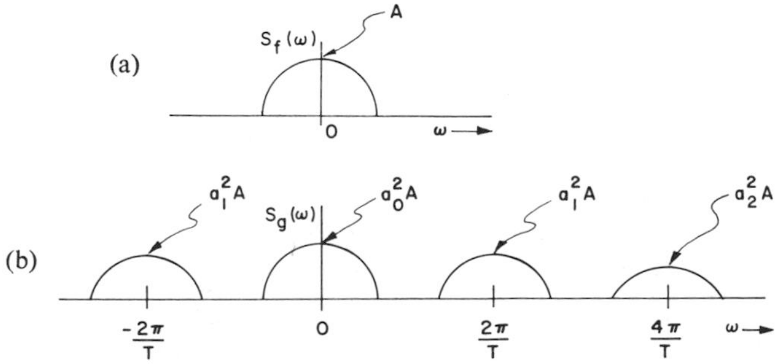

Figure 3.2 Pre- and post-sampling power spectral densities. (a) Representative signal spectrum; (b) sampled signal spectrum.

been lost in multiplying $f(t)$ by the function $b_\tau(t - \eta)$, even though $b_\tau(t - \eta)$ is identically zero for a large percentage of the time.

Since two quite different processes can have the same spectrum, this contention requires further support. To provide this support we need only consider the mean-squared difference between $f(t)$ and the filter output. Let $h(t)$ be the impulse response of the filter $H(j\omega)$. Then:

$$\begin{aligned} E\left\{f(t) - \int_{-\infty}^{\infty} g(\gamma)h(t-\gamma)\,d\gamma\right\}^2 &= R_f(0) - 2\int_{-\infty}^{\infty} R_{fg}(\gamma - t)h(t-\gamma)\,d\gamma \\ &\quad + \int_{-\infty}^{\infty}\int_{-\infty}^{\infty} R_g(\gamma_2 - \gamma_1)h(t-\gamma_1)h(t-\gamma_2)\,d\gamma_1\,d\gamma_2 \\ &= \int_{-\infty}^{\infty} S_f(\omega)[1 - 2H(j\omega)]\frac{d\omega}{2\pi} + \int_{-\infty}^{\infty} S_g(\omega)\,|H(j\omega)|^2\frac{d\omega}{2\pi} \end{aligned} \tag{3.2.6}$$

(Note: $R_{fg}(\alpha) = R_f(\alpha)E(b_\tau(t - \eta)) = R_f(\alpha)$.) This quantity is identically zero if $H(j\omega)$ is as defined above, and if $S_f(\omega) = S_g(\omega)$ for $|\omega| < \omega_m$ and $S_f(\omega) = 0$ for $|\omega| > \omega_m = 2\pi W$. The recovered process is indeed identical, in the mean-squared sense, to the original process under the conditions outlined.

Notice that this conclusion is independent of the pulse width τ. Letting $\tau \to 0$, we conclude that all of the information in the bandlimited process $f(t)$ is contained in the sample values

$$f(\eta + nT) \qquad n = 0, \pm 1, \pm 2, \ldots, \qquad T = 1/2W$$

for any η. The signal can therefore be recovered by passing the sampled sequence $g(t) = T \sum_{n=-\infty}^{\infty} f(t)\delta(t - nT)$ [$\delta(t)$ representing the Dirac delta function] through a filter with the impulse response

$$h(t) = \frac{1}{2\pi}\int_{-\infty}^{\infty} H(j\omega)\,e^{j\omega t}\,d\omega = \frac{\sin 2\pi W t}{\pi t}$$

Thus, if $f(t)$ has a power spectrum of width W or less,

$$\begin{aligned} f(t) &= \int_{-\infty}^{\infty} g(\alpha)h(t-\alpha)\,d\alpha \\ &= \sum_{n=-\infty}^{\infty}\int_{-\infty}^{\infty} Tf(\alpha)\delta(\alpha - nT)h(t-\alpha)\,d\alpha \\ &= \sum_{n=-\infty}^{\infty} f(nT)\frac{\sin 2\pi W(t - nT)}{2\pi W(t - nT)} \end{aligned} \tag{3.2.7}$$

This result is commonly referred to as the *sampling theorem*.

Unfortunately, most power spectral densities of interest are not zero outside any finite frequency range. No signal of finite duration, for example, can have a spectrum with this property. A more typical power spectral density is shown in Figure 3.3(a). However, while there.may be no frequency beyond

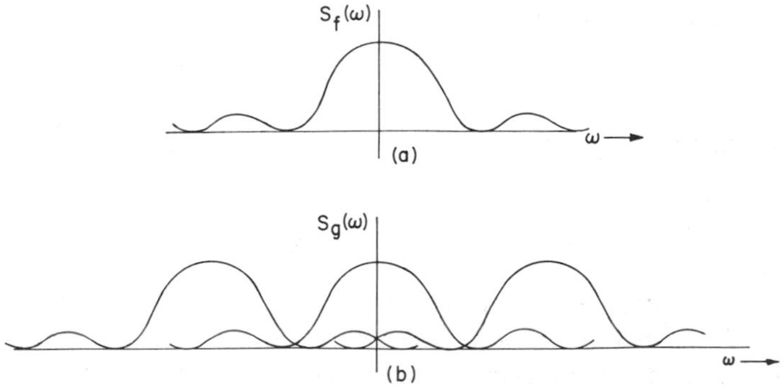

Figure 3.3 Aliasing. (a) Signal spectrum; (b) sampled signal spectrum.

which the spectral density is identically zero, the percentage of the total power represented by frequencies $|f| > f_0$ for some sufficiently large but finite f_0 is arbitrarily small for any process with finite power. It would then seem reasonable to sample the function every $T = 1/2f_0$ seconds yielding a function with the generic spectrum shown in Figure 3.3(b). The net result of such a procedure is to "fold back" the spectrum. If all energy at frequencies greater than f_0 is now filtered out, the spectrum is no longer that of the original process, and the original process cannot be recovered exactly from the periodic samples. The error due to this folding back phenomenon, often referred to as the *aliasing* error, is readily determined by using Equation (3.2.6.). The minimum acceptable sampling rate, in most practical situations, is constrained by the amount of aliasing error which is considered tolerable.

3.3 On the Evaluation of Pulse Modulation Systems

The conventional measure of fidelity in an analog communication system is the *signal-to-noise ratio* defined as follows: If the desired receiver output, the signal, is $s(t)$ and the actual output is $y(t)$, the noise power is, by definition

$$\sigma_n^2 = \lim_{T\to\infty} \frac{1}{2T} \int_{-T}^{T} [y(t) - s(t)]^2 \, dt \qquad (3.3.1a)$$

The signal power,† of course, is just

$$\sigma_s^2 = \lim_{T\to\infty} \frac{1}{2T} \int_{-T}^{T} [s(t) - \overline{s(t)}]^2 \, dt \qquad (3.3.1b)$$

†The average signal amplitude

$$\overline{s(t)} = \lim_{T\to\infty} \frac{1}{2T} \int_{-T}^{T} s(t) \, dt$$

is ideally zero. If $\overline{s(t)}$ were non-zero the same information could in principle be conveyed by transmitting the signal $s(t) - \overline{s(t)}$. The signal power in this case therefore, is still defined as the variance of $s(t)$.

and the signal-to-noise ratio is

$$\frac{S}{N} = \frac{\sigma_s^2}{\sigma_n^2} \tag{3.3.2}$$

By analogy, we define the signal-to-noise ratio in a pulse communication system as

$$\frac{S}{N} = \frac{\mathrm{Var}(s_i)}{E[(s_i - \hat{s}_i)^2]} \tag{3.3.3}$$

where s_i is the sample to be transmitted over the interval $iT_s < t < (i + 1)T_s$ and $\hat{s}_i$ is the estimate of this sample produced at the receiver output. (The sequences s_i and $\hat{s}_i$ are assumed stationary, so S/N is actually independent of i.) The measure of fidelity of a pulse communication system to be used in this chapter is the signal-to-noise ratio as defined in Equation (3.3.3). An equivalent measure, for a given signal variance, is the mean-squared error $E[(s_i - \hat{s}_i)^2]$.

One measure of the *cost* of achieving a particular level of fidelity is the bandwidth of the modulated signal. There are several meaningful ways of defining this quantity. The bandwidth can be defined, for example, as the width of an ideal bandpass filter necessary to pass, say, 90% of the signal power. This measure, of course, necessitates the determination of the actual signal spectrum.

A second, often useful, measure of the bandwidth occupancy of a signal is its *effective bandwidth*, defined as the reciprocal of the maximum density (in channels per unit bandwidth) with which channels conveying statistically identical signals can be frequency multiplexed without mutual interference, i.e., the output of the demodulator for any one channel is to be kept independent of the signals transmitted over any other channel.

As we shall be dealing exclusively with pulse communication systems, the following expression for the power spectral density of a random sequence of statistically independent pulses will be useful (cf. Reference 3.1):

$$S(\omega) = \frac{1}{T_s}\int p(\mu)\,|F_\mu(j\omega)|^2\,d\mu - \frac{1}{T_s}\left|\int p(\mu)F_\mu(j\omega)\,d\mu\right|^2 + \frac{2\pi}{T_s^2}\sum_{n=-\infty}^{\infty}\left|\int p(\mu)F_\mu(j\omega)\,d\mu\right|^2 \delta\!\left(\omega - \frac{2\pi n}{T_s}\right) \tag{3.3.4}$$

where $F_\nu(j\omega)$ is the Fourier transform of the time function $x(t, \nu)$. This latter function represents the transmitted waveform $x(t - iT_s, \nu_i)$ over any particular interval $iT_s < t < (i + 1)T_s$ and is assumed identically zero outside the interval $(0, T_s)$; i.e.,

$$F_\nu(j\omega) = \int_0^{T_s} x(t, \nu)\, e^{-j\omega t}\, dt$$

and $p(\nu)$ is the probability density function of the signal parameter ν. Equation (3.3.4) is valid when the successive parameters ν_i, $i = \ldots, -1, 0, 1 \ldots,$

in the transmitted sequence are all determined independently by a stationary source.

[To prove this expression, we first observe that the autocorrelation function (time average) of the transmitted pulse sequence $x(t - \gamma) \equiv x(t - iT_s - \gamma, \nu_i)$, $iT_s < t < (i + 1)T_s$, $i = \ldots, -1, 0, 1, \ldots,$ is given by:

$$R(\tau) = \lim_{T \to \infty} \frac{1}{T} \int_{-T/2}^{T/2} x(t)x(t + \tau)\, dt \tag{3.3.5}$$

We make the process $x(t - \gamma)$ stationary by defining γ as a random variable uniformly distributed in the interval $0 \leq \gamma < T_s$. If, in addition, the sequence of parameters ν_j is ergodic,† as we shall assume, $x(t - \gamma)$ is an ergodic process; the time and ensemble averages are identical. The integral in Equation (3.3.5) is then equal to the expected value of this same integral over any one interval $jT_s < t < (j + 1)T_s$. That is, letting $\tau = \eta + iT_s$, $i = \ldots, -2, -1, 0, \ldots, 0 \leq \eta < T_s$, and denoting by $p(\mu, \nu, i)$ the joint probability density function of the parameters $\nu_j = \mu, \nu_{j+i} = \nu$, we have, for $iT_s \leq \tau < (i + 1)T_s$,

$$\begin{aligned} R(\tau) &= \frac{1}{T_s} \iint d\mu\, d\nu\, p(\mu, \nu, i) \int_0^{T_s - \eta} x(t, \mu)x(t + \eta, \nu)\, dt \\ &\quad + \frac{1}{T_s} \iint d\mu\, d\nu\, p(\mu, \nu, i + 1) \int_{T_s - \eta}^{T_s} x(t, \mu)x(t + \eta - T_s, \nu)\, dt \end{aligned} \tag{3.3.6}$$

Then

$$\begin{aligned} S(\omega) &= \int_{-\infty}^{\infty} R(\tau)\, e^{-j\omega\tau}\, d\tau \\ &= \sum_{i=-\infty}^{\infty} \iint d\mu\, d\nu\, p(\mu, \nu, i) \frac{e^{-j\omega(iT_s)}}{T_s} \int_0^{T_s} \int_0^{T_s - \eta} x(t, \mu)x(t + \eta, \nu)\, e^{-j\omega\eta}\, dt\, d\eta \\ &\quad + \sum_{i=-\infty}^{\infty} \iint d\mu\, d\nu\, p(\mu, \nu, i + 1) \frac{e^{-j\omega(iT_s)}}{T_s} \\ &\qquad \times \int_0^{T_s} \int_{T_s - \eta}^{T_s} x(t, \mu)x(t + \eta - T_s, \nu)\, e^{-j\omega\eta}\, dt\, d\eta \\ &= \sum_{i=-\infty}^{\infty} \iint p(\mu, \nu, i) \frac{e^{-j\omega(iT_s)}}{T_s} \left[\int_0^{T_s} \int_0^{T_s} x(t, \mu)x(t + \eta, \nu)\, e^{-j\omega\eta}\, dt\, d\eta \right. \\ &\quad \left. + \int_0^{T_s} \int_0^{T_s} x(t, \mu)x(t + \eta - T_s, \nu)\, e^{-j\omega(\eta - T_s)}\, dt\, d\eta \right] d\mu\, d\nu \end{aligned} \tag{3.3.7}$$

This last expression results by substituting $i - 1$ for i in the second summation. The limits on the inner integrands can be changed with impunity because $x(t, \nu)$ is identically zero for $t > T_s$ and for $t < 0$. Now observe that

$$\begin{aligned} &\int_0^{T_s} \int_0^{T_s} x(t, \mu)x(t + \eta, \nu)\, e^{-j\omega\eta}\, dt\, d\eta + \int_0^{T_s} \int_0^{T_s} x(t, \mu)x(t + \eta - T_s, \nu)\, e^{-j\omega(\eta - T_s)}\, dt\, d\eta \\ &= \int_0^{T_s} x(t, \mu)\, e^{j\omega t} \int_t^{T_s + t} x(s, \nu)\, e^{-j\omega s}\, ds\, dt + \int_0^{T_s} x(t, \mu)\, e^{j\omega t} \int_{t - T_s}^{t} x(s, \nu)\, e^{-j\omega s}\, ds\, dt \\ &= \int_0^{T_s} x(t, \mu)\, e^{j\omega t} \int_0^{T_s} x(s, \nu)\, e^{-j\omega s}\, ds\, dt = F_\mu^*(j\omega)F_\nu(j\omega) \end{aligned} \tag{3.3.8}$$

---

†For the definition of ergodicity see, for example, Reference 3.2. Note, in particular, that the sequence $\{\nu_j\}$ is ergodic when the terms ν_j are all mutually independent, identically distributed random variables—the case of major interest here.

Consequently,

$$S(\omega) = \sum_{i=-\infty}^{\infty} \iint d\mu\, d\nu\, p(\mu, \nu, i) \frac{e^{-j\omega(iT_s)}}{T_s} F_\mu^*(j\omega)F_\nu(j\omega) \tag{3.3.9}$$

It the successive symbols are all mutually independent,

$$p(\mu, \nu, i) = \begin{cases} p(\mu)p(\nu) & i \neq 0 \\ p(\mu) & i = 0 \end{cases} \tag{3.3.10}$$

and

$$S(\omega) = \frac{1}{T_s}\int p(\mu)\,|F_\mu(j\omega)|^2\, d\mu - \frac{1}{T_s}\left|\int p(\mu)F_\mu(j\omega)\, d\mu\right|^2$$
$$+ \frac{1}{T_s}\left|\int p(\mu)F_\mu(j\omega)\, d\mu\right|^2 \sum_{i=-\infty}^{\infty} e^{-j\omega(iT_s)} \tag{3.3.11}$$

But $\sum_{i=-\infty}^{\infty} e^{-j\omega(iT_s)}$ is the Fourier series expansion of a periodic sequence of unit impulses:

$$\sum_{i=-\infty}^{\infty} e^{-j\omega(iT_s)} = \frac{2\pi}{T_s}\sum_{n=-\infty}^{\infty} \delta\left(\omega - \frac{2\pi n}{T_s}\right)$$

This can be verified by examining the limiting form as the pulse width approaches zero of the Fourier series expansion of a periodic sequence of rectangular pulses having height $1/W$ and width W. Using this identity then establishes the result stated in Equation (3.3.4). Clearly, if $x(t, \mu) = x_\mu(t)$ where μ is a discrete parameter, the probability density $p(\mu)$ is concentrated on a discrete set of points and the integrals over μ in Equation (3.3.4) become summations.]

3.4 Optimum Demodulation

The optimum continuous parameter pulse demodulator, according to the criterion of Section 3.3, minimizes the mean-squared difference between the transmitted parameter ν and the receiver output $\hat{\nu}$. Thus, the optimum demodulator is a parameter estimator as discussed in Section 2.9, with the cost of an error equal to its mean-squared value. In order to state the present problem in the discrete terminology of Section 2.9, we pass the received signal $y(t)$ through an ideal low-pass filter of bandwidth B. The filtered signal can be represented by the set of samples $Y = \{y_i\} = \{y_f(i\Delta t)\}$, $i = 1, 2, \ldots, k = 2BT_s$, with $\Delta t = 1/2B$ and the subscript f denoting the fact that $y(t)$ has been filtered prior to sampling.† Since the received signal is not confined to any finite bandwidth B, some information is, of course, lost in filtering. We will first determine the optimum demodulator based on the observables Y for some finite bandwidth B, and then consider the form of this demodulator as we let $B \to \infty$. In the limit, no information is lost and the resulting demodulator will be optimum without restriction.

---

†Again, since the statistics of the signal are independent of the particular interval in question, we can, without loss of generality, restrict our attention to the signal $y(t) = Ax(t, \mu) + n(t)$ for $0 < t < T_s$.

We have already determined in Chapter Two that the optimum estimator of the parameter ν based on the observables Y when the cost function is the mean-squared-error is the conditional mean

$$\hat{\nu} = E(\nu \mid Y) \tag{3.4.1}$$

To evaluate this estimate we need the a posteriori probability density function $p(\nu \mid Y)$:

$$p(\nu \mid Y) = \frac{p(Y \mid \nu)p(\nu)}{p(Y)} \tag{3.4.2}$$

where $p(\nu)$ is the a priori density function of the parameter ν.

Writing $y_i = y_f(i\Delta t) = Ax_f(i\Delta t, \nu) + n_f(i\Delta t) \triangleq Ax_i(\nu) + n_i$ with $x_f(t, \nu)$ and $n_f(t)$ representing the filtered signal and the filtered additive noise, respectively, we have

$$p(\nu \mid Y) = K_0 p[n_1 = y_1 - Ax_1(\nu), n_2 = y_2 - Ax_2(\nu), \ldots, n_k = y_k - Ax_k(\nu)]p(\nu) \tag{3.4.3}$$

where K_0 is a normalizing constant, independent of ν. Moreover, since $n(t)$ is a sample function of white Gaussian noise with the single-sided spectral density N_0 watts/Hz, $p(n_1, n_2, \ldots, n_k)$ is a multivariate Gaussian density function. The spectrum of this noise after it is passed through an ideal low-pass filter of bandwidth B is

$$S_n(f) = \begin{cases} \dfrac{N_0}{2} & |f| \leq B \\ 0 & \text{otherwise} \end{cases} \tag{3.4.4}$$

The autocorrelation function $R_n(\tau)$ is therefore

$$R_n(\tau) = N_0 B \frac{\sin 2\pi B\tau}{2\pi B\tau} \tag{3.4.5}$$

and $R_n(i\Delta t) = R_n(i/2B) = 0$ for $i = 1, 2, \ldots$, while $R_n(0) = N_0 B$. Thus

$$p(n_1, n_2, \ldots, n_k) = \frac{1}{(2\pi N_0 B)^{k/2}} \exp\left[-\frac{1}{2N_0 B}\sum_{i=1}^{k} n_i^2\right] \tag{3.4.6}$$

and

$$\begin{aligned} p(\nu \mid Y) &= \frac{K_0}{(2\pi N_0 B)^{k/2}} \exp\left[-\frac{1}{2N_0 B}\sum_{i=1}^{k}[y_f(i\Delta t) - Ax_f(i\Delta t, \nu)]^2\right]p(\nu) \\ &= K_1 p(\nu) \exp\left\{\frac{1}{N_0 B}\sum_{i=1}^{k}\left[y_f(i\Delta t)Ax_f(i\Delta t, \nu) - \frac{A^2}{2}x_f^2(i\Delta t, \nu)\right]\right\} \end{aligned} \tag{3.4.7}$$

with K_1 also independent of ν. Now, letting $B \to \infty$ $(\Delta t \to 0)$, we obtain

$$p(\nu \mid y(t)) = K_1' p(\nu) \exp\left\{\frac{1}{N_0}\int_0^{T_s}[2Ay(t)x(t, \nu) - A^2x^2(t, \nu)]\, dt\right\} \tag{3.4.8}$$

The optimum demodulator, therefore, determines the quantity

$$\hat{\nu} = \int \nu p(\nu \mid y(t))\, d\nu \tag{3.4.9}$$

with $p[\nu \mid y(t)] \triangleq p(\nu \mid y(t), 0 \leq t \leq T_s)$ as given in Equation (3.4.8).

It was observed in Section 2.9 that when $p(v \mid Y)$ is a unimodal symmetric function of v, the conditional expectation of v is equal to its conditional mode. Thus, an equivalent operation for the optimum demodulator, in this event, is to find the estimate $\hat{v}$ for which the a posteriori probability density $p[\hat{v} \mid y(t)]$ attains its maximum.

If the a priori probability density $p(v)$ is not known, other estimators, including the *minimax* and *maximum-likelihood* estimators (cf. Section 2.9) may be appropriate. Indeed, as shown in Appendix A, the maximum-likelihood estimate is efficient if an efficient estimate exists. When this is the case, it becomes the optimum estimate in the mean-squared sense when one imposes the additional constraint that the estimate be unbiased. Even when there is no efficient estimate, the maximum-likelihood estimate often has a small bias and a variance not greatly exceeding the theoretical minimum of Equation (2.9.5). Thus, it may well be the preferred estimate even when $p(v)$ is known.

One might be tempted to conclude that the maximum-likelihood estimate would always be efficient in the present context since the number k of independent observables y_i on which each estimate is based does approach infinity as the bandwidth B becomes infinite, and, as stated in Section 2.9, maximum-likelihood estimators are asymptotically efficient. Unfortunately, the variance $N_0 B$ of these observables becomes simultaneously infinite. The statement in Section 2.9, in contrast, assumed the statistics to be independent of k. The asymptotic properties of maximum-likelihood estimators will be exhibited, however, as we shall see, when the "input" signal-to-noise ratio $R = A^2 T_s/N_0$ becomes large. Here T_s is the direct analog of the number of independent samples when the observables are discrete.

The maximum-likelihood estimator exhibits another advantage over most other estimators, as we shall soon discover: it is often easier to mechanize. For these two reasons, because they are frequently the optimum unbiased estimators and because they are usually somewhat easier to implement than their competitors, maximum-likelihood (or maximum-likelihood-principle) estimators constitute the class of estimators most widely accepted for demodulation purposes.

The maximum-likelihood estimate $\hat{v}$ of the parameter v is, by definition, a solution of the likelihood equation

$$\frac{\partial}{\partial v} \log p[y(t) \mid v] = \frac{\frac{\partial}{\partial v} p[y(t) \mid v]}{p[y(t) \mid v]} = 0$$

Using Bayes' rule, and Equation (3.4.8), we conclude that the maximum-likelihood estimate is a solution of the equation

$$\frac{\partial}{\partial v} \log p[y(t) \mid v] = \frac{2A}{N_0} \int_0^{T_s} [y(t) - Ax(t, v)] \frac{\partial x(t, v)}{\partial v} \, dt = 0 \quad (3.4.10)$$

Regardless of the criterion used, optimum estimators all tend to have one important feature in common: they incorporate the received signal only through cross-correlation coefficients of the form

$$\int_0^{T_s} y(t)z(t, \nu)\, dt \tag{3.4.11}$$

where, for example, $z(t, \nu) = x(t, \nu)$ [Equation (3.4.8)] or $z(t, \nu) = \partial x(t, \nu)/\partial \nu$ [Equation (3.4.10)]. One way of determining these coefficients is to pass the received signal through a filter having the impulse response

$$h(t) = \begin{cases} z(T_s - t, \nu) & 0 < t < T_s \\ 0 & \text{otherwise} \end{cases} \tag{3.4.12}$$

Such a filter is called a *matched filter*. [It is "matched" to the signal $z(t, \nu)$.] It is always possible, of course, to construct a matched filter as a *cross-correlator*, i.e., by multiplying $y(t)$ by $z(t, \nu)$ and integrating, as shown in Figure 3.4(a).

Often $z(t, \nu)$ will be of the form $z(t, \nu) = s(t, \nu)\sqrt{2}\sin(\omega_c t + \phi)$, with $\sqrt{2}\sin(\omega_c t + \phi)$ representing a *carrier*. A generally more practical demodulator in this case is shown in Figure 3.4(b). The only difference between this device and the correlator of Figure 3.4(a) is in the inclusion of the low-pass filter. If $\omega_c/2\pi$ is large compared to the maximum significant frequency component in $s(t, \nu)$, then the output of the first multiplier consists of two relatively narrow-band terms, one centered about zero frequency and one about the frequency $2\omega_c/2\pi$. Because the multiplier-integrator combination acts as a filter with the impulse response $h(t) = s(T_s - t, \nu)$, it too is a matched filter, matched to the low frequency signal $s(t, \nu)$. As such, it is a low-pass filter with a bandwidth equal to the signal bandwidth, and hence small as compared to $2\omega_c/2\pi$. If the first filter has a wide bandwidth relative to that of the matched filter, it has no significant effect on the output of this filter and the demodulators of Figures 3.4(a) and (b) are indeed equivalent. Neverthe-

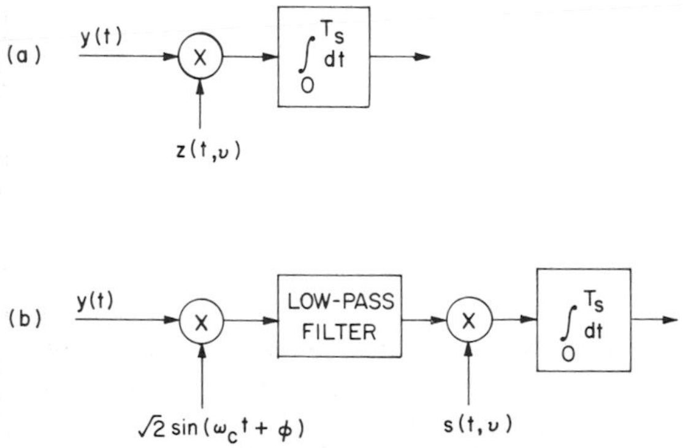

Figure 3.4 Matched filters. (a) Cross-correlator realization; (b) cross-correlator for modulated carrier.

less, if the bandwidth of this first filter is narrow as compared to $2\omega_c/2\pi$, then it will eliminate the double frequency term after the first product. Under the conditions outlined, therefore, it is irrelevant whether or not we include the carrier in the analysis of the communication system in question, so long as we incorporate a carrier product-demodulator into the receiver.

In general, regardless of how they are mechanized, a continuum of matched filters is required, one for each of the possible values of ν. If the demodulator is to be of any practical value, of course, the number of these devices must be reduced to some finite value. In some cases, this involves making an approximation to the optimum demodulator. Fortunately, the optimum demodulators for the modulation systems of this chapter will involve only one or two such matched filters, as we shall discover shortly.

In concluding this section, we should mention that the received signal may contain other parameters, such as the signal amplitude, or the carrier phase, which are also unknown. If these parameters can be assumed to remain constant over the entire transmission interval, they can generally be estimated (as discussed in Section 2.9) with sufficient accuracy to treat them as known. At the other extreme, when they can assume independent values over each T_s second interval, they may have to be estimated simultaneously with each data parameter ν_i even though they, in themselves, convey no information. Alternatively, if the a priori distributions of these "nuisance" parameters are known, the data parameter estimates can sometimes be averaged over these values, obviating the need to estimate them explicitly. That is, denoting the unknown parameters by the vector $\boldsymbol{\phi}$, we first determine the function $p[y(t)\,|\,\nu, \boldsymbol{\phi}]$ or $p[\nu\,|\,y(t), \boldsymbol{\phi}]$ or $E[\nu\,|\,y(t), \boldsymbol{\phi}]$ and then average it over the unknown parameters $\boldsymbol{\phi}$:

$$p[y(t)\,|\,\nu] = \int_{-\infty}^{\infty} p[y(t)\,|\,\nu, \boldsymbol{\phi}]p(\boldsymbol{\phi}\,|\,\nu)\,d\boldsymbol{\phi}$$

(cf. Section 2.4). A specific example of this procedure is first encountered in Section 3.6.

3.5 Coherent Pulse Amplitude Modulation (PAM)

The envelope of a pulse amplitude modulated signal consists of a sequence of pulses of fixed time duration T_s seconds, the amplitudes of which vary in accordance with the information ν_i to be transmitted. If the pulse envelopes are rectangular, the transmitted signal is of the form

$$x(t - iT_s, \nu_i) = \sqrt{2}\,\nu_i \sin \omega_c t \qquad iT_s < t < (i+1)T_s \tag{3.5.1}$$

In order to keep the signal statistics independent of the interval, we require that $\omega_c = \pi r/T_s$, where r is either an integer or else is large relative to unity.

Let $y(t)$ denote the received signal plus noise:

$$y(t) = Ax(t, v) + n(t) = \sqrt{2}\, Av \sin \omega_c t + n(t) \qquad 0 < t < T_s \quad (3.5.2)$$

From Equation (3.4.8), we find that the a posteriori distribution of v given $y(t)$, $0 < t < T_s$, is,

$$\begin{aligned} p[v \mid y(t)] &= Kp(v) \exp\left\{\frac{2A}{N_0}\int_0^{T_s} y(t)(\sqrt{2}\, v \sin \omega_c t)\, dt - v^2 \frac{\mathscr{E}}{N_0}\right\} \\ &= Kp(v) \exp\left\{2R\left(vI - \frac{v^2}{2}\right)\right\} \end{aligned} \quad (3.5.3)$$

where K is a normalizing constant, independent of v, $\mathscr{E} = A^2 T_s$ represents the received pulse energy when $v = 1$, $R = \mathscr{E}/N_0$, and $I = (1/AT_s)\int_0^{T_s} y(t)\sqrt{2}$ $\cdot \sin \omega_c t\, dt$. If the random variable v is uniformly distributed:

$$p(v) = \begin{cases} \dfrac{1}{2a} & -a \le v \le a \\ 0 & \text{otherwise} \end{cases} \quad (3.5.4)$$

then

$$\frac{\partial p[v \mid y(t)]}{\partial v} = \left[2R(I - v) + \frac{1}{p(v)}\frac{dp(v)}{dv}\right] p[v \mid y(t)] \quad (3.5.5)$$

and $p[v \mid y(t)]$ is a unimodal function of v with the mode $v = \hat{v}$ where

$$\hat{v} = \begin{cases} I & -a \le I \le a \\ -a & I < -a \\ a & I > a \end{cases} \quad (3.5.6)$$

The maximum a posteriori probability estimator of v is therefore $\hat{v}$ as defined in Equation (3.5.6). (Note that $p[v \mid y(t)]$ is not a symmetric function of v (unless $I = 0$) and $\hat{v}$ is not in general an optimum estimate of v in the mean-squared sense.)

The maximum-likelihood estimate $\hat{v}$ is, from Equation (3.4.10), a solution to the equation

$$\frac{\partial \log p[y(t) \mid v]}{\partial v} = \frac{2A}{N_0}\int_0^{T_s} [y(t) - \sqrt{2}\, Av \sin \omega_c t]\sqrt{2} \sin \omega_c t\, dt = 0 \quad (3.5.7)$$

which in this case has the unique solution

$$\hat{v} = I = \frac{1}{AT_s}\int_0^{T_s} \sqrt{2}\, y(t) \sin \omega_c t\, dt \quad (3.5.8)$$

for all I.

Regardless of which of these estimators is used, the demodulator is of the form shown in Figure 3.5. The estimator itself may be quite simple (a short circuit in the case of maximum-likelihood estimation) or relatively complex (as in the case of the estimator $\hat{v} = E[v \mid y(t)] = K \int vp(v)$ $\cdot \exp\{2R(vI - (v^2/2))\}\, dv$).

Although the output signal-to-noise ratio is not difficult to determine

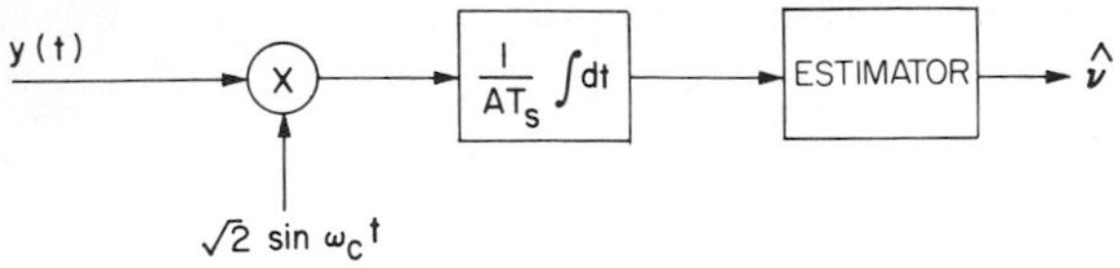

Figure 3.5 Phase-coherent PAM demodulator.

for any of these demodulators, we shall evaluate this quantity here only for the maximum-likelihood demodulator, where it assumes a particularly simple form. Since the noise is Gaussian, the maximum-likelihood estimate $\hat{\nu}$ of ν [Equation (3.5.8)] is Gaussianly distributed with the conditional mean

$$E(\hat{\nu}\mid \nu) = E\left\{\nu + \frac{\sqrt{2}}{AT_s}\int_0^{T_s} n(t)\sin\omega_c t\,dt \,\middle|\, \nu\right\} = \nu \tag{3.5.9}$$

and, hence, is unbiased. The conditional mean-squared error is

$$\sigma_n^2 = E[(\hat{\nu}-\nu)^2\mid \nu] = E\left\{\frac{2}{A^2T_s^2}\int_0^{T_s}\int_0^{T_s} n(t)n(u)\sin\omega_c t\sin\omega_c u\,du\,dt\right\} = \frac{1}{2R} \tag{3.5.10}$$

and is independent of ν. Since

$$E\left[\frac{\partial \log p[y(t)\mid \nu]}{\partial \nu}\,\middle|\, \nu\right]^2 = E[2R(\hat{\nu}-\nu)\mid \nu]^2 = 2R \tag{3.5.11}$$

the maximum-likelihood estimate of ν is efficient (cf. Section 2.9). Finally, the output signal-to-noise ratio is simply

$$\frac{S}{N} = \frac{\sigma_s^2}{\sigma_n^2} = \frac{2\bar{\mathscr{E}}}{N_0} \tag{3.5.12}$$

where $\bar{\mathscr{E}} = E(\nu^2)\mathscr{E}$ is average energy in the received signal.

For purposes of comparison, we now consider the optimum demodulators when the signal amplitude is Gaussianly distributed:

$$p(\nu) = \frac{1}{\sqrt{2\pi}\sigma_s}e^{-\nu^2/2\sigma_s^2} \tag{3.5.13}$$

In this case,

$$\begin{aligned} p[\nu\mid y(t)] &= K\exp\left\{2R\nu I - \left(R + \frac{1}{2\sigma_s^2}\right)\nu^2\right\} \\ &= K'\exp\left\{-\left(R + \frac{1}{2\sigma_s^2}\right)(\nu-\hat{\nu})^2\right\} \end{aligned} \tag{3.5.14}$$

where $\hat{\nu} = I/[1 + (1/2R\sigma_s^2)]$. Consequently, $p[\nu\mid y(t)]$ is also a Gaussian density function, and is symmetric and unimodal. Both the optimum mean-squared error estimate and the maximum a posteriori estimates of ν are therefore

$$\hat{\nu} = E[\nu\mid y(t)] = \frac{I}{1 + \dfrac{1}{2R\sigma_s^2}} = \frac{I}{1 + \left(\dfrac{2\bar{\mathscr{E}}}{N_0}\right)^{-1}} \tag{3.5.15}$$

where again $\bar{\mathscr{E}} = \mathscr{E}\sigma_s^2$ is the average received signal energy. Moreover,

$$\sigma_n^2 = E(\hat{\nu} - \nu)^2 = \frac{1}{2R + \dfrac{1}{\sigma_s^2}} \tag{3.5.16}$$

so that

$$\frac{S}{N} = \frac{\sigma_s^2}{\sigma_n^2} = 1 + \frac{2\bar{\mathscr{E}}}{N_0} \tag{3.5.17}$$

In contrast, the maximum-likelihood estimator, since it is independent of the a priori distribution, remains as before, yielding the signal-to-noise ratio given by Equation (3.5.12). The maximum-likelihood estimater, as already observed, is unbiased. The minimum mean-squared error estimate, Equation (3.5.15), on the other hand, is biased for any finite signal-to-noise ratio $\bar{\mathscr{E}}/N_0$:

$$E(\hat{\nu} \mid \nu) = \frac{E(I \mid \nu)}{1 + \left(\dfrac{2\bar{\mathscr{E}}}{N_0}\right)^{-1}} = \frac{\nu}{1 + \left(\dfrac{2\bar{\mathscr{E}}}{N_0}\right)^{-1}} \tag{3.5.18}$$

The spectrum, and hence the bandwidth occupancy, of a PAM signal can be readily determined with the aid of Equation (3.3.4). We will restrict our attention here, however, to the somewhat less practical but theoretically more interesting measure of bandwidth, the *effective bandwidth* introduced in Section 3.3. This measure, it will be recalled, is defined by the maximum possible density with which PAM channels can be frequency multiplexed without mutual interference. To determine this, consider the ideal situation in which all channels are in perfect synchronization, the lth channel transmitting signals of the form

$$\sqrt{2}\, x_l \sin \omega_l t$$

where x_l is constant for each interval $iT_s < t < (i + 1)T_s$ for all l. Then the output of the detector for the mth channel due to the pulse sent over the lth channel is

$$\frac{2}{T_s} x_l \int_0^{T_s} \sin \omega_l t \sin \omega_m t \, dt = 0 \qquad l \neq m$$

if both ω_l and ω_m are integer multiples of π/T_s. Since two adjacent channels need be separated in frequency by $1/2T_s$ Hz only, the effective bandwidth occupancy of a channel is evidently $B_{\text{eff}} = 1/2T_s$ Hz.

3.6 Phase-Incoherent Pulse Amplitude Modulation

In an incoherent communication system, the phase ϕ of the carrier is taken to be a random variable, uniformly distributed over the interval $(0,2\pi)$, and no attempt is made to estimate its true value. Incoherent communication is necessary when it is not practical to attempt to estimate ϕ directly. This situation may arise either because of equipment limitations or because ϕ is varying too rapidly due to oscillator instabilities, for example, to allow a precise estimate to be made. In any event, if ϕ is a slowly varying function of time relative to the interval T_s, it is reasonable (and convenient) to treat it as a uniformly distributed random variable, constant over any one pulse interval. Moreover, if T_s is large relative to the carrier period $2\pi/\omega_c$, there is no contradiction in the assumption, which will be made throughout this section, that the carrier phase is unknown, while the phase of the envelope is known exactly.

The transmitted signal must be altered when an incoherent decision is required to reflect the fact that the sign of ν can no longer be resolved. There is no way of determining whether ν was transmitted with the carrier phase ϕ or whether the amplitude was $-\nu$ and the carrier phase $\pi + \phi$. For this reason, ν is generally restricted to be positive when the demodulator is to be incoherent. The actual signal s is still assumed to have zero mean; therefore ν is of the form $s + a$, for example, when $p(s) = 1/2a$, $-a < s < a$.

With this qualification, the transmitted signal is as before [cf. Equation (3.5.2)], but now, so far as the demodulator is concerned the received signal is of the form $x(t, \nu, \phi) = \sqrt{2}\,\nu \sin(\omega_c t + \phi)$, and is a function of two random parameters, only one of which is to be estimated. As a consequence, the derivations of Section 3.4 must be modified to take this into account. We begin by observing that the optimum demodulator of the signal $y(t) = Ax(t, \nu, \phi) + n(t)$, if the carrier phase ϕ were known, involves the density function [see Equation (3.4.8)]

$$\begin{aligned} p[\nu \mid y(t), \phi] &= K' \exp\left[\frac{2R\nu}{AT_s}\int_0^{T_s} y(t)\sqrt{2}\sin(\omega_c t + \phi)\,dt - R\nu^2\right] p(\nu \mid \phi) \\ &= K' p(\nu) \exp\{-R\nu^2 + 2R\nu[X\cos\phi + Y\sin\phi]\} \qquad (3.6.1) \end{aligned}$$

where $X = \dfrac{1}{AT_s}\displaystyle\int_0^{T_s} y(t)\sqrt{2}\sin\omega_c t\,dt \qquad Y = \dfrac{1}{AT_s}\displaystyle\int_0^{T_s} y(t)\sqrt{2}\cos\omega_c t\,dt$

and ν is assumed to be independent of ϕ. The unconditional a posteriori probability density function $p[\nu \mid y(t)]$ is therefore

$$\begin{aligned} p[\nu \mid y(t)] &= \int_0^{2\pi} p[\nu \mid y(t), \phi]\, p[\phi \mid y(t)]\, d\phi \\ &= Kp(\nu)\int_0^{2\pi} p(\phi)\exp\{-R\nu^2 + 2R\nu M\cos(\theta - \phi)\}\,d\phi \qquad (3.6.2) \end{aligned}$$

with $M^2 = X^2 + Y^2$ and $\theta = \tan^{-1}[Y/X]$. If

$$p(\phi) = \frac{1}{2\pi}, \qquad 0 < \phi < 2\pi$$

then

$$p[\nu \mid y(t)] = Kp(\nu) \exp\{-R\nu^2\} \frac{1}{2\pi} \int_0^{2\pi} \exp\{2RM\nu \cos(\theta - \phi)\} \, d\phi$$
$$= Kp(\nu) \exp\{-R\nu^2\} I_0(2RM\nu) \tag{3.6.3}$$

where $I_n(x)$ is the modified nth order Bessel function of the first kind.

Since ν cannot be negative, $p[\nu \mid y(t)]$ will not be a symmetric function of ν regardless of the function $p(\nu)$. Thus, the optimum mean-squared error estimate, the maximum a posteriori estimate and the maximum-likelihood estimate will generally all be different. We consider only the latter estimate here. The maximum-likelihood estimate of ν is a solution of the equation

$$\frac{\partial \log p[y(t) \mid \nu]}{\partial \nu} = 2R\left(M \frac{I_1(2RM\nu)}{I_0(2RM\nu)} - \nu\right) = 0 \tag{3.6.4}$$

and hence is†

$$\hat{\nu} = M \frac{I_1(2RM\hat{\nu})}{I_0(2RM\hat{\nu})} \tag{3.6.5}$$

There are several difficulties with this result. First, the fact that the estimate of ν is a solution to a transcendental equation is inconvenient to say the least. Moreover, the solution is dependent on the signal-to-noise ratio $R = \mathscr{E}/N_0$ which therefore has to be known or estimated.

In the absence of noise, as $R \to \infty$, $I_1(2RM\nu)/I_0(2RM\nu) \to 1$ and the maximum-likelihood estimate of ν is

$$\hat{\nu} = M. \tag{3.6.6}$$

This estimate, although not optimum for arbitrary noise levels, is nevertheless meaningful even when the signal-to-noise ratio is not particularly large, and it is often convenient to use it instead of the true maximum-likelihood estimate. A block diagram of this approximation to the maximum-likelihood phase-incoherent PAM receiver is shown in Figure 3.6.‡

Interestingly, Equation (3.6.6) is precisely the estimate which would be used if both the amplitude and phase of the received signal were to be estimated simultaneously. The maximum-likelihood estimate of ν, for a given ϕ, [see Equation (3.4.10)] is one satisfying

$$\frac{\partial \log p[y(t) \mid \nu, \phi]}{\partial \nu} = 2R[X \cos \phi + Y \sin \phi - \nu] = 0 \tag{3.6.7a}$$

while the maximum-likelihood estimate of ϕ must be a solution of the equation

$$\frac{\partial \log p[y(t) \mid \nu, \phi]}{\partial \phi} = 2R\nu[Y \cos \phi - X \sin \phi] = 0 \tag{3.6.7b}$$

---

†This equation also has a solution $\nu = 0$ which, since it is independent of the received signal, can be ignored.

‡For an alternative implementation, see Section 4.3.

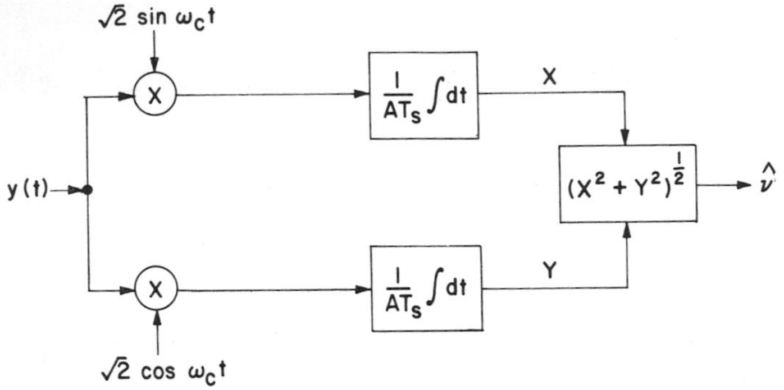

Figure 3.6 Phase-incoherent PAM demodulator.

Solving these two equations simultaneously, we find as the joint maximum-likelihood estimators of the two unknown parameters ν and ϕ

$$\hat{\phi} = \theta = \tan^{-1}\left[\frac{Y}{X}\right] \tag{3.6.8a}$$

and

$$\hat{\nu} = M = [X^2 + Y^2]^{1/2} \tag{3.6.8b}$$

(This solution is not unique. The other solution, obtained by equating $\hat{\nu} = 0$ and $\hat{\phi} = -\tan^{-1}[X/Y]$, involves an amplitude estimate which is independent of the received signal and hence is easily dismissed.)

While the estimation of ϕ is not attempted in incoherent reception, it can be estimated from the same observables used to estimate ν. Since ϕ must be assumed to be slowly varying with time before the detection scheme outlined here can be used, then $\hat{\phi}_i$, the estimate of the phase of the ith pulse, must provide some information concerning the phase of the $(i + 1)$st pulse. So in principle incoherent PAM reception could be improved by using the phase estimates which could conceivably be obtained with only slightly more effort.

To determine the mean and variance of the estimate $\hat{\nu} = M$, we observe that, when $y(t) = \sqrt{2}A\nu \sin(\omega_c t + \phi) + n(t)$, X and Y are jointly Gaussianly distributed with

$$E(X) = \nu \cos \phi$$

$$E(Y) = \nu \sin \phi$$

$$\sigma_n^2 = E(X^2) - E^2(X) = E(Y^2) - E^2(Y) = \frac{1}{2R} \tag{3.6.9}$$

and

$$E(XY) - E(X)E(Y) = 0$$

Thus

$$p(X, Y \mid \nu, \phi) = \frac{1}{2\pi\sigma_n^2} \exp\left\{-\frac{(X - \nu\cos\phi)^2 + (Y - \nu\sin\phi)^2}{2\sigma_n^2}\right\} \tag{3.6.10}$$

and, letting $(X^2 + Y^2)^{1/2} = M$ and $\tan^{-1}[Y/X] = \phi$, we obtain

$$p(M, \theta \mid v, \phi) = \frac{MR}{\pi} \exp\{-R[M^2 + v^2 - 2vM\cos(\theta - \phi)]\}$$

$$0 < M, \quad -\pi < \theta < \pi \qquad (3.6.11)$$

Then

$$p(M \mid v, \phi) = \int_{-\pi}^{\pi} p(M, \theta \mid \phi, v)\, d\theta$$

$$= 2RM \exp\{-R[M^2 + v^2]\} I_0(2MRv) = p(M \mid v)$$

$$0 < M \qquad (3.6.12)$$

and is independent of ϕ. Further (Reference 3.4)

$$E(M \mid v) = \int_0^\infty Mp(M \mid v)\, dM$$

$$= \frac{\sqrt{\pi}}{2R^{1/2}} e^{-Rv^2/2}\left[(1 + Rv^2) I_0\left(\frac{Rv^2}{2}\right) + Rv^2 I_1\left(\frac{Rv^2}{2}\right)\right]$$

$$E(M^2 \mid v) = \int_0^\infty M^2 p(M \mid v)\, dM = \frac{1}{R}(1 + Rv^2) \qquad (3.6.13)$$

Using the asymptotic expansion of the function $I_n(x)$

$$I_n(x) = \frac{e^x}{(2\pi x)^{1/2}} \sum_{k=0}^{\infty} \frac{(-1)^k \Gamma(n + k + \frac{1}{2})}{k!\,\Gamma(n - k + \frac{1}{2})(2x)^k}$$

we find

$$E(M \mid v) = v + \frac{1}{4Rv} + O\left(\frac{1}{R^2 v^3}\right)$$

and

$$E(M^2 \mid v) - E^2(M \mid v) = \frac{1}{2R} + O\left(\frac{1}{R^2 v^2}\right) \qquad (3.6.14)$$

Accordingly, at high signal-to-noise ratios R, this estimate is unbiased for any $v \neq 0$ and has a variance approaching that encountered in the coherent case. In addition, since for any v,

$$E\left\{\left(\frac{\partial}{\partial v} \log p[y(t) \mid v, \phi]\right)^2\right\} = 4R^2 E[X\cos\phi + Y\sin\phi - v]^2 = 2R$$

[cf. Equation (3.6.7)], the variance of the estimate asymptotically approaches the minimum possible variance as given in Equation (2.9.5). Thus, the estimate $\hat{v} = M$ is asymptotically efficient.

The signal parameter v must have some non-zero mean value $\bar{v}$ when the demodulator is phase-incoherent, as previously remarked. In particular, if v is uniformly distributed in the interval $(0, 2a)$, $\bar{\mathscr{E}} \triangleq \mathscr{E} E(v^2) = 4a^2\mathscr{E}/3$, and the phase-incoherent PAM signal-to-noise ratio when $\bar{\mathscr{E}}/N_0$ is large is approximately

$$\left(\frac{S}{N}\right)_{\substack{\text{incoherent}\\ \text{PAM}}} \approx \frac{\bar{\mathscr{E}}}{2N_0} = \frac{1}{4}\left(\frac{S}{N}\right)_{\substack{\text{coherent}\\ \text{PAM}}} \qquad (3.6.15)$$

[cf. Equation (3.5.12)].

The effective bandwidth for incoherent PAM is evaluated as in the coherent situation by determining the minimum separation between channels such that the mutual interference is zero. However, in the incoherent case the receiver does not know the phase of the transmitted signal. The output of the mth channel detector due to a pulse sent over the lth channel is equal to $(X^2 + Y^2)^{1/2}$ where

$$X \propto \int_0^{T_s} \sin(\omega_l t + \phi) \sin \omega_m t \, dt$$

and

$$Y \propto \int_0^{T_s} \sin(\omega_l t + \phi) \cos \omega_m t \, dt$$

If both X and Y are to be zero, $\omega_l - \omega_m$ must equal $2\pi k/T_s$ for some integer k, and the effective bandwidth occupancy is $1/T_s$, twice that of coherent PAM.

3.7 Phase Shift Keying (PSK)

If the phase of the carrier, rather than its amplitude, is varied in accordance with the (time discrete) information, the process is called phase shift keyed (PSK) modulation. The received signal has the form

$$y(t) = Ax(t, v) + n(t) = \sqrt{2}\, A \sin(\omega_c t + \phi_i) + n(t)$$
$$iT_s < t < (i+1)T_s \quad ((3.7.1)$$

where ϕ_i, $-\pi < \phi_i < \pi$, represents the signal. That is, $\phi_i = s_i\pi/a$ where s_i, $-a < s_i < a$, is the data sample to be transmitted over the ith pulse interval. As in the case of coherent PAM, ω_c is assumed to equal $r\pi/T_s$, with r either an integer or else large as compared to unity.

The a posteriori probability distribution of ϕ given $y(t)$, $0 \le t \le T_s$, is, from Section 3.4

$$p[\phi \mid y(t)] = Kp(\phi) \exp\left\{\frac{2A}{N_0}\int_0^{T_s} y(t)\sqrt{2}\sin(\omega_c t + \phi)\, dt\right\}$$
$$= Kp(\phi) \exp\{2R(X\cos\phi + Y\sin\phi)\} \quad (3.7.2)$$

where K is a constant independent of ϕ, $R = \mathscr{E}/N_0$. The terms X and Y are as defined in the previous section:

$$X = \frac{1}{AT_s}\int_0^{T_s} y(t)\sqrt{2}\sin\omega_c t\, dt \quad \text{and} \quad Y = \frac{1}{AT_s}\int_0^{T_s} y(t)\sqrt{2}\cos\omega_c t\, dt$$

The density function is clearly not a symmetric function of ϕ, except for certain special values of X and Y, regardless of the a priori distribution of ϕ. The mean and the mode of $p[\phi \mid y(t)]$ will not generally be equal so the optimum mean-squared error estimate and the maximum a posteriori probability estimate are not equivalent.

If the signal samples are uniformly distributed, $p(\phi) = 1/2\pi$, $-\pi < \phi < \pi$, and

$$\frac{\partial p[\phi \mid y(t)]}{\partial \phi} = 2R[-X \sin \phi + Y \cos \phi] p[\phi \mid y(t)] \tag{3.7.3}$$

The maximum a posteriori probability estimate of ϕ is, therefore, $\hat{\phi} = \theta = \tan^{-1}(Y/X)$. (This estimate yields a unique value in the range $-\pi < \phi < \pi$ if the signs of the terms X and Y are retained and used to determine the phase quadrant.) This is also the maximum-likelihood estimate of ϕ_i [cf. Equation (3.6.7b), with $\nu = 1$].† The resulting demodulator is illustrated in Figure 3.7.

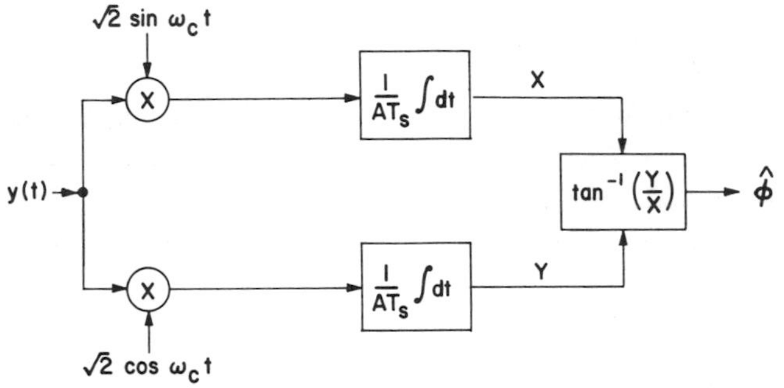

Figure 3.7 Maximum-likelihood PSK demodulator.

The optimum PSK receiver thus utilizes the same decision variables X and Y as the incoherent PAM receiver discussed in Section 3.6. It was

---

†The optimum mean-squared error estimate is also easily determined by expanding ϕ in a Fourier series:

$$\phi = \sum_{n=1}^{\infty} \frac{2}{n}(-1)^{n+1} \sin n\phi \qquad -\pi < \phi < \pi$$

Thus,

$$E[\phi \mid y(t)] = \frac{K}{2\pi} \sum_{n=1}^{\infty} \frac{2}{n}(-1)^{n+1} \int_{-\pi}^{\pi} \exp\{2RM \cos(\phi - \theta)\} \sin n\phi \, d\phi$$

$$= K \sum_{n=1}^{\infty} \frac{2}{n}(-1)^{n+1} \sin n\theta \, I_n(2RM)$$

where $M = (X^2 + Y^2)^{1/2}$, $\theta = \tan^{-1}(Y/X)$. Since

$$\int_{-\pi}^{\pi} p[\phi \mid y(t)] \, d\phi = 1 \qquad K = [I_0(2RM)]^{-1}$$

yielding

$$E[\phi \mid y(t)] = \sum_{n=1}^{\infty} \frac{2}{n}(-1)^{n+1} \sin n\theta \frac{I_n(2RM)}{I_0(2RM)}.$$

If $2RM \gg 1$, this estimate is approximately

$$\sum_{n=1}^{\infty} \frac{2}{n}(-1)^{n+1} \sin n\theta = \theta$$

and is equivalent to the maximum-likelihood estimate.

shown there that if $y(t) = \sqrt{2}\,\nu A \sin(\omega_c t + \phi) + n(t)$,

$$p(M, \theta \mid \phi, \nu) = \frac{M}{\pi} R \exp\{-R[M^2 + \nu^2 - 2\nu M \cos(\theta - \phi)]\}$$

$$0 < M, \quad -\pi < \theta < \pi \quad (3.7.4)$$

where $M = (X^2 + Y^2)^{1/2}$ and $\theta = \tan^{-1}(Y/X)$. Here, $\nu = 1$ is constant, and we are interested in the distribution of the estimate $\theta = \hat{\phi}$. But

$$p(\hat{\phi}) = \int_0^\infty p(M, \hat{\phi})\, dM = \frac{R}{\pi} \int_0^\infty M \exp\{-R[M^2 + 1 - 2M \cos\theta_e]\}\, dM$$

$$= \frac{e^{-R}}{2\pi} + \left(\frac{R}{\pi}\right)^{1/2} \cos\theta_e\, e^{-R\sin^2\theta_e} \frac{1}{\sqrt{2\pi}} \int_{-\infty}^{(2R)^{1/2}\cos\theta_e} e^{-x^2/2}\, dx = p(\theta_e) \quad (3.7.5)$$

where $\theta_e = \hat{\phi} - \phi$, $-2\pi \leq \theta_e \leq 2\pi$.

Because the density function of the error θ_e of the estimate of ϕ is symmetrical about $\theta_e = 0$, the expected value of this error is zero and the estimate is unbiased. The variance of the error is, in general, a rather complex function of the signal-to-noise ratio R. However, when R is large, θ_e will, with high probability, be small. As a result, the quantity $(2R)^{1/2} \cos\theta_e$ will be large, and

$$\frac{1}{\sqrt{2\pi}} \int_{-\infty}^{(2R)^{1/2}\cos\theta_e} e^{-x^2/2}\, dx \approx 1 - \frac{\exp[-R\cos^2\theta_e]}{\sqrt{2\pi}(2R)^{1/2}\cos\theta_e} \quad (3.7.6)$$

Hence,

$$p(\theta_e) \approx \left(\frac{R}{\pi}\right)^{1/2} \cos\theta_e \exp[-R\sin^2\theta_e] \quad (3.7.7)$$

Moreover, when θ_e is small, $\cos\theta_e \approx 1$ and $\sin\theta_e \approx \theta_e$, so that

$$p(\theta_e) \approx \left(\frac{R}{\pi}\right)^{1/2} \exp[-R\theta_e^2] \quad (3.7.8)$$

The error θ_e is asymptotically Gaussianly distributed with zero mean, and variance $\sigma_n^2 = 1/2R$. Again, as one can easily verify using Equation (3.6.7b), (with $\nu = 1$), this estimate is asymptotically efficient; it is unbiased and when R is large, has a minimum variance as given by Equation (2.9.5).

If the signal amplitude is uniformly distributed, the signal variance is

$$\sigma_s^2 = \frac{1}{2\pi} \int_{-\pi}^{\pi} \phi_\nu^2\, d\phi_\nu = \frac{\pi^2}{3} \quad (3.7.9)$$

The received signal energy is a constant $\mathscr{E} = A^2 T_s$. The output signal-to-noise ratio is therefore

$$\left(\frac{S}{N}\right)_{\text{PSK}} \approx \frac{2\pi^2}{3} \frac{\mathscr{E}}{N_0} = \frac{\pi^2}{3} \left(\frac{S}{N}\right)_{\text{PAM(coherent)}} \quad (3.7.10)$$

This is an approximate result, valid when the input signal-to-noise ratio $\mathscr{E}/N_0$ is sufficiently large.

The effective bandwidth requirement for PSK is precisely that of incoherent PAM. The arguments presented in Section 3.6 are directly applicable here. As a result, the effective bandwidth of a PSK signal is twice that required by a coherent PAM signal, although in practice the bandwidth which must be allocated to a single channel is essentially the same in both cases, as is evidenced by referring to their respective spectra.

3.8 Concluding Remarks

This chapter has been concerned with the salient features of continuous amplitude pulse communication systems: the form of the optimum demodulators (under various definitions of optimality), the attainable output signal-to-noise ratios, the signal spectra and their effective bandwidth requirements. Three specific systems, coherent PAM, incoherent PAM, and PSK were analyzed. When the signal samples were uniformly distributed, the large signal-to-noise ratio performance of PSK was found to be better than that of coherent PAM by a factor of $\pi^2/3$, while incoherent PAM exhibited a degradation by a factor of four relative to coherent PAM. The spectra of the three systems are nearly identical, while the effective bandwidth is $1/T_s$ Hz for PSK and incoherent PAM, and $1/2T_s$ Hz for coherent PAM.

It is interesting to note the amount of prior information which must be provided for the different maximum-likelihood demodulators to be operative. The coherent PAM demodulator [cf. Equation (3.5.8)] assumes knowledge of both the received signal power and the phase of the signal carrier, although the first of these requirements can often be waived when only the relative signal amplitude is of importance. Incoherent demodulation of PAM circumvents the need for a carrier phase reference although the true maximum-likelihood demodulator requires a knowledge of both the signal and noise power levels [cf. Equation (3.6.5)]. In the case of PSK it is not necessary to know either the signal amplitude or the noise level, but a phase reference is clearly needed since the magnitude of the modulating signal can be determined only from the difference between this reference phase and the phase of the received signal. (A form of "incoherent" PSK is discussed in the next chapter.) Finally, a PSK system possesses the important practical advantage of allowing the transmitted power to be constant. In contrast, coherent PAM with a uniform signal amplitude distribution requires a peak power of three times its average power.

Although the envelopes of the pulses hypothesized for each of the systems considered in this chapter were rectangular, essentially identical results would have been obtained for arbitrary envelopes $f(t)$. In particular, if the receiver filter is matched to the pulse $f(t) \sin \omega_c t$, $0 < t < T_s$, with $f(t)$ a slowly varying function relative to the carrier period $2\pi/\omega_c$, and if the received signal is $Af(t) \sin(\omega_c t + \phi) + n(t)$, the signal-to-noise ratio at the filter output is

$$\frac{\left(A\int_0^{T_s} f^2(t)\,dt \cos\phi\right)^2}{\frac{N_0}{2}\int_0^{T_s} f^2(t)\,dt} = 2\cos^2\phi\,\frac{\mathscr{E}}{N_0} \tag{3.8.1}$$

with $\mathscr{E}$ the received signal energy and N_0 the noise spectral density. This ratio is obviously independent of the pulse envelope $f(t)$.

Moreover, while the optimum system performance is achieved when the receiver filters are matched to the received pulse, the performance analyses of this chapter are equally applicable when the filters are not matched filters. When the received pulse is as described in the preceding paragraph and the receiver filter impulse response is $h(t) \sin \omega_c(T_s - t)$, the output signal-to-noise ratio becomes

$$\begin{aligned}
&\frac{\left(A\int_{-\infty}^{\infty} f(t)h(T_s - t)\,dt \cos\phi\right)^2}{\frac{N_0}{2}\int_{-\infty}^{\infty} h^2(T_s - t)\,dt} \\
&= 2\cos^2\phi\,\frac{\mathscr{E}}{N_0}\,\frac{\left(\int_{-\infty}^{\infty} f(t)h(T_s - t)\,dt\right)^2}{\int_{-\infty}^{\infty} f^2(t)\,dt \int_{-\infty}^{\infty} h^2(T_s - t)\,dt} \\
&\triangleq 2\gamma\cos^2\phi\,\frac{\mathscr{E}}{N_0}
\end{aligned} \tag{3.8.2}$$

The only effect of the non-matched filter is to reduce the output signal-to-noise ratio by the factor γ [which, by Schwarz's inequality, is less than unity, unless $h(T_s - t)$ is some multiple of $f(t)$]. With this modification, the earlier analyses are applicable even when the receiver filters are not matched.

This last statement must obviously be qualified when a sequence of pulses is transmitted and the filter impulse response $h(t)$ is not identically zero for $t > T_s$. When this is true, the filter output at each sampling instant is influenced by earlier pulses as well as the current one, and the average system performance is generally worse than that predicted on the basis of a single pulse alone. Moreover, the statement that the system performance is independent of the pulse shape must also be qualified when the communication channel is other than the infinite bandwidth, white Gaussian channel assumed here. If the channel bandwidth is not large relative to the signal bandwidth, for example, or if the channel exhibits multipath characteristics, with different transmission paths representing different delays, the performance will indeed be a function of pulse shape.

Notes to Chapter Three

The early applications of statistical techniques to the detection of signals and the estimation of signal parameters in the presence of noise were generally oriented toward radar (References 3.5, 3.6, and 3.7) although their relevance to communications problems were of course apparent. The most extensive work in applying statistical hypothesis testing and estimation theory to communications problems was done by Middleton and Van Meter [(References 3.8 and 3.9); see also Reference 3.4]. Sections 3.4 through 3.7 are largely specializations of these results. The sampling procedure discussed in Section 3.2 was first studied by Nyquist (Reference 3.10) and later by Shannon (Reference 3.11), Landau and Pollak (Reference 3.12), and many others. The derivation of the power spectral density of a random pulse sequence in Section 3.3 is due to Titsworth and Welch (Reference 3.1).

The channel model assumed throughout this discussion was a highly idealized one. The primary effect of relaxing some of the assumptions leading to the model (e.g., the white noise assumption) is to complicate the analysis somewhat, but to leave it basically unchanged (see Problem 3.1). Other modifications of the channel model (e.g., a channel transfer function which varies randomly with time, but which is still independent of frequency) can sometimes be accommodated by a simple extension of the more restrictive analysis (see Problem 4.1). More generally, however, a change in the channel model introduces a new set of problems which were absent in the original model. When the channel bandwidth cannot be assumed large relative to the signal spectrum, for example, the transmitted pulses disperse, creating potential inter-pulse interference problems at the receiver, and both the receiver structure and the pulse shapes should be modified accordingly (cf. References 3.13, 3.14). Other modifications may be called for when the channel characteristics can vary with time (Reference 3.15). The cardinal assumption leading to the results of this chapter, however, was the assumption of Gaussian noise. The optimum receiver, if indeed it can be determined, may be fundamentally different from those encountered here when other than Gaussian noise is postulated (cf. Reference 3.16).

Problems

3.1 (a) Show that when the (stationary Gaussian) noise is not necessarily white, Equation (3.4.8) becomes

$$p(\nu \mid y(t)) = Kp(\nu) \exp\left\{\int_0^{T_s} [2Ay(t)z(t, \nu) - A^2x(t, \nu)z(t, \nu)]\, dt\right\}$$

where $z(t, \nu)$ is defined by the integral equation

$$x(t, \nu) = \int_0^{T_s} R_n(t - \tau)z(\tau, \nu)\, d\tau$$

and $R_n(\tau)$ is the noise autocorrelation function.

(b) Let $y(t) = x(t, \nu) + n(t)$, $0 < t < T_s$, with $n(t)$ Gaussian noise with the power spectral density $S_n(\omega)$, and let $h(t)$ and $H(j\omega)$ be the impulse response and the transfer function of a *whitening filter* defined by the equation $|H(j\omega)|^2 S_n(\omega) = N_0/2$. Show that if $y(t)$ is impressed at the input of such a filter, the output is of the form $y_f(t) = \xi(t, \nu) + n_w(t)$ with $\xi(t, \nu) = \int_0^{T_s} x(\tau, \nu)h(t - \tau)\, d\tau$ and with $n_w(t)$ a white Gaussian process.

(c) Show that the receiver implied in part (a) can be realized as a whitening filter followed by the receiver which would have been used were a signal from the set $\{\xi(t, \nu)\}$, defined in part (b), received in the presence of white Gaussian noise.

3.2 (a) Determine the spectrum of the PAM signal defined in Equation (3.5.1), with the ν_i mutually independent random variables uniformly distributed in the interval (a, b). Compare this with spectrum of the signal $\sqrt{2}\,Am(t - \eta)$ $\sin(\omega_c t + \phi)$ with $m(t) = \nu_i$, $iT_s < t < (i + 1)T_s$, and with η and ϕ independent random variables uniformly distributed over the intervals $(0, T_s)$ and $(0, 2\pi)$ respectively.

(b) Repeat part (a) for the PSK signal defined in Equation (3.7.1), with ϕ_i uniformly distributed in the interval $(-\pi, \pi)$.

(c) Repeat parts (a) and (b) when the pulse envelopes are of the *raised-cosine* form rather than rectangular; i.e., when the signal consists of a sequence of amplitude- or phase-modulated pulses of the form $\sqrt{2}\,Ap(t) \sin(\omega_c t + \phi)$ with $p(t) = \sqrt{\tfrac{2}{3}}\,(1 - \cos 2\pi t/T_s)$.

3.3 (Reference 3.3). Show that the estimator of Equation (3.5.15) is the optimum *minimax* estimator of the amplitude of a PAM pulse under the mean-squared error cost criterion; i.e., show that this estimator minimizes the maximum possible mean-squared error when only the variance of the (zero *pt-pt* mean) signal amplitude is known. [*Hint*: Show that, of all zero-mean amplitude distributions having variance σ_s^2, the Gaussian distribution results in the greatest mean-squared error when this estimator is used.]

3.4 A PSK modulated signal $s_i(t) = \sqrt{2}\,A \sin(\omega_c t + \phi_i)$, $iT_s < t < (i + 1)T_s$, is received in the presence of white Gaussian noise. The receiver is designed to determine the optimum estimate $\hat{s}_i(t)$ of the received signal under the mean-squared error criterion, and to use the phase of this estimate as the estimate of θ_i. Show that the resulting estimate and the maximum-likelihood estimate of θ_i are identical.

References

3.1 R. C. Titsworth and L. R. Welch, "Power Spectra of Signals Modulated by Random and Pseudorandom Sequences," Tech. Rept. No. 32–140, Jet Prop. Lab., Calif. Inst. Tech., Pasadena, Calif., 1961.

3.2 J. L. Doob, *Stochastic Processes*, John Wiley & Sons, Inc., New York, N. Y., 1953.

3.3 J. M. Wozencraft and I. M. Jacobs, *Principles of Communication Engineering*, John Wiley & Sons, Inc., New York, N. Y., 1965.

3.4 D. Middleton, *Introduction to Statistical Communication Theory*, McGraw-Hill Book Co., New York, N. Y., 1960.

3.5 D. O. North, "Analysis of the Factors Which Determine Signal-to-Noise Discrimination in Radar," RCA Tech. Dept. PRT-6C, 1943.

3.6 J. I. Marcum, "A Statistical Theory of Target Detection by Pulsed Radar," RAND Corp. Repts. RM-15061, 1947; RM-753, R-113, 1948.

3.7 P. M. Woodward, *Probability and Information Theory with Radar Applications*, McGraw-Hill Book Co., New York, N. Y., 1953.

3.8 D. Van Meter and D. Middleton, "Modern Statistical Approaches to Reception in Communication Theory," *IRE Trans. Inf. Theory*, PGIT-4, p. 119, 1954.

3.9 D. Middleton and D. Van Meter, "Detection and Extraction of Signals in Noise from the Point of View of Statistical Decision Theory," *J. Soc. Ind. Appl. Math.*, **3,** p. 192, 1955; **4,** p. 86, 1956.

3.10 H. Nyquist, "Certain Topics in Telegraph Transmission Theory," *Trans. AIEE*, **47,** p. 617, 1928.

3.11 C. E. Shannon, "Communication in the Presence of Noise," *Proc. IRE*, **37,** p. 10, 1949.

3.12 H. J. Landau and H. O. Pollak, "Prolate Spheroidal Wave Functions, Fourier Analysis and Uncertainty—III: The Discussion of the Space of Essentially Time- and Band-limited Signals," *Bell Syst. Tech. J.*, **41,** p. 1295, 1962.

3.13 E. D. Sunde, "Theoretical Fundamentals of Pulse Transmission," *Bell Syst. Tech. J.*, Part I, p. 721; Part II, p. 987; 1954.

3.14 D. W. Tufts, "Nyquist's Problem—The Joint Optimization of Transmitter and Receiver in Pulse Amplitude Modulation," *Proc. IEEE*, **53,** p. 248, 1965.

3.15 T. Kailath, "Optimum Receivers for Randomly Varying Channels," *Proc. 4th London Symp. on Inf. Theory*, (C. Cherry, ed.) Butterworth, Washington, D. C., 1961.

3.16 H. Sherman and B. Reiffen, "An Optimum Demodulator for Poisson Processes: Photon Source Detectors," *Proc. IEEE*, **51,** p. 1316, 1963.

Supplementary Bibliography

Baghdady, E. J., Ed., *Lectures on Communication System Theory*, McGraw-Hill Book Co., New York, N. Y., 1960.

Bennett, W. R., and J. R. Davey, *Data Transmission*, McGraw-Hill Book Co., New York, N. Y., 1965.

Lucky, R. W., J. Salz, and E. J. Weldon, Jr., *Principles of Data Communication*, McGraw-Hill Book Co., New York, N. Y., 1968.

Schwartz, M., W. R. Bennett, and S. Stein, *Communication Systems and Techniques*, McGraw-Hill Book Co., New York, N. Y., 1966.

chapter four

Amplitude-Discrete, Time-Discrete Communication Systems

4.1 Introduction

The pulse modulation schemes of the preceding chapter involved signals which were continuous functions of some parameter. Although the optimum receiver for such a set of waveforms ostensibly requires a continuum of matched filters, one for each of the possible received signals, the systems considered involved only one or two such filters.

This is not generally the case. (Consider, for example, the optimum demodulator when the set of signals is $s_\omega(t) = \sqrt{2}\,A \sin \omega t,\ 0 < t < T_s$ with ω a continuous parameter.) In general, the optimum receiver does entail the determination of a function of the received signal for every possible value of the parameter to be estimated. Under these circumstances, optimum continuous parameter pulse communication receivers can be only approximated. A convenient approximation is often afforded by the observation that a filter matched to the signal $s_\alpha(t)$ will generally indicate a large output when a signal $s_{\alpha'}(t)$ is received, the parameter α' differing only slightly from the parameter α. This suggests that the receiver be "quantized" with a finite set of matched filters replacing the infinite set theoretically needed.

If quantizing the signal at the receiver simplifies the communication system, quantizing it at the transmitter results in even further simplification.

The fact that only a finite number of signals are available, of course, introduces a quantization error whenever the information to be transmitted is not inherently discrete in nature. This error may be completely insignificant, however, because in nearly every conceivable utilization of this information only a certain accuracy is required or even usable. And in any event, since the signal is inevitably perturbed by noise, the accuracy attainable is inherently limited, regardless of the modulation scheme.

If it can be argued that little is lost in quantizing the signal before transmitting it, it can be even more easily argued that much is gained. The advantages lie primarily in the ease with which digital information can be stored and transferred, as well as communicated from one location to another. Some of the less apparent advantages should become clearer in the succeeding chapters. The principal advantage to be exploited in this chapter is the relative simplicity of the analysis, as well as the mechanization, of such systems.

In its basic configuration, a *digital communication system* transmitter has as its input a sequence of digits (in some suitable numerical base). It converts these digits into a sequence of *symbols*, each selected from a finite set of *symbols* (which we shall call the *symbol alphabet*), and each corresponding uniquely to a specific sequence of input digits. Each of these symbols is then represented by a particular waveform† $s_\nu(t)$, $\nu = 1, 2, \ldots, N$, N indicating the number of symbols in the alphabet. One symbol is to be transmitted every T_s seconds; generally $s_\nu(t)$ will be assumed to be identically zero outside the interval $(0, T_s)$.

The measure of performance of the communication systems considered in the previous chapter was the output signal-to-noise ratio. It is possible to determine an output signal-to-noise ratio for digital systems as well (taking into account the mean-squared quantization error if so desired). However, a more natural, and, in most cases, a more satisfactory measure is the *probability of a symbol error*: the probability that the symbol at the receiver output differs from that transmitted. The mean-squared error in the case of analog systems is useful primarily because it is readily determined, not necessarily because this is the measure of greatest concern to the user. Indeed, as the comments of the previous paragraphs suggest, a persistent communication error which is always small relative to the accuracy required of the received information may be much less objectionable than an occasional, but large, error in this information. Yet the mean-squared error in the two cases could be identical.

The probability of a symbol error seems to be a meaningful basis for comparing two systems utilizing the same number of symbols. But one must

†We shall generally use the word symbol to refer both to the symbol as defined here and to the waveform which represents it. Similarly, the term symbol alphabet will denote both the set of symbols and the set of corresponding waveforms.

qualify this statement when systems employing a different number of symbols are to be compared. Specificially, if a signal alphabet contains N symbols, each received symbol conveys $\log_2 N$ *bits* of information (i.e., the same amount of information represented by $\log_2 N$ binary digits). Clearly, two systems having an unequal number of symbols can be compared in a meaningful way only if they use the same amount of energy to transmit each information bit. It is the total amount of energy needed to transmit the complete message that represents the cost of the transmission, not the amount of energy needed to transmit satisfactorily a particular symbol.

The figure of merit to be used in this chapter, then, is the probability of making a symbol error. In order to compare systems employing a different number of symbols, we shall determine this probability as a function of the average amount of energy expended for each bit of transmitted information. In the process of determining this symbol error probability, the probabilities of each kind of error [the probability of mistaking $s_\nu(t)$ for $s_\mu(t)$ for each μ and ν] will also generally be obtained. This information can always be used to evaluate the mean-squared error if so desired.

4.2 The Optimum Phase-Coherent Receiver and Its Performance

As Chapter Three was concerned with the statistical estimation of some parameter of the received signal, so here the problem becomes one of multiple hypothesis testing. This problem was discussed in Section 2.8 where it was remarked that the optimum decision, when the cost of an error is independent of the particular error committed, is to select the hypothesis having the largest a posteriori probability of being true. Since we wish to minimize the error probability, all errors are treated equally, and the optimum detector in the present situation is indeed the *maximum a posteriori probability detector*.

The waveform at the input to the receiver is $y(t) = As_{\nu_i}(t - iT_s) + n(t)$, $iT_s < t < (i+1)T_s$, for some value of $\nu_i = 1, 2, \ldots, N$.† Under the assumption of white Gaussian noise, the a posteriori probabilities $P[\nu \mid y(t)]$ are easily determined, by an argument identical to that of Section 3.4, to be of the form

$$P[\nu \mid y(t)] = kP(\nu)\exp\left\{\frac{2A}{N_0}\int_0^{T_s} y(t)s_\nu(t)\,dt - \frac{A^2}{N_0}\int_0^{T_s} s_\nu^2(t)\,dt\right\}$$

$$= k\exp\left\{\frac{2A}{N_0}\int_0^{T_s} y(t)s_\nu(t)\,dt - \frac{A^2K_\nu}{N_0}\right\} \tag{4.2.1}$$

†As in the preceding chapter, i will be taken to be zero, and the subscript on the ν dropped when it can be done without introducing ambiguities.

where k is a constant independent of ν and the terms

$$K_\nu = \int_0^{T_s} s_\nu^2(t)\,dt - \frac{N_0}{A^2}\log_e P(\nu) \tag{4.2.2}$$

are constants which are independent of the received signal $y(t)$. The optimum decision is to choose the signal $s_\mu(t)$, where μ is the value of ν which maximizes the expression:

$$z_\nu = \int_0^{T_s} y(t)s_\nu(t)\,dt - \frac{AK_\nu}{2} \tag{4.2.3}$$

The optimum detector thus consists of a bank of matched filters as shown in Figure 4.1.

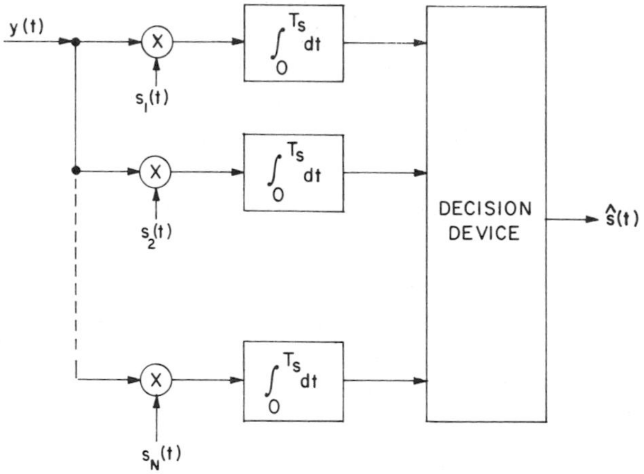

Figure 4.1 Optimum phase-coherent demodulator (N-symbol alphabet).

The additive noise $n(t)$ is assumed to be white and Gaussianly distributed. Then, if $y(t) = As_r(t) + n(t)$, the variables z_i are Gaussianly distributed for any given value of r. Since the noise is white, $E[n(t)] = 0$, and $E[n(u)n(t)] = (N_0/2)\,\delta(t - u)$, where N_0 is the single-sided noise spectral density. Consequently,

$$\eta_\nu(r) \triangleq E(z_\nu \mid r) = A\int_0^{T_s} s_r(t)s_\nu(t)\,dt - \frac{AK_\nu}{2} \tag{4.2.4}$$

while

$$\sigma_\nu^2(r) = \frac{N_0}{2}\int_0^{T_s} s_\nu^2(t)\,dt = \frac{N_0}{2A^2}\mathscr{E}_\nu = \sigma_\nu^2 \tag{4.2.5}$$

is the conditional variance of the random variable z_ν, with $\mathscr{E}_\nu$ the energy in the νth received symbol. The conditional covariance of z_ν and z_μ is

$$\rho_{\nu\mu}(r) = \frac{E\{[z_\nu - \eta_\nu(r)][z_\mu - \eta_\mu(r)] \mid r\}}{\sigma_\nu \sigma_\mu}$$

$$= \frac{1}{\sigma_\mu \sigma_\nu} \frac{N_0}{2} \int_0^{T_s} s_\mu(t) s_\nu(t)\, dt \triangleq \rho_{\nu\mu} = \rho_{\mu\nu} \tag{4.2.6}$$

and is also independent of the transmitted symbol.

If $s_r(t)$ is transmitted, the probability that it is correctly received is

$$P_c(N; r) = \Pr\left[z_r > \max_{k \neq r} z_k \mid r \right]$$

$$= \int_{-\infty}^{\infty} \int_{-\infty}^{w_r + \eta_r - \eta_1} \int_{-\infty}^{w_r + \eta_r - \eta_2} \cdots \int_{-\infty}^{w_r + \eta_r - \eta_N} \frac{\exp(-\frac{1}{2}\mathbf{w}^T \Lambda^{-1} \mathbf{w})}{(2\pi)^{N/2} |\Lambda|^{1/2}} dw_1\, dw_2 \ldots dw_N\, dw_r \tag{4.2.7}$$

where $w_i = z_i - \eta_i$, $\eta_i = \eta_i(r)$, $\mathbf{w}$ is the column vector $(w_1, w_2, \ldots, w_N)$, $\mathbf{w}^T$ its transpose, and Λ the (non-singular) covariance matrix

$$\Lambda = \{\sigma_\nu \sigma_\mu \rho_{\mu\nu}\}$$

The term Λ^{-1} denotes the inverse of Λ and $|\Lambda|$ its determinant. Finally, the error probability using the communication system in question is

$$P_e(N) = 1 - \sum P(r) P_c(N; r) \tag{4.2.8}$$

where again $P(r)$ is the a priori probability of the symbol $s_r(t)$.

While the Equation (4.2.7) is formidable, it can, in principle, be evaluated for any non-singular matrix Λ. If some restrictions are placed on the signal set $\{s_i(t)\}$, it is possible to obtain a somewhat more manageable expression. The first restriction is a practical as well as an analytically convenient one. We have argued that the optimum receiver bases its decisions on the terms defined in Equation (4.2.3) and hence involves knowledge of both A, the received signal amplitude, and N_0 the noise spectral density. If, however, K_ν is made independent of ν by requiring that both the signal energy $\mathcal{E}_\nu$ and the a priori probabilities $P(\nu)$ be independent of ν, neither of these parameters need be measured to implement the optimum receiver. In addition, the transmitter can operate more efficiently because it requires a relatively constant amount of energy. These restrictions, therefore, are of considerable practical advantage. Moreover, they are generally not constraints of significant consequence. The average information transmitted per symbol is obviously maximized when the a priori symbol probabilities are equal. In Chapter Ten methods are discussed whereby this situation can be approached for any source statistics. And if each symbol is a priori equally likely, each conveys the same amount of information. Hence each should be represented by equal energy unless this is precluded by other considerations.

When the N symbols all have an equal a priori probability of being selected by the source, and are all represented by waveforms of equal energy, the optimum receiver simply determines the quantities

$$z_\nu = \int_0^{T_s} y(t)s_\nu(t)\,dt \tag{4.2.9}$$

and selects the largest of these. In this case, the expression for the probability of a correct decision, Equation (4.2.7), still applies, but with

$$\eta_\nu(r) = A\int_0^{T_s} s_r(t)s_\nu(t)\,dt = A\mathscr{E}\rho_{\nu r}$$

and

$$\sigma_\nu^2 = (N_0/2)\mathscr{E} = \sigma^2 \tag{4.2.10}$$

and with

$$\rho_{ij} = \frac{1}{\mathscr{E}}\int_0^{T_s} s_i(t)s_j(t)\,dt$$

Equation (4.2.7) is only slightly simplified by these assumptions concerning the signal; its solution can still be accomplished only with considerable effort. For this reason we now consider several of the most important special cases.

4.2.1 Orthogonal Symbols

If $\rho_{ij} = 0$ for all $i \neq j$, the set of symbols is called, for obvious reasons, an orthogonal alphabet. In this case, Equation (4.2.7) can be simplified rather significantly. First, by symmetry, $P_c(N; r)$ is independent of r. The probability of a correct decision is equal to the probability of correctly determining, at the receiver, the transmitted signal, regardless of which specific signal was transmitted. Moreover, since $\rho_{ij} = 0$, $i \neq j$, the matrix $\mathbf{\Lambda}$ is just σ^2 times the identity matrix $\mathbf{I}$. Making this substitution in Equation (4.2.7) establishes

$$P_e(N) = 1 - P_c(N) = 1 - \int_{-\infty}^{\infty}\left[\int_{-\infty}^{\zeta}\frac{e^{-\xi^2/2}}{\sqrt{2\pi}}\,d\xi\right]^{N-1}\frac{e^{-[\zeta-(2R)^{1/2}]^2/2}}{\sqrt{2\pi}}\,d\zeta \triangleq \Phi_N(R_b) \tag{4.2.11}$$

where $R_b = R/\log_2 N = \mathscr{E}/N_0 \log_2 N$ is the ratio of the received signal energy per bit to the noise spectral density. Consequently, when $\rho_{ij} = 0$ for all $i \neq j$, the error probability is a function only of the number of signals N and of the input signal-to-noise ratio $R = \mathscr{E}/N_0$. This function $\Phi_N(R_b)$ has been evaluated numerically (cf. Reference 4.1). It is plotted here in Figure 4.2 as a function of R_b for various values of N.

When there are only two symbols (the binary orthogonal set), the error probability can be expressed in a somewhat more convenient form. For then

$$P_c(2) = \Pr\{z_1 > z_2 \mid 1\} = \Pr\{z_1 - z_2 > 0 \mid 1\}$$

Since z_1 and z_2 are independent Gaussian random variables, $z_1 - z_2 = w$ is

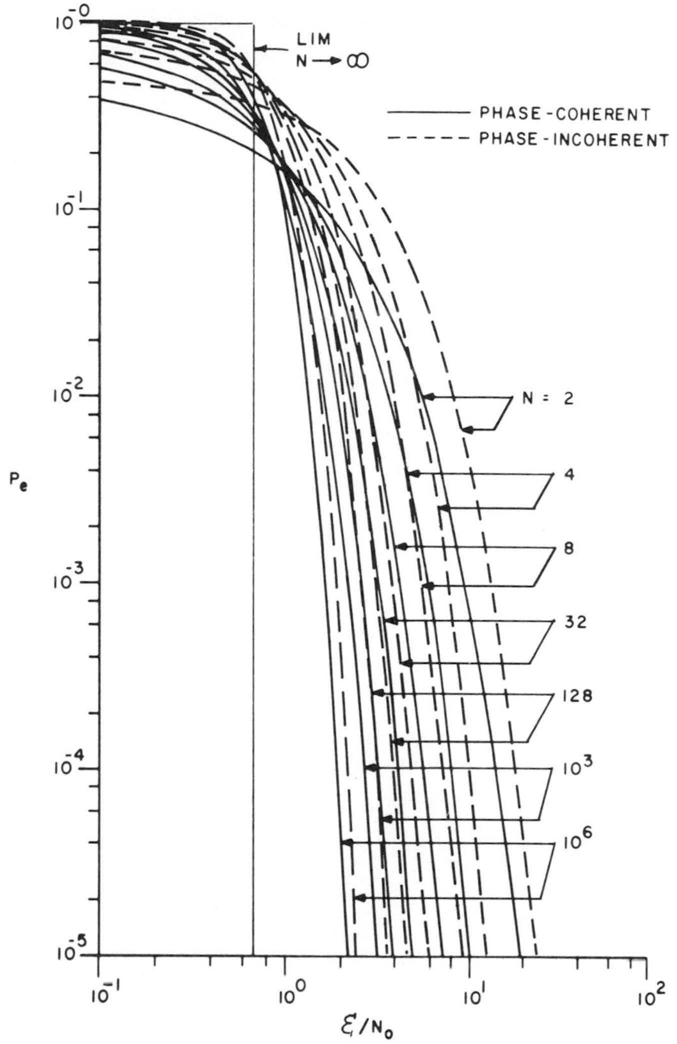

Figure 4.2 Orthogonal symbol error probabilities.

also a Gaussian variable with a mean equal to the difference of the means of z_1 and z_2 and a variance equal to the sum of their variances. Thus

$$P_e(2) = \Pr\{w \leq 0 \mid 1\} = \frac{1}{\sqrt{2\pi}} \int_{R_b^{1/2}}^{\infty} e^{-\xi^2/2}\, d\xi \triangleq \frac{1}{2}\left[1 - \operatorname{erf}\left(\frac{R_b}{2}\right)^{1/2}\right] \quad (4.2.12)$$

4.2.2 Error Probability when $\rho_{ij} = \rho$, all $i, j, i \neq j$.

When the cross-correlation coefficients $\rho_{ij} (i \neq j)$ are all equal to ρ and each symbol represents the same amount of energy $\mathscr{E}$, the error probability is again independent of the transmitted signal because of the inherent symmetry. While it is possible to introduce a transformation of variables into Equation (4.2.7) which reduces this situation to the orthogonal case just considered, a simpler method of determining the error probability is as follows: First suppose $\rho > 0$, and consider the signal set obtained by adding to each symbol $s_i(t)$, $0 < t < T_s'$, in an N-symbol orthogonal set, a fixed time function $r(t)$, $T_s' < t < T_s$. The correlation between any two of these extended symbols is

$$\int_0^{T_s'} s_i(t)s_j(t)\,dt + \int_{T_s'}^{T_s} r^2(t)\,dt = \begin{cases} \mathscr{E}'' & i \neq j \\ \mathscr{E}' + \mathscr{E}'' = \mathscr{E} & i = j \end{cases}$$

where $\mathscr{E}''$ is the energy in the $r(t)$ portion of the symbol, $\mathscr{E}'$ the energy in $s_i(t)$, and $\mathscr{E}$ the total energy in each of the extended symbols. If $\mathscr{E}''/\mathscr{E} = \rho$ the extended symbol set contains N symbols having the cross-correlation coefficients $\rho_{ij} = \rho$, $i \neq j$, as desired. But the performance of a communication system using this signal set is equivalent to one using orthogonal symbols. The last $T_s - T_s'$ seconds of correlation yield identical results for all N symbols and, consequently, contribute nothing to their distinguishability. Using the extended symbols defined here, therefore, is the same as using an orthogonal set of symbols, each having the energy

$$\mathscr{E}' = \mathscr{E} - \mathscr{E}'' = \mathscr{E}(1 - \rho)$$

Similarly, if $\rho < 0$, we can begin with the original signal set $\{s_i(t)\}$, $0 < t < T_s$ and form an orthogonal set by adding to each of these symbols the time function $r(t)$, $T_s < t < T_s'$, where

$$\int_{T_s}^{T_s'} r^2(t)\,dt = |\rho|\,\mathscr{E} = -\rho\mathscr{E}$$

Again, the performance of a system using the extended (orthogonal) symbols must be identical to that using the original non-orthogonal symbols. But the energy in the extended symbols is

$$\mathscr{E}' = \mathscr{E}(1 + |\rho|) = \mathscr{E}(1 - \rho)$$

Since the error probability is a function only of the coefficients ρ_{ij} and of the signal-to-noise ratio, it follows that any set of N symbols, each having the energy $\mathscr{E}$, with cross-correlation coefficients $\rho_{ij} = \rho$ for all $i, j, i \neq j$, results in the same error probability as a set of N orthogonal symbols having the energy $\mathscr{E}(1 - \rho)$:

$$P_e(N, \rho) = \Phi_N[R_b(1 - \rho)] \tag{4.2.13}$$

where Φ_N is defined as in Equation (4.2.11). This is true whether ρ is positive or negative.

4.2.3 On the Optimum Choice of ρ

It was concluded in the previous paragraphs that the effective signal energy of a signal set is $\mathscr{E}(1-\rho)$ where $\mathscr{E}$ is the actual energy of each symbol, and $\rho = \rho_{ij}$, the cross-correlation coefficient. It would seem that ρ should be made as highly negative as possible ($\rho = -1$) to minimize the error probability. Unfortunately, this is possible only when $N = 2$; otherwise ρ cannot be made so small. Specifically ρ is bounded by $-1/(N-1)$ when the alphabet contains N symbols.

Actually, we can prove the following more general result:

$$\begin{aligned}\rho_{\text{ave}} \stackrel{\Delta}{=} \operatorname*{ave}_{\substack{i,j\\ i\neq j}} \rho_{ij} &= \frac{1}{N(N-1)} \sum_{\substack{i,j=1\\ i\neq j}}^{N} \rho_{ij} = \frac{1}{N(N-1)}\left\{\sum_{i,j=1}^{N} \rho_{ij} - N\right\}\\ &= \frac{1}{N(N-1)}\left\{\sum_{i,j=1}^{N} \frac{1}{\sigma_i\sigma_j}\frac{N_0}{2}\int_0^{T_s} s_i(t)s_j(t)\,dt - N\right\} \\ &= \frac{1}{N(N-1)}\left\{\frac{N_0}{2}\int_0^{T_s}\left[\sum_{i=1}^{N}\frac{s_i(t)}{\sigma_i}\right]^2 dt - N\right\} \geq -\frac{1}{N-1}\end{aligned} \tag{4.2.14}$$

Further, since $\rho_{\max} \stackrel{\Delta}{=} \max_{\substack{i,j\\ i\neq j}} \rho_{ij} \geq \rho_{\text{ave}}$, both $\rho_{\max}$ and ρ_{ave} are at least as large as $-1/(N-1)$. In particular, when $\rho_{ij} = \rho$, for all $i, j, i \neq j$, it follows that $\rho \geq -1/(N-1)$. Note, too, if ρ is to equal $-1/(N-1)$, the signal set is *linearly dependent*; that is $\sum_{i=1}^{N} s_i(t)/\sigma_i = 0$ for all t, $0 < t < T_s$.

When N is large, $\rho_{\max}$ and ρ_{ave} are bounded by approximately zero, so orthogonal alphabets are nearly optimum in the sense of minimizing these values. When $\rho_{ij} = -1/(N-1)$ for all i and j, $i \neq j$, the error probability is equal to that of an N symbol orthogonal alphabet having the energy per symbol of $[N/(N-1)]\mathscr{E}$. Such an alphabet is called a *transorthogonal* or *regular simplex* alphabet.

It might be supposed that since the effective energy is proportional to $1-\rho$, the optimum signal sets would be those for which $\rho_{\max}$ (and ρ_{ave}) achieve the minimum value of $-1/(N-1)$. This is clearly true when all the correlations $\rho_{ij}(i \neq j)$ are constrained to be equal. But while the validity of this conclusion is intuitive even without the restriction of uniform cross-correlation coefficients ρ, it remains a conjecture (cf. Reference 4.2).

4.2.4 Biorthogonal Signal Sets

Another situation of some interest is the following: Suppose $s_1(t)$, $s_2(t), \ldots, s_{N/2}(t)$ comprise a set of $N/2$ orthogonal symbols, and consider the set of N symbols $s_1(t), s_2(t), \ldots, s_{(N/2)+1}(t) = -s_1(t), s_{(N/2)+2}(t) =$

$-s_2(t), \ldots, s_N(t) = -s_{N/2}(t)$. Then

$$\rho_{ij} = \begin{cases} 0 & i \neq j, j \pm N/2 \\ 1 & i = j \\ -1 & i = j \pm N/2 \end{cases}$$

and

$$\rho_{\text{ave}} = \frac{1}{N(N-1)} \sum_{\substack{i,j \\ i \neq j}} \rho_{ij} = -\frac{1}{N-1}$$

Since $\rho_{\max} = 0 > -1/(N-1)$, this is not an optimum signal set in the sense of the above discussion, but ρ_{ave} does attain its lower bound. In addition, as will become apparent shortly, such a set, called a *biorthogonal* alphabet, does have some significant advantages over the orthogonal alphabet. Since both $s_i(t)$ and $-s_i(t)$ are in the alphabet, the optimum receiver determines the quantities

$$z_i = \int_0^{T_s} y(t) s_i(t)\, dt \qquad i = 1, 2, \ldots, N/2$$

where all of the signals are again assumed to possess equal energy and occur with equal a priori probability. The z_i with the largest absolute value is selected, the sign of z_i indicating whether $+s_i(t)$ or $-s_i(t)$ was most likely received. It is here, incidentally, that one of the more important advantages of a biorthogonal set is found: only $N/2$ matched filters are needed in an N-symbol receiver.

The probability of a correct decision† $P_c(N)$ is again independent of which signal was transmitted and is just the probability that, when $s_{N/2}(t)$ was transmitted, $z_{N/2} > |z_i|$ for all $i < N/2$:

$$P_c(N) = \Pr\left[z_{N/2} > \max_{i < N/2} |z_i| \,\middle|\, \frac{N}{2} \right]$$

Since z_i is Gaussianly distributed with $E(z_i \mid N/2) = \begin{cases} 0, & i < N/2 \\ A\mathscr{E}, & i = N/2 \end{cases}$ and $\text{Var}(z_i \mid N/2) = \sigma_i^2 = (N_0/2A^2)\mathscr{E}$

$$\begin{aligned} P_e(N) &= 1 - P_c(N) \\ &= 1 - \int_0^{\infty} \left[\int_{-\zeta}^{\zeta} \frac{e^{-\xi^2/2}}{\sqrt{2\pi}}\, d\xi \right]^{(N/2)-1} \frac{e^{-[\zeta - (2R)^{1/2}]^2/2}}{\sqrt{2\pi}}\, d\zeta \\ &= \Phi_N'(R_b) \end{aligned} \tag{4.2.15}$$

where R_b is as previously defined. Equation (4.2.15) has also been solved numerically, and the probabilities $\Phi_N'(R_b)$ are plotted and tabulated in Reference 4.1, for example, as a function of R_b for various values of N. For any

---

†Note that the covariance matrix Λ is singular in this case so Equation (4.2.7) cannot be used directly.

value of N, $\Phi'_N(R_b) < \Phi_N(R_b)$ but the difference becomes insignificant for $N \geq 8$.

Again when $N = 2$, the error probability can be expressed in terms of the error function:

$$P_e(2) = 1 - \frac{1}{\sqrt{2\pi}} \int_{-(2R_b)^{1/2}}^{\infty} e^{-\zeta^2/2} d\zeta = \frac{1}{2}[1 - \text{erf}(R_b^{1/2})] \quad (4.2.16)$$

This result can also be obtained by noting that ρ_{12} is equal to -1 for a two-symbol biorthogonal alphabet. The biorthogonal symbol error probability, when $N = 2$, is just the orthogonal error probability with an effective signal-to-noise ratio $R(1 - \rho_{12}) = 2R$.

4.2.5 *Error Probability Bounds*

The expression for the probability of an error in an N-symbol communication system is quite unwieldy when no restrictions are placed on the correlation coefficients ρ_{ij} and it can be evaluated only numerically even in the special cases just considered. Fortunately, it is not difficult to obtain simple bounds on this expression, bounds which may be entirely satisfactory for many specific applications.

The simplest and most versatile upper bound on the error probability stems from the *union bound* used in Section 2.8. That is, if the outputs of the receiver correlators are denoted by z_i [Equation (4.2.9)],† an error is made if the rth symbol is transmitted and $z_i \geq z_r$ for some $i \neq r$. Since these events are not mutually exclusive

$$P_e(N; r) \leq \sum_{\substack{i=1 \\ i \neq r}}^{N} \Pr\{z_i \geq z_r\} \leq (N - 1) \max_{i \neq r} \Pr\{z_i \geq z_r\} \quad (4.2.17a)$$

Moreover, since the probability of at least one of these events is greater than the probability of any specific one of them

$$P_e(N; r) \geq \max_{\substack{i \\ i \neq r}} \Pr\{z_i \geq z_r\} \quad (4.2.17b)$$

It is only necessary to repeat the argument leading to Equation (4.2.12) to conclude that

$$\frac{1}{2}\left\{1 - \text{erf}\left(\frac{R^*}{2}\right)^{1/2}\right\} \leq P_e(N; r) \leq \frac{(N - 1)}{2}\left\{1 - \text{erf}\left(\frac{R^*}{2}\right)^{1/2}\right\} \quad (4.2.18)$$

where $R^* = \min_{i \neq r} R(1 - \rho_{ir})$. Finally, using the well known inequalities

$$\frac{e^{-x^2/2}}{\sqrt{2\pi}\, x}\left(1 - \frac{1}{x^2}\right) \leq \frac{1}{2}\left[1 - \text{erf}\left(\frac{x}{\sqrt{2}}\right)\right] \leq \frac{e^{-x^2/2}}{\sqrt{2\pi}\, x} \qquad x \geq 0$$

---

†If the a priori probabilities and signal energies are not equal for all symbols, then z_i is as defined in Equation (4.2.3).

we find

$$\frac{e^{-R^*/2}}{(2\pi R^*)^{1/2}}\left(1-\frac{1}{R^*}\right) \leq P_e(N; r) \leq \frac{1}{(2\pi R^*)^{1/2}} \exp\left\{-\frac{R^*}{2} + \log_e(N-1)\right\} \tag{4.2.19}$$

Note that the upper bound on the error probability asymptotically approaches zero as $N \to \infty$ if $R^*/\log_2 N = R_b^* > 2 \log_e 2$.

A generally tighter upper bound can be established with only slightly more effort. The discussion here will be limited to orthogonal symbol alphabets, although the results can be extended to arbitrary signal sets as well. From Equation (4.2.11) we have

$$P_e(N) = \int_{-\infty}^{\infty} \frac{e^{-(\zeta-\zeta_0)^2/2}}{\sqrt{2\pi}} \left\{1 - \left[\frac{1}{2}\left(1 + \text{erf}\left(\frac{\zeta}{\sqrt{2}}\right)\right)\right]^{N-1}\right\} d\zeta \tag{4.2.20}$$

where $\zeta_0^2 = 2R$. But for $\zeta \geq 0$

$$\begin{aligned}\frac{1}{2}\left[1 + \text{erf}\left(\frac{\zeta}{\sqrt{2}}\right)\right] &= 1 - \int_{\zeta}^{\infty} \frac{e^{-x^2/2}}{\sqrt{2\pi}}\, dx = 1 - \left[\int_{\zeta}^{\infty}\int_{\zeta}^{\infty} \frac{e^{-(x^2+y^2)/2}}{2\pi}\, dx\, dy\right]^{1/2} \\ &\geq 1 - \left[\int_{\sqrt{2}\zeta}^{\infty}\int_{0}^{\pi/2} \frac{e^{-r^2/2}}{2\pi}\, r\, d\theta\, dr\right]^{1/2} = 1 - \frac{1}{2} e^{-\zeta^2/2}\end{aligned} \tag{4.2.21}$$

The inequality follows because $x = r \sin \theta$ and $y = r \cos \theta$ are both greater than ζ only if $0 < \theta < \pi/2$ and $r \geq \sqrt{2}\zeta$; the converse is not true. Combining Equations (4.2.20) and (4.2.21), and noting that $1 + \text{erf}(\zeta/\sqrt{2}) \geq 0$ for all ζ, and that $(1-x)^{N-1} \geq 1 - Nx$ for all $x \geq 0$, we obtain, for any $\zeta_1 \geq 0$

$$P_e(N) \leq \int_{-\infty}^{\zeta_1} \frac{e^{-(\zeta-\zeta_0)^2/2}}{\sqrt{2\pi}}\, d\zeta + \frac{Ne^{-\zeta_0^2/4}}{2} \int_{\zeta_1}^{\infty} \frac{e^{-(\zeta-\zeta_0/2)^2}}{\sqrt{2\pi}}\, d\zeta \tag{4.2.22}$$

This upper bound is minimized when ζ_1 is defined by:

$$\frac{N}{2} e^{-\zeta_1^2/2} = 1 \tag{4.2.23}$$

since then the integrand of the first term is smaller than that of the second for all $\zeta < \zeta_1$ and larger for all $\zeta > \zeta_1$. Finally, using the bound of Equation (4.2.21) and Equation (4.2.23), we have

$$\begin{aligned}P_e(N) &\leq \int_{\zeta_0-\zeta_1}^{\infty} \frac{e^{-\xi^2/2}}{\sqrt{2\pi}}\, d\xi + e^{-(1/2)(\zeta_0^2/2-\zeta_1^2)} \int_{2(\zeta_1-\zeta_0/2)}^{\infty} \frac{e^{-\zeta^2/2}}{\sqrt{2\pi}}\, d\zeta \\ &\leq \begin{cases}\frac{1}{2}[e^{-(\zeta_0-\zeta_1)^2/2} + e^{-[(\zeta_1-\zeta_0/2)^2-(1/2)(\zeta_1^2-\zeta_0^2/2)]}] = e^{-(\zeta_0-\zeta_1)^2/2} & \zeta_1 < \zeta_0 < 2\zeta_1 \\ \frac{1}{2}e^{-(\zeta_0-\zeta_1)^2/2} + e^{-(1/2)(\zeta_0^2/2-\zeta_1^2)} \leq \frac{3}{2}e^{-(1/2)(\zeta_0^2/2-\zeta_1^2)} & 2\zeta_1 < \zeta_0\end{cases}\end{aligned} \tag{4.2.24}$$

where $\zeta_0^2 = 2R$ and $\zeta_1^2 = 2 \log_e(N/2)$.

The error probability therefore approaches zero asymptotically with N if $\zeta_0 > \zeta_1$ and hence if $R_b > \log_e 2$. (The union bound assured this to be the case only if $R_b > 2 \log_e 2$.) It is not difficult to verify that this is a necessary as well as a sufficient condition for an asymptotically zero error probability; that, in fact, $P_e(N)$ is asymptotically one if $R_b < \log_e 2$.

4.3 Phase-Incoherent Detection

For purposes of radio communications it is often convenient to represent the symbols to be transmitted in the form

$$\begin{aligned} s_i(t) &= s_i(t, \phi) = \sqrt{2}\, \xi_i(t) \sin(\omega_c t + \theta_j(t) + \phi) \\ &= \sqrt{2}\, \alpha_i(t) \sin(\omega_c t + \phi) + \sqrt{2}\, \beta_i(t) \cos(\omega_c t + \phi) \qquad 0 < t < T_s \end{aligned} \tag{4.3.1}$$

This representation has no bearing on any of the preceding remarks when both ω_c and ϕ are known exactly, since no assumptions were made concerning the form of the symbols to be used. The only concern was with their cross-correlation coefficients. If, however, as in Section 3.6, we are unwilling or unable to determine the carrier phase ϕ, the received symbol must be detected incoherently, and the more detailed representation of Equation (4.3.1) is needed to evaluate the system performance.†

If ϕ is assumed to be a random variable, uniformly distributed in the range $(0,2\pi)$, and $\alpha_i(t)$ and $\beta_i(t)$ are taken to be known functions of time [assumptions which are consistent only if $\alpha_i(t)$ and $\beta_i(t)$ are slowly varying relative to $\sin(\omega_c t + \phi)$], the optimum *phase-incoherent* detector is readily determined (cf. Section 3.6). The a posteriori probability that the ith symbol was transmitted when the signal $y(t) = As_r(t, \phi) + n(t)$ was received and the phase ϕ is known is

$$\begin{aligned} P(i \mid y(t), \phi) &= k \exp\left[\frac{2A}{N_0}\int_0^{T_s} y(t)s_i(t, \phi)\, dt - \frac{\mathscr{E}_i}{N_0}\right] P(i \mid \phi) \\ &= kP(i) \exp\left\{\frac{\mathscr{E}_i}{N_0} + \frac{2A}{N_0}[X_i \cos\phi + Y_i \sin\phi]\right\} \end{aligned} \tag{4.3.2}$$

where k is independent of i, $P(i)$ is the a priori probability of the ith symbol,

$$\mathscr{E}_i = \int_0^{T_s} A^2 s_i^2(t, \phi)\, dt \qquad X_i = \int_0^{T_s} s_i(t, 0) y(t)\, dt$$

and

$$Y_i = \int_0^{T_s} s_i\left(t, \frac{\pi}{2}\right) y(t)\, dt$$

It is concluded, as in Section 3.6, that

$$P(i \mid y(t)) = kP(i) \exp\left\{-\frac{\mathscr{E}_i}{N_0}\right\} I_0\left(\frac{2Az_i}{N_0}\right) \tag{4.3.3}$$

where $z_i = (X_i^2 + Y_i^2)^{1/2}$.

---

†The signal considered in Section 3.6 was of the same form as that in Equation (4.3.1) with $\beta_i(t) = 0$. This more general expression, with $\beta_i(t)$ not necessarily zero, allows us to include angle as well as amplitude modulated waveforms.

If $P(i)$ and $\mathscr{E}_i$ are independent of i, then [since $I_0(x)$ is a monotonically increasing function of x] $P(i \mid y(t))$ is maximized when z_i attains its maximum. Consequently, the maximum-likelihood (maximum a posteriori probability) decision rule is to accept the μth symbol as the one received if and only if $z_\mu > \max_{\nu \neq \mu} z_\nu$. This detector is shown in Figure 4.3.

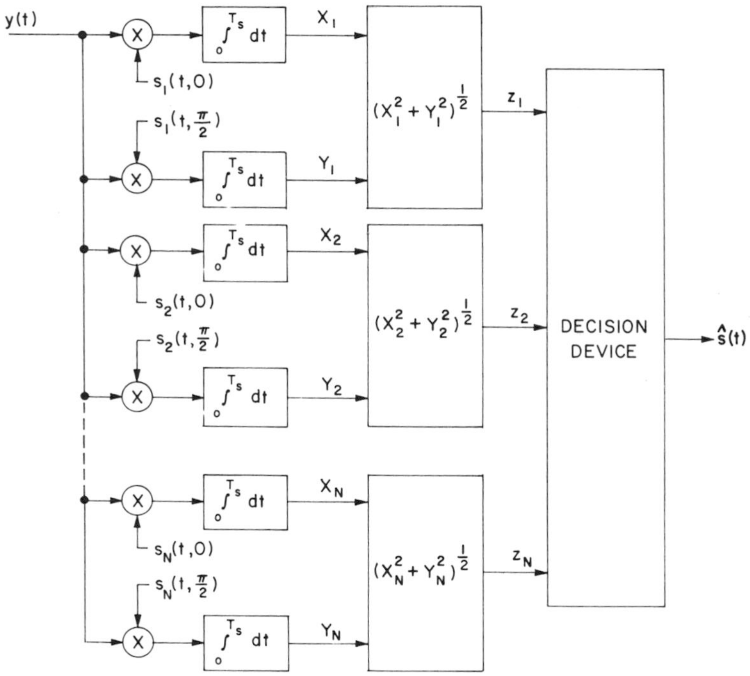

Figure 4.3 Optimum phase-incoherent demodulator.

An interesting and useful alternate implementation of the optimum phase-incoherent detector is suggested by considering the output of the filter having the impulse response

$$h_i(t) = \begin{cases} s_i(T_s - t, \phi) & 0 < t < T_s \\ 0 & \text{otherwise} \end{cases} \tag{4.3.4}$$

If the signal $y(t)$ is impressed at the input of such a filter, the output at time t is

$$\begin{aligned} \int_{-\infty}^{\infty} y(\tau) h_i(t - \tau)\, d\tau &= \int_{t-T_s}^{t} y(\tau) s_i(T_s + \tau - t, \phi)\, d\tau \\ &= Y_i(t) \cos(\omega_c t - \phi') - X_i(t) \sin(\omega_c t - \phi') \\ &= z_i(t) \cos[\omega_c t + \theta_i(t)] \end{aligned} \tag{4.3.5}$$

where

$$X_i(t) = \int_{t-T_s}^{t} y(\tau)\,[\sqrt{2}\,\alpha_i(T_s+\tau-t)\sin\omega_c\tau + \sqrt{2}\,\beta_i(T_s+\tau-t)\cos\omega_c\tau]\,d\tau$$

$$Y_i(t) = \int_{t-T_s}^{t} y(\tau)\,[\sqrt{2}\,\alpha_i(T_s+\tau-t)\cos\omega_c\tau + \sqrt{2}\,\beta_i(T_s+\tau-t)\sin\omega_c\tau]\,d\tau$$

$$\phi' = \omega_c T_s + \phi$$

$$z_i(t) = (X_i^2(t) + Y_i^2(t)^{1/2})$$

and

$$\theta_i(t) = \tan^{-1}\left(\frac{Y_i(t)}{X_i(t)}\right) - \phi'$$

The output of this filter then is a sinusoid with the envelope $z_i(t)$ and phase $\theta_i(t)$. Since $z_i(T_s)$ is precisely the decision statistic z_i defined in Equation (4.3.3), the optimum detector can evidently also be realized by a bank of filters of the form given in Equation (4.3.4), each filter followed by an *envelope detector*, a sampler, and a decision device, as shown in Figure 4.4. The filter $h_i(t)$, which we shall refer to as an *incoherent matched filter*, is matched to the signal $s_i(t, \phi)$ with ϕ arbitrary. The interpretation of $z_i(t)$ as the envelope of the function $z_i(t)\cos[\omega_c t + \theta_i(t)]$ is meaningful, and the realization of an envelope detector for extracting this envelope is possible, only if $z_i(t)$ and $\theta_i(t)$ are relatively constant over any $2\pi/\omega_c$ second interval. But since the time variations of $z_i(t)$ and $\theta_i(t)$ are due entirely to the time variation of the functions $\alpha_i(t)$ and $\beta_i(t)$ this is indeed the situation.

As a first step in determining the probability of a correct decision observe that X_i and Y_i are both Gaussian random variables with

$$\eta_x \triangleq E\{X_i | r, \phi\} = A\int_0^{T_s} s_r(t, \phi)s_i(t, 0)\,dt$$

$$= \frac{\mathscr{E}}{A}(\tilde{\rho}_{ir}\cos\phi - \hat{\rho}_{ir}\sin\phi) \qquad (4.3.6a)$$

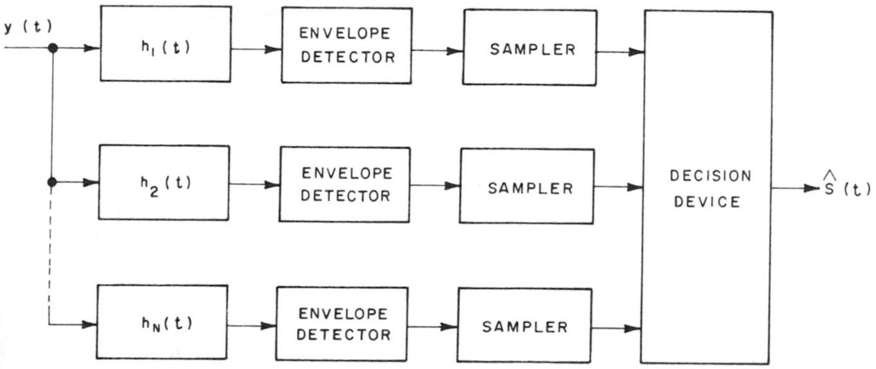

Figure 4.4 Matched filter realization of an optimum phase-incoherent demodulator.

and

$$\eta_y \triangleq E\{Y_i \mid r, \phi\} = \int_0^{T_s} s_r(t, \phi) s_i\left(t, \frac{\pi}{2}\right) dt = \frac{\mathscr{E}}{A}(\hat{\rho}_{ir} \cos\phi + \tilde{\rho}_{ir} \sin\phi) \tag{4.3.6b}$$

where

$$\tilde{\rho}_{ir} = \frac{1}{\mathscr{E}_0} \int_0^{T_s} [\alpha_i(t)\alpha_r(t) + \beta_i(t)\beta_r(t)]\, dt$$

$$\hat{\rho}_{ir} = \frac{1}{\mathscr{E}_0} \int_0^{T_s} [\alpha_i(t)\beta_r(t) - \beta_i(t)\alpha_r(t)]\, dt \tag{4.3.7}$$

and where $\mathscr{E} = \mathscr{E}_i$ is the received signal energy, assumed independent of i, and $\mathscr{E}_0 = \mathscr{E}/A^2$. The carrier frequency ω_c is assumed sufficiently large relative to the maximum significant modulation frequency to justify ignoring the integrals of the double frequency components. This assumption will be made throughout this discussion. Also, since $\tilde{\rho}_{ii} = 1$ and $\hat{\rho}_{ii} = 0$, we find

$$\operatorname{Var}\{X_i \mid r, \phi\} = \mathscr{E}_0 \frac{N_0}{2} = \operatorname{Var}\{Y_i \mid r, \phi\} \triangleq \sigma^2 \tag{4.3.8}$$

and

$$E\{X_i Y_i \mid r, \phi\} - E\{X_i \mid r, \phi\} E\{Y_i \mid r, \phi\} = 0$$

Consequently,

$$p(X_i, Y_i \mid r, \phi) = \frac{\exp\left[-\dfrac{(X_i - \eta_x)^2}{2\sigma^2}\right]}{\sqrt{2\pi}\,\sigma} \frac{\exp\left[-\dfrac{(Y_i - \eta_y)^2}{2\sigma^2}\right]}{\sqrt{2\pi}\,\sigma} \tag{4.3.9}$$

Letting $z_i = (X_i^2 + Y_i^2)^{1/2}$, $\theta_i = \tan^{-1}(Y_i/X_i)$, $a_i = \tilde{\rho}_{ir}\mathscr{E}_0$, and $b_i = \hat{\rho}_{ir}\mathscr{E}_0$, and assuming $p(\phi \mid r) = p(\phi) = 1/2\pi$, $0 \leq \phi < 2\pi$, we obtain, for $z_i \geq 0$,

$$\begin{aligned} p(z_i \mid r) &= \int_{-\infty}^{\infty} \int_{-\infty}^{\infty} p(z_i, \theta_i \mid r, \phi) p(\phi)\, d\phi\, d\theta_i \\ &= \frac{z_i}{\sigma^2} \exp\left[-\frac{z_i^2 + a_i^2 + b_i^2}{2\sigma^2}\right] I_0\left(\frac{(a_i^2 + b_i^2)^{1/2} z_i}{\sigma^2}\right) \end{aligned} \tag{4.3.10}$$

This function is sometimes called the Rician density function. When $a_i = b_i = 0$, $p(z_i \mid r)$ reduces to the Rayleigh density function:

$$p(z_i \mid r) = \frac{z_i}{\sigma^2} \exp\left(-\frac{z_i^2}{2\sigma^2}\right) \qquad z_i \geq 0 \tag{4.3.11a}$$

and

$$p(y_i = z_i^2 \mid r) = \frac{1}{2\sigma^2} \exp\left[-\frac{y_i}{2\sigma^2}\right] \qquad y_i \geq 0 \tag{4.3.11b}$$

where $p(y_i)$ will be recognized as the chi-square density function with two degrees of freedom.

The probability of a correct decision $P_c(N)$, of course, involves the joint density function $p(z_1, z_2, \ldots, z_N \mid r)$, not just the first-order density functions:

$$P_c(N) = \sum P(r) P_c(N; r) \tag{4.3.12}$$

where

$$P_c(N; r) = \Pr\{z_r > \max_{k \neq r} z_k \,|\, r\}$$

[cf. Equations (4.2.7) and (4.2.8)]. This expression, not surprisingly, is even more unwieldy than the corresponding expression in the coherent case. As before, restrictions are placed on the values of $\tilde{\rho}_{ij}$ (and $\hat{\rho}_{ij}$) in order to obtain a more manageable expression.

4.3.1 Orthogonal Symbols

If $\tilde{\rho}_{ij} = \hat{\rho}_{ij} = 0$ for all i, j, with $i \neq j$, the symbols are said to be orthogonal. The error probability in this case is quite easily determined. Since the Gaussian variables X_i and Y_i are all mutually uncorrelated, and hence, statistically independent, (i.e., since

$$E(X_i X_j) - E(X_i)E(X_j) = E(Y_i Y_j) - E(Y_i)E(Y_j) = \frac{N_0 \mathscr{E}_0}{2} \tilde{\rho}_{ij}$$

and

$$E(X_i Y_j) - E(X_i)E(Y_j) = \frac{N_0 \mathscr{E}_0}{2} \hat{\rho}_{ij}$$

for all i, j) the variables z_i, as functions of independent random variables, are also independent. Moreover, by symmetry, the error probability is again independent of the transmitted signal. As a result

$$P_c(N) = \Pr\left\{z_1 > \max_{k \neq 1} z_k \,|\, r = 1\right\}$$

$$= \int_{-\infty}^{\infty} p(z_1 | 1) \left[\int_{-\infty}^{z_1} p(z_i | 1)\, dz_i\right]^{N-1} dz_1 \tag{4.3.13}$$

From Equation (4.3.11) we have

$$\int_{-\infty}^{z_1} p(z_i | 1)\, dz_i = \frac{1}{2\sigma^2} \int_0^{z_1^2} e^{-y/2\sigma^2}\, dy = 1 - e^{-z_1^2/2\sigma^2} \qquad i \neq 1$$

and combining this with Equations (4.3.10) and (4.3.13) and defining $\zeta = z/\sigma$ and $\zeta_0 = (2R)^{1/2} = (2\mathscr{E}/N_0)^{1/2}$, we obtain

$$P_c(N) = \int_0^{\infty} \zeta \exp\left[-\frac{\zeta^2 + \zeta_0^2}{2}\right] I_0(\zeta\zeta_0)\left(1 - e^{-\zeta^2/2}\right)^{N-1} d\zeta \tag{4.3.14}$$

Using the binomial expansion,

$$(1 - e^{-\zeta^2/2})^{N-1} = \sum_{i=0}^{N-1} \binom{N-1}{i} (-1)^i e^{-i\zeta^2/2}$$

we find

$$P_c(N) = e^{-\zeta_0^2/2} \sum_{i=0}^{N-1} \binom{N-1}{i} (-1)^i \int_0^{\infty} \zeta e^{-\zeta^2(i+1)/2} I_0(\zeta\zeta_0)\, d\zeta \tag{4.3.15}$$

The solution to this last integral is known:

$$\int_0^\infty \zeta e^{-\zeta^2(i+1)/2} I_0(\zeta\zeta_0)\, d\zeta = \frac{1}{i+1} \exp\left\{\frac{\zeta_0^2}{2(i+1)}\right\} \tag{4.3.16}$$

so that

$$\begin{aligned} P_c(N) &= e^{-R} \sum_{j=1}^{N} \binom{N}{j} (-1)^{j-1} e^{R/j} \\ &= 1 - \frac{1}{N} \sum_{j=2}^{N} \binom{N}{j} (-1)^j e^{-[(j-1/j)R]} \end{aligned} \tag{4.3.17}$$

where we have used the identity

$$\binom{N-1}{i} \frac{1}{i+1} = \binom{N}{i+1} \frac{1}{N}$$

and let $j = i + 1$.

The incoherent detection error probability $P_e(N) = 1 - P_c(N)$ has also been evaluated numerically (e.g., Reference 4.3) and is plotted in Figure 4.2 as a function of $R_b = R/\log_2 N$ for various values of N. Notice when $N = 2$

$$P_e(2) = 1 - P_c(2) = \tfrac{1}{2} e^{-R/2} \tag{4.3.18}$$

4.3.2 Non-orthogonal Symbols

Expressions for the error probabilities in the phase-incoherent reception of non-orthogonal signals are generally more difficult to obtain than the corresponding phase-coherent results. The argument of Section 4.2.2, for example, concerning the method of regarding a set of symbols having uniform cross-correlation coefficients $\rho_{ij} = \rho$ as composed of part orthogonal and part identical segments is of little value in the incoherent situation since the identical segments still have bearing on the decision. Suppose $X_i = x_i + \alpha$ and $Y_i = y_i + \beta$, $i = 1, 2, \ldots, N$, where α and β are independent of i, and x_i, x_j, y_i and y_j are mutually independent for all i and j, $i \neq j$. The optimum decision is based on the maximum of the quantities $z_i = (X_i^2 + Y_i^2)^{1/2} = (x_i^2 + y_i^2 + 2\alpha x_i + 2\beta y_i + \alpha^2 + \beta^2)^{1/2}$. The cross-terms αx_i and βy_i preclude the conclusion that the decision is independent of α and β, even though these terms are common to all X_i and Y_i, respectively.

It is nevertheless possible to obtain a concise expression for the probability of a correct decision $P_c(N, \rho)$ in the incoherent reception of symbols having uniform cross-correlation coefficients $\tilde{\rho}_{ij} = \rho \geq 0$, when $\hat{\rho}_{ij} = 0$. Specifically,

$$\begin{aligned} P_c(N, \rho) = (1 - \rho)\, e^{-R} \int_0^\infty \int_0^\infty xy \exp[-\tfrac{1}{2}(x^2 + y^2)]\, I_0[(2R(1-\rho)^{1/2}x]\, I_0(\rho xy) \\ \times [1 - Q(\sqrt{\rho}\, y, x)]^{N-1}\, dx\, dy \end{aligned} \tag{4.3.19}$$

where

$$Q(\alpha, \beta) = \int_\beta^\infty x \exp(-\tfrac{1}{2}(x^2 + \alpha^2)) I_0(\alpha x)\, dx$$

is a tabulated function known as the Marcum Q function (Reference 4.4). (The proof of this result is somewhat lengthy and will not be presented here. See Reference 4.5 for details.) The error probability $P_e(N, \rho) = 1 - P_c(N, \rho)$ has been evaluated numerically and tabulated (Reference 4.12).

When $N = 2$, the phase-incoherent error probability is a function only of R and of $\rho \triangleq (\tilde{\rho}_{12}^2 + \hat{\rho}_{12}^2)^{1/2}$. A useful expression for this probability is found in Reference 4.6:

$$P_e(2, \rho) = \frac{1}{2} e^{-R/2} \left[I_0\left(\frac{\rho R}{2}\right) + 2 \sum_{n=1}^{\infty} \left(\frac{1 - \sqrt{1 - \rho^2}}{\rho} \right)^n I_n\left(\frac{\rho R}{2}\right) \right] \tag{4.3.20}$$

4.3.3 *Error Probability Bounds*

The inequalities in Equation (4.2.19), of course, are equally valid for phase-incoherent reception and establish that for arbitrary values of $\rho_{ij} = (\tilde{\rho}_{ij}^2 + \hat{\rho}_{ij}^2)^{1/2}$

$$\max_{i \neq r} P_e\{2, \rho_{ir}\} \leq P_e(N; r) \leq \max_{i \neq r} (N - 1) P_e\{2, \rho_{ir}\} \tag{4.3.21}$$

The expression $P_e\{2, \rho_{ir}\}$, as defined in Equation (4.3.20) is itself easily bounded. Since

$$0 \leq I_n(x) = \frac{1}{\pi} \int_0^{\pi} e^{x \cos\theta} \cos n\theta \, d\theta \leq \frac{1}{\pi} \int_0^{\pi} e^{x \cos\theta} \, d\theta = I_0(x) \tag{4.3.22a}$$

and since

$$1 \leq I_0(x) \leq \frac{1}{\pi} \int_0^{\pi} e^x \, d\theta = e^x \tag{4.3.22b}$$

it follows from Equation (4.3.20) that

$$\begin{aligned} \frac{1}{2} e^{-R/2} \leq P_e(2, \rho) &\leq \frac{1}{2} e^{-R/2} I_0\left(\frac{\rho R}{2}\right) \left[1 + 2 \sum_{n=1}^{\infty} \left(\frac{1 - \sqrt{1 - \rho^2}}{\rho} \right)^n \right] \\ &\leq \frac{1}{2} \left(\frac{1 + \rho}{1 - \rho} \right)^{1/2} e^{-R(1-\rho)/2} \end{aligned} \tag{4.3.23}$$

Since the upper bound on $P_e(2, \rho)$ is a monotonically increasing function of ρ, $P_e(N; r)$ can be bounded in terms of $\rho_r \triangleq \max_{i \neq r} (\tilde{\rho}_{ir}^2 + \hat{\rho}_{ir}^2)^{1/2}$. Indeed, it seems obvious that the error probability $P_e(N; r)$ will attain its minimum possible value when $\rho_r = 0$. Since the decision is based on the magnitudes of the decision variables, a negative value of either $\tilde{\rho}_{ij}$ or $\hat{\rho}_{ij}$ would presumably be equally as deleterious as a positive value. Intuitively, at least, the optimum phase-incoherent symbol set is the orthogonal set (cf. Reference 4.7).

It is interesting to compare this upper bound on the incoherent error probability with the comparable coherent probability bound [Equation (4.2.19)]. Combining Equations (4.3.21) and (4.3.23) yields for incoherent reception

$$P_e(N; r) \leq \frac{1}{2}\left(\frac{1+\rho_r}{1-\rho_r}\right)^{1/2} \exp\left[-\frac{R}{2}(1-\rho_r) + \log_e(N-1)\right] \quad (4.3.24)$$

The analogous result in the coherent case is just $(2/(1+\rho_r)\pi R)^{1/2}$ times this value. The terms ρ_r, however, are not necessarily equal in the two cases.

The argument leading to the bound in Equation (4.2.24) on the error probability for coherent reception of orthogonal symbols is also applicable here with only minor modification. Using Equation (4.3.14) and the upper bound in Equation (4.3.22) on $I_0(x)$, we obtain, with $\zeta_0^2 = 2R$,

$$P_e(N) \leq \int_0^\infty \zeta e^{-(\zeta-\zeta_0)^2/2}[1-(1-e^{-\zeta^2/2})^{N-1}]\, d\zeta$$

$$= \int_0^{\zeta_1} e^{-(\zeta-\zeta_0)^2}\, d\zeta + Ne^{-\zeta_0^2/4}\int_{\zeta_1}^\infty \zeta e^{-(\zeta-\zeta_0/2)^2}\, d\zeta \quad (4.3.25)$$

for any $\zeta_1 \geq 0$. Defining $e^{\zeta_1^2/2} = N/2$ and letting $P_0(N)$ represent the upper bound in Equation (4.2.24) on the coherent error probability, we have for all $\zeta_0 > \zeta_1$

$$P_e(N) - \sqrt{2\pi}\,\zeta_0 P_0(N) \leq \int_0^{\zeta_1} (\zeta - \zeta_0) e^{-(\zeta-\zeta_0)^2/2}\, d\zeta$$

$$+ 2e^{(1/2)(\zeta_1^2-(\zeta_0^2/2))}\int_{\zeta_1}^\infty \left(\zeta - \frac{\zeta_0}{2}\right) e^{-(\zeta-(\zeta_0/2))^2}\, d\zeta \leq e^{-(\zeta_0-\zeta_1)^2/2} \leq P_0(N) \quad (4.3.26)$$

so

$$P_e(N) \leq (2\sqrt{\pi R} + 1) P_0(N)$$

The error probability bounds for incoherent detection are evidently exponentially identical to the comparable coherent bounds, and the condition on R_b for the respective error bounds to be small when N is large is the same in the two cases. This actually should not be surprising since the carrier phase is assumed to remain constant during the entire incoherent detection interval. As $N \to \infty$, the detection interval also becomes infinite and the signal phase is, in effect, estimated perfectly.

4.4 On the Generation of Orthogonal Symbols

The previous several sections have been concerned with the performance of communication systems utilizing finite sets of symbols. The probability of an erroneous decision at the receiver was found to be a function only of the number of symbols N, their cross-correlation coefficients ρ_{ij}, and the ratio R_b of the symbol energy per bit to the noise spectral density. It was argued that the optimum values for the cross-correlation coefficients in the absence of other constraints was $\rho_{ij} = -1/(N-1)$ for phase-coherent systems and $\rho_{ij} = 0$ for phase-incoherent systems, for all i, j, with $i \neq j$. When N is large, an orthogonal set of symbols is therefore essentially optimum for both situations. It is clearly important to be able to construct such a symbol set which can be used in a practical system. In this section, we discuss two easily generated and commonly used orthogonal symbol sets, and investigate their effective bandwidths.

4.4.1 Frequency Shift Keyed Symbols (FSK)

The set of symbols used in a (discrete) *frequency shift keyed* (FSK) communication system is of the form

$$s_i(t) = \sqrt{2}\sin\left(\omega_c + \frac{\pi k_i}{T_s}\right)t \qquad i = 0, 1, 2, \ldots, N-1 \tag{4.4.1}$$

where $\omega_c = \pi l/T_s$, with l and the constants k_i all integers. The cross-correlation coefficients are

$$\rho_{ij} = \frac{2}{T_s}\int_0^{T_s}\sin\left(\omega_c + \frac{\pi k_i}{T_s}\right)t\,\sin\left(\omega_c + \frac{\pi k_j}{T_s}\right)t\,dt = \begin{cases}1 & k_i = k_j\\ 0 & k_i \neq k_j\end{cases} \tag{4.4.2}$$

and the set is orthogonal, regardless of the value of N. Orthogonality in the phase-incoherent sense, however, imposes an additional constraint on the integers k_i. For letting

$$\begin{aligned} s_i(t, \phi) &= \sqrt{2}\sin\left(\omega_c t + \frac{\pi k_i}{T_s}t + \phi\right) = \sqrt{2}\cos\frac{\pi k_i}{T_s}t\,\sin(\omega_c t + \phi) \\ &\quad + \sqrt{2}\sin\frac{\pi k_i}{T_s}t\,\cos(\omega_c t + \phi) \\ &= \sqrt{2}\,\alpha_i(t)\sin(\omega_c t + \phi) + \sqrt{2}\,\beta_i(t)\cos(\omega_c t + \phi) \end{aligned} \tag{4.4.3}$$

we find

$$\hat{\rho}_{ij} = \frac{1}{T_s}\int_0^{T_s}[\alpha_i(t)\beta_j(t) - \beta_i(t)\alpha_j(t)]\,dt = 0 \tag{4.4.4}$$

only if $k_i - k_j$ is an even integer.

A satisfactory set of orthogonal symbols for either phase-coherent or phase-incoherent reception, then, is defined by Equation (4.4.1). Since the signal frequencies need be separated by only $1/2T_s$ Hz to insure phase-coherent orthogonality, the *effective bandwidth* of such a set, according to the definition of that term in Section 3.3, is $N/2T_s$, N denoting the total number of symbols. If the orthogonal symbols are to be used in a phase-incoherent system, the effective bandwidth is twice this value.

A biorthogonal set can always be obtained from an orthogonal (phase-coherent) set by simply adjoining the symbols $-s_i(t)$ $i = 1, 2, \ldots, N$ to the original set. This procedure does not increase the bandwidth requirement over that of the orthogonal set, yet it doubles the number of symbols. Consequently the effective bandwidth for a biorthogonal set is $N/4T_s$, half that of an orthogonal set of the same size.

The power spectrum of a random sequence of any of these symbols can be calculated by using the method of Section 3.3.

4.4.2 *Pulse Position Modulation (PPM)*

A second commonly encountered orthogonal signal set is the following:

$$s_i(t) = \begin{cases} \sqrt{2N} \sin \omega_c t & \dfrac{iT_s}{N} < t < (i+1)\dfrac{T_s}{N} \\ 0 & \text{otherwise} \end{cases} \tag{4.4.5}$$

where generally, to keep the same energy per pulse, $\omega_c = l\pi N/T_s$ for some integer l, or for some $l \gg 1$. The symbols are pulses of energy, the position of the pulse defining the symbol. A communication system employing these symbols is called a *pulse position modulation* (PPM) system. Since the pulses can begin only at the instants of time $t = iT_s/N$, the system considered here is discrete PPM.

The pulses as defined in Equation (4.4.5) do not overlap in time; they are clearly orthogonal in both the phase-coherent and the phase-incoherent sense. The effective bandwidth is $N/2T_s$ Hz in the case of phase-coherent reception. That is

$$2N \int_{iT_s/N}^{(i+1)T_s/N} \sin \omega_{c_1} t \sin \omega_{c_2} t \, dt = 0$$

if ω_{c_1} and ω_{c_2} are separated by any multiple of $\pi N/T_s$ radians/second. As a result there will be no interference between any two identical PPM channels so long as their carrier frequencies are separated by some multiple of $N/2T_s$ Hz.

The same set of symbols, when detected incoherently, requires twice the effective bandwidth. This is because the cross-term

$$2N \int_{iT_s/N}^{(i+1)T_s/N} \sin \omega_{c_1} t \cos \omega_{c_2} t \, dt$$

must also vanish, requiring $\omega_{c_1} - \omega_{c_2}$ to be some multiple of $2\pi N/T_s$.

Again, by including the negative pulses $-s_i(t)$ $i = 1, 2, \ldots, N$, as well, the number of symbols can be doubled, yielding a phase-coherent biorthogonal set with an effective bandwidth half as great as that needed for an orthogonal set containing the same number of symbols.

There are, of course, many other practical methods for generating orthogonal signal sets. One technique of particular importance will be discussed in detail in Chapter Thirteen.

4.5 Phase-Shift Keying (PSK)—Phase-Coherent Detection

Another signal set of some interest is the following:

$$s_i(t) = \sqrt{2} \sin(\omega_c t + \phi_i) \qquad 0 < t < T_s \tag{4.5.1}$$

where $\phi_i = 2\pi i/N$, $i = 0, 1, \ldots, N - 1$, and $\omega_c = \pi l/T_s$ for some integer l. A system using this set of symbols differs from the PSK system discussed in Chapter Three only in the limitation here to a finite number of signal phases. The term PSK will be applied to both systems, with qualification only when there is danger of confusion.

The PSK cross-correlation coefficients (phase-coherent sense) are

$$\rho_{ij} = \cos\,(i - j)\frac{2\pi}{N} \tag{4.5.2}$$

When N is equal to 2, ρ_{ij} equals -1, for $i \neq j$, and the set is optimal in the sense of Section 4.2.3. When N is 3, ρ_{ij} equals $-\frac{1}{2}$, $i \neq j$, and again, the set is optimal. When N is 4, ρ_{ij} equals 0, for $|i - j| = 1$ or 3, and ρ_{ij} equals -1, for $|i - j| = 2$. This set is a biorthogonal set, and as such achieves the minimum average cross-correlation coefficient. The error probabilities in these cases have already been determined in Section 4.2. For larger values of N, however, the exact expressions for the error probabilities obtained in Section 4.2 are no longer applicable, since ρ_{ij} is not independent of i and j for $i \neq j$, nor is the set a biorthogonal set. The bounds of that section, of course, still apply. However, in the case of PSK modulation, it is possible to improve significantly upon those bounds.

The maximum a posteriori probability detector for PSK is easily determined. From Section 3.7, Equation (3.7.2), one finds

$$P[\phi_i\,|\,y(t)] = kP(\phi_i)\exp\left\{\frac{2A}{N_0}\int_0^{T_s} y(t)\sqrt{2}\,\sin\omega_c t\,dt\,\cos\phi_i \right.$$
$$\left. + \frac{2A}{N_0}\int_0^{T_s} y(t)\sqrt{2}\,\cos\omega_c t\,dt\,\sin\phi_i\right\} \tag{4.5.3}$$

where k is independent of ϕ_i. If $P(\phi_i)$ is also independent of i, the a posteriori probability $P[\phi_i\,|\,y(t)]$ is maximized for that ϕ_i for which

$$X\cos\phi_i + Y\sin\phi_i \tag{4.5.4}$$

is a maximum, where

$$X = \frac{1}{AT_s}\int_0^{T_s} y(t)\sqrt{2}\,\sin\omega_c t\,dt$$

and

$$Y = \frac{1}{AT_s}\int_0^{T_s} y(t)\sqrt{2}\,\cos\omega_c t\,dt$$

Defining $X = M\cos\phi$ and $Y = M\sin\phi$ ($\phi = \tan^{-1} Y/X$) we find that the a posteriori probability is maximized by that ϕ_i minimizing the value of $|\phi - \phi_i|$. For $X\cos\phi_i + Y\sin\phi_i = M\cos(\phi - \phi_i)$ is a maximum when the argument of the cosine is a minimum (modulo 2π).

Since there are N signals, equally spaced in phase, no error will be made at the receiver if, and only if, the difference θ_e between ϕ and the true phase ϕ_r of the received signal is bounded in magnitude by π/N. If $|\theta_e|$ is greater than this amount the difference $|\phi - \phi_{r+1}|$ or $|\phi - \phi_{r-1}|$ will be less than

$|\phi - \phi_r|$ and some other symbol will be selected as the received symbol. Then

$$P_e(N) = 1 - \int_{-\pi/N}^{\pi/N} p(\theta_e)\, d\theta_e \tag{4.5.5}$$

with $p(\theta_e)$ as defined in Equation (3.7.5).

This error probability is easily bounded. First note from Equation (4.5.5) that the probability of an error is independent of the transmitted signal. For a given signal-to-noise ratio, it depends only upon the number of equally spaced phase angles which are to be distinguished. Moreover, the error probability would clearly be unchanged were $\{\phi_i'\}$ rather than $\{\phi_i\}$ the set of possible signal phases, where

$$\phi_i' = \left(i - \frac{1}{2}\right)\frac{2\pi}{N} \qquad i = 0, 1, \ldots, N-1 \tag{4.5.6}$$

Now, if ϕ_0' is transmitted, and if Y is positive, an error will surely be made. For if $Y > 0$, $0 < \phi < \pi$, then some other phase ϕ_i' will be closer to ϕ than is ϕ_0'. Thus

$$P_e(N) \geq \Pr\{Y > 0 \,|\, \phi_0'\} = \Pr\left\{\frac{\pi}{N} < \phi < \frac{N+1}{N}\pi \,|\, \phi_0\right\} \tag{4.5.7}$$

But, in addition,

$$\begin{aligned} P_e(N) &= \Pr\left\{\frac{\pi}{N} < \phi < \pi \,|\, \phi_0\right\} + \Pr\left\{-\frac{\pi}{N} > \phi > -\pi \,|\, \phi_0\right\} \\ &\leq \Pr\left\{\frac{\pi}{N} < \phi < \frac{N+1}{N}\pi \,|\, \phi_0\right\} + \Pr\left\{-\frac{\pi}{N} > \phi > -\frac{N+1}{N}\pi \,|\, \phi_0\right\} \end{aligned}$$

and hence, by symmetry,

$$P_e(N) \leq 2\Pr\left\{\frac{\pi}{N} < \phi < \frac{N+1}{N}\pi \,|\, \phi_0\right\} = 2\Pr\{Y > 0 \,|\, \phi_0'\} \tag{4.5.8}$$

Since Y is a Gaussian random variable with

$$E\{Y \,|\, \phi_0'\} = -\sin\frac{\pi}{N}$$

and

$$\operatorname{Var}\{Y \,|\, \phi_0'\} = \frac{N_0}{2\mathcal{E}} \triangleq \frac{1}{2R}$$

$$\Pr\{Y > 0 \,|\, \phi_0'\} = \frac{1}{2}\left[1 - \operatorname{erf}\left(R^{1/2}\sin\frac{\pi}{N}\right)\right] \tag{4.5.9}$$

and

$$\frac{1}{2}\left[1 - \operatorname{erf}\left(R^{1/2}\sin\frac{\pi}{N}\right)\right] \leq P_e(N) \leq 1 - \operatorname{erf}\left(R^{1/2}\sin\frac{\pi}{N}\right) \tag{4.5.10}$$

The error probability, in this case, has been bounded much more tightly than is possible using the techniques of Section 4.2.5 [cf. Equation (4.2.19)].

When N is large, $R_b = R/\log_2 N$ must be of the order of $N^2/\pi^2 \log_2 N$ for $P_e(N)$ to be small. Quantized PSK therefore is generally competitive with the modulation systems discussed earlier only when N is relatively small.

The preceding argument can also be used to determine the exact error probability in the important case when $N = 2$, for then the lower bound on P_e is, in fact, attained. That is, if $\phi'_0 = -\pi/2 = 3\pi/2$, an error will be made if, *and only if,* Y is positive. If Y is negative $\pi < \phi < 2\pi$ and $|\phi - (3\pi/2)|$ will certainly be less than $|\phi - (\pi/2)|$. When $N = 2$, then,

$$P_e(2) = \tfrac{1}{2}[1 - \operatorname{erf}(R^{1/2})] \tag{4.5.11}$$

corroborating the earlier observations concerning biorthogonal symbol sets.

4.6 Phase-Shift Keying—Differentially Coherent Detection

It is apparent that ordinary PSK communication is useless without a phase reference, since the information is actually conveyed by the phase of the received signal. On the other hand, if the information resides not in the phases themselves, but rather in the difference between the phases of successive symbols, the phase reference requirement can be eliminated. In this case, it is necessary only to assume that the random phase component of the received signal remains essentially constant over a period of $2T_s$ seconds in order to be able to extract information from the received symbol sequence. Specifically, if the symbol $s_{i_1}(t) = \sqrt{2}\sin(\omega_c t + i_1(2\pi/N) + \alpha)$ is transmitted over the interval $0 < t < T_s$, and $s_{i_2}(t - T_s) = \sqrt{2}\sin(\omega_c(t - T_s) + i_2(2\pi/N) + \alpha)$ over the interval $T_s < t < 2T_s$, and if the absolute value of the change in α over this $2T_s$ second interval is small as compared to π/N radians, then the effect of this change on the probability of an erroneous decision concerning the phase difference $(i_1 - i_2)(2\pi/N)$ will be negligible. A communication system which relies on the ability to detect the difference in phase between successive PSK symbols is called a *differentially coherent PSK* or *DPSK* system.

To determine the form of the optimum (a posteriori probability) differentially coherent receiver, we begin with the usual Bayes' rule manipulation:

$$P[i_1, i_2 \,|\, y(t), \alpha] = \frac{p(y(t) \,|\, i_1, i_2, \alpha)\, P(i_1, i_2 \,|\, \alpha)}{p(y(t) \,|\, \alpha)} \tag{4.6.1}$$

where i_1 denotes the symbol received over time interval $0 < t < T_s$, i_2 the symbol received over the interval $T_s < t < 2T_s$, and $y(t)$ the received noisy signal over the interval $0 < t < 2T_s$. Subject to the usual assumption that

$P(i_1, i_2)$ is independent of the specific symbols i_1 and i_2, the optimum decision is to choose the pair of symbols i_1, i_2 maximizing the likelihood function

$$p[y(t)\,|\,i_1, i_2] = \int_0^{2\pi} p(n_1(t), n_2(t)\,|\,\alpha)\, p(\alpha)\, d\alpha \tag{4.6.2}$$

where, of course, $p(\alpha) = 1/2\pi$, $0 \leq \alpha < 2\pi$. The terms $n_1(t) = y_1(t) - s_{i_1}(t)$, $0 < t < T_s$, and $n_2(t) = y(t) - s_{i_2}(t)$, $T_s < t < 2T_s$, are as usual assumed to be white Gaussian random processes, and as they exist over disjoint time intervals, they are independent. As a result

$$\begin{aligned} p[y(t)\,|\,i_1, i_2] = \frac{K}{2\pi}\int_0^{2\pi} \exp\left\{\frac{2A}{N_0}\left[\int_0^{T_s} y(t)\sqrt{2}\sin\left(\omega_c t + i_1\frac{2\pi}{N} + \alpha\right) dt\right.\right. \\ \left.\left. + \int_{T_s}^{2T_s} y(t)\sqrt{2}\sin\left(\omega_c t + i_2\frac{2\pi}{N} + \alpha\right) dt\right]\right\} d\alpha \end{aligned} \tag{4.6.3}$$

This integral has already been encountered several times; proceeding as in Section 4.3, we find

$$\begin{aligned} p[y(t)\,|\,i_1, i_2] &= \frac{K}{2\pi}\int_{i_1(2\pi/N)}^{2\pi + i_1(2\pi/N)} \exp\{2R[X\cos\phi + Y\sin\phi]\}\, d\phi \\ &= KI_0[2R(X^2 + Y^2)^{1/2}] \end{aligned} \tag{4.6.4}$$

where

$$X = \frac{1}{AT_s}\int_0^{T_s} y(t)\sqrt{2}\sin\omega_c t\, dt + \frac{1}{AT_s}\int_{T_s}^{2T_s} y(t)\sqrt{2}\sin\left(\omega_c t + i\frac{2\pi}{N}\right) dt$$

$$Y = \frac{1}{AT_s}\int_0^{T_s} y(t)\sqrt{2}\cos\omega_c t\, dt + \frac{1}{AT_s}\int_{T_s}^{2T_s} y(t)\sqrt{2}\cos\left(\omega_c t + i\frac{2\pi}{N}\right) dt$$

and where $\phi = \alpha + i_1 2\pi/N$, and $i = i_2 - i_1$.

Since $I_0(x)$ is a monotonically increasing function of x, the optimum decision is to select that value of $i = i_2 - i_1$ maximizing the quantity $X^2 + Y^2$.

Writing

$$X = x_1 + x_2\cos i\frac{2\pi}{N} + y_2\sin i\frac{2\pi}{N}$$

and (4.6.5)

$$Y = y_1 - x_2\sin i\frac{2\pi}{N} + y_2\cos i\frac{2\pi}{N}$$

where

$$x_j = \frac{1}{AT_s}\int_{(j-i)T_s}^{jT_s} y(t)\sqrt{2}\sin\omega_c t\, dt$$

and

$$y_j = \frac{1}{AT_s}\int_{(j-1)T_s}^{jT_s} y(t)\sqrt{2}\cos\omega_c t\, dt$$

we have

$$X^2 + Y^2 = x_1^2 + y_1^2 + x_2^2 + y_2^2 + 2(x_1x_2 + y_1y_2)\cos\frac{2\pi i}{N} + 2(x_1y_2 - x_2y_1)\sin\frac{2\pi i}{N} \tag{4.6.6}$$

Defining $x_1x_2 + y_1y_2 = M\cos\phi$ and $x_1y_2 - x_2y_1 = M\sin\phi$, we find, analogously to the result obtained in the coherent case, that the optimum decision as to the differential phase $\Delta\phi = (2\pi/N)(i_2 - i_1) = \phi_2 - \phi_1$ is to select $\widehat{\Delta\phi}$ so as to minimize the value

$$|\phi - \widehat{\Delta\phi}|$$

Also note that the maximum-likelihood estimate of $\phi_1 \triangleq \alpha + i_1(2\pi/N)$ given the signal $y(t)$ is $\hat{\phi}_1 = \tan^{-1}(y_1/x_1)$ and the maximum-likelihood estimate of $\phi_2 = \alpha + i_2(2\pi/N)$ given $y(t)$ is $\hat{\phi}_2 = \tan^{-1}(y_2/x_2)$. But

$$\tan(\hat{\phi}_2 - \hat{\phi}_1) = \frac{\tan\hat{\phi}_2 - \tan\hat{\phi}_1}{1 + \tan\hat{\phi}_1\tan\hat{\phi}_2} = \frac{y_2x_1 - x_2y_1}{x_1x_2 + y_1y_2} = \tan\phi \tag{4.6.7}$$

Consequently, we have proved the following intuitively satisfying result: the optimum detector of the differential phase $\Delta\phi$ selects the value $\widehat{\Delta\phi}$ which minimizes the difference

$$|\widehat{\Delta\phi} - (\hat{\phi}_2 - \hat{\phi}_1)| \tag{4.6.8}$$

where $\hat{\phi}_2$ and $\hat{\phi}_1$ are the optimum estimates of the phases of the received symbols $s_{i_2}(t)$ and $s_{i_1}(t)$, respectively.

The detector suggested by this last observation is somewhat more complicated than it needs to be, however. A more practical implementation, shown in Figure 4.5, relies on the fact that the optimum decision requires only the two statistics $x_1x_2 + y_1y_2$ and $x_1y_2 - x_2y_1$ [see Equation (4.6.6)]. The first of these statistics is produced at the output of the upper filter in Figure 4.5 [cf. Equation (4.3.5)] and the second at the output of the lower filter. (Here $\omega_c T_s$ is assumed to be an even multiple of π.) The decision device then weights these outputs at the sampling instants $t = iT_s$ in accordance with Equation (4.6.6). In the binary case ($N = 2$), this detector is even simpler since only the upper statistic is needed and the decision depends only on the sign of this quantity.

Determining bounds on the error probability P_e in this case is rather difficult. It is possible to calculate the density function of the phase difference $\hat{\phi}_2 - \hat{\phi}_1$, but the resulting expression is cumbersome. Instead, we obtain an approximate expression for P_e as follows: The error $\theta_{e_1} = \phi_1 - \hat{\phi}_1$ in the estimate of ϕ_1, the phase of the symbol $s_{i_1}(t)$, is given by (cf. Equation 3.7.5)

$$p(\theta_{e_1}) = \frac{e^{-R}}{2\pi} + (2R)^{1/2}\cos\theta_{e_1} e^{-R\sin^2\theta_{e_1}}\frac{1}{2\pi}\int_{-\infty}^{(2R)^{1/2}\cos\theta_{e_1}} \exp\left(-\frac{x^2}{2}\right)dx$$
$$\approx \frac{(2R)^{1/2}}{\sqrt{2\pi}}\cos\theta_{e_1}\exp(-R\sin^2\theta_{e_1}) \tag{4.6.9}$$

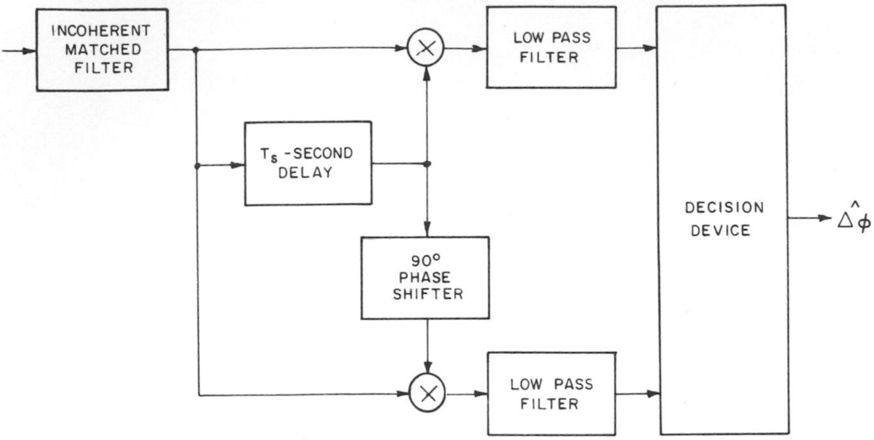

Figure 4.5 Differentially coherent PSK demodulator.

where again $R = (A^2T_s/N_0)$. One can argue that this last approximation is actually a lower bound on $p(\theta_{e_1})$ for $|\theta_{e_1}| \leq (\pi/N)$. In any event, because $|\theta_{e_1}|$ must with high probability be less than π/N if the system performance is to be acceptable, $(2R)^{1/2} \cos \theta_{e_1}$ will generally be large relative to unity, and this approximation is reasonably accurate. Similarly, defining $\theta_{e_2} = \phi_2 - \hat{\phi}_2$ where ϕ_2 is the phase of the symbol $s_{i_2}(t)$ and $\hat{\phi}_2$ its optimum estimate, we have

$$p(\theta_{e_2}) \approx \frac{(2R)^{1/2}}{\sqrt{2\pi}} \cos \theta_{e_2} \exp(-R \sin^2 \theta_{e_2}) \tag{4.6.10}$$

If, and only if, $|\theta_{e_2} - \theta_{e_1}|$ is less than $2\pi/2N$ will the decision concerning $\Delta\phi$ be correct, as only then will the function

$$|\widehat{\Delta\phi} - (\hat{\phi}_2 - \hat{\phi}_1)| = |\widehat{\Delta\phi} - \Delta\phi + \theta_{e_2} + \theta_{e_1}|$$

be minimized for $\widehat{\Delta\phi} = \Delta\phi$. Furthermore, since θ_{e_1} and θ_{e_2} must be small if a correct decision is to be made,

$$|\sin(\theta_{e_2} - \theta_{e_1})| = |\sin\theta_{e_2}\cos\theta_{e_1} - \sin\theta_{e_1}\cos\theta_{e_2}| \approx |\sin\theta_{e_2} - \sin\theta_{e_1}|$$

Therefore,

$$P_e(N) = 1 - \Pr\left\{|\theta_{e_2} - \theta_{e_1}| < \frac{\pi}{N}\right\} \approx 1 - \Pr\left\{|\sin(\theta_{e_2} - \theta_{e_1})| < \sin\frac{\pi}{N}\right\}$$

$$\approx 1 - \Pr\left\{|\sin\theta_{e_2} - \sin\theta_{e_1}| < \sin\frac{\pi}{N}\right\} \tag{4.6.11}$$

But

$$p(\alpha = \sin\theta_{e_1}) \approx \frac{(2R)^{1/2}}{\sqrt{2\pi}} \exp(-R\alpha^2)$$

and

$$p(\beta = \sin\theta_{e_2}) \approx \frac{(2R)^{1/2}}{\sqrt{2\pi}} \exp(-R\beta^2)$$

so that both α and β are approximately Gaussianly distributed, each with zero mean, and variance $1/2R$. As α and β are independent Gaussian variables, their difference is also a Gaussian random variable with $E(\alpha - \beta) = E(\alpha) - E(\beta) = 0$ and $\text{Var}(\alpha - \beta) = E(\alpha^2) + E(\beta^2) = (1/R)$. Thus

$$P_e(N) \approx 1 - \Pr\left\{|\alpha - \beta| < \sin\frac{\pi}{N}\right\} = 1 - \frac{(R)^{1/2}}{\sqrt{2\pi}} \int_{-\sin(\pi/N)}^{\sin(\pi/N)} \exp\left(-\frac{R}{2}\gamma^2\right) d\gamma$$
$$= 1 - \text{erf}\left\{\left(\frac{R}{2}\right)^{1/2} \sin\frac{\pi}{N}\right\} \tag{4.6.12}$$

A comparison of Equations (4.5.10) and (4.6.12) shows that the error probability for differentially coherent detection is essentially equal to that for coherent detection when the ratio of the signal energy to the noise spectral density in the latter situation is halved.

The validity of this expression for the error probability depends critically upon the approximation of Equation (4.6.11). This approximation in turn is dependent upon the assumption that $R\cos\theta_e$ is large. For large values of N, the condition that $|\theta_{e_1} - \theta_{e_2}| < \pi/N$ is sufficient to insure that θ_{e_1} and θ_{e_2} are individually small, and that the necessary assumptions are satisfied. Thus, the estimate of the error probability obtained here can be expected to be meaningful for reasonably large N. In the important case when $N = 2$, by the same token, the accuracy of this expression is suspect.

Fortunately, a different line of reasoning leads to an exact formula for the error probability when $N = 2$. For then there are only two possible symbols, $s_1(t) = \sqrt{2}\sin\omega_c t$ and $s_2(t) = -s_1(t)$. In attempting to ascertain the differential phase between two successive symbols, one needs to distinguish between only two events: either $s_i(t)$ was received [over the interval $(0, T_s)$] followed by $s_i(t - T_s)$ [over the interval $(T_s, 2T_s)$] or $s_i(t)$ was received followed by $-s_i(t - T_s)$. The situation is the same whether $i = 1$ or 2 since no attempt is ever made to determine i. Regardless of i, therefore, it is only necessary to distinguish between the two "extended-symbols" $u_i(t) = s_i(t), s_i(t - T_s)$ [indicating $s_i(t)$ followed by itself one symbol period later], and $u_2(t) = s_i(t), -s_i(t - T_s)$. This is entirely equivalent to making a phase-incoherent decision as to whether $u_1(t)$ or $u_2(t)$ was transmitted. Because $u_1(t)$ and $u_2(t)$ are obviously mutually orthogonal, each representing twice the energy in either of the symbols $s_1(t)$ and $s_2(t)$, the error probability is, from Equation (4.3.18)

$$P_e(2) = \tfrac{1}{2}\exp(-R) \tag{4.6.13}$$

In contrast, the approximation in Equation (4.6.12), for $N = 2$, gives

$$P_e(2) \approx \frac{2e^{-R/2}}{\sqrt{2\pi}\,R} \tag{4.6.14}$$

in this instance, a rather conservative estimate.

4.7 Bit versus Symbol versus Word Error Probabilities

The figure of merit for digital communication systems used throughout this chapter has been the probability of committing a symbol error. It should be realized, however, that even if two systems yield the same symbol error probability, their performances, from the user's viewpoint, may be quite different. The greater the number of bits per symbol, for example, the more these bit errors cluster together. If the symbol error probability is 10^{-3}, the expected number of symbols occurring between any two erroneous symbols is 1000. If each symbol represents one bit of information, the expected number of bits separating two erroneous bits is 1000, while the expected separation when there are 10 bits per symbol is 10,000 bits. Of course, a symbol error generally creates more bit errors in the second case, so the percentage of bit errors tends to be the same, as we shall see shortly. Nevertheless, this clustering effect may make one system more attractive than another, even at the same symbol error rate. Which one is preferable will depend upon the particular situation.

One rather common alternative figure of merit for digital systems is the probability of a bit error rather than that of a symbol error. But this measure suffers from the same disadvantage as the symbol error probability measure: it is not necessarily the measure of greatest interest to the user. Typically, the information bits can be grouped into information *words* of, say, k bits each. Each word may correspond to one experimental observation, for example, or to one data sample. When this is the case, it often follows that the significant figure of merit is the information *word* error probability. The user wants the received word to be correct; if any one bit in a word is in error, the information represented by the word may be useless to him. But to compare two systems on the basis of their word error probabilities is rather impractical, as a different comparison would result for each value of k, the number of bits per word.

In short, while the symbol error probability may not be the measure of concern to a particular user, any other measure would tend to be limited to a specific situation. Fortunately, both of the alternative measures mentioned above, i.e., the bit error probability and the word error probability, can generally be calculated quite easily from the symbol error probability. The next few paragraphs elaborate on this.

(1) Bit Error Probabilities; Uniform Cross-Correlation Symbol Sets ($\rho_{ij} = \rho$). To determine the bit error probability P_b in terms of the symbol error probability P_e in an N symbol communication system in which the cross-correlation coefficients $\rho_{ij} = \rho$, for all i, j, with $i \neq j$, we need only

observe that the probability of any one symbol being mistaken for any other is independent of the particular symbols involved. Given that a symbol error has been made, then, the transmitted binary n-tuple ($n = \log_2 N$) was equally likely to have been any one of the $2^n - 1$ possible n-tuples other than the one observed. Since there are $\binom{n}{i}$ n-tuples differing from a given n-tuple in exactly i bits, the expected fraction of the n bits which are in error, given that a symbol is in error, is

$$\eta_b = E(\text{fraction of bits in error} \mid \text{symbol in error})$$
$$= \frac{1}{2^n - 1}\sum_{i=1}^{n}\frac{i}{n}\binom{n}{i} = \frac{2^{n-1}}{2^n - 1} \tag{4.7.1}$$

and

$$P_b = \eta_b P_e = \frac{N}{2(N-1)} P_e \approx \frac{1}{2} P_e \tag{4.7.2}$$

(2) Bit Error Probabilities—Biorthogonal Symbol Sets. Denoting by P_{e_1} the probability of mistaking a symbol for any specified symbol other than its complement, by P_{e_2} the probability of mistaking a symbol for its complement, and by P_e the probability of a symbol error, we have

$$P_e = (N-2)P_{e_1} + P_{e_2} \tag{4.7.3}$$

Since $P_{e_2} < P_{e_1}$ the expected number of erroneous bits is obviously minimized when complementary n-tuples are represented by complementary symbols. When this is the case,

$$P_b = P_{e_1}\sum_{i=1}^{n-1}\frac{i}{n}\binom{n}{i} + P_{e_2} = (2^{n-1} - 1)P_{e_1} + P_{e_2} = \frac{1}{2}(P_e + P_{e_2}) \approx \frac{P_e}{2} \tag{4.7.4}$$

The approximation made in the last step of Equation (4.7.4) is valid except for small values of N. When $N = 2$, $P_e = P_{e_2}$ and P_b equals P_e not $P_e/2$. Interestingly, the *bit* error probabilities when $N = 2$ and $N = 4$ are identical when the two symbol sets represent the same signal energy per bit. To see this, let x denote the output of a filter matched to one of the $N = 2$ biorthogonal symbols, and x_1 and x_2 the outputs of filters matched to two of the orthogonal symbols in an $N = 4$ biorthogonal set. Suppose the mapping were such that when $x_1 > |x_2|$, the received symbol is identified with the message 00 and when $x_2 > |x_1|$ it is identified with 01. If the symbol 00 is actually transmitted, an error in the first bit occurs if and only if $x_1 + x_2 \leq 0$; for if $x_1 + x_2$ were positive, either $x_1 > |x_2|$ or $x_2 > |x_1|$. Thus, the probability of a bit error (by symmetry it is the same for either of the two bits) is the probability of the event $y \triangleq (x_1 + x_2)/2 \leq 0$. But since x and y are both Gaussian variables with $E(y \mid 00) = \frac{1}{2}E(x_1 \mid 00) = E(x \mid 0)$ and $\text{Var}(y) = \frac{1}{4}\text{Var}(x_1) + \frac{1}{4}\text{Var}(x_2) = \frac{1}{2}\text{Var}(x_1) = \text{Var}(x)$, $\Pr(y \leq 0 \mid 00) = \Pr(x \leq 0 \mid 0)$ and the bit error probabilities in the two cases are identical.

(3) Bit Error Probabilities—PSK Symbols. In this instance the probability of mistaking one symbol for either of the two "nearest" (in phase) symbols is clearly much greater, when the symbol error probability is acceptably small, than any other kind of symbol error. By mapping the binary n-tuples onto the symbols in such a way that the two n-tuples corresponding to any two adjacent symbols differ from each other in only one bit position, we can assure that by far the most probable number of bit errors, given a symbol error, is just one. Accordingly, subject to such a mapping,

$$P_b \approx \frac{P_e}{\log_2 N} \tag{4.7.5}$$

A mapping satisfying the constraint described in the preceding paragraph is called a *Gray code* (Reference 4.8). The existence of Gray codes of n-tuples for all integers n is easily established by induction. For let $g_1, g_2, \ldots, g_N$ be an ordering of binary n-tuples such that g_i and g_{i+1} differ in only one bit for all $i = 1, 2, \ldots, N$ (with $N + 1 \equiv 1$). Further, let νg_i denote the $(n + 1)$-tuple obtained by prefixing g_i with the binary digit ν. Then, clearly, $0g_1, 0g_2, \ldots, 0g_N, 1g_N, 1g_{N-1}, \ldots, 1g_2, 1g_1$, is an ordering of $(n + 1)$-tuples satisfying the same constraint. Since 0,1 is such an ordering for $n = 1$, Gray codes exist for all n.

(4) Data Word Error Probabilities. If k bits constitute a data word, and if one symbol is used to represent each bit, the probability of a word error P_w is

$$P_w = 1 - (1 - P_e)^k \tag{4.7.6}$$

where P_e is the probability of a symbol error. Figure 4.6 shows a comparison of the word error probabilities when (1) each bit of a 5-bit data word is transmitted as a biorthogonal symbol, and (2) when each word is transmitted as one of 32 biorthogonal symbols.

Perhaps the most significant conclusion to be drawn from the investigation of word error rather than bit or symbol error probabilities arises from the behavior of these probabilities at low signal-to-noise ratios R. Using either of the latter two measures, one finds it better for sufficiently low values of R to use as few symbols as possible. This situation quickly changes, however, as R is increased. If word error probabilities are of concern, in contrast, it appears to be always better to use a symbol to represent each word, rather than each bit (cf. Figure 4.6).

This latter phenomenon is actually quite easily explained. If the word error probability is of interest, the probability of mistaking the waveform representing one data *word* for any waveform representing another data word is the measure of concern, regardless of the number of bits corresponding to each symbol. If each bit is individually represented by one transorthogonal symbol ($\rho_{12} = -1$), for example, and k bits form a data word, the correlation between two such waveforms (representing data words) varies from

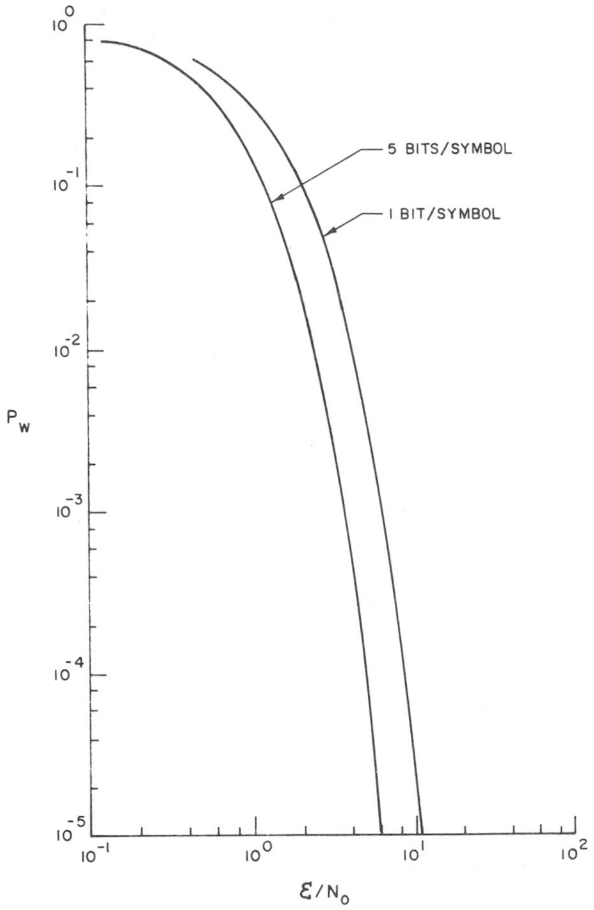

Figure 4.6 Comparison of word error probabilities ($k = 5$).

$\rho_{ij} = (k - 2)/k$ to -1. But, in a coherent system, transorthogonal codes, having $\rho_{ij} = -1/(2^k - 1)$ for all i and j, with $i \neq j$, are presumably optimum for all values of R (cf. Section 4.2.3). The superiority of mapping words rather than individual bits onto transorthogonal symbols must follow. Because the performance of both orthogonal and biorthogonal symbol sets is so similar to that of transorthogonal sets, we should expect the same conclusions to hold for the former alphabets as well.

Finally, when l data words are mapped onto one symbol, the word error probability can be expressed in terms of the symbol error probabilities in a manner similar to that used to obtain bit error probabilities from symbol error probabilities. As an example, suppose two words of k bits each are to be transmitted as one of 2^{2k} orthogonal symbols. The expected fraction of data words in error, given that a symbol is in error, is

$$\eta_w = \tfrac{1}{2}\text{Pr (1st word correct and second word in error)}$$
$$+ \tfrac{1}{2}\text{Pr (2nd word correct and first word in error)}$$
$$+ \text{Pr (both words in error)}$$
$$= \frac{2^k - 1}{2^{2k} - 1} + \frac{2^{2k} - 1 - 2(2^k - 1)}{2^{2k} - 1} = \frac{2^k}{2^k + 1} \tag{4.7.7}$$

Therefore,

$$P_w = \frac{2^k}{2^k + 1} P_e \tag{4.7.8}$$

4.8 Concluding Remarks

The probability of an erroneous reception in an equal-energy N-symbol communication system operating over a white Gaussian channel is a function only of the signal and noise levels at the outputs of the receiver filters and of the correlation (in either the coherent or the incoherent sense) between these various outputs. Comments of the sort made in Section 3.8, therefore, are equally pertinent here, subject to the same general conditions. In particular, the specific waveforms used in a communication system (of the sort hypothesized here) are irrelevant; all symbol alphabets having the same set of correlation coefficients perform equally well when operating at the same signal and noise levels. Moreover, the performance analyses of this chapter are not restricted to optimum detectors. The results are readily extended to other than matched filter detectors so long as the output signal and noise levels and the correlation between these various outputs can be determined. The results obtained using matched filters, however, provide lower bounds on the probability of error attainable by using any other set of filters.

The communication systems detailed in this and the preceding chapter will serve as the nucleus for subsequent discussions of the synchronization problem. This does not imply that all communication systems of interest have been included in these investigations. Indeed, some obvious modifications (e.g., discrete-amplitude PAM, continuous-amplitude PPM, FSK, and differentially coherent PSK) and hybridizations (e.g., PAM-PSK) of those systems immediately suggest themselves. Nevertheless, those systems which were discussed are sufficiently representative to enable us to broach the subject of synchronization with considerable generality.

Notes to Chapter Four

The first detailed investigation of an M-ary coherent channel was evidently that of Kotel'nikov (Reference 4.9; see also References 4.5 and 4.11).

Similar results for the incoherent channel were obtained in References 4.5 and 4.10. Error probabilities associated with this channel for various signal-to-noise ratios and correlation coefficients are tabulated in Reference 4.12. The error probability curves plotted in Figure 4.2 were taken from References 4.11 (coherent case) and 4.3 (incoherent case), and those in Figure 4.6 from Reference 4.13. The permission of the authors and publishers in each case is gratefully acknowledged. The error probability bounding arguments in Sections 4.2 and 4.3 are due to Wozencraft and Jacobs (Reference 4.14). The development in Sections 4.5 and 4.6 closely parallels the work of Arthurs and Dym (Reference 4.15).

Problems

4.1 (a) Suppose the signal

$$\sqrt{2}\,\gamma\xi(t)\sin(\omega_c t + \phi) + n(t), \qquad 0 \leq t \leq T_s$$

is received, with $\xi(t)$ one of two equally likely, equal energy, orthogonal signals, and $n(t)$ white Gaussian noise. The carrier phase is a random variable uniformly distributed over the interval $(0, 2\pi)$, and the signal amplitude γ is a Rayleigh-distributed random variable:

$$p(\gamma) = 2\gamma e^{-\gamma^2}, \qquad \gamma \geq 0$$

[If these random amplitude and phase characteristics are duc to the channel, the channel is called a *Rayleigh-fading* or *multipath* channel. Such a channel results, for example, when the signal energy travels over a large number of paths, with different paths subject to nearly equal attenuations but varying delays. When γ and ϕ can be assumed constant over any symbol interval, as here, the fading is said to be slow.] Show that the minimum attainable symbol error probability is

$$P_e = \frac{1}{2 + \bar{R}}$$

where $\bar{R}$ is the ratio of the average received energy per symbol to the noise spectral density.

(b) Now suppose the received signal is of the form

$$\sqrt{2}\,\xi(t)\sin\omega_c t + \sqrt{2}\,\gamma\xi(t)\sin(\omega_c t + \phi) + n(t)$$

with all terms as previously defined. [The first term, called the *spectral* component, might arise if one of the signal paths dominated all the rest.] Show that if the same receiver is used for this signal as was used in part (a) of this problem, the error probability becomes

$$P_e' = P_e \exp\{-P_e R_s\}$$

where P_e is as defined in part (a), and R_s is the ratio of the energy in the spectral component to the noise spectral density.

(c) Describe the optimum receiver for the signal defined in part (b).

4.2 (Reference 4.15). Show that $P_e(N)$, as defined in Equation (4.2.8), satisfies the inequalities

$$\frac{1}{2}\left\{1 - \operatorname{erf}\left(\frac{\bar{R}^*}{2}\right)^{1/2}\right\} \le P_e(N) \le \frac{N-1}{2}\left\{1 - \operatorname{erf}\left(\frac{R^{**}}{2}\right)^{1/2}\right\}$$

where

$$\bar{R}^* = R\left\{\frac{1}{N}\sum_{r=1}^{N}(1-\rho_r)^{1/2}\right\}^2$$

$$R^{**} = \min_r R(1-\rho_r)$$

and

$$\rho_r = \min_{\substack{i \\ i \neq r}} (1 - \rho_{ir})$$

[cf. Equation (4.2.18)].

4.3 Use the extreme-value results of Appendix A.3 to demonstrate that the orthogonal error probability bounds of Sections 4.2 and 4.3 are exact as $N \to \infty$.

References

4.1 S. W. Golomb, L. D. Baumert, M. F. Easterling, J. J. Stiffler, and A. J. Viterbi, *Digital Communications*, Prentice-Hall, Inc., Englewood Cliffs, N. J., 1964.

4.2 S. Farber, "On the Signal Selection Problem for Phase Coherent and Incoherent Communication Channels," Ph. D. Dissertation, Dept. of E. E., Calif. Inst. Tech., Pasadena, Calif. 1968.

4.3 W. C. Lindsey, "Coded Non-coherent Communications," *IEEE Trans. on Space Electronics and Telemetry*, SET-11, p. 6, 1965.

4.4 J. I. Marcum, *Tables of Q Functions*, The RAND Corp., RM-339, Santa Monica, Calif., 1950.

4.5 A. H. Nuttall, "Error Probabilities for Equicorrelated *M*-ary Signals Under Phase-coherent and Phase-incoherent Reception," *IRE Trans. Inf. Theory*, **1**, p. 305, 1962.

4.6 C. W. Helstrom, "The Resolution of Signals in White Gaussian Noise," *Proc. IRE*, **43**, No. 9, p. 1111, 1955.

4.7 R. A. Scholtz, and C. L. Weber, "Signal Design for Phase-Incoherent Communications," *IEEE Trans. Inf. Theory*, IT-12, p. 456, 1966.

4.8 W. M. Goodall, "Television by Pulse Code Modulation," *Bell System Tech. J.*, **30**, p. 33, 1951.

4.9 V. A. Kotel'nikov, *Theory of Optimum Noise Immunity*, McGraw-Hill Book Co., New York, N. Y., 1960.

4.10 S. Reiger, "Error Rates in Data Transmission," *Proc. IRE*, **46**, p. 919, 1958.

4.11 A. J. Viterbi, "On Coded Phase-coherent Communication," *IRE Trans. on Space Electronics and Telemetry*, SET-7, p. 3, 1961.

4.12 D. Shnidman, "Tabulation of the Probability of Detection and Correct Decision Among *M* Alternatives for Phase-incoherent Reception," *Air Force Cambridge Res. Lab. Res. Rep.* AFCRL-63-374, 1963.

4.13 A. J. Viterbi, *Principles of Coherent Communication*, McGraw-Hill Book Co., New York, N. Y., 1966.

4.14 J. M. Wozencraft, and I. M. Jacobs, *Principles of Communication Engineering*, John Wiley & Sons, Inc., New York, N. Y., 1965.

4.15 E. Arthurs, and H. Dym, "On the Optimum Detection of Digital Signals in the Presence of White Gaussian Noise," *IRE Trans. on Comm. Systems*, CS-10, p. 336, 1962.

Supplementary Bibliography

Balakrishnan, A. V., "A Contribution to the Sphere-packing Problem of Communication Theory," *J. Math. Anal. and Appl.*, **3**, p. 485, 1961.

Fano, R. M., *Transmission of Information*, The MIT Press, Cambridge, Mass., and John Wiley & Sons, Inc., New York, N. Y., 1961.

Lawton, J. G., "Comparison of Binary Data Transmission Systems," *2nd Nat. Conf. on Military Electron.*, 1958.

Turin, G. L., "The Asymptotic Behavior of Ideal *M*-ary Systems," *Proc. IRE*, (correspondence), **47**, p. 93, 1959.

Weber, C. L., "New Solutions to the Signal Design Problem for Coherent Channels," *IEEE Trans. Inf. Theory*, IT-12, p. 161, 1966.

chapter five

Optimum Filters, Matched Filters, and Phase-Locked Loops

5.1 Introduction

The problems which have been considered in the past two chapters have, in some form or other, been concerned with gleaning information from a noise-corrupted signal. In every case, the signal was a sequence of waveforms, each waveform a function of one or more parameters which assumed and retained unchanged a particular value for a known length of time, T_s seconds. The problem involved the determination, with as much accuracy as possible, of the parameter representing the desired information. This gave rise to the matched filters and related devices discussed in the preceding chapters.

We begin this chapter by investigating the form of the "optimum" filter for a considerably more general class of signals. Specifically, the received signal is a sample function of an arbitrary wide-sense stationary random process, described only in terms of its autocorrelation function (or, equivalently, its power spectral density). Let the filter input (the observables) be $y(t)$. [Generally $y(t)$ will be of the form $y(t) = x(t) + n(t)$, $x(t)$ representing the information and $n(t)$ denoting the noise.] Let the output be denoted $\hat{z}(t)$ and the desired output $z(t)$. Then the optimum filter is defined as the one producing the best approximation in the mean-squared error sense to the desired signal $z(t)$; i.e., the optimum filter is to be chosen so as to minimize the error

$$\sigma_e^2 = E[\hat{z}(t) - z(t)]^2 \tag{5.1.1}$$

The random processes $y(t)$ and $z(t)$ are assumed stationary and the filter time invariant so σ_e^2 will be independent of the time t.

In order to determine the optimum filter it is generally necessary to place restrictions on the class of devices which are to be considered. To this end, we specify that the filter be linear; i.e., that $\hat{z}(t)$ be a linear function of the observables $y(t)$. This restriction has a twofold advantage: it not only results in a relatively simple optimization procedure, but also yields a device which can be realized physically (or, at least, closely approximated). The transfer function of the optimum linear filter will be derived in the following section.

It should be acknowledged that some error criterion other than the mean-squared error might be as satisfactory, or, in some situations, more so. The mean absolute error

$$e_a = E|\hat{z}(t) - z(t)| \tag{5.1.2}$$

or the maximum absolute error

$$e_M = \max_t |\hat{z}(t) - z(t)| \tag{5.1.3}$$

for example, could well be of greater concern to the user than the mean-squared error. The mean error

$$\eta = E[\hat{z}(t) - z(t)] \tag{5.1.4}$$

is, of course also significant. However, if η is other than zero, the mean-squared error could easily be reduced, as is readily shown. For suppose $\hat{z}(t)$ is the output of the optimum mean-squared filter, and consider the function

$$\sigma_e^2(a) = E[\hat{z}(t) - a - z(t)]^2 = E[\hat{z}(t) - z(t)]^2 - 2a\eta + a^2$$

Since $\sigma_e^2(a)$ is a monotonically increasing function of $|a - \eta|$, it is possible to decrease the mean-squared error of the filter output by subtracting from it the constant η. For all practical purposes, then, the optimum linear filter in the mean-squared sense would not be altered were a condition also imposed on the mean error. In all cases of interest, in fact, the mean error at the output of such a filter will indeed be zero.†

The reason for selecting the mean-squared error criterion over the other possibilities is primarily expedience. The mathematical manipulations are tractable and in fact a solution to the problem can be obtained, a statement which is untrue in general for almost any of the other non-trivial criteria one could choose. Moreover, the mean-squared error is a rather intuitively satisfying measure. It is, after all, the "power" in, or, from a statistical point of view, the variance of, the error signal. Finally, if further support is needed

†This is not to say that the estimate $\hat{z}(t)$ of $z(t)$ is unbiased (cf. Section 2.9). Rather, the expected value of the bias is zero.

for accepting the mean-squared error criterion, we mention that in many important cases, the resulting filter is optimum under a considerably more general class of criteria. The problem outlined, after all, is just an estimation problem of the kind discussed in Section 2.9. (As in previous chapters, we can first filter and sample the received signal, to restrict the set of observables to a discrete set Y, and then let the prefilter bandwidth become infinite.) The task is to estimate the variable $z(t_0)$ at some time t_0, on the basis of the observables $y(t)$, $t_1 < t < t_2$. (In general, t_1 will be $-\infty$ and $t_2 \leq t_0$ in the problems to be considered here.) But, as was stated in Section 2.9, the optimum estimate $\hat{z}(t_0)$ of $z(t_0)$ under a wide variety of criteria, including the mean-squared error criterion, is $\hat{z}(t_0) = E[z(t_0) \mid Y]$. In particular, this is true if the cost of an error $e(t_0) = \hat{z}(t_0) - z(t_0)$ is any monotonically non-decreasing function of $e(t_0)$ and if the density function $p[z(t_0) \mid Y)]$ is a symmetric, unimodal function of $z(t_0)$. This situation arises, for example, when $y(t)$ and $z(t)$ are joint Gaussian random processes. It is also interesting to note that $E[z(t_0) \mid Y]$ is a linear function of the observables $Y = \{y_i\}$ when the random variables $z(t_0)$, $y_1, y_2, \ldots$, are jointly Gaussianly distributed. Thus, when $y(t)$ and $z(t)$ are joint Gaussian processes, the restriction that $\hat{z}(t_0)$ be a linear function of $y(t)$ is vacuous. The same result follows without this restriction for any cost function which is a monotonically non-decreasing function of the error $e(t_0) = \hat{z}(t_0) - z(t_0)$.

5.2 The Optimum Linear Filter

The optimum filter, as defined in the preceding paragraphs, is the linear filter which minimizes the mean-squared difference σ_e^2 between some desired signal $z(t)$ and the filter output $\hat{z}(t)$. Thus, we wish to find the filter with the impulse response $h(t)$ such that

$$\begin{aligned}\sigma_e^2 &= E[\hat{z}(t) - z(t)]^2 = E\left\{\int_{-\infty}^{\infty} y(t-\tau)h(\tau)\,d\tau - z(t)\right\}^2 \\ &= R_z(0) - 2\int_{-\infty}^{\infty} R_{yz}(\tau)h(\tau)\,d\tau + \int_{-\infty}^{\infty}\int_{-\infty}^{\infty} R_y(\tau-\eta)h(\tau)h(\eta)\,d\tau\,d\eta\end{aligned} \tag{5.2.1}$$

attains its minimum possible value. The terms $R_{\alpha\beta}(\tau)$, of course, are the correlation functions: $R_{\alpha\beta}(\tau) = E(\alpha(t)\beta(t+\tau))$ and $R_\alpha(\tau) = R_{\alpha\alpha}(\tau)$.

As we shall see, the optimum filter is readily determined when no additional constraints are placed on it. One constraint which must usually be invoked, however, if the solution is to be of any practical value, is that the filter be *causal*, that it not produce an output before it receives an input, and, hence, that its impulse response $h(t)$ be identically zero for all $t < 0$. Although this constraint somewhat complicates the analysis, the optimum causal filter can be determined with only slightly more effort.

It will also be convenient in the following derivations to assume the existence of the transfer function $H(j\omega)$ of the optimum filter. This assumption is easily seen to be justified if the filter is to be *stable*; i.e., if every bounded input $x(t)$ into the filter is to produce a bounded output $y(t)$. To verify this statement, we show that a filter is stable if and only if its impulse response is absolutely integrable:

$$\int_{-\infty}^{\infty} |h(t)|\, dt < \infty \tag{5.2.2}$$

This is a sufficient condition for the existence of the Fourier transform $H(j\omega)$, since

$$|H(j\omega)| = \left| \int_{-\infty}^{\infty} h(t) e^{-j\omega t}\, dt \right| \leq \int_{-\infty}^{\infty} |h(t)|\,|e^{-j\omega t}|\, dt = \int_{-\infty}^{\infty} |h(t)|\, dt$$

To prove the equivalence between the stability of a filter and the absolute integrability of its transfer function, we note that

$$|y(t)| = \left| \int_{-\infty}^{\infty} h(\tau)x(t-\tau)\, d\tau \right| \leq \int_{-\infty}^{\infty} |h(\tau)|\,|x(t-\tau)|\, d\tau$$

and if $|x(t)| \leq X$ for all t

$$|y(t)| \leq X \int |h(\tau)|\, d\tau.$$

On the other hand, if, for some value of t, the input is

$$x(t-\tau) = \operatorname{sgn} h(\tau) = \begin{cases} 1 & h(\tau) > 0 \\ -1 & h(\tau) < 0 \end{cases}$$

for all τ, then, for this input,

$$y(t) = \int_{-\infty}^{\infty} |h(\tau)|\, d\tau$$

and the statement is proved.

It is not difficult to show that the optimum filter has an impulse response satisfying the following integral equation, known as the *Wiener-Hopf equation:*

$$\int_{a}^{\infty} h(\tau) R_y(\tau - \eta)\, d\tau = R_{yz}(\eta) \tag{5.2.3}$$

for all $\eta > a$. If the filter is to be causal, $a = 0$; otherwise $a = -\infty$. The mean-squared error when the filter impulse response satisfies this equation is

$$\sigma_e^2 = R_z(0) - \int_{a}^{\infty}\int_{a}^{\infty} h(\tau)h(\eta)R_y(\tau-\eta)\, d\tau\, d\eta = R_z(0) - \int_{a}^{\infty} R_{yz}(\tau)h(\tau)\, d\tau \tag{5.2.4}$$

as can be verified by substituting Equation (5.2.3) into Equation (5.2.1).

To demonstrate the optimality of a filter satisfying Equation (5.2.3), we show that the error difference $\sigma_e^2\,[g(t)] - \sigma_e^2[h(t)]$ is non-negative, $\sigma_e^2[g(t)]$ representing the mean-squared error when an arbitrary filter $g(t)$ is used and $\sigma_e^2[h(t)]$ the same quantity when the filter satisfying Equation (5.2.3) is used. If $h(t)$ is to be causal, then, of course, so must be $g(t)$. Otherwise, $g(t)$ is unconstrained. Using both Equations (5.2.1) and (5.2.4), we have

$$\begin{aligned}
&\sigma_e^2[g(t)] - \sigma_e^2[h(t)] \\
&= -2\int_a^\infty g(\tau)R_{yz}(\tau)\,d\tau + \int_a^\infty\int_a^\infty [g(\tau)g(\eta) + h(\tau)h(\eta)]R_y(\tau-\eta)\,d\tau\,d\eta \\
&= -2\int_a^\infty\int_a^\infty g(\tau)h(\eta)R_y(\tau-\eta)\,d\tau\,d\eta \\
&\quad + \int_a^\infty\int_a^\infty [g(\tau)g(\eta) + h(\tau)h(\eta)]R_y(\tau-\eta)\,d\tau\,d\eta \qquad (5.2.5)\\
&= \int_a^\infty\int_a^\infty [g(\tau)-h(\tau)][g(\eta)-h(\eta)]R_y(\tau-\eta)\,d\tau\,d\eta \\
&= \frac{1}{2\pi}\int_{-\infty}^\infty |G(j\omega) - H(j\omega)|^2 S_y(\omega)\,d\omega \geq 0
\end{aligned}$$

In the last expression, $G(j\omega)$ and $H(j\omega)$ are the transfer functions of the two filters and $S_y(\omega)$ the power spectral density of the process $y(t)$. Note, too, that the solution $h(t)$ is unique in that the error difference is zero only if $G(j\omega) = H(j\omega)$ for all ω for which $S_y(\omega) \neq 0$.

This result is of little practical value, however, unless we somehow obtain an explicit solution for $h(t)$, or equivalently, for $H(j\omega)$. To accomplish this, we rewrite Equation (5.2.1) in the frequency domain. Making the substitutions

$$R(\tau) = \frac{1}{2\pi}\int_{-\infty}^\infty S(\omega)e^{j\omega\tau}\,d\omega$$

and interchanging the order of integration, we find

$$\begin{aligned}
\sigma_e^2 &= \frac{1}{2\pi}\int_{-\infty}^\infty [S_z(\omega) - S_{yz}(\omega)H(-j\omega) - S_{zy}(\omega)H(j\omega) + S_y(\omega)|\,H(j\omega)|^2]\,d\omega \\
&= \frac{1}{2\pi}\int_{-\infty}^\infty \left\{\left|\Phi(\omega)H(j\omega) - \frac{S_{yz}(\omega)}{\Phi^*(\omega)}\right|^2 + \left[S_z(\omega) - \frac{|S_{yz}(\omega)|^2}{S_y(\omega)}\right]\right\}d\omega \qquad (5.2.6)
\end{aligned}$$

where $\Phi(\omega)$ is any function satisfying the relationship $|\Phi(\omega)|^2 = S_y(\omega)$ and $\Phi^*(\omega)$ is its complex conjugate. The terms $S_z(\omega)$ and $S_y(\omega)$, of course, denote the power spectral densities of the processes $z(t)$ and $y(t)$, respectively, and $S_{zy}(\omega)$ their cross-spectral density.

If it is not required to be causal, the optimum filter is easily determined from Equation (5.2.6). Since only the first term of the last integrand is a function of $H(j\omega)$, and since this term cannot be negative, the mean-squared error is minimized by equating it to zero. Thus

$$H(j\omega) = \frac{S_{yz}(\omega)}{S_y(\omega)} \qquad (5.2.7)$$

is the transfer function of the optimum non-causal filter. The resulting mean-squared error is

$$\sigma_0^2 = \frac{1}{2\pi}\int_{-\infty}^{\infty} \frac{S_y(\omega)S_z(\omega) - |S_{yz}(\omega)|^2}{S_y(\omega)}\,d\omega \tag{5.2.8}$$

Since additional restrictions on $H(j\omega)$ can only increase the error, the σ_0^2 of Equation (5.2.8) is often referred to as the *irreducible* error.

Unfortunately, the impulse response of the filter defined by the transfer function of Equation (5.2.7) will not be zero for negative values of t, and hence is not causal. In order to impose the causality condition, it is convenient to make an assumption concerning the spectrum of the input process $y(t)$: $S_y(\omega)$ will be assumed to be a rational function of ω. The optimization does not require a rational spectrum, but the arguments are considerably simplified when such is the case. Moreover, the spectrum of $y(t)$ can generally be approximated as closely as desired by a rational function, even if it is not strictly rational.

If $S_y(\omega)$ is assumed rational, it can be written in the form

$$S_y(\omega) = a^2\frac{(\omega - \alpha_1)\ldots(\omega - \alpha_m)}{(\omega - \beta_1)\ldots(\omega - \beta_n)} \tag{5.2.9}$$

Because $S_y(\omega)$ represents the spectral density of the process $y(t)$, it must be a real, non-negative, symmetric function of ω. From this it is readily concluded that in the above expression a^2 must be real, any poles and zeros with non-zero imaginary parts must occur as complex conjugate pairs, and all real poles and zeros must occur with even multiplicity.

As a consequence of these conditions on the spectral density of $y(t)$, $S_y(\omega)$ can be factored into two parts:

$$\begin{aligned} S_y(\omega) &= a\frac{(\omega - \alpha_1)\ldots(\omega - \alpha_{m/2})}{(\omega - \beta_1)\ldots(\omega - \beta_{n/2})}\cdot a\frac{(\omega - \alpha_1^*)\ldots(\omega - \alpha_{m/2}^*)}{(\omega - \beta_1^*)\ldots(\omega - \beta_{n/2}^*)} \\ &\triangleq \Psi(\omega)\Psi^*(\omega) \end{aligned} \tag{5.2.10}$$

where (*) indicates the complex conjugate. The factorization is such that all poles and zeros of $\Psi(\omega)$ are in the upper half and those of $\Psi^*(\omega)$ in the lower half complex ω-plane. (In subsequent manipulations it will be convenient to assume that $S_y(\omega)$, and therefore $\Psi(\omega)$ and $\Psi^*(\omega)$, have no singularities on the real ω-axis. If $S_y(\omega)$ actually does have such singularities they must occur with even multiplicity. Hence, they can be moved off the axis as conjugate pairs through substitutions of the form $\omega_i^2 \to \omega_i^2 + \epsilon_i^2$. The optimum filter can be determined for this revised spectrum, and the desired solution then obtained by allowing the terms ϵ_i to approach zero.)

It was mentioned that a solution to the filtering problem can be found even if the spectrum $S_y(\omega)$ is not rational. In fact, if

$$\int_{-\infty}^{\infty}\left|\frac{\log S_y(\omega)}{1 + (\omega/2\pi)^2}\right|d\omega < \infty \tag{5.2.11}$$

$S_y(\omega)$ can still be factored and a solution obtained. If, on the other hand, the integral in Equation (5.2.11) is not finite, the stationary process $y(t)$ is *deterministic*; the future behavior of the function can be exactly predicted by a linear operation on its past (Reference 5.1).

In the subsequent derivation we will find it convenient to introduce the following notation. Let $F(\omega)$ and $f(t)$ be a Fourier transform pair:

$$f(t) = \int_{-\infty}^{\infty} F(\omega)e^{j\omega t}\frac{d\omega}{2\pi}, \quad F(\omega) = \int_{-\infty}^{\infty} f(t)e^{-j\omega t}\,dt$$

We define

$$f_+(t) = \begin{cases} f(t) & t \geq 0 \\ 0 & t < 0 \end{cases} \tag{5.2.12a}$$

and

$$f_-(t) = \begin{cases} 0 & t \geq 0 \\ f(t) & t < 0 \end{cases} \tag{5.2.12b}$$

so that $f(t) = f_+(t) + f_-(t)$. Now, let $F_+(\omega)$ and $f_+(t)$ be a Fourier transform pair, and similarly for $F_-(\omega)$ and $f_-(t)$. Then $F(\omega) = F_+(\omega) + F_-(\omega)$, with the transform of the first term non-zero only for $t \geq 0$, and the transform of the second term non-zero only for $t < 0$.

Returning to the problem at hand, we wish to find the transfer function $H(j\omega)$ of the *causal* filter minimizing the mean-squared error as given in Equation (5.2.6). Defining

$$F(\omega) = \Psi(\omega)H(j\omega) - \frac{S_{yz}(\omega)}{\Psi^*(\omega)} = F_+(\omega) + F_-(\omega) \tag{5.2.13}$$

we can rewrite Equation (5.2.6), with $\Phi(\omega) = \Psi(\omega)$, in the form

$$\begin{aligned}\sigma_e^2 &= \frac{1}{2\pi}\int_{-\infty}^{\infty} |F(\omega)|^2\,d\omega + \sigma_0^2 \\ &- \frac{1}{2\pi}\int_{-\infty}^{\infty} [|F_+(\omega)|^2 + F_+(\omega)F_-^*(\omega) + F_+^*(\omega)F_-(\omega) + |F_-(\omega)|^2]\,d\omega + \sigma_0^2\end{aligned} \tag{5.2.14}$$

where σ_0^2 is as defined in Equation (5.2.8) and is independent of $H(j\omega)$. But

$$\begin{aligned}\frac{1}{2\pi}\int_{-\infty}^{\infty} F_+(\omega)F_-^*(\omega)\,d\omega &= \frac{1}{2\pi}\int_{-\infty}^{\infty} F_-^*(\omega)\int_{-\infty}^{\infty} f_+(t)e^{-j\omega t}\,dt\,d\omega \\ &= \int_{-\infty}^{\infty} f_+(t)\frac{1}{2\pi}\int_{-\infty}^{\infty} F_-^*(\omega)e^{-j\omega t}\,d\omega\,dt = \int_{-\infty}^{\infty} f_+(t)f_-^*(t)\,dt = 0\end{aligned} \tag{5.2.15}$$

and similarly for the term $\frac{1}{2\pi}\int_{-\infty}^{\infty} F_+^*(\omega)F_-(\omega)\,d\omega$. Furthermore,

$$F_-(\omega) = [\Psi(\omega)H(j\omega)]_- - \left[\frac{S_{yz}(\omega)}{\Psi^*(\omega)}\right]_- = -\left[\frac{S_{yz}(\omega)}{\Psi^*(\omega)}\right]_- \tag{5.2.16}$$

and is independent of $H(j\omega)$. This follows because

$$\frac{1}{2\pi}\int_{-\infty}^{\infty} \Psi(\omega)H(j\omega)e^{j\omega t}\,d\omega = \int_{-\infty}^{\infty} \phi(\eta)h(t-\eta)\,d\eta \qquad (5.2.17)$$

where $\phi(t)$, the Fourier transform of $\Psi(\omega)$, and $h(t)$ are both zero for all $t < 0$.† Thus, the convolution in Equation (5.2.17) is zero for all $t < 0$, $[\Psi(\omega)H(j\omega)]_- = 0$, and Equation (5.2.16) holds. The only term in Equation (5.2.14) dependent on $H(j\omega)$ is

$$F_+(\omega) = [\Psi(\omega)H(j\omega)]_+ - \left[\frac{S_{yz}(\omega)}{\Psi^*(\omega)}\right]_+ = \Psi(\omega)H(j\omega) - \left[\frac{S_{yz}(\omega)}{\Psi^*(\omega)}\right]_+ \qquad (5.2.18)$$

this last step also following directly from Equation (5.2.17) (i.e., the Fourier transform of $\Psi(\omega)H(j\omega)$ is non-zero only for $t \geq 0$). The optimum causal filter, then, is specified by equating this term to zero, and has the transfer function

$$H(j\omega) = \frac{1}{\Psi(\omega)}\left[\frac{S_{yz}(\omega)}{\Psi^*(\omega)}\right]_+ \qquad (5.2.19)$$

Any other causal filter will necessarily yield a larger mean-squared error.

If the term inside the brackets in Equation (5.2.19) is rational, $H(j\omega)$ can be determined quite easily by making a partial fraction expansion of this term and retaining only the upper-half ω-plane poles. If it is not rational, one can first find its Fourier transform, and then retain only the positive time portion of this transform, obtaining

$$H(j\omega) = \frac{1}{\Psi(\omega)}\int_0^{\infty} e^{-j\omega t}\left[\frac{1}{2\pi}\int_{-\infty}^{\infty} e^{j\omega' t}\frac{S_{yz}(\omega')}{\Psi^*(\omega')}\,d\omega'\right]dt \qquad (5.2.20)$$

Thus far we have said little about the desired signal $z(t)$. We have only assumed that $S_z(\omega)$ and $S_{yz}(\omega)$ could be determined. The most common situation is when $y(t) = x(t) + n(t)$, with the signal $x(t)$ and noise $n(t)$ orthogonal random processes ($R_{xn}(\tau) = 0$) and $z(t) = x(t + t_0)$ for some constant t_0. (If $t_0 = -\infty$, representing an infinite delay, the optimum filter is once again the non-causal filter [Equation (5.2.7)]; an infinite future as well as an infinite past is then available.) Under these conditions, we have:

$$S_y(\omega) = S_x(\omega) + S_n(\omega), \text{ and } S_{yz}(\omega) = S_x(\omega)e^{j\omega t_0}$$

If $t_0 > 0$, the resulting filter is called a *prediction* filter.

The filter of greatest concern to us, however, is the *smoothing* filter in which $z(t) = x(t)$ and $y(t) = x(t) + n(t)$. Equation (5.2.19) can be somewhat simplified when $t_0 = 0$ and the spectra of the (orthogonal) signal and noise

---

†Remember that $f(t)$ is non-zero for $t > 0$ only if its (rational) transform $F(\omega)$ has poles in the upper-half ω-plane (the left-half s-plane, where $s = j\omega$) and is non-zero for $t < 0$ only if $F(\omega)$ has poles in the lower-half ω-plane.

are both rational (cf. Reference 5.2). For, in this event

$$H(j\omega) = \frac{1}{\Psi(\omega)}\left[\frac{S_x(\omega)}{\Psi^*(\omega)}\right]_+ = \frac{1}{\Psi(\omega)}\left[\frac{\Psi(\omega)\Psi^*(\omega) - S_n(\omega)}{\Psi^*(\omega)}\right]_+$$

$$= 1 - \frac{1}{\Psi(\omega)}\left[\frac{S_n(\omega)}{\Psi^*(\omega)}\right]_+ \qquad (5.2.21)$$

If, in addition, the noise is white, $S_n(\omega) = N_0/2$, and the signal spectral density approaches zero for large ω, this expression can be further simplified. Since

$$\lim_{\omega\to\infty} \Psi(\omega)\Psi^*(\omega) = \lim_{\omega\to\infty}\left[S_x(\omega) + \frac{N_0}{2}\right] = \frac{N_0}{2}$$

then

$$\lim_{\omega\to\infty} \Psi(\omega) = \lim_{\omega\to\infty} \Psi^*(\omega) = \left(\frac{N_0}{2}\right)^{1/2}$$

and the term $S_n(\omega)/\Psi^*(\omega)$ has a non-zero residue at $\omega = \infty$. However,

$$\left[\frac{S_n(\omega)}{\Psi^*(\omega)}\right]_+ = \left[\left(\frac{N_0}{2}\right)^{1/2} - \left(\frac{N_0}{2}\right)^{1/2} \frac{\Psi^*(\omega) - \left(\frac{N_0}{2}\right)^{1/2}}{\Psi^*(\omega)}\right]_+$$

$$= \left(\frac{N_0}{2}\right)^{1/2} - \left(\frac{N_0}{2}\right)^{1/2}\left[\frac{\Psi^*(\omega) - \left(\frac{N_0}{2}\right)^{1/2}}{\Psi^*(\omega)}\right]_+ \qquad (5.2.22)$$

and this last term does have a zero residue at $\omega = \infty$. Hence its partial fraction expansion involves terms of the form $K/(\omega + \alpha)^k$ only (with k a positive integer), and since all of the zeros of $\Psi^*(\omega)$ are in the lower-half plane

$$\left[\frac{\Psi^*(\omega) - \left(\frac{N_0}{2}\right)^{1/2}}{\Psi^*(\omega)}\right]_+ = 0$$

Thus,

$$H(j\omega) = 1 - \frac{1}{\Psi(\omega)}\left[\frac{S_n(\omega)}{\Psi^*(\omega)}\right]_+ = 1 - \frac{\left(\frac{N_0}{2}\right)^{1/2}}{\Psi(\omega)} \qquad (5.2.23)$$

when the noise is white and orthogonal to the signal and when the signal spectral density $S_x(\omega)$ is a rational function of ω, with $\lim_{\omega\to\infty} S_x(\omega) = 0$.

The mean-squared error using the optimum filter is as given in Equation (5.2.4). The equivalent expressions in the frequency domain are also of interest:

$$\sigma_e^2 = \frac{1}{2\pi}\int_{-\infty}^{\infty} [S_z(\omega) - S_y(\omega)\,|\,H(j\omega)\,|^2]\, d\omega$$

$$= \frac{1}{2\pi}\int_{-\infty}^{\infty} [S_z(\omega) - S_{yz}(\omega)\,H(-j\omega)]\, d\omega \qquad (5.2.24)$$

When $z(t) = x(t)$, $y(t) = x(t) + n(t)$, and $x(t)$ and $n(t)$ are orthogonal processes, these equations become:

$$\sigma_e^2 = \frac{1}{2\pi}\int_{-\infty}^{\infty} \{S_x(\omega)[1 - |H(j\omega)|^2] - S_n(\omega)|H(j\omega|^2\}\, d\omega$$
$$= \frac{1}{2\pi}\int_{-\infty}^{\infty} S_x(\omega)[1 - H(j\omega)]\, d\omega \tag{5.2.25}$$

This last expression is particularly convenient when either Equations (5.2.21) or (5.2.23) apply. In the latter case, we have

$$\sigma_e^2 = \left(\frac{N_0}{2}\right)^{1/2}\int_{-\infty}^{\infty} \frac{S_x(\omega)}{\Psi(\omega)}\frac{d\omega}{2\pi} \tag{5.2.26}$$

Still another useful expression for the mean-squared error under these same conditions follows from Equation (5.2.6)

$$\sigma_e^2 = \frac{1}{2\pi}\int_{-\infty}^{\infty} [S_x(\omega)|1 - H(j\omega)|^2 + S_n(\omega)|H(j\omega)|^2]\, d\omega \tag{5.2.27}$$

This equation, unlike the previous error expressions, is valid even when $H(j\omega)$ does not represent the optimum filter.

5.3 The Matched Filter Revisited

The parameter estimators and detectors of Chapters Three and Four were found without exception to involve *matched filters*, filters with impulse responses of the form

$$h(t) = f(t_0 - t) \tag{5.3.1}$$

where $f(t)$ $(0 < t < t_0)$ was related to the expected signal. In this section, we pause to show an interesting relationship between optimum mean-squared error filters and matched filters. To this end we pose another problem which also has as a solution the matched filter of Equation (5.3.1), but which is solved by the techniques developed in the previous section. The problem is as follows: The received signal is $y(t) = f(t) + n(t)$ with $f(t)$ a known function of time and $n(t)$, the noise, a wide-sense stationary random process. The signal is to be passed through a linear filter with the impulse response $h(t)$. We denote by $g(t)$ the output of this filter which would be observed were there no noise, and by σ_n^2 the mean-squared filter output due to noise. The task is to find the filter $h(t)$ which maximizes the ratio

$$\frac{g^2(t_0)}{\sigma_n^2} \tag{5.3.2}$$

with t_0 some specific instant of time.† Since the signal-to-noise ratio is of

†Since $f(t)$ is a known time function, $y(t)$ is no longer stationary. Consequently, in contrast to the filters of the previous sections, the optimum filter here will in general depend on the specific time instant t_0.

concern, and not the signal level itself, this maximization can be accomplished by requiring the signal portion of the output to have some fixed value at time $t = t_0$ and finding the filter which minimizes the mean-square noise level. Accordingly, since

$$g(t_0) = \int_{-\infty}^{\infty} h(t_0 - t) f(t)\, dt = \frac{1}{2\pi} \int_{-\infty}^{\infty} e^{j\omega t_0} H(j\omega) F(j\omega)\, d\omega$$

where

$$F(j\omega) = \int_{-\infty}^{\infty} f(t) e^{-j\omega t}\, dt$$

we wish to find the transfer function $H(j\omega)$ which minimizes the quantity

$$\sigma_n^2 + \lambda g(t_0) = \frac{1}{2\pi} \int_{-\infty}^{\infty} \left\{ |H(j\omega)|^2 S_n(\omega) + \frac{\lambda}{2} e^{j\omega t_0} H(j\omega) F(j\omega) + \frac{\lambda}{2} e^{-j\omega t_0} H(-j\omega) F(-j\omega) \right\} d\omega \tag{5.3.3}$$

This expression is identical to Equation (5.2.6) when we substitute the terms $S_n(\omega)$, $-(\lambda/2)e^{-j\omega t_0} F(-j\omega)$, and 0, for $S_y(\omega)$, $S_{yz}(\omega)$, and $S_z(\omega)$, respectively. The only condition placed on these quantities in order to derive Equation (5.2.20) for the optimum causal filter was that $S_y(\omega)$, or in this case $S_n(\omega)$, be factorable. Thus, if $S_n(\omega) = \Psi(\omega)\, \Psi^*(\omega)$, with all the singularities of $\Psi(\omega)$ contained in the upper-half ω-plane, the optimum causal filter in the present problem has the transfer function

$$H(j\omega) = \frac{k}{\Psi(\omega)} \int_0^{\infty} e^{-j\omega t} \frac{1}{2\pi} \int_{-\infty}^{\infty} \frac{e^{j\omega'(t-t_0)} F(-j\omega')}{\Psi^*(\omega')}\, d\omega'\, dt \tag{5.3.4}$$

where $k = -\lambda/2$ is simply a gain constant and does not affect the signal-to-noise ratio. If the noise is white, $S_n(\omega) = N_0/2$,

$$\frac{1}{2\pi} \int_{-\infty}^{\infty} \frac{e^{j\omega'(t-t_0)} F(-j\omega')}{\left(\frac{N_0}{2}\right)^{1/2}}\, d\omega' = \frac{1}{\left(\frac{N_0}{2}\right)^{1/2}} f(t_0 - t)$$

and

$$H(j\omega) = k_1 \int_0^{\infty} f(t_0 - t) e^{-j\omega t}\, dt \tag{5.3.5}$$

with $k_1 = 2k/N_0$ an arbitrary constant. Thus, setting $k_1 = 1$, we have

$$h(t) = \begin{cases} f(t_0 - t) & t > 0 \\ 0 & t < 0 \end{cases} \tag{5.3.6}$$

If $f(t)$ is identically zero for $t > t_0$, this is indeed the impulse response of the matched filter as defined in Chapter Three.

5.4 Tracking Phase Estimators

Let $Ax(t)$, a known signal periodic with period T, be subjected to an unknown τ second time delay and received in the presence of additive white Gaussian noise $n(t)$ (with single-sided spectral density N_0). Suppose it were desired to estimate τ on the basis of the received signal $y_l(t) = Ax(t + \tau) + n(t)$, $lT < t < (l + 1)T$. (Since τ can only be estimated modulo T, it can be either negative or positive.) The maximum-likelihood estimate of τ, under these circumstances, is a solution to the equation (cf. Section 3.4)

$$\frac{\partial}{\partial\hat{\tau}} \log_e p(y_l(t)\,|\,\hat{\tau}) = \frac{\partial}{\partial\hat{\tau}}\left\{\frac{1}{N_0}\int_{lT}^{(l+1)T} [2Ay_l(t)x(t+\hat{\tau}) - A^2x^2(t+\hat{\tau})]\right\} dt$$
$$= \frac{2A}{N_0}\int_{lT}^{(l+1)T} y_l(t)\frac{\partial x(t+\hat{\tau})}{\partial\hat{\tau}}\,dt = 0 \qquad (5.4.1)$$

In some cases, when $x(t)$ is a sinusoid, for example, Equation (5.4.1) leads to an explicit solution for the optimum estimate $\hat{\tau}$ of the delay τ (cf. Section 3.7). More generally, it does not. In any event, the problem to be investigated here differs somewhat from those of concern in Chapter Three. Here τ, rather than assuming independent values over each successive T second interval, is a continuous function of time, relatively constant over any particular interval, but slowly varying from one interval to the next.

Since τ is assumed to be nearly constant over any T second interval, one could make an independent maximum-likelihood estimate of it over each of these intervals using the estimator defined by Equation (5.4.1). But since τ is a slowly varying function of time, its values over successive intervals will be highly correlated. To ignore this correlation is to discard much useful information. If it were possible to specify the density function $p(\tau_l\,|\,\hat{\tau}_{l-1}, \hat{\tau}_{l-2}, \ldots,)$ of the variable τ during the interval $lT < t < (l + 1)T$ conditioned on the estimates of τ over the preceding intervals, a maximum a posteriori probability estimator could in principle be defined.

A generally more practical approach achieving much the same result is suggested by the fact that, subject to certain regularity conditions (see Appendix A.2),

$$E\left[\left(\frac{\partial \log_e p[y_l(t)\,|\,\hat{\tau}]}{\partial\hat{\tau}}\right)_{\hat{\tau}=\tau}\,\middle|\,\tau\right] = 0 \qquad (5.4.2a)$$

and

$$E\left[\left(\frac{\partial^2 \log_e p[y_l(t)\,|\,\hat{\tau}]}{\partial^2\hat{\tau}}\right)_{\hat{\tau}=\tau}\,\middle|\,\tau\right]$$
$$= -E\left[\left(\frac{\partial \log_e p[y_l(t)\,|\,\hat{\tau}]}{\partial\hat{\tau}}\right)^2_{\hat{\tau}=\tau}\,\middle|\,\tau\right] \qquad (5.4.2b)$$

Thus, the expected value of the integral

$$e_l = \int_{lT}^{(l+1)T} y_l(t) \frac{\partial x(t + \hat{\tau})}{\partial \hat{\tau}}\, dt \tag{5.4.3}$$

is a monotonically decreasing function of τ in the vicinity $\hat{\tau} \approx \tau$, positive for $\hat{\tau} < \tau$ and negative for $\hat{\tau} > \tau$. Suppose, therefore, that e_l were used to effect a change in the phase $\hat{\tau}$ of the locally generated signal, causing $\hat{\tau}$ to decrease when it (e_l) is negative and to increase when it is positive. In principle, then, if $\hat{\tau} = \tau$ is the only value of $\hat{\tau}$ for which Equations (5.4.2) hold, $\hat{\tau}$ should converge to, and track, τ. Actually, to exploit the fact that $\hat{\tau}$ is a slowly varying function of time, $\hat{\tau}$ should be controlled by a weighted average of the past indicators of the sign of the error $\tau - \hat{\tau}$; i.e., by the quantity $\epsilon_{l+1} \triangleq \sum_{i=-\infty}^{l} \alpha_{l-i} e_i$, rather than by e_l alone.

The block diagram of a device for accomplishing the task outlined in the preceding paragraph is shown in Figure 5.1. If $h(0) = 0$ and

$$h_\alpha(t) = \alpha_i \qquad iT < t < (i+1)T \tag{5.4.4}$$

the input to the phase shifter over the interval $(l+1)T < t < (l+2)T$ can be expressed in the form:

$$\epsilon_{l+1} = \sum_{i=-\infty}^{l} \alpha_{l-i} e_i \triangleq \epsilon((l+1)T) \tag{5.4.5}$$

where

$$\epsilon(t) \triangleq \int_{-\infty}^{t} y(\eta) x'(\eta + \hat{\tau}) h_\alpha(t - \eta)\, d\eta$$

Moreover, if τ is relatively constant over any T second interval, as has been assumed, the difference $\alpha_i - \alpha_{i-1}$ should be small for all i. Consequently, $h_\alpha(t)$ can be closely approximated by a continuous (and even differentiable) function of time, say $h_a(t)$, and the phase shifter inputs ϵ_l can be replaced by

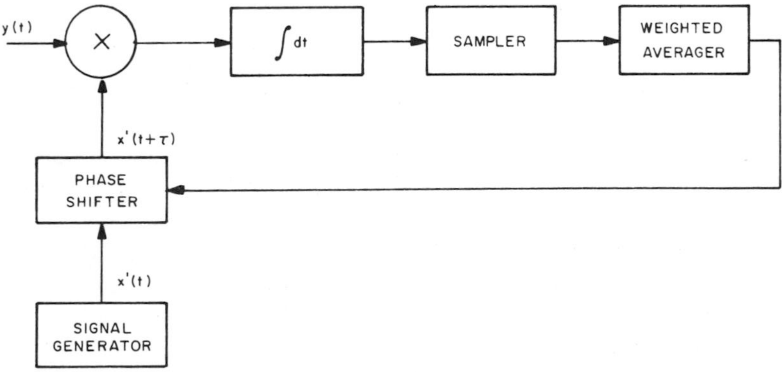

Figure 5.1 Tracking phase estimator.

the input signal $\epsilon(t)$ with virtually no effect on the performance of the device. Finally, we remark that rather than relating $\hat{\tau}$ directly to the function $\epsilon(t)$, we could equivalently have related the rate of change of $\hat{\tau}$ to its derivative

$$\epsilon'(t) = \int_{-\infty}^{t} y(\eta)x'(\eta + \hat{\tau})h_a'(t - \eta)\, d\eta \tag{5.4.6}$$

Now consider the device represented by the block diagram in Figure 5.2. The voltage controlled oscillator (VCO) oscillates at a frequency which, ideally, is linearly proportional to its input voltage. The VCO output is used to clock the signal generator producing the local signal $x'(t + \hat{\tau})$, thereby controlling the rate of change of $\hat{\tau}$. In view of the observations made in the preceding paragraph, therefore, we conclude that, if the impulse response of the loop filter is $h(t) = h_a'(t)$, the devices of Figures 5.1 and 5.2 are functionally equivalent. The latter device, known as a *phase-locked loop*, represents a generally more practical implementation.

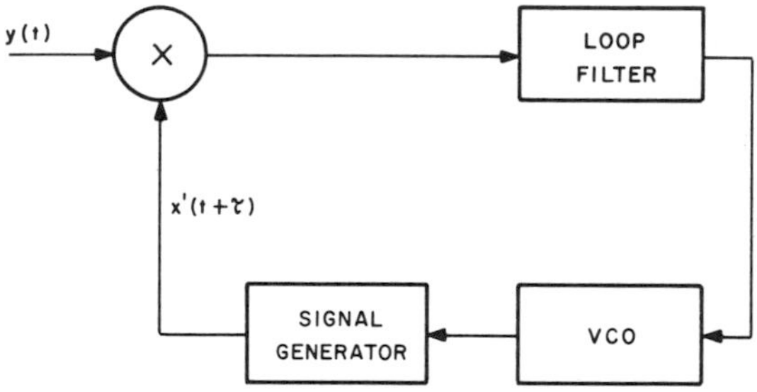

Figure 5.2 Phase-locked loop.

The phase-locked loop, then, is in essence a feedback device designed to force a solution to the likelihood Equation (5.4.1). The as yet unspecified loop filter is needed because the received signal phase τ varies with time; the past therefore should generally not be weighed uniformly in estimating the present phase. The remaining problem is to determine the optimum loop filter as a function of the statistics of $\tau(t)$. Before this can be accomplished, however, it is necessary to develop a linear model of the phase-locked loop. This is the subject of the following section. In order to avoid complicating the subsequent discussion unnecessarily, $x(t)$ will be assumed to be a sinusoid. The results obtained are extended, in Section 5.8, to include non-sinusoidal functions as well.

It might be well, before proceeding, to re-emphasize that one of the fundamental differences between the smoothing filter and the matched filter approaches to the estimation problem is in the amount of a priori knowledge assumed about the signal. The matched filter necessitates a complete description of the expected signal, except perhaps for an unknown but constant parameter or parameters; the smoothing filter requires knowledge only of its second-order statistics. In the present situation, both approaches will be found applicable to the same problem. A matched filter can be used because the functional form of the expected signal is known except for an unknown parameter, but now this parameter is a random process, not a random variable. A smoothing filter will therefore be required to provide as accurately as possible a continuous estimate of this time varying parameter.

5.5 The Linear Model

The block diagram of a phase-locked loop designed to track the phase of a sinusoidal signal is shown in Figure 5.3. Since the VCO output is itself a sinusoid, the signal generator of Figure 5.2 can be replaced by a 90° phase-shifter, or even omitted entirely. (The 90° shifter is needed only if the VCO output is to serve as a replica of the received signal.) The exact analysis of the tracking error associated with this device is difficult. Because the differential

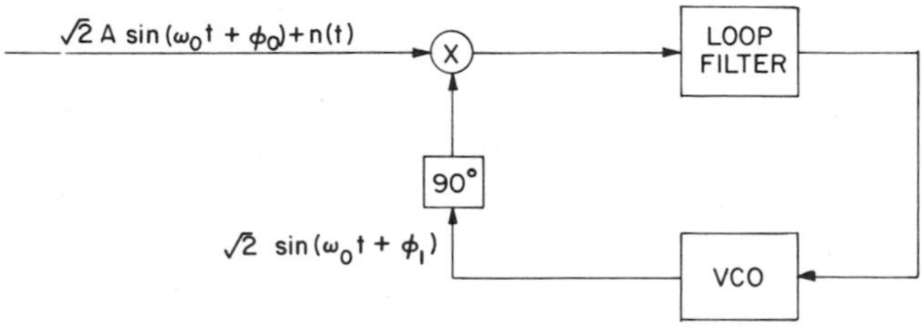

Figure 5.3 Sinusoidal phase-locked loop.

equation relating its output to its input is non-linear, it can be solved exactly only in certain special cases. In order to obtain more tractable results concerning the loop performance, as well as to specify the form of the optimum loop filter, it is necessary to make an approximation which has the effect of linearizing the system differential equation. The validity of this linear approximation depends upon the assumption of a relatively large signal-to-noise ratio with a consequent small tracking error. In most applications with which we shall be concerned, this condition will be necessarily satisfied.

If the received signal is $y(t) = \sqrt{2}\,A \sin[\omega_0 t + \phi_0(t)] + n(t)$, and if the phase-shifted VCO output is $\sqrt{2}\cos[\omega_0 t + \phi_1(t)]$, then subject to the above condition $\phi_0(t) \approx \phi_1(t)$. The low-frequency component of the product of these terms, $A \sin[\phi_0(t) - \phi_1(t)] \approx A[\phi_0(t) - \phi_1(t)]$, therefore, is an approximately linear function of the phase difference. The double-frequency component of this product, on the other hand, can be ignored entirely. This follows because the phase-locked loop is essentially equivalent to the device of Figure 5.1, and hence responds only to the full-period integral of this product. The rate of the change of phase (i.e., the frequency) of the output of the VCO is then (approximately) linearly proportional to a filtered and noise-corrupted version of the tracking error

$$\phi_0(t) - \phi_1(t) \triangleq \phi_e(t) \tag{5.5.1}$$

Consequently, so long as the tracking error is small, the phase-locked loop can be represented as shown in Figure 5.4.

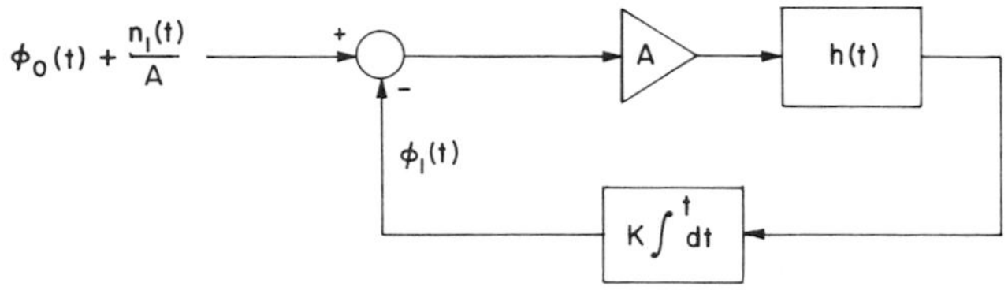

Figure 5.4 Linear model of a phase-locked loop.

In order to proceed, it is necessary to determine the statistics of the input noise, $n_1(t) \triangleq \sqrt{2}\cos[\omega_0 t + \phi_1(t)]n(t) \triangleq \alpha(t)n(t)$. The analysis seems considerably complicated by the dependence between this noise and the phase output of the VCO. But this phase is dependent, not upon $n(t)$ itself, but upon the integral of a filtered version of the product $n(t)\alpha(t)$. If the noise bandwidth is B_n, $n(t)$ and $n(t-\tau)$ will be essentially independent for any $\tau > k/B_n$, with k a generally small number. Since the phase $\phi_0(t)$ is assumed to be slowly varying relative to the signal period $2\pi/\omega_0$, the optimum phase estimate $\phi_1(t)$ will be influenced by the last $2\pi K/\omega_0$ seconds of the input signal with $K \gg 1$. Accordingly, if $\omega_0/2\pi K \triangleq B_l \ll B_n$, and this, of course, is always the sort of inequality needed in order to call the noise "white," the last k/B_n seconds of received noise will have a negligibly small effect on $\phi_1(t)$. Thus, $\phi_1(t)$ and, as a consequence, $\alpha(t)$ will be virtually independent of $n(t)$ and

$$E\{n_1(t)\} \approx E\{n(t)\}E\{\alpha(t)\} = 0 \tag{5.5.2a}$$

while

$$E\{n_1(t_1)n_1(t_2)\} \approx \begin{cases} E\{n(t_1)\alpha(t_1)\alpha(t_2)\}E\{n(t_2)\} = 0 & t_2 > t_1 \\ E\{n^2(t_1)\}E\{\alpha^2(t_1)\} & t_2 = t_1 \end{cases} \tag{5.5.2b}$$

But, if the input noise is "white," (i.e., $B_n \gg B_l$)

$$E\{n(t)n(t+\tau)\} \approx \frac{N_0}{2}\delta(\tau)$$

so that

$$E\{n_1(t)n_1(t+\tau)\} \approx \frac{N_0}{2}E\{\alpha^2(t)\}\delta(\tau) \tag{5.5.3}$$

If $\phi_0(t)$ is uniformly distributed, as it presumably is, and if the phase error is independent of $\phi_0(t)$, as it clearly is in a phase-locked loop, then $\phi_1(t)$

is also uniformly distributed in the interval $(0, 2\pi)$. Consequently, since $\alpha(t) = \sqrt{2}\cos[\omega_0 t + \phi_1(t)]$, $E[\alpha^2(t)] = 1$. The noise at the input to the linear model of the phase-locked loop is therefore essentially white and wide-sense stationary.

The input noise spectrum $S_{n_1}(\omega)$ has just been determined. Because the cross-spectral density $S_{n_1\phi_0}(\omega)$ is also needed in the determination of the optimum filter, it too must be investigated. But since $n(t_1)$ is effectively independent of $\alpha(t_1)$ and strictly independent of the signal phase $\phi_0(t_2)$ for any t_2,

$$E\{n_1(t_1)\phi_0(t_2)\} \approx E\{\alpha(t_1)\phi_0(t_2)\}E\{n(t_1)\} = 0 \tag{5.5.4}$$

and $S_{n_1\phi_0}(\omega)$ is negligibly small.

Lastly, because not only the variance but also the distribution of the phase error $\phi_e(t) = \phi_0(t) - \phi_1(t)$ will be of subsequent interest, it is useful to consider the complete statistics of the input noise process $n_1(t) = n(t)\alpha(t)$. To do this, it is convenient to suppose, for the moment, that the loop is preceded by a bandpass filter of bandwidth $B \leq 2f_0 = 2\omega_0/2\pi$ symmetrically centered about the frequency f_0. The noise variance σ_n^2 is then $N_0 B$. Let $n_f(t)$ represent this filtered noise and express it in the form:

$$n_f(t) = n_a(t)\cos\omega_0 t - n_b(t)\sin\omega_0 t \tag{5.5.5}$$

If $n(t)$ is a zero mean Gaussian process, $n_a(t)$ and $n_b(t)$ are mutually independent Gaussian processes, both having means equal to zero and variances equal to that of $n_f(t)$.† Since $\alpha(t) = \sqrt{2}\cos[\omega_0 t + \phi_1(t)]$ the low frequency components of the product $n_f(t)\alpha(t)$ can be written as

$$[n_f(t)\alpha(t)]_{lf} = \frac{1}{\sqrt{2}}\{n_a(t)\cos\phi_1(t) + n_b(t)\sin\phi_1(t)\} \tag{5.5.6}$$

Now define

$$\sqrt{2}\,[n_f(t)\alpha(t)]_{lf} = n'(t) = \{n_a(t)\cos\phi_1(t) + n_b(t)\sin\phi_1(t)\} \tag{5.5.7}$$

and

$$n''(t) = \{n_a(t)\sin\phi_1(t) - n_b(t)\cos\phi_1(t)\}$$

and consider the joint density function of the random variables n', n'', ϕ_1 at any particular instant of time. The Jacobian of the transformation from n_a, n_b, and ϕ_1 to n', n'', and ϕ_1 is unity. Moreover,

$$p(n_a, n_b, \phi_1) = \frac{1}{2\pi\sigma_n^2}e^{-(n_a^2+n_b^2)/2\sigma_n^2}p(\phi_1)$$

The last step follows because n_a and n_b are independent Gaussian variables and, as has already been argued, $n_f(t)$ and hence $n_a(t)$ and $n_b(t)$ are essentially independent of $\phi(t)$. But $n_a^2 + n_b^2 = (n')^2 + (n'')^2$ and so

†See, for example, Reference 5.3.

$$p(n', n'', \phi_1) = \frac{1}{2\pi\sigma_n^2} e^{-[(n')^2+(n'')^2]/2\sigma_n^2} p(\phi_1)$$

Finally,

$$p(n') = \int_{-\infty}^{\infty} \int_0^{2\pi} p(n', n'', \phi_1)\, d\phi_1\, dn'' = \frac{1}{\sqrt{2\pi}\,\sigma_n} e^{-(n')^2/2\sigma_n^2} \qquad (5.5.8)$$

and the first-order statistics of the noise at the loop filter input, $[n_f(t)x(t)]_{lf} = (1/\sqrt{2})\,n'(t)$, are Gaussian, with zero mean and variance $\frac{1}{2}\sigma_n^2$. Since the bandwidth of the low-frequency component is half that of the bandpass filter, the two-sided spectral density of this component is $(\frac{1}{2}\sigma_n^2)/(\frac{1}{2}B) = N_0/2$, the same as that of $n(t)$.

Although it was convenient in the above argument to precede the loop with a bandpass filter of bandwidth $B \leq 2f_0$, the presence of such a filter is actually irrelevant to the conclusions. Any noise power outside the frequency range $0 < f < 2f_0$ will be far outside the bandwidth of the loop after being multiplied by $\alpha(t)$. Thus, B can be arbitrarily large without affecting the loop output, and the filter can in fact be eliminated entirely. Accordingly, so far as the loop is concerned, $[n_1(t)]_{lf}$ is a white Gaussian random process with a spectral density $N_0/2$. Again, the high-frequency components of $n_1(t)$ will have no effect on the loop performance and can be ignored.

5.6 Optimization of the Phase-Locked Loop Filter

There are several approaches to the optimization of the phase-locked loop, depending upon the assumptions made concerning the signal. There is inevitably a random variation due to oscillator instabilities between the VCO output phase and the received signal phase even in the absence of noise. Although both oscillators will contribute to this random fluctuation, it is generally convenient to treat the VCO as a perfect oscillator, and the transmitter oscillator as the sole source of this variation. For most analytical purposes, the actual source is irrelevant since only the phase difference is of concern. If the spectrum of this phase fluctuation is known or if it can be approximated, the phase-locked loop filter can readily be optimized so as to minimize the mean-squared phase difference.

Defining the Laplace transforms

$$\theta_i(s) = \int_0^{\infty} \phi_i(t) e^{-st}\, dt \qquad i = 0,1$$

and

$$H(s) = \int_0^{\infty} h(t) e^{-st}\, dt$$

where $s = \alpha + j\omega$, $\alpha \geq 0$, we have, in the absence of noise (cf. Figure 5.4; $\phi_1(t)$, $\phi_2(t)$, and $h(t)$ are as defined there)

$$\theta_1(s) = Y(s)\theta_0(s) \tag{5.6.1}$$

where $Y(s)$, the loop transfer function, is (with K denoting the VCO gain)

$$Y(s) = \frac{A\,KH(s)}{s + A\,KH(s)} \tag{5.6.2}$$

This relationship is illustrated in Figure 5.5. To determine the optimum loop filter $H(s)$, it is sufficient to find the optimum transfer function $Y(s)$ and then solve for $H(s)$:

$$H(s) = \frac{sY(s)}{A\,K[1 - Y(s)]} \tag{5.6.3}$$

Since the loop must be stable if it is to be of any utility, $Y(s)$ will represent a stable filter (cf. Section 5.2). Thus, the Fourier transform $Y(j\omega)$ of the im-

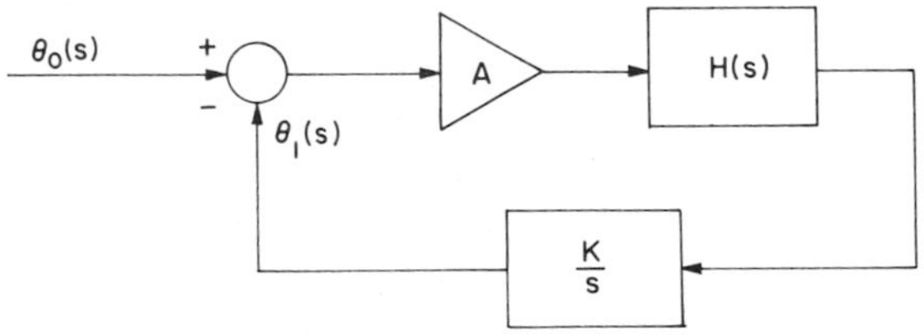

Figure 5.5 Linear model in the frequency domain.

pulse response of this filter also exists, and the optimization procedure of Section 5.2 is directly applicable here. The optimum loop filter $H(s)$ will not necessarily be stable, however, as we shall see.

As an example, suppose the received signal is of the form

$$\sqrt{2}\,A_0 \sin[(\omega_0 + 2\pi f)t] + n(t) \tag{5.6.4}$$

where f is a random frequency component with the spectrum

$$S_f(\omega) = \frac{2\sigma_f^2\beta}{\omega^2 + \beta^2}$$

and $\omega_0/2\pi$ is the VCO center frequency. The phase spectrum is then

$$S_\phi(\omega) = \frac{2\sigma_f^2\beta}{\omega^2(\omega^2 + \beta^2)}$$

The optimum filter $Y(j\omega)$ is the smoothing filter when the signal has the spectrum $S_\phi(\omega)$ and the noise is white with the spectral density

$$S_n(\omega) = \frac{N_0}{2A_0^2}$$

Since the noise is white and the signal spectrum rational (and asymptotically zero) the optimum filter is as defined in Equation (5.2.23):

$$Y(j\omega) = 1 - \frac{\left(\frac{N_0}{2A_0^2}\right)^{1/2}}{\Psi(\omega)} \tag{5.6 5}$$

Here

$$|\Psi(\omega)|^2 = S_\phi(\omega) + S_n(\omega) = \frac{N_0}{2A_0^2}\left[\frac{B_0^2 + aj\omega - \omega^2}{j\omega(j\omega + \beta)}\right]\left[\frac{B_0^2 - aj\omega - \omega^2}{j\omega(j\omega - \beta)}\right]$$

where $a = (2B_0^2 + \beta^2)^{1/2}$ and $B_0^4 = 4A_0^2\sigma_f^2\beta/N_0$. Therefore,

$$Y(j\omega) = \left[\frac{B_0^2 + (a - \beta)j\omega}{B_0^2 + aj\omega - \omega^2}\right] \tag{5.6.6}$$

The mean-squared error due to the noise is [cf. Equation (5.2.26)]

$$\sigma_n^2 = \frac{1}{2\pi}\int_{-\infty}^{\infty} \frac{N_0}{2A_0^2}|Y(j\omega)|^2\, d\omega = \frac{N_0 B_L}{A_0^2} \tag{5.6.7}$$

where the *loop bandwidth* $B_L = [B_0^2 + (a - \beta)^2]/4a$, is the width of an ideal rectangular filter which would pass the same noise power. The error due to the frequency jitter is

$$\sigma_s^2 = \frac{1}{2\pi}\int_{-\infty}^{\infty} S_0(\omega)\,|1 - Y(j\omega)|^2\, d\omega = \frac{\sigma_f^2\beta}{aB_0^2} \tag{5.6.8}$$

It is interesting to observe the relationship between the phase error here due to the additive noise [Equation (5.6.7)] and the corresponding term using an optimum phase estimator when the received phase is assumed constant (see Section 3.7). The respective error terms in these two situations

$$\sigma_n^2 = \frac{N_0 B_L}{A_0^2} \quad \text{and} \quad \sigma_n^2 = \frac{N_0}{2A_0^2 T}$$

are identical if B_L and $1/2T$ are equated. The term T, it will be recalled, is bounded by the length of time the received sinusoid can be assumed to retain the same phase when making a maximum-likelihood estimation.

In most practical situations, there are transient phenomena as well as random oscillator fluctuations causing relative variation between the received and VCO signal phases. One such transient occurs, for example, when the signal is first received and the loop is brought into lock. Another one might be caused by a Doppler shift in the received signal frequency moving it away from the center frequency of the loop. Often, these transient phenomena constitute a much grosser effect than the random phase variation, and loop filters are frequently selected solely to counteract tracking errors due to anticipated transients. With this approach, the loop filter is designed to minimize the mean-squared error caused by the additive noise subject to a condition on the allowable transient error. As in the matched filter analysis of Section 5.3, the optimization procedure using this criterion of optimality is identical to that for the smoothing filter as soon as one makes the proper identifications.

The optimization, subject to a transient constraint, proceeds as follows. The mean-squared phase error caused by the additive noise is

$$\sigma_n^2 = \frac{1}{2\pi}\int_{-\infty}^{\infty} |Y(j\omega)|^2 S_n(\omega)\, d\omega \tag{5.6.9}$$

where $S_n(\omega) = N_0/2A_0^2$ is, as before, the normalized noise spectral density. The transient error is defined as the integral of the square of the difference between the input and output phases, in the absence of noise, due to a transient beginning at time $t = 0$:

$$e_T^2 = \int_0^{\infty} [\phi_1(t) - \phi_0(t)]^2\, dt = \int_0^{\infty} \phi_e^2(t)\, dt \tag{5.6 10}$$

Since this integral is to be bounded, there exists, by Plancherel's theorem, a function $\theta_e(j\omega)$ such that

$$\frac{1}{2\pi}\int_{-\infty}^{\infty} |\theta_e(j\omega)|^2\, d\omega = \int_{-\infty}^{\infty} \phi_e^2(t)\, dt \tag{5.6.11}$$

Moreover, if $\phi_e(t)$ is absolutely integrable, as we would also like to require of the transient error, then

$$\theta_e(j\omega) = \int_{-\infty}^{\infty} \phi_e(t) e^{-j\omega t}\, dt = \lim_{\alpha\to 0} \theta_e(s) = \lim_{\alpha\to 0} |1 - Y(s)|\theta_0(s) \tag{5.6.12}$$

where, again, $s = \alpha + j\omega$.

It is desired to minimize σ_n^2 subject to a constraint on the transient error; that is, the loop filter is sought which minimizes the expression

$$E^2 = \sigma_n^2 + \lambda^2 e_T^2 \tag{5.6.13}$$

(The Lagrange multiplier λ is determined by specifying the acceptable transient error as will be shown presently.) Thus we wish to find the transfer function $Y(s)$ minimizing

$$E^2 = \frac{1}{2\pi}\int_{-\infty}^{\infty} \{|Y(j\omega)|^2 S_n(\omega) + \lambda^2 |Y(j\omega) - 1|^2 |\theta_0(j\omega)|^2\}\, d\omega \tag{5.6.14}$$

But Equations (5.2.6) and (5.6.14) are identical when the term $\lambda^2|\theta_0(j\omega)|^2$ in the latter is substituted for $S_z(\omega)$ in the former and when $S_{zn}(\omega) = 0$. Accordingly, the optimum filters of Section 5.2 are also optimum here under this identification of terms.

Clearly one of the most important transients to consider is a frequency step $\theta_0(s) = \Delta\omega/s^2$ since the center frequencies of the transmitter and VCO oscillators will never be exactly equal. The loop must be designed to keep the tracking error due to this difference within reasonable bounds. But since

$$\lambda^2 |\theta_0(j\omega)|^2 = \lim_{\beta\to 0} \frac{\lambda^2(\Delta\omega)^2}{\omega^2(\omega^2 + \beta^2)} \tag{5.6.15}$$

the desired solution is immediately obtainable from Equation (5.6.6) by substituting $\lambda^2(\Delta\omega)^2$ for $\sigma_f^2\beta$ and letting $\beta \to 0$. The optimum loop transfer function, therefore, is just

$$Y(j\omega) = \frac{B_0^2 + \sqrt{2}\,B_0 j\omega}{B_0^2 + \sqrt{2}\,B_0 j\omega - \omega^2} \tag{5.6.16}$$

where $B_0^4 = [4A_0^2(\lambda\Delta\omega)^2]/N_0$. As before, the mean output noise power can be written

$$\sigma_n^2 = \frac{1}{2\pi}\int_{-\infty}^{\infty} \frac{N_0}{2A_0^2}\,|Y(j\omega)|^2\,d\omega = \frac{N_0 B_L}{A_0^2}\ \text{rad}^2 \tag{5.6.17}$$

where $B_L = (3B_0/4\sqrt{2})$ is the loop bandwidth.

Using Equation (5.6.3), with $A = A_0$, we have

$$H(s) = \frac{B_0^2 + \sqrt{2}\,B_0 s}{A_0 K s} \tag{5.6.18}$$

As a result, the optimum filter depends upon B_0 and A_0 and hence on the signal amplitude, the signal-to-noise ratio, the magnitude of the frequency step, and the value of λ which in turn controls the magnitude of the transient error. To verify this last statement, it is only necessary to observe that

$$e_T^2 = \frac{1}{2\pi}\int_{-\infty}^{\infty} |\theta_0(j\omega)|^2\,|1 - Y(j\omega)|^2\,d\omega = \frac{(\Delta\omega)^2}{\left(\frac{8B_L}{3}\right)^3}\ \text{rad}^2\ \text{sec} \tag{5.6.19}$$

and is inversely proportional to $\lambda^{3/2}$.

If the signal power is A_1^2 rather than the design value A_0^2 volts², and the noise spectral density is N_1 rather than N_0 watts/Hz, the output phase jitter becomes

$$\sigma_n^2 = \left(\frac{B_L N_1}{A_1^2}\right)\frac{1}{3}\left[2\frac{A_1}{A_0} + 1\right]\ \text{rad}^2 \tag{5.6.20a}$$

and the transient error

$$e_T^2 = \frac{(\Delta\omega)^2}{\left(\frac{8B_L}{3}\right)^3}\left(\frac{A_0}{A_1}\right)^2\ \text{rad}^2\ \text{sec} \tag{5.6.20b}$$

with B_L as previously defined.

It is, of course, possible to design the loop to compensate for different transient behavior. The loop analyzed above is generally referred to as a *second-order loop* since $Y(s)$ involves two poles. A *first-order loop* results when the transient due to a phase step

$$\theta_0(s) = \frac{\Delta\theta}{s} \tag{5.6.21}$$

is to be limited. In this case, the optimum loop transfer function is easily shown to be

$$Y(s) = \frac{4B_L}{s + 4B_L} \tag{5.6.22a}$$

while

$$H(s) = \frac{4B_L}{A_0 K} \tag{5.6.22b}$$

where

$$B_L = \frac{\sqrt{2}\,\lambda(\Delta\theta)\,A_0}{4N_0^{1/2}}$$

The mean-square phase error due to the additive noise, and the transient error, for an arbitrary signal amplitude A_1 and noise spectral density $N_1/2$, are, respectively,

$$\sigma_n^2 = \frac{N_1 B_L}{A_1^2}\left(\frac{A_1}{A_0}\right) \tag{5.6.23a}$$

and

$$e_T^2 = \frac{(\Delta\theta)^2}{8B_L\left(\frac{A_1}{A_0}\right)} \tag{5.6.23b}$$

For the *third-order loop*, designed to track frequency ramps,

$$\theta_0(s) = \frac{\Delta\alpha}{s^3}$$

the optimum loop filter is found to be

$$H(s) = \frac{\left(\frac{6}{5}B_L\right)^3 + 2\left(\frac{6}{5}B_L\right)^2 s + \frac{12}{5}B_L s^2}{A_0 K s^2} \tag{5.6.24}$$

where

$$B_L^3 = \left(\frac{5}{6}\right)^3 \frac{\sqrt{2}\,\lambda(\Delta\alpha)A_0}{(N_0)^{1/2}}$$

The corresponding errors for arbitrary signal and noise levels are

$$\sigma_n^2 = \frac{N_1 B_L}{A_1^2}\frac{3}{5}\left[\frac{4\left(\frac{A_1}{A_0}\right) + 1}{4\left(\frac{A_1}{A_0}\right) - 1}\right] \tag{5.6.25a}$$

and

$$e_T^2 = \frac{\left(\frac{5}{6}\right)^3 (\Delta\alpha)^2}{\left(\frac{A_1}{A_0}\right)\left[4\left(\frac{A_1}{A_0}\right) - 1\right]B_L^3} \tag{5.6.25b}$$

The advantages of a second-order over a first-order loop can be easily demonstrated. If the input to a first-order loop is a frequency step of magnitude $\Delta\omega$, the phase error is

$$\phi_e(t) = \frac{1}{2\pi j}\int_{-j\infty}^{j\infty} (1 - Y(s))\frac{\Delta\omega}{s^2} e^{st}\, ds = \frac{\Delta\omega}{4B_L}\{1 - e^{-4B_L t}\}$$

and $e(\infty) = \Delta\omega/4B_L$ leaving an objectionable steady-state error. The second-order loop, of course, responds to the same input with a zero steady-state error. On the other hand, since the frequency difference between two oscillators does not generally increase (or decrease) indefinitely, a third-order loop will seldom be called for in synchronization problems.

In summary, it has been shown that by proper design techniques it is possible to construct a *tracking* matched filter. The filter can be designed to compensate for random phase fluctuations or for transient behavior (or for both simultaneously) equally well. Unfortunately, the optimum filter in general depends upon the signal-to-noise ratio. In particular, the optimum loop bandwidth in each of the three phase-locked loops just considered was found to be directly proportional to some fractional power of the signal-to-noise ratio. Yet, if the loop filter parameters are not varied as the input signal-to-noise ratio changes, the effective loop bandwidth is a function of the signal amplitude only [cf. Equations (5.6.20), (5.6.23), and (5.6.25)]. As a result, although both the mean-squared phase error and the transient error decrease in every case as the signal power increases from its design value, they do not decrease proportionately, as they would were the loop filter optimized at every signal level.

In practice, phase-locked loops are usually preceded by bandpass limiters or automatic gain control devices. Such devices tend to force the loop bandwidth to vary as a function of the signal-to-noise ratio rather than as a function of the signal level only, and hence to exhibit a more nearly optimum dependence on these parameters. This last statement is an immediate consequence of the following two facts concerning bandpass limiters and automatic gain control devices.

(1) The signal-to-noise ratio $S_1'/N_1'B_1$ at the output of such devices is directly proportional to the ratio S_1/N_1B_1 at their input; i.e.,

$$\frac{S_1'}{N_1'B_1} = k\frac{S_1}{N_1B_1}$$

where $\pi/4 < k < 2$ for bandpass limiters with bandwidth B_1,† and, presumably, $k = 1$ for automatic gain control devices.

(2) The total power at the output of such devices is a constant, independent of the input signal-to-noise ratio; i.e.,

$$S_1' + N_1'B_1 = K$$

†cf. Reference 5.3; see also Problem 5.6.

Thus,

$$S_1' = K \frac{S_1}{S_1 + \dfrac{N_1 B_1}{k}}$$

and

$$N_1' = \frac{K}{k} \frac{N_1}{S_1 + \dfrac{N_1 B_1}{k}}$$

Accordingly, an increase in the received signal level is partially realized, at the loop input, as a decrease in the noise level. The loop bandwidth, therefore, increases less rapidly (and hence more nearly optimally) with the signal level than it would in the absence of the limiter. Similarly, an increase in noise level is translated in part into a decrease in signal power at the loop input. This in turn causes the loop bandwidth to decrease thereby conforming more closely to the ideal situation.

In any event, it is usually acceptable to design the loop to operate optimally under the worst conditions expected. While the performance will not be optimum at other than the design level, it will still be better than at that extreme, which in itself is hopefully satisfactory.

In general, random phase fluctuations (oscillator noise) will be present even when the loop is designed to compensate solely for the additive noise and expected transients. This oscillator noise will increase the mean-squared error beyond that due to additive noise only. Nevertheless, in most cases the loop bandwidth needed to provide satisfactory transient behavior will be sufficiently wide to render the effect of this oscillator noise relatively small compared to that of the additive noise. This will be presumed in subsequent analyses involving phase-locked loops.

The analysis of this section began with the assumption that the linear model of the phase-locked loop provided a valid approximation to its actual behavior. Experimental evidence indicates this to be the case when the mean-squared loop tracking error does not exceed $\frac{1}{4}$ rad^2. The non-linear analysis of the first-order loop produces results which are in very close agreement with those of the linear analysis over this same range. Other approximate non-linear analyses also support this conclusion for higher-order loops. In all the applications with which we shall be concerned, a mean-squared phase error of greater than $\frac{1}{4}$ rad^2 would be intolerable. Accordingly, the linear model will be generally acceptable for our purposes. The major exception to this statement is in the initial acquisition process, when the loop is not yet tracking the received signal. This aspect of the problem is discussed in the next section.

5.7 Acquisition Time

In this section we derive estimates for the amount of time needed to bring a phase-locked loop into lock† when the initial difference between the VCO and received signal frequencies is Δf. Since two independent oscillators will never be operating at exactly the same frequency, this situation inevitably occurs in the initial lock-up phase of operation. The second-order loop, as suggested by the preceding discussion, is capable of adjusting to such a frequency offset with no steady-state error. Higher-order loops also have this property, but, because they exhibit stability problems as the signal parameters move away from their design values, second-order loops are generally preferred. Moreover, as already noted, the kinds of transients which require higher-order loops are typically of limited duration. For these reasons, we will limit our attention here to the lock-up capability of the second-order loop only, when the initial frequency offset is a constant Δf.

The time necessary to achieve lock with a phase-locked loop, although certainly related to the in-lock behavior of the loop, is inherently more difficult to analyze. This is because it is no longer possible to use a linear model of the loop; clearly, the loop must generally operate outside the linear range before lock is achieved. In the analysis to follow, some approximate answers are obtained by, among other things, neglecting the presence of noise at the loop input. Since phase estimates must be quite precise if they are to be useful in synchronous systems, the high signal-to-noise ratio situations are the ones of fundamental interest. The noise-free lock-up time of the loop, therefore, should provide a reasonable estimate of its performance in most synchronization applications.

The analysis of the lock-up time begins by returning to the non-linearized model of the phase-locked loop (Figure 5.6). The high frequency terms following the multiplier can be ignored since they will still be virtually eliminated by the filter. The filter has been designated $H(p)$, p indicating the operator d/dt, so that the analysis can be performed in the time domain. The output frequency of the VCO consists of a constant term ω_0, the center frequency, plus a term which is proportional to the input voltage $H(p)A_0 \sin \phi_e(t)$. Consequently,

$$\phi_1(t) = \frac{K_0 H(p)}{p} A_0 \sin \phi_e(t)$$

where K_0 is the VCO gain constant, and

$$\phi_0(t) - \phi_1(t) = \phi_0(t) - \frac{H(p) A_0 K_0}{p} \sin \phi_e(t) = \phi_e(t) \tag{5.7.1}$$

†A phase-locked loop is said to be "in lock" when the phase difference between the VCO and received signals drops below, and remains strictly less than, π radians.

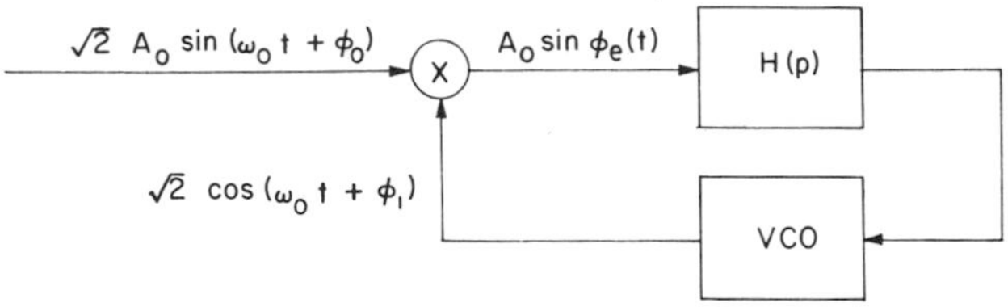

Figure 5.6 Phase-locked loop (non-linear model).

Differentiating, we find

$$\frac{d\phi_e(t)}{dt} + H(p)A_0K_0 \sin \phi_e(t) = \frac{d\phi_0(t)}{dt} = \Delta\omega \tag{5.7.2}$$

the last equality following because the initial difference between the VCO and received signal frequencies is, by hypothesis, $\Delta\omega$ radians/sec; $\phi_0(t) = (\Delta\omega)t + \phi_0$. The second-order loop filter encountered in Section 5.6 was of the form

$$H(p) = K_1\frac{1 + \tau_0 p}{p} \tag{5.7.3}$$

(The optimum filter, according to the criterion of Section 5.6, has $K_1 = B_0^2/A_0K_0$ and $\tau_0 = \sqrt{2}/B_0$.) Substituting for $H(p)$ in Equation (5.7.2) and again differentiating yields

$$\frac{d^2\phi_e}{dt^2} + \frac{d\phi_e}{dt}\tau_0 K \cos \phi_e + K \sin \phi_e = \frac{d^2\phi_0(t)}{dt^2} \tag{5.7.4}$$

where $K = A_0K_0K_1$. The input frequency is assumed constant so $d^2\phi_0(t)/dt^2 = 0$. It is convenient to make the substitution $\tau = \tau_0 Kt$. Then Equation (5.7.4) becomes, more simply,

$$\frac{d^2\phi_e}{d\tau^2} + \cos \phi_e \frac{d\phi_e}{d\tau} + \frac{1}{K\tau_0^2} \sin \phi_e = \dot{f} + f \cos \phi + \frac{1}{K\tau_0^2} \sin \phi = 0 \tag{5.7.5}$$

where $\phi = \phi_e(t)$, $\quad f = \dfrac{d\phi_e(t)}{d\tau}$ $\quad$ and $\quad \dot{f} = \dfrac{d^2\phi_e(t)}{d\tau^2}$.

Thus

$$\frac{\dot{f}}{f} = \frac{\dfrac{df}{d\tau}}{\dfrac{d\phi}{d\tau}} = \frac{df}{d\phi} = -\cos \phi - \frac{\sin \phi}{4\zeta^2 f} \tag{5.7.6}$$

where

$$\zeta^2 = \frac{K\tau_0^2}{4}$$

Equation (5.7.6) can be used to derive an approximate expression for the time needed to achieve lock. To begin, we multiply both sides of Equation

(5.7.6) by $\sin\phi\, d\phi$ and integrate between the limits $\phi = 2\pi k$ and $\phi = 2\pi(k+1)$, where k is an arbitrary integer, obtaining

$$\int_{\phi=2\pi k}^{\phi=2\pi(k+1)} \sin\phi\, df = -\int_{2\pi k}^{2\pi(k+1)} \frac{\sin^2\phi}{4\zeta^2 f}\, d\phi = -\int_{2\pi k}^{2\pi(k+1)} f\cos\phi\, d\phi \qquad (5.7.7)$$

the last integral gotten by integrating the first integral by parts. Similarly, multiplying both sides of Equation (5.7.6) by $f d\phi$, integrating between the same limits, and substituting from Equation (5.7.7) yields

$$\frac{1}{2}[f^2(2\pi(k+1)) - f^2(2\pi k)] = -\int_{2\pi k}^{2\pi(k+1)} f\cos\phi\, d\phi = -\int_{2\pi k}^{2\pi(k+1)} \frac{\sin^2\phi}{4\zeta^2 f}\, d\phi \qquad (5.7.8)$$

If f is positive, this last integral is positive and $f[2\pi(k+1)] < f(2\pi k)$ indicating that, regardless of the value of the frequency difference f, each complete cycle results in a net decrease in this difference. If f is negative, corresponding to a phase transversal in the opposite direction, from $2\pi(k+1)$ to $2\pi k$ radians, the integral on the right is negative and $|f(2\pi k)| < |f[2\pi(k+1)]|$. The frequency difference thus approaches zero, regardless of the initial difference. The theoretical frequency lock-in range is infinite.

To estimate the amount of time necessary to achieve lock, we recall that $f = d\phi/d\tau$ and consequently

$$t = \frac{\tau}{\tau_0 K} = \frac{1}{\tau_0 K}\int \frac{d\phi}{f} \qquad (5.7.9)$$

By summing Equation (5.7.8) over n periods we find

$$f^2(2\pi n) - f^2(0) = -\frac{1}{2\zeta^2}\int_0^{2n\pi} \frac{\sin^2\phi}{f}\, d\phi$$

For some value of n, $f^2(2n\pi) \approx 0$ and

$$f^2(0) - f^2(2\pi n) = \frac{1}{2\zeta^2}\int_0^{2\pi n} \frac{\sin^2\phi}{f}\, d\phi = \frac{\alpha}{2\zeta^2}\int_0^{2\pi n} \frac{d\phi}{f} = \frac{2T_{\Delta f}\alpha}{\tau_0} \qquad (5.7.10)$$

where $T_{\Delta f}$ is the time necessary for the frequency difference to decay to approximately zero when the initial difference is $\Delta f = [\tau_0 K f(0)/2\pi]$ Hz, and where α is some constant in the interval $0 < \alpha < 1$. Consequently,

$$T_{\Delta f} = \frac{\pi^2}{8\alpha\zeta}\left(\frac{1+4\zeta^2}{4\zeta}\right)^3 \frac{(\Delta f)^2}{B_L{}^3} \text{ secs} \qquad (5.7.11)$$

where $B_L = (1/4\tau_0)(1 + 4\zeta^2)$ is the loop bandwidth when the loop filter is as given in Equation (5.7.3). (If the optimum filter of Section 5.6 is used, $\zeta^2 = \frac{1}{2}$ and $\tau_0 = \sqrt{2}/B_0$. It is interesting to note that the lock-up time of Equation (5.7.11) is actually minimized when $\zeta^2 = \frac{1}{2}$.) A generally good estimate of $T_{\Delta f}$ results when $\sin^2 x$ in Equation (5.7.10) is replaced by its average value of $\frac{1}{2}$. Then $\alpha = \frac{1}{2}$ and

$$T_{\Delta f} \approx \frac{\pi^2}{4\zeta}\left(\frac{1+4\zeta^2}{4\zeta}\right)^3 \frac{(\Delta f)^2}{B_L{}^3} \text{ secs} \qquad (5.7.12)$$

In some situations it may be possible, and desirable, to increase B_L during the lock-up mode to decrease the time required, and then to decrease B_L after an in-lock situation is indicated. The amount by which B_L can be changed, however, is limited by the noise, since a larger loop bandwidth increases the effect of the additive noise and consequently increases the probability of altogether failing to attain lock.

The fact remains, therefore, that the time needed to achieve lock grows as the square of the difference between the VCO and the received signal frequencies. When the initial frequency uncertainty is large the lock-up time can be excessive. When this is the case the method generally adopted is to add a linearly increasing (or decreasing) voltage to the VCO input, thereby sweeping the VCO frequency over the range of uncertainty. The previous discussion can be extended to obtain some limited results in this situation. Now, since

$$\phi_1(t) = \frac{K_0 H(p)}{p} A_0 \sin \phi_e(t) - \frac{1}{2}\beta t^2 \tag{5.7.13}$$

where β is the sweep rate in radians per second per second, Equation (5.7.2) becomes

$$\frac{d\phi_e(t)}{dt} + A_0 K_0 H(p) \sin \phi_e(t) = \frac{d\phi_0(t)}{dt} + \beta t \tag{5.7.14}$$

Substituting for $H(p)$ and differentiating, as before, yields

$$\frac{d^2\phi_e}{dt^2} + \frac{d\phi_e}{dt}\tau_0 K \cos \phi_e + K \sin \phi_e = \frac{d^2\phi_0(t)}{dt^2} + \beta \tag{5.7.15}$$

in place of Equation (5.7.4). Equation (5.7.6) becomes

$$\frac{df}{d\phi} = -\cos \phi + \frac{\dfrac{\beta}{K} - \sin \phi}{4\zeta^2 f} \tag{5.7.16}$$

Integrating both sides of this equation between the limits $2\pi k$ and $2\pi(k + 1)$ shows

$$f(2\pi(k+1)) - f(2\pi k) = \frac{1}{4\zeta^2}\int_{2\pi k}^{2\pi(k+1)} \frac{\dfrac{\beta}{K} - \sin \phi}{f}\, d\phi \tag{5.7.17}$$

But now, if $\beta > K$, a positive value of f over any one cycle indicates $f[2\pi(k + 1)]$ to be greater than $f(2\pi k)$, while a negative frequency difference establishes that $f(2\pi k) < f(2\pi(k + 1))$ when $\beta < -K$. In either of these instances, the frequency difference grows in absolute magnitude over any one cycle. Furthermore, when f is small in absolute magnitude, Equation (5.7.16) shows f to be an increasing function of ϕ when $f > 0$ and $\beta > K$, and when $f < 0$ and $\beta < -K$. Accordingly, the absolute value of f is an increasing function of time for small values of f. These two factors strongly suggest an unstable situation when β exceeds K in absolute magnitude with the consequent inability of the loop to come into lock. Experimental evidence and computer simulation support this conclusion, as well as the slightly weaker converse that when $|\beta| < K/2$, lock can be assured. (At lower signal-to-noise ratios $|\beta|$ must be still smaller relative to K. However, at the signal-to-noise ratios generally encountered in coherent communication and synchronization problems, $|\beta|$ can be nearly as large as $K/2$.)

Sweeping across the region of uncertainty, Δf Hz, at the rate of $\beta/2\pi$ Hz /sec requires $2\pi\Delta f/\beta$ seconds. The lock-up time using this method is on the order of

$$T'_{\Delta f} = \frac{2\pi\Delta f}{\beta} \approx \frac{4\pi\Delta f}{K} = \frac{\pi\Delta f}{\left(\frac{4\zeta}{1+4\zeta^2}\right)^2 B_L^2} \tag{5.7.18}$$

Comparing Equations (5.7.18) with Equation (5.7.12), we find sweeping the VCO across the region of uncertainty beginning to be effective if

$$\Delta f > \frac{4}{\pi}\frac{4\zeta^2}{1+4\zeta^2} B_L \tag{5.7.19}$$

These lock-up time estimates have ignored the time needed to reduce the phase error to some satisfactorily small value once the frequency error has been so reduced. To estimate this quantity, consider the situation in which the loop is initially in lock, but in which the received phase undergoes a step change of magnitude $\Delta\theta$ at the time $t = 0$. If $\Delta\theta$ is not too large, the linear model of the loop is again applicable. Since we are considering a second-order loop, we have

$$\theta_e(t) = \frac{1}{2\pi j}\int_{-j\infty}^{j\infty} \frac{\Delta\theta}{s}[1 - Y(s)]e^{st}\, ds = \sqrt{2}\,\Delta\theta\, e^{-B_0 t/\sqrt{2}} \cos\left(\frac{\sqrt{2}\,B_0 t}{2} + \frac{\pi}{4}\right)$$

The time needed to reduce the envelope of this transient error to approximately $\frac{1}{4}$ of its original value is therefore $T_{\Delta\theta} = 1/B_L$. This, of course, provides only a rough indication of the magnitude of the quantity of interest. Nevertheless, if B_L is small relative to the frequency uncertainty Δf, this final portion of the lock-up time is presumably negligible relative to the time needed to overcome the initial frequency offset [Equation (5.7.12) or (5.7.18)].

5.8 Phase-Locked Loops and Non-Sinusoidal Signals

The discussion leading to the concept of the phase-locked loop in Section 5.4 was in no sense limited to sinusoidal signals. The same approach is equally applicable to any periodic signal $x(t)$. Since the output of the signal generator (Figures 5.1 and 5.2) is denoted by $x'(t)$, the preceding statement might be qualified to include only differentiable periodic functions. This is actually not necessary, however. The same device could be postulated for tracking the phase of the signal $x(t)$ using some locally generated periodic signal other than $x'(t)$. The argument in Section 5.4 leading to the phase-locked loop of Figure 5.2 is equally applicable when the local signal is any periodic function $z(t)$ such that

$$\int_0^T x(t)z(t + \tau)\, dt\,|_{\tau=0} = 0 \tag{5.8.1}$$

and

$$\frac{\partial}{\partial\tau}\int_0^T x(t)z(t + \tau)\, dt\,|_{\tau=0} < 0$$

and such that these two conditions hold uniquely at the point $\tau = 0$. Indeed, superior tracking performance could conceivably result using some local signal $z(t)$ not equal to $x'(t)$. We will return to this point shortly.

For the present, in order to keep the discussion as general as possible, the local signal $z(t)$ will be left as an otherwise unspecified periodic function. Both $z(t)$ and $x(t)$, however, will be normalized to represent unity power. Since the amplitude A of the received signal is arbitrary, and since increasing the amplitude of $z(t)$ is equivalent, for example, to increasing the otherwise unspecified loop gain, this normalization imposes no loss of generality.

To the extent that the conditions leading to the phase-locked loop are applicable (i.e., to the extent that both τ and $\hat{\tau}$ are relatively slowly varying functions of time) the phase-locked loop and the tracking device of Figure 5.1 are equivalent. Under these same conditions, then, it follows that the influence of the received signal $x(t + \tau)$ on the loop performance is exclusively dependent upon the correlation function

$$\rho_{xz}(\hat{\tau} - \tau) = \frac{1}{T}\int_0^T x(t + \tau)z(t + \hat{\tau})\, dt \tag{5.8.2}$$

Furthermore, if the loop can be said to be operating in a sufficiently small region about the point $\hat{\tau} = \tau$ with sufficiently high probability,

$$\rho_{xz}(\tau - \hat{\tau}) \approx (\tau - \hat{\tau})\lambda \tag{5.8.3}$$

where

$$\lambda = \frac{d}{d\tau}\rho_{xz}(\tau)\,|_{\tau=0} \triangleq \rho'_{xz}(0)$$

(This, of course, was the approximation which lead to the linear loop in Section 5.5.) It would appear therefore that, so long as the tracking error remains within this restricted region, the effect of the functional form of the received signal can be measured solely in terms of the quantity $\rho'_{xz}(0)$, that all signals $x(t)$ and $z(t)$ giving rise to the same parameter $\rho'_{xz}(0)$ are equivalent.

To support this contention that only the correlation function $\rho_{xz}(\tau)$, or in the linear region the slope of this correlation function, is of significance, however, it is also necessary to re-examine the product $z(t + \hat{\tau})n(t)$ representing the noise at the input to the loop filter. The phase $\hat{\tau} = \hat{\tau}(t)$ is dependent upon the noise $n(t_1)$, but, as in Section 5.5, only for $t_1 < t$. Thus, the argument that the spectrum of this product is "white" if $n(t)$ occupies a band broad relative to the loop bandwidth remains inviolate. If the two-sided power spectral density of $n(t)$ is $N_0/2$, the spectral density of $z(t + \hat{\tau})n(t)$ is, from Section 5.5,

$$\frac{N_0}{2} E[z^2(t + \hat{\tau})] = \frac{N_0}{2} \tag{5.8.4}$$

We next consider the distribution of the amplitude, at any instant in time, of the low-frequency components of the product $z(t + \hat{\tau})n(t)$. (The high-frequency components will again be excluded by the filtering action of the loop.) To do this, we represent the periodic signal $z(t + \hat{\tau})$ by its Fourier series expansion

$$z(t + \hat{\tau}) = \sum_{i=0}^{\infty} a_i \cos(\omega_i t + \theta_i) \qquad \omega_i = \frac{2\pi i}{T}$$

We then temporarily precede the loop with a bank of ideal bandpass filters, one centered about each of the harmonic frequencies $\omega_i = 2\pi i/T$, $i \geq 0$, with each filter having the same bandwidth $B \leq 1/T$ (except the zero frequency bandwidth which is half this value). As in Section 5.5, the noise through the ith filter can be written

$$n_i(t) = n_{c_i}(t) \cos \omega_i t - n_{s_i}(t) \sin \omega_i t \tag{5.8.5}$$

Then

$$\begin{aligned} \{z(t + \hat{\tau})n(t)\}_{lf} &= \left\{\sum_{i,j} a_i \cos(\omega_i t + \theta_i)(n_{c_j}(t) \cos \omega_j t - n_{s_j}(t) \sin \omega_j t)\right\}_{lf} \\ &= \sum_{i=1}^{\infty} \frac{a_i}{2}(n_{c_i}(t) \cos \theta_i + n_{s_i}(t) \sin \theta_i) + a_0 n_{c_0}(t) \end{aligned} \tag{5.8.6}$$

This is just a weighted sum of terms of the form found in Equation (5.5.7), terms which were shown to be zero-mean Gaussianly distributed random variables with time independent second moments. The purpose of preceding the loop with bandpass filters was to provide a convenient representation for the noise. But the argument leading to Equation (5.8.6) is valid even when $B = 1/T$, that is, when there is actually no filter at all. We conclude, as in Section 5.5, that the low-frequency noise component at the input to the loop filter is a sample function of a white (i.e., broadband relative to the loop bandwidth) Gaussian random process.

When $z(t)$ is a square wave assuming only the amplitudes plus one and minus one, the fact that $n(t)z(t)$ has the same distribution as does $n(t)$ when $n(t)$ is white and symmetrically distributed is apparent. The distributions of $n(t)$ and $-n(t)$ are exactly the same and, since $n(t)$ and $n(t + \epsilon)$ are independent for all $\epsilon \neq 0$, it is impossible to tell whether $n(t)$ or $-n(t)$ is being observed. The noise statistics obviously remain unaltered in this case. The argument of the preceding paragraphs shows this same conclusion to be true for the low frequency components of the product $n(t)z(t)$ where $z(t)$ is any periodic waveform. Intuitively, $\{z(t)n(t)\}$ simply represents a frequency translated version of white noise, which, because it is white, does not change any of its properties.

Combining these two facts (i.e., (1) the statistics of the noise $n_1(t) = \{n(t)z(t+\hat{\tau})\}_{lf}$ are virtually independent of $z(t)$, and (2) the error signal in the linear model of the loop is dependent not on the signal $z(t)$ itself, but on the slope of the correlation function $\rho_{xz}(\tau)$), we come to the following conclusion. The performance of a phase-locked loop designed to track an arbitrary periodic signal $Ax(t)$ of period T is identical to that of a loop having a sinusoidal input (as predicted by the linear model) when the rms amplitude of the sinusoid is

$$A_e = \frac{AT}{2\pi}\,|\rho'_{xz}(0)| \tag{5.8.7}$$

This conclusion is valid so long as the loop is operating within the region over which the slope of the correlation function $\rho_{xz}(\tau)$ is relatively constant.

To return to the question as to whether or not $z(t)$ should be equated to $x'(t)$, as suggested by the argument in Section 5.4, we note that the "best" choice for $z(t)$, in view of Equation (5.8.7), is one yielding a correlation function with the maximum possible slope in the neighborhood of $\tau = 0$. The optimum choice for $z(t)$ therefore is indeed $x'(t)$. This follows because

$$|\rho'_{xz}(0)| = \left|\frac{\partial}{\partial\tau}\frac{1}{T}\int_0^T x(t)z(t+\tau)\,dt\right|_{\tau=0} = \left|\frac{1}{T}\int_0^T x(t)z'(t)\,dt\right| \tag{5.8.8}$$

Integrating this last expression by parts and using Schwarz's inequality, we find

$$|\rho'_{xz}(0)| = \left|\frac{1}{T}\int_0^T x'(t)z(t)\,dt\right| \leq \left[\frac{1}{T}\int_0^T [x'(t)]^2\,dt\right]^{1/2}\left[\frac{1}{T}\int_0^T z^2(t)\,dt\right]^{1/2} \tag{5.8.9}$$

But since $z(t)$ is constrained to represent unity power, the right side of the above inequality is independent of this function. And since this inequality is an equality if, and only if, $z(t) = kx'(t)$ for some constant k, the stated conclusion follows. Actually, this result is directly related to the efficiency of the maximum-likelihood estimator of τ [Equation (5.4.1)].

This is not the complete story, however, since the loop will not necessarily always operate in the linear region about the point $\tau = 0$, particularly if this region is small. The preceding paragraph does establish that the optimum local signal $z(t)$ for a given received signal $x(t)$ is $z(t) = x'(t)$, *provided* the correlation function $\rho_{xx'}(\tau)$ is an approximately linear function of τ over a sufficiently large neighborhood of the point $\tau = 0$.†

To be somewhat more quantitative about the significance of the extent of the linear region, suppose $\rho_{xz}(\tau)$ were a strictly linear function of τ for $|\tau| < L/2$, and that the phase of the received signal were constant. Then, according to the linear model, the probability of the loop tracking error $\tau_e = \tau - \hat{\tau}$ remaining within this region would be‡

$$\Pr\left\{|\tau_e| < \frac{L}{2}\right\} = \operatorname{erf}\left(\frac{\pi L}{2\sigma_\phi T}\right) \tag{5.8.10}$$

with $\sigma_\phi^2 = N_0 B_L / A_e$. If $\rho_{xz}(\tau)$ is not a strictly linear function of τ over the region $|\tau| \leq L/2$, but rather is of the generic form shown in Figure 5.7, it still follows that

$$\Pr\left(|\tau_e| < \frac{L}{2}\right) = \operatorname{erf}\left(\frac{\pi L k(L)}{\sqrt{2}\,\sigma_\phi T}\right) \tag{5.8.11}$$

for some $k(L)$ in the interval $\lambda(L)/\lambda_0 \leq k(L) \leq 1$, and with $\lambda(L)$ as defined in Figure 5.7. The magnitude of this probability as a function of L provides a measure of the adequacy of the linear model of the phase-locked loop for the particular application of interest.

[This same argument can be used to estimate the loss of lock probability. Since, by definition, loss of lock occurs whenever $|\tau_e| > T/2$,

$$p \triangleq \Pr(\text{loss of lock}) = \Pr\left\{|\tau_e| > \frac{T}{2}\right\} = 1 - \operatorname{erf}\left(\frac{\pi k}{\sqrt{2}\,\sigma_\phi}\right) \tag{5.8.12}$$

for some k, $\max\limits_{L \leq T} L\lambda(L)/\lambda_0 T \leq k \leq 1$. If independent samples of the tracking error are taken until a loss of lock is detected, the expected number of samples required is

$$E(n) = \sum_{n=1}^{\infty} n(1-p)^n p = \frac{1}{p} \tag{5.8.13}$$

And since the minimum time interval separating independent samples at the loop output is on the order of $1/2B_L$, the expected time to loss of lock is approximately

$$\bar{T} = \frac{E(n)}{2B_L} = \frac{1}{2B_L\left[1 - \operatorname{erf}\left(\dfrac{\pi k}{\sqrt{2}\,\sigma_\phi}\right)\right]} \tag{5.8.14}$$

---

†Since the derivative of a sinusoid is reasonably linear over a sizable fraction of its period, the local signal, when $x(t)$ is a sinusoid, should presumably also be a sinusoid, as was assumed in the preceding sections. Even here, however, other criteria would lead to different conclusions. If it were deemed more important to extend the range of linearity of the function $\rho_{xz}(\tau)$, for example, than to maximize $\rho'_{xz}(0)$, $z(t)$ might be chosen differently (cf. Reference 5.5).

‡Since, under the assumptions here, the tracking error is a linear function of a Gaussian random process (cf. Section 5.5), it is itself a Gaussian random process.

$$\lambda(L) = \begin{cases} \lambda_0 & |\tau| \leq \dfrac{L_1}{2} \\[2mm] \dfrac{\rho_{xz}(\tau)}{\tau} & \dfrac{L_1}{2} \leq |\tau| \leq \dfrac{L_2}{2}, \dfrac{L_3}{2} \leq |\tau| \leq \dfrac{T}{2} \\[2mm] \lambda_1 & \dfrac{L_2}{2} \leq |\tau| \leq \dfrac{L_3}{2} \end{cases}$$

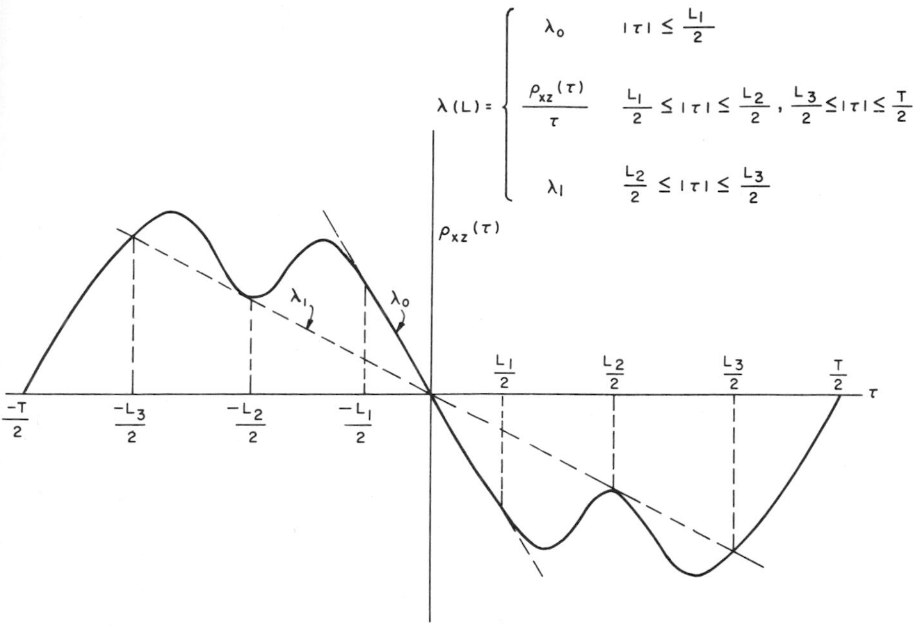

Figure 5.7 Representative correlation function and associated bounds.

While this argument is entirely heuristic, its validity can be confirmed, for first-order loops, using the non-linear model. If $\rho_{xz}(\tau)$ is sinusoidal, for example, the mean time to loss of lock in a first-order loop is well approximated for all $\sigma_\phi < \frac{1}{2}$ rad by Equation (5.8.14), with $k = 2/\pi$ (Reference 5.6).]

Notes to Chapter Five

The theory of optimum linear filtering and prediction was originally formulated by Wiener (Reference 5.1) and Kolmogoroff (Reference 5.7). The approach presented here loosely parallels that of Bode and Shannon (Reference 5.8). The method in Section 5.6 for determining the optimum loop filter is due to Jaffe and Rechtin (Reference 5.4). The lock-up time analysis in Section 5.7 is taken from Reference 5.9. The recent book by Viterbi (Reference 5.6) is highly recommended for a thorough discussion of the performance of phase-locked loops, both as tracking devices and as demodulators. See Reference 5.6 also for an exact analysis of the steady-state performance of a first-order loop. Significant progress has recently been made in extending the nonlinear analysis to higher-order loops (Reference 5.10).

Problems

5.1 The signal $y(t) = x(t) + n(t)$ is passed through two smoothing filters, designed to produce the minimum mean-squared estimates, $\hat{x}(t)$ and $\hat{n}(t)$, of $x(t)$ and $n(t)$, respectively.
(a) Compare the two estimates $\hat{x}(t)$ and $y(t) - \hat{n}(t)$ of $x(t)$.
(b) Express $\hat{y}(t) \triangleq \hat{x}(t) + \hat{n}(t)$ in terms of $x(t)$ and $n(t)$.

5.2 Verify Equations (5.6.22) through (5.6.25).

5.3 The second-order loop filter defined in Equation (5.6.18) is of the form $H(s) = a + (b/s)$ and, hence, involves a perfect integrator. In practice, the integrator will not be perfect and the filter will have the form $H(s) = (as + b)/(s + \delta)$, with δ small, but not zero. Determine the noise bandwidth of a loop having such a filter and find the steady-state response of the loop to a frequency step input.

5.4 Show, using for example the root-locus method, that the linear phase-locked loop models of both the first- and second-orders (with and without a perfect integrator) are stable regardless of the deviation of the signal and noise levels from their design values. Show that this is not true of third-order loops.

5.5 The exact analysis of the non-linear model of the first-order phase-locked loop yields, as the distribution of the phase error, the *Tikhonov* distribution:

$$p(\phi) = \exp [R_L g(\phi)] \Big/ \int_{-\pi}^{\pi} \exp [R_L g(\phi)] d\phi, \qquad |\phi| \leq \pi$$

where $g(\phi) = \int_{-\pi}^{\phi} \rho_{xz}(T\eta/2\pi)\, d\eta$ with $\rho_{xz}(\tau)$ as defined in Equation (5.8.2), and where $R_L = 4A/N_0K$ with A^2 the received signal power, N_0 the noise spectral density, and K the VCO gain constant. [This result is valid when the received signal phase is constant, the noise is white and Gaussian, and the phase error is interpreted modulo 2π; cf. Reference 5.6.] Show that the phase error distribution is indeed approximately Gaussian, for sufficiently large values of R_L, with a variance σ_ϕ^2 as predicted by the linear model [with an effective signal amplitude as given in Equation (5.8.7)]. Show further that if $\rho_{xz}(\tau)$ is an odd function of τ, and if its first three derivatives all exist, the Gaussian approximation is valid for all R_L for which

$$\sigma_\phi^2 \ll \frac{48\pi^2}{T^2} \left| \frac{\rho'_{xz}(0)}{\rho'''_{xz}(0)} \right|$$

5.6 The signal $y(t) = \sqrt{2}\, A \sin (\omega_c t + \theta) + n(t)$, $n(t) =$ white Gaussian noise, is passed through a bandpass filter with noise bandwidth B_1 centered about ω_c. The filter output is in turn passed through an odd νth-law device [i.e., a device which converts an input $(x)t$ into an output $|x(t)|^\nu \operatorname{sgn}(x(t))$] followed by a second bandpass filter with noise bandwidth B_2 centered about ω_c. Show that if $B_2 \ll B_1 \ll \omega_c$ and the input signal-to-noise ratio is sufficiently large, the output signal-to-noise ratio is inversely proportional to $1 + \nu^2$ and hence is maximized for $\nu = 0$. (The signal-to-noise ratio is here defined as the ratio of the mean-squared value of the signal which would be observed in the absence of noise to that of the difference between this signal and the one actually observed.) Show that the accuracy with which this signal can be tracked with a phase-locked loop under these conditions is, in contrast, independent of ν. [Hint: express the input to the νth-law device in the form $B \sin(\omega_c t + \psi)$ and find the in-phase and quadrature components of the fundamental at its output.] What does this imply, so far as a phase-locked loop is concerned, about the 3 db improvement in signal-to-noise ratio attainable (when this ratio is large) bypassing such a signal through a bandpass limiter? [See page 136].

References

5.1 N. Wiener, *Extrapolation, Interpolation, and Smoothing of Stationary Time Series*, John Wiley & Sons, New York, N. Y., 1949.

5.2 M. C. Yovits, and J. L. Jackson, "Linear Filter Optimization with Game Theory Considerations," *IRE Nat. Conv. Record*, Part 4, p. 193, 1955.

5.3 W. B. Davenport, Jr., and W. L. Root, *Random Signals and Noise*, McGraw-Hill Book Co., New York, N. Y., 1958.

5.4 R. Jaffe, and E. Rechtin, "Design and Performance of Phase-Lock Circuits Capable of Near-Optimum Performance over a Wide Range of Input Signal and Noise Levels," *IRE Trans. on Inf. Theory*, IT-1, p. 66, 1955.

5.5 L. M. Robinson, "Tanlock: A Phase-Locked Loop of Extended Tracking Capability," *Conf. Proc. 1962 Nat. Winter Conv. on Military Electron.*, p. 396, 1962.

5.6 A. J. Viterbi, *Principles of Coherent Communication*, McGraw-Hill Book Co., New York, N. Y., 1966.

5.7 A. N. Kolmogoroff, "Interpolation and Extrapolation," *Bull. Acad. Sci. U.S.S.R., Ser. Math*, **5,** p. 3, 1944.

5.8 H. W. Bode, and C. E. Shannon, "A Simplified Derivation of Linear Least-square Smoothing and Prediction Theory," *Proc. IRE*, **38,** p. 417, 1950.

5.9 A. J. Viterbi, "Acquisition and Tracking Behavior of Phase-Locked Loops," *Proc. Symposium on Active Networks and Feedbacks*, **10,** Polytech. Inst. Brooklyn, N. Y., 1960.

5.10 W. C. Lindsey, "Nonlinear Analysis and Synthesis of Generalized Tracking Systems," Univ. So. Calif., Los Angeles, Calif., Part I, Rep. 317, 1968; Part II, Rep. 342, 1969.

Supplementary Bibliography

Gardner, F. M. *Phaselock Techniques*, John Wiley & Sons, New York, N. Y., 1966.

Lee, Y. W., *Statistical Theory of Communication*, John Wiley & Sons, Inc., New York, N. Y., 1960.

Tausworthe, R. C., "Cycle Slipping in Phase-Locked Loops," *IEEE Trans. Comm. Tech.*, Com.-15, p. 417, 1967.

Tikhonov, V. I., "The Effects of Noise on Phase-Lock Oscillator Operation," *Avtomat. i Telemakh.*, **20**, p. 1160, 1959; "Phase-Lock Automatic Frequency Control Application in the Presence of Noise," *Avtomat. i Telemakh.*, **21**, p. 209, 1960.

Van Trees, H. L., *Detection, Estimation, and Modulation Theory, Vol. I*, John Wiley & Sons, Inc., New York, N. Y., 1968.

PART TWO

SYNCHRONIZATION

chapter six

Separate Channel Synchronization

6.1 Introduction

This chapter, and the succeeding two, are concerned with techniques for establishing the synchronization needed for the efficient operation of communication systems of the type described in Chapters Three and Four. Perhaps the simplest and most straightfoward way to provide this information is to utilize a separate communication channel, or channels, solely for synchronization purposes. Some of the aspects of this approach are investigated in this chapter.

As discussed in Chapter One, the first step in the synchronization procedure is usually to slave the receiver and transmitter clocks, thereby establishing a common clock (carrier) reference throughout the system. The second step is then to increase the unambiguous time interval to an acceptable T_0 seconds.

Slaving the receiver to the transmitter clock entails transmitting a periodic signal along with the message and tracking its phase at the receiver. Typically, the clock also serves as a carrier or subcarrier, and therefore is most generally a sinusoid. Regardless of the periodic signal used, however, the optimum receiver is, in essence, a phase-locked loop, as the preceding chapter attempted to demonstrate. The following two sections investigate the relationship between the clock signal and the attainable tracking accuracy.

6.2 Sinusoidal Clock Signals

If the clock signal is to be a sinusoid, the receiver should presumably involve a phase-locked loop using a sinusoidal reference, as discussed in Chapter Five. Under these circumstances, the only choices remaining as to either the signal or the receiver structure concern the signal frequency and the loop filter. Moreover, these two choices are not independent. The optimum loop filter was found, in Chapter Five, to be completely defined in terms of the signal-to-noise ratio, and of the phase statistics or the anticipated phase transients. The signal-to-noise ratio is prescribed by external constraints. Similarly, once both the transmitter and receiver oscillator stabilities and their relative motion are specified, the phase characteristics are also determined. The only uncertainty in these characteristics is one of magnitude since, as will be shown momentarily, this is a function of the oscillator frequency. Thus, in effect, the single choice remaining is the choice of frequency.

If the variance of the loop phase error is σ_0^2 rad^2 and the signal frequency is $\omega_0 = 2\pi/T_0$ rad/sec, the variance of the tracking error in *seconds* squared is $\sigma_\tau^2 = \sigma_0^2/\omega_0^2$. The synchronization error could apparently be decreased simply by increasing the clock frequency from ω_0 rad/sec to, say, $\omega > \omega_0$ rad/sec. This is only part of the story, however, since the loop filter, and hence the phase error variance σ_0^2, are also functions of ω. Specifically, the filter used in a phase-locked loop is the result of a compromise between the effect of a randomly varying or transient signal phase and the effect of the additive noise. To minimize the tracking error due to additive noise, the loop bandwidth should be made as small as possible; to minimize the transient error it should be as large as possible. Since the magnitude of the transient is presumably directly proportional to the signal frequency, the higher the frequency, the greater the required loop bandwidth. Thus, when the signal power and the noise spectral density are both fixed, the phase tracking error can be expected to increase as the design frequency of the loop is increased.

This increase in phase error variance will be at most directly proportional to the frequency increase, however. Indeed, the presence of additive noise and the consequent necessity to compromise between the two sources of error will usually cause the bandwidth of the optimum loop and hence the phase error variance to increase less than proportionately with the signal frequency. In general, therefore, the phase error variance can be written in the form $\sigma_0^2(\omega/\omega_0)^r$ rad^2 for some r in the interval (0, 1), and $\sigma_\tau^2 = (\sigma_0^2/\omega_0^r)(1/\omega^{2-r})$ sec^2. Thus σ_τ is inversely proportional to $\omega^{1-r/2}$ and does decrease as ω is increased.

Of course, ω cannot be made arbitrarily large. The rms phase error must be small as compared to $\pi/2$ radians if the loop is to remain in lock over long periods of time; i.e., the condition

$$\sigma_0\left(\frac{\omega}{\omega_0}\right)^{r/2} = \frac{\pi}{2k} \tag{6.2.1}$$

must hold for some suitably large constant k. (The value of k must be of the order of 3 or greater if the loss of lock probability is to be acceptably small; cf. Section 5.8.) Accordingly, the maximum acceptable signal frequency is

$$\omega = \omega_0\left(\frac{\pi}{2k\sigma_0}\right)^{2/r} \tag{6.2.2}$$

and the variance of the phase-error of a loop operating at this frequency would be

$$\sigma_\phi^2 = \sigma_\tau^2\omega_0^2 = \left(\frac{4k^2}{\pi^2}\right)^{(2/r)-1}\sigma_0^{4/r}\ \text{rad}^2 \tag{6.2.3}$$

Here, for comparative purposes, we have expressed the variance in terms of the original period $T_0 = 2\pi/\omega_0$ (i.e., 2π radians correspond to T_0 seconds). The phase-error variance of a loop using the maximum acceptable frequency is seen to be proportional to the $(2/r)$th power of the variance σ_0^2 of a loop operating at the frequency ω_0 provided, of course, this variance is small compared to unity.

As an example consider the second-order loop designed to track a frequency step of magnitude $\Delta\omega/2\pi$. From Section 5.6

$$\sigma_0^2 = \frac{N_0 B_L}{A_0^2} = \frac{3}{4(2)^{1/4}}\left(\frac{N_0}{A_0^2}\right)^{3/4}(\lambda\Delta\omega)^{1/2}$$

Since the magnitude of the frequency step is presumably directly proportional to the frequency, r is in this case equal to $\frac{1}{2}$.† (Typically r will be equal to the reciprocal of the order of the loop; see Section 5.6.) Thus

$$\sigma_\phi^2 = \left(\frac{2k}{\pi}\right)^6\sigma_0^8 = \left(\frac{2k}{\pi}\right)^6\left(\frac{N_0 B_L}{A_0^2}\right)^4 \tag{6.2.4}$$

an impressive improvement when $N_0 B_L/A_0^2$ is small as compared to unity.

6.3 Non-Sinusoidal Clock Signals

The performance of a phase-locked loop in tracking the signal $y(t) = Ax(t) + n(t)$ by using the locally generated signal $z(t)$ (see Figure 5.2) was found, in Section 5.8, to be identical to that of the same loop were $x(t)$ and $z(t)$ both sinusoids with

†The parameter λ is independent of the frequency; it merely represents the relative weight attached to the noise and transient errors. To verify this observe that if λ is constant σ_n^2 and e_T^2 both exhibit the same dependence on $\Delta\omega$ (cf. Section 5.6).

$$A_e = \frac{AT_0}{2\pi}\,|\rho'_{xz}(0)| \tag{6.3.1}$$

the received signal rms amplitude and T_0 the signal period.

As an example of the tracking performance attainable using nonsinusoidal signals, let $x(t)$ be a periodic pulse of width ΔT and repetition rate T_0. Since the actual pulse shape is of secondary importance in the sequel (cf. Problem 6.2), it will be assumed for convenience to be rectangular:

$$x(t) = \begin{cases} \sqrt{T_0/\Delta T} & -\frac{\Delta T}{2} < t < \frac{\Delta T}{2} \quad \text{modulo } T_0 \\ 0 & \text{otherwise} \end{cases} \tag{6.3.2a}$$

Let the local signal $z(t)$ be defined by:

$$z(t) = \begin{cases} \sqrt{T_0/2\Delta t} & -\frac{\Delta T + \Delta t}{2} < t < -\frac{\Delta T - \Delta t}{2} \quad \text{modulo } T_0 \\ -\sqrt{T_0/2\Delta t} & \frac{\Delta T - \Delta t}{2} < t < \frac{\Delta T + \Delta t}{2} \quad \text{modulo } T_0 \\ 0 & \text{otherwise} \end{cases} \tag{6.3.2b}$$

with $\Delta t \leq \Delta T$. These functions are shown in Figure 6.1 along with their

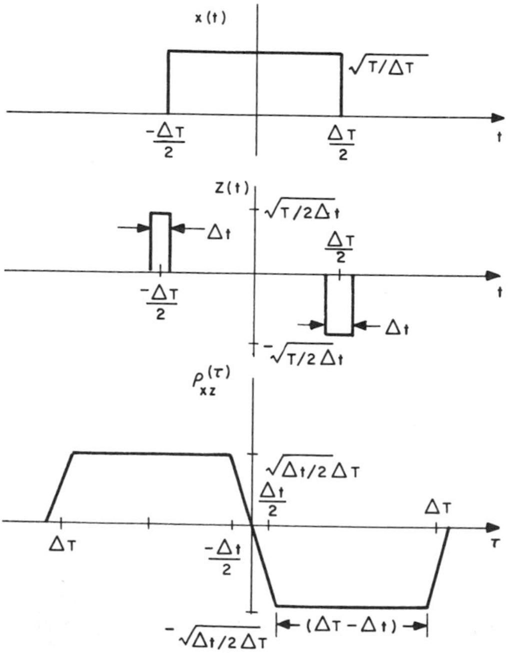

Figure 6.1 Functions defined in Equation (6.3.2).

cross-correlation function $\rho_{xz}(\tau)$, which assumes the form

$$\rho_{xz}(\tau) = -\tau\sqrt{2/\Delta T \cdot \Delta t} \qquad |\tau| < \Delta t/2 \tag{6.3.2c}$$

over the region of primary interest.

The slope of the correlation function $\rho_{xz}(\tau)$ in the range $|\tau| < \Delta t/2$ is inversely proportional to $(\Delta t)^{1/2}$. This would suggest letting Δt approach zero to minimize the tracking error. Indeed, as $\Delta t \to 0$, $z(t) \to x'(t)$, the optimum local signal so far as maximizing $\rho'_{xz}(0)$ is concerned. Clearly, however, Δt cannot be made arbitrarily small since the maximum magnitude of $\rho_{xz}(\tau)$ is *directly* proportional to $(\Delta t)^{1/2}$. Determining the best value for Δt requires taking into account the fact that the optimum loop bandwidth B_e is a function of the effective signal amplitude A_e [Equation (6.3.1)]. Specifically,

$$B_e = B_L\left(\frac{A_e}{A}\right)^r$$

where B_L is the bandwidth of the optimum loop designed to track a sinusoid of amplitude A, and r depends on the order of the loop. (Again, r is typically the reciprocal of the order of the loop.) The mean-squared phase error due to the additive noise, therefore, when $x(t)$ and $z(t)$ are as defined in Equation (6.3.2), is

$$\sigma_\phi^2 = \frac{N_0 B_e}{A_e^2} = \sigma_0^2\left(2\pi^2 \frac{\Delta T}{T_0}\frac{\Delta t}{T_0}\right)^{1-(r/2)} \tag{6.3.3}$$

where

$$\sigma_0^2 = \frac{N_0 B_L}{A^2}$$

would be the tracking error were the signal a sinusoid of period T_0.

If Equation (6.3.3) is to be valid, the loop must operate in the linear region $|\tau| \leq \Delta t/2$ with high probability. To insure that this be the case, and to avoid the difficulty when Δt approaches zero, we impose the additional constraint that

$$\sigma_\phi = \frac{\pi}{k}\frac{\Delta t}{T_0} \tag{6.3.4}$$

for some sufficiently large k (cf. Section 6.2). Combining Equations (6.3.3) and (6.3.4), and solving for Δt, we find

$$\frac{\Delta t}{T_0} = \left(2\pi^2 \frac{\Delta T}{T_0}\right)^{(2-r)/(2+r)}\left(\frac{k\sigma_0}{\pi}\right)^{4/(2+r)}$$

and, using this value of Δt,

$$\sigma_\phi^2 = \left(2k\pi \frac{\Delta T}{T_0}\right)^{2[(2-r)/(2+r)]} \sigma_0^{8/(2+r)} \tag{6.3.5}$$

The mean-squared error in tracking a rectangular pulse with a second-order loop ($r = \frac{1}{2}$), for example, would be proportional to $(N_0 B_L/A^2)^{8/5}$ when the local signal is defined as in Equation (6.3.2) and Δt is chosen in accordance with the preceding discussion. This represents a potential improvement relative to the error in tracking a sinusoid of the same period, although a considerably less spectacular one than was found to be attainable using higher frequency sinusoids (Section 6.2). This scheme does have the advantage of not introducing multiple lock-in points in the original T_0 second interval.

The parameter ΔT, the received pulse width, was tacitly assumed in the preceding discussion to be a fixed constant. If this is not the case, it too can be optimized in much the same way as was Δt. Again, combining Equations (6.3.3) and (6.3.4), but now solving for ΔT, we find

$$\frac{\Delta T}{T_0} = \frac{2^{(2-r)/2r} k^{2/r} \sigma_0^{2/r}}{\pi\left(\dfrac{\Delta t}{\Delta T}\right)^{(2+r)/2r}}$$

where, for the moment, we have kept the ratio $\Delta t/\Delta T$ constant. Substituting this result into Equation (6.3.3) or (6.3.4) yields

$$\sigma_\phi^2 = \left(\frac{2k^2}{\Delta t/\Delta T}\right)^{(2/r)-1} \sigma_0^{4/r} \tag{6.3.6}$$

Now the tracking error is inversely proportional to Δt. Since Equation (6.3.2) is valid only for $\Delta t \leq \Delta T$ (and there is clearly no advantage in making $\Delta t > \Delta T$) the optimum value for Δt is here ΔT. Thus, we come to the following conclusions: if the received pulse is sufficiently wide, it is advantageous to make the width of the local signal as narrow as possible consistent with an acceptable loss-of-lock probability. If, however, it is possible to optimize the pulse width of the transmitted signal, it should be made as narrow as possible subject to the same loss-of-lock constraint. When this capability exists, no further advantage can be realized by narrowing the local signal pulse width. While only moderate improvement is possible in optimizing the local pulse width [Equation (6.3.5)], the advantage to be realized in optimally selecting the width of the transmitted pulse is essentially the same as that attainable by optimizing the frequency of a transmitted sinusoid [compare Equations (6.3.6) and (6.2.3)].

Occasionally, a square-wave rather than a sinusoid is used as a clock signal, with the local signal $z(t)$ either a sinusoid or another square-wave. In this event, the frequency of the square-wave can be readily optimized by using the approach suggested in Section 6.2. When $z(t)$ is a sinusoid,

$$\rho_{xz}(\tau) = \frac{2\sqrt{2}}{\pi} \sin \frac{2\pi\tau}{T_0}$$

and the effective signal amplitude is $(2\sqrt{2}/\pi)A$ representing a slight degradation relative to the situation when the transmitted signal is also a sinusoid. Similarly, when $z(t)$ is a square-wave, the effective signal amplitude is further decreased to the value $2A/\pi$.

The optimum correlation function $\rho_{xz}(\tau)$ would evidently be a unit amplitude square-wave, having infinite slope at the point $\tau = 0$, and absolute magnitude one for all $\tau \neq 0$, modulo T_0. Unfortunately, no cross-correlation function between two finite energy signals can be of this form. The specification of an "optimum" realizable correlation function is formidable due, among other things, to the difficulty in ascertaining what deviations from the ideal square-wave correlation function are least objectionable.

If only the slope $|\rho'_{xz}(0)|$ of the correlation function at the point $\tau = 0$ were of concern, it could be made arbitrarily large by any of the techniques discussed in this and the previous section. The cost of doing so is in the narrowing of the linear region and in the introduction of multiple lock-in points or, in the case of a narrow pulse, in the potentially increased lock-up time due to the long intervals of τ in which $\rho_{xz}(\tau)$, and hence the error signal, are identically zero. An alternative approach, therefore, might be to attempt to maximize the quantity $|\rho'_{xz}(0)|$ subject to a condition on the overall function $\rho_{xz}(\tau)$, the condition, for example, that $\rho_{xz}(\tau)$ be a monotonically non-decreasing function of τ over the region $|\tau| \leq T_0/4$.

This last condition would be satisfied for the class of signals discussed in this section, for instance, if ΔT in Equation (6.3.1) were equated to $T_0/2$. The width Δt of the local signal pulse could then be minimized (in order to maximize $|\rho'_{xz}(0)|$) as discussed above. The somewhat limited improvement obtained in this manner, relative to using, say, a sinusoid of frequency $f = 1/T_0$, is indicative of the penalty to be paid in placing global rather than local constraints on the correlation function $\rho_{xz}(\tau)$. In order to maximize the effectiveness of the available power, therefore, it is often necessary to use a high-frequency clock, and to eliminate the resulting ambiguities by other means. This is particularly true when peak power or bandwidth or acquisition time restrictions limit the effectiveness of the narrow pulse approach.

6.4 Epoch Search

If, in the interest of decreasing the power needed for synchronization, the clock signal period ΔT is chosen to be smaller than the basic T_0 second time ambiguity, the synchronization procedure is not completed with the acquisition of the clock. Instead, there remain $N \triangleq T_0/\Delta T$ points, or *epochs*, in each T_0 second interval, any one of which could represent the correct sync reference point. The determination of the correct one of these epochs can be accomplished by transmitting a second synchronization signal $x(t)$, periodic with minimum period T_0, and identifying its phase at the receiver.

At first glance, this approach would not seem to offer any advantage over using a T_0-second period signal for the clock itself. The existence of a clock reference, however, insures the needed tracking accuracy. The purpose of the second signal is only to resolve the remaining N-fold timing ambiguity.

As a consequence considerably less power need be relegated to this signal than would be needed were it to serve as a clock also. The total required power, therefore, can be sizably reduced using two signals rather than one.

The task at the receiver, then, is to determine which of the signals $Ax_\nu(t) = Ax(t - \nu\Delta T)$, $\nu = 0, 1, \ldots, N-1$, is actually being received in the presence of the usual additive, white Gaussian noise. This is generically equivalent to the detection problems considered in Chapter Four. The maximum-likelihood solution† involves the formulation of the correlations

$$z_\nu = \sum_{j=0}^{M-1} \int_{jT_0}^{(j+1)T_0} y(t)x(t - \nu\Delta T)\, dt \triangleq \sum_{j=0}^{M-1} z_{\nu j} \qquad \nu = 0, 1, \ldots, N-1 \tag{6.4.1}$$

and the selection of the largest of these. The observation period MT_0 here, in contrast to the situation in Chapter Four, is not necessarily limited by the duration of the received signal, which may well be transmitted indefinitely. Moreover, if the local signal $x(t)$ is timed by the recovered clock, the often troublesome random drift in phase between the received and local signals will be essentially non-existent. Nevertheless, it is usually desirable to make a decision concerning the epoch as quickly as possible. Accordingly, the task of this section is to estimate the minimum value of M such that a decision can be made with a predetermined acceptably small error probability.

The situation is further altered relative to that in Chapter Four when, as is frequently the case, equipment limitations preclude simultaneous processing of the observables pertaining to each of the epochs. In this event, a search may be in order. Both the fixed-sample-size and sequential search algorithms of Chapter Two, therefore, will also be applied to the problem of concern in this section.

6.4.1 Simultaneous Observation

Since the noise is white and Gaussian, the variables $z_{\nu j}$ (the *decision variables*)‡ are Gaussianly distributed with means

$$\eta_\nu(\mu) \triangleq E(z_{\nu j} \mid \mu) = \int_{jT_0}^{(j+1)T_0} Ax_\mu(t)x_\nu(t)\, dt \triangleq AT_0\rho_i \triangleq \eta_i \qquad \nu = 0, 1, \ldots, \tag{6.4.2a}$$

and covariances

$$\sigma_{\mu\nu}^2 \triangleq E(z_{\mu j}z_{\nu j}) - E(z_{\mu j})E(z_{\nu j}) = \frac{N_0}{2}\int_{jT_0}^{(j+1)T_0} x_\mu(t)x_\nu(t)\, dt = \frac{N_0}{2} T_0\rho_i \triangleq \sigma^2\rho_i \tag{6.4.2b}$$

†The a priori probability $P(\nu)$ of the νth epoch being the correct one will be assumed independent of ν throughout this chapter; $P(\nu) = 1/N$, $\nu = 0,1, \ldots, N-1$. Thus the terms maximum-likelihood and maximum a posteriori probability synchronization can be used interchangeably.

‡The term decision variables is often used to denote the statistics upon which the decision is actually based.

where μ represents the true epoch and $i = \mu - \nu$, modulo N. (With no loss of generality, we can stipulate that $(1/T_0)\int_0^{T_0} x_\mu^2(t)\,dt = 1$ so that $|\rho_i| \leq 1$.)

The situation is identical to that in Section 4.2. The bounds of Equation (4.2.19), in particular, are directly applicable here. The probability P_e of an erroneous decision is bounded by

$$\frac{e^{-r^2/2}}{(2\pi)^{1/2}r}\left(1 - \frac{1}{r^2}\right) \leq P_e \leq \frac{e^{-[(r^2/2)-\log_e(N-1)]}}{(2\pi)^{1/2}r} \tag{6.4.3}$$

where

$$r^2 \triangleq \min_{\substack{\mu,\nu \\ \mu\neq\nu}} \frac{E^2(z_\nu - z_\mu \mid \nu)}{\operatorname{Var}(z_\nu - z_\mu \mid \nu)} = MR(1-\rho)$$

with $\rho \triangleq \max_{i\neq 0} \rho_i$ and $R = A^2T_0/N_0$. If P_e is to be small, r^2 must be large compared to unity. Thus, taking the logarithm of the inequalities (6.4.3), we obtain

$$2\log_e(1/P_e) \leq r^2 \leq 2\log_e(1/P_e) + 2\log_e(N-1) \tag{6.4.4a}$$

or, more conveniently,

$$r^2 = 2\kappa_N \log_e(1/P_e) \tag{6.4.4b}$$

where $1 \leq \kappa_N \leq 1 + [\log_e(N-1)/\log_e(1/P_e)]$. The terms on the order of $\log_e r$ have been omitted since they will be small relative to r^2. Typically P_e will be small compared to $1/N$ so these bounds on r^2 will be quite tight.

As a consequence of Equation (6.4.4), we have

$$M = \frac{2\kappa_N \log_e(1/P_e)}{R(1-\rho)} \tag{6.4.5}$$

and the required observation time is MT_0 seconds. If $\rho_i = \rho$ for all $i \neq 0$, the exact expression for the error probability derived in Section 4.2 is, of course, equally applicable here. One need only substitute the term $(1 - \rho)MR/\log_2 N$ for the R_b of Figure 4.2, to adapt those results to the present problem.

6.4.2 Fixed-Sample-Size Search

To apply the search algorithm discussed in Section 2.5 to the situation here, we test at the νth stage of the search the null hypothesis H_ν (that the received signal epoch is the νth epoch, modulo N) against some alternative hypothesis $\bar{H}_\nu$. The observables are to be the statistics z_ν as defined in Equation 6.4.1.†

†These statistics are clearly not sufficient; all of the statistics $z_1, z_2, \ldots, z_N$ are needed for an optimum decision. But the purpose of considering a search in the first place is to avoid the necessity of simultaneously processing all of these quantities.

Although the alternative hypothesis $\bar{H}_\nu$ is ostensibly composite, it is reasonable, and clearly expedient, to replace it with any of the simple hypotheses H_μ for any $\mu \neq \nu$. For consider the likelihood ratio logarithm

$$\log_e \frac{p(z_\nu \mid \nu)}{p(z_\nu \mid \mu)} = \log_e \Lambda_{\nu\mu} = \frac{\eta_0(1-\rho_i)}{\sigma^2} z_\nu - \frac{M\eta_0^2(1-\rho_i^2)}{2\sigma^2} \tag{6.4.6}$$

where η_0, σ^2, and ρ_i are as defined in Equation (6.4.2). If there were only two possible epochs, the most powerful test of hypothesis H_ν at any level α_ν would be to accept it if the decision variable z_ν exceeds some threshold a_ν and to reject it otherwise. But, since $\rho_i < 1$ for all $i \neq 0$, the acceptance region of H_ν for any specific α_ν will be independent of μ. The test in Equation (6.4.6) of hypothesis H_ν is therefore uniformly most powerful against all of the alternative hypotheses H_μ. Regardless of the alternative hypothesis, H_ν will be accepted if z_ν exceeds the threshold a_ν and will be rejected otherwise.

For reasons mentioned in Section 2.5, we will restrict our attention to the most straightforward of the fixed-sample-size search algorithms. Each hypothesis is to be tested at the same level and is to be tested sufficiently long to assure a small probability of an erroneous decision. The threshold a_ν will therefore be kept constant, independent of ν, and will be chosen to keep the probability of both kinds of error small. An error will be assumed to occur if no decision has been made by the time all N states have been tested.

To estimate the number of observables needed to complete the search, we denote by α the probability of erroneously rejecting H_ν, and by $\beta_\nu(\mu)$ the probability of accepting it when the μth epoch is actually the correct one. Further, let $a \triangleq M\eta_0[\rho + \gamma(1-\rho)]$ be the decision threshold where $\rho \triangleq \max_{i\neq 0} |\rho_i|$ and η_0 is as previously defined. (Since γ has not been specified, this does not imply any constraint on a. However, if α and $\beta_\nu(\mu)$ are to be small, γ must clearly fall in the interval $0 \leq \gamma \leq 1$.) Then

$$\alpha_\nu = \alpha = \Pr\{z_\nu < a \mid \nu\} = \tfrac{1}{2}\{1 - \operatorname{erf}[(1-\gamma)r]\} \tag{6.4.7a}$$

where $r^2 = M(1-\rho)^2\eta_0^2/2\sigma^2$. Similarly,

$$\beta_\nu(\mu) = \Pr\{z_\nu > a \mid \mu\} \leq \tfrac{1}{2}[1 - \operatorname{erf}\gamma r] \triangleq \beta \tag{6.4.7b}$$

the equality holding if $\rho_i = \rho$ for all $i \neq 0$.

If the ρ_i were identical for all $i \neq 0$, we would have, under the conditions outlined (cf. Section 2.5)

$$P_e \approx \alpha + \frac{N-1}{2}\beta \tag{6.4.8}$$

As α and β are both functions of γ, it is possible to specify the value of γ which minimizes P_e. When the approximation (6.4.8) is applicable, the optimum value of γ is determined by the equation:

$$\frac{d\alpha}{d\gamma} + \frac{N-1}{2}\frac{d\beta}{d\gamma} = 0$$

which is satisfied if

$$0 \leq \gamma = \frac{1}{2}\left(1 + \frac{1}{r^2}\log_e \frac{N-1}{2}\right) \leq 1 \tag{6.4.9}$$

Using this value of γ in Equation (6.4.8) and the asymptotic expansion of the error function, we obtain

$$P_e = \frac{2\exp\left\{-\left(1 - \dfrac{\log_e \dfrac{N-1}{2}}{r^2}\right)^2 \dfrac{r^2}{4}\right\}}{\sqrt{\pi}\, r\left[1 - \left(\dfrac{\log_e \dfrac{N-1}{2}}{r^2}\right)^2\right]}\left[1 + O\left(\frac{1}{r^2}\right)\right] \triangleq P_0(N) \tag{6.4.10}$$

Again, because r^2 must be large if P_e is to be small, we find, by taking the logarithm of both sides of Equation (6.4.10) and neglecting terms on the order of $\log_e r$, that

$$r^2 = 4\kappa'_N \log_e (1/P_e) \tag{6.4.11}$$

where

$$\kappa'_N = \frac{1}{2} + \frac{\log_e [(N-1)/2]}{4\log_e (1/P_e)} + \frac{1}{2}\left(1 + \frac{\log_e [(N-1)/2]}{\log_e (1/P_e)}\right)^{1/2}$$

When the ρ_i are not necessarily equal, P_e can still be bounded as follows: Clearly, to the extent that Equation (6.4.8) is valid,

$$P_e \leq \alpha + \frac{N-1}{2}\beta \tag{6.4.12}$$

where β is defined as in Equation (6.4.7). Moreover, because at least one epoch will be erroneously accepted with the probability β, and since with probability $\frac{1}{2}$ it will be tested before the correct epoch,

$$P_e \geq \tfrac{1}{2}(\beta + \alpha(1-\beta)) + \tfrac{1}{2}\alpha \approx \alpha + \tfrac{1}{2}\beta \tag{6.4.13}$$

Using the optimal decision thresholds, as defined in Equation (6.4.9), for both the upper and lower bounds on P_e, we obtain

$$P_0(2) \leq P_e \leq P_0(N) \tag{6.4.14}$$

[see Equation (6.4.10)] and, consequently,

$$r^2 = 4\kappa''_N \log_e (1/P_e) \tag{6.4.15}$$

where $\kappa'_2 \leq \kappa''_N \leq \kappa'_N$.

The expected search time is $E(\nu)MT_0 \triangleq \bar{M}_T T_0$ where ν is the number of epochs tested before the search is concluded. When $P_e \ll 1$, it follows from equation (2.5.9) that $E(\nu) \approx (N+1)/2$, and from Equation (6.4.15) that

$$\bar{M}_T \approx \frac{2(N+1)\kappa''_N \log_e (1/P_e)}{R(1-\rho)^2} \tag{6.4.16}$$

where R is as previously defined.

6.4.3 Deferred Decision

An interesting variation on the fixed-sample-size method involves making a *deferred decision* as to the correct epoch. That is, the decision is deferred until all of the observables z_ν, $\nu = 0, 1, 2, \ldots, N-1$ are processed, even though the correlations yielding these observables are made serially. Undoubtedly the most straightforward decision rule is to select as the correct epoch that value of ν maximizing the variable z_ν itself. (This is obviously a sub-optimum decision; the optimum deferred decision is discussed presently.)

The probability of an error in making such a decision satisfies the same bounds [cf. Equation (6.4.3)] as does the error probability in a parallel decision. The only difference is in the fact that the observations producing the statistics z_ν are no longer made simultaneously. Thus, $\sigma_{\nu\mu}^2 = 0$, $\mu \neq \nu$, and the term r^2 of Equation (6.4.3) must be redefined as in Equation (6.4.7). Equation (6.4.5) therefore becomes in the present context

$$M_T = \frac{2N\kappa_N \log_e (1/P_e)}{R(1-\rho)^2} \tag{6.4.17}$$

It is interesting to compare Equations (6.4.16) and (6.4.17), especially when N is large and $P_e \ll 1/N$, as it typically will be. In this event, κ_N, κ_2', κ_N' and hence κ_N'' are all approximately unity and the search times in the two cases are essentially identical. It makes little difference in performance whether all epochs are tested serially and a (sub-optimum) deferred decision made, or whether the fixed-sample-size search algorithm is used. The most important difference is that the search time in the latter case is a random variable and is equally likely to assume any value from $[2/(N+1)]\bar{M}_T T_0$ to $[2N/(N+1)]\bar{M}_T T_0$.

To translate these results into more concrete form and to expose one of the weaknesses of the fixed-sample-size serial search algorithm, let the synchronization signal be of the form

$$x(t) = \begin{cases} \sqrt{n} & 0 < t < \dfrac{T_0}{n} \\ 0 & \dfrac{T_0}{n} < t < T_0 \end{cases} \tag{6.4.18}$$

for some integer n, $2 \leq n \leq N$, when T_0/N seconds represents the maximum acceptable time uncertainty. Then

$$\rho\left(\frac{\nu T_0}{N}\right) \triangleq \rho_\nu = \begin{cases} 1 - \left|\dfrac{\nu n}{N}\right| & |\nu| \leq \dfrac{N}{n} \\ 0 & \text{otherwise} \end{cases}$$

and $\rho = \max\limits_{\nu \neq 0} \rho_\nu = 1 - (n/N)$. Substituting this into Equations (6.4.5) and (6.4.17) [or (6.4.16)] yields

$$M_T = \frac{2N\kappa_N \log_e (1/P_e)}{nR} \tag{6.4.19a}$$

for simultaneous processing, and

$$M_T = \frac{2N^3\kappa_N \log_e (1/P_e)}{n^2 R} \tag{6.4.19b}$$

for serial processing. When N is large, the serial processing constraint can represent a significant degradation in performance.†

One would strongly suspect that this result could be improved upon when n is small as compared to N. The serial search as outlined is so obviously inefficient in this case. All of the (independent) statistics z_ν [Equation (6.4.1)] must be processed (at least with the deferred decision method). Yet only one of them, the largest, influences the decision. The fact that z_1 and z_{N-1} must be nearly as large as z_0 (when $\nu = 0$ represents the true epoch) and that $z_{N/2}$ should be much smaller than either of these quantities is ignored in the decision.

Indeed, given the statistics $z_0, z_1, \ldots, z_{N-1}$, the optimum (maximum-likelihood) decision is to accept that epoch maximizing the likelihood function $p(z_0, z_1, \ldots, z_{N-1} \mid \nu)$. Since the statistics z_ν are independent when the observations are made serially,

$$p(z_0, z_1, \ldots, z_{N-1} \mid \nu) = \frac{1}{(2\pi M)^{N/2}\sigma^N} \exp\left\{-\frac{1}{2}\sum_{i=0}^{N-1}\frac{(z_{i+\nu} - M\eta_i)^2}{M\sigma^2}\right\} \tag{6.4.20}$$

where the subscripts of $z_{i+\nu}$ are to be taken modulo N. This function is maximized by maximizing the variable

$$v_\nu = \sum_{i=0}^{N-1} \eta_i z_{i+\nu} \tag{6.4.21}$$

When $\eta_i = \eta_1$ for all $i \neq 0$

$$v_\nu = (\eta_0 - \eta_1)z_\nu + \eta_1 \sum_{i=0}^{N-1} z_i$$

and v_ν is maximized for the same ν maximizing z_ν. But for any other set of means η_i, basing the decision on z_ν alone is sub-optimal. We shall refer to a decision based on the statistics v_ν as opposed to the statistics z_ν as an *optimum deferred decision*.

The statistics v_ν are sums of Gaussian random variables and hence are also Gaussianly distributed. Accordingly, the probability P_e of accepting the wrong epoch when an optimum deferred decision is made can be bounded in the usual way (letting $\nu = 0$ correspond to the correct epoch) in terms of the ratio

---

†When N is large relative to n, a synchronization error may be less deleterious than when they are more nearly equal. This is because in the first situation the most common error would be to select some epoch close to the true epoch. In some cases, such errors may be tolerable. But since the search time depends only logarithmically on P_e this is a small consolation.

$$r^2 = \min_{\mu \neq 0} \frac{E^2(v_\mu - v_0)}{\text{Var}(v_\mu - v_0)} = \frac{M\left(\sum_{i=0}^{N-1} (\eta_{i+\mu} - \eta_i)\eta_i\right)^2}{\sigma^2 \sum_{i=0}^{N-1} (\eta_{i+\mu} - \eta_i)^2} \tag{6.4.22}$$

[cf. Equation (6.4.3)]. When the signal is of the form given in Equation (6.4.18), the minimum of the ratios $E^2(v_\mu - v_0)/\text{Var}(v_\mu - v_0)$ is attained when $\mu = 1$. Thus, combining Equations (6.4.4) and (6.4.22), and to simplify the result, assuming N/n to be an integer, we find the required search time to be $M_T T_0$ seconds, where

$$M_T = \frac{2N^2 \kappa_N \log_e (1/P_e)}{nR} \tag{6.4.23}$$

The optimum deferred decision search is decidedly superior to the earlier serial search methods when N is large relative to n; that is, when the desired timing resolution is small as compared to the width of the pulse being used over the sync channel. When N equals n, of course, the optimum method reduces to the deferred search described earlier.

6.4.4 *Sequential Search*

To apply the sequential search algorithm to the present situation, it is only necessary to compare the running sum

$$Z_m = \sum_{j=1}^{m} \zeta_{\nu j} \triangleq \sum_{i=1}^{m} \log_e \frac{p(z_{\nu j} \mid \nu)}{p(z_{\nu j} \mid \bar{\nu})} \tag{6.4.24}$$

to the two thresholds $\log_e A$ and $\log_e B$. The hypothesis H_ν (that the νth epoch is the correct one) is accepted if this ratio exceeds $\log_e A$ and rejected if it drops below $\log_e B$. If neither of these events takes place, the test continues, the new statistic $\zeta_{\nu, m+1}$ is added to the running sum, and Z_{m+1} is compared to these same thresholds.

Since the likelihood ratio test is here a uniformly most powerful one, the composite alternative hypothesis will be replaced by the simple hypothesis $H_{\bar{\nu}}$ for some $\bar{\nu} \neq \nu$. The probability of erroneously accepting a particular epoch, of course, is a function of the epoch being tested. Specifically, (cf. Appendix A.1) the probability of accepting the μth epoch is

$$P_\mu = \frac{1 - B^{h(\mu)}}{A^{h(\mu)} - B^{h(\mu)}} \tag{6.4.25}$$

where A and B are the decision thresholds, and $h(\mu)$ the non-zero solution to the equation $E(e^{h\zeta_{\nu j}} \mid \mu) \triangleq g(h) = 1$.

In the situation here $p(z_{\nu j} \mid \nu) = (1/\sqrt{2\pi}\,\sigma) \exp\{-[(z_{\nu j} - \eta_0)^2/2\sigma^2]\}$, $p(z_{\nu j} \mid \bar{\nu}) = (1/\sqrt{2\pi}\,\sigma) \exp\{-[(z_{\nu j} - \eta_i)^2/2\sigma^2]\}$, $i = \nu - \bar{\nu}$, modulo N, and for the moment let $p(z_{\nu j} \mid \mu) = (1/\sqrt{2\pi}\,\sigma') \exp\{-[(z_{\nu j} - \eta)^2/2(\sigma')^2]\}$. Then $h(\mu)$

is the non-zero solution to

$$1 = g(h) = \frac{e^{f(h)}}{\sqrt{2\pi}\,\sigma'} \int_{-\infty}^{\infty} e^{-[(\xi-\eta')^2/2(\sigma')^2]}\, d\xi = e^{f(h)}$$

where $\eta' = \eta + h(\sigma'/\sigma)^2(\eta_0 - \eta_i)$ and $f(h) = [(\eta_0 - \eta_i)/2\sigma^2]h\{2\eta + (\eta_0 - \eta_i)(\sigma'/\sigma)^2 h - (\eta_0 + \eta_i)\}$. Since the last equality in this equation holds for any finite value of h, the necessary condition for $g(h)$ to equal unity is for $f(h)$ to equal zero and hence, if $h \neq 0$ and $\eta_i \neq \eta_0$, for

$$h = h(\mu) = \frac{(\eta_0 + \eta_i) - 2\eta}{\left(\dfrac{\sigma'}{\sigma}\right)^2 (\eta_0 - \eta_i)} \tag{6.4.26}$$

If $\sigma' \leq \sigma$ and $\eta \leq \eta_i$, then $h(\mu) \geq 1$. Accordingly, if α and β are the parameters defining the thresholds A and B (see Section 2.4.2) $P_\mu \leq \beta$. Furthermore, if $\sigma' \leq \sigma$ and $\eta \geq \eta_0$, $h(\mu) \leq -1$, and in this case, $P_\mu \geq 1 - \alpha$.

The conditions leading to Equations (2.6.8) and (2.6.12) are therefore satisfied here if $\eta_i = \max_{l \neq 0} \eta_l = \eta_0 \rho$ and $\sigma' = \sigma$.† Since, in this case,

$$\zeta_{\nu j} = \frac{\eta_0(1-\rho)}{\sigma^2} z_{\nu j} - \frac{\eta_0^2(1-\rho^2)}{2\sigma^2}$$

then

$$E(\zeta_{\nu j} \mid \mu) = R(1-\rho)[2\rho_i - (1+\rho)] \tag{6.4.27}$$

where $i = \nu - \mu$, modulo N, $\rho = \max_{i \neq 0} \rho_i$, and R and ρ_i are as previously defined. The expected time needed to complete the search, therefore, is $\bar{M}_T T_0$ where

$$\bar{M}_T = \frac{C}{R(1-\rho)} \sum_{i=1}^{N-1} \frac{1}{(1+\rho-2\rho_i)} + \frac{\left(\log_e \dfrac{N-1}{P_e} - c\right)}{R(1-\rho)^2} \tag{6.4.28}$$

and where

$$C = c = 1 \qquad \alpha \to 1, \quad \beta \to 0 \qquad (P_e \ll 1)$$

and

$$C = \frac{1}{2}\log_e \frac{2}{P_e}, \quad c = 0 \qquad \alpha = \frac{P_e}{2}, \quad \beta = \frac{P_e}{N-1} \qquad (P_e \ll 1)$$

[cf. Equations (2.6.8) and (2.6.12)].

The sequential search should demonstrate a significant advantage vis-à-vis the fixed-sample-size search, for example, when the sync signal is of the form given in Equation (6.4.18) with N large as compared to n. For, in this event, some epochs are much easier to dismiss than others. The fixed-sample-size search must allow the same amount of time for testing each epoch,

†Note that if the estimates of η_0, ρ, and σ are conservative, if the actual value of η_0 is at least as great as its assumed value, if $\rho_i < \rho$ for all $i \neq 0$, and if the actual variance is not greater than σ, then the test is at least as reliable as it is claimed to be.

whereas the sequential search is able to spend less time on those epochs having statistics differing significantly from those of the correct one. Equation (6.4.28) becomes, in this case, (assuming N/n to be an integer)

$$\bar{M}_T \approx \frac{2C}{R}\left(\frac{N}{n}\right)^2\left[1+\log_e\left(\frac{2N}{n}-1\right)+\frac{\left(N-\frac{2N}{n}-1\right)}{4\left(\frac{N}{n}-\frac{1}{2}\right)}\right]$$
$$+\frac{1}{R}\left(\frac{N}{n}\right)^2\left(\log_e\frac{N-1}{P_e}-c\right) \tag{6.4.29}$$

Thus, when n is small, the sequential search with α small is actually inferior to the optimum deferred decision method. As $\alpha \to 1$, the sequential search demonstrates a possible advantage for all $n \geq 2$, although for small n this advantage is marginal. At the other extreme, when $n = N$, the sequential search is approximately four times faster than the fixed-sample-size search when $\alpha = P_e/2$ and faster by a factor of $2 \log_e (2/P_e)$ when $\alpha \to 1$.

6.5 Epoch Estimation

Although the synchronization problem of the preceding section was introduced as a detection problem, it can be advantageous, when equipment limitations preclude parallel processing, to rephrase it as an estimation problem; i.e., to treat $\tau = \nu\Delta T$ as a continuous parameter. Indeed, in some cases, when N is very large, for example, or when a common clock is not required, it is somewhat artificial to restrict the number of contending epochs to be finite.

When τ is continuous, its maximum-likelihood estimate satisfies the likelihood equation

$$\frac{\partial \log_e p[y(t)\,|\,\tau]}{\partial \tau} = \frac{2A}{N_0}\int_0^{MT_0} y(t)\frac{\partial}{\partial \tau}x(t+\tau)\,dt = 0 \tag{6.5.1}$$

If this equation cannot be solved explicitly for τ, processing limitations would generally restrict us either to using a phase-locked loop type device to force its solution or to considering only a finite set of epochs τ. Both of these approaches have already been investigated in the preceding sections of this chapter.

For some signals $x(t)$, however, Equation (6.5.1) can be solved for τ. Suppose, for example, $x(t)$ were a sinusoid: $x(t) = \sqrt{2}\sin \omega_0 t$ with $\omega_0 = 2\pi/T_0$. Then

$$\frac{\partial \log_e p[y(t)\,|\,\tau]}{\partial \tau} = 2R\omega_0[-X\sin \omega_0\tau + Y\cos \omega_0\tau] \tag{6.5.2}$$

where

$$\begin{Bmatrix} X \\ Y \end{Bmatrix} = \frac{1}{AT_0} \sum_{i=0}^{M-1} \int_{iT_0}^{(i+1)T_0} y(t)\sqrt{2} \begin{Bmatrix} \sin \omega_0 t \\ \cos \omega_0 t \end{Bmatrix} dt \triangleq \sum_{i=0}^{M-1} \begin{Bmatrix} X_i \\ Y_i \end{Bmatrix}$$

and $R = A^2T_0/N_0$. The maximum-likelihood estimate $\hat{\tau}$ is therefore

$$\hat{\tau} = \frac{T_0}{2\pi} \tan^{-1}\left(\frac{Y}{X}\right) \tag{6.5.3}$$

The mean and variance of this estimate can readily be determined from the corresponding results in Section 3.7. For our purposes it is sufficient to recall that $\hat{\tau}$, as a maximum-likelihood estimate, is asymptotically unbiased and with a variance

$$E[(\hat{\tau} - \tau)^2 \mid \tau) = \frac{1}{4R^2\omega_0^2 ME[-X_i \sin \omega_0\tau + Y_i \cos \omega_0\tau]^2} = \frac{T_0^2}{(2\pi)^2 2RM} \tag{6.5.4}$$

It is tempting to look for other synchronization signals which, like a sinusoidal signal, afford a reduction in equipment complexity, but hopefully yield superior performance. Such a reduction in complexity can evidently be accomplished if $z(t + \tau) = \partial x(t + \tau)/\partial\tau$ is *separable* in the sense that it can be expressed in the form

$$z(t + \tau) = \sum_{i=1}^{n} f_i(t) g_i(\tau) \tag{6.5.5}$$

If this can be done, only n integrals need be determined, the estimate of τ obtained by solving the equation

$$\sum_{i=1}^{n} I_i g_i(\tau) = 0 \tag{6.5.6}$$

where

$$I_i = \int_0^{MT_0} y(t) f_i(t)\, dt$$

[cf. Equation (6.5.1)]. If this is to represent an actual reduction in complexity, n should be small and Equation (6.5.6) relatively easy to solve.

In any event, since $z(t + \tau)$ is a periodic function of time, it can be expanded in a Fourier series:

$$\begin{aligned} z(t + \tau) &= \sum_{l=1}^{n/2} a_l \sin[\omega_l(t + \tau) + \theta_l] \\ &= \sum_{l=1}^{n/2} [a_l \sin \omega_l t \cos(\omega_l\tau + \theta_l) + a_l \cos \omega_l t \sin(\omega_l\tau + \theta_l)] \end{aligned} \tag{6.5.7}$$

where $\omega_\nu = 2\pi k_\nu/T_0$ with T_0 the period of $z(t)$ and k_ν an integer. The $f_\nu(t)$ and $g_\nu(\tau)$ of Equation (6.5.5) can therefore be defined as

$$f_\nu(t) = \begin{cases} \sin \omega_{(\nu+1)/2} t & \nu \text{ odd} \\ \cos \omega_{\nu/2} t & \nu \text{ even} \end{cases} \tag{6.5.8}$$

and

$$g_\nu(\tau) = \begin{cases} a_{(\nu+1)/2} \cos(\omega_{(\nu+1)/2}\tau + \theta_{(\nu+1)/2}) & \nu \text{ odd} \\ a_{\nu/2} \sin(\omega_{\nu/2}\tau + \theta_{\nu/2}) & \nu \text{ even} \end{cases}$$

respectively. Since the I_i of Equation (6.5.6) are statistically independent when the functions $f_\nu(t)$ are as defined in Equation (6.5.8), at least n correlations are required with this approach. For assume the contrary, that the integral in Equation 6.5.1 could be expressed in the form $\sum_{j=1}^{m} J_j g_j'(\tau)$ for some set of functions $g_j'(\tau)$ and with $m < n$. Then

$$\sum_{i=1}^{n} I_i g_i(\tau) = \sum_{j=1}^{m} J_j g_j'(\tau)$$

and, since the functions $g_i(\tau)$ are orthogonal over the interval $(0, T_0)$,

$$I_\nu = \sum_{j=1}^{m} \alpha_{\nu j} J_j \qquad \nu = 1, 2, \ldots, n$$

with $\alpha_{\nu j} \triangleq \int_0^{T_0} g_j'(\tau) g_i(\tau)\, d\tau \Big/ \int_0^{T} g_i^2(\tau)\, d\tau$. But if this were the case, the variables J_j could be expressed as linear combinations of the first m of the I_ν (unless these variables were linearly dependent). This in turn would imply a linear relationship among some of the variables I_ν, and would hence lead to a contradiction.

We have already commented on the performance of the synchronizer when $z(t)$ is defined by Equation (6.5.7) with $n = 2$. The preceding paragraph suggests that this result cannot be improved without using more correlators. When n is greater than two, the resulting maximum-likelihood equation (6.5.6) would seem to be of sufficient complexity to limit the utility of this approach. This limitation can be overcome by solving each of the equations

$$I_\nu g_\nu(\tau) + I_{\nu+1} g_{\nu+1}(\tau) = 0 \qquad \nu \text{ odd}$$

separately and using the solution to the equation corresponding to the maximum of the frequencies ω_ν to define the estimate $\hat{\tau}$ of τ. The solution to the rest of these equations can then be used to eliminate the remaining ambiguities.† The resulting estimate, while sub-optimum, will generally not differ substantially from the optimum estimate. A slight modification of this scheme, which leads to a particularly simple implementation, is discussed in detail in Section 6.9.

Actually, rather effective estimators can sometimes be devised by using statistics other than those suggested by Equation (6.5.1). As an example, suppose the sync signal is a periodic rectangular pulse of width $T_0/2$. The autocorrelation function of such a signal is shown in Figure 6.2. Observe that every value of $\rho(\tau)$ in the range $\{0, 1\}$ corresponds to only two possible

---

†A method for determining the optimum relationship between the different frequencies ω_ν can be found in Reference 6.1.

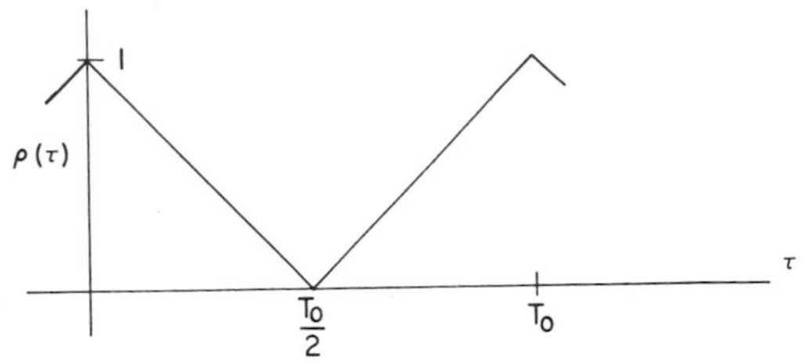

Figure 6.2 Periodic autocorrelation function of a rectangular pulse of width $T_0/2$.

values of τ in the interval $\{0, T_0\}$. (When $\rho(\tau) = 1$, the value of τ is uniquely determined modulo T_0.) Accordingly, a good estimate of $\rho(\tau)$ could be used to estimate τ and from this the correct epoch (the point $\tau = 0$). Since

$$\rho(\tau) = 1 - \frac{2|\tau|}{T_0} \qquad |\tau| < \frac{T_0}{2} \tag{6.5.9a}$$

solving for $|\tau|$ yields

$$|\tau| = \frac{T_0}{2}[1 - \rho(\tau)] \tag{6.5.9b}$$

The suggestion then is to estimate $\rho(\tau)$ by

$$\hat{\rho}(\tau) = \frac{1}{MAT_0}\int_0^{MT_0} y(t)x(t)\,dt \tag{6.5.10}$$

where $x(t)$ is as defined in Equation (6.4.18) (with $n = 2$) and $y(t) = Ax(t - \tau) + n(t)$, and from this to estimate $|\tau|$:

$$\widehat{|\tau|} = \frac{T_0}{2}[1 - \hat{\rho}(\tau)] \tag{6.5.11}$$

When it exists, the maximum-likelihood estimator of $|\tau|$, given the observable $\hat{\rho}(\tau)$, is indeed that of Equation (6.5.11). For, except at the points of discontinuity of $d\rho(\hat{\tau})/d\hat{\tau}$,

$$0 = \frac{\partial \log p[\hat{\rho}(\tau)\,|\,\hat{\tau}]}{\partial\hat{\tau}} = \pm\, k[\hat{\rho}(\tau) - \rho(\hat{\tau})] \tag{6.5.12}$$

with k a constant independent of $\hat{\tau}$. This equation has the unique solution $\hat{\rho}(\tau) = \rho(\hat{\tau})$. If $|\hat{\rho}(\tau)| \leq 1$, the maximum-likelihood estimate of $|\tau|$ is consequently $\widehat{|\tau|}$ where

$$\rho(\hat{\tau}) = 1 - \frac{2\widehat{|\tau|}}{T_0} = \hat{\rho}(\tau) \tag{6.5.13}$$

Since $\hat{\rho}(\tau)$ is Gaussianly distributed, so also is $\widehat{|\tau|}$, the mean and variance of which are given by

$$\eta_{\hat{\tau}} = E(\widehat{|\tau|}) = (T_0/2)[1 - E(\hat{\rho}(\tau)] = (T_0/2)[1 - \rho(\tau)] = |\tau| \tag{6.5.14a}$$

and

$$\sigma_{\hat{\tau}}^2 = \text{Var}(\widehat{|\tau|}) = (T_0/2)^2 \text{ Var}[\hat{\rho}(\tau)] = \frac{T_0^2}{8MR} \tag{6.5.14b}$$

respectively.

This estimation technique is quite comparable, both in the complexity of the equipment required, and in the results obtained, with the maximum-likelihood estimation of the epoch of a sinusoidal signal [compare Equations (6.5.4) and (6.5.14b)]. This is perhaps surprising since the latter is a maximum-likelihood estimate based on the observables $y(t)$, whereas the former is maximum-likelihood only in terms of the statistic $\hat{\rho}(\tau)$. Indeed, if the observables $y(t)$ are used optimally in conjunction with a signal having the autocorrelation function of Equation (6.5.9a) much better results are realized. The time needed to detect the correct one of the N contending epochs can be reduced by a factor of approximately N relative to the estimation time (cf. Section 6.4).

To relate these results more effectively to the detection results of the previous sections, we stipulate that a synchronization error is committed if the estimate $\widehat{|\tau|}$ differs from $|\tau|$ by more than $\Delta T/2 = T_0/2N$ seconds. Then

$$P_e = 1 - \text{erf}\,(M^{1/2}R^{1/2}/N)$$

and the search time is on the order of $M_T T_0$ where

$$M_T = \frac{N^2}{R} \log_e \frac{1}{P_e} \tag{6.5.15}$$

The search is not finished, however, since so far we have only estimated $|\tau|$ not τ. Additional observation is needed to determine whether $\hat{\tau} = \widehat{|\tau|}$ or $\hat{\tau} = -\widehat{|\tau|}$ is the better estimate. Which of the two estimates is better can be decided, for example, by making fixed-sample-size tests at the epoch $\tau = \widehat{|\tau|}$ and at the epoch $\tau = -\widehat{|\tau|}$ and accepting the one yielding the larger correlation. The number of observables needed for this decision when the two epochs are investigated serially can be as large as $M' = (N^2/R) \log_e (1/P_e)$, [cf. Equation (6.4.17)] a number ironically as great as that needed for the original estimation. On the other hand, if the two contending epochs can be observed simultaneously, requiring only two correlators, the required number of observables is, from Equation (6.4.5), on the order of at most $M'' = M'/N$ which, for large N, represents a negligibly small increase in the total estimation time.

It should be acknowledged that the epoch estimation method depends more crucially on an accurate knowledge of the signal amplitude than the other methods heretofore considered. If the signal amplitude is not known it must be estimated either prior to the search, or simultaneously with it. In either case, the inevitable inaccuracies of this estimate will increase the probability of an error in the final decision. This liability may be somewhat mitigated if a clock reference is available. If the amplitude of the received clock signal bears some known relationship to that of the sync signal, the amplitude reference may well be a by-product of the clock reference. Thus, this precondition for the successful application of the epoch estimation procedure may be less severe than might be supposed.

6.6 Phase-Incoherent Synchronization

It may happen that synchronization is to be achieved without prior knowledge of the synchronization channel carrier phase. This would be true either when the system constraints are such that coherent detection is impractical, or when it is for some reason more efficient first to synchronize and then to establish phase-coherence. It should be noted that the assumed clock reference does not assure phase-coherence. When the carrier frequency is much greater than the clock frequency, for example, the clock provides virtually no information concerning the carrier phase. In some situations it may even be possible to synchronize without any common clock reference, the transmitter clock uncertainty being small enough to ignore during the synchronization process. In this section, some of the synchronization techniques of the previous sections are re-examined in regard to the phase-incoherent channel.

The maximum-likelihood phase-incoherent receiver was discussed in Chapters Three and Four. If the phase can be assumed to remain essentially constant, at least over any particular T_0 second duration, the maximum-likelihood phase-incoherent synchronizer involves the determination of the quantities:

$$\int_0^{2\pi} \cdots \int_0^{2\pi} p(\phi_0, \phi_1, \ldots, \phi_{M-1}) \times \left[\prod_{l=0}^{M-1} \exp \frac{2A}{N_0} \int_{lT_0}^{(l+1)T_0} y(t)s_\nu(t, \phi_l)\, dt\right] d\phi_0\, d\phi_1 \ldots d\phi_{M-1} \tag{6.6.1}$$

where $\{s_\nu(t, \phi) = \sqrt{2}\,\xi(t - \nu\Delta T) \sin [\omega_c(t - \nu\Delta T) + \theta(t - \nu\Delta T) + \phi]\}$ represents the set of possible received synchronization signals and $y(t) = As_r(t, \phi) + n(t)$ is the received signal plus noise. This term is proportional to the a posteriori probability of the νth epoch, given the signal $y(t)$, $0 < t < MT_0$.

Expression (6.6.1) would be rather formidable even if the Mth order density function $p(\phi_0, \phi_1, \ldots, \phi_{M-1})$ could be specified. This density function reflects, among other things, the rate at which the phase of the received carrier drifts with respect to the local reference. In most cases it can only be approximated. To avoid this difficulty we shall limit the discussion here to the two extreme cases:

(1) $p(\phi_0, \phi_1, \ldots, \phi_{M-1}) = p(\phi_0)\delta(\phi_0 - \phi_1)\delta(\phi_0 - \phi_2) \ldots \delta(\phi_0 - \phi_{M-1})$

and

$$(2)\ p(\phi_0, \phi_1, \ldots, \phi_{M-1}) = p(\phi_0)p(\phi_1)\ldots p(\phi_{M-1})$$

where $p(\phi_i) = (1/2\pi)$, $0 \leq \phi_i < 2\pi$. In the first, the phase is constant throughout the synchronization period; in the second, it assumes independent values from one T_0 second interval to the next. Phase-incoherent synchronization, in effect, involves a simultaneous estimation of the epoch of the received signal and of the phase of its carrier, although the latter is not estimated explicitly. The longer the phase remains constant, the better it can be estimated, and, hence, the better the estimate of the epoch. The two cases to be considered, therefore, represent two extremes in potential synchronizer performance.†

Expression (6.6.1) under these respective assumptions becomes (cf. Section 4.3; as there we assume $\omega_c \gg 2\pi/T_0$):

Case (1):

$$\frac{1}{2\pi}\int_0^{2\pi} \exp \frac{2A}{N_0}\int_0^{MT_0} y(t)s_\nu(t, \phi)\, dt\, d\phi = I_0(2MRw_\nu) \tag{6.6.2}$$

where $w_\nu = (a_\nu^2 + b_\nu^2)^{1/2}$, $R = \mathscr{E}/N_0$, $\mathscr{E} = A^2\int_0^{T_0} s_\nu^2(t, \phi)\, dt \triangleq A^2\mathscr{E}_0$ and

$$a_\nu = \frac{1}{M}\sum_{l=0}^{M-1} a_{\nu l} \qquad a_{\nu l} = \frac{A}{\mathscr{E}}\int_{lT_0}^{(l+1)T_0} y(t)s_\nu(t, 0)\, dt$$

$$b_\nu = \frac{1}{M}\sum_{l=0}^{M-1} b_{\nu l} \qquad b_{\nu l} = \frac{A}{\mathscr{E}}\int_{lT_0}^{(l+1)T_0} y(t)s_\nu\left(t, \frac{\pi}{2}\right) dt$$

Case (2):

$$\prod_{l=0}^{M-1}\frac{1}{2\pi}\int_0^{2\pi}\exp\left(\frac{2A}{N_0}\int_{lT_0}^{(l+1)T_0} y(t)s_\nu(t, \phi_l)\, dt\right) d\phi_l = \prod_{l=0}^{M-1} I_0(2Rz_{\nu l}) \triangleq z_\nu \tag{6.6.3}$$

where

$$z_{\nu l} = (a_{\nu l}^2 + b_{\nu l}^2)^{1/2}$$

In case (1), since $I_0(2MRw_\nu)$ is a monotonic function of w_ν, w_ν itself can be used as the decision variable. The discussion of Section 4.3 established

$$p(w_\nu \mid \mu) = \begin{cases} 2MRw_\nu \exp\left[-MR(w_\nu^2 + \rho_i^2)\right] I_0(2MR\rho_i w_\nu) & w_\nu > 0 \\ 0 & w_\nu < 0 \end{cases} \tag{6.6.4}$$

where $i = \mu - \nu$, $\rho_i^2 = \tilde{\rho}_i^2 + \hat{\rho}_i^2$, $\tilde{\rho}_i = (1/\mathscr{E}_0)\int_0^{T_0} s_0(t, 0)s_i(t, 0)\, dt$ and $\hat{\rho}_i =$

---

†The phase could vary even during the T_0 second intervals, of course, thereby degrading the performance even further. This situation is fairly unlikely to occur in a practical system, however, at least when T_0 represents the symbol period. If T_0 corresponds to a higher-order synchronization interval, it might be well to use a known sequence of symbols for the synchronization signal, using the method discussed here for symbol synchronization and employing the word synchronization techniques of the later chapters to resolve the remaining ambiguities. If the sync signal is a pulse of width T_0/n, of course, it is only necessary for the phase to be relatively constant over each T_0/n second interval for condition (2) to be satisfied.

$(1/\mathscr{E}_0)\int_0^{T_0} s_0(t, 0)s_i(t, \pi/2)\,dt$. The results of that section are directly applicable here. When $\tilde{\rho}_i$ and $\hat{\rho}_i$ are both zero for all $i \neq 0$, for example, the probability of a correct decision is given by Equation (4.3.17) (with R replaced by MR). The other results of Section 4.3, including the bounds for arbitrary ρ_i can also be used here.

The fixed-sample-size search involves probability ratios of the form

$$\frac{p(w_\nu \mid \nu)}{p(w_\nu \mid \mu)} = \frac{e^{-MR} I_0(2MRw_\nu)}{e^{-MR\rho_i^2} I_0(2MR\rho_i w_\nu)} \tag{6.6.5}$$

The ratio $I_0(ax)/I_0(bx)$ can be shown to be a monotonically increasing function of x for $a > b$. Since $\rho_i^2 < 1$, for all $i \neq 0$, the probability ratio test is a one-sided test of w_ν; the νth epoch is accepted if, and only if, w_ν exceeds some specific value. The test is therefore uniformly most powerful, and the selection of the alternative hypothesis is irrelevant. The probability of an error of the first kind is, from Section 4.3 (with $\nu = 0$ the correct epoch),

$$\alpha = 1 - \int_\theta^\infty p(w_0)\,dw_0 = 1 - Q[(2MR)^{1/2}, (2MR)^{1/2}\theta] \tag{6.6.6}$$

where $Q(x, y)$ is defined as in Equation (4.3.19), and θ is such that when $w_\nu > \theta$, the ratio in Equation (6.6.5) exceeds the desired threshold. Similarly, the probability of an error of the second kind is

$$\beta_\nu = \int_\theta^\infty p(w_\nu)\,dw_\nu = Q[(2MR)^{1/2}\rho_\nu, (2MR)^{1/2}\theta] \tag{6.6.7}$$

When $\tilde{\rho}_\nu$ and $\hat{\rho}_\nu = 0$ for all $\nu \neq 0$, (the orthogonal case), then

$$\beta_\nu = Q[0, (2MR)^{1/2}\theta] = e^{-MR\theta^2} \tag{6.6.8}$$

We can thus equate P_e to $\alpha + [(N-1)/2]\beta$, for example, and select θ so as to minimize the error probability as in Section 6.5.

Unfortunately, the conditions necessary for the validity of the conclusions concerning the duration of a sequential probability ratio test (Section 2.3) are not satisfied here. Specifically the ratio $p(w_\nu \mid \nu)/p(w_\nu \mid \mu)$ cannot be expressed as the product of ratios of the form $p(w_{\nu l} \mid \nu)/p(w_{\nu l} \mid \mu)$. To do so would be to ignore the dependence of the carrier phase over the successive intervals. If this dependence were non-existent, the decision variables would be those of case (2) discussed above. In analyzing the sequential search in this latter case, of course, we are simultaneously underbounding the performance of a sequential search based on the decision variables w_ν. This same statement is also true of the other search procedures. Moreover, the coherent synchronization analyses of the preceding sections provide upper bounds on the performance attainable using the statistics w_ν. Consequently, we shall turn directly to the analysis of the synchronizer performance when the situation of case (2) is operative.

In case (2) the search is to be based on the statistics z_ν as defined in Equation (6.6.3), or more conveniently, on the statistics

$$\log z_\nu = \sum_{l=0}^{M-1} \log I_0(2Rz_{\nu l}) \tag{6.6.9}$$

Since $\log z_\nu$ is the sum of a (presumably) large number of independent random variables, thereby satisfying the conditions necessary for the applicability of the central limit theorem,† it is approximately Gaussianly distributed. This fact can be used to simplify the analysis when the statistics z_ν are the decision variables. But since it is generally impractical to evaluate Bessel functions in a communications receiver, an expeditious as well as a mathematically convenient approximation to $\log z_\nu$ results from the following observation. The most difficult situation obviously arises when the signal-to-noise ratio $R = \mathscr{E}/N_0$ is small. If this ratio is large almost any synchronization scheme will work satisfactorily. But if R is small compared to unity, we have

$$\log_e [I_0(2Rz_{\nu l})] \approx \log_e (1 + R^2 z_{\nu l}^2) \approx R^2 z_{\nu l}^2 \tag{6.6.10}$$

and the decision could be made on the basis of the sum of the squares of the variables $z_{\nu l}$ rather than on the sum of Bessel functions of these variables, a significant simplification.‡

Define

$$\zeta_\nu = \sum_{l=0}^{M-1} z_{\nu l}^2 \tag{6.6.11}$$

Again, ζ_ν is the sum of a large number of independent random variables, and, from the central limit theorem, is approximately Gaussianly distributed. It will be recalled that most of the bounds on the decision times developed in the previous sections of this chapter depended solely on ratios of the form

$$r^2 = \min_{\nu \neq 0} \frac{E^2(x_0 - x_\nu)}{\operatorname{Var}(x_0 - x_\nu)}$$

where the x_ν were Gaussianly distributed random variables corresponding to the νth epoch (with the zeroth epoch the correct one). In the present situation§

†The existence of the second moments of the identically distributed random variables $\log I_0(2Rz_{\nu l})$ is sufficient to justify this statement (cf. Reference 6.2).

‡At the other extreme, when R is large as compared to unity, $\zeta_{\nu l} = \log_e I_0(2Rz_{\nu l}) \approx 2Rz_{\nu l}$ and the sum of the $z_{\nu l}$ themselves could be used. But, as we shall see, the performance of a system using the sum of the squares of the variables $z_{\nu l}$ is only slightly inferior to that of a coherent system when R is large. Thus the possible advantage in ever using the large signal-to-noise approximation to the ideal synchronizer is presumably marginal.

§For future reference, we also note that when ϕ is known to have one of the two values $\phi = 0$ or $\phi = \pi$, the decision should be based on the statistic $\zeta'_\nu = \sum_{l=0}^{M} a_{\nu l}^2$ rather than on the ζ_ν of Equation (6.6.11). Further,

$$E(\zeta'_\nu) = M\left(\rho_\nu^2 + \frac{1}{2R}\right)$$

and

$$\operatorname{Var}(\zeta'_\nu - \zeta'_\mu) = M\left\{\frac{1}{R^2}(1 - \rho_i^2) + \frac{2}{R}(\rho_\nu^2 + \rho_\mu^2 - 2\rho_i\rho_\mu\rho_\nu)\right\}$$

$$E(\zeta_\nu) = M\left(\rho_\nu^2 + \frac{1}{R}\right) \tag{6.6.12a}$$

$$\text{Var}(\zeta_\nu) = M\left[\frac{1}{R^2} + \frac{2}{R}\rho_\nu^2\right]$$

and, if the μth and νth epochs are tested simultaneously,

$$\text{Var}(\zeta_\nu - \zeta_\mu) = 2M\left\{\frac{1}{R^2}(1 - \rho_i^2) + \frac{1}{R}[\rho_\nu^2 + \rho_\mu^2 - 2\tilde{\rho}_i(\hat{\rho}_\nu\hat{\rho}_\mu + \tilde{\rho}_\nu\tilde{\rho}_\mu) - 2\hat{\rho}_i(\hat{\rho}_\nu\tilde{\rho}_\mu - \tilde{\rho}_\nu\hat{\rho}_\mu)]\right\} \tag{6.6.12b}$$

where $\rho_j^2 = \tilde{\rho}_j^2 + \hat{\rho}_j^2$, and $i = \nu - \mu$, modulo N. Using the bounds of Equation (6.4.4), for example, we find that the time needed for a parallel observation decision is on the order of $M_T T_0$ seconds, where

$$M_T = \frac{4\kappa_N \log_e (1/P_e)}{(1 - \rho^2)R^2/(R + 1)} \tag{6.6.13}$$

and where $\rho^2 = \max_{\nu \neq 0} \rho_\nu^2$. Similarly the time required for a fixed-sample-size search is approximately $\bar{M}_T T_0$ seconds, with

$$\bar{M}_T = \frac{4N\kappa_N'' \log_e (1/P_e)}{(1 - \rho^2)^2\{R^2/[R(1 + \rho^2) + 1]\}} \tag{6.6.14}$$

[see Equation (6.4.15)].

Other results of the previous section can also be extended to the phase-incoherent case. The sequential search algorithm, for example, can be most easily applied to the present problem by treating the statistics $z_{\nu l}^2$ as though they were Gaussianly distributed with the mean $E(z_{\nu l}^2 | \nu) = \eta_0$ and the variance $\sigma^2 = \max_\mu \text{Var}(z_{\nu l}^2 | \mu)$ and by choosing as the alternative hypothesis epoch the epoch $\bar{\nu}$ having the mean $\eta_1 = E(z_{\nu l}^2 | \bar{\nu}) = \max_\mu E(z_{\nu l}^2 | \mu)$ and the variance σ^2. The search would be based on the decision variables

$$\zeta_{\nu l} = \frac{\eta_0 - \eta_1}{\sigma^2}\left\{z_{\nu l}^2 - \frac{\eta_0 + \eta_1}{2}\right\}$$

The expected search time could then be estimated by using Equations (2.6.8) and (2.6.12). The error probability occurring in these expressions, of course, does not represent the actual error probability since the observables $z_{\nu l}^2$ are not distributed as hypothesized. The results of Section 2.6 concerning the effect on the level and the power of the test when the distribution of the observables is neither that of the null hypothesis nor of the alternative hypothesis, however, can be used to determine the true probability of an error. [The distribution of the observables $z_{\nu l}$ is given in Equation (4.3.10)].

Although, the decision variables ζ_ν are optimum only for very low signal-to-noise ratios R, it is encouraging to note that when R is large, the resulting expression for M_T [Equations (6.6.13) and (6.6.14)] when $\rho = 0$ are only twice the corresponding expression in the coherent case. Moreover, when R is large, the required search times in the two situations become identical as $\rho \to 1$. Since the performance of a synchronizer relying on the variables w_ν [Equation (6.6.2)] is bounded from above by the coherent synchronizer and from below by the synchronizer using the variables ζ_ν, it too must yield only marginally better performance than the ζ_ν synchronizer, under the conditions just mentioned, even when the phase can be assumed constant over the full MT_0 seconds of the search. Finally, we can conclude that the degradation in performance sustained in using ζ_ν rather than $\log z_\nu$ is small, regardless of the signal-to-noise ratio R. When R is large, the preceding comments apply; when R is small ζ_ν and $\log z_\nu$ are essentially equal and any advantage afforded by the latter variable must be negligible.

The dependence of the search time on the number of epochs N is the same in both coherent and incoherent synchronization searches. The dependence of the search time on the signal-to-noise ratio R in the two situations is also essentially the same when R is large relative to unity. When R is small, however, the phase-incoherent synchronizer tends to exhibit a $1/R^2$ dependence on this parameter rather than the $1/R$ dependence of the coherent synchronizer.

Before concluding this section, we should also remark that phase-locked loops can sometimes be used to advantage even when synchronization is to be accomplished incoherently. A rectangular pulse of the sort discussed in Section 6.3, for example, can be tracked by a phase-locked loop even in the absence of a carrier reference. The performance analysis of such a loop will be deferred for the moment, however, since the same device will be encountered again in a more general context in Chapter Eight (Section 8.6).

6.7 Synchronization Signal Design—Pseudo-Noise Sequences

All of the epoch detection schemes considered in the preceding sections yield search times, or at least bounds on these search times, which are monotonically increasing functions of the maximum correlation coefficient ρ (or, in the phase-incoherent case, of ρ^2). Accordingly, if we are free to choose the synchronization channel signal, we should presumably attempt to minimize this parameter. The problem is to find a periodic signal $x(t)$, with period T_0, yielding a set of N signals $x_\nu(t) = x[t - (\nu T_0/N)]$, $\nu = 0, 1, \ldots, N-1$, such that the maximum of the correlation coefficients

$$\rho_{\mu\nu} = \frac{1}{\mathscr{E}} \int_0^{T_0} x_\mu(t) x_\nu(t)\, dt \tag{6.7.1}$$

is as small as possible. As was shown in Section 4.2.3, the inequality

$$\max_{\substack{\mu,\nu \\ \mu\neq\nu}} \rho_{\mu\nu} \geq -\frac{1}{N-1} \tag{6.7.2}$$

holds for any set of N signals. This lower bound obviously applies here. A class of signals which essentially meets this bound will be exhibited shortly. But first note that when N is large an orthogonal set (with $\rho_{\mu\nu} = 0$) is nearly optimum (and presumably is optimum in the incoherent case). One class of orthogonal signals, the PPM signals discussed in Section 4.4, immediately suggests itself: if $x(t) = \sqrt{2N} \sin \omega_c t$, $0 \leq t \leq T_0/N$, $\rho_{\mu\nu} = 0$ for all $\mu \neq \nu$, as desired.

Thus a nearly optimum synchronization signal, when any of the search procedures of the previous sections are to be used, is just a pulse of energy of width T_0/N transmitted repetitively every T_0 seconds. The fundamental disadvantage of such a signal lies in the fact that the signal amplitude is proportional to $\sqrt{N}$, and hence the peak transmitted power is proportional to N, if the total transmitted power is to remain constant. This severely limits the value of N in a practical system and thereby restricts the amount by which the original time uncertainty T_0 can be reduced.

Fortunately, it is possible not only to generate a signal having essentially optimum correlation coefficients $\rho_{\nu\mu}$ without the severe peak power requirement of a single pulse, but to do so under the additional constraint that its envelope be constant, independent of time.

Consider the following function†

$$x(t) = \begin{cases} 1 & 0 < t < T_0/3 \\ 1 & T_0/3 > t < 2T_0/3 \\ -1 & 2T_0/3 < t < T_0 \end{cases} \tag{6.7.3}$$

This function, shown in Figure 6.3(a) can be represented by the sequence $\{1, 1, -1\}$. The autocorrelation function $\rho(\eta)$ is depicted in Figure 6.3(b). Notice that $\rho(\eta)$ varies linearly between the values it assumes at the points $\eta = 0, T_0/3, 2T_0/3, \ldots,$; the function need be determined only at these points in order to be completely specified. Clearly, this will always be the case when $x(t)$ is a sequence of pulses of constant amplitude and uniform duration ΔT seconds; e.g., whenever $x(t)$ can be represented by a sequence of ones and minus ones. The function defined in Equation (6.7.3) is actually slightly better (for phase-coherent applications) than a rectangular pulse with pulse width $\Delta T = T_0/3$. But its most important advantage is in its constant power level.

To achieve a narrower effective pulse width entails the construction of

†The actual transmitted signal will usually be of the form $x(t)\sqrt{2}A \sin \omega_c t$. It is assumed here that carrier demodulation has already been effected.

(a)

(b)

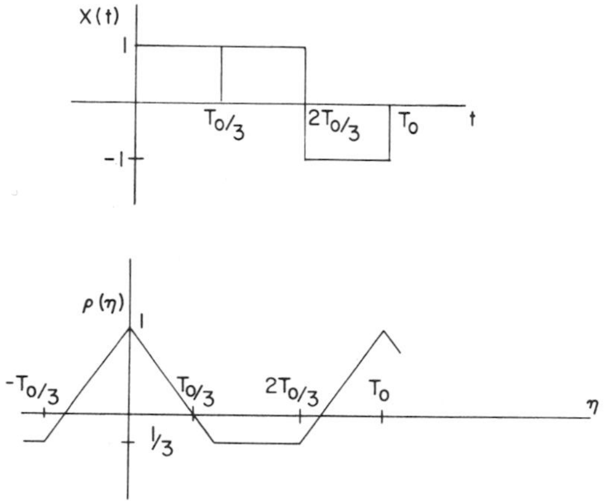

Figure 6.3 (a) The function $x(t)$ [Equation (6.7.3)]; (b) periodic autocorrelation function of $x(t)$.

longer sequences of *ones* and *minus ones* having autocorrelation functions approaching the lower bound of Equation (6.7.2). Note the effective pulse width ΔT of a signal of this type cannot be less than T_0/N where N is the number of symbols in (one period of) the two-level sequence representing $x(t)$. The goal, therefore, is to specify a two-level sequence of period N having a uniformly small autocorrelation function $\rho(\eta)$ for all η in the interval $iT_0 + T_0/N < \eta < (i + 1)T_0 - T_0/N$. We assert this can be done for any N of the form $N = 2^k - 1$, where k is an integer.

To illustrate the procedure for accomplishing this, we point out a special feature of the sequence $\{1, 1, -1\}$ which is the key to its generalization. Replacing the *minus-one* term in that sequence by zero, we find the modulo-two sum of the first two terms is equal to the third; the modulo-two sum of the second and third term is equal to the first, and the modulo-two sum of the

(a)

(b)

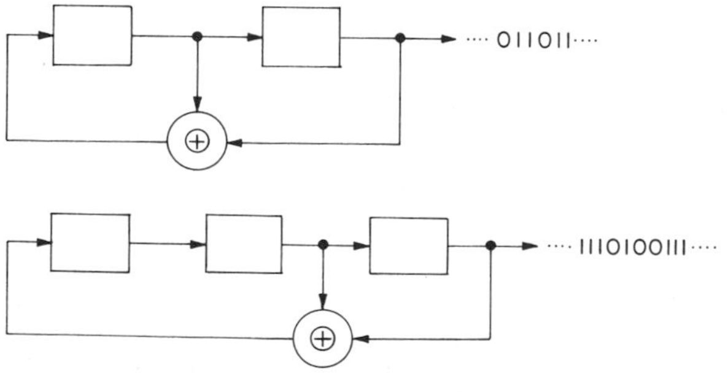

Figure 6.4 Both (a) and (b) show shift-register sequence generators.

third and the first term is equal to the second. The sequence could be generated by a *linear feedback shift-register* as shown in Figure 6.4(a). The square boxes represent storage elements which retain a *one* or a *zero* for one unit of time ($T_0/3$ seconds, in this example) and then shift it in the direction of the output arrow. The circle containing the sign $\oplus$ denotes a modulo-two adder; it instantaneously takes the modulo-two sum of its inputs and shifts it in the direction of its output arrow. That this shift-register does generate, repetitively, the sequence . . . . . . 110110 . . . . . is easily verified.

Thus, a two-stage (i.e., two storage units) shift-register can generate a sequence of period $3 = 2^2 - 1$. A three-stage shift-register is shown in Figure 6.4(b) along with a sequence which it generates. (The sequence should be read from right to left. The symbol on the far right was the first output from the register.) Whenever the same three-tuple is encountered for the second time the sequence begins to repeat itself again, since any three consecutive symbols determine the remainder of the sequence. This three-stage shift-register has generated a sequence of period $7 = 2^3 - 1$. In fact, the following statement is true: There exists some *linear* feedback arrangement (i.e., one involving modulo-two sums only) whereby a k-stage shift-register can generate a binary sequence of period $2^k - 1$ for any integer k. The proof of this statement follows directly from results to be developed in Chapter Thirteen and will not be presented here. It should be emphasized that not all linear feedback connections result in sequences of period $2^k - 1$. Indeed, only a small fraction of the possible feedback arrangements produce the desired results.

A sequence of period $2^k - 1$ produced by a linear k-stage shift-register is often denoted a *maximal-length linear shift-register sequence*. The rationale for this term becomes apparent upon observing that the period of the sequence is equal to the minimum number of digits separating any two identical k-tuples in it. This is because the contents of the k storage elements uniquely determine the subsequent output. Further, if the storage elements all contained *zeros* at some point, the output would be *zeros* only from then on. Thus, the longest k-stage shift-register sequence can contain each non-zero k-tuple once and only once before it begins to repeat itself. Since there are exactly $2^k - 1$ such (binary) k-tuples, and since any sequence of period N contains N l-tuples for any $l \leq N$, the period of a linear k-stage shift-register sequence cannot exceed $N = 2^k - 1$.

It has not yet been demonstrated that the desired autocorrelation properties of the maximal-length sequence $\{1, 1, 0\}$ are retained in the longer sequences. This is easily accomplished upon observing the following two properties of maximal-length, linear shift-register sequences:

(1) The modulo-two sum of a maximal-length sequence with any cyclic permutation of itself by any number of positions l, where $l \neq 0$, modulo N, is some third cyclic permutation of the same sequence.

(2) Any linear sequence of maximal-length $2^k - 1$ must contain exactly 2^{k-1} *ones* and $2^{k-1} - 1$ *zeros.*
We will prove both of these statements presently. Before doing so, let us first examine their consequence in relation to the autocorrelation function of these sequences.

The autocorrelation function of the time function $x(t)$ represented by a sequence of *ones* and *minus ones*, as already noted, is determined by its value at the points $\tau_j = jT_0/N$. If x_i is the value of the ith term of this sequence (with $x_i = +1$ or -1) then

$$\rho_j = \rho\left(j\frac{T_0}{N}\right) = \frac{1}{N}\sum_{i=1}^{N} x_i x_{i+j} = \frac{A_j - D_j}{N} \tag{6.7.4}$$

where $x_{N+l} \equiv x_l$, and where A_j denotes the number of times $x_i = x_{i+j}$ and D_j is the number of times $x_i \neq x_{i+j}$. This last expression on the right of Equation (6.7.4) is true whether the *one-minus-one* or the *one-zero* representation of the sequence is used. But the number of times $x_i = x_{i+j}$ in the *one-zero* representation is equal to the number of *zeros* in the N-tuple $\{x_i \oplus x_{i+j}\}$. The first of the two properties of a maximal-length linear sequence listed above states that the sum sequence $\{x'_i = x_i \oplus x_{i+j}\}$, where $j \neq 0$, modulo N, is just a cyclic permutation of the sequence $\{x_i\}$, $\{x'_i = x_{i+l}\}$ for some integer l. The second property, in turn, asserts that the number of *zeros* in the sequence $\{x_{i+l}\}$ and hence in the N-tuple $\{x_i \oplus x_{i+j}\}$ is exactly $2^{k-1} - 1$. Consequently,

$$\rho_j = \frac{A_j - D_j}{N} = \frac{(2^{k-1} - 1) - (2^{k-1})}{2^k - 1} = -\frac{1}{N} \tag{6.7.5}$$

for any $j \neq 0$, modulo N. When $j = 0$, modulo N, of course, $A_j = N$ and $\rho_j = 1$. If these two statements concerning maximal-length, linear sequences are true, and they will be proved shortly, the time function $x(t)$ corresponding to such a sequence has the autocorrelation function shown in Figure 6.5. Thus, the correlation coefficients $\rho_{\mu\nu} = \rho_{|\mu-\nu|} = -(1/N)$, $\mu \neq \nu$, and come very close to achieving the lower bound $-[1/(N-1)]$ as discussed earlier. In fact, no two-level sequence of length N greater than two can improve upon this result. This follows because, from Equation (6.7.4), $\rho_j = l/N$ for some integer

Figure 6.5 Periodic autocorrelation function of maximal-length sequence.

l, and since $\max_j \rho_j \geq -[1/(N-1)]$, we have $l \geq -[N/(N-1)]$. But since l is an integer, l must be greater than or equal to -1 for all $N > 2$.

We now prove the two statements upon which the validity of Equation (6.7.5) depends. The second statement, that a linear sequence of maximal length $2^k - 1$ contains exactly $2^{k-1} - 1$ *zeros* and 2^{k-1} *ones*, follows almost immediately from the earlier observation that each k-tuple must occur once and only once in a maximum-length sequence of length $2^k - 1$. Each sequence symbol is contained in exactly k k-tuples (it is the first digit in one, the second in another, etc.). The number M of *ones* in the sequence is therefore

$$M = \frac{1}{k}\sum_{i=1}^{2^k} m_i = \frac{1}{k}\left(2^k \frac{k}{2}\right) = 2^{k-1} \tag{6.7.6}$$

where m_i is the number of *ones* in the ith non-zero k-tuple. The number of *zeros*, of course, is just $2^k - 1 - M = 2^{k-1} - 1$.

The first of the two statements needed to obtain the autocorrelation function of $x(t)$ (that the sequence $\{x_i \oplus x_{i+j}\}$ is a sequence of the form $\{x_{i+l}\}$ for some integer l) is a consequence of the linearity of the shift-register generating these sequences. Any term in the sequence is a linear combination of the k preceding terms:

$$x_i = \sum_{\nu=1}^{k} a_\nu x_{i-\nu} \tag{6.7.7}$$

where the summation is taken modulo-two. The coefficients a_ν are either *one* or *zero* and are determined by whether or not the corresponding storage element output is an input to the modulo-two adder. The sum $x_i \oplus x_{i+j}$ can be written as:

$$(x_i \oplus x_{i+j}) = \sum_{\nu=1}^{k} a_\nu (x_{i-\nu} \oplus x_{i+j-\nu}) \tag{6.7.8}$$

Accordingly, the sequence $\{x'_i = x_i \oplus x_{i+j}\}$ can also be generated by the same feedback logic as the original sequence $\{x_i\}$. Since any k-tuple uniquely determines the remainder of this sequence, all non-zero k-tuples must occur somewhere in it if any of them do. The sequence either contains only *zeros* or else is some cyclic permutation of any other maximal-length sequence generated by the same logic. If the two k-tuples $x_i, x_{i+1}, \ldots, x_{i+k-1}$ and $x_{i+j}, x_{i+j+1}, \ldots, x_{i+j+k-1}$ are not equal (and they will be equal only if $j = 0$, modulo N), the sequence generated by their sum is not the sequence of *zeros* only. Therefore, unless $j = 0$, modulo N, the sum of any two cyclic permutations of the same maximal-length, linear shift-register sequence is a third permutation of the same sequence. This proves the first statement.

Because these sequences have certain properties in common with random binary sequences (cf. Chapter Thirteen), they are commonly referred to in the literature as *pseudo-random* or *pseudo-noise* sequences. This latter designation (often abbreviated *p-n* sequences) will be used in subsequent references to maximal-length shift-register sequences. It should be remarked that there also exist binary sequences, other than pseudo-noise sequences, having desirable two-level autocorrelation functions (References 6.3, 6.4). Although these sequences do supplement the pseudo-noise sequences (they can have periods not of the form $2^k - 1$) they are generally more difficult to generate and hence are less frequently used.

A pseudo-noise sequence, then, provides a means of realizing a signal with an effective pulse width of $\Delta T = T_0/N$ seconds (with $N = 2^k - 1$) while allowing the radiated power to be constant. Accordingly, it can be used for synchronization purposes in much the same way as a rectangular pulse of the same effective width. The discussion of Section 6.3, in particular, concerning the tracking of a periodic rectangular pulse with a phase-locked loop is equally applicable to *p-n* sequence tacking. The error signal resulting in the phase-error variance of Equation (6.3.6) (with $\Delta t = \Delta T$), for example, can be realized, when the received signal is the *p-n* sequence $Ap(t)$, by using the local (unity power) signal $p^*(t) = \sqrt{N/2(N+1)}[p(t + \Delta T/2) - p(t - \Delta T/2)]$. (The effective signal power here is actually greater by a factor of $(N+1)/N$ than that of a rectangular pulse of width ΔT.)

6.8 Synchronization Signal Design—Rapid Acquisition Sequences

The past sections of this chapter have been concerned with the estimation of the amount of time needed to attain the desired timing accuracy using various search algorithms and various synchronization signals. Although the choice of either the search method or the sync signal or both are often circumscribed by external constraints, in many cases the designer may have a wide latitude in specifying the system to be used. The question then arises as to the optimum (in the sense of minimizing the search time) synchronization method under these reduced constraints.

If there are no constraints at all (other than the usual power limitation) the optimum method is evidently to transmit a pseudo-noise sequence (or its equivalent) over the sync channel and to make a maximum-likelihood decision as to the correct epoch (cf. Section 4.2).

As we have already remarked, one of the most common constraints is an equipment limitation forcing the epochs to be observed serially rather than simultaneously. What is the best class of signals to use in this event? Frequently, pseudo-noise sequences are used in this case too, for obvious reasons. Indeed, as a quick glance at the previous sections of this chapter will verify, the searches considered thus far are most efficient when $\rho = -1/N$ (or, in the incoherent case, when $\rho = 0$). If a pseudo-noise sequence is used, the conditions delineated in Section 2.7 are satisfied, and the *optimum* serial search is a sequential search with $\alpha \to 1$. The minimum possible expected search time attainable with a *p-n* sequence is therefore directly proportional to the length of the sequence [cf. Equation (6.4.29) with $n = N$].

There is reason to believe the serial search time could be a much less rapidly increasing function of N. One is led to this conclusion upon observing the rather obvious inefficiencies in the serial search procedure when pseudo-noise sequences are used. Regardless of the algorithm used, considerably more time is spent in rejecting erroneous epochs, when N is large, than in accepting the correct one. Yet a rejection eliminates only one epoch from contention while an acceptance eliminates $N - 1$. The return is not proportional to the effort. The epoch estimation scheme of Section 6.5 avoids this difficulty. The observation provides the same amount of information regardless of which epoch is actually tested. Although the required decision time is greater using this technique than the expected search time using the sequential approach, at least when the sync signal is a pseudo-noise sequence, the estimation algorithm is in some sense more efficient.

The key to reducing the search time lies in exploiting this feature of the epoch estimation algorithm, in combining an algorithm which extracts the same amount of information from each decision with a class of signals which can be used efficiently with this algorithm. Suppose the sync signal were such that the observables pertaining to any particular epoch assumed, in the absence of noise, one of m possible values depending upon which epoch were the correct one. If m were equal to N, and each possible relationship between the correct and the observed epoch resulted in a unique expected value for the observable, the search would be completed after only one decision. If $m = N^{1/2}$ two such decisions would be needed. In general, for any value of m, up to m^r different epochs could be resolved with r decisions. Thus, ideally, r could be as small as the smallest integer equal to or greater than $\log_m N$.

In order to minimize the number of decisions needed, m should be made as large as possible. On the other hand, the variance of the magnitude of the decision variable must be inversely proportional to the square of the number of levels which are to be distinguished if a reliable decision is to be assured. Accordingly, the amount of time needed for decision must be roughly proportional to m^2 and the total search time proportional to $m^2 \log_m N = (m^2/\log_e m) \log_e N$. Differentiating $m^2/\log_e m$ with respect to m, we find that it is an increasing function of m for $m > e^{1/2}$ and a decreasing function for $m < e^{1/2}$. Since m must be an integer, its optimum value is apparently 2.

Thus, the search should proceed on the basis of a binary decision made at each epoch investigated as do most of the search routines discussed earlier, but with one important difference. Each binary decision should eliminate half of the epochs still in contention at that stage of the search, regardless of which epoch is being obesrved. If this can be accomplished, the total search time would be proportional to $\log_2 N$, rather than N, an impressive improvement when N is large.

It is easy to select a signal such that the first binary decision would eliminate half of the contending epochs. In particular, suppose the sync signals $s_1(t)$ were a unity power sinusoid or square wave with period $2T_0/N$. (Any function of time $f(t)$ having a minimum period $2T_0/N$ and satisfying the condition $f[t + (T_0/N)] = -f(t)$, i.e., any *anti-periodic* function, would work as well.) The autocorrelation coefficients $\rho(\tau = iT_0/N) \triangleq \rho_i$ of such a signal are just

$$\rho_i = (-1)^i \tag{6.8.1}$$

The maximum-likelihood decision as to whether the epoch being observed is an odd or an even numbered epoch (relative to the correct epoch) is made on the basis of a sum of integrals of the form

$$z_1(j) = \int_{jT_0}^{(j+1)T_0} y_1(t)s_1(t)\,dt \tag{6.8.2}$$

where the received signal $y_1(t)$ equals $As_1[t - (iT_0/N)] + n(t)$, for some integer i. If $\sum_{j=0}^{M-1} z_1(j)$ is positive, the tested and the correct epochs presumably have the same parity (i.e., an odd number of epochs must be separating them); if this sum is negative they must have opposite parity. In either event, only $N/2$ epochs will remain in contention after the decision, as was desired.

But after making this decision, the situation is changed only in that there are now $N/2$ evenly spaced epochs rather than the original N. Thus, a second sync channel could be utilized in exactly the same way as the first one to reduce again by a factor of two the number of contending epochs. One need only transmit over this second channel an anti-periodic signal $s_2(t)$ of period $4T_0/N$ and make a decision on the basis of integrals of the form

$$z_2(j) = \int_{jT_0}^{(j+1)T_0} y_2(t)s_2(t)\,dt \tag{6.8.3}$$

And, in general, if N is a power of 2, the number of contending epochs can be reduced from $N/2^{i-1}$ to $N/2^i$ by transmitting over the ith channel the anti-periodic signal $s_i(t)$ with the period $2^iT_0/N$ and formulating the integrals

$$z_i(j) = \int_{jT_0}^{(j+1)T_0} y_i(t)s_i(t)\,dt \tag{6.8.4}$$

Only $\log_2 N$ such channels and hence $\log_2 N$ such decisions are required to identify the correct epoch (assuming all of these decisions are made correctly).

Although this procedure has been described as though $\log_2 N$ separate sync channels were available, only one channel is actually needed. This is because the signals $s_i(t)$ are orthogonal:

$$\frac{1}{T_0}\int_0^{T_0} s_\nu(t)s_\mu(t)\,dt = 0 \qquad \mu \neq \nu \tag{6.8.5}$$

Thus the transmitted signal can be

$$s(t) = \frac{1}{\sqrt{\log_2 N}} \sum_{i=1}^{\log_2 N} s_i(t) \tag{6.8.6}$$

and each of the decision variables $z_i(j)$ can be formed without being affected by the presence of any of the other signals $s_l(t)$, $l \neq i$. If the amplitude of each of the component signals $s_i(t)$ is $1/\sqrt{\log_2 N}$ as indicated in Equation (6.8.6), the total received signal power is independent of N.

At the ith stage of the search, then, a decision must be made on the basis of the statistics $z_i(j)$ as to whether the received signal epoch and the epoch of the signal $s_i(t)$ have the same parity modulo 2^i (the hypothesis H_i) or whether they have different parities (the hypothesis $\bar{H}_i$). Since the two kinds of errors are identical so far as their effect on the outcome of the search is concerned, the corresponding error probabilities α and β should be equal. The optimum decision at each stage, in the sense of minimizing the expected time needed, is of course a sequential decision. The expected number of observables required is, since $\alpha = \beta$,

$$E(M \mid H_i) = E(M \mid \bar{H}_i) = \frac{(1 - 2\alpha)\log_e \dfrac{1-\alpha}{\alpha}}{E(\zeta_i \mid H_i)} \tag{6.8.7}$$

where ζ_i is the logarithm of the probability ratio

$$\frac{p_0(z_i)}{p_1(z_i)} = \exp\left\{\frac{2\eta}{\sigma^2} z_i\right\} \tag{6.8.8}$$

and where $\eta \triangleq |E(z_i)| = AT_0/\sqrt{n}$, $\sigma^2 \triangleq \mathrm{Var}(z_i) = N_0T_0/2$ and $n \triangleq \log_2 N$. Thus, the expected search time is $\bar{M}_T T_0$ seconds, where

$$\bar{M}_T = (\log_2 N)E(M \mid H_i) = \frac{1-2\alpha}{4R}\log_e\left(\frac{1-\alpha}{\alpha}\right)(\log_2 N)^2 \tag{6.8.9}$$

with $R = A^2T_0/N_0$. Finally, since the probability P_e of accepting an erroneous epoch is approximately $n\alpha$ when $n\alpha \ll 1$ (cf. Section 2.6; $n\alpha\{1 - [(n-1)/2]\alpha\} \leq P_e = 1 - (1-\alpha)^n \leq n\alpha$) we have

$$\bar{M}_T \approx \frac{1}{4}\log_e\left(\frac{\log_2 N}{P_e}\right)\frac{(\log_2 N)^2}{R} \tag{6.8.10}$$

A fixed-sample-size decision could also have been made at each stage. The resulting search time, when P_e is small, is larger by a factor of approximately four than the sequential search time under the same conditions.

Although this result [Equation (6.8.10)] is somewhat worse than the $\log_2 N$ proportionality suggested in the preceding paragraphs,† it nevertheless

†Note that if it is possible to alter the transmitted signal at the appropriate instants during the course of the synchronization search, the signals $s_i(t)$ could be transmitted serially. The power in each component signal would then be effectively increased by the factor of $\log_2 N$, and the search time would indeed be proportional to $\log_2 N$.

represents a sizable reduction in search time relative to the time needed using pseudo-noise sequences. Moreover, it seems apparent that no further reduction is possible with a serial search, at least if the decisions are to be binary. The signal used over each "channel" was optimum in the sense that no other signal of the same energy could make the two possible hypotheses more distinguishable. Since each decision was equally important to the final outcome, all of the "channels" should represent the same energy, as indeed they did. No decision routine is more rapid on the average than the sequential decision, so the time spent making each decision is minimized. And, finally, no fewer than $\log_2 N$ binary decisions, on the average, could have determined the correct epoch.

One disadvantage these signals have vis-à-vis the pseudo-noise sequences is in the non-uniformity of the transmitted power. The peak power [see Equation (6.8.6)] is $\log_2 N$ times as great as the average power. While this is a considerable improvement over the ratio of N to 1 characterizing the signal pulse of width T_0/N, it still may be objectionable. In any event, these signals can be replaced by constant-power binary sequences with only a minimal deterioration in performance.

To see how this can be accomplished we let the signals $s_i(t)$ be square-waves and note that they can be unambiguously represented by a sequence $s_i = (\sigma_1^i, \sigma_2^i, \ldots, \sigma_N^i)$ of *ones* and *minus ones*, σ_j^i corresponding to the amplitude of the signal $s_i(t)$ over the interval $(j-1)T_0/N < t < jT_0/N$. The signal $s(t)$, then, is represented by the sequence

$$s = \left(\sum_{i=1}^{n} \sigma_1^i, \sum_{i=1}^{n} \sigma_2^i, \ldots, \sum_{i=1}^{n} \sigma_N^i\right) \tag{6.8.11}$$

The goal is to replace this sequence by a sequence $s' = (\theta_1, \theta_2, \ldots, \theta_N)$ of *ones* and *minus ones* corresponding to a signal $s'(t)$ in such a way as to minimize the increase in the expected synchronization search time. From Equation (6.8.9), this search time is determined solely by the terms η and σ, with the latter independent of the received signal. Thus we wish to select s' in such a way as to maximize

$$\eta_i' = A\int_0^{T_0} s'(t)s_i(t)\,dt = \frac{AT_0}{N}\sum_{j=1}^{N}\theta_j\sigma_j^i$$

for all $i = 1, 2, \ldots, n = \log_2 N$. But

$$\begin{aligned}\min_i \eta_i' \le \operatorname*{ave}_i \eta_i' &= \frac{AT_0}{Nn}\sum_{i=1}^{n}\sum_{j=1}^{N}\theta_j\sigma_j^i \\ &= \frac{AT_0}{Nn}\sum_{j=1}^{N}\theta_j\sum_{i=1}^{n}\sigma_j^i \le \frac{AT_0}{Nn}\sum_{j=1}^{N}\left|\sum_i \sigma_j^i\right|\end{aligned} \tag{6.8.12}$$

the last inequality satisfied with equality if, and only if,

$$\theta_j = \operatorname{sgn}\left\{\sum_i \sigma_j^i\right\} \triangleq \begin{cases} 1 & \sum_i \sigma_j^i \ge 0 \\ -1 & \sum_i \sigma_j^i < 0 \end{cases} \tag{6.8.13}$$

The definition of sgn $\{x\}$ here when $x = 0$ is arbitrary so far as this last statement is concerned. For convenience, we have defined sgn $\{0\} = 1$ although this is at variance with the usual definition which has sgn $\{0\} = 0$. As we shall see, when θ_j is defined as in Equation (6.8.13), the terms $\min_i \eta'_i$ as well as $\operatorname{ave}_i \eta'_i$ attain the upper bound of Equation (6.8.12). Consequently, the sequence s', which we denote as a *rapid acquisition sequence,* is optimum under the constraints in effect here. The rapid acquisition sequence s' of length $N = 32$ is shown in Figure 6.6, along with the component sequences s_i.

The key to the proof of the optimality of the sequence s' lies in the observation that the set of n-tuples $\{\sigma_j^1, \sigma_j^2, \ldots, \sigma_j^n\}$ contains each of the N binary n-tuples (of *ones* and *minus ones*) once and only once. Moreover, since θ_j is a symmetric function of σ_j^i and σ_j^k,

$$\eta'_k - \eta'_i = \frac{AT_0}{N}\sum_{j=1}^{N}(\sigma_j^k - \sigma_j^i)\theta_j = \frac{AT_0}{N}\sum_{j=1}^{N}(\sigma_j^i - \sigma_j^k)\theta_j = \eta'_i - \eta'_k$$

and hence, $\eta'_i = \eta'_k$ for all i and k. Accordingly, $\operatorname{ave}_i \eta'_i = \min_i \eta'_i$ and all of the inequalities in Equation (6.8.12) are indeed equalities when θ_j is as defined above. Finally,

$$\eta' \triangleq \eta'_i = \frac{AT_0}{Nn}\sum_{j=1}^{N}\left|\sum_{i=1}^{n}\sigma_j^i\right|$$

$$= \frac{2AT_0}{Nn}\sum_{\nu=0}^{[n/2]}\binom{n}{\nu}(n-2\nu) = \begin{cases} \dfrac{AT_0}{2^{n-1}}\dbinom{n-1}{\frac{n-1}{2}} & n \text{ odd} \\[2ex] \dfrac{AT_0}{2^n}\dbinom{n}{\frac{n}{2}} & n \text{ even} \end{cases} \tag{6.8.14}$$

with the square brackets denoting the integer part of the enclosed fraction. The last summation again follows from the fact that each of the N binary n-tuples occurs exactly once in the set $\{\sigma_j^1, \sigma_j^2, \ldots, \sigma_j^n\}$.

When n is large an application of Sterling's approximation to the binomial coefficients yields

$$\eta' \approx \sqrt{\frac{2}{\pi}}\frac{AT_0}{(\log_2 N)^{1/2}} \tag{6.8.15}$$

The only cost of replacing $s(t)$ by $s'(t)$ as the transmitted signal is to increase the search time by a factor of approximately $\pi/2$ [see Equation (6.8.9) with η replaced by η'].

It is interesting to note that rapid acquisition sequences contain their own clock. Specifically, if such a sequence is impressed at the input to a phase-locked loop, if the local signal $z(t)$ is a square-wave with a period equal to that of the highest-frequency component defining the sequence, and if the

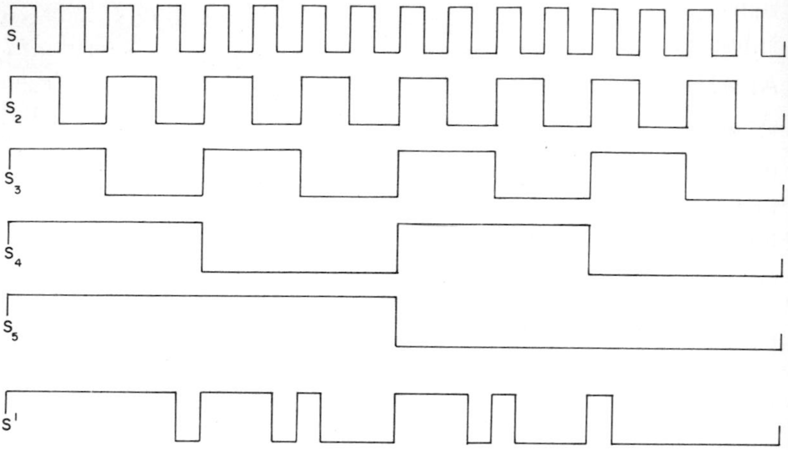

Figure 6.6 Rapid acquisition sequence and component sequences ($N = 32$).

loop bandwidth is on the order of $1/2T_0$ or less, this component will be tracked. So far as the loop is concerned, the sequence appears to be a pure square-wave with the amplitude $A\eta'$ with η' as defined in Equation (6.8.14). The tracking error is therefore $\sigma_\phi^2 = (\pi/2)^2(N_0B_L/A^2(\eta')^2)$ (see Section 6.3).

Generally, considerably more power should be relegated to the clock component than to any of the other components defining the composite sequence since the clock component determines the ultimate synchronization accuracy. This can be accomplished by simply attaching more weight to the highest frequency sequence in determining the composite sequence. Thus, for example, when $N = 16$, one could take the signs of the sums of the corresponding terms in the four component sequences

2	−2	2	−2	2	−2	2	−2	2	−2	2	−2	2	−2	2	−2
1	1	−1	−1	1	1	−1	−1	1	1	−1	−1	1	1	−1	−1
1	1	1	1	−1	−1	−1	−1	1	1	1	1	−1	−1	−1	−1
1	1	1	1	1	1	1	1	−1	−1	−1	−1	−1	−1	−1	−1

obtaining the composite sequence

1	1	1	−1	1	−1	1	−1	1	−1	1	−1	1	−1	−1	−1

This sequence has the correlation $\eta_{cl} = \frac{3}{4}$ with the clock component and $\eta = \frac{1}{4}$ with each of the other component sequences. In general, if the clock is represented by the sequence $a, -a, a, -a, \ldots$, while all the other components are represented by sequences of *plus ones* and *minus ones*, the correlation between the composite sequence thus formed and the clock is easily seen to be

$$\eta_{cl} = \frac{2A - N}{N} \tag{6.8.16}$$

where

$$A = 2 \sum_{i=0}^{[(n-1+a)/2]} \binom{n-1}{i} - \binom{n-1}{\frac{n-1+a}{2}}$$

with $n = \log_2 N$. The second binomial coefficient is taken to be zero if $(n - 1 + a)/2$ is not an integer. Similarly, the correlation with each of the other component sequences is

$$\eta = \frac{2B - N}{N} \tag{6.8.17}$$

where

$$B = 2 \sum_{i=0}^{[(n-1+a)/2]} \binom{n-2}{i} - \binom{n-2}{\frac{n-1+a}{2}} + 2 \sum_{i=0}^{[(n-1-a)/2]} \binom{n-2}{i} - \binom{n-2}{\frac{n-1-a}{2}}$$

Note that $\eta_{cl} = 1$, $\eta = 0$, when $a > n - 1$ and that $\eta_{cl} = \eta$ when $a = 1$. If n is relatively large, almost any desired relationship between η_{cl} and η can be obtained with some a in the range $1 \leq a \leq n - 1$.

6.9 Concluding Remarks

This chapter has been concerned with methods for establishing synchronization, both as to the choice of the signal to use, and as to techniques for processing it efficiently, when a separate channel is available solely for that purpose. It should be remarked that a separate channel does not imply a separate transmitter and receiver. The sync signal can be multiplexed with the information bearing signal in a number of ways: by utilizing separate subcarriers, for example, or by interleaving the two signals in time. In some cases, the communication system can operate in two modes, the initial mode establishing the synchronization needed for the remainder of the transmission. Synchronization is retained either by continuing to track a clock reference or by relying on sufficiently stable oscillators in both the transmitter and the receiver. Regardless of the multiplexing scheme used, if the sync channel can be treated as a separate entity the subject matter of this chapter is applicable.

The first step in the synchronization process generally, although not always, is to establish a common clock reference throughout the communication system. This involves transmitting a periodic signal and tracking its phase at the receiver. If the period of this clock signal is equal to (or greater than) the minimum acceptable T_0-second unambiguous synchronization interval, no further effort is needed. If, however, in order to reduce the acquisition time, a clock signal with period $\Delta T < T_0$ is used, there remain $N = T_0/\Delta T$ ambiguous synchronization epochs, and a signal may be required to eliminate this N-fold ambiguity.

The optimum strategy for eliminating this ambiguity is a function of the signal to be used for this purpose and of the equipment constraints. The major conclusions of this chapter in this regard can be summarized as follows:

(1) If there are no equipment or sync signal constraints except perhaps for a peak power constraint, the optimum signal is a pseudo-noise signal or its equivalent (a periodic orthogonal signal for an incoherent channel) and the optimum detector a bank of matched-filters, one for each signal epoch. The decision time is then proportional to $\log N$, where N is the number of contending signal epochs.

(2) If equipment limitations force serial observations of the epochs the optimum signal is as discussed in Section 6.8. If there is a severe peak power restriction, these sequences can be converted into binary rapid acquisition sequences with only a slight increase in search time. In either case, the search time is of the order of $(\log N)^2$. If pseudo-noise sequences are used under this constraint, the search time will be of the order of N. (Methods for combining several pseudo-noise sequences to produce a composite signal are discussed in References 6.3 and 6.11. The search time using such sequences is less than that required when only one *p-n*-sequence component is used but greater than that needed with rapid acquisition sequences.)

(3) If the sync signal is also restricted, the optimum search algorithm may be either a deferred decision, a sequential search, or an estimation, depending on the specific signal involved. In any event, the search time will be typically proportional to N^2.

(4) If severe equipment limitations preclude all but the simplest search routines, (e.g., the fixed-sample-size serial search) the search time may exhibit an N^3 proportionality.

It has been tacitly assumed, in arriving at that portion of these conclusions pertaining to signal design, that the synchronization channel is signal-to-noise ratio limited rather than bandwidth limited. Indeed, if the pulse-width-bandwidth product is of the order of 3 or 4 or greater, the finite bandwidth effect is entirely negligible and these signal design conclusions need no qualification. If this is not the case, however, it may no longer be advantageous to make the pulse width as narrow as would otherwise be dictated by signal-to-noise considerations alone. Nevertheless, it may still be possible, by altering the pulse shape, to narrow the *effective* pulse width within the constraints imposed by the band-limited channel. The most efficacious pulse shape, though, is intimately related to the channel pass-band characteristics, and it is difficult to make any general conclusions in this regard. This same qualification applies to the choice of the frequency of a sinusoidal subcarrier serving as a clock. Obviously, the clock signal must be kept within the channel pass band, a fact which may limit its frequency to some value less than that suggested in Section 6.2.

A second assumption underlying the above conclusions should also be noted. Namely, it was assumed that the correlations producing the decision variables involved an integral number of periods of the synchronizing signal. If the signal-to-noise ratio is sufficiently large and the synchronization signal period sufficiently long, it might be possible to make a reliable decision concerning the signal phase before a complete period has been observed. In this case, the above results are no longer valid. The *partial-correlation coefficients*, obtained by cross-correlating a pseudo-noise sequence with its local replica, or a rapid acquisition sequence with one of its component sequences, over only part of its period, will have non-zero variances even in the absence of noise. Pseudo-noise sequences are superior to rapid acquisition sequences in this regard, and, when the signal-to-noise ratio is large enough for partial correlations to suffice, will undoubtedly outperform them. See Problem 6.6 for one method of exploiting this capability.

This chapter, like the preceding ones, has concentrated on optimum or near-optimum solutions to the problems to which it was addressed, an optimum solution corresponding to the most effective utilization of the available signal power. As in the preceding chapters, the extension of the analyses of this chapter to situations in which the optimum configurations can only be approximated is generally straightforward.

To one concerned primarily with equipment complexity, the most vulnerable element in the equipment needed to implement the techniques of this chapter is undoubtedly the tracking loop. Indeed, synchronization is obviously possible without such devices, although their elimination is generally quite costly in terms of the signal power required to obtain a specified accuracy.

If the received signal is a sinusoid, the phase-locked loop functions as a narrow-band filter. Because a loop is capable of tracking the signal frequency, its bandwidth B_L can be much narrower than the dynamic frequency range Δf of the received signal. Nevertheless, the received signal could be filtered with a passive filter having a bandwidth $B \geq \Delta f$. The filtered signal would then be of the form (cf. Section 5.4)

$$\begin{aligned} y_f(t) &= A \sin(\omega_c t + \phi) + n_f(t) \\ &= [A + n_1(t)] \sin(\omega_c t + \phi) + n_2(t) \cos(\omega_c t + \phi) \\ &\approx A \sin\left(\omega_c t + \phi + \frac{n_2(t)}{A}\right) \end{aligned} \tag{6.9.1}$$

The last approximation holds when $|n_f(t)| \ll A$, as it must be if the recovered signal is to provide an acceptable clock reference. Thus, the phase-error variance is approximately

$$\sigma_\phi^2 = \frac{E[n_2^2(t)]}{A^2} = \frac{N_0 B}{A^2} \tag{6.9.2}$$

where N_0 is the noise spectral density. To obtain the same accuracy with a filter of bandwidth B as is attainable with a phase-locked loop of bandwidth B_L, therefore, necessitates an increase in the signal power by a factor of B/B_L. Since $B \geq \Delta f$ while, generally, $B_L \ll \Delta f$, the required power increase can be appreciable.

The same general conclusion as to the cost of eliminating the tracking loop holds when the received signal is non-sinusoidal. When the synchronization signal consists of periodically transmitted narrow pulses, for example, the tracking loop of Section 6.3 is sometimes replaced by a filter followed by a threshold device, the estimated epoch determined by the instant at which the filter output exceeds some threshold.† Again, the cost of eliminating the loop is a considerably increased synchronizer bandwidth (cf. Problem 6.3) and hence a corresponding increase in the signal power needed for satisfactory results.

Notes to Chapter Six

Much of the literature relating to the subject matter of this chapter has been oriented toward ranging rather than synchronization (cf. References 6.3, 6.6). One exception is the paper by Selin and Tuteur (Reference 6.7). Detailed discussions of pseudo-noise and other sequences having two-level autocorrelation functions can be found in References 6.3 and 6.4. The material on rapid acquisition sequences originally appeared in Reference 6.8.

Problems

6.1 (Reference 6.9). (a) A signal $x(t - \tau)$ is received in the presence of additive white Gaussian noise with single-sided spectral density N_0. The function $x(t)$ is known but the time delay τ is a random variable, uniformly distributed over the interval $(-T_0/2, T_0/2)$. Show that, regardless of the method used to estimate

---

†The synchronization problem in this guise is closely related to the problem of estimating the time of arrival of a radar pulse for ranging purposes. This problem is examined in detail in Reference 6.5.

τ, the mean-squared estimation error is underbounded by

$$\sigma_\tau^2 \geq \max_{0 \leq u \leq T_0/2} \frac{u^2}{8}\left\{1 - \operatorname{erf}\left[\frac{(1-\rho(u))R}{2}\right]^{1/2}\right\}$$

where $R = \mathscr{E}/N_0$, $\mathscr{E} = \int_{-\infty}^{\infty} x^2(t)\,dt$, and $\rho(u) = (1/\mathscr{E})\int_{-\infty}^{\infty} x(t)x(t+u)\,dt$.

[*Hint*: Consider the following problem: One of two equally likely signals $x_1(t) = x[t - (u/2)]$ and $x_2(t) = x[t + (u/2)]$ is received. To determine which, the signal time delay τ is estimated and the decision based on the sign of this estimate $\hat{\tau}$. The probability of an erroneous decision is clearly

$$P_e = \frac{1}{2} Pr\left\{\tau_e > \frac{u}{2}\middle| x_1\right\} + \frac{1}{2} Pr\left\{\tau_e < -\frac{u}{2}\middle| x_2\right\}$$

where $\tau_e = \tau - \hat{\tau}$ is the estimation error. To obtain the stated result, use Equations (4.2.12) and (4.2.13) to underbound P_e, and use the Chebycheff inequality, $Pr\,(|x| \geq k) \leq E(x^2)/k^2$, to overbound it.]

(b) Show that if

$$x(t) = \begin{cases} \left(\dfrac{\mathscr{E}}{\Delta T}\right)^{1/2}, & 0 < t < \Delta T \\ 0, & \text{otherwise} \end{cases}$$

then

$$\sigma_\tau^2 \geq \frac{k(\Delta T)^2}{R^2}$$

where

$$k = \max_{0 \leq \xi^2 \leq R/2} \left[\frac{\xi^4}{2}(1 - \operatorname{erf} \xi)\right] = 0.0925$$

Compare this estimation error variance with that resulting when the same signal is transmitted periodically, with period T_0, and tracked with a phase-locked loop having a bandwidth $B_e = 1/2T_0$ (cf. Section 6.3).

6.2 The *root-mean-squared* (rms) *bandwidth* B_0 of a periodic sequence of pulses $x(t)$ is defined by

$$B_0^2 = \frac{\int_{-\infty}^{\infty} (\omega/2\pi)^2\,|X(\omega)|^2\,d\omega}{\int_{-\infty}^{\infty} |X(\omega)|^2\,d\omega}$$

with $X(\omega)$ the Fourier transform of $x(t)$.

(a) Show that if such a pulse sequence is to be tracked with a phase-locked loop as discussed in Section 6.3, the phase-error variance can approach, at sufficiently large signal-to-noise ratios, the value

$$\sigma_\phi^2 = \frac{N_0 B_L}{A^2 B_0^2 T^2}$$

where T is the period of the pulse sequence, A^2 the received signal power, N_0 the noise spectral density, B_L the loop bandwidth, and B_0 the (finite) rms signal bandwidth.

(b) Using the Cramér-Rao inequality (Appendix A.2), determine a lower bound on the variance of any unbiased estimate of the phase of such a signal and compare with the preceding result.

(c) Show that the pulse-width optimization technique of Section 6.3 applies equally well to pulses having finite rms bandwidths and leads to essentially the same results.

6.3 One common method for establishing synchronization is to transmit a periodic sequence of narrow pulses and to note the instants at which the amplitude of a suitably filtered version of the received signal exceeds some threshold. Show that at sufficiently large signal-to-noise ratios, the mean-squared synchronization error is identical to that attainable by tracking the same signal with a phase-locked loop having a bandwidth $B_L = 1/2T$, with $1/T$ the pulse repetition rate. Show that this is true for both rectangular pulses and for pulses having a finite rms bandwidth. [*Hint*: At high signal-to-noise ratios, the relationship between the rms timing error σ_τ and the rms noise σ_n at the filter output can be approximated as shown in Figure P6.3.] What is the analog here to the loss-of-lock probability associated with phase-locked loops?

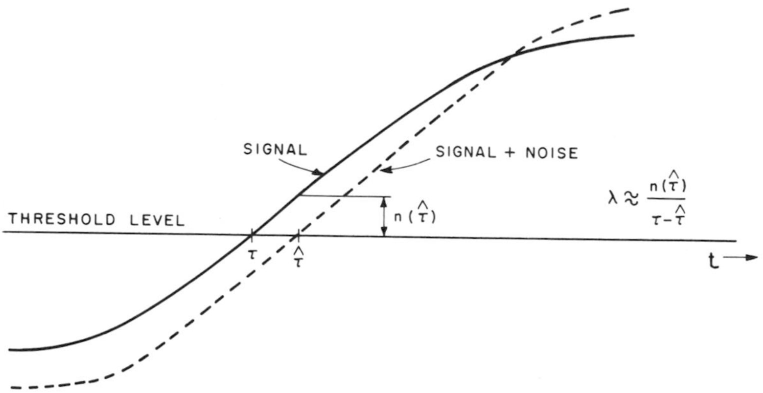

Figure P6.3 Approximate relationship between noise level and timing error.

6.4 (Reference 6.10). The phase of a white Gaussian noise-corrupted square-wave of period T is to be estimated by correlating it simultaneously with two locally generated square-waves of the same period, one delayed by an amount $T/4$ relative to the other. If after MT seconds of correlation the normalized output of one correlator is x and of the second correlator is y, show that the maximum-likelihood-principle estimator of the delay of the received square-wave is

$$\hat{\tau} = \frac{T}{8}[2 \operatorname{sgn} y - \operatorname{sgn} x \operatorname{sgn} y - \operatorname{sat}(x \operatorname{sgn} y - y \operatorname{sgn} x)]$$

where

$$\text{sat } u = \begin{cases} u, & |u| < 1 \\ \text{sgn } u, & |u| \geq 1 \end{cases}$$

and

$$\text{sgn } u = \begin{cases} 1, & u \geq 0 \\ -1, & u < 0 \end{cases}$$

[*Hint*: Let $u = (x - y)/\sqrt{2}$, $v = (x + y)/\sqrt{2}$ and note that the maximum-likelihood-principle estimate minimizes $\{[x - E(x \mid \hat{\tau})]^2 + [y - E(y \mid \hat{\tau})]^2\}^{1/2}$.] Determine the mean-squared error in making such an estimate when the signal-to-noise ratio is large. Compare this with the corresponding parameter when sinusoids rather than square-waves are used for the local references.

6.5 Suppose a sync signal is used to frequency-modulate a carrier; i.e., suppose the received signal is of the form $\sqrt{2}A \sin\left[\omega_c t + \Delta f \int s(t)\,dt\right] + n(t)$, with $n(t)$ broadband Gaussian noise. It is well-known and easily demonstrated that, at signal-to-noise ratios exceeding the FM demodulator threshold, the demodulator output assumes the form

$$\Delta f\, s(t) + \frac{n'(t)}{A}$$

where $n'(t)$ is the derivative of $n(t)$. This signal is to be tracked with a phase-locked loop with a fixed bandwidth B_L narrow relative to Δf. Show that: (a) If $s(t)$ is a sinusoid, the attainable synchronization accuracy is independent of its frequency. (This statement, of course, is subject to the usual assumption that the loop bandwidth is narrow relative to the reciprocal of the sync signal period.) (b) If $s(t)$ is a pulse sequence, periodic with period T, and if $\max_t s(t) = 1$, the timing error variance is

$$\sigma_\tau^2 = \frac{N_0 B_L}{A^2(2\pi\Delta f)^2}\frac{TS_2}{S_1^2}$$

with $S_i \triangleq \int_0^T (d^i s(t)/dt^i)^2\,dt$ assumed finite for both $i = 1$ and $i = 2$.

6.6 Consider the following scheme for determining the phase of a noise-corrupted pseudo-noise sequence of length $2^k - 1$: First, k consecutive received bits are detected (using the methods of Chapter Four) and read into a shift-register wired to generate the same sequence as the one being received. Successive received bits are then detected and compared to the corresponding shift-register output. The number of times the shift-register and detector outputs agree is counted and, on the basis of this count, a sequential test is conducted to determine whether or not the local and received sequences are in phase. If they are determined not to be in phase, a new set of k detected bits is read into the shift-register and a new sequential test begun. This procedure is continued until some sequential test ends with the acceptance of the hypothesis that the two sequences are in phase. Estimate the number of received bits required to conclude the sync search, with an error probability not exceeding P_e, under the following assumptions: (a) Bit synchronization has already been established, (b) If the two sequences are out-of-phase a bit comparison is equally likely to produce an agreement as a disagreement, with successive agreements and disagreements statistically independent events. At what signal-to-noise ratios does this method outperform the methods of Sections 6.7 and 6.8?

6.7 Show that a pseudo-noise sequence can be tracked with a phase-locked loop using either of the local signals

$$z_1(t) = p(t) - p(t - \Delta T)$$

or

$$z_2(t) = \begin{cases} p(t) & l\Delta T < t < (l + \frac{1}{2})\,\Delta T \\ -p(t) & (l + \frac{1}{2})\,\Delta T < t < (l + 1)\,\Delta T \end{cases} \qquad l = \ldots -1, 0, 1, \ldots$$

where $p(t)$ denotes the sequence and ΔT its bit period. [Such devices are called *delay-locked loops* (Reference 6.11).] Compare the resulting error signals.

References

6.1 J. J. Stiffler, "Phase-Locked Loop Synchronization with Sinusoidal Signals," *Jet Prop. Lab. Space Prog. Sum.* 37–27, Vol. 4, 1964.

6.2 H. Cramér, *Mathematical Methods of Statistics*, Princeton University Press, Princeton, N. J., 1946.

6.3 S. W. Golomb, *et al.*, *Digital Communications*, Prentice-Hall, Inc., Englewood Cliffs, N. J., 1964.

6.4 S. W. Golomb, *Shift-Register Sequences*, Holden-Day, Inc., San Francisco, 1967.

6.5 C. W. Helstrom, *Statistical Theory of Signal Detection*, Pergamon Press, Oxford, England, 1960.

6.6 G. C. Kronmiller, Jr., and E. J. Baghdady, "The Goddard Range and Range Rate Tracking System: Concept Design, and Performances," *Proc. of Int. Space Elec. Symp.*, Miami Beach, Fla., 1965.

6.7 I. Selin and F. Tuteur, "Synchronization of Coherent Detectors," *IEEE Trans. on Comm. Systems*, CS-11, p. 100, 1963.

6.8 J. J. Stiffler, "Rapid Acquisition Sequences," *IEEE Trans. Inf. Theory*, IT-14, p. 213, 1968.

6.9 J. Ziv and M. Zakai, "Some Lower Bounds on Signal Parameter Estimation," *IEEE Trans. Inf. Theory*, IT-15, p. 386, 1969.

6.10 S. Butman, "On Estimating the Phase of a Square Wave in White Noise," *Jet Prop. Lab. Space Prog. Sum.* 37–53, Vol. 3, p. 200, 1968.

6.11 Titsworth, "Optimal Ranging Codes," *IEEE Trans. on Space Elec. and Telemetry*, SET-10, p. 19, 1964.

6.12 Spilker, J. J., Jr., and D. T. Magill, "The Delay-Lock Discriminator—an Optimum Tracking Device," *Proc. IRE*, **49**, p. 1403, 1961.

Supplementary Bibliography

Bussgang, J. J., and D. Middleton, "Optimum Sequential Detection of Signals in Noise," *IRE Trans. Inf. Theory*, IT-1, **5**, p. 5, 1955.

Layland, J. W., "On Optimal Signals for Phase-Locked Loops," *IEEE Trans. on Comm. Tech.*, COM-17, p. 526, 1969.

Reed, I. S., and I. Selin, "A Sequential Test for the Presence of A Signal in One of *K* Possible Positions," *IRE Trans. Inf. Theory*, IT-9, p. 286, 1963.

Shaft, P. D., "Optimum Design of the Nonlinearity in Signal Tracking Loops," Stanford Res. Inst., Menlo Park, Calif., 1968.

Slepian, D., "Estimation of Signal Parameters in the Presence of Noise," *IRE Trans. Inf. Theory*, IT-3, p. 68, 1954.

Spilker, J. J., Jr., "Delay-Lock Tracking of Binary Signals," *IEEE Trans. on Space Elect. and Tele.*, SET-9, p. 1, 1963.

Stiffler, J. J., "On the Selection of Signals for Phase-locked Loops," *IEEE Trans. on Comm. Tech.*, COM-16, p. 239, 1968.

Stiffler, J. J., "Coding and Synchronization—the Signal Design Problem," *Advances in Communication Systems* (A. V. Balakrishnan, ed.), Vol. 3, Academic Press Inc., New York, N. Y., 1968.

Swerling, P., "Parameter Estimation Accuracy Formulas," *IEEE Trans. Inf. Theory*, IT-10, p. 302, 1964.

Ward, R. B., "Acquisition of Pseudonoise Signals by Sequential Estimation," *IEEE Trans. on Comm. Tech.*, COM-13, p. 475, 1965.

chapter seven

Maximum-Likelihood Symbol Synchronization

7.1 Introduction

Chapter Six investigated techniques for providing both a clock reference and higher-order synchronization through the use of auxiliary channels. Most of the remaining chapters will be concerned, in one way or another, with methods for gleaning this information directly from the received message. The ultimate goal is to be able to dispense entirely with all auxiliary synchronization channels.

It should be apparent that the information needed to establish symbol synchronization in particular is actually present in the message-bearing signal itself. There clearly must be a discernible difference between the different received symbols if any information is to be transmitted. And since the symbols must be distinguishable, it should be possible to determine directly from the received sequence just when, in time, the transition from one symbol to the next can take place. The purpose of this chapter is to investigate methods for exploiting this capability.

7.2 The Maximum-Likelihood Symbol Synchronizer

In this section, the maximum-likelihood approach is applied to the symbol synchronization problem. The maximum-likelihood decision as to the symbol epoch is to accept that epoch† ν maximizing the function $p[y(t)\,|\,\nu]$, thereby identifying the symbol epoch most likely to account for the observed signal $y(t)$. In the development of this approach, the symbols will all be assumed to have equal and known time durations, T_s seconds. Any device which attempts to ascertain the symbol epoch will be referred to as a *symbol synchronizer;* a synchronizer which functions by finding the maximum over ν of the density functions $p[y(t)\,|\,\nu]$ will be called a *maximum-likelihood symbol* synchronizer.

The derivation of the mathematical form of a maximum-likelihood synchronizer parallels that of the maximum-likelihood detector in Chapter Three. Let the received signal $y(t) = x_j(t - \nu\Delta T) + n(t)$ be passed through an ideal low pass filter with bandwidth B, and sampled at the rate $2B$ samples/second (cf. Section 3.4). The output is the sequence of samples

$$y_i = x^j_{i-\nu k} + n_i \qquad \begin{array}{l} i = 0,1,\ldots, ML + (N-1)k \\ j = 1,2,\ldots, n \end{array} \tag{7.2.1}$$

The terms n_i represent independent samples from a Gaussian random process having zero mean and variance $\sigma^2 = N_0 B$, N_0, as usual, denoting the single-sided noise spectral density. The superscript j indicates the jth of the set of n possible symbols $x_j(t)$, L is the number of samples per symbol, $k = L\Delta T/T_s = L/N$ is the number of samples in the interval separating any two consecutive symbol epochs, and M indicates the total number of symbols observed. Finally, the epoch ν is some integer in the interval $\{0, N-1\}$.

Since the set of observables $Y = \{y_i\}$ are independent Gaussian variables when j and ν are known

$$\begin{aligned} p(Y\,|\,\nu) &= \prod_{l=0}^{M-1} \sum_{j=1}^{n} \frac{1}{(\sqrt{2\pi}\,\sigma)^L} \exp\left\{-\frac{1}{2\sigma^2} \sum_{i=lL+\nu k}^{(l+1)L+\nu k-1} (y_i - x^j_{i-\nu k})^2\right\} P(j) \\ &= K_1 \prod_{l=0}^{M-1} \sum_{j=1}^{n} \exp\left\{\frac{1}{\sigma^2} \sum_{i=lL+\nu k}^{(l+1)L+\nu k-1} y_i x^j_{i-\nu k} - \frac{1}{2}(x^j_{i-\nu k})^2\right\} P(j) \end{aligned} \tag{7.2.2}$$

where $P(j)$ is the probability of the jth symbol, and where

$$K_1 = \frac{1}{(\sqrt{2\pi}\,\sigma)^{ML}} \exp\left\{-\frac{1}{2\sigma^2} \sum_{i=\nu k}^{ML+\nu k-1} y_i^2\right\}$$

It will be noted that the end samples y_i for $0 \leq i < \nu k$ and $ML + \nu k \leq i < (M+1)L$ have been omitted in Equation (7.2.2). The set of observables Y necessarily includes all the samples y_i, $0 \leq i < (M+1)L$, because the terms $p(Y\,|\,\nu)$ must be evaluated for all ν. To ignore these end samples is to disregard some information which could have bearing on the decision. Nevertheless, if the number $(M+1)L$ of samples required for a reliable decision is large, these end samples represent a negligibly small contribution to the total information available and can be safely ignored. At very high

†As in Chapter Six, it is convenient to restrict our attention initially to the case in which there are only a finite number of contending epochs.

signal-to-noise ratios this will not be true, as we shall shortly discover. With this exception, however, the inclusion of these end samples in the subsequent manipulations only complicates the results without significantly altering them. With the same qualification, the term K_1 in Equation (7.2.2) is effectively independent of ν and can therefore be treated as a constant.

Taking the limit of the summation in Equation (7.2.2) as the bandwidth B becomes infinite and the sampling interval approaches zero, we find that the maximum-likelihood decision is to select as the symbol epoch that value of ν for which

$$\prod_{l=0}^{M-1}\left(\sum_{j=1}^{n} \exp\left\{\frac{1}{N_0}\int_{lT_s+\nu\Delta T}^{(l+1)T_s+\nu\Delta T} [2y(t)x_j(t-\nu\Delta T) - x_j^2(t-\nu\Delta T)]\,dt\right\}P(j)\right) \tag{7.2.3}$$

is a maximum. This is the mathematical operation to be performed by the maximum-likelihood synchronizer.

A frequently useful alternative approach is to make a simultaneous maximum-likelihood decision concerning the epoch *and* the symbols $x_j(t-\nu\Delta T)$. If this is done, Equation (7.2.3) would evidently be replaced by

$$\prod_{l=0}^{M-1}\left[\max_{j_l} \exp\left\{\frac{1}{N_0}\int_{lT_s+\nu\Delta T}^{(l+1)T_s+\nu\Delta T} [2y(t)x_{j_l}(t-\nu\Delta T) - x_{j_l}^2(t-\nu\Delta T)]\,dt\right\}\right] \tag{7.2.4}$$

Since some information is inevitably lost in substituting Equation (7.2.4) for Equation (7.2.3), the resulting synchronizer will be somewhat inferior to the maximum-likelihood synchronizer. As we shall see, however, Equation (7.2.4) often leads to a much more practical device. Moreover, such an approach may well be in order when the a priori probabilities $P(j)$ are unknown. [Still another alternative is to make a maximum a posteriori probability decision concerning the symbols and a maximum-likelihood decision as to the epoch. The only difference between this and the method just discussed is in the inclusion of the a priori probabilities $P(j)$ when finding the maximum over j of the factors in Equation (7.2.4).]

In obtaining Equations (7.2.3) and (7.2.4) we have implicitly assumed the functional form of the symbols $x_j(t-\tau)$ to be known exactly except for the random time delay τ. This assumption will often be valid. If a phase-coherent clock reference is available, for example, this assumed ability to reproduce the symbols (except for the time delay) at the receiver may well exist. Not infrequently, however, there may be other unknown parameters even while symbol synchronization is being attempted.

One of the parameters which is most frequently unknown is the phase of the carrier. In this event the distribution of the set of observables Y is a function of the carrier phase as well as of the epoch ν. If the carrier phase is assumed to remain constant over each symbol interval, this assumption being consistent with that made in the previous discussions of phase-incoherent reception, the maximum-likelihood phase-incoherent symbol synchronizer involves determination of the quantities

$$\int \cdots \int p(Y \mid \nu, \phi_1, \phi_2, \ldots, \phi_M) p(\phi_1, \phi_2, \ldots, \phi_M) d\phi_1 d\phi_2 \ldots d\phi_M \tag{7.2.5}$$

where ϕ_i is the phase during the ith interval $iT_s + \nu\Delta T < t < (i+1)T_s + \nu\Delta T$. As in Section 6.6, the form of the synchronizer depends upon the assumptions made concerning the density function $p(\phi_1, \phi_2, \ldots, \phi_M)$. The discussion here will be limited to the synchronizer resulting under the approximation

$$p(\phi_1, \phi_2, \ldots, \phi_M) = p(\phi_1)p(\phi_2) \ldots p(\phi_M) \tag{7.2.6}$$

(Clearly, the phase statistics can also be assumed to be independent of the symbol epoch ν.) If there is some phase dependence between successive symbols (as there invariably is) this approximation yields a suboptimum synchronizer in that some phase information which could be used is ignored. Nevertheless, the performance of this suboptimum synchronizer provides a lower bound on that of the optimum synchronizer were all the phase information used. An upper bound on this performance, of course, is established by the phase-coherent synchronizer (cf. Section 6.6).

Utilizing Equation (7.2.6), we find the maximum-likelihood symbol epoch to be the value of ν maximizing the expression

$$\prod_{l=0}^{M-1} \left(\sum_{j=1}^{n} \int p(\phi) \left\{ \exp \frac{1}{N_0} \int_{lT_s+\nu\Delta T}^{(l+1)T_s+\nu\Delta T} [2y(t)x_j(t-\nu\Delta T, \phi) - [x_j(t-\nu\Delta T, \phi)]^2]\, dt \right\} d\phi P(j) \right) \tag{7.2.7}$$

Again, a sometimes expeditious variation on the maximum-likelihood synchronizer is afforded by making joint maximum-likelihood decisions as to epoch, symbols, and perhaps even phases. The corresponding synchronizer determines the epoch ν maximizing the function

$$\prod_{l=0}^{M-1} \max_{j_l} \int p(\phi) \left\{ \exp \frac{1}{N_0} \int_{lT_s+\nu\Delta T}^{(l+1)T_s+\nu\Delta T} [2y(t)x_{j_l}(t-\nu\Delta T, \phi) - [x_{j_l}(t-\nu\Delta T, \phi)]^2]\, dt \right\} d\phi \tag{7.2.8}$$

or the function

$$\prod_{l=0}^{M-1} \max_{j_l, \phi_l} \exp \frac{1}{N_0} \int_{lT_s+\nu\Delta T}^{(l+1)T_s+\nu\Delta T} [2y(t)x_{j_l}(t-\nu\Delta T, \phi_l) - [x_{j_l}(t-\nu\Delta T, \phi_l)]^2]\, dt \tag{7.2.9}$$

Other signal parameters besides the time delay and the carrier phase may be unknown. The symbol amplitude, for example, or the noise spectral density are not always known precisely. Generally, however, these parameters can be estimated with sufficient accuracy to justify treating them as known, and the associated uncertainties can usually be ignored in specifying the appropriate synchronizer.

7.3 PAM Symbol Synchronization

In this section, we shall consider the maximum-likelihood phase-coherent and phase-incoherent symbol synchronizers for the pulse amplitude modulation systems described in Sections 3.5 and 3.6. We begin with the phase-coherent situation in which the carrier phase has already been determined.

The received signal is of the form $y(t) = \sqrt{2}\,Ax\sin(\omega_c t + \phi) + n(t)$ where A and ϕ are known constants, and x is a random variable, constant over each interval $iT_s + \nu\Delta T < t < (i+1)T_s + \nu\Delta T$, and having the known a priori probability density function $p(x)$. The sum over j in Equation (7.2.3) becomes here an integral and the probability $P(j)$, a probability density. The maximum-likelihood epoch is that ν maximizing

$$\prod_l \int_{-\infty}^{\infty} \exp\left\{\frac{1}{N_0}\int_{lT_s+\nu\Delta T}^{(l+1)T_s+\nu\Delta T} [2Axy(t)\sqrt{2}\sin(\omega_c t + \phi) - A^2x^2]\,dt\right\} p(x)\,dx \tag{7.3.1}$$

This product can be written as

$$\prod_l e^{Ra_{\nu l}^2} \int_{-\infty}^{\infty} e^{-R(x-a_{\nu l})^2} p(x)\,dx \tag{7.3.2}$$

where $R = A^2T_s/N_0$ and

$$a_{\nu l} \triangleq \frac{1}{AT_s}\int_{lT_s+\nu\Delta T}^{(l+1)T_s+\nu\Delta T} y(t)\sqrt{2}\sin(\omega_c t + \phi)\,dt$$

The integral

$$\frac{R^{1/2}}{\sqrt{\pi}}\int_{-\infty}^{\infty} e^{-R(x-a_{\nu l})^2} p(x)\,dx \tag{7.3.3}$$

is the convolution of the density functions $p(x)$ and

$$p(a_{\nu l}) = \frac{R^{1/2}}{\sqrt{\pi}} e^{-Ra_{\nu l}^2} \tag{7.3.4}$$

It is an expression of the fact that the exponential term $\exp(Ra_{\nu l}^2)$ should be given less weight when $a_{\nu l}$ differs significantly from those values predicted by the a priori distribution of its expected value x. Generally, however, this term will not significantly affect the synchronizer decision.

To support this contention when x is uniformly distributed, we consider the integral in Expression (7.3.3) in the two extremes when $R \gg 1$ and when $R \ll 1$. In the first instance, when $R \gg 1$, $e^{-R(x-a_{\nu l})^2}$ is negligible, except when

$x \approx a_{\nu l}$. If $p(x) = 1/(x_0 - x_1)$ for $x_1 < x < x_0$, we have

$$\frac{R^{1/2}}{\sqrt{\pi}} \int_{x_1}^{x_0} e^{-R(x-a_{\nu l})^2} p(x)\, dx \approx \begin{cases} \dfrac{1}{x_0 - x_1} & x_1 < a_{\nu l} < x_0 \\ 0 & \text{otherwise} \end{cases}$$

But when R is large $a_{\nu l}$ will be within the interval x_1, x_0 with high probability, so the weighting function is nearly always a constant. Similarly, if $R \ll 1$, $e^{-R(x-a_{\nu l})^2} \approx 1$ unless $a_{\nu l}$ is significantly outside the interval x_1, x_0. In this case,

$$\frac{R^{1/2}}{\sqrt{\pi}} \int_{x_1}^{x_0} e^{-R(x-a_{\nu l})^2} p(x)\, dx \approx \frac{R^{1/2}}{\sqrt{\pi}}$$

and is again essentially constant.

When x is Gaussianly distributed, $p(x) = (1/\sqrt{2\pi}\sigma_s)e^{-x^2/2\sigma_s^2}$, the situation is even more straightforward. In this event,

$$\frac{R^{1/2}}{\sqrt{\pi}} \int_{-\infty}^{\infty} e^{-R(x-a_{\nu l})^2} p(x)\, dx = \frac{1}{\sqrt{\pi}} \left(\frac{R}{1 + 2R\sigma_s^2} \right)^{1/2} e^{-[R/(1+2R\sigma_s^2)]a_{\nu l}^2}$$

and Equation (7.3.2) becomes

$$K \prod_l e^{[(2R\sigma_s^2)/(1+2R\sigma_s^2)]a_{\nu l}^2}$$

But since the ν maximizing the expression $K_1 \prod_l \exp(K_2 a_{\nu l}^2)$ is independent of both K_1 and K_2, the decision remains the same whether the factor in Equation (7.3.3) is included or not.

As a final motivation for equating the term in Equation (7.3.3) to a constant, we observe that when a joint maximum-likelihood symbol and epoch decision is to be made [Equation (7.2.4)] this term is replaced by the constant

$$\max_x \left[\frac{R^{1/2}}{\sqrt{\pi}} e^{-R(x-a_{\nu l})^2} \right] = \frac{R^{1/2}}{\sqrt{\pi}} \tag{7.3.5}$$

With this justification we proceed to ignore the integral in Equation (7.3.2), treating it as a constant, realizing that the resulting synchronizer will possibly be suboptimum. The relative simplicity of the statistic

$$z'_\nu = \prod_l e^{Ra_{\nu l}^2} \tag{7.3.6}$$

will generally more than compensate for its possible suboptimality. As usual, it will be more convenient to work with the logarithm of this statistic rather than the statistic itself. To this end we define

$$z_\nu = \sum_l a_{\nu l}^2 \tag{7.3.7}$$

If the number of terms involved in each of these summations is large, we can use the central limit theorem to argue that z_ν is approximately Gaussianly distributed.† The probability that the true symbol epoch (which will be assumed without loss of generality to be the zeroth epoch, $\nu = 0$) can be correctly distinguished from all other epochs can then be calculated in terms of the first and second moments of the variables z_ν.

Let M be the total number of terms in the summation of Equation (7.3.7), and let $m = E(x)$ be the expected value of the signal amplitude, $\mu_i = E[(x - m)^i]$ its ith central moment, and $R = A^2T_s/N_0$, the input signal-to-noise ratio (see Section 3.5). Further, let $\lambda_\nu = \nu\Delta T/T_s = \nu/N$. The first and second moments of the variables z_ν can be expressed in terms of these parameters as follows:

$$\eta_\nu = E(z_\nu) = M\left\{[(1 - \lambda_\nu)^2 + \lambda_\nu^2]\mu_2 + m^2 + \frac{1}{2R}\right\} \tag{7.3.8a}$$

and

$$\sigma_\nu^2 = M\sigma_{sxs}^2 + M\sigma_{sxn}^2 + M\sigma_{nxn}^2 \tag{7.3.8b}$$

where

$$\begin{aligned}\sigma_{sxs}^2 &= [\lambda_\nu^2 + (1 - \lambda_\nu)^2]^2(\mu_4 - \mu_2^2) + 4[\lambda_\nu^2 + (1 - \lambda_\nu)^2]m\mu_3 \\ &\quad + 4\lambda_\nu^2(1 - \lambda_\nu)^2\mu_2^2 + 4m^2\mu_2 \\ \sigma_{sxn}^2 &= \{[(1 - \lambda_\nu)^2 + \lambda_\nu^2]\mu_2 + m^2\}\frac{2}{R}\end{aligned}$$

and

$$\sigma_{nxn}^2 = \frac{1}{2R^2}$$

The subscripts on the last three terms denote, respectively, "signal-cross-signal" (the variance due to the signal alone), "signal-cross-noise" (the variance due to the interaction of the signal and noise) and "noise-cross-noise" (the variance caused by the noise alone).‡ Finally,

$$\begin{aligned}\Delta_{\nu\mu}^2 &\triangleq E[(z_\nu - z_\mu)^2] - E^2(z_\nu - z_\mu) \\ &= M\Delta_{sxs}^2 + M\Delta_{sxn}^2 + M\Delta_{nxn}^2\end{aligned} \tag{7.3.9}$$

---

†Again, the easily verified existence of the second moments of the identically distributed random variables $a_{\nu l}^2$ justifies this statement. The variables $a_{\nu l}^2$ and $a_{\nu,l+i}^2$ are statistically independent for all $i \neq 0, 1$. Consequently, the sums $\sum_{\text{odd } l} a_{\nu l}^2$ and $\sum_{\text{even } l} a_{\nu l}^2$ both are asymptotically Gaussianly distributed, and, since it is the sum of two Gaussian variables, so too is z_ν.

‡In all of these expressions, ω_c is assumed to be either some integer multiple of $2\pi/\Delta T$ or else sufficiently large relative to the signal bandwidth that the carrier can be eliminated by a product demodulator (cf. Section 3.4). This assumption only serves to simplify the resulting expressions; it does not substantially alter the conclusions. If the carrier did not satisfy one of these conditions the synchronization time would presumably decrease because some contending epochs would be more obviously different from the true epoch.

where

$$\Delta_{sxs}^2 = 4(\lambda_\mu - \lambda_\nu)^2[1 - (\lambda_\mu + \lambda_\nu)]^2 \mu_4$$

$$\Delta_{sxn}^2 = 4(\lambda_\mu - \lambda_\nu)[1 - (2\lambda_\nu + \lambda_\mu) + 3\lambda_\nu\lambda_\mu]\frac{\mu_2}{R}$$

$$\Delta_{nxn}^2 = 2(\lambda_\mu - \lambda_\nu)[1 - (\lambda_\mu - \lambda_\nu)]\frac{1}{R^2}$$

and where $\mu \geq \nu$. We again emphasize that these expressions have been derived by assuming M, the number of symbols observed, to be sufficiently large to justify omitting from consideration the partial symbol intervals at the beginning and end of the observation interval.

If all the epochs are tested simultaneously, the probability P_e of an erroneous decision is bounded, in the usual way, by

$$\max_{i\neq 0} \Pr\{z_0 - z_i < 0\} \leq P_e = 1 - \Pr\left\{z_0 > \max_{i\neq 0} z_i\right\}$$
$$\leq \sum_{i=1}^{N-1} \Pr\{z_0 - z_i < 0\} \leq (N-1) \max_{i\neq 0} \Pr\{z_0 - z_i < 0\} \quad (7.3.10)$$

When M is large enough to justify the use of the Gaussian approximation to the distribution of the random variables z_i, the variable $y_i = z_0 - z_i$ is also Gaussianly distributed with the mean

$$m_i = E(y_i) = \eta_0 - \eta_i \quad (7.3.11a)$$

and variance

$$\sigma_i^2 = E(y_i^2) - E^2(y_i) = \Delta_{0i}^2 \quad (7.3.11b)$$

defined in Equations (7.3.8) and (7.3.9), respectively. Paralleling the discussion of Section 6.4, we conclude that if the error probability is to be small, the smallest of the ratios

$$\frac{m_i^2}{\sigma_i^2} = \frac{2M\lambda_i(1-\lambda_i)\mu_2^2}{2\lambda_i(1-\lambda_i)\mu_4 + \dfrac{2\mu_2}{R} + \dfrac{1}{R^2}} \quad (7.3.12)$$

must be of the order of $2\kappa_N \log_e 1/P_e$. [The term κ_N, it will be recalled, is bounded from below by 1 and from above by $1 + \log_e(N-1)/\log_e(1/P_e)$; if, as is usually the case, $P_e \ll 1/N$, κ_N is essentially unity.] This ratio is an increasing function of λ_i for $\lambda_i < \frac{1}{2}$ and a decreasing function of λ_i for $\lambda_i > \frac{1}{2}$. It thus attains its minimum value at $\lambda_i = 1/N$ (and $\lambda_i = 1 - 1/N$). The total number of symbols M_T needed for a decision is consequently

$$M_T = 2N\kappa_N \log_e \frac{1}{P_e}\left[\frac{\mu_4}{N\mu_2^2} + \frac{N}{N-1}\left(\frac{1}{R\mu_2} + \frac{1}{2R^2\mu_2^2}\right)\right] \quad (7.3.13)$$

It is interesting to compare this result with the analogous result of Chapter Six. The dependence of the search time upon N is the same in the two cases when the width of the sync signal autocorrelation function is of the order of $T_s/2$. The presence of both the $1/R$ and $1/R^2$ terms in Equation (7.3.13) recalls the phase-incoherent situation considered in Chapter Six. This is not surprising since in both cases the decision is based on the square of the integral of a signal plus Gaussian noise.

As R approaches infinity the ratio in Equation (7.3.12) becomes $M\mu_2^2/\mu_4$ which suggests that synchronization errors are possible even in the absence of noise. This happens because the end effects become less and less ignorable as $R \to \infty$. Nevertheless, a correct decision can always be made on the basis of the statistic z_ν [Equation (7.3.7)] even after only one symbol interval has been observed when there is no noise. To verify this, it is only necessary to observe that the quantity

$$\int_{\tau_1}^{\tau_2} y^2(t)\,dt - \frac{\left[\int_{\tau_1}^{\tau} \sqrt{2}\,y(t)\sin\omega_c(t-\tau)\,dt\right]^2}{\int_{\tau_1}^{\tau} 2\sin^2\omega_c(t-\tau)\,dt} - \frac{\left[\int_{\tau}^{\tau_2} \sqrt{2}\,y(t)\sin\omega_c(t-\tau)\,dt\right]^2}{\int_{\tau}^{\tau_2} 2\sin^2\omega_c(t-\tau)\,dt} \qquad \tau_1 < \tau < \tau_2$$

is positive unless the signal $y(t)$ is a constant amplitude sinusoid in each of the two intervals of integration; i.e., unless $y(t) = x_1 \sin\omega_c(t-\tau)$, $\tau_1 < t < \tau$, and $y(t) = x_2 \sin\omega_c(t-\tau)$, $\tau < t < \tau_2$.† Thus, the sum of the second two terms attains its maximum value when $\tau = \nu\Delta T$ with ν the correct symbol epoch. This maximum will be unique unless $y(t)$ is constant over the total interval $\tau_1 < t < \tau_2$ which can happen only if two successive symbols are equal (an event of zero probability under the assumptions here) or if the region of integration is totally contained in one symbol interval. Accordingly, a completely reliable decision can be made on the basis of these last two terms (in the absence of noise) so long as the interval of observation is at least T_s seconds. Yet only the numerators of these terms are functions of the received signal, and these numerators are indeed statistics of the form (7.3.7).

When the different epochs cannot be tested simultaneously, the same decision variables z_ν can be used in conjunction with any of the serial tests discussed in Chapter Six. The analyses of these various serial synchronization schemes parallel in all essential details the corresponding analyses in Chapter Six and lead to analogous results.

The maximum-likelihood synchronizer for phase-incoherent PAM, from Equation (7.2.7), determines the epoch ν maximizing

---

†This statement is an immediate consequence of Schwarz's inequality.

$$\prod_l \frac{1}{2\pi}\int_0^{2\pi}\int_{-\infty}^{\infty} \exp\left\{\frac{1}{N_0}\int_{lT_s+\nu\Delta T}^{(l+1)T_s+\nu\Delta T} [2Ax\, y(t)\sqrt{2}\sin(\omega_c t + \phi) - A^2x^2]\, dt\right\} \times p(x)\, dx\, d\phi \quad (7.3.14)$$

[We have made the usual assumptions for incoherent PAM; the received signal is $\sqrt{2}Ax\sin(\omega_c t + \phi)$ with x having the a priori distribution $p(x)$, and with ϕ uniformly distributed in the interval $(0,2\pi)$. Also, $\omega_c = 2\pi k/T_s$ where k is either an integer, or else is large compared to unity.] If we define

$$a_{\nu l} = \frac{1}{AT_s}\int_{lT_s+\nu\Delta T}^{(l+1)T_s+\nu\Delta T} \sqrt{2}\, y(t)\sin\omega_c t\, dt \quad (7.3.15)$$

and

$$b_{\nu l} = \frac{1}{AT_s}\int_{lT_s+\nu\Delta T}^{(l+1)T_s+\nu\Delta T} \sqrt{2}\, y(t)\cos\omega_c t\, dt$$

and proceed as in the phase-coherent case, we can rewrite Equation (7.3.14) in the form

$$\prod_l \frac{1}{2\pi}\int_0^{2\pi} \exp[Rc_{\nu l}^2(\phi)]\int_{-\infty}^{\infty} \exp[-R[x - c_{\nu l}(\phi)]^2]\, p(x)\, dx\, d\phi \quad (7.3.16)$$

where $c_{\nu l}(\phi) = a_{\nu l}\cos\phi + b_{\nu l}\sin\phi$, and R is as previously defined. Again, either by arguing that the integral with respect to x is essentially constant, or by making a joint maximum likelihood decision concerning both the epoch and the symbols [Equation (7.2.8)], we obtain as the quantity from which to determine the epoch

$$K\prod_l \frac{1}{2\pi}\int_0^{2\pi} \exp\left[\frac{R}{2}\zeta_{\nu l}^2(1 + \cos 2(\theta - \phi))\right] d\phi \quad (7.3.17)$$

where $a_{\nu l} = \zeta_{\nu l}\cos\theta$, $b_{\nu l} = \zeta_{\nu l}\sin\theta$, and K is (at least approximately) constant. Carrying out the integration, we obtain (neglecting the constant terms)

$$\prod_l \exp\left(\frac{R}{2}\zeta_{\nu l}^2\right) I_0\left(\frac{R}{2}\zeta_{\nu l}^2\right) \quad (7.3.18)$$

Both $\exp[(R/2)\zeta_{\nu l}^2]$ and $I_0[(R/2)\zeta_{\nu l}^2]$ are monotonic functions of $\zeta_{\nu l}^2$ suggesting that $\zeta_{\nu l}^2$ itself might constitute a satisfactory decision variable. In fact, if $R \ll 1$,

$$e^{(R/2)\zeta_{\nu l}^2} I_0\left(\frac{R}{2}\zeta_{\nu l}^2\right) \sim e^{(R/2)\zeta_{\nu l}^2}\left(1 + \frac{1}{4}\left(\frac{R}{2}\right)^2 \zeta_{\nu l}^4 + \dots\right)$$
$$\sim e^{(R/2)\zeta_{\nu l}^2} \quad (7.3.19)$$

And when $R \gg 1$

$$e^{(R/2)\zeta_{\nu l}^2} I_0\left(\frac{R}{2}\zeta_{\nu l}^2\right) \sim e^{R\zeta_{\nu l}^2}\left(1 + \frac{1}{\frac{R}{2}\zeta_{\nu l}^2} + \dots\right)$$
$$\sim e^{R\zeta_{\nu l}^2} \quad (7.3.20)$$

This last expression would also result if a maximum-likelihood decision were made as to all of the unknown parameters, the x's and ϕ's as well as ν [see Equation (7.2.9)]. Taking the logarithm of either of these expressions, (7.3.19) or (7.3.20), yields as the epoch decision variable

$$z_\nu = \sum_l \zeta_{\nu l}^2 = \sum_l (a_{\nu l}^2 + b_{\nu l}^2) \tag{7.3.21}$$

Although this decision variable is generally suboptimum when the a priori distribution of either x or ϕ, or both, is known, the relative simplicity with which it can be determined in a practical communication system easily justifies its use. Moreover, the preceding arguments suggest that the resulting synchronizer can be expected to perform nearly as well as synchronizers which do presume knowledge of these distributions.

The performance analysis of this synchronizer parallels that of the phase-coherent synchronizer. The only difference, in fact, is that the noise term in the mean, and the noise-cross-noise term in the variance of Equations (7.3.8) and (7.3.9) are doubled over their corresponding values in the phase-coherent case. For future reference we rewrite these equations here:

$$\begin{aligned}
\eta_\nu &= E(z_\nu) = M\left\{[(1-\lambda_\nu)^2 + \lambda_\nu^2]\mu_2 + m^2 + \frac{c_0}{2R}\right\} \\
\sigma_\nu^2 &= M(\sigma_{sxs}^2 + \sigma_{sxn}^2 + \sigma_{nxn}^2) \\
\sigma_{sxs}^2 &= [(1-\lambda_\nu)^2 + \lambda_\nu^2]^2(\mu_4 - \mu_2^2) + 4[\lambda_\nu^2 + (1-\lambda_\nu)^2]m\mu_3 \\
&\quad + 4\lambda_\nu^2(1-\lambda_\nu)^2\mu_2^2/c_1 + 4m^2\mu_2 \\
\sigma_{sxn}^2 &= \{[(1-\lambda_\nu)^2 + \lambda_\nu^2]\mu_2 + m^2\}\frac{2}{R} \\
\sigma_{nxn}^2 &= \frac{c_0}{2R^2} \\
\Delta_{\nu\mu}^2 &= E[(z_\nu - z_\mu)^2] - E^2(z_\nu - z_\mu) \\
&= M(\Delta_{sxs}^2 + \Delta_{sxn}^2 + \Delta_{nxn}^2) \\
\Delta_{sxs}^2 &= 4(\lambda_\mu - \lambda_\nu)^2[1 - (\lambda_\mu + \lambda_\nu)]^2\mu_4/c_1 \\
\Delta_{sxn}^2 &= 4(\lambda_\mu - \lambda_\nu)[1 - (2\lambda_\nu + \lambda_\mu) + 3\lambda_\mu\lambda_\nu]\frac{\mu_2}{R} \\
\Delta_{nxn}^2 &= 2c_0(\lambda_\mu - \lambda_\nu)[1 - (\lambda_\mu - \lambda_\nu)]\frac{1}{R^2}
\end{aligned} \tag{7.3.22}$$

where $\mu \geq \nu$. The term c_0 is equal to 1 in the coherent case and to 2 in the incoherent case; c_1 is 1 in both cases. The reason for introducing this latter term will become apparent shortly.

7.4 Phase Shift Keying (PSK) Symbol Synchronization

The system to be considered here is the one discussed in Section 4.5 (and Section 3.7) in which any of the n waveforms $x_i(t) = \sqrt{2}A\sin(\omega_c t + \phi + \theta_i)$, with ϕ known and $\theta_i = 2\pi i/n$, $i = 1, 2, \ldots, n$, is equally likely to be transmitted over any given T_s second symbol interval. From Equation (7.2.3) we conclude that the most likely symbol epoch corresponds to that value of ν maximizing the expression

$$\prod_l \frac{1}{n}\sum_{j=1}^{n} \exp\left\{\frac{2A}{N_0}\int_{lT_s+\nu\Delta T}^{(l+1)T_s+\nu\Delta T} y(t)\sqrt{2}\sin(\omega_c t + \phi + \theta_j)\,dt\right\}$$
$$= \prod_l \frac{1}{n}\sum_{j=1}^{n} \exp\{2R(a_{\nu l}\cos\theta_j + b_{\nu l}\sin\theta_j)\} \qquad (7.4.1)$$

where

$$a_{\nu l} = \frac{1}{AT_s}\int_{lT_s+\nu\Delta T}^{(l+1)T_s+\nu\Delta T} y(t)\sqrt{2}\sin(\omega_c t + \phi)\,dt$$
$$b_{\nu l} = \frac{1}{AT_s}\int_{lT_s+\nu\Delta T}^{(l+1)T_s+\nu\Delta T} y(t)\sqrt{2}\cos(\omega_c t + \phi)\,dt$$

and $R = A^2T_s/N_0$.

Again, because of the difficulties in implementing the synchronizer suggested by Equation (7.4.1) (to say nothing of the difficulties in analyzing its performance), it is of interest to develop some approximations to this ideal. First consider the low signal-to-noise ratio approximation. Expanding the exponential term in Equation (7.4.1) in a power series yields

$$\prod_l \frac{1}{n}\sum_{j=1}^{n} \exp\{2R(a_{\nu l}\cos\theta_j + b_{\nu l}\sin\theta_j)\}$$
$$= \prod_l \frac{1}{n}\sum_{j=1}^{n} [1 + 2Ra_{\nu l}\cos\theta_j + 2Rb_{\nu l}\sin\theta_j + 2R^2 a_{\nu l}^2\cos^2\theta_j$$
$$+ 4R^2 a_{\nu l}b_{\nu l}\sin\theta_j\cos\theta_j + 2R^2 b_{\nu l}^2\sin^2\theta_j + O(R^3)] \qquad (7.4.2)$$

Since $\theta_j = 2\pi j/n, j = 1, 2, \ldots, n$,

$$\sum_{j=1}^{n}\cos k\theta_j = \sum_{j=1}^{n}\sin k\theta_j = 0$$

for any integer k which is not a multiple of n. Equation (7.4.2) therefore becomes

$$\prod_l \left[1 + \frac{1+\delta_{n2}}{2}2R^2 a_{\nu l}^2 + \frac{1-\delta_{n2}}{2}2R^2 b_{\nu l}^2 + O(R^3)\right]$$
$$\approx \begin{cases} K_1 \sum_l (a_{\nu l}^2 + b_{\nu l}^2) & n > 2 \\ K_2 \sum_l a_{\nu l}^2 & n = 2 \end{cases} \qquad (7.4.3)$$

where $\delta_{n2} = 1$ when $n = 2$ and zero otherwise, and where K_1 and K_2 are independent of ν. This approximation is valid for continuous PSK as well, the summation over j becoming an integration over θ.

The performance analysis of this approximation, Equation (7.4.3), to the maximum-likelihood synchronizer is entirely analogous to the preceding analysis for PAM. In fact, when $n = 2$, ($z_\nu = \sum_l a_{\nu l}^2$), the first two moments of z_ν are identical to those of the phase-coherent PAM decision variable [Equation (7.3.22), with $m = 0$, $\mu_2 = \mu_4 = c_0 = c_1 = 1$]. When $n > 2$ ($z_\nu = \sum_l (a_{\nu l}^2 + b_{\nu l}^2)$), the first two moments of z_ν are also as given in Equation 7.3.22, but now with $m = 0$, $\mu_2 = \mu_4 = 1$, and $c_0 = c_1 = 2$.† Thus, a reliable (parallel observation) decision concerning the correct epoch will be made if the number of symbols observed is of the order of

$$M_T \approx 2NK_N \log_e \frac{1}{P_e}\left(\frac{1}{N} + \frac{1}{R} + \frac{1}{2R^2}\right) \tag{7.4.4a}$$

when $n = 2$, and

$$M_T \approx 2NK_N \log_e \frac{1}{P_e}\left(\frac{1}{2N} + \frac{1}{R} + \frac{1}{R^2}\right) \tag{7.4.4b}$$

when $n > 2$ [cf. Equation (7.3.13)].

Interestingly, the optimum low signal-to-noise ratio synchronizer for differentially coherent PSK signals is equivalent (except when $n = 2$) to the one just derived for the coherent case. In the absence of a phase reference, the expression in Equation (7.4.1) is replaced by

$$\prod_l \frac{1}{n} \sum_{j=1}^{n} \frac{1}{2\pi} \int_0^{2\pi} \exp\{2R(a_{\nu l}\cos(\theta_j + \phi) + b_{\nu l}\sin(\theta_j + \phi))\}\, d\phi$$
$$= \prod_l I_0(2R(a_{\nu l}^2 + b_{\nu l}^2)^{1/2}) \tag{7.4.5}$$

where $a_{\nu l}$ and $b_{\nu l}$ are as previously defined [Equation (7.4.1)] but with the local phase ϕ now arbitrary. Using the small argument expansion of the Bessel function, as in Section 7.3, we obtain as a low signal-to-noise ratio approximation to Equation (7.4.5),

$$\prod_l [1 + R^2(a_{\nu l}^2 + b_{\nu l}^2) + O(R^4)] \tag{7.4.6}$$

which suggests that the decision as to the correct symbol epoch should be based on the terms

$$z_\nu = \sum_l (a_{\nu l}^2 + b_{\nu l}^2) \tag{7.4.7}$$

regardless of the value of n. Except when $n = 2$, this is identical with the conclusion reached in the coherent case. And, except for $n = 2$, the corresponding synchronizers will yield identical performance. The performance of the binary differentially coherent synchronizer is clearly equivalent to that of

---

†The successive phases are here assumed statistically independent and uniformly distributed. See Problem 7.1 for the modification in these results when this latter restriction is removed.

the binary phase-incoherent PAM synchronizer discussed in the previous section, with $m = 0$, and $\mu_2 = \mu_4 = 1$.

Actually, it should not be surprising that the phase-coherent and differentially coherent PSK synchronizers are identical. In either situation, the symbol epoch information is conveyed by the phase changes from one symbol to the next; the synchronizer performance depends upon the phase differences, not upon the specific phase itself.

At the other extreme, when the signal-to-noise ratio R is large compared to unity, the term in the summation of Equation (7.4.1) for which the quantity $(a_{vl} \cos \theta_j + b_{vl} \sin \theta_j)$ attains its maximum can be expected to dominate the remainder of this summation. Under this assumption, the logarithm of Equation (7.4.1) is proportional to

$$\sum_l \max_{j_l} \{a_{vl} \cos \theta_{j_l} + b_{vl} \sin \theta_{j_l}\} \tag{7.4.8}$$

This expression would also have been obtained had we attempted a joint decision concerning the epoch and the symbols [Equation (7.2.4)]. The θ_j maximizing the term $a_{vl} \cos \theta_j + b_{vl} \sin \theta_j$ is θ_k where

$$|\theta - \theta_k| \leq \frac{\pi}{n}$$

$$\theta = \tan^{-1} \frac{b_{vl}}{a_{vl}}$$

(see Section 4.5). If n is large $\theta \approx \theta_k$ for some k and

$$\max_j \{a_{vl} \cos \theta_j + b_{vl} \sin \theta_j\} = (a_{vl}^2 + b_{vl}^2)^{1/2} \cos(\theta - \theta_k) \approx (a_{vl}^2 + b_{vl}^2)^{1/2} \tag{7.4.9}$$

In contrast, if $n = 2$ (with $\theta_j = 0$ or π)

$$\max_j \{a_{vl} \cos \theta_j + b_{vl} \sin \theta_j\} = |a_{vl}| \tag{7.4.10}$$

The differentially coherent synchronizer of Equation (7.4.5) can be approximated, when R is large, by using the asymptotic expansion of the Bessel function:

$$\prod_l I_0[2R(a_{vl}^2 + b_{vl}^2)^{1/2}] \sim \prod_l \exp\{2R(a_{vl}^2 + b_{vl}^2)^{1/2}\} \tag{7.4.11}$$

This same expression results when a joint maximum-likelihood estimate is made of the phase ϕ and the symbol j:

$$\prod_l \max_{j_l \phi_l} \exp\{2R[a_{vl} \cos(\theta_{j_l} + \phi_l) + b_{vl} \sin(\theta_{j_l} + \phi_l)]\}$$

$$= \prod_l \exp\{2R(a_{vl}^2 + b_{vl}^2)^{1/2}\} \tag{7.4.12}$$

In either event the decision should be based on the terms

$$z'_v = \sum (a_{vl}^2 + b_{vl}^2)^{1/2} \tag{7.4.13}$$

The best approximation to the maximum-likelihood synchronizer evidently does depend upon both the signal-to-noise ratio and the number of possible phases. As in Section 6.6, however, we can quite reasonably justify limiting our attention to the low signal-to-noise ratio approximations. This, after all, represents the most difficult synchronization situation. Moreover, the performance of such a synchronizer provides a lower bound on the performance of the ideal synchronizer. Finally, and perhaps most important, the high signal-to-noise ratio performance of these approximations to the ideal synchronizer is, so far as the effects of the noise are concerned, virtually identical to that of the optimum synchronizers in the absence of any modulation. There is little margin left for improvement. To support this contention we need only compare Equations (7.4.4) and (7.3.13) with Equation (6.4.5), noting that, in the present situation, $\max_i \rho(iT_s/N) = 1 - 1/N$.

7.5 Synchronization of Orthogonal Symbols

In the preceding several sections, the approach to the symbol synchronization problem outlined in the introduction to this chapter was applied specifically to PAM and PSK communication systems. This same approach is equally applicable to the orthogonal (and biorthogonal) communication systems discussed in Chapter Four. The practicality of this approach depends upon the particular orthogonal symbol set being used. In later chapters, orthogonal symbol constructions will be presented which lend themselves to efficient synchronization techniques much more readily than either of the two constructions (PPM and FSK) considered in Chapter Four. Consequently, the discussion of the synchronization problem as applied to those two classes of symbol sets will be kept brief.

7.5.1 FSK Symbol Synchronization

When the set of symbols is

$$x_i(t) = \sqrt{2}\sin\omega_i t \qquad 0 < t < T_s \qquad i = 1, 2, \ldots, n \tag{7.5.1}$$

where $\omega_i = \pi k_i/T_s$, for some integer k_i (cf. Section 4.4), the maximum-likelihood phase-coherent synchronizer statistics [Equation (7.2.3)] take the form

$$\prod_l \sum_{j=1}^{n} \exp\,[2Rz_{\nu l}(j)]\,P(j) \tag{7.5.2}$$

where

$$z_{\nu l}(j) = \frac{1}{AT_s}\int_{lT_s+\nu\Delta T}^{(l+1)T_s+\nu\Delta T} \sqrt{2}\,y(t)\sin\omega_j(t - lT_s - \nu\Delta T)\,dt$$

and $R = A^2T_s/N_0$. Again, those terms independent of ν have been omitted.

Presumably, the communication system will operate at a signal-to-noise ratio R sufficiently large to yield a small synchronous error probability. Consequently, one (or two, when $\nu \neq 0$) of the terms of the summation in Equation (7.5.2) will tend to dominate. This suggests making a joint maximum-likelihood estimate of the epoch and the symbols and hence basing the decision on the sum

$$z_\nu \triangleq \sum_l z_{\nu l} \triangleq \sum_l \max_{j_l} z_{\nu l}(j_l) \tag{7.5.3}$$

As in the preceding sections, we can estimate the number of observables needed for a decision by requiring the ratio

$$r^2 = \min_\nu \frac{E^2(z_0 - z_\nu | 0)}{\mathrm{Var}(z_0 - z_\nu | 0)} \tag{7.5.4}$$

to be of the order of $2\kappa_N \log_e (1/P_e)$.

Letting $y(t) = \sqrt{2}A \sin \omega_r(t - lT_s) + n(t)$, $lT_s < t < (l+1)T_s$, and $y(t) = \sqrt{2}A \sin \omega_s[t - (l+1)T_s] + n(t)$, $(l+1)T_s < t < (l+2)T_s$, and neglecting the double-frequency terms, yield

$$z_{\nu l}(j) = (-1)^{k_r} \frac{N-\nu}{N} \frac{\cos \frac{\omega_r + \omega_j}{2}(N-\nu)\Delta T \sin \frac{\omega_r - \omega_j}{2}(N-\nu)\Delta T}{\frac{\omega_r - \omega_j}{2}(N-\nu)\Delta T}$$

$$+ (-1)^{k_j} \frac{\nu}{N} \frac{\cos \frac{\omega_s + \omega_j}{2}\nu\Delta T \sin \frac{\omega_s - \omega_j}{2}\nu\Delta T}{\frac{\omega_s - \omega_j}{2}\nu\Delta T} + n_j \tag{7.5.5}$$

with

$$n_j = \frac{1}{AT_s} \int_{lT_s + \nu\Delta T}^{(l+1)T_s + \nu\Delta T} \sqrt{2}\, n(t) \sin \omega_j(t - lT_s - \nu\Delta T)\, dt$$

Consider first the special case in which $k_j = 2N(k+j)$ with k some arbitrary integer. This, of course, requires the frequencies to be separated by a considerably greater amount than that needed simply to assure orthogonality. There are situations in which such a condition would be quite acceptable. In any case, this restriction will provide an indication of the synchronizability of FSK signals under more general conditions. When $k_j = 2N(k+j)$

$$z_{\nu l}(j) = \left(1 - \frac{\nu\Delta T}{T_s}\right)\delta_{rj} + \frac{\nu\Delta T}{T_s}\delta_{sj} + n_j \tag{7.5.6}$$

where δ_{ij} is the Kronecker delta function

$$\delta_{ij} = \begin{cases} 1 & i = j \\ 0 & i \neq j \end{cases}$$

We further simplify the analysis by assuming the signal-to-noise ratio to be sufficiently large that the largest correlator output will, with high probability, come from the correlator having the largest *expected* output. Under this assumption, when $\nu \leq N/2$, and when Equation (7.5.6) is applicable, the numerator of Equation (7.5.4) is

$$M^2\left(\frac{n-1}{n}\right)^2 \frac{\nu^2}{N^2} \tag{7.5.7}$$

while the denominator is

$$M\frac{\nu}{N}\frac{n-1}{n}\left[\frac{1}{R}+\frac{1}{n}\frac{\nu}{N}^2\right] \tag{7.5.8a}$$

when the epochs are tested simultaneously, and

$$M\left[\frac{1}{R}+\frac{n-1}{n^2}\left(\frac{\nu}{N}\right)^2\right] \tag{7.5.8b}$$

when they are tested serially. (The successive symbols are assumed independent and uniformly distributed; when $\nu > N/2$, modulo N, replace ν in these expressions by $N-\nu$.) The worst case again occurs for $\nu = 1$ (or $N-1$), so the number of symbols needed for a parallel observation decision is

$$M_T \approx 2\kappa_N \log_e \frac{1}{P_e}\left[\frac{N}{R}\left(\frac{n}{n-1}\right)+\frac{1}{n-1}\right] \tag{7.5.9}$$

Similarly, in the case of fixed-sample-size serial search, the total number of symbols which must be observed is approximately

$$M_T \approx 2\kappa_N'' \log_e \frac{1}{P_e}\left[\left(\frac{n}{n-1}\right)^2\frac{N^3}{R}+\frac{N}{n-1}\right] \tag{7.5.10}$$

The order of magnitude performance resulting from this approach, therefore, is the same as already encountered with the PAM and PSK synchronizers. (The $1/R^2$ terms are not present here, but then we began by assuming a large signal-to-noise ratio R.) The other methods discussed in the previous sections (estimation, optimum deferred decision, etc.) are also applicable here, with, as is readily verified, the same relative performance as previously observed.

The same conclusions hold as well for the large signal-to-noise ratio phase-incoherent FSK synchronizer. If a joint maximum-likelihood decision is made as to both the received symbol and its phase, the resulting synchronizer will base its decision on the function [see Equation (7.2.9)]

$$\begin{aligned}\log_e \prod_l \max_{j_l, \phi_l} \exp\{2R(a_{j_l}(\nu)\cos\phi_l + b_{j_l}(\nu)\sin\phi_2)\} \\ = 2R\sum_l \max_{j_l}(a_{j_l}^2(\nu)+b_{j_l}^2(\nu))^{1/2} \triangleq 2R\zeta_\nu\end{aligned} \tag{7.5.11}$$

where

$$\left.\begin{matrix}a_{j_l}(\nu)\\ b_{j_l}(\nu)\end{matrix}\right\} = \frac{1}{AT_s}\int_{lT_s+\nu\Delta T}^{(l+1)T_s+\nu\Delta T} y(t)\sqrt{2}\begin{Bmatrix}\sin\omega_j(t-\nu\Delta T)\\ \cos\omega_j(t-\nu\Delta T)\end{Bmatrix} dt$$

[This same expression also results upon taking the large signal-to-noise ratio approximation to the synchronizer described by Equation (7.2.8).] Such a device can be implemented, for example, by a bank of phase-incoherent matched filters, one centered at each of the frequencies $\omega_i/2\pi$, and each followed by an envelope detector (cf. Section 4.3).

Under the large signal-to-noise ratio assumption being made here, both the expected value of the output of any of the envelope detectors, and the variance of this output, are approximately equal to these same quantities at the output of the corresponding phase-coherent matched filter [cf. Equation (3.6.14)]. Accordingly, the performance of the phase-incoherent FSK synchronizer is nearly identical to that of the phase-coherent synchronizer of the preceding paragraphs, under the same conditions.

The dependence of this discussion on the assumption that the detector with the largest expected output will, with high probability, be the detector with the largest actual output should be emphasized. For values of ν near zero or N, this assumption is equivalent to a small synchronous error probability requirement. However, as ν approaches $N/2$, two phenomena alter this situation: (1) There are two (or more) detectors rather than one having relatively large expected outputs [see Equation (7.5.6)]; and (2) the probability increases that the output of one of the detectors with a zero expected value actually exceeds one of those having a non-zero expected value. Both of these phenomena tend to increase the expected value of the maximum output beyond the maximum of the expected values of these outputs. The effect of these biases is to decrease the reliability with which some of the epochs can be dismissed. This will not greatly alter the search time in the parallel or fixed-sample-size serial search modes, since these must be of sufficient duration to dismiss reliably the most difficult epoch (which will still presumably be the epoch $\nu = 1$), regardless of the epoch actually being observed. Those search procedures which realize savings because some epochs can be more easily rejected than others, however, do lose their effectiveness when these biases are taken into account.

These same observations hold, but with even more severe limitations on the signal-to-noise ratio, when the integers k_j [Equation (7.5.1)] are not necessarily of the form $k_j = 2N(k + j)$. In this case, the absolute value of the expected output of the jth (phase-coherent) detector is bounded by [see Equation (7.5.5)]

$$\frac{N-\nu}{N}\left|\frac{\sin\frac{\omega_r - \omega_j}{2}(N-\nu)\Delta T}{\frac{\omega_r - \omega_j}{2}(N-\nu)\Delta T}\right| + \frac{\nu}{N}\left|\frac{\sin\frac{\omega_s - \omega_j}{2}\nu\Delta T}{\frac{\omega_s - \omega_j}{2}\nu\Delta T}\right| \leq 1 \qquad (7.5.12)$$

the strict inequality holding for any $\nu \neq 0$ unless $j = r = s$. Consequently, for sufficiently high signal-to-noise ratios, the zeroth epoch can still be distinguished from all other epochs. Now, however, because more of the detectors will exhibit non-zero expected outputs, the performance of this synchronization scheme will deteriorate even more rapidly as the signal-to-noise ratio is decreased. Although this performance can be bounded we shall limit the discussion here to the above observations concerning the qualitative behavior of these techniques for synchronizing orthogonal FSK signals. As we have already remarked, when synchronization is to be accomplished from properties of the symbols themselves, considerably more efficacious orthogonal symbol sets exist. These will be discussed in detail in later chapters.

7.5.2 PPM Symbol Synchronization

The symbol set of interest here is comprised of symbols of the form

$$x_i(t) = \sqrt{2n}\,A \sin \omega_c t \qquad \frac{(i-1)T_s}{n} < t < \frac{iT_s}{n} \qquad i = 1, 2, \ldots, n \tag{7.5.13}$$

with $\omega_c = 2\pi kn/T_s$ for some integer k. Each symbol is of T_s seconds duration while the non-zero amplitude interval, which we will denote the *sub-symbol* interval, lasts T_s/n seconds. One synchronization strategy is therefore to limit the task initially to the identification of the n epochs at which the symbol amplitude can change from a zero to a non-zero value, or vice versa. The second step is then to determine which of these n remaining epochs is actually the correct symbol epoch. If equipment limitations do not preclude the possibility, it is of course potentially faster to combine these two operations. Nevertheless, since the two-stage strategy is clearly the more practical of the two approaches, the discussion here is restricted to the problem of identifying the sub-symbol epoch; i.e., of establishing sub-symbol synchronization. The second stage of the procedure will not be examined here since it belongs to the class of problems to be considered in Chapter Fourteen.

To determine the form of the maximum-likelihood sub-symbol synchronizer, we treat the sub-symbols as though they, themselves, were symbols. In so doing we shall ignore the dependence between successive sub-symbols. The information, then, which is to be provided the synchronizer is that the lth received sub-symbol has the form

$$y(t) = \alpha_l \sqrt{2n}\,A \sin \omega_c (t - \nu\Delta T) + n(t)$$
$$l\frac{T_s}{n} + \nu\Delta T < t < (l+1)\frac{T_s}{n} + \nu\Delta T \tag{7.5.14}$$

where $n(t)$ is white Gaussian noise. The density function of the random variable α_l is

$$p(\alpha_l) = \frac{1}{n}\delta(\alpha_l - 1) + \frac{n-1}{n}\delta(\alpha_l) \tag{7.5.15}$$

$\delta(x)$ denoting the Dirac delta function. The maximum-likelihood synchronizer, Equation (7.2.3), determines the maximum over $\nu(\nu = 0, 1, \ldots, (N/n) - 1)$ of the function

$$\prod_l \int_{-\infty}^{\infty} \exp\,[2Rz_{\nu l}\alpha_l - R\alpha_l^2]p(\alpha_l)\,d\alpha_l \tag{7.5.16}$$

where

$$z_{\nu l} = \frac{1}{AT_s}\int_{l(T_s/n)+\nu\Delta T}^{(l+1)(T_s/n)+\nu\Delta T} y(t)\sqrt{2n}\sin\omega_c(t - \nu\Delta T)\,dt$$

and of course, $R = A^2T_s/N_0$. Substituting from Equation (7.5.15) and integrating yields

$$\prod_l\left[\frac{n-1}{n} + \frac{1}{n}e^{2R(z_{\nu l}-1/2)}\right] = \prod_l\left\{\frac{1}{n}\left[e^{2R(z_{\nu l}-1/2)} - 1\right] + 1\right\} \tag{7.5.17}$$

as the desired statistic.

If n and R are both large as compared to unity, and again R must be of the order of $\log_2 n$ if the error probability is to be small, the terms $z_{\nu l}$ are significant, so far as the synchronizer is concerned, only if they exceed $\frac{1}{2}$. This suggests first deciding whether or not a sub-symbol is even present before allowing the corresponding statistic to influence the decision concerning the epoch. If a joint maximum-likelihood decision were to be made as to both the epoch and the sub-symbol, for example, the synchronizer would determine the value of ν maximizing the equation

$$\prod_l \max_{\alpha_l} \exp\,[2Rz_{\nu l}\alpha_l - R\alpha_l^2] = \prod_{l\in L}\exp\left[2R\left(z_{l\nu} - \frac{1}{2}\right)\right] \tag{7.5.18}$$

where L is the set of subscripts l for which $z_{\nu l} > \frac{1}{2}$. If the decision as to the presence or absence of a sub-symbol is made instead on the basis of the a posteriori probability of this event, the same expression results but with L redefined so as to include only those subscripts for which $z_{\nu l} > \frac{1}{2} + (1/2R)\log_e(n-1)$. In any event, the decision can be based on a set of statistics of the form

$$z_\nu = \sum_{l\in L}\left(z_{\nu l} - \frac{1}{2}\right) \tag{7.5.19}$$

If the synchronous error probability is small, the term $z_{\nu l}$ will generally not contribute to the sum z_ν unless a non-zero sub-symbol pulse is actually present. Under this assumption, the number of symbols required for a decision is easily seen to be essentially that needed to identify the correct one of N/n contending epochs when a pulse of width T_s/n is transmitted every T_s seconds over a separate sync channel, as discussed in Chapter Six. Similar conclusions hold, subject to the same conditions, for phase-incoherent PPM synchronizers.

7.6 Tracking Symbol Synchronizers

The investigation up to now has been concerned with the initial determination of the symbol epoch. Since all of the suggested schemes for accomplishing this tacitly assume knowledge of the received symbol period, either this period must be quite stable, or else the receiver clock must be slaved to that of the transmitter (cf. Chapter Six). In the latter case particularly, therefore, symbol synchronization is in principle completed once the symbol epoch has been correctly ascertained; any subsequent fluctuations in the symbol epoch will be reflected in the receiver clock. Nevertheless, it is often advantageous to be able to track variations in the symbol epoch directly, without relying on an auxiliary clock. This is the subject of interest here.

The maximum-likelihood decision concerning the symbol epoch $\tau \triangleq \nu\Delta T$, given the observation $y(t)$, was found in Section 7.2 to involve density functions typically of the form

$$p[y_l(t)\,|\,\tau] = K\left(\sum_{j=1}^{n} \exp\left\{\frac{1}{N_0}\int_{lT_s+\tau}^{(l+1)T_s+\tau} [2y(u)x_j(u-\tau) - x_j^2(u-\tau)]\,du\right\}P(j)\right) \tag{7.6.1}$$

The epoch detectors of the past sections of this chapter functioned by finding the maximum over τ of the quantity

$$p[y(t)\,|\,\tau] = \prod_{l=0}^{M-1} p[y_l(t)\,|\,\tau] \tag{7.6.2}$$

The alternative to be considered in this section is to use a weighted sum of the derivatives

$$\frac{\partial \log_e p[y_l(t)\,|\,\tau]}{\partial\tau} \tag{7.6.3}$$

to control the current estimate of τ. Since this expression will be zero when τ equals the true symbol epoch τ_0, and since it is a monotonically decreasing function of τ in the vicinity of $\tau = \tau_0$, a feedback device of the sort discussed in Chapter Five should be able to force τ to converge to the maximum-likelihood estimate of τ_0 (cf. Section 5.4). The description and analysis of such symbol tracking devices are the subjects of this section. We begin with the coherent PAM and bi-phase PSK situation.

7.6.1 *Coherent PAM and Bi-phase PSK*

The function $\log_e p(y_l(t)\,|\,\tau)$ was approximated in Section 7.3 by the quantity

$$Ka_l^2(\tau) = K\left[\frac{1}{AT_s}\int_{lT_s+\tau}^{(l+1)T_s+\tau} y(u)\sqrt{2}\,\sin(\omega_c u + \phi)\,du\right]^2 \triangleq Kz[(l+1)T_s + \tau] \tag{7.6.4}$$

This suggests approximating the derivative $\partial \log_e p[y_i(t) \mid \tau]/\partial\tau$ by the function

$$\frac{z(t) - z(t - \Delta t)}{\Delta t} \tag{7.6.5}$$

The resulting symbol-tracking device is as shown in Figure 7.1. The output of the box labeled "signal generator" is the periodic sampling function $s(t) = \sum_i [\delta(t - \Delta t/2 - iT_s) - \delta(t + \Delta t/2 - iT_s)]$ with $\delta(t)$ the Dirac delta function. The loop filter then serves to weight the successive inputs $z(t) - z(t - \Delta t)$, $t = \tau + l'T_s$, $l' = l, l - 1, l - 2, \ldots$, thereby establishing the error signal over the interval $lT_s < t < (l + 1)T_s$, as discussed in Section 5.4. (The term $1/\Delta t$ is absorbed in the loop gain constant.)

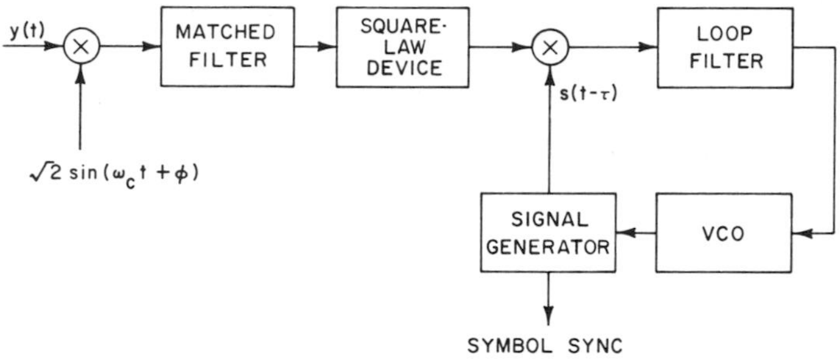

Figure 7.1 Symbol synchronizer–coherent PAM and bi-phase PSK.

Analysis of the performance of this device is facilitated upon recognizing its equivalence to the generalized phase-locked loop discussed in Section 5.8. It was shown there that the mean-squared tracking error associated with such a device in the absence of any transient phase perturbations or oscillator jitter is a function only of the effective signal amplitude, the zero-frequency noise spectral density $S_n(0)$ at the loop filter input (assuming, as usual, that the loop bandwidth is narrow relative to the pre-loop bandwidth), and the loop noise bandwidth B_L.

The signal† at the loop input is

$$E[z(t)] \triangleq \eta(t) \tag{7.6.6}$$

(With no loss of generality, we let $\tau_0 = 0$ denote the true symbol epoch.) The effective amplitude, therefore, is

$$A_e = \left| \frac{1}{2\pi} \int_0^{T_s} \eta(t) s'(t)\, dt \right| = \left| \frac{1}{2\pi} \int_0^{T_s} \eta'(t) s(t)\, dt \right| \tag{7.6.7}$$

†The signal amplitude at any instant of time is now a random variable. Again, however, if the loop bandwidth is narrow relative to $1/T_s$ the VCO output will be determined by a weighted sum of a large number of input pulses. Thus, to a good approximation the signal at the loop input can be replaced by its expected value.

(cf. Section 5.8. Since it does not significantly complicate the subsequent analysis and since the resulting generalization is of some practical interest, the feedback signal $s(t)$ will for the present be allowed to be any periodic function of t such that $\int_0^{T_s} s(t)\,dt = 0$.) Note that $z(t)$ is identical to the z_ν of Equation (7.3.7), with $\nu\Delta T = \tau$ and with $M = 1$. Thus, $\eta(t)$ equals the η_ν of Equation (7.3.8) under the same conditions.

The remainder of the input

$$\zeta(t) = z(t) - \eta(t) \tag{7.6.8}$$

represents noise. The noise at the input to the loop filter is therefore

$$n(t) = \zeta(t)s(t - \tau) \tag{7.6.9}$$

Further,

$$\begin{aligned} S_n(0) &= \int_{-\infty}^{\infty} R_n(u)\,du = \sum_{i=-\infty}^{\infty} \int_0^{T_s} R_n(u + iT_s)\,du \\ &= \frac{1}{T_s}\sum_{i=-\infty}^{\infty} \int_0^{T_s}\int_0^{T_s} R_n(u + iT_s, t)\,dt\,du \end{aligned} \tag{7.6.10}$$

where

$$R_n(u, t) \triangleq E[n(t)n(t + u)]$$

with the operator $E(\cdot)$ representing an ensemble average. Since both $R_\zeta(u, t) \triangleq E[\zeta(t)\zeta(t + u)]$ and $s(t)$ are periodic functions of t, with period T_s, $S_n(0)$ can be rewritten in the form

$$S_n(0) = \frac{1}{T_s}\sum_{i=-\infty}^{\infty} \int_0^{T_s}\int_0^{T_s} R_\zeta(u + iT_s, t)s(t - \tau)s(t + u - \tau)\,dt\,du \tag{7.6.11}$$

The signal $s(t)$ was treated as deterministic in this last manipulation, although its phase is randomly varying. This is justified, however, if the loop bandwidth is narrow relative to the bandwidth of the received signal (cf. Section 5.8).

Now consider the function

$$\begin{aligned} \Delta^2(u, t) &\triangleq \lim_{M\to\infty} \frac{1}{M} E\Big(\sum_{i=1}^{M} [\zeta(t + iT_s) - \zeta(t + u + iT_s)]\Big)^2 \\ &= \lim_{M\to\infty} \frac{1}{M}\sum_{i=1}^{M}\sum_{j=1}^{M} \{R_\zeta[(j - i)T_s, t] - 2R_\zeta[u + (j - i)T_s, t] \\ &\quad + R_\zeta[(j - i)T_s, t + u]\} \\ &= \lim_{M\to\infty} \sum_{\nu=-(M-1)}^{M-1} \Big(1 - \frac{|\nu|}{M}\Big)[R_\zeta(\nu T_s, t) - 2R_\zeta(u + \nu T_s, t) \\ &\quad + R_\zeta(\nu T_s, t + u)] \\ &= \sum_{\nu=-\infty}^{\infty} [R_\zeta(\nu T_s, t) - 2R_\zeta(u + \nu T_s, t) + R_\zeta(\nu T_s, t + u)] \end{aligned} \tag{7.6.12}$$

This last expression is valid so long as $\lim_{\nu\to\infty} R_\zeta(\nu T_s, t) \to 0$. Combining

Equations (7.6.11) and (7.6.12), and again recalling that both $R_\zeta(\nu T_s, t)$ and $s(t)$ are periodic in t, and that $\int_0^{T_s} s(t)\,dt = 0$, we obtain

$$S_n(0) = -\frac{1}{2T_s}\int_0^{T_s}\int_0^{T_s} \Delta^2(u, t)s(t-\tau)s(t+u-\tau)\,dt\,du \tag{7.6.13}$$

The utility of this expression for $S_n(0)$ becomes apparent upon observing that

$$\Delta^2(u, t) = \lim_{M\to\infty}\frac{1}{M}\Delta^2_{\nu\mu} \tag{7.6.14}$$

where $\Delta^2_{\nu\mu}$ is as defined in Equation (7.3.9) with $\nu\Delta T = t$, $\mu\Delta T = t + u$, for $0 \leq t \leq u + t \leq T_s$ and $\nu\Delta T = t + u - T_s$, $\mu\Delta T = t$, for $0 \leq t < T_s < u + t < T_s + t$. Note that $\Delta^2(u, t)$ is periodic in t.

Returning to the symbol tracking device of Figure 7.1, in which $s(t) = \sum_i [\delta(t - (\Delta t/2) - iT_s) - \delta(t + (\Delta t/2) - iT_s)]$, we find, from Equation (7.3.8), that the expected value of the error signal at the loop filter input has the form

$$\eta\left(t + \frac{\Delta t}{2}\right) - \eta\left(t - \frac{\Delta t}{2}\right) = \begin{cases} -2\mu_2\left(1 - \dfrac{\Delta t}{T_s}\right)\dfrac{2\tau}{T_s} & 0 \leq |\tau| \leq \dfrac{\Delta t}{2} \\ -2\mu_2\dfrac{\Delta t}{T_s}\left(1 - \dfrac{2\tau}{T_s}\right) & \dfrac{\Delta t}{2} \leq \tau \leq T_s - \dfrac{\Delta t}{2} \end{cases} \tag{7.6.15}$$

with $t = \tau$, modulo T_s. This signal is plotted in Figure 7.2. From Equation (7.6.7)

$$A_e = \frac{2\mu_2}{\pi T_s}\left(1 - \frac{\Delta t}{T_s}\right) \tag{7.6.16}$$

From Equation (7.6.13)

$$S_n(0) = \frac{\Delta^2\left(\Delta t, \tau - \dfrac{\Delta t}{2}\right)}{T_s} \tag{7.6.17}$$

Although the noise spectral density is dependent on the sampling instant τ, it is a relatively slowly varying function of this parameter. In any event, if the loop is tracking properly, $\tau \approx 0$ and, from Equation 7.6.14,

$$\begin{aligned} S_n(0) &\approx \frac{\Delta^2\left(\Delta t, -\dfrac{\Delta t}{2}\right)}{T_s} \\ &= \frac{2\Delta t}{T_s^2}\left(1 - \frac{\Delta t}{T_s}\right)\left[\left(2 - \frac{3\Delta t}{2T_s}\right)\frac{\mu_2}{R} + \frac{1}{R^2}\right] \qquad 0 \leq \Delta t \leq \frac{T_s}{2} \end{aligned} \tag{7.6.18}$$

Finally, the mean-squared tracking error in radians squared (with 2π

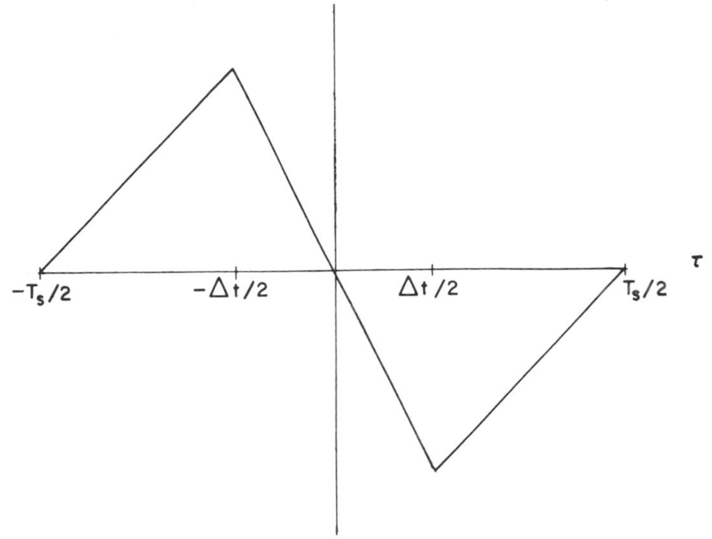

Figure 7.2 Error signal for synchronizer of Figure 7.1 with sampling function feedback.

radians corresponding to T_s seconds) is

$$\begin{aligned}\sigma_\phi^2 &= \frac{2S_n(0)B_L}{A_e^2}\\ &= 2\pi^2\left(\frac{\dfrac{\Delta t}{T_s}}{1-\dfrac{\Delta t}{T_s}}\right)\left[\left(1-\frac{3\Delta t}{4T_s}\right)\frac{1}{\mu_2 R}+\frac{1}{2\mu_2^2 R^2}\right]B_L T_s\\ &\triangleq \left(\frac{a}{\mu_2 R}+\frac{b}{\mu_2^2 R^2}\right)B_L T_s \qquad (7.6.19)\end{aligned}$$

The parameters a and b are plotted in Figure 7.3 as functions of $\Delta t/T_s$. Again we encounter the by now familiar situation in which both the tracking error variance and the linear region of the error signal are directly proportional to the same parameter, in this case Δt. As before (cf. Section 6.3, for example), Δt can be optimized as a function of the loop signal-to-noise ratio. Since the linear region extends over only a Δt second interval, Δt cannot become arbitrarily small [at least if Equation (7.6.19) is to remain valid]. On the other hand, Δt can often be considerably less than $T_s/2$ seconds.

The symbol synchronizer of Figure 7.1 is also amenable to a number of alternative implementations. Two alternatives are suggested by the following observation: Let $\alpha_l(\tau)$ denote the integral

$$\alpha_l(\tau) = \frac{1}{AT_s}\int_{lT_s+\tau+(\Delta t/2)}^{(l+1)T_s+\tau-(\Delta t/2)} \sqrt{2}\, y(t)\sin(\omega_c t+\phi)\,dt$$

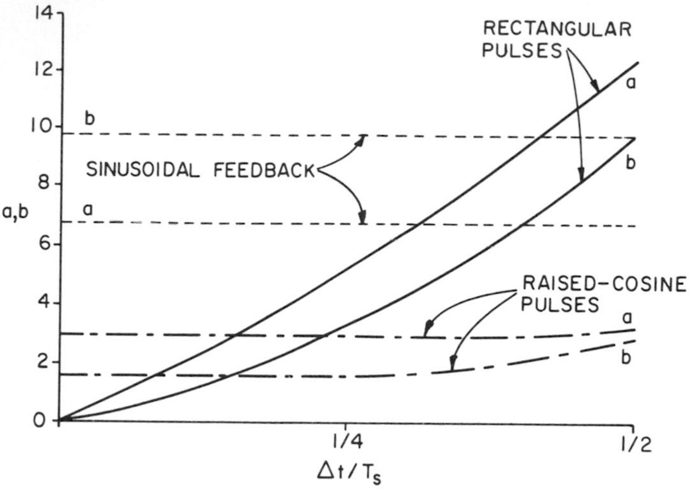

Figure 7.3 Parameters of Equations (7.6.19), (7.6.22), and (7.7.4).

and $\beta_l(\tau)$ the integral

$$\beta_l(\tau) = \frac{1}{AT_s} \int_{lT_s+\tau-(\Delta t/2)}^{lT_s+\tau+(\Delta t/2)} \sqrt{2}\, y(t) \sin(\omega_c t + \phi)\, dt$$

The error signal at the VCO input (Figure 7.1) can then be expressed as a weighted sum of the form

$$\sum_l w_l\{[\alpha_l(\tau) + \beta_{l+1}(\tau)]^2 - [\alpha_l(\tau) + \beta_l(\tau)]^2\}$$
$$\approx \sum_l w_l \beta_l(\tau)[\alpha_{l-1}(\tau) - \alpha_l(\tau)] \qquad (7.6.20)$$

These expressions for the error signal are valid if the loop filter is narrow relative to the reciprocal of the symbol period T_s; i.e., if the terms w_l are slowly varying functions of l.

The two synchronizer configurations suggested by these manipulations are shown in Figures 7.4 and 7.5. The first of these, known as an *early-late gate* or *split gate* synchronizer, represents an alternative, and possibly simpler, implementation of the more general device depicted in Figure 7.1.

The modified early-late gate synchronizer shown in Figure 7.5 suggests a further modification. When the loop is properly tracking $\tau \approx 0$; the upper integrator in Figure 7.5 therefore integrates over the transition region, while the lower integrator integrates over the interval during which a transition does not occur. In the case of binary PSK, for example, the upper integrator output is, ideally, multiplied by zero when no transition takes place (such an event provides no synchronization information) and by $+1$ or -1, the sign of the transition, when one does take place. When the signal-to-noise ratio is sufficiently large to enable reliable decisions, this last interpretation suggests

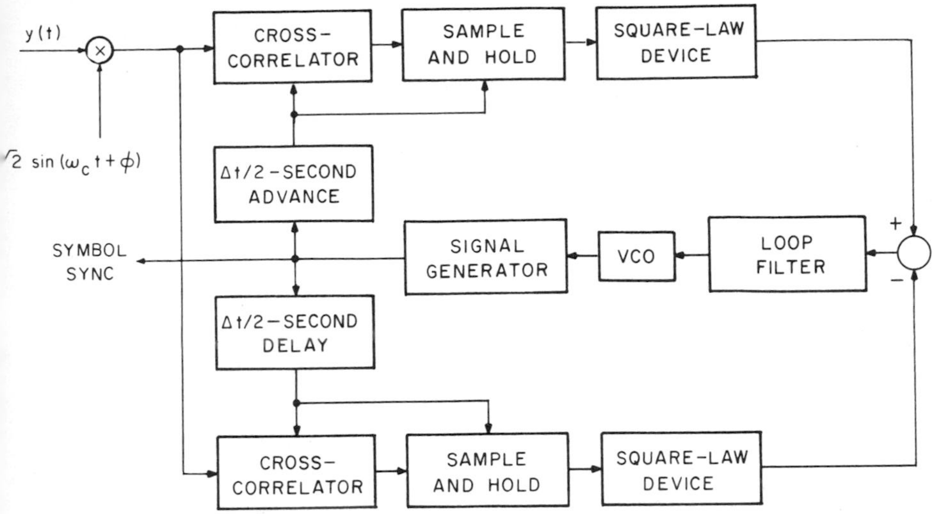

Figure 7.4 Early-late gate synchronizer.

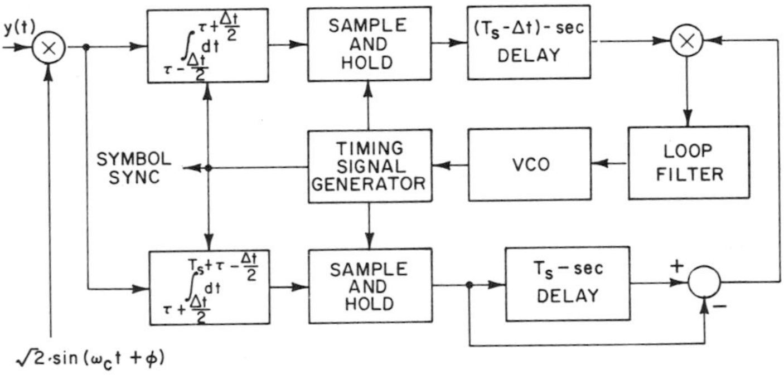

Figure 7.5 Modified early-late gate synchronizer.

making a hard decision as to whether or not a transition took place and weighting the integral across the transition region accordingly; e.g., following the integrator in the lower loop of Figure 7.5 by a hard limiter. Essentially this same device, incidentally, would have resulted directly had we begun with the large signal-to-noise ratio approximation to the maximum-likelihood synchronizer statistic [Equation (7.4.10)]. If every decision concerning the presence or absence of a transition were correct, the tracking-error variance obtainable with such a device would be

$$\sigma_\phi^2 = 2\pi^2 \frac{\Delta t}{T_s} \frac{B_L T_s}{R} \tag{7.6.21}$$

as is easily verified. The improvement possible in making a hard decision, therefore, is limited at least at large signal-to-noise ratios R. [Compare Equations (7.6.19) and (7.6.21); see also Problem 7.3.] Nevertheless, the potential simplification in implementing this modified synchronizer may well make it worth considering.

The symbol-tracking device shown in Figure 7.1 can also be used in conjunction with other local signals $s(t)$. If $s(t)$ is a sinusoid with period T_s, for example, Equation (7.6.7) yields

$$A_e = \frac{2\mu_2}{\pi^2}$$

while Equation (7.6.13) becomes (when the tracking error is small)

$$S_n(0) = \frac{2T_s}{\pi^2}\left[\frac{\mu_2}{2R}\left(1 + \frac{15}{4\pi^2}\right) + \frac{1}{R^2}\right]$$

so that

$$\sigma_\phi^2 = \frac{2S_n(0)B_L}{A_e^2} = \pi^2\left[\frac{1}{2\mu_2 R}\left(1 + \frac{15}{4\pi^2}\right) + \frac{1}{\mu_2^2 R^2}\right]B_L T_s \qquad (7.6.22)$$

The mean-squared tracking error is again of the form $(a/\mu_2 R + b/\mu_2^2 R^2)B_L T_s$. For comparison, these parameters a and b are also indicated in Figure 7.3.

Although the performance of the sinusoidal-feedback synchronizer is decidedly inferior to that of the sampling-function synchronizer, its simpler implementation may more than compensate for this, at least when R is sufficiently large and $B_L T_s$ sufficiently small. This is particularly true when the sinusoidal-feedback phase-locked loop can be replaced by a narrow-band filter as discussed in Section 6.9. The filter bandwidth B_f will generally have to be considerably larger than the bandwidth of the loop it replaces, but the mean-squared tracking error, now given by Equation (7.6.22) with B_L replaced by B_f (at least if $B_f \ll 1/T_s$), may still be within acceptable bounds.

7.6.2 *Incoherent PAM, Non-Binary PSK, and Differentially Coherent PSK*

The symbol synchronizers of the preceding paragraphs were limited to coherent PAM and binary PSK signals. The generalization needed to enable them to operate when the modulation is incoherent PAM, non-binary PSK, or differentially coherent PSK is obvious in view of the results obtained in Sections 7.3 and 7.4.

The only change needed in the above derivations to accommodate such modulations is to redefine the $z(t)$ of Equation (7.6.4) as

$$z(t) = \left[\frac{1}{AT_s}\int_{t-T_s}^{t} y(u)\sqrt{2}\sin\omega_c u\,du\right]^2 + \left[\frac{1}{AT_s}\int_{t-T_s}^{t} y(u)\sqrt{2}\cos\omega_c u\,du\right]^2 \qquad (7.6.23)$$

[cf. Equations (7.3.21) and (7.4.7)]. With this modification, the arguments proceed exactly as before. The tracking error is as given in Equation (7.6.19) or (7.6.22), except that the coefficient of the $1/R^2$ term is doubled. [Again, see Sections 7.3 and 7.4, and Equation (7.3.22).]

The modification in the implementation of the resulting synchronizer is correspondingly minor. The coherent matched filters in the previous synchronizers need only be replaced by incoherent matched filters followed by envelope detectors (cf. Section 4.3). Everything else remains the same. The synchronizer shown in Figure 7.1, for example, now assumes the form illustrated in Figure 7.6.

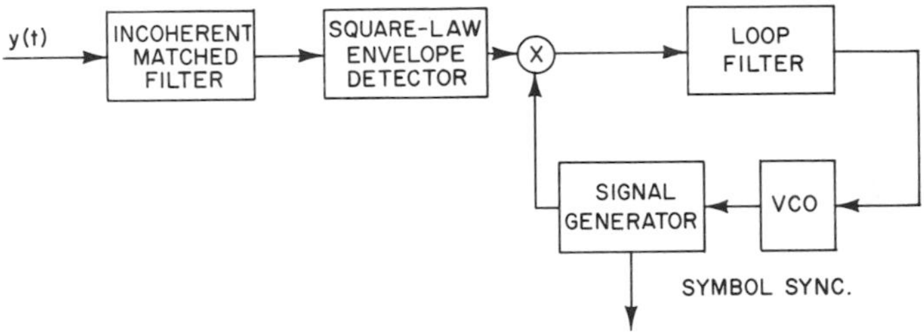

Figure 7.6 Symbol synchronizer—incoherent PAM and non-binary PSK.

7.6.3 Orthogonal Symbols

The symbol synchronization techniques of Section 7.5 are also easily adaptable to a symbol tracking configuration. Consider the difference $d(\tau)$ between the output at the time $\tau + \Delta t/2$ and that at the time $\tau - \Delta t/2$ of a filter matched to an FSK symbol or to a PPM sub-symbol. If the true symbol (or sub-symbol) epoch is τ_0, if $|\tau - \tau_0| < \Delta t/2$, and if the symbol corresponding to the observed matched filter was actually received, the expected value of $d(\tau)$ is easily scen to be a linear function of $\tau - \tau_0$. The quantity $d(\tau)$ can therefore be used as an error signal to control the sampling instant τ. Moreover, if $\tau \approx \tau_0$, a reliable decision can be made on the basis of the filter output at the time τ as to whether or not the particular symbol or sub-symbol in question was received, and $d(\tau)$ can be accepted as an error signal only when this decision is made in the affirmative.

The tracking devices suggested by the preceding paragraph are shown in Figures 7.7, for incoherent FSK, and 7.8, for coherent PPM signals. The decision apparatus decides which symbol was received (in the FSK case) or whether or not a sub-symbol pulse is present (in the PPM case) and gates the error signal acccordingly. These devices are quite similar to the large signal-to-noise ratio binary PSK symbol tracking device discussed earlier. The tracking error variance, assuming the decisions concerning the

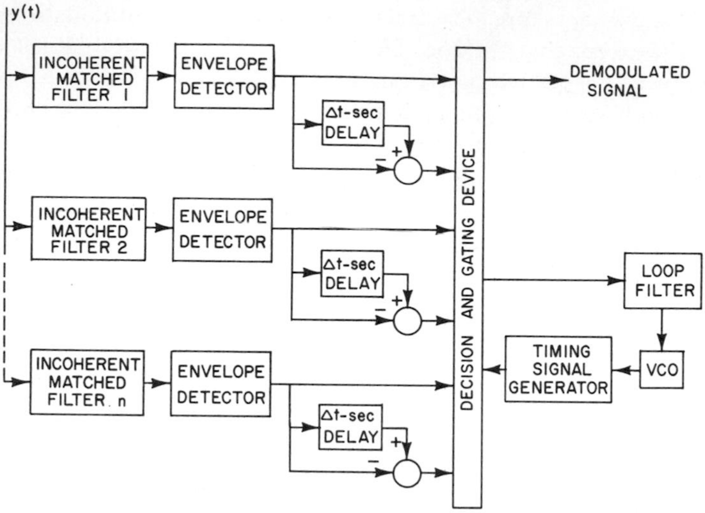

Figure 7.7 Symbol synchronizer—FSK.

presence or absence of a symbol are always correct, is easily seen to be

$$\sigma_\phi^2 = 4\pi^2 \frac{\Delta t}{T_s} \frac{B_L T_s}{pR} \tag{7.6.24}$$

where

$$p = \begin{cases} \Pr\{\text{two successive symbols differ}\} = \dfrac{n-1}{n} & \text{FSK} \\ \Pr\{\text{sub-symbol pulse followed by no pulse}\} = \dfrac{n^2-1}{n^2} & \text{PPM} \end{cases}$$

The FSK symbol tracking error is here defined in terms of a symbol period (2π radians corresponding to T_s seconds) while the PPM error is defined in terms of a sub-symbol period (2π radians corresponding to T_s/n seconds).

This last symbol tracking scheme is clearly applicable to any pulse communication system. Once a sufficiently accurate estimate of the symbol

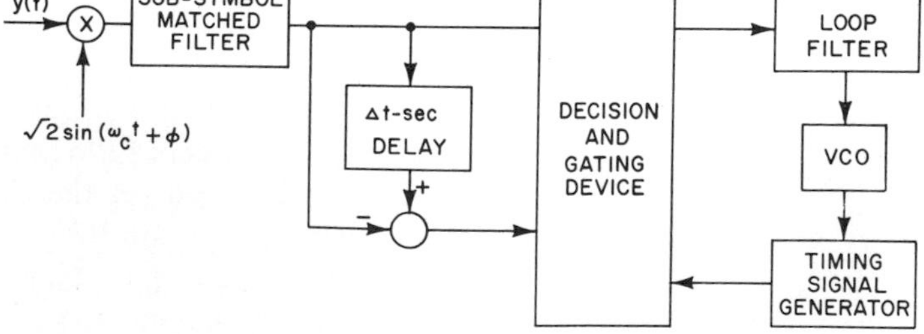

Figure 7.8 Symbol synchronizer—PPM.

epoch is available a reliable decision can presumably be made as to the symbol just received. Since both the received symbol and the transition region are known, the latter can be processed to refine the estimate of the symbol epoch. The applicability of this approach depends upon the not always valid assumption of a relatively large signal-to-noise ratio and on the ability to obtain the initial epoch estimate in a satisfactorily short period of time.

7.7 Non-Rectangular Symbol Synchronization

In contrast to the synchronous situation, the performance of a pulse communication system in the synchronization mode is dependent upon the pulse shape even in the absence of bandwidth limitations. As in Chapter Six, for instance, the tracking accuracy can be increased (or the search time decreased) by narrowing the effective width of the pulses being used. On the other hand, the narrower the pulse width the more rapidly the magnitude of the output of a filter matched to this pulse decreases from its maximum value. Thus, the accuracy *required* increases accordingly, and the situation tends to remain unchanged.

This is not to say that a synchronization advantage cannot result from careful pulse shaping. One potential advantage in pulse shaping, for example, is the possibility of simplifying the equipment needed for synchronization. *Return-to-zero* modulation, in which energy is transmitted during only a part of each symbol period, is sometimes used for this purpose. This is equivalent to multiplying the symbols by an on-off rectangular wave (not necessarily symmetrical) and hence introduces a frequency component at the symbol frequency which can be filtered out or tracked. If energy is transmitted over only the first half of each symbol period in an FSK system, for example, and if the symbol frequencies are sufficiently separated, the outputs of the matched filters (or of the envelope detectors) of Figure 7.7 are non-overlapping triangular pulses. The sum of these outputs is a (noisy) triangular wave which can be tracked with a phase-locked loop (or filtered with a narrow-band filter) to establish the desired symbol synchronization, thus allowing the sampler and decision device in Figure 7.7 to be replaced by a simple summing device. Except for the n-fold increase in the noise level (only one output contains a signal component at any instant while all n contain noise) the performance of such a device is just that of a phase-locked loop tracking a triangular-wave signal. At large signal-to-noise ratios, this n-fold noise increase may be an acceptable price to pay for the reduced equipment complexity.

Another motive for modifying the signal pulse shape for synchronization purposes is to nullify the stochastic properties of the signal. The analyses of this chapter all assumed statistical independence between successive symbols, an assumption which is rarely strictly satisfied. If, in actuality, successive symbols tended to be equal, the number of symbol transitions could be considerably less than assumed in the above analyses, and the synchronizer performance considerably poorer. Again, return-to-zero modulation can be used to overcome this problem. A related technique in the case of binary PSK is to use a *Manchester code* in which one message symbol is represented by

$$x(t) = \begin{cases} \sqrt{2} \sin \omega_c t & 0 < t < T_s/2 \\ -\sqrt{2} \sin \omega_c t & T_s/2 < t < T_s \end{cases}$$

and the other symbol by $-x(t)$. Such modifications in symbol structure make the synchronizer performance relatively independent of the signal statistics. The cost is in the increased bandwidth required to transmit these modified symbols.

The question to which a discussion of this sort inevitably leads, the question as to the "optimum" signal under, in this case, a given set of communication and synchronization requirements, will not be pursued here. Clearly, the answer, if it could be obtained, would be highly dependent upon the constraints. Moreover, the effect of symbol synchronization errors on the overall comunication system performance can generally be kept to negligible levels through the use of efficient synchronization techniques, regardless of the symbol pulse shape (cf. Chapter Nine). The symbol synchronization requirement, therefore, would tend to have only minor influence on the signal design problem in any event.

The results of this chapter can be generalized to include non-rectangular pulses as well. When the pulse envelope $p(t)$, $0 \leq t \leq T_s$, is arbitrary, Equations (7.3.22), in particular, are altered as follows: Let $\lambda_i = \frac{1}{T_s}\int_0^{T_s} p(t - T_s + i\Delta T)p(t)\,dt$ with $\lambda_N = 1$, and define $\gamma_1 = \lambda_\nu, \gamma_2 = \lambda_{N-\nu}, \gamma_3 = \lambda_\mu, \gamma_4 = \lambda_{N-\mu}$, $\gamma_5 = \lambda_{(\mu-\nu)}$, $\gamma_6 = \lambda_{N-(\mu-\nu)}$, $a = \mu_4 + 4m\mu_3 + 4m^2\mu_2 - \mu_2^2$, $b = m\mu_3 + 2m^2\mu_2$, and $d = \mu_2^2 + 4m^2\mu_2$. Then

$$\begin{aligned}
\eta_\nu &= \mu_2(\gamma_1^2 + \gamma_2^2) + m^2(\gamma_1 + \gamma_2)^2 + c_0/2R \\
\sigma_{sxs}^2 &= a(\gamma_1^2 + \gamma_2^2)^2 + 8b\gamma_1\gamma_2(\gamma_1^2 + \gamma_2^2) + 4d\gamma_1^2\gamma_2^2/c_1 \\
\sigma_{sxn}^2 &= 2\mu_2(\gamma_1^2 + \gamma_2^2)/R + 2m^2(\gamma_1 + \gamma_2)^2/R \\
\sigma_{nxn}^2 &= c_0/2R^2 \\
\Delta_{sxs}^2 &= a[(\gamma_1^2 + \gamma_2^2) - (\gamma_3^2 + \gamma_4^2)]^2 + 8b(\gamma_1\gamma_2 - \gamma_3\gamma_4)[(\gamma_1^2 + \gamma_2^2) - (\gamma_3^2 + \gamma_4^2)] \\
&\quad + 4d(\gamma_1\gamma_2 - \gamma_3\gamma_4)^2/c_1 \\
\Delta_{sxn}^2 &= 2m^2[(\gamma_1 + \gamma_2)^2 + (\gamma_3 + \gamma_4)^2 - 2(\gamma_1 + \gamma_2)(\gamma_3 + \gamma_4)(\gamma_5 + \gamma_6)]/R \\
&\quad + 2\mu_2[\gamma_1^2 + \gamma_2^2 + \gamma_3^2 + \gamma_4^2 - 2\gamma_1\gamma_3\gamma_6 - 2\gamma_2\gamma_4\gamma_6 \\
&\quad - 2\gamma_2\gamma_3\gamma_5]/R \\
\Delta_{nxn}^2 &= c_0(1 - \gamma_5^2 - \gamma_6^2)/R^2
\end{aligned} \tag{7.7.1}$$

The synchronizers of section 7.6, and the arguments leading to their consideration, apply equally well to any pulse shape so long as the associated matched filters or cross-correlators are matched to the specific pulse shape being used. The mean-squared tracking error corresponding to the device depicted in Figure 7.1 is still of the form

$$\sigma_\phi^2 = \frac{2S_n(0)B_L}{A_e^2} \tag{7.7.2}$$

with A_e and $S_n(0)$ as defined in Equations (7.6.7) and (7.6.13), respectively. Now, however, the terms η_ν and $\Delta_{\mu\nu}^2$ in these equations are modified as indicated in Equation (7.7.1).

As already remarked, the appropriateness of a particular pulse shape is strongly dependent on the conditions under which it is to be used. One of the most commonly used pulse shapes, other than rectangular, is the so-called *raised-cosine* pulse defined by†

$$p(t) = \sqrt{2/3}[1 - \cos(2\pi t/T_s)] \qquad 0 \le t \le T_s \tag{7.7.3}$$

Thus, to illustrate the effect of the pulse shape on the performance of a tracking synchronizer, we choose the raised-cosine pulse as a representative, practical alternative to the rectangular pulse.

Consider the synchronizer shown in Figure 7.1 with $s(t)$ the sampling function $s(t) = \sum_i [\delta(t - (\Delta t/2) + iT_s) - \delta(t + (\Delta t/2) + iT_s)]$, and with the received signal a sequence of PAM or PSK modulated raised-cosine pulses (i.e., $y(t) = \sqrt{2}\,Ax(t - iT_s)\sin(\omega_c t + \theta) + n(t)$, $iT_s < t < (i+1)T_s$, with either x or θ representing the information). The mean-squared tracking error is again of the form

$$\sigma_\phi^2 = \left(\frac{a}{\mu_2 R} + \frac{c_0 b}{\mu_2^2 R^2}\right) B_L T_s \tag{7.7.4}$$

where $a = a(\Delta t/T_s)$ and $b = b(\Delta t/T_s)$ can be determined from Equations (7.6.13) and (7.7.1). These parameters are also plotted in Figure 7.3. (The mean PAM amplitude is here assumed to be zero.) It will be noted that, in contrast to the case when the pulses are rectangular, the tracking error is virtually independent of Δt, especially for $\Delta t < T_s/4$. Similarly, the shape of the error signal is much less dependent on Δt. (See Figure 7.9. The amplitude of each error signal has been normalized so that the corresponding noise levels in all cases are identical.)

If the shape of the error signal is not adversely affected, it is presumably advantageous to allow Δt to approach zero in Equation (7.6.5) and to design the synchronizer accordingly. [Figure 7.10 suggests more practical implementations for such a synchronizer than the one indicated in Figure 7.1. The feedback signal here is of the form $s(t) = \sum_i \delta(t - iT_s)$.] The mean-

---

†One obvious advantage of this pulse shape vis-à-vis the rectangular pulse is in its narrower spectrum; cf. Problem 3.2.

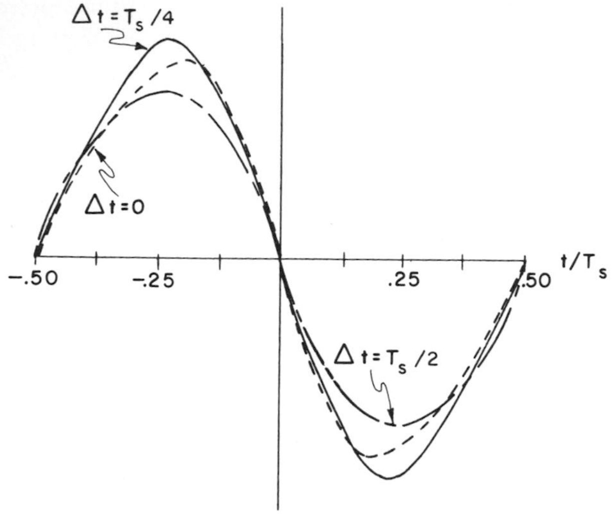

Figure 7.9 Error signal for raised-cosine symbol synchronizer.

squared tracking error of the resulting PAM or PSK symbol synchronizer can be readily determined with the aid of Equations (7.7.1) and (7.7.2). Specifically,

$$\sigma_\phi^2 = \frac{2\pi^2\left\{\dfrac{2}{R'} + \dfrac{c_0}{(R')^2}\right\} B_L}{\displaystyle\int_0^{T_s} (p'(t))^2\, dt} \tag{7.7.5}$$

where $R' = A^2T_s(\mu_2 + m^2)/N_0$, and $p'(t)$ denotes the derivative of the pulse envelope $p(t)$ (see Problem 7.1).

It is interesting to note that the synchronizer of Figure 7.1 is able to track

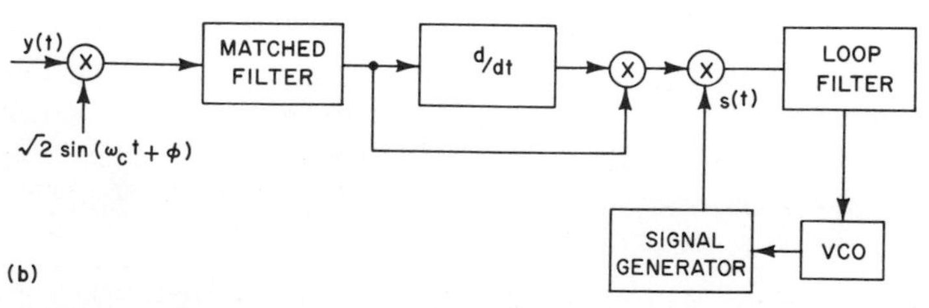

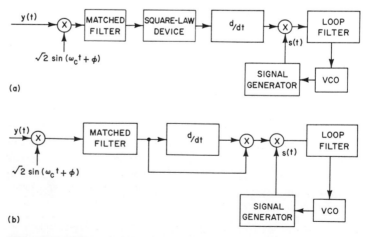

Figure 7.10 Differentiating symbol synchronizers (phase coherent case).

raised-cosine pulses more accurately than rectangular pulses for $\Delta t/T_s$ greater than approximately 0.1 while the situation reverses for $\Delta t/T_s$ smaller than this value. In general, one would expect that the more abrupt the transition between successive pulses, the more accurately could this point be located and tracked. This is supported by the preceding results, but only to the extent that the signal-to-noise ratio is sufficiently large to allow this property to be fully exploited. That is, only when the loop signal-to-noise ratio is relatively large can Δt be small without incurring an unacceptably large loss-of-lock probability (cf. Section 6.3). And only in this case is the abruptness of the transition between successive rectangular pulses turned to advantage.

We should also note, in assessing the relative merits of rectangular and non-rectangular pulses from the synchronization point of view, that synchronization information can be gleaned from a sequence of non-rectangular pulses even in the absence of any modulation (cf. Problem 7.1). This of course is not true when the pulses are rectangular. Indeed, in the latter case, it may be necessary to constrain the modulation to assure a minimum number of transitions and thereby avoid possible loss of synchronization.

7.8 Concluding Remarks

This chapter has been concerned with methods of obtaining symbol synchronization directly from the received sequence of information-bearing symbols. Two types of synchronizers were examined, the first designed to identify the correct one of a finite set of symbol epochs, the second to track this epoch. Investigation into this first category of synchronizers centered on the so-called maximum-likelihood synchronizers. The performance of this class of synchronizers was found to be quite comparable with that attainable using the two-channel approach of Chapter Six when the effective pulse width of the synchronizing signal there is of the order of the symbol period here. Since the two-channel approach necessitates dividing the available power between the data and synchronization channels, while here all of the power is used simultaneously for both communication and synchronization, the single-channel method may outperform the two-channel approach even when the synchronizing pulse width in the latter case is relatively narrow.

The motivation for considering the class of tracking synchronizers postulated in Section 7.6 was threefold. First, this class is sufficiently general to include many symbol synchronizers of practical interest. Second, it includes as a subclass synchronizers designed to force a solution to the maximum-likelihood equation (cf. Section 5.4), at least in the most difficult situation when the signal-to-noise ratio R is small. Consequently, the resulting estimate of the symbol epoch is asymptotically optimum in the mean-squared sense (Section 2.8) as the loop signal-to-noise ratio approaches infinity; i.e., as the loop bandwidth B_L approaches zero. [This, of course, does not guarantee optimality under any other set of conditions. Furthermore, the proof of the asymptotic optimality of the maximum-likelihood estimator depends on certain regularity conditions (Appendix A.2) which do not hold unless constraints are placed on the first several derivatives of the pulse envelope. In a physical system, of course, channel bandwidth limitations preclude rectangular pulses in any event, so this qualification is not of serious consequence. The conclusions based on rectangular pulses remain unaltered so long as such pulses are recognized as idealizations. Third, and probably most important, relatively simple analytical expressions can be derived relating the mean-squared tracking error to the received signal-to-noise ratio and other physical parameters.

Notes to Chapter Seven

Except for a brief summary of some of the results in Sections 7.6 and 7.7 (Reference 7.1), and a portion of Sections 7.2 and 7.3 (Reference 7.2) the material in this chapter has not been previously published. Other publications relating to the subject matter of this chapter include Reference 7.3 (concerning the performance of symbol trackers of the sort shown in Figure 7.4, but with the square-law non-linearity replaced by an absolute-value non-linearity); Reference 7.4 (in which synchronizers of the sort shown in Figure 7.1, but with the phase-locked loop replaced by a low-pass filter, are digitally simulated); Reference 7.5 (regarding the use of symbol tracking devices on telephone channels); Reference 7.6 (concerning optimum binary symbol synchronization based on a restricted set of statistics); and References 7.7 and 7.8 (in which still other bit synchronization estimators are investigated).

Problems

7.1 (a) Let $p(t)$ be an arbitrary pulse shape as defined in Equation (7.7.1) and assume its first and second derivatives exist and are absolutely integrable over the interval $0 \leq t \leq T_s$. Show that, when $\mu = \nu + 1$ in Equation (7.7.1)

$$\lim_{\Delta T \to 0} \frac{\Delta^2_{sxs}}{(\Delta T)^2} = 4a(\gamma_1\gamma_1' - \gamma_2\gamma_2')^2 - 16b(\gamma_1\gamma_1' - \gamma_2\gamma_2')(\gamma_1\gamma_2' - \gamma_2\gamma_1') + 4d(\gamma_1\gamma_2' - \gamma_2\gamma_1')^2/c_1$$

$$\lim_{\Delta T \to 0} \frac{\Delta^2_{sxn}}{(\Delta T)^2} = \frac{2m^2}{R}\{(\gamma_1' - \gamma_2')^2 - (\gamma_1 + \gamma_2)^2\gamma_3''\} + \frac{2\mu_2}{R}\{(\gamma_1')^2 + (\gamma_2')^2 - (\gamma_1^2 + \gamma_2^2)\gamma_3''\}$$

$$\lim_{\Delta T \to 0} \frac{\Delta^2_{nxn}}{(\Delta T)^2} = -\frac{c_0}{R^2}\gamma_3''$$

where $\gamma(t) = \frac{1}{T_s}\int_0^{T_s} p(u - T_s + t)p(u)\,du$, $\gamma'(t) = d\gamma(t)/dt, \gamma''(t) = d^2\gamma(t)/dt^2$, and the subscripts 1, 2, and 3 denote the arguments $\tau \triangleq \lim_{\Delta T \to 0} \nu\Delta T$, $T_s - \tau$, and T_s, respectively.

(b) Use this result to show that, if the signal-to-noise ratio is sufficiently large to insure a small tracking error, the tracking error variance is as given in Equation (7.7.5). Note that, under these conditions, the mean-squared error in tracking PSK modulated symbols is independent of the modulation.

7.2 In the derivations in Sections 7.6 and 7.7 the dependence of the noise spectral density $S_n(0)$ at the loop input on the tracking error τ was ignored [cf. Equation (7.6.17)]. (a) Show that, when $B_L T_s$ is small and when σ_ϕ^2 is sufficiently small to justify the linear approximation to the phase-locked loop, to a good approximation

$$\sigma_\phi^2 = \int_{-\infty}^{\infty} \sigma_\phi^2(\tau)p(\tau)\,d\tau$$

where $\sigma_\phi^2(\tau)$ is the variance which would theoretically result were $S_n(0)$ evaluated at the point τ and where $p(\tau) = (1/\sqrt{2\pi}\,\sigma_\tau)\exp(-\tau^2/2\sigma_\tau^2)$, with $\sigma_\tau = (T_s/2\pi)\sigma_\phi$. (b) Using this result, show that when the dependence of $S_n(0)$ on τ is taken into account the tracking-error variance as given in Equation (7.6.19) is increased by the factor

$$\left[1 - \left(\frac{\mu_4}{c_1\mu_2^2} + \frac{3}{2\mu_2 R}\right)\frac{B_L T_s}{1 - (\Delta t/T_s)}\right]^{-1}$$

7.3 Consider the tracking device shown in Figure 7.5, but with the lower integrator followed by a hard-limiter. (a) Show that when the envelopes of the received binary PSK pulses are rectangular the (linear model) variance of the tracking error of such a device is

$$\sigma_\phi^2 = 2\pi^2\left(\frac{\Delta t}{T_s}\right)\frac{B_L T_s}{(1 - 2p)^2 R}$$

where p denotes the synchronous symbol error probability at the signal-to-noise ratio $R' = R[1 - (\Delta t/T_s)]$.

(b) Show that, for small signal-to-noise ratios R, the presence of the hard-limiter increases the phase-error variance by a factor of $\pi/2$ while, at the other extreme, when $R \to \infty$, it decreases the variance by the factor $[1-(\Delta t/T_s)]/[1-(\frac{3}{4}\Delta t/T_s)]$. (c) Derive the analogous results when the pulse envelopes are raised-cosine rather than rectangular in shape.

7.4 Estimate the time needed to establish PAM and PSK symbol synchronization directly from the modulation using the various serial search techniques examined in Chapter 6 [cf. Equation (7.3.22)].

7.5 Show that when the successive pulses in a PSK modulated signal have statistically independent, but not necessarily uniformly distributed, phases, Equation (7.7.1) still applies but with $m^2 = E(\cos\Delta\theta_l)$, $\mu_2 = 1 - m^2$, $a = b = 0$, and $d/c_1 = E(\cos^2\Delta\theta_l) + 2E(\cos\Delta\theta_l\cos\Delta\theta_{l+1}) - 3E^2(\cos\Delta\theta_l)$, and with $\Delta\theta_l = \theta_l - \theta_{l+1}$, θ_l denoting the phase of the pulse over the interval $lT_s \le t \le (l+1)T_s$.

References

7.1 Stiffler, J. J., "On the Performance of a Class of PCM Bit Synchronizers," *Proc. of the Nat. Tele. Conf.*, Washington, D. C., p. 67, 1969.

7.2 Stiffler, J. J., "Maximum-Likelihood Symbol Synchronization," *Jet Propulsion Lab., Space Programs Summary* 37–35, Vol. 4, Calif. Inst. Tech., Pasadena, Calif., p. 349, 1965.

7.3 Simon, M. K., "Nonlinear Analysis of an Absolute Value Type of Early-Late Gate Bit Synchronizer," *IEEE Trans. on Comm. Tech.*, COM-18, 1970.

7.4 Wintz, P. A., and E. J. Luecke, "Performance of Optimum and Suboptimum Synchronizers," *IEEE Trans. on Comm. Tech.*, COM-17, p. 380, 1969.

7.5 Saltzberg, B. R., "Timing Recovery for Synchronous Binary Data Transmission," *Bell Syst. Tech. Jour.*, **45**, p. 593, 1966.

7.6 Pitcher, T. S., and H. Rumsey, "A Bayes Estimate for Synchronization," *Proc. of the IEEE*, **56**, p. 1095, 1968.

7.7 McBride, A. L., and A. P. Sage, "On Discrete Sequential Estimation of Bit Synchronization," *IEEE Trans. on Comm. Tech.*, COM-18, 1970.

7.8 Lee, G. M., and J. J. Komo, "PCM Bit Synchronization by Non-Linear Theory" (to be published).

chapter eight

Maximum-Likelihood Carrier Synchronization

8.1 Introduction

In the previous chapter, methods were investigated for establishing symbol synchronization directly from the received message. In this chapter, some of these same techniques are shown to be equally applicable to the problem of obtaining a carrier (or subcarrier) reference from the modulated carrier. Since the carrier phase will inevitably vary with time, the problem is one of tracking a periodic signal rather than one of detecting its phase, and the pertinent techniques used are those of Chapter Five.

The problem of extracting a carrier reference from the data channel when the modulated carrier spectrum contains a discrete component at that frequency has already been treated in Chapter Six. It is only necessary to determine the ratio of the power in the carrier component to the noise spectral density in that frequency region to estimate the accuracy with which it can be tracked. This chapter, therefore, is concerned only with data channels having no discrete carrier frequency component, or, at least, none of sufficient strength to be tracked with any degree of reliability.

In the first several sections of this chapter, carrier tracking devices are developed and analyzed for various types of modulation under the assumption that symbol synchronization is known. Since symbol synchronization must be established before demodulation can begin, and since the carrier is to be tracked throughout the communication mode, this assumption is entirely justified. In the acquisition mode, symbol synchronization may or may not be available prior to carrier acquisition. (It could be established using one of the phase-incoherent techniques of Chapter Seven, for example.) If it is not, the methods to be developed here will still be operative, although at a somewhat reduced level of efficiency. In later sections, this assumption of prior symbol synchronization will be removed.

The situation, then, is the following: The received signal $y(t)$ is of the form $y_l(t) = x_{j_l}(t - \tau, \phi_0) + n(t)$, $\tau + lT_s < t < \tau + (l+1)T_s$, $(l = \ldots, -1, 0, 1, 2, \ldots; j_l = 1, 2, \ldots, n)$, with $n(t)$ white Gaussian noise, and with ϕ_0 denoting the carrier phase and τ the symbol epoch, initially assumed known. (Without loss of generality, therefore, let $\tau = 0$.) The problem is to track the carrier phase ϕ_0. The approach will be similar to that taken in Sections 5.4 and 7.6; viz., a feedback arrangement will be used to force a solution to a weighted sum of the likelihood equations

$$\frac{\partial \log_e p[y_l(t) \mid \phi]}{\partial \phi} = 0 \tag{8.1.1}$$

where

$$p[y_l(t) \mid \phi] = K\left(\sum_{j=1}^{n} \exp\left\{\frac{1}{N_0}\int_{lT_s}^{(l+1)T_s} [2y(t)x_j(t, \phi) - x_j^2(t, \phi)]\,dt\right\} P(j)\right) \tag{8.1.2}$$

(This last equation follows immediately from the analogous result in Section 7.2.) The resulting device will be called a *maximum-likelihood carrier synchronizer*. It is an optimum device in the same sense that a phase-locked loop is optimum for tracking an unmodulated carrier.

In taking the maximum-likelihood approach, we are of course ignoring any information concerning the carrier phase which might be implied by the presumed symbol synchronization. This is justified either if the carrier phase and the symbol epoch are in fact independent or if the signal spectrum is narrow relative to its center frequency. In the latter case, a synchronization reference could well be treated as exact so far as symbol timing is concerned and yet be quite coarse relative to a carrier period. If neither of these situations holds, there is in effect no carrier; rather the carrier should be regarded as part of the symbol structure and the techniques of Chapter Seven applied.

8.2 PAM Carrier Synchronization

If the received signal is of the form† $y_l(t) = \sqrt{2}Ax_l \sin(\omega_c t + \phi_0) + n(t)$, $lT_s < t < (l+1)T_s$ (with x_l denoting the information, $n(t)$ white noise

†The pulse envelopes here will be assumed rectangular. Their actual shape, however, is irrelevant to the results which follow, so long as the resulting signal spectrum is narrow relative to its center frequency. The modifications required in the ensuing tracking devices to accommodate non-rectangular pulses are obvious.

having the single-sided spectral density N_0, and ϕ_0 the carrier phase) we can argue, as in Section 7.3, that a good approximation to the function $p[y_l(t) \mid \phi]$ is

$$p[y_l(t) \mid \phi] = Ke^{Ra_l^2(\phi)} \tag{8.2.1}$$

with

$$a_l(\phi) = \frac{1}{AT_s} \int_{lT_s}^{(l+1)T_s} y(t)\sqrt{2} \sin(\omega_c t + \phi)\, dt$$

[Equation (8.2.1), it will be recalled, is exact when the amplitudes of the PAM pulses are Gaussianly distributed, and is a reasonable approximation at all signal-to-noise ratios $R = A^2T_s/N_0$ for other distributions as well.] Using Equation (8.2.1) we find

$$\begin{aligned}\frac{\partial \log_e p[y_l(t) \mid \phi]}{\partial \phi} &= \frac{2R}{A^2T_s^2} \int_{lT_s}^{(l+1)T_s} y(t)\sqrt{2} \sin(\omega_c t + \phi)\, dt \\ &\quad \times \int_{lT_s}^{(l+1)T_s} y(t)\sqrt{2} \cos(\omega_c t + \phi)\, dt \end{aligned} \tag{8.2.2}$$

To analyze the performance of the resulting carrier tracking device (Figure 8.1) let $y_l(t) = \sqrt{2}\, Ax_l \sin[\omega_c t + \phi_0(t)] + n(t)$ be the received signal, and let $z(t) = \sqrt{2} \sin[\omega_c t + \phi_1(t)]$ be the VCO output. Then the input to the loop filter over the interval $(l+1)T_s < t < (l+2)T_s$ is

$$\begin{aligned}&[AT_s x_l \sin\phi_e(l) + n_1(l)][AT_s x_l \cos\phi_e(l) + n_2(l)] \\ &\quad = \tfrac{1}{2}A^2T_s^2 x_l^2 \sin 2\phi_e(l) + AT_s x_l \sin\phi_e(l) n_2(l) \\ &\qquad + AT_s x_l \cos\phi_e(l) n_1(l) + n_1(l) n_2(l)\end{aligned} \tag{8.2.3}$$

where $\phi_e(l) = \phi_1(t) - \phi_0(t)$, $lT_s < t < (l+1)T_s$, and is assumed constant over each T_s-second interval, and where

$$n_1(l) = \int_{lT_s}^{(l+1)T_s} n(t)\sqrt{2} \cos[\omega_c t + \phi_1(t)]\, dt$$

and (8.2.4)

$$n_2(l) = \int_{lT_s}^{(l+1)T_s} n(t)\sqrt{2} \sin[\omega_c t + \phi_1(t)]\, dt$$

Only the first term on the right of Equation (8.2.3) conveys any information regarding the phase of the received signal. The remaining terms constitute noise. If the loop bandwidth is small as compared to $1/T_s$ [and this will be the case, if, as we have assumed, $\phi_0(t)$, and hence $\phi_1(t)$ and $\phi_e(t)$ are slowly varying functions of time], the VCO output will be determined by the weighted sum of a large number of input pulses. As in Section 7.6, the signal at the loop filter input can be approximated by the expectation over x_l of the first term on the right side of Equation (8.2.3). The effective signal amplitude is therefore

$$A_e = \frac{d}{d\phi_e}\left[\frac{1}{2}A^2T_s^2 E(x^2) \sin 2\phi_e\right]_{\phi_e = 0} = A^2T_s^2 E(x^2) \tag{8.2.5}$$

where $E(x^2) \triangleq E(x_l^2)$ is assumed independent of l [cf. Equation (5.8.7)].

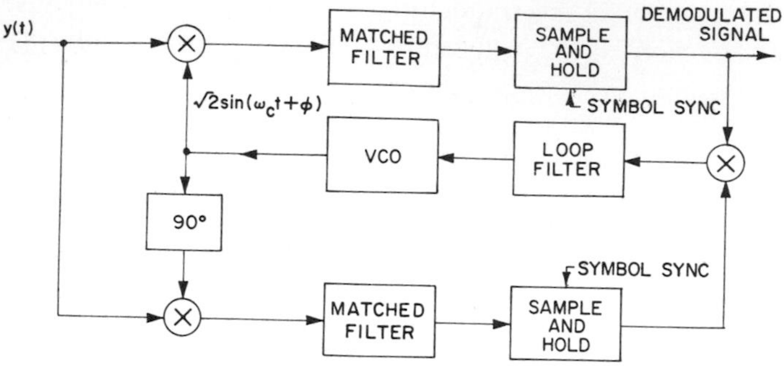

Figure 8.1 PAM carrier synchronizer.

Since the noise $n(t)$ at the input to the device is assumed white, the noise terms at the loop filter input corresponding to any two different intervals, $lT_s < t < (l+1)T_s$ and $mT_s < t < (m+1)T_s$, $l \neq m$, are independent. And since

$$E[n_1^2(l)] = E[n_2^2(l)] = \frac{N_0}{2} T_s \tag{8.2.6a}$$

and

$$E[n_1(l)n_2(l)] = 0 \tag{8.2.6b}$$

the autocorrelation function of the noise is

$$R_n(\tau) = \begin{cases} \sigma_n^2\left(1 - \dfrac{|\tau|}{T_s}\right) & |\tau| \leq T_s \\ 0 & |\tau| > T_s \end{cases} \tag{8.2.7}$$

where

$$\begin{aligned}\sigma_n^2 &= E\{AT_s x_l \sin\phi_e(l) n_2(l) + AT_s x_l \cos\phi_e(l) n_1(l) + n_1(l)n_2(l)\}^2 \\ &= A^2 T_s^2 E(x^2) \frac{N_0}{2} T_s + \left(\frac{N_0}{2}\right)^2 T_s^2\end{aligned}$$

Thus,

$$S_n(\omega) = \sigma_n^2 \int_{-T_s}^{T_s} \left(1 - \frac{|\tau|}{T_s}\right) e^{-j\omega\tau}\, d\tau = \sigma_n^2 T_s \left(\frac{\sin \dfrac{\omega T_s}{2}}{\dfrac{\omega T_s}{2}}\right)^2 \tag{8.2.8}$$

As in the case of the ordinary phase-locked loop, then, the tracking device of Figure 8.1 is well approximated, for small tracking errors, by the linear model shown in Figure 8.2. The bandwidth of the noise $n'(t)$ is of the order of $1/T_s$, and because the loop bandwidth is assumed narrow relative to this quantity, the noise is effectively white with the two-sided spectral density $S_n(0) = \sigma_n^2 T_s$. Thus, the results of Section 5.6 are again applicable here.

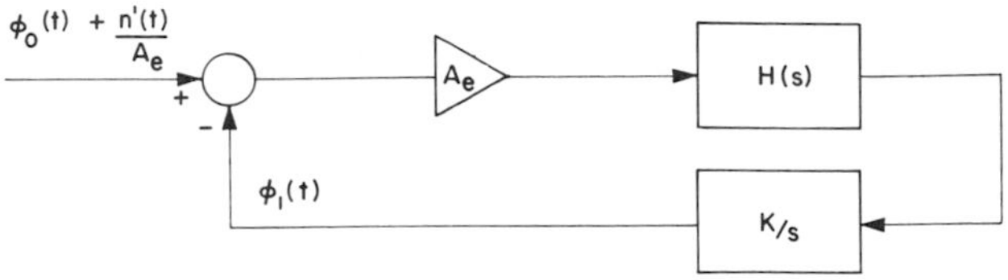

Figure 8.2 Linear model of synchronizer in Figure 8.1.

The phase error variance due to the input noise is

$$\sigma_\phi^2 \approx \frac{2S_n(0)B_L}{A_e^2} = B_L T_s\left(\frac{1}{R'} + \frac{1}{2R'^2}\right) \tag{8.2.9}$$

where $R' = A^2E(x^2)T_s/N_0$ is the ratio of the average signal energy to the spectral density of the additive noise, and B_L is the noise bandwidth of the equivalent loop of Figure 8.2.

It will be noted that the signal at the output of the tracking device in Figure 8.1 can either be in phase with the received carrier or 180° out-of-phase with it. This 180° phase ambiguity is inevitable; the carrier tracker clearly has no way, in the absence of an absolute reference, of discerning between the signals $\sqrt{2}\,Ax\sin(\omega_c t + \phi_0)$ and $\sqrt{2}\,A(-x)\sin(\omega_c t + \phi_0 + \pi)$. There are, of course, a number of ways of resolving this ambiguity. Most generally either the natural redundancy in the information sequence or the redundancy deliberately inserted for error control, frame synchronization, or other purposes will be sufficient to enable the user to distinguish between the true message and its inverse. In any event, while such ambiguities can be bothersome, they seldom present a problem of serious consequence.

8.3 PSK Carrier Phase Tracking

Now let the received signal be a noise-corrupted, PSK modulated sinusoid, $y_l(t) = \sqrt{2}\,A\sin[\omega_c t + \theta_{j_l} + \phi_0(t)] + n(t)$, $lT_s < t < (l+1)T_s$, with $\theta_j = 2\pi j/n$, $j = 1, 2, \ldots, n$, representing the information and $\phi_0(t)$ the phase which is to be tracked. At first glance, it might appear meaningless to attempt to track $\phi_0(t)$ when θ_{j_l} is also a random variable. But if the local phase reference were perfect, ϕ_0 would be zero and θ_{j_l} could assume only one of n precisely known values. The imperfect phase reference causes a discrepancy between the values θ_{j_l} is known to be able to have and the values it apparently does assume. In other words, $\phi_0(t)$ simply causes a bias in the observed phases. If $\phi_0(t)$ varies slowly with time, remaining relatively constant over a number of T_s-second intervals, this bias can presumably be tracked.

The function $p[y_l(t)\,|\,\phi]$ can now be expressed in the form (cf. Section 7.4)

$$p[y_l(t)\,|\,\phi] = \frac{K}{n}\sum_{j=1}^{n}\exp\{2R[a_l(\phi)\cos\theta_j + b_l(\phi)\sin\theta_j]\} \tag{8.3.1}$$

where

$$\left.\begin{matrix} a_l(\phi) \\ b_l(\phi) \end{matrix}\right\} = \frac{1}{AT_s} \int_{lT_s}^{(l+1)T_s} y(t)\sqrt{2} \left\{\begin{matrix} \sin(\omega_c t + \phi) \\ \cos(\omega_c t + \phi) \end{matrix}\right\} dt$$

The maximum-likelihood estimate of ϕ, given $y_l(t)$, $lT_s < t < (l+1)T_s$, is a solution of the equation

$$\frac{\partial \log_e p[y_l(t) \mid \phi]}{\partial \phi} = 0 \tag{8.3.2}$$

We contend, however, that any $\phi = \hat{\phi}$ satisfying the equation

$$\frac{d}{d\phi} \sum_{j=1}^{n} [a_l(\phi) \cos \theta_j + b_l(\phi) \sin \theta_j]^n = 0 \tag{8.3.3}$$

will also be a solution to equation (8.3.2).

To prove this statement, we first expand the exponential of Equation (8.3.1) in a power series, take the logarithm of the resulting expression, and differentiate with respect to ϕ, obtaining

$$\frac{\partial \log_e p[y_l(t) \mid \phi]}{\partial \phi} = \frac{\dfrac{K}{n} \displaystyle\sum_{\nu=0}^{\infty} \frac{(2R)^\nu}{\nu!} \frac{dF_\nu(\phi)}{d\phi}}{p[y_l(t) \mid \phi]} \tag{8.3.4}$$

where

$$F_\nu(\phi) = \sum_{j=1}^{n} [a_l(\phi) \cos \theta_j + b_l(\phi) \sin \theta_j]^\nu = \sum_{j=1}^{n} [c_l(\phi) e^{i\theta_j} + c_l^*(\phi) e^{-i\theta_j}]^\nu$$

with $c_l(\phi) \triangleq [a_l(\phi) - ib_l(\phi)]/2$ and $c_l^*(\phi)$ its complex conjugate. Expanding the summand in a binomial series, interchanging the order of summation, and noting that $\sum_{j=1}^{n} e^{i\theta_j \mu} = \begin{cases} 0 & \mu \neq 0, \bmod n \\ n & \mu = 0, \bmod n \end{cases}$, we obtain

$$F_\nu(\phi) = n \binom{\nu}{\frac{\nu}{2}} |c_l(\phi)|^\nu + n \sum_{\mu=1}^{[\nu/n]} \binom{\nu}{\frac{\nu - \mu n}{2}} \{[c_l(\phi)]^{\mu n} + [c_l^*(\phi)]^{\mu n}\} \tag{8.3.5}$$

where $[\nu/n]$ denotes the largest integer not exceeding ν/n and the binomial coefficients $\binom{\nu}{j}$ are taken to be zero if j is not an integer. But, as is easily verified, $|c_l(\phi)|$ is independent of ϕ, and

$$\frac{d}{d\phi} F_\nu(\phi) = in^2 \sum_{\mu=1}^{[\nu/n]} \mu \binom{\nu}{\frac{\nu - \mu n}{2}} \{[c_l(\phi)]^{\mu n} - [c_l^*(\phi)]^{\mu n}\} \tag{8.3.6}$$

This summation will certainly be zero if $c_l^n(\phi) = [c_l^*(\phi)]^n$; that is, if $(d/d\phi)F_\nu(\phi) = 0$ for $\nu = n$. Thus, if

$$\frac{d}{d\phi} F_n(\phi) = \frac{n^2}{2^{n-1}} \sum_{\substack{\nu=1 \\ \nu \text{ odd}}}^{n} (-1)^{(\nu-1)/2} \binom{n}{\nu} a_l^{n-\nu}(\phi) b_l^\nu(\phi) = 0 \tag{8.3.7}$$

the numerator of Equation 8.3.4 will be identically zero and the likelihood equation satisfied, as was to be shown.

If $y_l(t) = \sqrt{2}\,AT_s \sin(\omega_c t + \theta_r + \phi_0) + n(t)$ is the signal received over the interval $lT_s < t < (l+1)T_s$, then

$$\begin{aligned} a_l(\phi) &= \cos\phi' + n_1 \\ b_l(\phi) &= \sin\phi' + n_2 \end{aligned} \tag{8.3.8}$$

where $\phi' = \phi_0 - \phi + \theta_r \triangleq \phi_e + \theta_r$ and

$$\left.\begin{matrix} n_1 \\ n_2 \end{matrix}\right\} = \frac{1}{AT_s}\int_{lT_s}^{(l+1)T_s} n(t)\sqrt{2}\begin{Bmatrix} \sin(\omega_c t + \phi) \\ \cos(\omega_c t + \phi) \end{Bmatrix} dt$$

Thus, $c_l(\phi) = (e^{-i\phi'}/2) + m_1$, $c_l^*(\phi) = (e^{i\phi'}/2) + m_2$, where $m_1 = (n_1 - in_2)/2$, $m_2 = m_1^*$, and

$$in^2 E\{c_l^n(\phi) - [c_l^*(\phi)]^n\} = \frac{n^2}{2^{n-1}} \sin n\phi' = \frac{n^2}{2^{n-1}} \sin n\phi_e \tag{8.3.9}$$

From Equation (8.3.9) we conclude that one of the roots of the function in Equation (8.3.3) is indeed an unbiased estimate of ϕ_0. In addition, the sign of the expected value of this function will be positive when the estimate ϕ of ϕ_0 is less than ϕ_0 and negative when ϕ exceeds ϕ_0 so long as $|\phi_0 - \phi| \leq \pi/n$. Accordingly, if this function itself, after suitable filtering, is used as the input to a VCO, the resulting device will be able to track the received carrier phase. These PSK carrier tracking loops are shown in Figure 8.3 for $n = 2$, 3, and 4.

There are, of course, $2n - 1$ other values of ϕ_e for which the expected value of Equation (8.3.3) is zero. The expected value of the second derivative of this equation is negative for $n - 1$ of these solutions, indicating a total of n possible lock-in points. This would also be the situation if the actual maximum-likelihood estimator were used, since, as we have already observed, any solution to Equation (8.3.3) is also a solution to the maximum-likelihood equation. Indeed, this n-fold ambiguity is clearly inevitable if the carrier is to be recovered directly from the modulation.†

These carrier tracking loops are sub-optimum only because the expected values of the derivatives of the expressions in Equations (8.3.2) and (8.3.3) are not necessarily equal. The true maximum-likelihood estimate might therefore have a smaller variance. Nevertheless, when the signal-to-noise ratio $R = A^2T_s/N_0$ is small, the expansion in Equation (8.3.4) is well approximated by retaining only the first non-zero term. In this case, then, the two estimators are essentially identical. The estimator based on the function of Equation (8.3.3) is therefore asymptotically optimum as $R \to 0$. Moreover,

†The comments made concerning the analogous problem in the preceding section are equally applicable here. Still another method for overcoming this problem in the case of digital systems is to encode the information *differentially*; i.e., to represent the information in terms of the changes between successive symbols rather than by the symbols themselves (cf. Section 12.6).

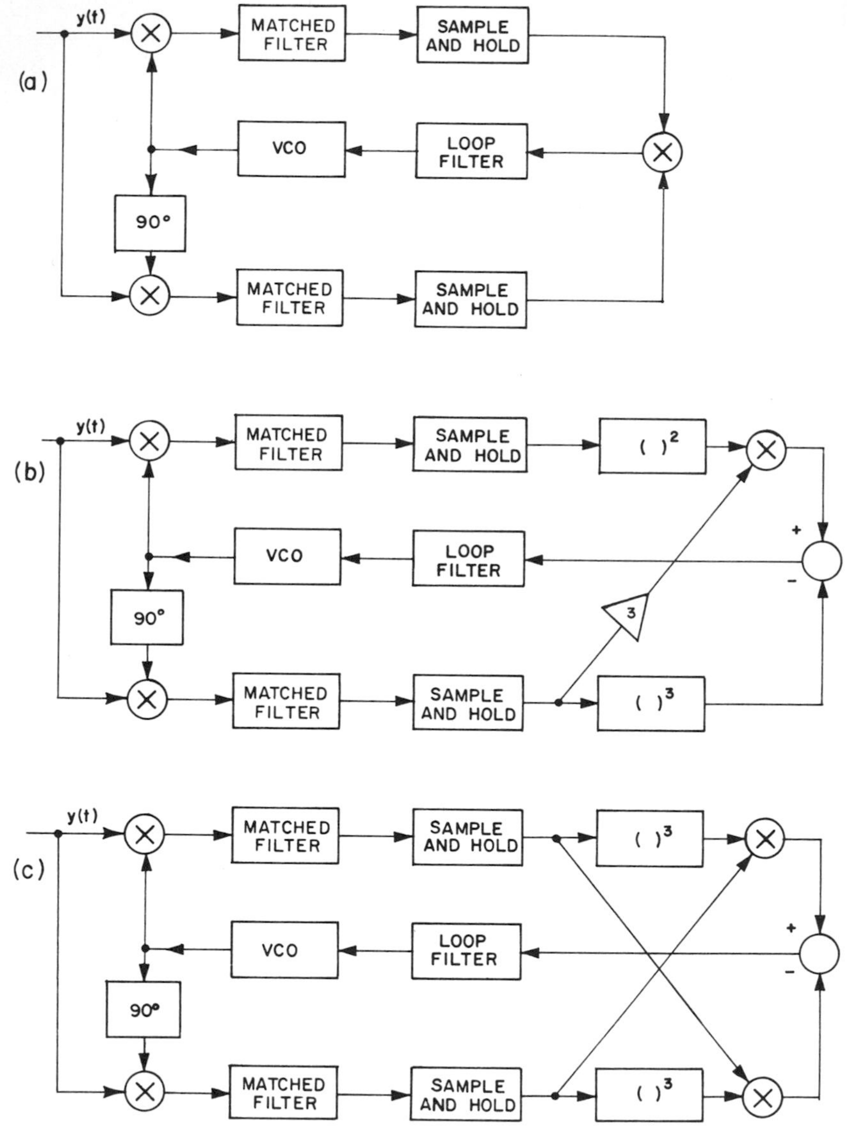

Figure 8.3 PSK carrier synchronizers. (a) Bi-phase; (b) three phase; (c) four phase.

as we shall see, this same estimator is also asymptotically optimum at the other extreme, when $R \to \infty$.

The performance analysis of these loops is identical to that of the PAM carrier tracking loop. The phase variance, when the equivalent loop noise bandwidth B_L is small as compared to $1/T_s$ is

$$\sigma_\phi^2 = \frac{2\sigma_n^2 B_L T_s}{A_e^2} \tag{8.3.10}$$

with $\sigma_n^2 = -n^4[E\{c_l^n(\phi) - [c_l^*(\phi)]^n\}^2 - E^2\{c_l^n(\phi) - [c_l^*(\phi)]^n\}]$ the variance of the signal at the loop filter input, and with A_e the effective signal amplitude. The noise variance σ_n^2 can be shown to be, after some manipulation,

$$\sigma_n^2 = \frac{n^4}{2^{2n-3}} \sum_{k=1}^{n} \binom{n}{k}^2 k! \frac{1}{R^k} \tag{8.3.11}$$

Also, from Equation (8.3.9), the effective signal amplitude is

$$A_e = \frac{d}{d\phi_e}\left\{\frac{n^2}{2^{n-1}} \sin n\phi_e\right\}_{\phi_e=0} = \frac{n^3}{2^{n-1}} \tag{8.3.12}$$

Thus

$$\sigma_\phi^2 = \frac{B_L T_s}{n^2} \sum_{k=1}^{n} \binom{n}{k}^2 k! \frac{1}{R^k} \tag{8.3.13}$$

When R is large,

$$\sigma_\phi^2 \sim \frac{N_0 B_L}{A^2} \tag{8.3.14}$$

Since this would be the phase variance in the absence of any modulation whatsoever, it clearly represents the optimum performance attainable from any carrier phase tracking device. This validates the statement made above that the loops described here are asymptotically optimum as $R \to \infty$. They are also optimum when R is small (if B_L is also sufficiently small) in which case

$$\sigma_\phi^2 \sim B_L T_s \frac{n!}{n^2} \frac{1}{R^n} \tag{8.3.15}$$

Clearly, such loops are practical only for small values of n, unless R is also relatively large.

8.4 Large Signal-to-Noise Ratio Carrier Phase Tracking

Although the tracking loops just discussed are asymptotically optimum as $R \to \infty$, other devices can, in some cases, yield better performance when R is large but finite. One method is to use the analog of Equation (7.2.4) as the function on which to base a decision, i.e., to use the function

$$\max_j \exp\left\{\frac{1}{N_0}\int_{lT_s}^{(l+1)T_s} [2y(t)x_j(t, \phi) - x_j^2(t, \phi)]\, dt\right\} \tag{8.4.1}$$

This approach, called *decision-directed feedback*, involves first a determination of the symbol most likely received and then an attempt to estimate, or more specifically to update the estimate of the carrier phase.

As an example, consider again the PSK system of the previous section. Then, when R is large and j_l is the value of j maximizing (8.4.1)

$$\frac{N_0}{2A}\frac{\partial \log_e p[y_l(t)\mid\phi]}{\partial\phi} \approx \int_{lT_s}^{(l+1)T_s} y(t)\sqrt{2}\cos(\omega_c t + \theta_{j_l} + \phi)\,dt \quad (8.4.2)$$

suggesting the device shown in Figure 8.4. This device differs from an ordinary phase-locked loop only in the presence of a phase-shifter which is controlled by the output of the symbol detector. The T_s second delay is needed because a complete symbol is to be received before a decision is made as to its phase. This decision then alters the VCO output so as to form the function defined in Equation (8.4.2). (The integrator can be omitted since it has virtually no effect if, as has been assumed, the loop bandwidth B_L is small as compared to $1/T_s$.)

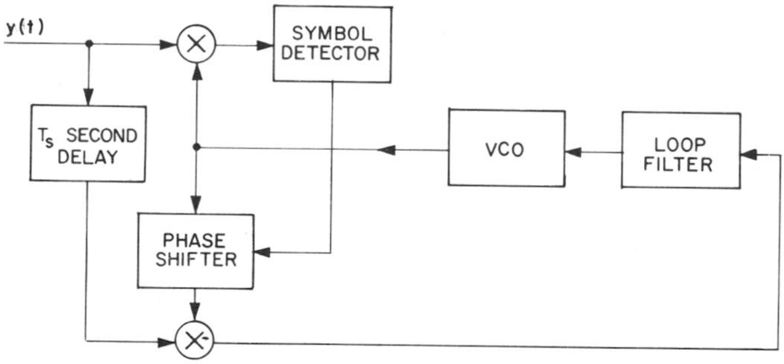

Figure 8.4 Decision-directed feedback PSK carrier synchronizer.

If the symbol detector always made the correct decision, the performance of this device would be identical to that of a phase-locked loop having the same input, but without the modulated phase, and having the same loop filter. The effect of an incorrect symbol decision is to reduce the input signal power, and to increase the noise power. Suppose the received signal is $y_l(t) = \sqrt{2}A\sin(\omega_c t + \theta_r + \phi_0)$ and that the detector decides on the phase θ_s. Then, the low-frequency component of the loop filter input due to the signal will be, over the next T_s seconds,

$$A\cos\theta_e \sin\phi_e + A\sin\theta_e\cos\phi_e \quad (8.4.3)$$

where $\phi_e = \phi_0 - \phi$, ϕ is the current VCO phase, and $\theta_e = \theta_r - \theta_s$. This first term is the desired signal, reduced in amplitude by the factor $\cos\theta_e$; the second term constitutes noise. This noise term is easily shown to be uncorrelated with the additive noise term

$$n_1(t) = \sqrt{2}\,n(t)\cos(\omega_c t + \theta_s + \phi) \quad (8.4.4)$$

[The output of the largest correlator in the symbol detector is influenced by the noise solely through the term $\sqrt{2}\,n(t)\sin(\omega_c t + \theta_s + \phi)$ which is independent of $n_1(t)$. The other correlator outputs are not independent of

$n_1(t)$, of course, but the contribution to these outputs due to $n_1(t)$ could equally well have had either sign without altering the result.] Moreover, the noise pulses $A \sin \theta_e \cos \phi_e$ corresponding to any two different symbol intervals are also uncorrelated. Thus, when $B_L \ll 1/T_s$ this noise simply adds to the low-frequency noise spectral density the factor

$$\sigma^2 T_s \tag{8.4.5}$$

where

$$\sigma^2 = A^2 E\{\sin^2 \theta_e \cos^2 \phi_e\} \approx A^2 E\{\sin^2 \theta_e\}$$

Since the average signal amplitude is reduced by the factor $E(\cos \theta_e)$, the phase error variance at the VCO output is, approximately,

$$\sigma_\phi^2 = \frac{N_0 B_L}{A^2 E^2(\cos \theta_e)} [1 + RE(\sin^2 \theta_e)] \tag{8.4.6}$$

By far the most prevalent error, at high signal-to-noise ratios R, is to select the phase nearest to the correct one. Thus, for n-phase PSK, $|\theta_e| = 2\pi/n$ with high probability, and

$$E^2(\cos \theta_e) \approx \left[1 - \left(1 - \cos \frac{2\pi}{n}\right) P_e\right]^2$$

$$E(\sin^2 \theta_e) \approx \left(\sin^2 \frac{2\pi}{n}\right) P_e \tag{8.4.7}$$

where P_e is the probability of a symbol error.

This same technique is also applicable, for example, when the modulation is orthogonal FSK or PPM. The phase-shifter in Figure 8.4 would be replaced by a frequency-shifter and an "and-gate," respectively. Since the signals are orthogonal, the effect of an erroneous symbol decision would be to leave the noise unaltered, and to eliminate the signal entirely. Thus, the phase error variance of such a loop would be

$$\sigma_\phi^2 \approx \frac{N_0 B_L}{A^2(1 - P_e)^2} \tag{8.4.8}$$

where, again, P_e is the probability of a symbol error. This, of course, assumes the existence of nearly perfect symbol synchronization enabling the symbol modulation to be eliminated before the signal reaches the loop filter.

One possible drawback to the approach outlined here is in the required T_s second delay. If the symbol detector output can be used to obtain an estimate of the currently most probable received symbol at every instant of time, this delay can be omitted entirely. In the case of binary PSK, for example, the maximum-likelihood estimate of the symbol being received at any instant of time t is provided by the sign of the correlator output at time t, i.e., by the function

$$s(t) \triangleq \operatorname{sgn} \int_{lT_s}^{t} y(t)\sqrt{2} \sin (\omega_c t + \phi)\, dt \qquad lT_s < t < (l + 1)T_s \tag{8.4.9}$$

Thus, when the modulation is binary PSK, the delay can be eliminated in the device shown in Figure 8.4 and the phase-shifter replaced by a multiplier having as one input the VCO output and as the other input the function $s(t)$.

Obviously, this modified device will be inferior to the device having the delay since the early estimates $s(t)$ will be wrong with a probability of nearly $\frac{1}{2}$. The phase variance in both cases will be given by Equation (8.4.6) with $E(\sin^2 \theta_e) = 0$, $E^2(\cos \theta_e) = (1 - 2P_0)^2$. In the loop having the delay, P_0 is simply the probability of a symbol error. In the modified loop, P_0 is the percentage of time $s(t)$ has the wrong sign:

$$\begin{aligned} P_0 &= \frac{1}{2} - \frac{1}{T_s}\frac{1}{\sqrt{2\pi}}\int_0^{T_s}\int_0^{[2R(t/T_s)]^{1/2}} e^{-x^2/2}\,dx\,dt \\ &= \frac{1}{2} - \frac{1}{\sqrt{2\pi}}\int_0^{(2R)^{1/2}}\left(1 - \frac{x^2}{2R}\right)e^{-x^2/2}\,dx \\ &= \frac{1}{2} - \frac{1}{2}\left(1 - \frac{1}{2R}\right)\operatorname{erf}(R^{1/2}) - \frac{1}{2\pi^{1/2}R^{1/2}}e^{-R} \end{aligned} \tag{8.4.10}$$

Consequently

$$(1 - 2P_0)^2 = \left[\left(1 - \frac{1}{2R}\right)\operatorname{erf}(R^{1/2}) + \frac{1}{R^{1/2}\pi^{1/2}}e^{-R}\right]^2 \cong \begin{cases} 1 & R \gg 1 \\ \dfrac{16}{9\pi}R & R \ll 1 \end{cases} \tag{8.4.11}$$

and

$$\sigma_\phi^2 \approx B_L T_s\left\{\frac{1}{R(1 - 2P_0)^2}\right\} \cong \begin{cases} B_L T_s \dfrac{1}{R} & R \gg 1 \\ \dfrac{9\pi}{16} B_L T_s \dfrac{1}{R^2} & R \ll 1 \end{cases} \tag{8.4.12}$$

In the binary PSK tracking loop with delay,

$$P_e = \frac{1}{2} - \frac{1}{\sqrt{2\pi}}\int_0^{(2R)^{1/2}} e^{-x^2/2}\,dx = \frac{1}{2}(1 - \operatorname{erf} R^{1/2}) \tag{8.4.13}$$

and

$$\sigma_\phi^2 \approx B_L T_s\left\{\frac{1}{R(1 - 2P_e)^2}\right\} \cong \begin{cases} B_L T_s \dfrac{1}{R} & R \gg 1 \\ \dfrac{\pi}{4} B_L T_s \dfrac{1}{R^2} & R \ll 1 \end{cases} \tag{8.4.14}$$

Both of these derivations have ignored a second-order effect which will actually cause the system performance to deteriorate more rapidly as the signal-to-noise ratio R decreases. In the determination of both P_0 and P_e the carrier reference was assumed exactly known. Yet, as R decreases, the carrier phase-reference will become less and less precise thereby reducing the effective input signal-to-noise ratio and hence further increasing both P_e and P_0. Nevertheless, if σ_ϕ^2 is satisfactorily small, this effect will indeed remain a secondary effect and the approximate expressions in Equations (8.4.12) and (8.4.14) will be indicative of the true performance of these systems.

8.5 The Costas Loop and the Squaring Loop

The phase tracking techniques of the previous sections all presumed prior symbol synchronization. This information is actually not that critical to the performance of the tracking loop. At worst, when symbol synchronization is in error by $T_s/2$ seconds, the average signal amplitude at the input to the loop filter is reduced by a factor of only two, for example, for both PSK and PAM (when the symbol amplitude distribution is symmetric about zero). The phase error variance is therefore increased by at most a factor of four in the absence of symbol synchronization.

In this section, however, we show that it is possible to suffer even less degradation in performance when symbol synchronization is not known, provided the tracking loop is designed without assuming this information. The approach is simply to replace the integrators in the loop by low-pass filters. Since these integrators alone require symbol synchronization, the altered loop will be completely independent of this information. Although the same approach is applicable to all of the loops of Section 8.3, the present discussion will be limited to the PAM and binary PSK loops.

These two loops, which are identical, take the form depicted in Figure 8.5 when the integrators are replaced by arbitrary filters. This device is generally referred to as a *Costas loop* (Reference 8.1).

Again this loop can be analyzed in much the same way as an ordinary phase-locked loop. Let the received signal be $y(t) = \sqrt{2}\,A(t) \sin [\omega_c t + \phi_0(t)] + n(t)$. Then the signal at the output of the upper filter (Figure 8.5) can be written in the form $A_1(t) \sin \phi_e(t) + n_1(t)$ where $A_1(t)$ is the filtered version of $A(t)$, $\phi_e(t) = \phi_0(t) - \phi_1(t)$, and where $n_1(t)$ is the filtered product $n(t)\sqrt{2} \cos [\omega_c t + \phi_1(t)]$. Similarly, the signal at the output of the lower filter is $A_1(t) \cos \phi_e(t) + n_2(t)$, with $n_2(t)$ the filtered product $n(t) \sqrt{2} \sin [\omega_c t + \phi_1(t)]$. If these filters are wideband relative to the loop bandwidth B_L (and since the bandwidth of these filters will be of the order of $1/T_s$, this is the same assumption made in the preceding section), the conclusions of Section 5.5 are equally applicable here. In particular, identifying $\sqrt{2}\,n_1(t)$ with the $n'(t)$ of Equation (5.5.7), and $\sqrt{2}\,n_1(t)$ with the $n''(t)$ of the same equation, we conclude that $n_1(t)$ and $n_2(t)$ are mutually independent Gaussian processes, and, so far as the remainder of the loop is concerned, essentially white. (That is, $E\{n_1(t)n_1(t + \tau)\} = E\{n_2(t)n_2(t + \tau)\} \approx 0$ for $\tau \gg 1/B_n$, B_n denoting the noise bandwidth of the two filters.) Moreover, since $\phi_e(t)$ is a slowly varying function of time (i.e., since the loop bandwidth is narrow) $n_1(t_2)$ is effectively independent of $\phi_e(t_1 + \tau)$ and hence also of $A_1(t_1 + \tau)$,

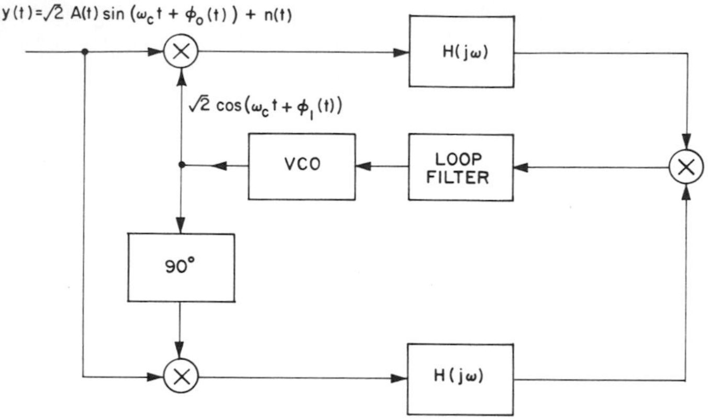

Figure 8.5 Costas loop.

for $t_1 \leq t_2$ and $\tau \ll 1/B_L$ (and similarly for $n_2(t_2)$; see Section 5.5]. Thus, we find, for example, that

$$E\{n_1(t)n_1(t+\tau)A_1(t)\sin\phi_e(t)A_1(t+\tau)\sin\phi_e(t+\tau)\}$$
$$\approx \begin{cases} E\{n_1(t)n_1(t+\tau)\}E\{A_1(t)A_1(t+\tau)\sin^2\phi_e(t)\} & \tau \ll 1/B_L \\ E\{n_1(t)A_1(t)\sin\phi_e(t)A_1(t+\tau)\sin\phi_e(t+\tau)\}E[n_1(t+\tau)] = 0 & \\ & \tau \gg 1/B_n \end{cases} \tag{8.5.1}$$

and, if $B_n \gg B_L$, we have

$$E\{n_1(t)n_1(t+\tau)A_1(t)\sin\phi_e(t)A_1(t+\tau)\sin\phi_e(t+\tau)\}$$
$$\approx R_{n_1}(\tau)E\{A_1(t)A_1(t+\tau)\sin^2\phi_e(t)\} \tag{8.5.2}$$

for all τ.

Using Equation (8.5.2) and analogous approximations to the other expectations of this same form, we come to the following conclusions: (1) The signal $A_1^2(t)\sin\phi_e(t)\cos\phi_e(t) = \frac{1}{2}A_1^2(t)\sin 2\phi_e(t)$ at the loop filter input is uncorrelated with the noise $n'(t) = n_1(t)A_2(t)\cos\phi_e(t) + n_2(t)A_2(t)\sin\phi_e(t) + n_1(t)n_2(t)$ at this point. (2) The noise autocorrelation function $R_{n'}(\tau)$ can be expressed in the form

$$R_{n'}(\tau) = R_{n_i}(\tau)R_{A_1}(\tau) + R_{n_i}^2(\tau) \tag{8.5.3}$$

where

$$R_{n_i}(\tau) = R_{n_1}(\tau) = R_{n_2}(\tau)$$

and, of course,

$$R_{A_1}(\tau) = E\{A_1(t)A_1(t+\tau)\}$$

Finally, again because $B_L \ll 1/T_s$, the signal magnitude at the loop filter input can be approximated by its average value as in Section 8.2. Accordingly, if $\phi_e(t)$ remains small, the tracking loop of Figure 8.5 is well approximated by the conventional linear phase-locked loop model (cf.

Figure 8.2) with the input $\phi_0(t) + n'(t)/E[A_1^2(t)]$. The phase-error variance due to the additive noise is therefore approximately

$$\sigma_\phi^2 = \frac{2B_L S_{n'}(0)}{E^2[A_1^2(t)]} \tag{8.5.4}$$

where B_L is the noise bandwidth of the linear model of the loop, and from Equation (8.5.3),

$$S_{n'}(0) = \int_{-\infty}^{\infty} R_{n'}(\tau)\, d\tau = \int_{-\infty}^{\infty} S_{A_1}(f) S_{n_i}(-f)\, df + \int_{-\infty}^{\infty} S_{n_i}(f) S_{n_i}(-f)\, df \tag{8.5.5}$$

But

$$A_1(t) = \int_0^{\infty} h(\tau) A(t - \tau)\, d\tau \tag{8.5.6}$$

where $h(t)$ is the impulse response of the filters preceding the multiplier. Consequently, if $S_A(f)$ is the spectral density of $A(t)$

$$E[A_1^2(t)] = \int_{-\infty}^{\infty} S_A(f) | H(j2\pi f) |^2\, df \tag{8.5.7}$$

Similarly when $n(t)$ is white with the single-sided spectral density N_0,

$$S_{n_i}(f) = \frac{N_0}{2} | H(j2\pi f) |^2 \tag{8.5.8}$$

and

$$S_{n'}(0) = \frac{N_0}{2} \int_{-\infty}^{\infty} S_A(f) | H(j2\pi f) |^4\, df + \left(\frac{N_0}{2}\right)^2 \int_{-\infty}^{\infty} | H(j2\pi f) |^4\, df \tag{8.5.9}$$

Thus,

$$\sigma_\phi^2 = N_0 B_L \left\{ \frac{\int_{-\infty}^{\infty} S_A(f) | H(j2\pi f) |^4\, df + \frac{N_0}{2} \int_{-\infty}^{\infty} | H(j2\pi f) |^4\, df}{\left[\int_{-\infty}^{\infty} S_A(f) | H(j2\pi f) |^2\, df\right]^2} \right\} \tag{8.5.10}$$

The optimum filter $H(j\omega)$ is presumably the filter minimizing σ_ϕ^2. But minimizing σ_ϕ^2 is equivalent to minimizing the numerator of Equation (8.5.10) subject to a constraint on the denominator, or, more conveniently, on the square root of the denominator. Thus, we wish to find the transfer function $H(j\omega)$ minimizing

$$\begin{aligned} &\int_{-\infty}^{\infty} \left\{ \left[S_A(f) + \frac{N_0}{2}\right] | H(j2\pi f) |^4 - \lambda S_A(f) | H(j2\pi f) |^2 \right\} df \\ &= \int_{-\infty}^{\infty} \left\{ \left[S_A(f) + \frac{N_0}{2}\right] \left[| H(j2\pi f) |^2 - \frac{\lambda S_A(f)}{2[S_A(f) + (N_0/2)]} \right]^2 \right. \\ &\qquad \left. - \frac{\lambda^2 S_A^2(f)}{4[S_A(f) + (N_0/2)]} \right\} df \end{aligned} \tag{8.5.11}$$

This last integral clearly attains its minimum value when

$$| H(j2\pi f) |^2 = \frac{(\lambda/2) S_A(f)}{S_A(f) + (N_0/2)} \tag{8.5.12}$$

This is the transfer function of the optimum filter. Since it is a condition on $|H(j\omega)|^2$ only, $H(j\omega)$ can obviously represent a causal filter. The constant $\lambda/2$ is arbitrary, as it does not affect the ratio of Equation (8.5.10). When $H(j\omega)$ is as defined in Equation (8.5.12), this ratio becomes

$$\sigma_\phi^2 = \frac{N_0 B_L}{\displaystyle\int_{-\infty}^{\infty} \frac{S_A^2(f)}{S_A(f) + (N_0/2)}\, df} \tag{8.5.13}$$

For either PAM or binary PSK modulation with rectangular-shaped pulses the spectrum of $A(t)$ is of the form

$$S_A(f) = P_s T_s \left(\frac{\sin \pi f T_s}{\pi f T_s}\right)^2 \tag{8.5.14}$$

where P_s is the average received signal power. Although the general expression for σ_ϕ^2 is rather unwieldy when $S_A(f)$ is as given by Equation (8.5.14), its asymptotic values at both large and small signal-to-noise ratios are easily found. When $N_0/2 \ll S_A(f)$ [for all f for which the contribution of $S_A(f)$ to the total signal power is of significance] we have

$$\sigma_\phi^2 \approx \frac{N_0 B_L}{P_s} = \frac{B_L T_s}{R} \tag{8.5.15}$$

with $R = P_s T_s / N_0$. This result, which incidentally is independent of the signal spectrum, represents the phase variance which would be observed in the absence of any modulation whatsoever. At the other extreme, when $N_0/2 \gg S_A(f)$, for all f,

$$\sigma_\phi^2 \approx \frac{N_0^2 B_L}{2\displaystyle\int_{-\infty}^{\infty} S_A^2(f)\, df} = \frac{3N_0^2 B_L}{4P_s^2 T_s} = \frac{3}{4}\frac{B_L T_s}{R^2} \tag{8.5.16}$$

The phase-error variance resulting from use of the loops of the previous section was found to be, for both PAM and binary PSK,

$$\sigma_\phi^2 = B_L T_s \left(\frac{1}{R} + \frac{1}{2R^2}\right) \tag{8.5.17}$$

Thus, when R is large, the performance in the two cases is nearly identical; when R is small the Costas loop variance is larger by a factor of $\frac{3}{2}$.

As a final comparison, and to gauge the sensitivity of the loop to the choice of the filter $H(j\omega)$, suppose we choose

$$H(j2\pi f) = \begin{cases} 1 & |f| < \dfrac{1}{T_s} \\ 0 & |f| > \dfrac{1}{T_s} \end{cases} \tag{8.5.18}$$

Then, since

$$P_s T_s \int_{-1/T_s}^{1/T_s} \left(\frac{\sin \pi f T_s}{\pi f T_s}\right)^2 df \approx P_s$$

$$\sigma_\phi^2 \approx \frac{B_L N_0\left(P_s + \frac{N_0}{T_s}\right)}{P_s^2} = B_L T_s\left(\frac{1}{R} + \frac{1}{R^2}\right) \tag{8.5.19}$$

An interesting variation on this Costas loop phase tracking method is suggested by the following manipulation: Let $y(t)$ be the input to the Costas loop and let $h(t)$ be the impulse response of the two pre-multiplier filters. Then the output of the upper filter is

$$\int_{-\infty}^{\infty} y(\tau)\sqrt{2} \cos [\omega_c \tau + \phi_1(\tau)] h(t - \tau)\, d\tau \tag{8.5.20}$$

But since this filter is assumed wide relative to the loop bandwidth, $\phi_1(\tau) \approx \phi(t)$ for all τ for which $h(t - \tau)$ is not negligibly small. Thus, the output of the upper filter can be rewritten in the form

$$\begin{aligned} &\left(\int_{-\infty}^{\infty} y(\tau) \cos \omega_c \tau\, h(t - \tau)\, d\tau\right)\sqrt{2} \cos [\phi_1(t)] \\ &\qquad - \left(\int_{-\infty}^{\infty} y(\tau) \sin \omega_c \tau\, h(t - \tau)\, d\tau\right)\sqrt{2} \sin [\phi_1(t)] \\ &\triangleq Y(t)\sqrt{2} \cos \phi_1(t) - X(t)\sqrt{2} \sin \phi_1(t) \end{aligned} \tag{8.5.21}$$

Similarly, the output of the lower filter is

$$Y(t)\sqrt{2} \sin \phi_1(t) + X(t)\sqrt{2} \cos \phi_1(t) \tag{8.5.22}$$

so that

$$[Y^2(t) - X^2(t)] \sin 2\phi_1(t) + 2X(t)Y(t) \cos 2\phi_1(t) \tag{8.5.23}$$

represents the input to the loop filter.

Now consider the square of the output at time t of a filter having the impulse response $h(t) \cos \omega_c t$:

$$\begin{aligned} &\left(\int_{-\infty}^{\infty} y(\tau) h(t - \tau) \cos [\omega_c (t - \tau)]\, d\tau\right)^2 \\ &\quad = \tfrac{1}{2}[Y^2(t) + X^2(t)] + \tfrac{1}{2}[Y^2(t) - X^2(t)] \cos 2\omega_c t + X(t)Y(t) \sin 2\omega_c t \end{aligned} \tag{8.5.24}$$

If this signal is multiplied by the term $4 \sin 2[\omega_c t + \phi_1(t)]$, the resulting low frequency term is precisely as given in Equation (8.5.23).

Thus, an alternative mechanization of the Costas loop, called the *squaring loop*, is shown in Figure 8.6. Since the error signals in the two loops are identical they will produce identical carrier references. The choice between the two devices can be solely on the basis of their relative ease of implementation. This in turn is a function of the frequencies and bandwidths involved. Note that this same argument holds when $h(t)$ is a matched filter,

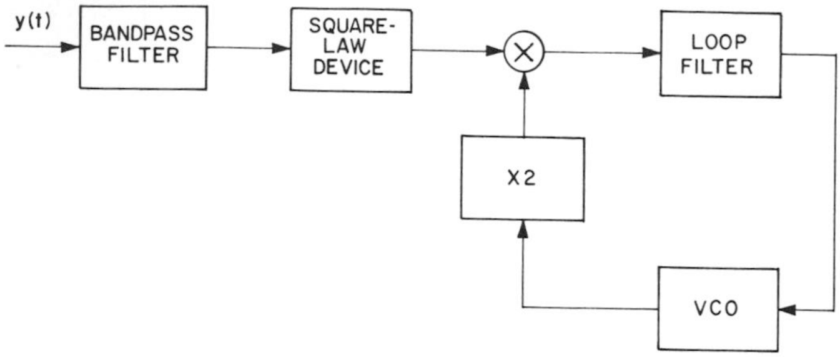

Figure 8.6 Squaring loop.

$h(t) \cos \omega_c t$ then representing an incoherent matched filter. The optimum PAM and binary PSK carrier tracking loops of Sections 8.2 and 8.3 can therefore also be realized as squaring loops. Finally, an obvious extension of this same argument can be used to show an equivalence between the n-ary PSK carrier tracking loop of Section 8.3 and an *nth power loop*. In the latter, a phase-incoherent matched filter is followed by an nth power device, the output of which is tracked by an ordinary phase locked loop (with a times-n frequency multiplier following the VCO).

8.6 Tracking Symbols Having a "Fine-Structure"

A fundamental difference between the tracking error variances associated with the symbol tracking loops of Chapter Seven and those derived for the carrier tracking loops in this chapter is in the latter's relative independence of the symbol statistics. This independence derives from the fact that the carrier frequency was assumed to be large as compared to the reciprocal of the symbol period. Thus each symbol represented many carrier cycles, and the correlation between the local reference and the received carrier was largely unaffected by the relatively rare symbol transitions. The symbol tracking loops, in contrast, depend partially (or exclusively in the case of rectangular pulses) on the occurrence of symbol transitions to produce an error signal.

The advantage in not having to rely on symbol transitions is apparent, particularly when successive symbols are not statistically independent. The key to realizing this advantage in symbol tracking loops is to endow the symbols with a "fine-structure" similar to that exhibited by the carrier. Specifically, the νth symbol might assume the form $x_\nu(t)\sigma(t)$ where $x_\nu(t)$ is as defined in Chapters Three and Four, depending upon the nature of the modulation, and $\sigma(t)$ represents the fine-structure. The signal $\sigma(t)$, for the moment, will be assumed to have a period T_s equal to the symbol period. If the period of $\sigma(t)$ were less than T_s, the fine-structure could be tracked without uniquely establishing symbol synchronization, and the purpose of introducing this structure would be largely lost.

The tracking loops introduced in Section 7.6 are equally applicable when the symbols are of the form $x_\nu(t)\sigma(t)$. The matched filters, of course, must now be matched to these altered symbols. One method for effecting this modification is shown in Figure 8.7(a) for coherent PAM and binary PSK and in Figure 8.7(b) for incoherent PAM and non-binary PSK symbol tracking loops. Since the signal is multiplied by $\sigma(t)$ before it passes through the boxes labeled "matched filter," these filters are matched to the pulse which would be observed were the fine-structure absent.

The analysis in Section 7.6 of the performance of these loops is also applicable here. The variance (in radians squared) of the tracking error is approximately

$$\sigma_\phi^2 = \frac{2S_n(0)B_L}{A_e^2} \tag{8.6.1}$$

where

$$A_e = \frac{1}{2\pi}\left|\frac{\partial}{\partial\tau}E\left\{\zeta\left(\tau+\frac{\Delta t}{2}\right)-\zeta\left(\tau-\frac{\Delta t}{2}\right)\right\}\right|_{\tau=0}$$

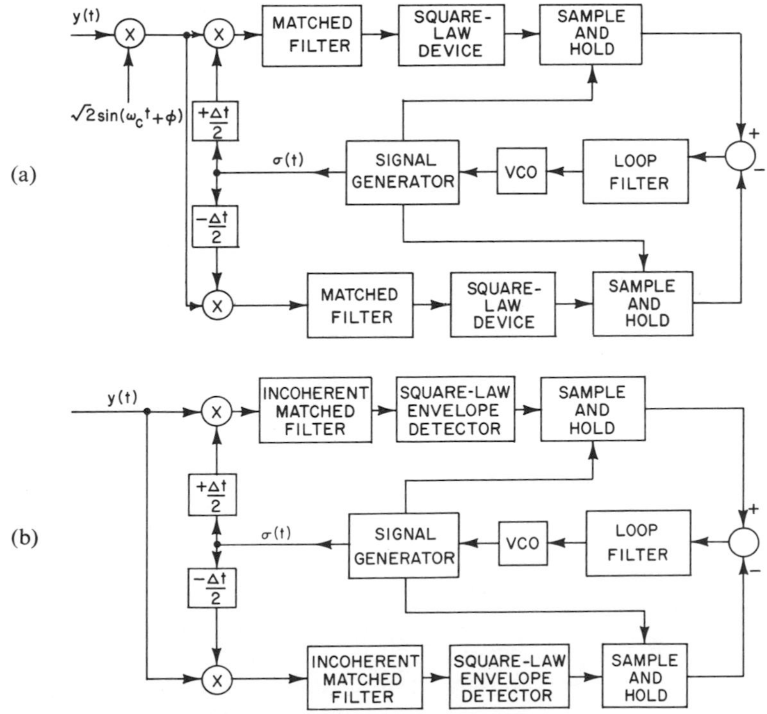

Figure 8.7 Tracking devices for symbols having a fine-structure. (a) Coherent; (b) incoherent.

and

$$S_n(0) = \frac{\Delta^2(\Delta t, -\Delta t/2)}{T_s}$$

with

$$\Delta^2(u, t) = \lim_{M \to \infty} \frac{1}{M} \operatorname{Var}[\zeta(t) - \zeta(t + u)]$$

[cf. Equations (7.6.7) and (7.6.17)]. The function $\zeta(t)$ represents the (normalized) output of the loop square-law device (Figure 8.7) at time t.

Since the purpose of introducing the fine-structure was to make the error signal effectively independent of the symbol statistics, $\sigma(t)$ can be assumed to be such that this goal is accomplished. Accordingly, the mean and variance of the variable $\zeta(t)$ are as determined in Section 6.6 [Equation (6.6.12) with $\zeta_\nu = \zeta(t = \nu \Delta T)$], and are functions only of the signal-to-noise ratio and the autocorrelation function $\rho(\tau)$ of $\sigma(t)$.

To convert this expression [Equation (8.6.1)] into something more manageable, let $\sigma(t)$ represent a pseudo-noise sequence of period T_s and with bit period $\Delta t = T_s/n$. Then

$$\rho(\tau) = \int_0^{T_s} \sigma(t)\sigma(t + \tau)\, dt = \begin{cases} 1 - \dfrac{n+1}{n}\dfrac{\tau}{\Delta t} & |\tau| < \Delta t \\ -\dfrac{1}{n} & |\tau| \geq \Delta t \\ & \text{modulo } T_s \end{cases} \tag{8.6.2}$$

and, for large n,

$$\sigma_\phi^2 = 2\pi^2 \left(\frac{\Delta t}{T_s}\right)^2 \left(\frac{1}{R} + \frac{c_0}{R^2}\right) B_L T_s \tag{8.6.3}$$

with

$$R = \begin{cases} A^2 E(x^2) T_s / N_0 & \text{PAM [cf. Equation (8.2.9)],} \\ A^2 T_s / N_0 & \text{PSK} \end{cases}$$

and

$$c_0 = \begin{cases} 1 & \text{coherent PAM and binary PSK} \\ 2 & \text{incoherent PAM and non-binary PSK} \end{cases}$$

This variance is in radians squared with 2π radians corresponding to T_s seconds.

It will be noted that not only is the rms tracking error independent of the symbol statistics [except for the term $E(x^2)$ in the PAM case], but that it is also multiplied by the factor $(\Delta t/T_s)$. This apparent advantage in using a long sequence is entirely counteracted, however, by the fact that the acceptable tracking error is decreased by this same factor due to the inclusion of the sequence. Moreover, since the effective signal amplitude A_e is increased when the sequence is added, the loop bandwidth B_L presumably must be increased because of its presence (see Section 6.3). Often the most fundamental constraint on the length of the sequence, however, is a bandwidth constraint; the bandwidth required to accommodate the modified symbols is increased by the factor $T_s/\Delta t$. Since this ratio must be moderately large for the fine-structure to be effective, the bandwidth expansion aspect of this approach generally represents its most serious drawback.

This technique of introducing a fine-structure into the symbol stream for synchronization purposes is easily extended to provide higher-order synchronization as well. If $\sigma(t)$ were a pseudo-noise sequence of period kT_s, for example, tracking this sequence at the receiver would identify a unique point in each sequence of k symbols. If the ratio $T_s/\Delta t$ is kept large, the tracking error will still be approximately as given in Equation (8.6.3). The reason this result is only approximate in this case is because now the filters preceding the square-law devices are matched to only a segment of the full pseudo-noise sequence. Thus, the error signal is determined not by the autocorrelation function of Equation (8.6.2) (since T_s is no longer the period of the p-n sequence) but rather by the set of functions

$$\rho_i(\tau) = \int_{iT_s}^{(i+1)T_s} \sigma(t)\sigma(t+\tau)\,dt \qquad (8.6.4)$$

These *partial autocorrelation functions*, however, are still approximately of the form of Equation (8.6.2) (with proper normalization) *if* Δt is sufficiently small relative to T_s.

Since the phase-error variance of Equation (8.6.3) is independent of the symbol modulation, it must be valid even in the absence of any modulation. Consequently, this expression also represents the phase error in tracking incoherently a pseudo-noise sequence, or equivalently a sequence of rectangular pulses of width Δt (cf. Section 6.6). In the latter case, the sampler in Figure 8.7(b) may still be required. This would be the situation if the carrier phase, while relatively constant over any Δt second interval, could not be assumed constant for T_s seconds. If, however, the carrier phase is essentially constant over any T_s second interval, and this assumption is implicit if $\sigma(t)$ is non-zero over the full T_s seconds and if the received symbols are not otherwise modulated, the sampler in Figure 8.7(b) is clearly unnecessary; the output of the envelope detector represents a valid input to the tracking loop at every instant of time. Were it not for the non-linear element preceding the sampler, its omission would not alter the tracking performance. Because of this non-linearity, however, the system performance can actually be improved, when there is no symbol modulation, by eliminating the sampler.

The analysis of the device in Figure 8.7(b) when the sampler is omitted, and when the matched filter is replaced by an arbitrary filter, is similar to the analysis of the Costas loop in Section 8.5. As one can verify by paralleling the development of that section when $\sigma(t)$ is a pseudo-noise sequence with bit period Δt, the phase-error variance at the output of this modified tracking loop is (when the pre-loop to loop bandwidth ratio is large)

$$\sigma_\phi^2 = 2\pi^2\left(\frac{\Delta t}{T_s}\right)^2\left\{\frac{1}{R} + \frac{2}{R^2}\frac{T_s}{2\pi}\int_{-\infty}^{\infty} |H(j\omega)|^4\,d\omega\right\}B_L T_s \qquad (8.6.5)$$

Here $H(j\omega)$ is the Fourier transform of $h(t)$, with $h(t) \cos \omega_c t$ representing the impulse response of the filter preceding the square-law device. If $h(t)$ is a rectangular filter of bandwidth $B = 1/2T_s$, the phase error variance here is seen to be identical to that using a matched filter followed by a sampler. If $h(t)$ represents a filter matched to a rectangular pulse of width T_s, the omission of the sampler does indeed result in a slight improvement.

This last statement would seem to be at odds with the conclusions of Section 6.6 concerning the optimum phase-incoherent synchronizer. The conflict, however, is with the model of the phase statistic leading to those conclusions, not with the conclusions themselves. Specifically, the inclusion of the sampler is a consequence of the stipulation that the carrier phase $\phi(t)$ is constant over each interval $iT_s < t < (i + 1)T_s$, but not over the interval $iT_s + \tau < t < (i + 1)T_s + \tau$ for any $\tau \neq 0$, modulo T_s. As pointed out earlier, this model is sometimes unrealistic.

8.7 Concluding Remarks

The most direct method for providing the synchronization needed in a communication system is to use three (or more) separate channels, one conveying a clock (and carrier or subcarrier) reference, one higher-order synchronization, and the third the information itself. There are two drawbacks to this approach. First, the three channels must somehow be multiplexed, generally either in time or in frequency. Not only does this increase the system complexity, it often presents interchannel interference problems which may degrade the system performance. Second, the power relegated to the synchronization channels represents power which could have been used to transmit data were some other means of synchronization available.

Accordingly, methods were investigated in the past two chapters for eliminating the necessity of one or both of these auxiliary channels. Chapter Seven formulated and analyzed procedures for achieving symbol synchronization directly from the sequence of received symbols, thereby eliminating, at least in part, the need for one of the synchronization channels. (Methods for obtaining higher-order synchronization from the data stream are explored in later chapters; see also Section 8.6.) This chapter continued this trend by examining methods for abstracting a clock reference directly from the modulated carrier.

It should not be surprising that in every case the accuracy with which the sync information could be determined directly from the modulation was diminished relative to the separate channel approach when the two channels represented the same signal power. Indeed, one would probably have expected the contrast between the two situations to be more pronounced. But since the percentage of power relegated to the synchronization channel should hopefully be small, the synchronization estimates obtained directly from the modulation may well be the superior of the two. To verify this last conjecture, however, we must determine the optimum allocation of power between the information channel and the synchronization channels. This is one of the topics to be investigated in Chapter Nine.

Notes to Chapter Eight

The Costas loop (Reference 8.1) was first proposed as a device for demodulating suppressed carrier amplitude modulation. Most of the early work dealing with carrier extraction devices for pulse modulated systems concentrated on the basically equivalent squaring-loop approach (References 8.2, 8.3 and 8.4). The form of the optimum filter for the non-symbol-synchronized squaring loop (cf. Section 8.5) was originally determined by Layland (Reference 8.5). Some of the decision-directed feedback schemes discussed in Section 8.4 are investigated in Reference 8.6. (See also Reference 8.7 in this regard.) The fine-structure tracking loops of Section 8.6 are examined in References 8.8 and 8.9. The material in the first three sections of this chapter has been published previously only in brief summary (cf. Reference 8.10).

Problems

8.1 Derive Equation (7.7.5) using the approach introduced in Section 8.2.

8.2 Show that the n-ary PSK carrier tracking loop of Section 8.3 is equivalent, under the conditions imposed in the squaring loop discussion in Section 8.5, to an nth power loop; i.e., a phase-locked loop preceded by a bandpass filter and an nth-law device.

8.3 Consider the carrier tracking loop suggested in the previous problem for quadraphase PSK, but with the matched filters replaced by rectangular bandpass filters with bandwidth $2/T_s$. Show that, in this case, the tracking error variance is approximately

$$\sigma_\phi^2 = \left(\frac{1}{R} + \frac{36}{R^2} + \frac{18}{R^3} + \frac{8}{R^4}\right) B_L T_s$$

Compare this with the result when matched filters are used.

8.4 Repeat Problem 5.6 with $y(t)$ a noisy, binary PSK signal, [$y(t) = \sqrt{2}\, A \sin(\omega_c t + \theta_l) + n(t)$, $lT_s < t < (l+1)T_s$, $\theta_l = 0$ or π], with the odd νth law device replaced by an even νth law device ($x(t) \rightarrow |x(t)|^\nu$), and with B_2 centered about $2\omega_c$. Show, in particular, that under the conditions delineated in Problem 5.6 the output signal-to-noise ratio is now inversely proportional to $l + (\nu/2)^2$ but that the accuracy with which the carrier can be reconstructed from this output signal is again independent of ν.

8.5 Show from Equation (8.3.1) that the ideal *binary* PSK carrier tracking loop is as shown in Figure 8.3a but with the upper loop containing the non-linearity tanh $[2R(\cdot)]$. Compare this loop to the loops discussed in Sections 8.3 and 8.4. In particular show that when $R \rightarrow \infty$ this loop is identical to the loop discussed in Section 8.4 with the delay.

8.6 The variance of all the carrier tracking loops examined in this chapter approached $B_L T_s / R$ at large signal-to-noise ratios R. This same variance would have resulted were the carrier unmodulated and the synchronizer an ordinary phase-locked loop. Should the loop bandwidth B_L be the same in the two cases? Is the loss-of-lock probability the same?

References

8.1 Costas, J.P., "Synchronous Communications," *Proc. IRE*, Vol. 44, p. 1713, 1956.

8.2 Van Trees, H., "Optimum Power Division in Coherent Communication Systems," MIT Lincoln Lab. Tech. Rept. No. 301, Lexington, Mass., 1963.

8.3 Stiffler, J. J., "On the Allocation of Power in a Synchronous Binary PSK Communication System," *Proc. of the National Telemetering Conf.*, p. 51, 1964.

8.4 Chang, S. S. L., and B. Harris, "An Optimum Self-Synchronized System," *AIEE Trans. on Comm. and Elec.*, No. 60, p. 110, 1962.

8.5 Layland, J. W., "An Optimum Squaring-Loop Filter," Jet Propulsion Lab., Space Programs Summary 37–37, Vol. 4, Calif. Inst. Tech., Pasadena, Calif., p. 250, 1967.

8.6 Proakis, J. G., P. R. Drouilhet, Jr., and R. Price, "Performances of Coherent Detection Systems Using Decision-Directed Channel Measurement," *IEEE Trans. on Comm. Syst.*, Vol. CS-12, p. 54, 1964.

8.7 Heller, J. A., "Sequential Decoding for Channels with Time Varying Phase," Ph.D. Thesis, Mass. Inst. Tech., Cambridge, Mass., 1967.

8.8 Gill, W. J., "A Comparison of Binary Delay-Lock Tracking-Loop Implementations," *IEEE Trans. on Aero. and Elec. Syst.*, Vol. AES-2, p. 415, 1966.

8.9 Ward, R. B., "Digital Communications on a Pseudo-noise Tracking Link Using Sequence Inversion Modulation," *IEEE Trans. on Comm. Tech.*, Vol. COM-15, p. 69, 1967.

8.10 Stiffler, J. J., "On Tracking PSK Modulated Carriers," *Proc. Hawaii International Conf. on Syst. Science*, Honolulu, Hawaii, p. 800, 1968.

chapter nine

The Power Allocation Problem

9.1 Introduction

Whatever the method used to provide the required synchronization, this information will invariably be imperfect. Obviously, these inaccuracies will cause some degradation in system performance. The estimation of this effect is one of the tasks of this chapter. This knowledge is necessary in the evaluation of the overall communication system and in answering, among other things, the question as to whether the synchronization method contemplated is satisfactory for the situation under consideration. Having this information, moreover, enables one to determine if it is advantageous to relegate more power for synchronization purposes, or if the system is in fact already "over-synchronized" in the sense that some of the power used for synchronization could be used more effectively to transmit information. This question of power allocation between synchronization and information signals will also be investigated in this chapter.

If the statistics of the reference errors are known, their effect on the system performance is, in principle, easily determined. When the output signal-to-noise ratio is a measure of the system performance, for example, and the carrier, subcarrier,† and symbol synchronization errors are ϕ_1, ϕ_2, and ϕ_3, respectively, it is first necessary to examine the consequent output signal-to-noise ratio as a function of these errors:

$$(S/N) = f(\phi_1, \phi_2, \phi_3)$$

Knowing this, we have, as the average performance of the system

$$(S/N)_{\text{ave}} = \iiint f(\phi_1, \phi_2, \phi_3) p(\phi_1, \phi_2, \phi_3)\, d\phi_1\, d\phi_2\, d\phi_3 \tag{9.1.1}$$

where $p(\phi_1, \phi_2, \phi_3)$ is the joint density function of the three reference errors.

Similarly, in discrete systems in which the probability of an error is significant, this probability can be evaluated as a function of the timing errors, and integrated over these errors to determine the average error probability:

$$P_e = \iiint P_e(\phi_1, \phi_2, \phi_3) p(\phi_1, \phi_2, \phi_3)\, d\phi_1\, d\phi_2\, d\phi_3 \tag{9.1.2}$$

The function $P_e(\phi_1, \phi_2, \phi_3)$ denotes the probability of an error, given the reference errors ϕ_1, ϕ_2, and ϕ_3. It is important to note that Equation (9.1.2) is valid, so far as carrier and subcarrier errors are concerned, only if these errors remain relatively constant over a symbol interval. If this is not the case, the symbol error probability is a function of the entire history of the reference error over the symbol interval.

Both of these parameters, the average signal-to-noise ratio and the average error probability, are functions of the *distribution* $p(\phi_1, \phi_2, \phi_3)$ of the phase errors. The major effort in the preceding three chapters has been to evaluate the variances of these errors using different estimation schemes. Fortunately, the phase errors are, to a good approximation, Gaussian in nearly all cases of practical interest, regardless of the estimation procedure. Moreover, the different phase errors are invariably either statistically independent (if they are derived from different signals) or directly proportional to each other (if they are derived from the same signal). Thus, we are concerned only with the single variate distributions $p(\phi_i)$, and, if ϕ_i is a Gaussian random variable, these distributions are completely defined in terms of the variances of the ϕ_i. [Generally, $E(\phi_i)$ will be zero. If the symbol epoch is *detected*, as discussed in Chapters Six and Seven, rather than tracked, an erroneous decision would, of course, result in a constant non-zero error.]

†In previous chapters, no explicit distinction was made between carriers and subcarriers. The synchronization techniques considered were, within limitations, equally applicable whether one or more carriers were involved. (In a multiple carrier coherent system, of course, some method must be provided for obtaining reference signals for all the carriers.) However, each imperfectly known reference signal does contribute to the degradation of the system performance and must be taken into account in its evaluation. Even here though, one of the carriers will usually be dominant so far as its effect on the performance is concerned. Because of this, and because of the impossibility of considering all contingencies anyway, most of the subsequent discussion will be limited to the single carrier situation.

To support this contention that the phase errors are approximately Gaussianly distributed, we first observe that it is obviously true when the phase estimate is a linear function of the Gaussian noise corrupted, received signal. Thus, the Gaussian approximation is valid when the estimate is produced at the output of a phase-locked loop so long as the phase-error variance is sufficiently small (less than about $\frac{1}{4}$ rad^2 in the case of sinusoidal signals; cf. Section 5.5). In most situations of interest here, the phase-error variance will indeed be small enough to justify this assumption; if it were not, the loss-of-lock probability, for one thing, would be unacceptably large.

Further justification for the assumption of a Gaussianly distributed phase error is needed when the tracking loop involves, or is preceded by, non-linear elements, as was the case in Chapters Seven and Eight. Because of non-linearities, the noise at the loop filter input is no longer Gaussian as assumed in Section 5.5. Nevertheless, if the loop bandwidth is narrow relative to the bandwidth of the filters preceding the non-linear devices, an assumption consistently made in the discussions of Chapters Seven and Eight, the phase error at any instant is due to the weighted average of a large number of effectively independent noise contributions. The central limit theorem, therefore, can be used to argue that the phase error is approximately Gaussianly distributed even in this case. This contention is given further support in Reference 9.1.

In the ensuing discussion, the synchronization and information channels will be assumed independent, and any effect the signal or noise of one channel might have on the other will be ignored. In some cases, this assumption is easily justified, as, for instance, when the two channels are sufficiently separated in frequency. Often, however, it is valid only under additional constraints. If the synchronization is derived directly from the modulation as discussed in the past two chapters, for example, or if the synchronization and information channels are not relegated to mutually exclusive portions of the spectrum, the noise contributions to the outputs of the two channels may not be independent. These noise terms, however, do become essentially independent if they are produced at the output of filters having widely disparate bandwidths. For then the noise at the output of the narrow filter represents a negligible segment of the output of the wide filter, and hence is virtually uncorrelated with it. Since the demodulator bandwidth is generally on the order of $1/T_s$, this condition is satisfied if the synchronizer noise bandwidth B_L is small compared to $1/T_s$.† This constraint, it will be recalled, was consistently imposed for other reasons on the synchronizers of the preceding two chapters. Even when the noise contributions to the two channels can be said to be independent, the signals may still mutually interfere. In the sequel, we will ignore this effect. It should be cautioned, however, that in some cases

†It is also satisfied in the rather unlikely condition that $B_L \gg 1/T_s$.

it must be taken into account if the analysis of the system performance is to be meaningful (cf. Problem 9.3).

It should be acknowledged that the average signal-to-noise ratio and the average error probability are somewhat arbitrary measures. A better measure in some cases, for example, might be the percentage of time the error probability exceeds a certain value or the signal-to-noise ratio falls below some prescribed level. Nevertheless, because the average of these quantities is the most frequently used measure, the analysis of the following sections will be concerned solely with the average signal-to-noise ratio, or the average error probability, whichever is applicable. The general approach can readily be extended to other measures if desired.

9.2 The Effect of Inexact References—Pulse Amplitude Modulation

We first consider the effect of a carrier (or subcarrier) phase error on the observed signal-to-noise ratio in a PAM system. If the received signal is

$$y(t) = \sqrt{2}\, Ax \sin \omega_c t + n(t) \tag{9.2.1}$$

and the local estimate of the carrier is $\sqrt{2} \sin (\omega_c t - \phi)$, the detector (Chapter Three) forms the integral

$$I = \frac{1}{AT_s} \int_0^{T_s} y(t) \sqrt{2} \sin (\omega_c t - \phi)\, dt \tag{9.2.2}$$

As in Chapter Three, we identify the output signal-to-noise ratio with the ratio of the signal variance to the mean-squared estimation error. Accordingly, for any given ϕ,

$$\left(\frac{S}{N}\right)_\phi = \frac{\text{Var}(x)}{E(I - x)^2} = \frac{2R}{1 + 2R'(1 - \cos \phi)^2} \tag{9.2.3}$$

where $R = (A^2 T_s/N_0)\text{Var}(x)$ and $R' = (A^2 T_s/N_0) E(x^2)$.

If the carrier phase reference error is Gaussianly distributed

$$\left(\frac{S}{N}\right)_{\text{ave}} = \frac{2R}{\sqrt{2\pi}\, \sigma_\phi} \int_{-\infty}^{\infty} \frac{\exp(-\phi^2/2\sigma_\phi^2)}{1 + 2R'(1 - \cos \phi)^2}\, d\phi \tag{9.2.4}$$

Since $(1 + x)^{-1} \geq 1 - x$ for all real x,

$$\begin{aligned}
\left(\frac{S}{N}\right)_{\text{ave}} &\geq \frac{2R}{\sqrt{2\pi}\, \sigma_\phi} \int_{-\infty}^{\infty} \exp\left(-\frac{\phi^2}{2\sigma_\phi^2}\right)[1 - 2R'(1 - \cos \phi)^2]\, d\phi \\
&= 2R\left[1 - 3R' + 4R' \exp\left(-\frac{\sigma_\phi^2}{2}\right) - R' \exp(-2\sigma_\phi^2)\right] \\
&\approx 2R\left[1 - \frac{3}{2} R' \sigma_\phi^4\right]
\end{aligned} \tag{9.2.5}$$

This bound is actually a good estimate of the average signal-to-noise ratio when σ_ϕ is sufficiently small that $2R'(1 - \cos \phi)^2 \ll 1$ for $|\phi| \leq 3\sigma_\phi$. The last approximation in Equation (9.2.5) is also valid under this assumption.

To determine the effect of an error in the choice of the (rectangular)† symbol interval, we observe that the integral in Equation (9.2.2) becomes, in the presence of a symbol timing error of τ seconds (assuming the carrier frequency is large compared to $1/T_s$ so the carrier can be ignored)

$$I = \frac{1}{AT_s} \int_{\tau}^{T_s} x_1(t)\, dt + \frac{1}{AT_s} \int_{T_s}^{T_s+\tau} x_2(t)\, dt \tag{9.2.6a}$$

for $\tau \geq 0$ and

$$I = \frac{1}{AT} \int_{\tau}^{0} x_0(t)\, dt + \frac{1}{AT} \int_{0}^{T_s+\tau} x_1(t)\, dt \tag{9.2.6b}$$

for $\tau < 0$, where

$$x_i(t) = Ax_i + n(t) \qquad iT_s < t < (i+1)T_s$$

Thus, if x_0, x_1, and x_2 are assumed to be identically and independently distributed,

$$\left(\frac{S}{N}\right)_\tau = \frac{\operatorname{Var}(x)}{E(I-x)^2} = \left[\frac{2R}{1 + 4R'\left(\dfrac{\tau}{T_s}\right)^2}\right] \tag{9.2.7}$$

If the reference error $\xi \triangleq \tau/T_s$ is Gaussianly distributed with zero mean and variance σ_ξ^2, Equation (9.1.1) yields

$$\begin{aligned}\left(\frac{S}{N}\right)_{\text{ave}} &= \frac{2R}{\sqrt{2\pi}\,\sigma_\xi} \int_{-\infty}^{\infty} \frac{\exp\left(-\dfrac{\xi^2}{2\sigma_\xi^2}\right)}{1 + 4R'\xi^2}\, d\xi \\ &= \sqrt{\frac{\pi}{2}}\, \frac{R \exp\left(\dfrac{1}{8R'\sigma_\xi^2}\right)}{(R')^{1/2}\sigma_\xi} \left\{1 - \operatorname{erf}\left[\frac{1}{2\sigma_\xi(2R')^{1/2}}\right]\right\} \\ &\approx 2R\{1 - 4R'\sigma_\xi^2\}\end{aligned} \tag{9.2.8}$$

Thus, the effect of imperfect symbol synchronization is to reduce the signal-to-noise ratio by the factor $(1 - 4R'\sigma_\xi^2)$. An analogous derivation yields this same conclusion, when R is large, for phase-incoherent PAM (cf. Section 3.6).

Finally, combining both carrier and symbol timing-error effects, we have

$$I = \frac{1}{AT_s} \int_{\tau}^{T_s+\tau} y(t)\sqrt{2}\, \sin(\omega_c t - \phi)\, dt \tag{9.2.9}$$

If $\omega_c \gg 1/T_s$

$$\left(\frac{S}{N}\right)_{\phi\xi} = \left[\frac{2R}{1 + 2R'(1-\cos\phi)^2 + 4R'\,|\xi|\cos\phi(1-\cos\phi) + 4R'\xi^2\cos^2\phi}\right] \tag{9.2.10}$$

---

†The most common motive for using other than rectangular pulses is to narrow the signal spectrum. As a result, the non-rectangular pulses generally used will have broader autocorrelation functions than rectangular pulses, and a communication system using such pulses will be correspondingly less adversely affected by symbol timing inaccuracies.

where the notation is as previously defined. If both σ_ϕ and σ_ξ are small, we obtain

$$\left(\frac{S}{N}\right)_{\text{ave}} \approx 2R\left[1 - 4R'\sigma_\xi^2 - 2R'\sigma_\phi^2\sigma_\xi\left(\sqrt{\frac{2}{\pi}} - 2\sigma_\xi\right) - \frac{3}{2}R'\sigma_\phi^4\right] \quad (9.2.11)$$

The result just derived was under the assumption that the symbol timing error and the carrier phase-error were independent. Frequently, they are both obtained from the same reference. In this case, $\xi = \phi/\pi m$ where m is the number of carrier half-cycles per symbol ($\omega_c = \pi m/T_s$). Then

$$\left(\frac{S}{N}\right)_\phi = \left\{\frac{1}{2R'} + (1 - \cos\phi)^2 + 2f(\phi)[f(\phi) + (1 - \cos\phi)]\right\}^{-1}\frac{R}{R'} \quad (9.2.12)$$

where

$$f(\phi) = \frac{1}{\pi m}|\phi\cos\phi - \sin\phi|$$

Since the last term in Equation (9.2.12) is of the order of $|\phi^5|/3\pi m$ for small ϕ while the second term is on the order of $\phi^4/4$, the last term can be ignored for reasonably small values of σ_ϕ, regardless of the value of m. This leads us to conclude that the degradation in performance is dominated by the degradation due to the carrier phase error alone when both the carrier and symbol timing are obtained from the same reference [cf. Equation (9.2.3)].

To summarize, the PAM signal-to-noise ratio reduction factor (when the carrier phase-error variance is σ_ϕ^2 and the symbol timing-error variance is σ_ξ^2) is approximately

$$\left[1 - 4R'\sigma_\xi^2 - 2R'\sigma_\phi^2\sigma_\xi\left(\sqrt{\frac{2}{\pi}} - 2\sigma_\xi\right) - \frac{3}{2}R'\sigma_\phi^4\right] \quad (9.2.13)$$

If both references are derived from the same signal, the reduction factor is essentially that due to the carrier phase error alone [obtained from Equation (9.2.13) by setting $\sigma_\xi = 0$].

9.3 The Effect of Inexact References—Phase Shift Keying

In this section, we investigate the effects of inexact references on the performance of the various PSK systems discussed in Chapters Three and Four. We begin with the binary case. If the local carrier reference is in error by ϕ radians, the expected output of the detector correlator is (cf. Section 4.5)

$$E(I|\phi) = \frac{1}{AT_s}\int_0^{T_s}(\pm\sqrt{2}A\sin\omega_c t)\sqrt{2}\sin(\omega_c t - \phi)\,dt = \pm\cos\phi \quad (9.3.1)$$

where it is assumed $\omega_c = \pi m/T_s$ for m either an integer or else large as compared to unity. Since the variance of this output is independent of ϕ, the sole consequence of the carrier reference error is to reduce the effective signal amplitude by the factor $\cos \phi$. Thus, when the carrier reference phase is in error by ϕ radians, the conditional probability of a symbol error is

$$P_e(\phi) = \tfrac{1}{2}[1 - \operatorname{erf}(R^{1/2} \cos \phi)] \tag{9.3.2}$$

where $R = A^2 T_s / N_0$ [cf. Equation (4.5.11)]. If ϕ is Gaussianly distributed with variance σ_ϕ^2, the symbol error probability is

$$P_e = \int_{-\infty}^{\infty} p(\phi) P_e(\phi)\, d\phi = \frac{1}{\pi \sigma_\phi} \int_0^{\infty} \exp\left(-\frac{\phi^2}{2\sigma_\phi^2}\right) \int_{\sqrt{2} R^{1/2} \cos \phi}^{\infty} \exp\left(-\frac{y^2}{2}\right) dy\, d\phi \tag{9.3.3}$$

This probability has been evaluated numerically and is plotted in Figure 9.1 as a function of R for various values of σ_ϕ. Note that as $R \to \infty$, P_e approaches a non-zero constant, sometimes called the *irreducible error probability*, for any non-zero variance σ_ϕ^2. An increase in the signal-to-noise ratio becomes less and less effective without a corresponding decrease in the reference error.

The irreducible error probability, of course, is the probability that the reference phase error ϕ falls outside the interval $(-\pi/2, \pi/2)$ modulo 2π. Since the Gaussian approximation to the distribution of ϕ is clearly weakest for values of ϕ exceeding $\pi/2$ radians, the error probability indicated in Equation (9.3.3) as $R \to \infty$ will be a correspondingly less precise estimate of the true error probability than it is for small values of R. In general, the error introduced by the Gaussian approximation starts to be significant for values of R exceeding $1/(2\sigma_\phi^2)$ since the probability of a phase error greater than $\pi/2$ radians then ceases to be negligible relative to the probability of an error induced by the additive noise (cf. Problem 9.1). This is a situation of little concern in practice since, as we see, σ_ϕ^2 will almost always be less than $1/(2R)$ in a well-designed system. Nevertheless, some of the curves in Figure 9.1 have been extended beyond the point $R = 1/(2\sigma_\phi^2)$ to provide a rough indication, generally an optimistic one, of the error probability which could be expected were the system to operate under these conditions.†

The effect of a *symbol* synchronization error on the error probability depends on the presence or absence of a symbol transition. If two successive (rectangular) symbols are identical, an incorrect symbol reference will have no effect whatsoever on the error probability. If the two successive symbols differ, however, the magnitude of the expected correlator output is reduced by a factor of $1 - (2|\tau|/T_s)$ where τ is the timing error [cf. Equation (9.2.6) with $x_1 = 1$, $x_0 = x_2 = -1$]. The probability of an error, given a timing

†When the reference signal is produced at the output of a phase-locked loop, the Tikhonov distribution (cf. Problem 5.5) or an appropriate generalization could be substituted for the Gaussian distribution in Equation (9.3.3) to provide a generally more accurate estimate of the error probability (Reference 9.9). However, except in special cases (e.g., a first-order loop with no frequency offset), the computational effort required for a solution is increased manyfold and, in any event, the results agree quite well with those obtained using the Gaussian approximation for all $\sigma_\phi^2 \leq 1/(2R)$.

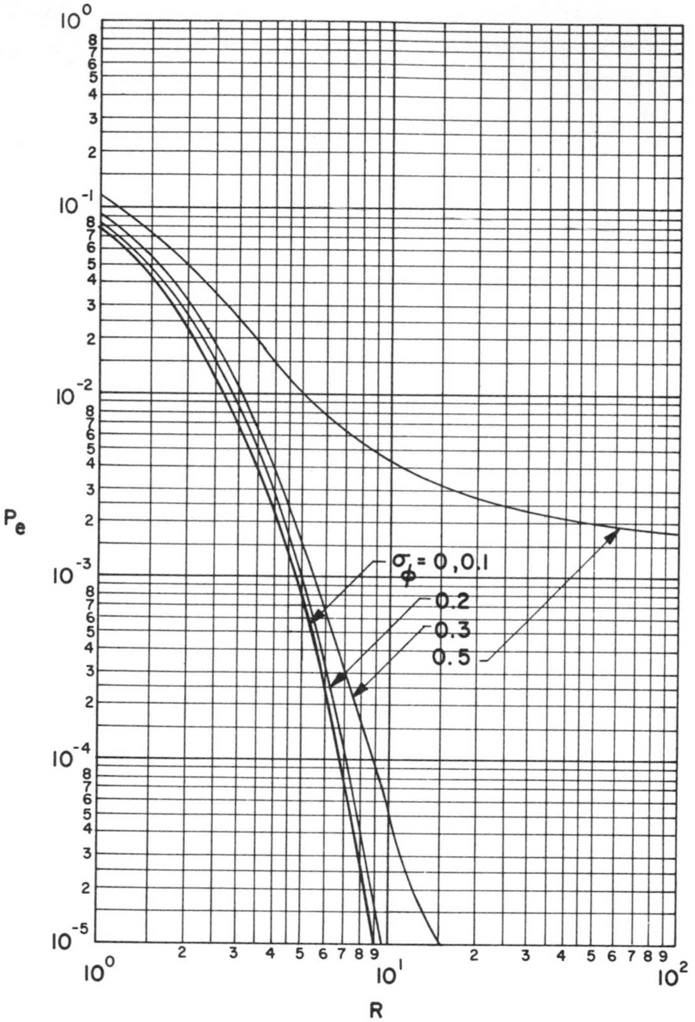

Figure 9.1 Binary PSK symbol error probability with imperfect carrier synchronization

error of τ seconds, is consequently

$$P_e(\tau) = \Pr(\text{error} \mid \text{transition}, \tau)\Pr(\text{transition} \mid \tau) + \Pr(\text{error} \mid \text{no transition}, \tau)\Pr(\text{no transition} \mid \tau) \qquad (9.3.4)$$

If the successive symbols are independent and equally likely to be either of the two binary symbols, the probability of a transition is one-half, and, if the timing error $\xi = \tau/T_s$ is Gaussianly distributed,

$$P_e = \int_{-\infty}^{\infty} P_e(\tau)p(\tau)\,d\tau$$
$$= \frac{1}{4\pi\sigma_\xi}\int_{-\infty}^{\infty} \exp\left(-\frac{\xi^2}{2\sigma_\xi^2}\right)\int_{(2R)^{1/2}(1-2|\xi|)}^{0} \exp\left(-\frac{x^2}{2}\right)dx\,d\xi$$
$$+ \frac{1}{2\sqrt{2\pi}}\int_{(2R)^{1/2}}^{\infty} \exp\left(-\frac{x^2}{2}\right)dx \tag{9.3.5}$$

Figure 9.2 shows this error probability as a function of R for several different values of σ_ξ.

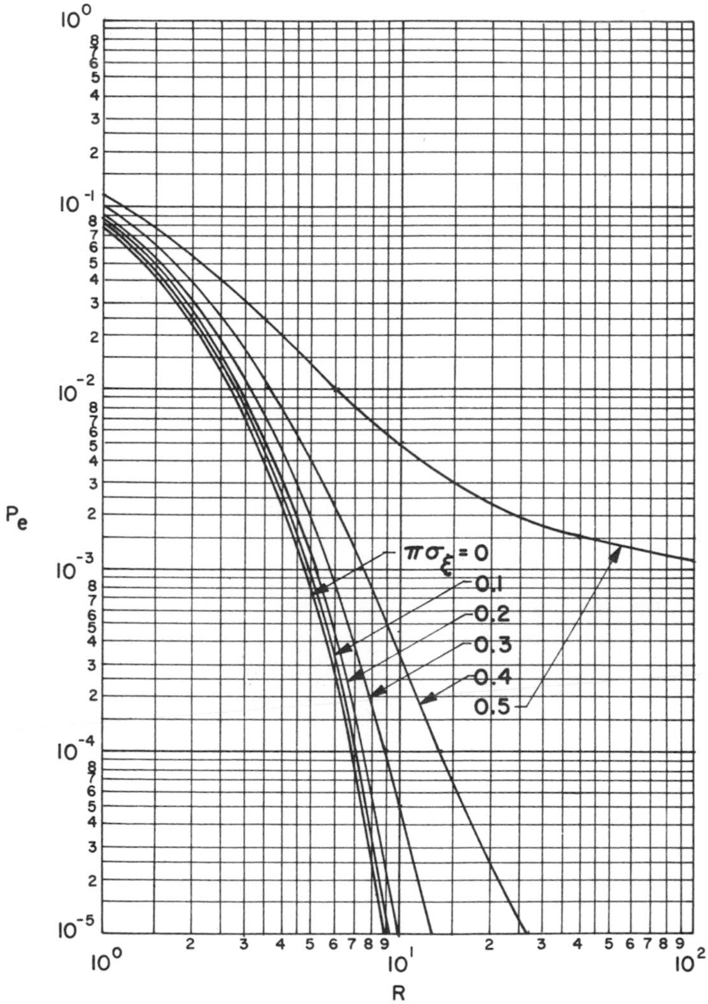

Figure 9.2 Binary PSK symbol error probability with imperfect symbol synchronization

To determine the effect of simultaneous phase and symbol sync errors, let the received signal be $\pm\sqrt{2}A \sin \omega_c t$, the local reference $\sqrt{2} \sin(\omega_c t - \phi)$, and the symbol timing error τ seconds. Then as is easily verified the expected correlator output is

$$E(I \mid \phi, \xi) = \cos\phi - 2\delta\left\{\xi \cos\phi - \frac{1}{2\pi m}[\sin\phi + \sin(2\pi m\xi - \phi)]\right\} \operatorname{sgn}(\xi) \tag{9.3.6}$$

where $\delta = 1$ when there is, and $\delta = 0$ when there is not, a symbol transition. Here again $\omega_c = m\pi/T_s$ with m either an integer or else $m \gg 1$. If $m \gg 1$, then

$$E(I \mid \phi, \xi) = (1 - 2\delta \mid \xi \mid) \cos\phi = E(I \mid \phi) E(I \mid \xi) \tag{9.3.7}$$

In contrast, suppose the same reference is used to establish the carrier phase and the symbol timing. Then $\xi = \phi/\pi m$ and, from Equation (9.3.6),

$$E(I \mid \phi, \xi) = \cos\phi - \frac{2\delta}{\pi m}[\phi\cos\phi - \sin\phi] \operatorname{sgn}(\phi) \tag{9.3.8}$$

Since

$$\sin\phi \geq \phi\cos\phi \qquad \text{for} \qquad 0 \leq \phi \leq \phi_0 > \pi$$

and

$$\sin\phi \leq \phi\cos\phi \qquad \text{for} \qquad -\pi > -\phi_0 < \phi \leq 0$$

the degradation due to the combined symbol and carrier phase errors is always less than that due to the carrier phase error alone. An upper bound on the combined effect is therefore afforded by considering only the carrier phase error. For large values of m, of course, the difference in the two cases is negligible.

Another situation of some interest is when a square-wave carrier (subcarrier) is used in place of a sinusoid. In this case, the decrease in the signal level due to a timing error of τ seconds is (when both symbol and carrier timing are derived from the same reference)

$$E(I \mid \xi) = \left[1 - 4 \mid \xi \mid + \frac{4\delta \mid \xi \mid}{m}\right] \qquad \xi \leq \frac{1}{2} \tag{9.3.9}$$

where $\xi = \tau/T_0$ with $T_0 = 2T_s/m$ the carrier period. The negative term is due to the carrier phase error alone. Again, the effect of the combined synchronization errors can be bounded by ignoring the symbol timing error. This bound rapidly approaches an exact expression as m becomes large.

The derivation of analogous results for q-ary PSK for $q > 2$ involves a straightforward, although somewhat tedious, extension of the arguments of the preceding paragraphs. The analysis is simplified by using the error probability bounding approach of Section 4.5. Adapting those bounds to the present discussion, we find, for example, that the probability of a symbol error in a q-ary PSK system subject to a ϕ radian carrier reference phase error is bounded by

$$\tfrac{1}{2}\{P_0(q;\phi)+P_0(q;-\phi)\}\le\max\{P_0(q;\phi),P_0(q;-\phi)\}\le P_e(q;\phi)$$
$$\le P_0(q;\phi)+P_0(q;-\phi) \qquad (9.3.10)$$

with $P_0(q;\phi)=\frac{1}{2}-\frac{1}{2}\operatorname{erf}(R^{1/2}\sin(\pi/q-\phi))$. The average error probability is therefore well approximated by Equation (9.3.3) with $\cos\phi$ replaced by $\sin[(\pi/q)-|\phi|]$. Similarly, if the symbol synchronization is in error by τ seconds, and if the difference in the phases of the two symbols affecting a particular decision is $\Delta\phi$ radians, then

$$\tfrac{1}{2}[P_0(q;\tau,\Delta\phi)+P_0(q;\tau,-\Delta\phi)]\le P_e(q;\tau,\Delta\phi)$$
$$\le P_0(q;\tau,\Delta\phi)+P_0(q;\tau,-\Delta\phi) \qquad (9.3.11)$$

where

$$P_0(q;\tau,\Delta\phi)=\frac{1}{2}-\frac{1}{2}\operatorname{erf}\left\{R^{1/2}\left[(1-|\xi|)\sin\frac{\pi}{q}+|\xi|\sin\left(\frac{\pi}{q}-\Delta\phi\right)\right]\right\}$$

and $\xi=\tau/T_s$. The average error probability can be bounded by averaging over ξ and $\Delta\phi$ [cf. Equation (9.3.5)]. Note that the lower inequalities in Equations (9.3.10) and (9.3.11) are both equalities when $q=2$. Moreover, when $q=4$, these same lower bounds represent the *bit* error probabilities when the carrier reference phase and the symbol synchronization errors are ϕ and τ, respectively. This conclusion follows directly from the discussion in Section 4.7.

An even simpler, although approximate, measure of the effect of timing errors in a PSK communication system results upon using the approximation to the distribution of the phase estimation error introduced in Section 3.7. In particular, if the received signal is $\sqrt{2}A\sin(\omega_c t+\phi+\phi_e)+n(t)$, with ϕ_e the reference phase error, the distribution of the estimate $\hat{\phi}$ of the received phase can be approximated by

$$p(\hat{\phi}\mid\phi_e)\approx\sqrt{\frac{R}{\pi}}\exp[-R(\hat{\phi}-\phi-\phi_e)^2] \qquad (9.3.12)$$

Consequently,

$$p(\hat{\phi})\approx\frac{1}{\sqrt{2\pi}\sigma}\int_{-\infty}^{\infty}\exp\left[-\frac{\phi_e^2}{2\sigma_\phi^2}\right]p(\hat{\phi}\mid\phi_e)\,d\phi_e$$
$$\approx\frac{1}{\sqrt{2\pi}\sigma_0}\exp\left[-\frac{(\hat{\phi}-\phi)^2}{2\sigma_0^2}\right] \qquad (9.3.13)$$

where $\sigma_0^2=(1+2\sigma_\phi^2R)/2R$ and where $\sigma_\phi^2=E(\phi_e^2)$. The effect of an imprecise phase reference, therefore, is to increase the estimation error variance from $1/2R$ to $\sigma_0^2=(1+2\sigma_\phi^2R)/2R$ and hence to increase the error probability from

$$P_e(\sigma_\phi=0)\approx 1-\operatorname{erf}\left(\frac{\pi}{q}R^{1/2}\right) \qquad (9.3.14)$$

to

$$P_e(\sigma_\phi) \approx 1 - \text{erf}\left[\frac{\pi}{q}\frac{R^{1/2}}{(1 + 2\sigma_\phi^2 R)^{1/2}}\right] \tag{9.3.15}$$

This approximation is particularly useful when q is large.

It also follows from Equation (9.3.13) that the average signal-to-noise ratio in a continuous-amplitude PSK system subject to a reference phase error is

$$\left(\frac{S}{N}\right)_{\text{ave}} = \frac{\pi^2}{3\sigma_0^2} = \frac{\pi^2}{3}\frac{2R}{(1 + 2\sigma_\phi^2 R)} \tag{9.3.16}$$

with σ_ϕ^2 the variance of this reference error.

A differentially-coherent PSK system, of course, does not require a phase-coherent carrier reference. This statement can be misleading, however. If the demodulation is accomplished by estimating the successive symbol phases, for instance, and observing the phase changes [cf. Equation (4.6.8)] an accurate carrier reference is clearly presupposed. While the steady-state phase of this reference is irrelevant, its short-term phase variation must be small or a bias will be introduced in the estimated phase differences. Equivalently, if the delay technique shown in Figure 4.5 is used, this delay must be quite precise. If it were in error by ΔT seconds, an $\omega_c \Delta T$ radian bias error would be introduced into every phase comparison. The effect of these errors, and of any symbol synchronization errors as well, can be estimated by making the appropriate modifications in the corresponding results for coherent PSK. The increased error probability in a binary differentially-coherent system, for example, due to a constant phase bias error, or due to symbol synchronization jitter, can be determined by replacing the function $1 - \text{erf}(x)$ with e^{-x^2} in Equations (9.3.2) or (9.3.5), respectively. When $q > 2$ it is only necessary to halve the signal-to-noise ratio in the phase-coherent results to find the analogous differentially-coherent expressions (cf. Section 4.6).

9.4 The Effect of Inexact References—Orthogonal Symbols

As in the communication systems discussed above, the dominant source of performance degradation in an orthogonal symbol system is generally the carrier phase error. This is particularly true when the carrier frequency is large relative to the reciprocal of the symbol period and when all references, including the carrier reference, are derived from the same source. Because of this, and because the effect of the other timing errors is highly dependent on the specific orthogonal alphabet being used, we limit consideration here to the effect of an imperfect carrier reference.

Suppose the signal alphabet consists of the set of unit energy waveforms

$$s_i(t, \phi) = \sqrt{2}\,\alpha_i(t) \sin(\omega_c t + \phi) + \sqrt{2}\,\beta_i(t) \cos(\omega_c t + \phi)$$
$$0 < t < T_s \tag{9.4.1}$$

[see Equation (4.3.1)]. The detector, upon receiving the signal $y(t) = As_r(t, \phi_0) + n(t)$, $0 < t < T_s$, determines the correlations

$$I_i = \frac{1}{AT_s} \int_0^{T_s} y(t) s_i(t, \phi_1) \, dt \tag{9.4.2}$$

where ϕ_1 is the estimated carrier phase. (The symbol timing error is assumed negligible.) The expected value of I_i, for a given $\phi = \phi_0 - \phi_1$, is

$$E(I_i \mid \phi) = \tilde{\rho}_{ri} \cos \phi + \hat{\rho}_{ri} \sin \phi \tag{9.4.3}$$

[cf. Equation (4.3.6); as usual, the carrier frequency ω_c is assumed sufficiently large to enable us to ignore the double-frequency components]. If the signal set is orthogonal in the phase-incoherent sense,

$$E(I_i \mid \phi) = \begin{cases} \cos \phi & i = r \\ 0 & i \neq r \end{cases} \tag{9.4.4}$$

and the effect of the phase error is to reduce the signal amplitude by the factor $\cos \phi$.

In most practical situations in which the carrier reference error is significant, it is generally well worth the possible increased bandwidth to choose a signal set which is orthogonal in the phase-incoherent sense. If the phase error ϕ is always sufficiently small to assure that the term $\hat{\rho}_{ri} \sin \phi$ is negligible, phase-coherent orthogonality would probably be satisfactory. In either case, the expected correlator outputs are essentially as stated in Equation (9.4.4).

The variance of these outputs is clearly unaltered by the reference phase error, so the only change in the system performance is due to the effective reduction of the signal power by a factor of $\cos^2 \phi$. Thus, referring to Equation (4.2.11), we find, as the average error probability,

$$P_e = 1 - \frac{1}{(2\pi)^{(N+1)/2}\sigma_\phi} \int_{-\infty}^{\infty} \exp\left(-\frac{\phi^2}{2\sigma_\phi^2}\right) \times \int_{-\infty}^{\infty} \exp\left(-\frac{x^2}{2}\right) \left[\int_{-\infty}^{x+\sqrt{2}R^{1/2}\cos\phi} \exp\left(-\frac{y^2}{2}\right) dy\right]^{N-1} dx \, d\phi \tag{9.4.5}$$

This error probability is plotted in Figure 9.3 as a function of $R_b = R/\log_2 N$ for various N and σ_ϕ. These curves (like those in Figures 9.1 and 9.2) tend to underestimate P_e when the probability of the event $\{|\phi| \geq \pi/2\}$ is not small relative to P_e; that is, when σ_ϕ^2 exceeds $1/R$. Again, however, σ_ϕ^2 will nearly always be less than $1/R$ in a properly designed system.

This same approach is clearly applicable to biorthogonal as well as orthogonal signal sets. Because of the similarity in performance of biorthogonal and orthogonal systems when more than four symbols are involved, the effect of an imprecise carrier phase reference will be nearly identical in the two situations.

It should be re-emphasized that the results concerning the effect of timing

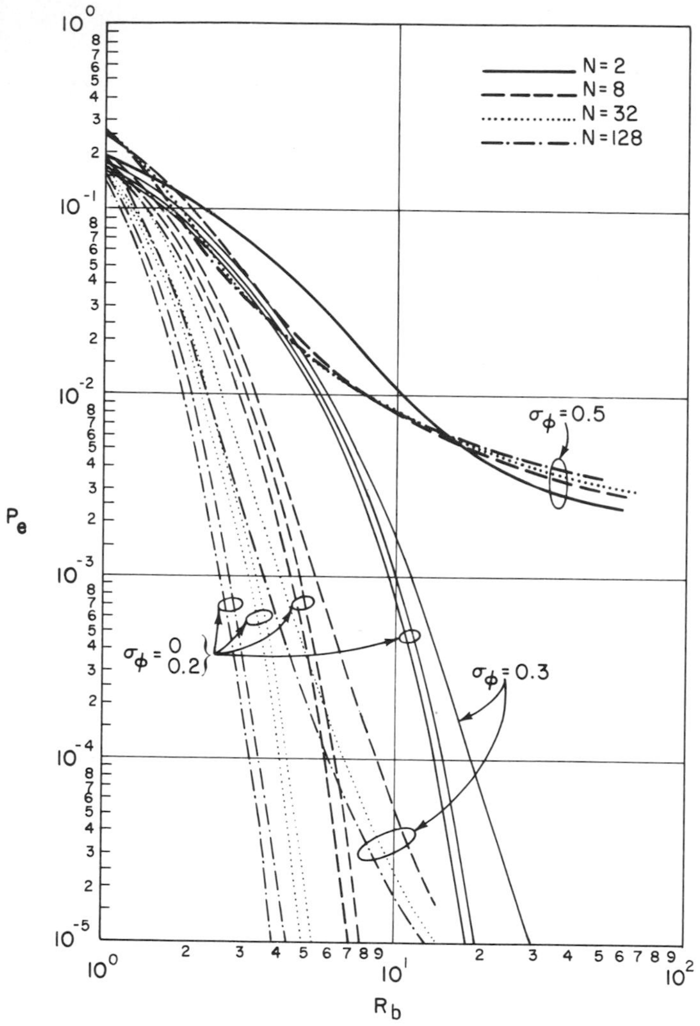

Figure 9.3 Orthogonal symbol error probability with imperfect carrier synchronization.

errors on the symbol error probabilities are based on the assumption of constant timing errors during any one correlation interval. At the other extreme, when the timing errors can vary considerably during an integration period, it may be more meaningful to determine the average effective signal power and to use this average to evaluate the error probability. For example, in the situation just considered, the average amplitude of the output of the correct correlator is

$$E(I_r) \approx \frac{1}{\sqrt{2\pi}\sigma_\phi} \int_{-\infty}^{\infty} \exp\left(-\frac{\phi^2}{2\sigma_\phi^2}\right) \cos\phi \, d\phi = \exp\left(-\frac{\sigma_\phi^2}{2}\right) \quad (9.4.6)$$

A measure of the deterioration in performance is then the increase in the symbol error probability when R is decreased from A^2T_s/N_0 to $(A^2T_s/N_0)\exp(-\sigma_\phi^2)$. This effect will be significantly less than that predicted by Equation (9.4.5) for the same value of σ_ϕ. Nevertheless, it more accurately reflects the change in the error probability when ϕ is varying rapidly.

9.5 Optimum Power Allocation

Often practical considerations dictate the amount of power which can be transmitted in the reference signal. When this is the case, the analyses of the previous sections enable one to determine the pertinent parameters (signal-to-noise ratios, or error probabilities) of the communication system. When the other constraints are less severe, however, it may be possible to vary the percentage of the total power which is relegated to the synchronization channel. Clearly, if all the power is used to transmit information and no attempt is made to obtain synchronization, the system will perform very badly. Similarly, if all the power is used to synchronize, none is left to transmit information. Consequently, there must be some point between these extremes at which the system performance is superior to, or at least as good as, that given any other division of power between the information and the synchronization signals.

The derivation of relatively simple expressions for this condition of optimum power allocation is the task of this section. Since, as we have observed, the most pronounced effect on the system performance is generally due to the carrier reference error, we shall limit the discussion to the problem of the allocation of power between the modulated and unmodulated portions of the carrier. The carrier reference is to be extracted from the unmodulated carrier by tracking it with a phase-locked loop having the noise bandwidth B_L Hz ($\sigma_\phi^2 = N_0B_L/P_s$ where P_s is the power in the unmodulated carrier). The extension of the techniques used here to other power allocation problems is straightforward. If the system involves a subcarrier as well as a carrier, for example, and the subcarrier and carrier references are independent (if they were derived from the same source, the subcarrier reference would presumably be nearly perfect), the power can generally be satisfactorily allocated between the three channels by superposition. That is, the power needed in the unmodulated carrier can be determined, from the results to be derived here, as though the subcarrier reference were perfect. Similarly, the same results can be used to determine the subcarrier reference power by neglecting the perturbation due to the imperfect carrier. If both references are reasonably accurate, as they must be if the system is to perform properly, the second-order effects due to the simultaneous presence of both reference errors are usually negligible so far as the optimum power allocation is concerned.

We illustrate the general method by investigating the carrier power allocation problem in conjunction with each of the systems discussed in the previous sections:

9.5.1 Pulse Amplitude Modulation

Expressing the reciprocal of the phase variance as

$$\frac{1}{\sigma_\phi^2} = \frac{P_s T_s}{N_0} Q \tag{9.5.1}$$

where $Q = 1/T_s B_L$, we have

$$R + \frac{1}{Q\sigma_\phi^2} = \frac{(P_s + P_m)}{N_0} T_s = \frac{P_0 T_s}{N_0} \triangleq R_0 \tag{9.5.2}$$

where P_s is the power in the unmodulated carrier, P_m the average power in the modulation, and P_0 the total, presumably limited, available power. Then

$$1 + \frac{1}{Q} \frac{d\left(\frac{1}{\sigma_\phi^2}\right)}{dR} = 0 \tag{9.5.3}$$

and, differentiating Equation (9.2.5) (with $R' = R$), subject to this constraint, we find the condition for an extremum (which is easily shown to be a maximum) in the average signal-to-noise ratio to be, approximately,

$$R = \frac{1}{2Q\sigma_\phi^2}\left\{\left(1 + \frac{4}{3}\frac{Q}{\sigma_\phi^2}\right)^{1/2} - 1\right\} \tag{9.5.4}$$

When $Q^2 R_0 \gg 1$ the optimum ratio of the carrier power to the total power is

$$\frac{P_s}{P_0} \approx \left(\frac{3}{Q^2 R_0}\right)^{1/3} \tag{9.5.5}$$

The ratio of the power in the synchronization signal to the total power is inversely proportional to Q. This result is typical of the relationships which we shall find in all of the power allocation problems.

9.5.2 Phase Shift Keying

Equating to zero the derivative of Equation (9.3.3) with respect to R under the constraint that $R + 1/Q\sigma_\phi^2$ is a constant yields

$$\int_0^\infty \exp\left(-\frac{x^2}{2}\right) \exp\left[-R\cos^2(\sigma_\phi x)\right] \cos(\sigma_\phi x)\, dx$$
$$= R\sigma_\phi^3 Q \int_0^\infty \exp\left(-\frac{x^2}{2}\right) \exp\left[-R\cos^2(\sigma_\phi x)\right] x \sin(\sigma_\phi x)\, dx \tag{9.5.6}$$

This equation has also been solved numerically. The solution expressed as the ratio P_s/P_0 is plotted in Figure 9.4 as a function of R_0 for various values of Q.

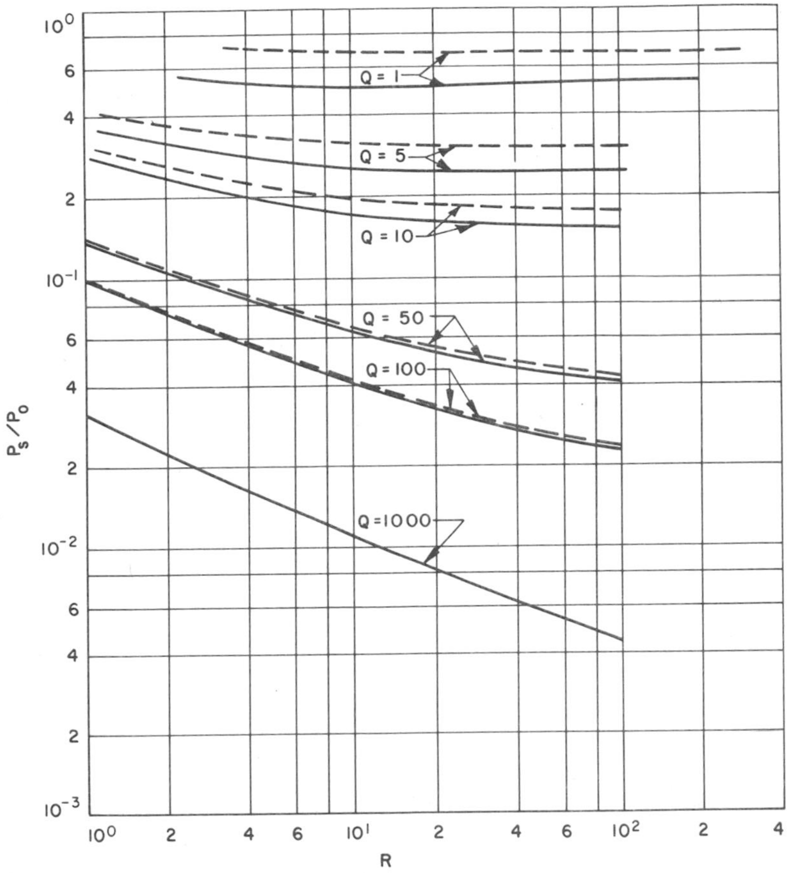

Figure 9.4 Optimum power allocation between binary PSK modulated signal and carrier

An approximate solution to Equation (9.5.6) results when $\sigma_\phi \ll 1$, so that $\cos \sigma_\phi x \approx 1 - (\sigma_\phi^2 x^2/2)$ and $x \sin \sigma_\phi x \approx \sigma_\phi x^2$. Defining $\sigma_x = (1 - 2\sigma_\phi^2 R)^{-1/2}$, we obtain under these approximations the condition

$$\frac{2}{\sqrt{2\pi}\sigma_x}\int_0^\infty \exp\left(-\frac{x^2}{2\sigma_x^2}\right) dx = \sigma_\phi^2(1 + 2RQ\sigma_\phi^2)\frac{1}{\sqrt{2\pi}\sigma_x}\int_0^\infty x^2 \exp\left(-\frac{x^2}{2\sigma_x^2}\right) dx$$

or

$$1 = \frac{\sigma_\phi^2}{2}\frac{(1 + 2RQ\sigma_\phi^2)}{(1 - 2\sigma_\phi^2 R)} \tag{9.5.7}$$

This last equation can be rewritten in the form

$$\frac{P_s}{P_0} = \frac{1}{\left(R_0 - \frac{1}{4}\right)\left[\left(1 + \frac{R_0(Q+2)}{(R_0 - \frac{1}{4})^2}\right)^{1/2} - 1\right]} \tag{9.5.8}$$

This expression is shown as the dashed line in Figure 9.4. The symbol error probability is plotted as a function of R_0 in Figure 9.5 with the power optimally divided between the carrier and data channels, the symbol synchronization assumed perfect. Also shown for comparison are the symbol error probabilities resulting from differentially coherent and from phase-incoherent detection again assuming perfect symbol synchronization.

Note that if the two waveforms under consideration are not antipodal, but have the cross-correlation coefficient ρ, the effective signal-to-noise ratio is reduced by a factor of $(1 - \rho)/2$ vis-à-vis the antipodal case (cf. Chapter Four). If R is replaced by $R(1 - \rho)/2 = R'$ in the preceding discussion, and Q by $Q' = 2Q/(1 - \rho)$, the above results are easily seen to hold for arbitrary ρ.

This same approach can be taken in conjunction with the error probabil-

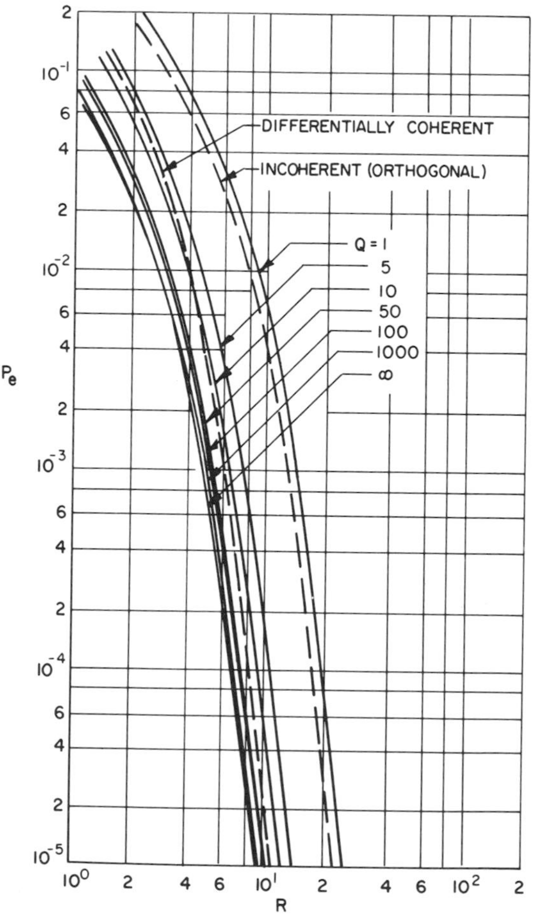

Figure 9.5 Binary PSK symbol error probability with optimum power allocation

ity expressions for $q > 2$ [e.g., Equation (9.3.10)], but the algebra is tedious and the results not particularly instructive. Instead we proceed directly to the expressions resulting from the approximation in Equation (9.3.13). For either discrete- or continuous-amplitude PSK, the goal of the power allocation strategy, under this approximation, is to minimize the variance σ_0^2 [Equation (9.3.13)]. Differentiating this quantity with respect to R, subject to the constraint that $R + (1/Q\sigma_\phi^2)$ is a constant, yields as the condition for an extremum in the average signal-to-noise ratio

$$2\sigma_\phi^4 QR^2 = 1 \tag{9.5.9}$$

Consequently, the ratio of the carrier to the total power should be

$$\frac{P_s}{P_0} = \frac{1}{1 + \left(\frac{Q}{2}\right)^{1/2}} \tag{9.5.10}$$

Notice that when Q is large as compared to R_0, the optimum power allocation in the binary case results in a carrier power of

$$P_s \approx \frac{P_0}{R_0^{1/2} Q^{1/2}} \tag{9.5.11a}$$

[cf. Equation (9.5.8)]. In contrast, when the number of phases q is large, the carrier power under the same conditions should be

$$P_s \approx \frac{P_0}{(Q/2)^{1/2}} \tag{9.5.11b}$$

[cf. Equation (9.5.10)] representing an increase by the factor $(2R_0)^{1/2}$.

9.5.3 Orthogonal Symbols

Differentiation of Equation (9.4.5) subject to the condition that $R + (1/Q\sigma_\phi^2)$ is constant yields

$$\frac{dP_e}{dR} = -\frac{1}{\sqrt{2\pi}} \int_{-\infty}^{\infty} \exp\left(-\frac{z^2}{2}\right) \frac{1}{(2R)^{1/2}} [\cos \sigma_\phi z - R\sigma_\phi^3 Qz \sin \sigma_\phi z] \times \left[\left\{\frac{d}{da} P_N(a)\right\}_{a=\sqrt{2R}\cos \sigma_\phi z}\right] dz \tag{9.5.12}$$

where

$$P_N(a) = \frac{1}{(2\pi)^{N/2}} \int_{-\infty}^{\infty} \exp\left(-\frac{x^2}{2}\right) \left[\int_{-\infty}^{x+a} \exp\left(-\frac{y^2}{2}\right) dy\right]^{N-1} dx$$

The optimum power allocation is in principle determined by equating this expression to zero and solving for σ_ϕ^2 in terms of R. Although this could be done numerically, a considerable amount of computation would be involved.†

---

†An alternative numerical method would be to fix the total power and evaluate P_e for various values of R and σ_ϕ^2 subject to this constraint. This is the approach taken in Reference 9.2. This could be done graphically by plotting on the curves of Figure 9.3 the locus of points corresponding to a fixed total power, a relatively easy task for any particular set of parameters N, R_0, and Q.

By making some approximations, however, we will be able to obtain a relatively simple analytical expression for the optimum power allocation. The results will be found to agree quite closely with the more precise numerical solution for all signal-to-noise ratios of interest.

We begin by arguing that for reasonably large values of a,

$$\frac{d}{da}P_N(a) \approx (N-1)\frac{d}{da}P_2(a) \tag{9.5.13}$$

This conclusion is supported by a cursory glance at Figure 4.2 which indicates the rate of change of the error probabilities to be, except for a multiplicative constant, independent of N. To support this analytically, we observe

$$\begin{aligned} P_N(a) &= \frac{1}{\sqrt{2\pi}}\int_{-\infty}^{\infty}\exp\left(-\frac{x^2}{2}\right)\left[1-\frac{1}{2}\operatorname{erfc}\frac{x+a}{\sqrt{2}}\right]^{N-1}dx \\ &\approx \frac{1}{\sqrt{2\pi}}\int_{-\infty}^{\infty}\exp\left(-\frac{x^2}{2}\right)\left[1-\frac{N-1}{2}\operatorname{erfc}\frac{x+a}{\sqrt{2}}\right]dx \end{aligned} \tag{9.5.14}$$

where

$$\operatorname{erfc}\frac{x+a}{\sqrt{2}} \triangleq 1-\operatorname{erf}\frac{x+a}{\sqrt{2}} = 1-\frac{1}{\sqrt{2\pi}}\int_{-(x+a)}^{x+a}\exp\left(-\frac{y^2}{2}\right)dy$$

The approximation of Equation (9.5.14) is valid when a is large and hence $\operatorname{erfc}[(x+a)/2]$ small for all x for which $\exp(-x^2/2)$ is significantly greater than zero. Then

$$\begin{aligned} \frac{dP_N(a)}{da} &\approx (N-1)\frac{d}{da}\left\{-\frac{1}{2\sqrt{2\pi}}\int_{-\infty}^{\infty}\exp\left(-\frac{x^2}{2}\right)\operatorname{erfc}\left(\frac{x+a}{\sqrt{2}}\right)dx\right\} \\ &= (N-1)\frac{dP_2(a)}{da} \end{aligned} \tag{9.5.15}$$

and, from Equation (9.5.12),

$$\begin{aligned} \frac{dP_e}{dR} \approx &-\frac{(N-1)}{\sqrt{2\pi}}\int_{-\infty}^{\infty}\exp\left(-\frac{z^2}{2}\right)\frac{1}{(2R)^{1/2}}[\cos\sigma_\phi z - R\sigma_\phi^3 Qz\sin\sigma_\phi z] \\ &\times\left[\left\{\frac{d}{da}P_2(a)\right\}_{a=\sqrt{2R}\cos\sigma_\phi z}\right]dz \end{aligned} \tag{9.5.16}$$

But the condition under which this expression is zero is independent of N. Since, when $N = 2$, the optimum power allocation occurs when Equation (9.5.6) is satisfied (with R replaced by $R/2$ and Q by $2Q$ since here $\rho = 0$), this is a good approximation to the optimum allocation for all values of N. And since Equation (9.5.8) is an approximate solution to Equation (9.5.6), valid when $\sigma_\phi \ll 1$, it is also applicable for all N under this same condition. (Here, too, R_0 must be replaced by $R_0/2$ and Q by $2Q$.)

It should be observed that the parameter R is the ratio of the signal energy per *symbol* to the noise spectral density. As noted in Chapter Four, R must increase proportionally with $\log_2 N$ if the error probability is to remain constant as N is increased. Furthermore, if the bit rate $1/T_b$ is held constant, $Q = 1/B_L T_s = 1/B_L T_b \log_2 N$ is inversely proportional to $\log_2 N$. Since Q must be large if the assumption of a constant phase reference during any integration interval is to be valid, the restriction on B_L becomes more severe as N is increased; B_L must be small as compared to $1/T_s = 1/T_b \log_2 N$.

In concluding, we observe that the preceding argument can be used, essentially unchanged, when a biorthogonal rather than an orthogonal set of symbols is used. The optimum power allocation remains virtually the same.

9.6 Single-Channel versus Separate-Channel Synchronization

This chapter has been concerned with the effect of imprecise synchronization references on the performance of various communication systems, and with the related problem of the allocation of power between the synchronization and information channels. In Chapters Seven and Eight, on the other hand, several techniques were investigated for obtaining this synchronization information directly from the information-modulated signal. The efficacy of those techniques can now be determined by using the results of this chapter. Two questions immediately suggest themselves in this regard: (1) Are synchronization estimates obtained directly from the modulated signal inferior to those provided by separate channels under an optimum division of power? (2) Is it beneficial to supplement these estimates with separate channel estimates if it is possible to do so? In general, the answer to both questions is no.

To support this statement, we need only to compare the variances of the two estimates. Let σ_1^2 be the variance of a synchronization estimate obtained directly from the modulated signal, and σ_2^2 the variance of the corresponding estimate provided by a second channel under an optimal division of power. We note that

$$\frac{1}{\sigma_2^2} = \frac{P_s}{P_0} QR_0 \tag{9.6.1}$$

and since the optimum ratio P_s/P_0 was found, in every case, to be of the order of Q^{-r} with $0 < r < 1$, $1/\sigma_2^2$ will be of the order of Q^s with $0 < s < 1$. In contrast, $1/\sigma_1^2$ was shown, in Chapter Eight, invariably to be directly proportional to Q. Evidently if $Q = 1/B_L T_s$ is sufficiently large, the estimate obtained directly from the data will be superior.

In practice, the term "sufficiently large" will generally encompass all the values of Q of interest. For example, the reciprocal of the variance of the separate channel PAM carrier reference is, optimally,

$$\frac{1}{\sigma_2^2} = (3QR_0^2)^{1/3} \tag{9.6.1a}$$

[Equation (9.5.5)]. The Costas or squaring-loop estimate reciprocal variance is, approximately,

$$\frac{1}{\sigma_1^2} = \frac{QR_0^2}{R_0 + 1} \tag{9.6.1b}$$

[Equation (8.5.19)]. Thus, the latter estimate is superior if

$$Q > \sqrt{3}\,\frac{(R_0 + 1)^{3/2}}{R_0^2} \tag{9.6.2}$$

The term on the right of this inequality approaches zero as R_0 becomes large, and is only $2\sqrt{6}$ when $R_0 = 1$. The analysis of the Costas loop and related devices all made use of the presumed inequality $B_L \ll 1/T_s$. Accordingly, their evident superiority exists at least so far as a PAM carrier reference is concerned, for all values of Q for which their analysis is valid in the first place.

Similarly, in the case of binary PSK carrier synchronization, the Costas loop will provide a more exact reference if

$$Q > \frac{5}{2R_0} + \frac{3}{2R_0^2} + \frac{1}{R_0^3} \tag{9.6.3}$$

If an error probability of 10^{-3} is desired, for example, R_0 must be of the order of 5, and the Costas loop approach is superior for any Q greater than approximately 0.57. To make the comparison between the two methods more graphic, the symbol error probability, when a Costas loop is used to derive the carrier reference, is plotted in Figure 9.6. This should be contrasted with the corresponding curves in Figure 9.5. Note that, except at low signal-to-noise ratios, the Costas loop approach yields results equal or superior to the differentially coherent scheme for all $Q \geq 2$. Interestingly, the performance attainable by using the optimum PSK carrier tracking loop of Section 8.3 rather than the Costas loop is virtually identical, when $Q = 2$, to that resulting from differentially coherent detection for all signal-to-noise ratios. When $Q = 2$, $B_L = 1/2T_s$ and the tracking loop bandwidth equals that of a T_s second integrator. The near equivalence of the two results therefore should not be surprising. It must be cautioned, however, that the analysis of Section 8.3, as well as the assumption of a Gaussianly distributed phase error, is at best only approximate when $Q = (B_L T_s)^{-1}$ is not large. In any event, the performance degradation using a Costas loop to provide the carrier reference is negligible even for relatively small values of Q.

Comparable results obtain for the other modulation systems and the other synchronization problems considered in the preceding analyses. Deterioration in the performance of a bi-phase PSK system due to imperfect (rectangular) symbol synchronization, for example, is seen to be small (for $R < 10$) if the synchronization error standard deviation $\sigma_\phi \triangleq 2\pi\sigma_\xi$ is 0.2 radian or less (Figure 9.2). The tracking error variance of the symbol tracking loop of Section 7.6 was found to be [Equation (7.6.19)]

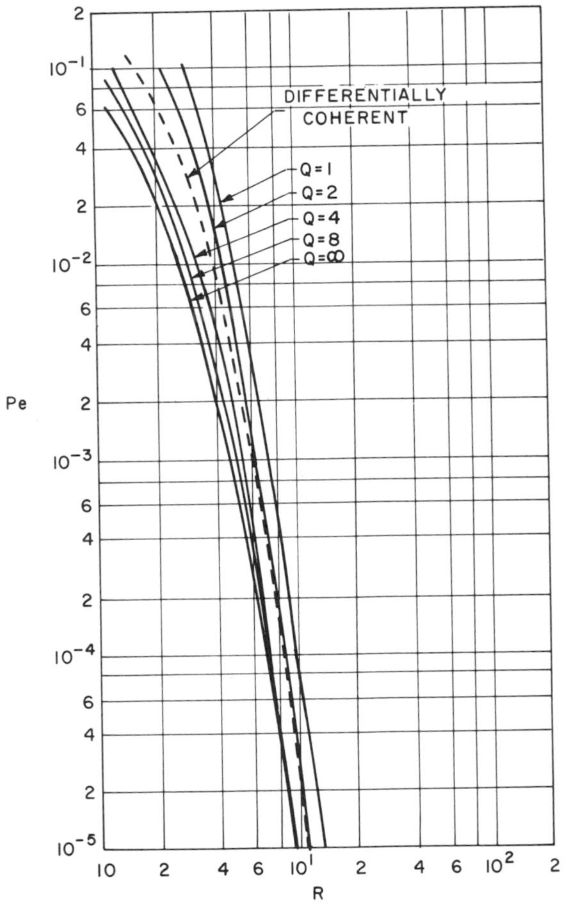

Figure 9.6 Binary PSK symbol error probability with a Costas loop derived carrier reference

$$\sigma_\xi^2 = \frac{1}{2Q}\left(\frac{\Delta t/T_s}{1-\Delta t/T_s}\right)\left[\left(1-\frac{3\Delta t}{4T_s}\right)\frac{1}{R}+\frac{1}{2R^2}\right] \tag{9.6.4}$$

Since the error signal is linear over the region $|\tau| \leq \Delta t/2$, $\Delta t/T_s$ can be of the order of 1/6 when $\sigma_\xi = 0.1/\pi$ without violating the conditions of Section 7.6. Thus, if $R = 5$, the synchronization error standard deviation can be as small as 0.2 radian with a Q of about 20.

In answer to the second question posed above, the question as to whether or not a separate synchronization channel should be used in a supplementary category, we observe that if the single-channel reference is more precise than the separate-channel reference when the power is divided optimally, then no power at all should be relegated to the synchronization channel. Suppose both methods (the single-channel and the separate-channel methods) were used simultaneously. If the combined estimate were better than the separate-channel estimate under an optimum power allocation, the system performance would obviously improve by allocating more power to the information channel. This follows because the synchronization estimate deteriorates under such a redistribution of power even less rapidly than when it is obtained only from the synchronization channel. And since the performance would improve by so re-allocating the power when only the one estimate is available, this conclusion must also be true when the two estimates are used. Thus, the system performance would continue to improve upon re-allocating power from the synchronization channel to the information channel, at least until the synchronization reference is no more precise than it would be under an optimum separate-channel-only allocation. But if the single-channel reference itself is actually more precise than this value, no power at all should be left in the synchronization channel. A separate synchronization channel, therefore, could not be used profitably in such a situation.

It should be emphasized that the preceding comments have concerned the relative quality of separate-channel and single-channel synchronization estimates. The measure of interest, of course, is usually the output signal-to-noise ratio or the symbol error probability. Since all available power is used to transmit information in the single-channel case, the overall performance of such a system may well be superior to that of a separate-channel system, even when more precise synchronization results using the latter approach. In any event, when Q is large both methods of synchronization result in nearly optimum performance and the decision as to which to use should be based on other considerations. The single-channel method, for example, avoids the multiplexing problem and may therefore be easier to implement. On the other hand, the absolute phase reference resulting from the separate-channel approach might make it more attractive, since without it some other technique is needed to resolve possible ambiguities in the information sequences (cf. Sections 8.2 and 8.3).

9.7 Concluding Remarks

The major conclusions suggested by the results of this chapter can be summarized as follows:

(1) In a coherent system in which the recovered carrier also serves as the receiver clock, the carrier phase error is the dominant source of performance degradation. In an incoherent system, lacking a high-frequency clock reference, or in a coherent system in which symbol synchronization is not related to the carrier reference, the symbol synchronization error can also be significant. Indeed, the availability of a relatively accurate clock reference can be a major advantage of a coherent system vis-à-vis an incoherent system.

(2) For even moderately large values of the parameter $Q = 1/B_L T_s$, however, the effect of inexact synchronization references will be relatively

minor. This is particularly true when the synchronization is derived directly from the modulation, but the statement also holds when a separate synchronization channel is used so long as the available power is properly allocated between the two channels.

(3) Except for quite small values of the parameter Q, synchronization obtained directly from the modulated signal is decidedly superior to that provided by a separate channel under an optimum power allocation strategy.

The analyses of this chapter have all assumed demodulators of the type introduced in Chapters Three and Four. These demodulators were shown to be optimum, assuming perfect symbol synchronization, and in the case of coherent reception, perfect carrier synchronization as well. An interesting generalization of these results would be the determination of the form of the optimum detector taking into account the inevitable synchronization inaccuracies. In general, this problem seems quite formidable, and, in view of the results of this chapter, of somewhat academic interest. In some special cases, however, answers can be obtained which are both tractable and instructive (cf. Problems 9.2, 9.5, and Reference 9.13).

Notes to Chapter Nine

Much of the material in this chapter was originally published in Reference 9.4. Other early publications in this area are due to Kaneko (Reference 9.5) and Van Trees (Reference 9.6). More recent work concerning the effect of an imperfect carrier reference on the performance of PSK systems includes several papers by Lindsey (see for example References 9.2 and 9.7) and papers by Aein (Reference 9.8), Lindsey and Simon (Reference 9.9), Oberst and Schilling (Reference 9.11), Bussgang and Leiter (Reference 9.14), and Mazo and Salz (Reference 9.10). The latter paper also investigates this problem in conjunction with vestigial sideband modulated signals.

Problems

9.1 [Cf. References 9.8 and 9.10]. (a) The carrier reference used to demodulate a binary PSK signal is obtained from a phase-locked loop. If the tracking-error variance is $\sigma_\phi^2 = 1/2cR$, with $R = ST_b/N_0$ the ratio of the signal energy per bit to the noise spectral density in the information channel, show that the probability of a bit error is bounded by

$$P_e \leq \min \begin{cases} \dfrac{\exp[(3c-1)R/2]I_0[(c-1)R/2]}{8I_0(2cR)} \\ \qquad \times \{F(\sqrt{\pi R}) + (1/4)G[(c-1)R/2]\} + \dfrac{1-e^{-2cR}}{4cR\, I_0(2cR)} \\ \dfrac{\exp[c^2R]}{8I_0(2cR)}\{F(\sqrt{\pi R}) + \pi^2/12\} + \dfrac{1-e^{-2cR}}{4cR\, I_0(2cR)} \end{cases}$$

where $F(x) = x[(1 + (4/x^2))^{1/2} - 1]$, $G(x) = \dfrac{1}{\pi I_0(x)} \int_0^\pi \phi^2 e^{x \cos\phi} d\phi$, and $I_0(x)$ denotes the zeroth-order modified Bessel function of the first kind. [*Hint*: The following inequalities will be useful: $[(x^2 + 2)^{1/2} - x]e^{-x^2}/\sqrt{\pi} \le \text{erfc}(x) \le (\pi/2)[(x^2 + (4/\pi))^{1/2} - x]e^{-x^2}/\sqrt{\pi}$; $F(x) \le x[(\cos^2\phi + (4/x^2))^{1/2} - \cos\phi] \le F(x) + \phi^2$; $(2\phi/\pi) \le \sin\phi \le \phi\,(|\phi| \le \pi/2)$; $\cos\phi \ge 1 - (\phi^2/2)$; $1 - \cos\phi \ge \frac{1}{2}(1 - \cos^2\phi)$.]

(b) Show that when R is large and $c > 1$, the effect of the imperfect carrier reference is to increase the bit error probability by a factor of at most $\sqrt{c/(c-1)}$.

(c) Compare the bound of part (a) when $c = 1$ and R is large with the error probability for differentially-coherent demodulation. [Note that when the signal and noise levels are the same in the two channels (i.e., the information channel and the synchronization channel), $c = 1$ corresponds to a loop bandwidth of $B_L = 1/2T_b$.]

(d) Show that, if the Tikhonov rather than the Gaussian distribution is used to describe the carrier phase error, the same upper bounds apply.

(e) Derive lower bounds on P_e using both the Tikhonov and Gaussian distributions and compare with the upper bounds. In particular, show that when $c \ge 1$ both distributions lead to bounds which are exponentially identical to the upper bounds, but that when $c < 1$, only the Tikhonov distribution has this property.

9.2 (References 9.10, 9.14). Let $y_1(t) = \sqrt{2}A \sin[\omega_c t + \phi + (2\pi\nu/n)] + n_0(t)$, $-T_s \le t < 0$, ν some integer in the interval $(0, n-1)$, and $y_2(t) = \sqrt{2}B \sin(\omega_c t + \phi) + n_1(t)$, $-T_c < t \le 0$, be two received signals with $n_0(t)$ and $n_1(t)$ sample functions of mutually independent white Gaussian processes having single-sided spectral densities N_0 and N_1 respectively.

(a) Show that the optimum (maximum-likelihood) detector for ν is identical to the optimum phase-coherent detector but with $r(t) = \sqrt{2}\sin(\omega_c t + \hat{\phi})$ serving as the reference, with $\hat{\phi}$ the maximum-likelihood estimate of ϕ based on the signal $y_2(t)$ (cf. Section 4.6).

(b) What is the probability of an error using this detector when $n = 2$? [*Hint*: (Cf. Reference 9.12). (1) If x_1, x_2, y_1, y_2 are mutually independent Gaussian variables with $\text{Var}(x_i) = \text{Var}(y_i) = \sigma_i^2$, $i = 1, 2$, then $u_1 = (x_1 + kx_2)/2$, $u_2 = (x_1 - kx_2)/2$, $v_1 = (y_1 + ky_2)/2$, and $v_2 = (y_1 - ky_2)/2$ are all Gaussian and are also mutually independent when $k = \sigma_1/\sigma_2$. (2) $\Pr(x_1x_2 + y_1y_2 < 0) = \Pr(u_1^2 + v_1^2 < u_2^2 + v_2^2)$. (3) If r_1 and r_2 are two independent Rician-distributed random variables,

$$p(r_i) = r_i \exp\left[-\frac{a_i^2 + r_i^2}{2}\right] I_0(a_i r_i), \qquad r_i > 0,\ i = 1, 2,$$

then

$$\Pr(r_1^2 < r_2^2) = \frac{1}{2}\left(1 - Q\left(\frac{a_1}{\sqrt{2}}, \frac{a_2}{\sqrt{2}}\right) + Q\left(\frac{a_2}{\sqrt{2}}, \frac{a_1}{\sqrt{2}}\right)\right),$$

with $Q(a, b)$ the Marcum Q function.]

(c) Let $R = A^2T_s/N_0$, $cR = B^2T_c/N_1$, and use the fact that when $\alpha\beta \gg 1$ and $\alpha \gg \alpha - \beta > 0$, $Q(\alpha, \beta) \approx \frac{1}{2}[1 + \text{erf}((\alpha - \beta)/\sqrt{2})]$ and $Q(\beta, \alpha) \approx \frac{1}{2}\sqrt{\alpha/\beta}\,[1 - \text{erf}((\alpha - \beta)/\sqrt{2})]$ to derive an approximate expression for the error probability when R is large. Compare this result with the bounds of Problem 9.1. Why does the phase-locked loop approach result in performance superior to the "optimum" method used here when $c < 1$?

9.3 A PAM signal is to be transmitted along with an unmodulated sinusoid which is to be tracked with a phase-locked loop for carrier reference purposes. Four signal configurations are under consideration:

(1) $\sqrt{2}A_0s(t)\sin\omega_ct + \sqrt{2}A_1\sin\omega_ct$
(2) $\sqrt{2}A_0s(t)\sin\omega_ct + \sqrt{2}A_1\cos\omega_ct$
(3) $\sqrt{2}A_0s(t)\sin\omega_ct + A_1\sin(\omega_c+\omega_0)t + A_1\sin(\omega_c-\omega_0)t$
(4) $\sqrt{2}A_0s(t)\sin\omega_0t\sin\omega_ct + \sqrt{2}A_1\sin\omega_ct$

with $s(t)$ denoting the amplitude modulated pulse sequence and $\omega_0 = 2\pi/T_s$. Compare the relative advantages and disadvantages of these four configurations and estimate the value of $Q = 1/B_LT_s$ required if the effect of the mutual interference between the signal and the carrier is to be negligible.

9.4 Examine the effect of an inexact symbol sync reference on the error probability in demodulating a sequence of binary PSK symbols having a raised-cosine pulse shape. Compare with the results of Section 9.3. Estimate the minimum value of Q for which the single-channel symbol sync reference is at least as accurate as that provided by a separate channel under an optimum division of power.

9.5 (Reference 9.3). A signal, consisting of a sequence of pulses of the form

$$s_i(t) = \sqrt{2}\,\alpha_i(t)\sin(\omega_ct+\phi) + \sqrt{2}\,\beta_i(t)\cos(\omega_ct+\phi) \qquad \begin{cases} 0 < t < T_s \\ i = 0, 1 \\ \omega_c \gg 1/T_s \end{cases}$$

is received in the presence of additive white Gaussian noise. The local carrier reference

$$c(t) = \sqrt{2}\sin(\omega_ct+\hat{\phi})$$

is produced at the output of a (first-order) phase-locked loop with a phase error $\phi_e = \phi - \hat{\phi}$ satisfying the Tikhonov distribution

$$p(\phi_e) = \frac{1}{2\pi I_0(2cR)}\exp[2cR\cos\phi_e]$$

(As in Problem 9.1, the loop signal-to-noise ratio is denoted $2cR$, with R the ratio of the symbol energy to the noise spectral density and c an arbitrary constant.) Show that the maximum-likelihood detector (called a *partially coherent detector*) in this case includes both a phase-coherent and a phase-incoherent demodulator. Show further that the optimum decision is based on a weighted average of the outputs of the two demodulators, with the two outputs given the respective weights $4cR$ and 1.

References

9.1 Kac M., and A. J. F. Siegert, "On the Theory of Noise in Radio Receivers with Square-Law Detectors," *Journ. App. Phys.*, 8, p. 583, 1947.

9.2 Lindsey, W. C., "Phase-Shift-Keyed Detection with Noisy Reference Signals,"

IEEE Trans. on Aerospace and Electronics Systems, AES-2, p. 393, 1968.

9.3 Viterbi, A. J., "Optimum Detection and Signal Selection for Partially Coherent Binary Communication," *IEEE Trans. Inf. Theory*, IT-11, p. 239, 1965.

9.4 Stiffler, J. J., "On the Allocation of Power in a Synchronous Binary PSK Communication System," *Proc. National Telemetry Conf.*, Los Angeles, Calif., 1964.

9.5 Kaneko, H., "A Statistical Analysis of the Synchronization of Digital Receivers," M. S. Thesis, Dept. Elec. Eng., Univ. of Calif., Berkeley, 1962.

9.6 Van Trees, H., "Optimum Power Division in Coherent Communication Systems," Lincoln Lab. Tech. Rept. No. 301, Mass. Inst. Tech., Lexington, Mass., Feb. 1963.

9.7 Lindsey, W. C., "Optimal Design of One-Way and Two-Way Coherent Communication Links," *IEEE Trans. on Comm. Tech.*, COM-14, p. 418, 1966.

9.8 Aein, J. M., "Coherency for the Binary Symmetric Channel," *IEEE Trans. on Comm. Tech.*, COM-18, p. 344, 1970.

9.9 Lindsey, W. C., and M. K. Simon, "The Effect of Loop Stress on the Performance of Phase-Coherent Communication Systems," *IEEE Trans. on Comm. Tech.*, COM-18, 1970.

9.10 Mazo, J. E., and J. Salz, "Carrier Acquisition for Coherent Demodulation of PAM," *IEEE Trans. on Comm. Tech.*, COM-18, p. 353, 1970.

9.11 Oberst, J. F., and D. L. Schilling, "Performance of Self-Synchronized Phase-Shift Keyed Systems," *IEEE Trans. on Comm. Tech.*, COM-17, p. 664, 1969.

9.12 Stein, S., "Unified Analysis of Certain Coherent and Non-Coherent Binary Communications Systems," *IEEE Trans. Inf. Theory*, IT-10, p. 43, 1964.

9.13 Tufts, D. W., and T. Berger, "Optimum Pulse Amplitude Modulation, Part II: Inclusion of Timing Jitter," *IEEE Trans. Inf. Theory*, IT-13, p. 209, 1967.

9.14 Bussgang, J. J., and M. Leiter, "Phase Shift Keying with a Transmitted Reference," *IEEE Trans. on Comm. Tech.*, COM-14, p. 14, 1966.

PART THREE

CODING

chapter ten

Encoding the Source—Decodability

10.1 Introduction

The digital communication systems of Chapter Four presumed the capability of representing the information as a sequence of discrete-amplitude symbols. The most straightforward method for rendering the information in this form if it is not inherently digital is simply to sample the source output and to quantize these samples, the sampling rate and the fineness of the quantization being determined by the magnitude of the acceptable error. There are several limitations to this approach. The number of quantization levels, for example, and hence the resulting number of symbols, may be greater than the number of waveforms which are to be used in the communication system. Then, too, such a representation of the message may not be especially efficient. Accordingly, it is often expedient to encode the information, i.e., to map it onto some more convenient set of symbols, which we shall refer to as the (source) *code symbols*.

One common method for decreasing, or in some cases increasing, the number of different symbols needed to represent the information is called *pulse code modulation* (*PCM*). (The term modulation is somewhat misleading in this context since nothing is implied about the particular waveforms used to represent the information.) Suppose the information samples are quantized into N levels and the modulator output is to be one of r waveforms. Then each sample can be represented as a sequence of $\log_r N$, or the smallest integer larger than $\log_r N$, r-ary symbols. (If r is greater than N, it may be possible to map several samples onto each code symbol.) The *source alphabet* (i.e., the ensemble of symbols representing the information) is thereby *matched* to the *channel alphabet* (cf. Chapter One). An interesting alternative to this approach, called *delta modulation*, is to increase the sampling rate to the point at which it is possible to represent each sample by only one r-ary symbol (in this case with $r = 2$) indicating the sign of the difference between the current sample and the one preceding it (Reference 10.1). Numerous variations on both of these techniques are clearly possible.

Even after the source has been matched to the channel, or perhaps prior to this event, further processing may be in order. It was generally assumed in Chapter Four that the successive symbols at the modulator input were independent and equally likely to be any one of the symbols in the alphabet. Unfortunately, sources seldom operate that way. Yet the greater the deviation of the source from this ideal the less efficient the overall system, unless some effort is made to compensate for this deviation.

One factor causing a disparity between an actual source and an ideal one is the tendency of most sources to produce different outputs with different probabilities. As an illustration of the resulting inefficiencies, suppose the source alphabet consists of four symbols a, b, c, and d occurring with the respective probabilities $P(a) = \frac{1}{2}$, $P(b) = \frac{1}{4}$, $P(c) = P(d) = \frac{1}{8}$. These symbols could of course be represented by the four binary two-tuples 00, 01, 10, and 11, each source symbol thereby requiring two binary code symbols. In contrast, if the mapping were from a to 1, b to 01, c to 001, and d to 000, half of the source symbols would require three code symbols, but the average number of code symbols needed to represent a source symbol would be

$$\tfrac{1}{2}\cdot 1 + \tfrac{1}{4}\cdot 2 + \tfrac{1}{8}\cdot 3 + \tfrac{1}{8}\cdot 3 = 1\tfrac{3}{4}$$

a reduction of $12\frac{1}{2}\%$ over the first method. The suggestion, then, is to effect a relationship between the probability of each symbol and the number of code symbols used to represent it, the more probable the source symbol, the shorter the corresponding sequence of code symbols. An algorithm for actually minimizing the average number of code symbols needed to represent a source symbol is presented in Section 10.3.

A second defect of real, as opposed to ideal, sources is the almost inevitable dependence between the successive source outputs. Often, the source can be modeled as a finite order *Markov source*, an nth order Markov source defined as one in which the probability distribution of the current output, given all the past outputs, is a function of the previous n outputs only. If an nth order Markov source is capable of producing N different outputs, N^n conditional distributions are needed to describe completely the statistics of the current output, and the source is said to have N^n possible *states*. Regardless of the form of this dependence among the successive source outputs, however, the effect is to introduce redundancy in the message sequence. Each symbol conveys less information than it is capable of conveying since it is to some extent predictable from the preceding symbols. Numerous schemes are available for reducing the effect of this redundancy in the source output, and hence for decreasing the average number of symbols needed to represent a message. These schemes are usually referred to as *data compression* or *data compaction* techniques.

Other constraints, such as equipment limitations, ease of encoding and decoding, and the desire to preserve a particular ordering in the mapping between the source and the code symbols, all affect to a greater or lesser extent the encoding chosen. Whatever the criterion for selecting a particular encoding, however, the result is that each source output, or each sequence of outputs, is represented by a sequence of code symbols.† Since generally more than one code symbol is needed for each source output, the code symbols are necessarily grouped into *words*, each word consisting of several symbols. The ensemble of allowable words will be referred to as the (source) *code dictionary*.

Any dictionary must satisfy one important condition if it is to have any practical value: it must be *uniquely decodable*. A sequence of words from a dictionary D is uniquely decodable, or more simply, *decodable*, if, and only if, it is impossible to interpret it as some other sequence of words from the same dictionary. A dictionary D is decodable if, and only if, every finite sequence of words from D is decodable. This concept is best illustrated by a dictionary lacking this property, as, for example, the dictionary $\{w_1 = 0, w_2 = 10, w_3 = 1001\}$. It is impossible to determine whether the symbol sequence 10010 was formed from the word sequence $w_3 w_1$, or from $w_2 w_1 w_2$. This dictionary is not uniquely decodable, and, consequently, of no practical use. (It might be argued that a "space" could always be left between words assuring their decodability even if the dictionary were not uniquely decodable. But if the decoder is able to distinguish a space from the other symbols, the space too could be used as a symbol. Using this space symbol only to separate words is not particularly efficient in general, as we shall see.)

Although any finite sequence of words from a uniquely decodable dictionary can be decoded in only one way, it does not necessarily follow that a *continuing* sequence of words has this property. As a consequence, there may be some delay before a word sequence can be decoded. Consider, for example, the dictionary $D = \{101, 00110, 10111, 11001\}$. The sequence of words

$$10111, 00110, 11001$$

---

†When, as is assumed here, the source outputs are grouped into blocks prior to encoding, with each block containing a predetermined number of outputs, the resulting code is sometimes called a *block code*. We shall reserve this term for the more restricted class of codes, however, in which each code word also contains the same number of code symbols.

cannot be distinguished from the sequence

$$101,\ 11001,\ 10110,\ 00110$$

until the fifteenth symbol is received. The dictionary D is decodable, but fifteen (or more) symbols may be needed before the first word can be uniquely identified. The number d_c of symbols in the longest sequence of words from a decodable dictionary needed to distinguish uniquely the first word of that sequence is called the *decoding delay*.

The decoding delay of a decodable dictionary is not necessarily finite, as the dictionary $D' = \{101, 00111, 10111, 11001\}$ demonstrates. This dictionary can be shown to be decodable. Nevertheless, the sequence

$$101110011100111\ldots$$

can be decoded as

$$101,\ 11001,\ 11001,\ldots$$

or as

$$10111,\ 00111,\ 00111,\ldots$$

The length of the longest ambiguous sequence in this case is unbounded.

There are many possible conditions which can be imposed on a source encoding, but the resulting dictionaries must in every case be decodable if they are to be of any utility. Accordingly, most of this chapter is devoted to the investigation of the decodability of arbitrary dictionaries rather than to an examination of specific dictionaries. The only exception to this is in Section 10.3 where, as we have already mentioned, an algorithm is presented to minimize the average number of code symbols needed to represent a source symbol when the relative probabilities of these symbols are known.

10.2 The Kraft Inequality and Exhaustive Dictionaries

We now prove a necessary and sufficient condition for the existence of a decodable dictionary having N words with the respective lengths (number of code symbols) $l_1, l_2, \ldots, l_N$. We begin with the necessary condition.

Theorem 10.1. If a dictionary of N r-symbol words of lengths $l_1, l_2, \ldots, l_N$ is uniquely decodable, then

$$\sum_{i=1}^{N} r^{-l_i} \leq 1 \tag{10.2.1}$$

(This inequality is generally referred to as the *Kraft* or *Kraft-Szilard inequality*.)

Proof. Consider the expression

$$\left(\sum_{i=1}^{N} r^{-l_i}\right)^n = \sum_{\nu=nl_{\min}}^{nl_{\max}} N_\nu r^{-\nu}$$

where N_ν is the number of sequences of exactly n code words containing exactly ν symbols. The equality follows from the fact that the nth power of $\sum_{i=1}^{N} r^{-l_i}$ contains only terms of the form $r^{-\sum_{j=1}^{n} l_{i_j}} = r^{-\nu}$, where ν can assume values within the range $nl_{\min}$ to $nl_{\max}$ ($l_{\max} \triangleq \max_i l_i$, $l_{\min} \triangleq \min_i l_i$). By definition, there are exactly N_ν such terms for any ν. Clearly, if the dictionary is to be uniquely decodable, no two different combinations can yield the same sequence. Since there are r^ν sequences of ν r-ary symbols, $N_\nu \leq r^\nu$, and

$$\left(\sum_{i=1}^{N} r^{-l_i}\right)^n \leq \sum_{\nu=nl_{\min}}^{nl_{\max}} 1 = n(l_{\max} - l_{\min}) + 1$$

for any n. Since for finite length code words, $l_{\max} - l_{\min}$ is finite, it is always possible to select n large enough so that this equality is not satisfied for any $\sum_{i=1}^{N} r^{-l_i} > 1$. Thus, if the dictionary is to be decodable this sum cannot exceed unity.

That this inequality can be an equality for some decodable dictionaries is immediately apparent since a dictionary for which $l_i = l (i = 1, 2, \ldots, N)$ is decodable if all the words are distinct. The condition

$$\sum_{i=1}^{N} r^{-l} = Nr^{-l} = 1$$

simply limits N to the value r^l. As there are exactly r^l different l-tuples of r symbols, this condition can be met with a decodable dictionary. That code dictionaries exist for *all* sets of l_i satisfying the inequality may be more surprising.

A sequence p of r-ary symbols is called a *prefix* of a word w if $w = ps$ for some sequence s containing one or more symbols. In like manner, the sequence s is referred to as a *suffix* of the word w. We define a code dictionary with the *prefix property* as one in which no code word is a prefix of any other code word. If a dictionary has the prefix property, each word can be unambiguously decoded as soon as it is received in its entirety. Such dictionaries are called *instantaneously decodable*. (To require even shorter delays, i.e., to demand that the longest words be decodable even before they have been entirely received, would clearly be impractical: if this were possible the unnecessary ending of these words could have been omitted in the first place.) Obviously, if a dictionary is to be instantaneously decodable, it must have the prefix property. The two terms are synonymous. We now proceed to prove that dictionaries with the prefix property and, a fortiori, decodable dictionaries exist with words of length $\{l_i\}$, $(i = 1, 2, \ldots, N)$ for all sets $\{l_i\}$ satisfying condition (10.2.1).

Theorem 10.2. There exists a dictionary with the prefix property defined over an r-symbol alphabet with code words of length $l_1, l_2, \ldots, l_N$, for all sets $\{l_i\}$ satisfying

$$\sum_{i=1}^{N} r^{-l_i} \leq 1$$

Proof. Let n_j be the number of code words in the dictionary which are to have length j $(0 \leq n_j \leq N)$. If $n_1 \leq r$ we can select the necessary number of unique 1-tuples for the dictionary. Suppose we have selected the n_1 1-tuples and begin choosing 2-tuples. If the dictionary is to have the prefix property, none of these 1-tuples can be the prefix of any of the chosen 2-tuples. Accordingly, there remain $r - n_1$ allowed initial symbols for the 2-tuples, for a total of $(r - n_1)r$ eligible 2-tuples from which n_2 must be selected. This is possible if $n_2 \leq (r - n_1)r$. Similarly, the necessary number of 3-tuples is available if $n_3 \leq [(r - n_1)r - n_2]r = r^3 - n_1 r^2 - n_2 r$ and, in general, the necessary number of i-tuples is available if

$$n_i \leq \{\ldots\{[(r - n_1)r - n_2]r - n_3\}r \ldots - n_{i-1}\}r$$
$$= r^i - n_1 r^{i-1} - n_2 r^{i-2} \ldots - n_{i-1} r$$

If the longest code word has length L, a dictionary exists having the prefix property if

$$\sum_{j=1}^{L} n_j r^{L-j} \leq r^L$$

or

$$\sum_{j=1}^{L} n_j r^{-j} = \sum_{i=1}^{N} r^{-l_i} \leq 1$$

which was to be proved.

As an example, suppose $r = 2$, $l_1 = 1$, $l_2 = 2$, $l_3 = 3$, $l_4 = l_5 = 4$. Then $\sum_{i=1}^{5} 2^{-l_i} = 1$ and a dictionary having the prefix property exists. All of the dictionaries (10.2.2), in fact, have the prefix property and satisfy the necessary condition on the word lengths.

1	1	1	1
01	01	00	00
001	000	011	010
0001	0010	0100	0110
0000	0011	0101	0111

(10.2.2)

0	0	0	0
10	10	11	11
110	111	100	101
1110	1101	1011	1001
1111	1100	1010	1000

Interestingly, not only are decodable dictionaries not necessarily unique for any set of attainable word lengths, but, even with the added prefix property constraint, dictionaries can generally be realized in many ways. There are, in fact, $\binom{r}{n_1}$ ways of selecting the first n_1 words, $\binom{r^2 - rn_1}{n_2}$ ways of selecting the next n_2 words, etc., yielding a total number $N\{l_i\}$ of instantaneously decodable dictionaries where

$$N\{l_i\} = \binom{r}{n_1}\binom{r^2 - rn_1}{n_2}\binom{r^3 - r^2n_1 - rn_2}{n_3} \cdots \binom{r^L - r^{L-1}n_1 - \cdots - rn_{L-1}}{n_L} \tag{10.2.3}$$

Even when the lengths l_i of the code words are specified and the dictionary is required to be instantaneously decodable, it will still generally be underconstrained. Further requirements can be imposed.

The significance of the summation $\sum_{i=1}^{N} r^{-l_i}$ becomes more evident upon examining the concept of an *exhaustive* dictionary. An exhaustive dictionary is defined as one containing at least one word which is the prefix of any semi-infinite sequence of symbols over which the code is defined. As a consequence, all finite sequences of symbols can be interpreted as a sequence of code words, ending either with a code word or a code word prefix.

Theorem 10.3. If an N-word dictionary is exhaustive, then

$$\sum_{i=1}^{N} r^{-l_i} \geq 1$$

where l_i is the length of the ith code word.

Proof. Let s_L represent any sequence of length $L \geq l_{\max} = \max_i l_i$. Then the dictionary is exhaustive only if there is a code word which is the prefix of every such sequence s_L. A code word of length l_i is the prefix of exactly r^{L-l_i} sequences of length L. There are a total of r^L sequences of that length, so, if the code is exhaustive

$$\sum_{i=1}^{N} r^{L-l_i} \geq r^L$$

which proves the theorem.

Theorem 10.4. Any two of the properties (1) exhaustive property, (2) prefix property, (3) $\sum_{i=1}^{N} r^{-l_i} = 1$, of a code dictionary imply the third.

Proof. If the dictionary is exhaustive, $\sum_{i=1}^{N} r^{-l_i} \geq 1$ by Theorem 10.3; if it has the prefix property $\sum_{i=1}^{N} r^{-l_i} \leq 1$ by Theorem 10.2. Clearly, (1) and (2) imply (3). If a dictionary is exhaustive and if $\sum_{i=1}^{N} r^{-l_i} = 1$ then, in the terminology of Theorem 10.3, each of the r^L sequences of length L must occur exactly once as one of the $\sum_{i=1}^{N} r^{L-l_i}$ sequences prefixed by a code word. Consequently, no two of these sequences can be identical and no code word can be the prefix of any other word. Thus (1) and (3) imply (2). Finally, if the dictionary has the prefix property and

$\sum_{i=1}^{N} r^{-l_i} = 1$, then, again in the terminology of Theorem 10.3, all $\sum_{i=1}^{N} r^{L-l_i}$ sequences of length L prefixed by a code word must be unique. If this sum is equal to r^L, all possible sequences are so represented and the code is exhaustive; (2) and (3) imply (1).

Corollary. If a dictionary is exhaustive and decodable, it has the prefix property.

Proof. Since the dictionary is exhaustive, $\sum r^{-l_i} \geq 1$; since it is decodable $\sum r^{-l_i} \leq 1$. Conditions (1) and (3) of Theorem 10.4 apply, and consequently so does condition (2).

The following theorems concerning exhaustive dictionaries will be useful in later investigations.

Theorem 10.5. If D is an exhaustive, decodable dictionary, p the prefix of some word in D, and $D(p)$ the set of all suffixes s_i such that ps_i is a word in D, then the *suffix set* $D(p)$ is also an exhaustive, decodable dictionary.

Proof. First, since D is exhaustive and decodable, it has the prefix property. Suppose c'_i were both a prefix and a code word in $D(p)$. Then pc'_i would be both a prefix and a code word in D. Thus, $D(p)$ must have the prefix property and hence is decodable. Now consider all code words in D having the prefix p. Since the dictionary is exhaustive, every infinite sequence has some code word c_i as a prefix, and, in particular, every infinite sequence beginning with p has some code word $c_i = pc'_i$ for a prefix, where c'_i is a word in $D(p)$. [If p were a code word, then, by the prefix property, $D(p)$ would be empty.] Consequently, every infinite sequence has some code word c'_i in $D(p)$ as a prefix, and $D(p)$ is exhaustive.

Theorem 10.6. If an N-word decodable dictionary D defined over an r-symbol alphabet is exhaustive, the number of distinct prefixes of words in D is equal to $(N - r)/(r - 1)$.

Proof. If $N = r$, the theorem is certainly true since the word lengths must all be 1 if the dictionary is to be exhaustive and decodable. Suppose the theorem is true for any exhaustive dictionary containing N' words for all N' in the range $r \leq N' < N$, and consider the N-word dictionary D. Divide D into r suffix sets, one set for each single-symbol prefix. By Theorem 10.5 every such set is itself a decodable, exhaustive dictionary. If the ith suffix set D_i has a total of n_i distinct words *and* word prefixes, the sum of the number of words and prefixes in D is $\sum_{i=1}^{r} n_i + r$. This follows because all distinct words and prefixes in any set D_i give rise to distinct words and prefixes in D; no two words or prefixes in D corresponding to different suffix sets can be the same. Each set D_i contributes its n_i words and prefixes plus the ith single-symbol prefix. Note this is true even when D_i is empty since D is exhaustive. But each of the dictionaries D_i has fewer than N words; by hypothesis $n_i = [(N_i - r)/(r - 1)] + N_i$ where N_i is the number of words in D_i. Thus

$$\sum_{i=1}^{r} r\frac{N_i - 1}{r - 1} + r = \frac{N - r}{r - 1} + N$$

and the theorem is true by induction.

Corollary. The longest word in an N-word exhaustive, decodable dictionary D has length

$$L \leq \frac{N - 1}{r - 1}$$

Proof. If D has a word of length L then it must have at least $L - 1$ prefixes. The prefix property of D assures that none of these prefixes are themselves code words, so, from Theorem 10.6, $L - 1 \leq [(N - r)/(r - 1)]$, as was to be shown.

Corollary. An exhaustive, decodable, N-word dictionary D exists only if $N = r + \lambda(r - 1)$ for some non-negative integer λ.

Proof. There must be an integer number λ of distinct prefixes in D. But from Theorem 10.6, $\lambda = (N - r)/(r - 1)$.

10.3 Huffman Codes

In this section, we describe an algorithm for the construction of instantaneously decodable dictionaries which, of all such dictionaries, minimize the average number of symbols needed to encode a message, given that each source output is to be encoded as an individual entity, without regard to its predecessors or successors. (This, of course, does not preclude the possibility that prior processing has already been effected to reduce the redundancy in these outputs as discussed in Section 10.1.)

It is convenient in describing the encoding algorithm to introduce the concept of an r-ary *tree*. An r-ary tree is as shown in Figure 10.1 (for $r = 3$), each *node* or *branch point* giving rise to r *branches*, each branch in turn ending in a node. These nodes are referred to as *immediate descendents* of the original node. If one node can be reached from another through a sequence of immediate descendents, then it is called a *descendent* of that node. The nodes are labeled as follows: The first node is unlabeled; its r immediate descendents are labeled with the r symbols of the alphabet. In general, the immediate descendents of the node labeled $\alpha_1 \alpha_2 \ldots \alpha_i$ are in turn labeled $\alpha_1 \alpha_2 \ldots \alpha_i\beta$ where β takes on the r different alphabet symbols. Clearly all l-tuples defined over an r-symbol alphabet are represented as a node in a tree of this sort. A dictionary is defined by identifying some of these nodes as code words and ignoring the others.

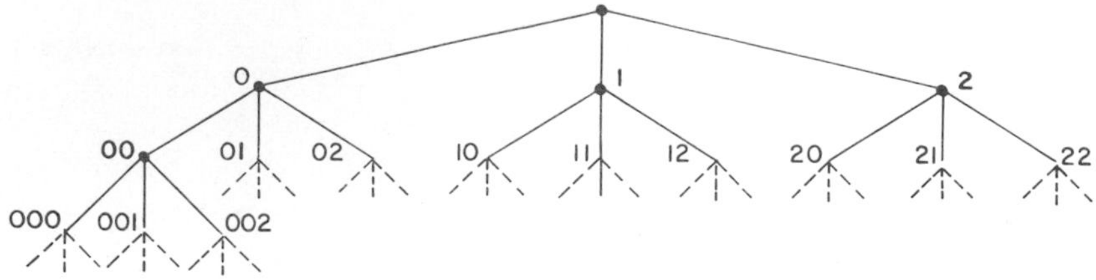

Figure 10.1 A ternary tree

If the dictionary is to have the prefix property, no descendent of a code word can also be a code word. If a node of such a dictionary is a code word, all of its descendent nodes can be omitted from the tree; each code word is then a *terminal node* producing no descendents. A tree of this sort is called a *code tree*. The code tree corresponding to the dictionary {0, 1, 20, 21, 220, 221, 222} is shown in figure 10.2. Since all dictionaries which can be represented by a code tree are obviously instantaneously decodable, we have proved:

Theorem 10.7. A dictionary is instantaneously decodable if, and only if, it can be represented as a code tree.

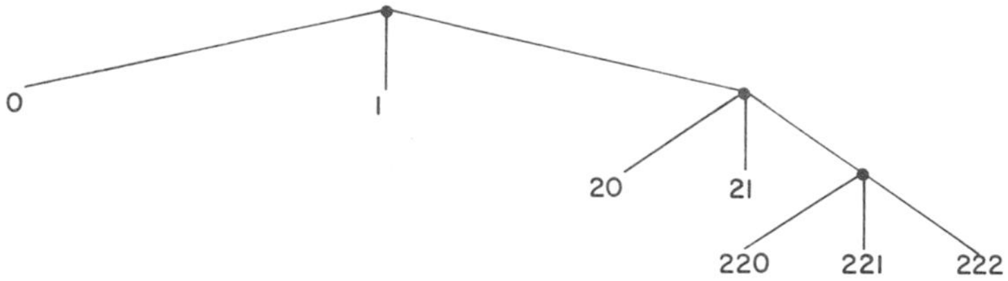

Figure 10.2 A code tree

Note that some of the terminal nodes of the tree need not be in the dictionary; i.e., they may correspond to "source symbols" occurring with zero probability.

Theorem 10.8. The set of word lengths $\{l_i\}$ of any dictionary which can be represented by a code tree satisfies the bound (10.2.1) with equality. (Again, in order to truncate the tree, it may be necessary to include some zero probability code words.)

Proof. It is only necessary to demonstrate that such a dictionary is exhaustive, that any semi-infinite sequence of r-ary symbols can be separated into code words. Consider an arbitrary sequence of symbols. If the first symbol is a code word it corresponds to a terminal node of the code tree. If it is not, it corresponds to a non-terminal node in the tree. In the latter case, the first two symbols correspond to a node at the end of one of the branches leading from this non-terminal node. The process can be continued; each new node, determined by each successive symbol, must be either a terminal or a non-terminal node. In either case, it is in the tree. Thus, it is always possible to move from node to node within the tree until a terminal node is encountered. When this happens, a word has been decoded and the process begins anew. Since every possible sequence can be decoded in this way, this dictionary is exhaustive.

A procedure for constructing an instantaneously decodable dictionary which minimizes the average number of code symbols needed to encode a source symbol is due to Huffman (Reference 10.7). The method is simply stated: Let S_0 denote the set of N source symbols, each of which is to be assigned to some node in a code tree. Take the r nodes corresponding to the r source symbols having the least probability of occurring and connect them via r branches to a single composite node. The $N - r$ remaining nodes plus the composite node can be identified with a new source S_1, with the composite node representing a new "symbol" having a probability equal to the sum of the probabilities of the nodes connected to it. Similarly, connect the r least probable nodes of the source S_1 to one node defining still another source S_2 of $N - 2(r - 1)$ symbols. Continue this procedure until only one node remains. This defines the code tree. The number of nodes under consideration is reduced by $r - 1$ at each step. There will be exactly r nodes left at the next to the last step if, and only if,

$$N = r + \lambda(r - 1) \tag{10.3.1}$$

for some integer λ. (Note this includes the case $N = r^{\alpha}$ for any integer α.) If N is not of this form, "dummy" source symbols, each having zero probability of occurring, must be defined until the total number of symbols plus dummy symbols does satisfy Equation (10.3.1). The example in Figure 10.3 should help illustrate the procedure.

The code tree defined by the Huffman algorithm is not unique. In the example of Figure 10.3(b), the symbols A_4, A_5, and A_6, for example, could obviously be interchanged without affecting the average message length. The symbols A_5, A_6, and A_7 could also be interchanged, since they all occur with the same probability.

Suppose the symbols of an N-word source S are represented by the words or some dictionary D. Then the average code word length is defined as

$$L = \sum_{i=1}^{N} p_i l_i \tag{10.3.2}$$

where l_i is the length of the code word representing the ith source symbol, and p_i is the probability of that symbol occurring at the source output. Clearly,

Message Units (S^0)	Probabilities					
	S^0	S^1	S^2	S^3	S^4	S^5
A^1	0.40	0.40	0.40	0.40	0.50	1.00
A^2	0.20	0.20	0.20	0.20	0.40	
A^3	0.10	0.10	0.12	0.18	0.20	
A^4	0.08	0.08	0.10	0.12		
A^5	0.05	0.05	0.08	0.10		
A^6	0.05	0.05	0.05			
A^7	0.05	0.05	0.05			
A^8	0.03	0.04				
A^9	0.02	0.03				
A^{10}	0.01					
A^{11}	0.01					

(a)

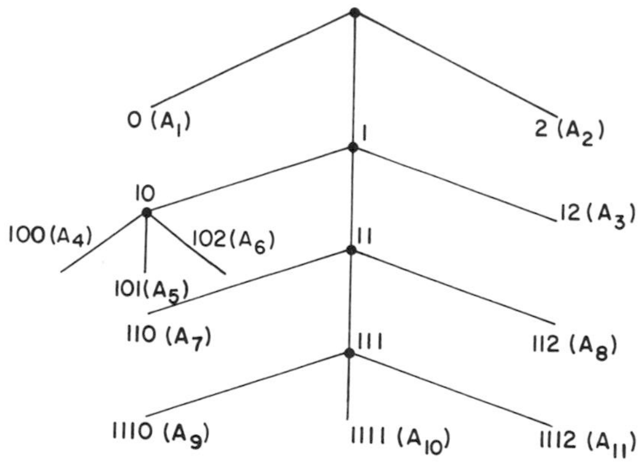

(b)

Figure 10.3 (a) A set of source symbols and their probabilities; (b) a Huffman code tree for the message set of (a) ($r = 3$)

at least when each source symbol is encoded without regard to its predecessors or successors, the average number of symbols needed to represent a sequence of source symbols is minimized by minimizing L. If the dictionary is to be uniquely decodable, the set of code word lengths $\{l_i\}$ must also satisfy the bound (10.2.1). We now prove that no uniquely decodable dictionary produces a shorter average message length for a given set of probabilities $\{p_i\}$ than do the Huffman codes.

Theorem 10.9. No instantaneously decodable dictionary has a shorter average word length to represent a source symbol than the appropriate Huffman dictionary.

Proof. Let S be an N-symbol source with symbol probabilities $\{p_i\}$, $1 \geq p_1 \geq p_2 \geq \cdots \geq p_N \geq 0$, and let D be an instantaneously decodable dictionary which, when used to encode S, attains the minimum possible average code word length L. (Since zero probability code words do not affect the average word length, N can be assumed with no loss in generality to be of the form $N = r + \lambda(r - 1)$, with λ an integer.) Clearly, each of the r least probable source symbols (including any zero probability symbols) will be represented by a code word having length at least as great as that of any other word in D. Moreover, since D can be represented by a code tree, it is obviously possible by a proper assignment of code words to represent these r least probable symbols by nodes which are all immediate descendants of a common node. Replacing these r-nodes by this common node yields a tree having $N' = N - r + 1$ terminal nodes. This tree represents a dictionary D' which could be used to encode an N'-symbol source S' having symbol probabilities $\{p'_i\}$ with $p'_i = p_i$, $1 \leq i \leq N - r$, $p'_{N'} = \sum_{i=N-r+1}^{N} p_i \triangleq p_0$. Thus, $L = L' + p_0$ with L' the average length of a word in D' (the averaging, of course, with respect to the probabilities $\{p'\}$). Accordingly, D is optimum for S if and only if D' is optimum for S'. Moreover, if D' is a Huffman code so also is D. Since all r-word Huffman codes are trivially optimum, all N-word Huffman codes are therefore optimum by induction for any N.

Corollary. No uniquely decodable dictionary has a shorter average word length than the appropriate Huffman dictionary.

Proof. All uniquely decodable dictionaries must satisfy the bound $\sum r^{-l_i} \leq 1$. Suppose there were such a dictionary with a set of word lengths $\{l_i\}$ yielding a shorter average word length than the optimum Huffman dictionary. Then, from Theorem 10.2, there must be an instantaneously decodable dictionary with this property. But from Theorem 10.9, no instantaneously decodable dictionary is superior to a Huffman dictionary, thus leading to a contradiction.

The Huffman algorithm, then, is ideally suited to the situation in which prior processing has reduced to a minimum the dependence between successive information symbols. When this is the case the Huffman code uses each of the r-ary code symbols approximately $(1/r)$th of the time and each symbol in an encoded sequence provides a minimum of information concerning its successor. This can be seen quite easily from the corresponding code tree. Each immediate descendent of any one node represents an event of, as nearly as possible, equal probability. Accordingly, any one symbol is, again as nearly as possible, equally likely to follow any other, and any one symbol is used approximately as often as any other.

10.4 On Testing an Arbitrary Dictionary for Unique Decodability

A useful tool in the analysis of the decodability properties of a particular dictionary is the *suffix decomposition* of the dictionary. The suffix decomposition of a dictionary consists of identifying the sets S_i defined as follows: The set S_0 is the dictionary itself, $S_0 = D$. The set S_i consists of all symbol sequences s_i (which we will call *segments*) such that either $s_{i-1}s_i = c$ or $s_{i-1} = cs_i$ where c is some word in D and s_{i-1} is any segment in S_{i-1}. Note every segment in every set S_i, $i \geq 1$, is a code word suffix. To illustrate, let $D = \{0, 10, 12, 21, 112, 1122\}$. Then the suffix decomposition of D can be written

$S_0 = D$	S_1	S_2	S_3	S_4	S_5	...
0	2	1	0	1	0	
10			2	2	2	
12			12		12	...
21			122		122	
112					1	
1122						

and the sets S_i are identical for all $i \geq 5$. It is important to remember that there are two reasons for including a sequence of symbols in a particular set. The set S_4, for example, includes the term 1 because 21 is in D and 2 is in S_3. The term 2 is contained in S_4 since 12 is in D and 122 in S_3.

We now prove the following theorem relating unique decodability to a condition on the suffix decomposition of D. This theorem was first proved by Sardinas and Patterson (Reference 10.2).

Theorem 10.10. A code dictionary is uniquely decodable if, and only if, no set S_i contains a code word.

Proof. We will first prove the necessity of this condition for unique decodability by showing that if a set contains a code word it is possible to construct a symbol sequence which is decomposable into two different word sequences. For notational convenience, s_i will designate any one of the segments in the ith set S_i. We begin by selecting a term s_m from the mth set. By definition, there is a segment s_{m-1} such that either $s_{m-1}s_m = c_{m-1}$ or $c_{m-1}s_m = s_{m-1}$ where c_{m-1} is a code word (the subscript simply identifying it as a code word pertaining to the $m-1$st set). If $s_{m-1}s_m = c_{m-1}$, we form the sequence $s_{m-1}s_m$; if $c_{m-1}s_m = s_{m-1}$, we form the sequence $c_{m-1}s_m$. Continuing we must find a term in the $m-2$nd set such that $s_{m-2}s_{m-1} = c_{m-2}$ or $c_{m-2}s_{m-1} = s_{m-2}$. In the first situation, we adjoin s_{m-2} as the next term on the left of the sequence formed thus far, in the second we adjoin c_{m-2}. We continue this procedure through all $m+1$ sets (the set S_0 containing only code words) obtaining the sequence

$$c_0x_1x_2 \ldots x_{m-1}s_m \tag{10.4.1}$$

where x_i is either a code word c_i or a code word suffix s_i.

Now consider any portion of the above sequence of the form

$$s_i c_{i+1} c_{i+2} \ldots c_{i+l-2} c_{i+l-1} s_{i+l} \tag{10.4.2}$$

Since $c_{i+l-1}s_{i+l} = s_{i+l-1}$, this sequence can be written as

$$s_i c_{i+1} c_{i+2} \dots c_{i+l-2} s_{i+l-1}$$

and continuing to combine the last two terms, we obtain

$$\begin{aligned} & s_i c_{i+1} c_{i+2} \dots s_{i+l-2} \\ &= s_i c_{i+1} c_{i+2} \dots s_{i+l-3} \\ &= \dots\dots\dots\dots \\ &= s_i s_{i+1} = c_i \end{aligned}$$

Similarly,

$$s_{i-k} c_{i-k} \dots c_{i-1} s_i = s_{i-k} s_{i-k+1} = c_{i-k} \tag{10.4.3}$$

Moreover, the sub-sequence

$$\begin{aligned} & c_0 c_1 c_2 \dots s_j \\ &= c_0 s_1 = s_0 \end{aligned} \tag{10.4.4}$$

and s_0, a term in the set S_0, is a code word. Thus, every term s_i in the sequence (10.4.1) represents both the end of one code word and the beginning of another. One decomposition of the sequence (10.4.1) into code words is the following: the first code word has the prefix c_0 and the suffix s_i where i is the first integer j such that $x_j = s_j$. If $x_{i+1} = c_{i+1}$, this term is the second word. If $x_{i+1} = s_{i+1}$, then the second word begins with s_{i+1} and ends with s_{i+l} where $i + l$ is the third integer such that $x_{i+l} = s_{i+l}$. In general, the remaining words begin with the term representing the 2νth occurrence of an "s" as opposed to a "c" and end with the $2\nu + 1$st such term. All those terms between the $2\nu + 1$st and the $2\nu + 2$nd "s" terms are "c" terms and hence are themselves code words. The second decomposition into code words is to take $c_0, c_1, \dots, c_{i-1}$ as individual code words and then continue by using the first, and in general the $(2\nu + 1)$st, "s" term as the prefix and the 2nd, and in general the $(2\nu + 2)$nd "s" term as the suffix of a code word. The "c" terms between the $(2\nu + 2)$nd and the $(2\nu + 3)$rd "s" terms are read as individual code words. There are, therefore, two different decompositions, one ending with the term s_m, the other ending with the term x_{m-1}. The complete sequence then has two decompositions whenever $s_m = c_m$ for some code word c_m. This completes the proof of the necessity of the stated condition.

To prove the sufficiency of the condition for unique decodability, we show that any sequence which can be decomposed into two word sequences must be representable in the form (10.4.1) with s_m a code word. To accomplish this we define the integer sequences

$$p_0 p_1 p_2 \dots p_l$$

and

$$q_0 q_1 q_2 \dots q_k$$

where $p_0 = q_0 = 0$ and p_i and q_i $(i > 0)$ are defined so that the ith word in one decoding of the ambiguous code word sequence ends with the p_ith symbol while the ith word in a second decomposition ends with the q_ith symbol. By hypothesis $p_l = q_k$ and, without loss of generality p_i and q_j can be assumed unequal ($i \neq 0, l$, $j \neq 0, k$). (If this is not true we simply have two or more shorter ambiguous sequences.) Now arrange the p_i's and q_i's into one sequence

$$r_0 r_1 r_2 \dots r_{m+1}$$

where $r_0 = p_0 = q_0 = 0$, $r_{m+1} = r_{l+k-1} = p_l = q_k$, $r_\nu > r_{\nu-1}$, and each term r_ν is one of the terms p_i or q_j. If $r_\nu = p_i$ and $r_{\nu+1} = p_{i+1}$ or if $r_\nu = q_j$ and $r_{\nu+1} = q_{j+1}$, the subsequence beginning with the $(r_\nu + 1)$st symbol and ending with the $r_{\nu+1}$st symbol is a code word. If, on the other hand, $r_\nu = p_i$ and $r_{\nu+1} = q_j$ or, if $r_\nu = q_j$ and $r_{\nu+1} = p_i$, then this same subsequence is not a code word but the prefix of one code word and the suffix of another. In either event, by identifying this subsequence with the term y_ν in the sequence

$$y_0 y_1 \dots y_m$$

we obtain a sequence of the form (10.4.1). The subsequence $y_i y_{i+1} \dots y_{i+l}$ where y_{i+l} is the first suffix in the sequence following y_{i-1} is consequently a term in the set S_i. Further, y_m is a suffix since $p_l = q_k$. Hence the mth set contains y_m. Finally, y_m is a code word as r_{m-1} equals either p_{l-1} or q_{k-1}. Thus, any ambiguous symbol sequence results in a suffix decomposition violating the conditions of the theorem. This completes the proof.

In order to illustrate the method more concretely, consider the code dictionary $D = 11, 010, 100, 110, 101, 0011$. The suffix decomposition of the dictionary begins:

$S_0 = D$	S_1	S_2	S_3	S_4	S_5
11	0	10	0	10	0
010		011	1	011	1
100				1	00
110				00	10
101				01	01
0011					11

and S_5 contains the code word 11. To reconstruct an ambiguous sequence, we pick $x_5 = s_5 = 11$, $x_4 = s_4 = 00$ (so $s_4 s_5$ is a code word), $x_3 = s_3 = 1$ ($s_3 s_4 = 100$), $x_2 = s_2 = 10$ ($s_2 s_3 = 101$), $x_1 = s_1 = 0$ ($s_1 s_2 = 010$) and $x_0 = s_0 = 11$ ($s_0 s_1 = 110$). The sequence $s_0 s_1 s_2 s_3 s_4 s_5 = 1101010011$ can be decoded as

$$11, 010, 100, 11$$

and also as

$$110, 101, 0011$$

Let D^* denote the dictionary obtained by reversing (reading from right to left) the words of the dictionary D (e.g., if $D = \{1, 01, 02, 012\}$ then $D^* = \{1, 10, 20, 210\}$). Then:

Theorem 10.11. The dictionary D is uniquely decodable if, and only if, the dictionary D^* is uniquely decodable.

Proof. If D^* is not uniquely decodable there is some sequence of symbols which can be decoded into at least two sequences of words from D^*. The reverse of this sequence can clearly be decoded into at least two sequences of words from D. Accordingly, D is uniquely decodable only if D^* is, and conversely.

Let N_s be the number of distinct suffixes of words in the dictionary D which are not themselves in D and let N_s^* be the corresponding number for the dictionary D^*. Then:

Theorem 10.12. No more than n_0 sets, excluding the set $S_0 = D$, need be investigated to determine whether a given dictionary D is decodable, where

$$n_0 \leq 1 + \min [N_s, N_s^*]$$

Proof. Each set $S_i(i > 0)$ contains only code word suffixes and the contents of S_i are determined exclusively by those of S_{i-1}. If S_j is a subset of the union of the sets S_i, $i = 1, 2, \ldots, j - 1$, so also must be S_{j+1}. If this is the case, and none of these sets contain a code word, neither does S_{j+1}. Unless each new set contains at least one suffix that has not previously occurred the investigation need proceed no further. There are at most N_s such sets. Accordingly, if no code word is included in any of the first $N_s + 1$ sets of the suffix decomposition, no code word will occur in any of them.

If a dictionary D is not uniquely decodable, there is a sequence of words from it which can be decoded in at least two ways. The reverse of this sequence is a sequence of code words from D^* which can also be decoded in at least two ways (Theorem 10.11). Moreover, the two sequences are decomposed by the two decodings into the same number of segments. Since at most $N_s^* + 1$ sets of the suffix decomposition of D^* need be investigated to determine its decodability, this also represents an upper bound on the number of sets of the decomposition of D that need to be considered. Thus

$$n_0 \leq \min [N_s + 1, N_s^* + 1]$$

and the theorem is proved.

As an example, consider the code dictionary 10, 01, 011, 111. The suffixes are 0, 1, 11 and $N_s = 3$. The reverse dictionary 01, 10, 110, 111 has suffixes 1, 0, 10, 11, so $N_s^* = 4$. It is necessary to investigate only four sets.

$S_0 = D$	S_1	S_2	S_3	S_4
10 01 011 111	1	0 11	1 11	0 11 1

Since none of the four sets contains a code word, the dictionary is uniquely decodable. Actually, it was not even necessary to examine the fourth set; all of the suffixes occur in the first two sets. A code word would have been exhibited in one of the first three sets if it were ever to occur. Or even simpler, since only suffixes appear in the sets S_i and since no suffix in D is a code word in D, none of these sets could possibly contain a code word. The suffix decomposition is not even necessary.

It is interesting to observe that even though $N_s^* > N_s$ the reverse dictionary D^* demonstrates its unique decodability even more quickly. When $S_0 = D^*$, S_1 is empty.

$S_0 = D^*$	S_1
01 10 110 111	

All remaining sets are empty and D^* (and hence D) is uniquely decodable. Notice that D^* has the prefix property and is obviously uniquely decodable without further investigation. All sets in the suffix decomposition of any dictionary having the prefix property will clearly be empty since no suffix can be adjoined to any code word to yield another code word. The reverse of a dictionary having the prefix property is one having what might be called the *suffix property:* no suffix of a code word is itself a code word. It follows either from Theorem 10.11, or from the argument in the preceding paragraph, that a dictionary having the suffix property is always uniquely decodable.

The number N_s of distinct suffixes of code words in a dictionary D has been used to establish a bound on the number of sets in the suffix decomposition of D which need be observed before it can be declared uniquely decodable. We will presently find other uses for this number N_s (e.g., in bounding the decoding delay of a code dictionary). It will be useful to have an upper bound on N_s in terms of the dictionary itself without having to detail its suffixes and to count those which are distinct. The following theorem provides such a bound.

Theorem 10.13. Let D be an N word dictionary over an r-symbol alphabet and denote by D_0 the subset of D containing the N_0 words $(1 \leq N_0 \leq N)$ in D which are not suffixes of any other word in D. Then the number N_s of distinct suffixes in D which are not also words in D is bounded by

$$N_s \leq \sum_i im_i - N_0[\log_r N_0] + \frac{r}{r-1}(r^{[\log_r N_0]-1} - 1)$$

where m_i is the number of words in D_0 of length i. The square brackets denote the integral part of the quantity they enclose.

Proof. The number N_s is obviously bounded by the number of distinct suffixes in D_0. Let $k = [\log_r N_0]$ and suppose all of the words in D_0 have length at least k. Then at most r suffixes of D_0 of length 1, r^2 of length 2, r^3 of length 3, etc., are distinct. Hence a total of at most

$$\sum_{i=1}^{k-1} r^i = \frac{r}{r-1}(r^{k-1} - 1)$$

of the $N_0(k-1)$ suffixes of D_0 of length $k-1$ and less are distinct. At least in this special case, N_s cannot exceed

$$\sum_i im_i - N_0 k + \frac{r}{r-1}(r^{k-1} - 1)$$

where $k = [\log_r N_0]$. This proves the theorem when the words in D_0 have minimum length k.

To extend the proof to include all dictionaries, suppose D_0 contains at least one word w_0 of length $l_0 \leq k-1$. Suppose further that there are $i(i \leq r)$ words of maximum length $l_m > k$ having the same suffix s_0 of length l_m^{-1}. Let D_0' be the set of words obtained by removing from D_0 these i words and w_0, and by adding to it the common suffix s_0 and the i words formed by adjoining i one-symbol prefixes to w_0. Then: (1) Set D_0' contains N_0 distinct words. This follows because the shortest word in D_0 was not a suffix of any other word; all the words formed by adjoining a prefix to this word are therefore unique in D_0'. (2) The number of distinct suffixes in D_0' is equal to the corresponding number in D_0, and the total number of suffixes (distinct or not) has not increased. (3) No word in D_0' is a suffix of any other word. We can continue in this manner decreasing the length of the longest words and increasing that of the shortest words of D_0 until we obtain a set of words $D_0^{(s)}$ all of which have length at least k, and the above arguments are applicable. And since the total number of suffixes in $D_0^{(s)}$ does not exceed, and the number of distinct suffixes in $D_0^{(s)}$ is equal to, the corresponding numbers in D_0, the same bound holds regardless of the initial word lengths.

It is also possible, using the suffix decomposition of a code dictionary, to obtain a necessary and sufficient condition for a bounded decoding delay.

Theorem 10.14. A code dictionary D is decodable with a bounded delay if, and only if, the set S_{n_0} in the suffix decomposition of D is empty for some $n_0 \leq N_s + 1$.

Proof. If no set in the suffix decomposition is empty, one can obviously construct an ambiguous sequence of infinite length. Let $s_i s_{i+1} \ldots s_j$ be a *chain* of segments where s_ν is in S_ν *if* $s_{\nu-1}$ is in $S_{\nu-1}$. If such a chain contains any one segment twice, then it is clearly possible to construct an infinite chain. This in turn implies that no set is empty. Accordingly no chain can contain the same segment more than once if the decoding delay is to be bounded. Since each segment is a suffix, and since there are N_s distinct suffixes which are not code words (no code word can appear in any set if the dictionary is decodable) a set must be empty for some $n_0 \leq N_s + 1$. This proves the necessity of the condition of the theorem.

That the condition is also sufficient follows immediately because all chains, and hence all ambiguous sequences, must be finite if the set S_{n_0} is empty for some finite integer n_0.

Theorem 10.15. If a code dictionary is decodable with finite delay d_c then

$$\left[\frac{n_0}{2}\right] l_{\min} < d_c \leq \left[\frac{n_0 + 1}{2}\right] l_{\max}$$

where S_{n_0} is the first empty set of the dictionary suffix decomposition and $l_{\max}$ and $l_{\min}$ are the number of symbols in the maximum and minimum length code words respectively.

Proof. Since the first empty set is S_{n_0}, there must be a sequence of the form (10.4.1) containing n_0 terms ($m = n_0 - 1$). Any two consecutive terms in this sequence contain at least $l_{\min}$ and at most $l_{\max}$ symbols. This follows because either $x_i x_{i+1} = s_i s_{i+1} = c_j$ or $x_i x_{i+1} = c_i s_{i+1} = s_i$ where c_j is a code word with length $l_{\min} \leq l_j \leq l_{\max}$ and s_i is a code word suffix. Two different decodings of this sequence are possible at least through the term x_{m-1}, but only one decoding can proceed beyond the term s_m. Accordingly, if n_0 is even, there exists an ambiguous sequence involving at least $(n_0 - 2)/2$ pairs $x_i x_{i+1}$ plus the term c_0 while at most $n_0/2$ pairs $x_i x_{i+1}$ are needed to resolve the decoding ambiguity. Similarly, if n_0 is odd, at least $(n_0 - 1)/2$ pairs $x_i x_{i+1}$ and at most $(n_0 - 1)/2$ pairs plus c_0 are so involved. Since the number of symbols in any pair $x_i x_{i+1}$ and in c_0 is bounded by $l_{\min}$ and $l_{\max}$, the statement of the theorem follows.

The term n_0 in Theorem 10.15, incidentally, is bounded by $N_s + 1$, not the minimum of $N_s + 1$ and $N_s^* + 1$. While the reverse of a decodable code is decodable, the reverse of a code with bounded decoding delay does not necessarily have a bounded delay. As an example of this, consider the dictionary $D = 10$, 01, 011, and 111 discussed above. While both D and D^* are decodable, the sequence

$$0110111011101110111\ldots$$

could have resulted from sending the code words 01, 10, 111, 011, 10, 111, 011 . . . or the code words 011, 011, 10, 111, 011, 10, 111, 01. This sequence of words from D could continue indefinitely before this ambiguity is resolved. The reverse code D^*, in contrast, has a decoding delay of three since it has the prefix property.

Notes to Chapter Ten

A proof that the Kraft inequality is both a sufficient condition for unique decodability and a necessary condition for instantaneous decodability can be found in Reference 10.3. McMillan (Reference 10.4) showed it to be a necessary condition for unique decodability as well, although the proof presented here is due to Karush (Reference 10.5). The properties of exhaustive codes are discussed in Reference 10.6. Huffman first described the encoding algorithm of Section 10.3 in Reference 10.7. An extension of this algorithm to the case in which the code symbols are of unequal time duration (the criterion being to minimize the duration of a code sequence needed to transmit an average message) has been made by Karp (Reference 10.8). See also Reference 10.9 in this connection. The original proof of the Sardinas and Patterson algorithm is found in Reference 10.2, although numerous alternative proofs have been presented subsequently (References 10.10, 10.11, and 10.12). A number of the other results in Section 10.4 are due to Levenshtein (Reference 10.13).

Problems

10.1 Which of the following dictionaries are decodable, which are decodable with unbounded delay, and which are not decodable? Identify a sequence having the maximum decoding delay in each case.

(a)	(b)
1	1
011	001
0011	101
01011	0001
00011	00001
001011	10101
000011	00101
010011	

(c)	(d)
1	10
001	001
101	0111
0001	0101
00001	1110
10101	1100
00101	
01000	

(e)	(f)
01	10
011	000
010	001
0110	011
0111	111
0100	0100
01110	0101
01100	1100
01000	1101
01111	

(g)	(h)
11	11
000	110
001	1100
010	1101
011	11000
1000	11001
1001	11010
1010	110000
1011	110001
	110010
	110100
	110101

(i)	(j)
11	011
001	110
011	0000
0001	0010
0101	0001
1001	0100
1011	1000
01001	1010
10001	1100
10101	1101
010001	1111
101001	
1010001	

Repeat the above for the reverses of these dictionaries.

10.2 Suppose the successive outputs of a source S are grouped into n-symbol blocks. If these outputs are symbols in an N-symbol alphabet A_1, the grouped symbols can be regarded as the outputs of a new source S^n, called the nth *extension* of the source S, whose outputs are symbols in an N^n symbol alphabet A_n (i.e., each of the possible N^n n-tuples appearing at the output of S is identified as one symbol in the alphabet A_n). Let L_n be the expected number of code symbols needed to represent a symbol in S^n.

(a) Show that if both S^n and S^{in} are coded optimally (on a symbol-by-symbol basis), with i some positive integer, then $L_{in} \leq iL_n$.

(b) Let A_1 be a binary alphabet with $\Pr(0) = \frac{9}{10}$, $\Pr(1) = \frac{1}{10}$. Compare L_1, $L_2/2$, and $L_3/3$ using a binary Huffman code in each case.

(c) Let S be a binary first-order Markov source with $\Pr(0 \mid 0) = .9$, $\Pr(1 \mid 0) = .1$, $\Pr(0 \mid 1) = \Pr(1 \mid 1) = .5$ (i.e., the probability that the ith output is 0 given that the $(i-1)$st output is 0 is .9, etc.). Compare the minimum average number of binary code bits needed per source output when: (1) the extended source S^2 is encoded; (2) S^3 is encoded; (3) S^2 is encoded but with different codes permitted for different source states.

(d) Repeat part (c) when $\Pr(0 \mid 0) = \Pr(1 \mid 1) = .9$, $\Pr(0 \mid 1) = \Pr(1 \mid 0) = .1$.

10.3 (Shannon's first theorem; see Reference 10.14).

(a) Let S be an N-symbol source with alphabet $\{s_i\}$ and let p_i be the probability that s_i appears at the source output. Define the *entropy* of S as $H_r(S) =$

$-\sum_{i=1}^{N} p_i \log_r p_i$. Show that if the symbols $\{s_i\}$ are to be represented as words in a uniquely decodable code over an r-ary alphabet, then the average number L of code symbols per source symbol is underbounded by $H_r(S)$. [*Hint*: Show that $\sum_{i=1}^{N} p_i \log(1/p_i) \leq \sum_{i=1}^{N} p_i \log(1/q_i)$ for any set of non-negative numbers q_i such that $\sum_{i=1}^{N} q_i = 1$. Then let $q_i = r^{-l_i}/\sum_{i=1}^{N} r^{-l_i}$, with l_i the length of the code word representing s_i.]

(b) Let s_i, $i = 1, 2, \ldots, N$, be encoded as an l_i-symbol r-ary word with l_i the unique integer in the range $\log_r(1/p_i) \leq l_i < \log_r(1/p_i) + 1$. Show that a uniquely decodable dictionary does indeed exist with these word lengths. Show further that the average length L of a word in such a dictionary is overbounded by $H_r(S) + 1$.

(c) Show that if S is a *zero-memory* source (i.e., if the successive outputs of S are statistically independent), the best code for its nth extension S^n has an average word length L_n bounded by

$$H_r(S) \leq \frac{L_n}{n} < H_r(S) + \frac{1}{n}$$

and hence that $\lim_{n\to\infty} (L_n/n) = H_r(S)$. Evaluate the codes of Problem 10.2(b) in terms of this result. [*Hint*: Show that $H_r(S^n) = nH_r(S)$.]

(d) Now let S be an (ergodic) mth order Markov source and let $\{\sigma_j(\nu); j = 1, 2, \ldots, N^\nu\}$ denote the set of ν-tuples defined over the N-symbol source alphabet. Define $H_r(S) = -\sum_{j=1}^{N^{m+1}} P(\sigma_j(m+1)) \log_r P(\sigma_j(m+1) \mid \sigma_j(m))$ with $P(\sigma_j(m+1))$ the probability that the $(m+1)$-tuple $\sigma_j(m+1)$ occurs as $m+1$ consecutive source outputs, and $P(\sigma_j(m+1) \mid \sigma_j(m))$ the probability of the same event conditioned on the first m of those outputs. (Note that this definition of $H_r(S)$ degenerates to the earlier definition when $m = 0$. Further, define $H_r(\overline{S^\mu}) = -\sum_{j=1}^{N^\mu} P(\sigma_j(\mu)) \log_r P(\sigma_j(\mu))$. Show that the best r-ary dictionary for the nth extension of this source for any $n \geq m$ has an average word length bounded by

$$H_r(S) + \frac{H_r(\overline{S^m}) - mH_r(S)}{n} \leq \frac{L_n}{n} \leq H_r(S) + \frac{H_r(\overline{S^m}) - mH_r(S) + 1}{n}.$$

Use this result to evaluate the codes of Problem 10.2, parts (c) and (d).

10.4 Another method for encoding a binary source, called *run-length coding*, can be effective either when one of the two symbols is much more likely than the other to appear at the source output or when dependence between successive outputs tends to cause long runs of the same symbol. The method is to represent an output sequence in terms of the numbers of symbols separating successive occurrences of the less likely symbol, or in terms of the lengths of the successive runs of either symbol. This information is mapped onto n-tuples. If the code alphabet is also binary, runs of length l for $0 \leq l \leq 2^n - 2$ and a continuing run (i.e., a run not yet concluded) of length $2^n - 1$ are so represented (e.g., if the code is designed for a source which produces mainly *zeros* the sequence 0010000000110000001 might be represented as 010,111,000,000,

110; if the code were designed for a source which tends to repeat symbols, the same sequence might be represented by 010,000,110,001,101. The encoding algorithms in the two instances differ only in that in the first case *zero* is always treated as the dominant symbol while in the second, the last source output is considered to be dominant). Compare run-length coding using n-tuples with Huffman coding of the nth extension of the source for the sources described in the previous problem. How is the efficiency of the run-length encoding scheme related to n?

10.5 Still another algorithm (the Shannon-Fano algorithm) for matching a binary code to the source statistics is described as follows. The source symbols are listed in order of their associated probabilities. The first step in the algorithm is to divide the set S of these symbols into two disjoint subsets, S_0, S_1, such that the probability of any symbol in S_0 is at least as large as that of any symbol in S_1, and with the sum of the symbol probabilities in each set as nearly equal as possible. Each subset in turn is further subdivided, $S_i \to S_{i0}, S_{i1}$, in exactly the same way. The procedure continues until each subset contains only one symbol. Each symbol is then labeled with the subscripts attached to its subset [e.g., if the source has four possible outputs a, b, c, and d, occurring with the respective probabilities 0.6, 0.2, 0.1, and 0.1, then $S_0 = (a)$, $S_1 = (b, c, d)$, $S_{10} = (b)$, $S_{11} = (c, d)$, $S_{110} = (c)$, $S_{111} = (d)$.] Compare the codes obtained using this algorithm with the appropriate Huffman codes for the following three sets of source symbol probabilities:

Symbols / Source	a	b	c	d	e
A	0.4	0.3	0.1	0.1	0.1
B	0.3	0.2	0.2	0.2	0.1
C	0.35	0.17	0.17	0.16	0.15

10.6 One occasionally useful method of characterizing a variable-length code is in terms of an associated code graph. A *code graph G* consists of a network of nodes and branches, with each branch labeled by a symbol in the code alphabet and directed from one node to another (not necessarily different) node. One node, labeled S_0, is called the *initial node*, a second node, S_0^*, is called the terminal node, and all others are called intermediate nodes. Each path through G, from S_0 to S_0^*, defines a code word in the associated code dictionary D, the code word symbols determined by the labels on the successive branches of the path. As an example, the code words of length 4 or less defined by the (binary) graph in Figure P10.6 are $\{01, 001, 011, 111, 0001, 0101, 1101\}$. Prove the following statements concerning the relationship between a graph G and its associated code dictionary D.

(a) D is finite if and only if G is finite and contains no loops (i.e., no paths passing through any node more than once).

(b) (Reference 10.15). Let X be the $(\nu + 2) \times (\nu + 2)$ matrix $\{n_{ij}x\}$ with n_{ij} the number of branches leading directly from the $(i - 1)$st to the $(j - 1)$st node of the $(\nu + 2)$-node graph G $(S_0, S_1, \ldots, S_\nu, S_0^* \triangleq S_{\nu+1})$ (e.g., in the example above, $n_{11} = 0$, $n_{12} = 2$, $n_{33} = 1$, $n_{34} = 1$, $n_{43} = 0$, etc.) Further, let $F(x) = \{f_{ij}(x)\} = (I - X)^{-1}$, the inverse of the matrix $I - X$, with I the

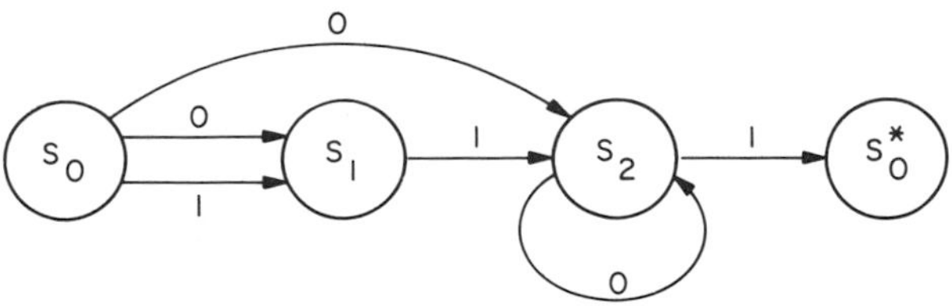

Figure P10.6

identity matrix. Then the number of words of length n in the dictionary defined by G is the coefficient of x^n in the power series expansion of $f_{1,\nu+2}(x)$. More generally, the number of distinct paths in G, beginning with the $(i-1)$st node and ending with the $(j-1)$st and involving exactly n branches, is the coefficient of x^n in the power series expansion of $f_{ij}(x)$. [*Hint*: Note that, formally, $(I - X)^{-1} = I + X + X^2 + \cdots$.]

(c) There exists a (generally not unique) code graph G corresponding to every code dictionary D. At least one such graph exists containing not more than $\min\{N_\mu, N_\mu^*\} + 2$ nodes, with N_μ and N_μ^* the number of distinct suffixes (including those which are also code words) in D and D^*, respectively.

(d) If D has the prefix property it can be represented by a code graph G with at most one branch having a given label leaving any one node (and with no branches leaving the node S_0^*). If, in addition, D is exhaustive, each node (excluding S_0^*) of G has exactly one branch leaving it labeled by each symbol in the code alphabet. Conversely, if G is a code graph satisfying this constraint (or these constraints), D has the prefix property (and is exhaustive). This last statement is true even if D is infinite, i.e., even if G contains loops.

References

10.1 de Jager, F. "Delta Modulation, A Method of PCM Transmission Using the One-Code," *Phillips Res. Rept.* 7, 1952.

10.2 Sardinas, A. A., and G. W. Patterson, "A Necessary and Sufficient Condition for the Unique Decomposition of Coded Messages," *IRE Conv. Record*, Part 8, p. 104, 1953.

10.3 Kraft, L. G., "A Device for Quantizing, Grouping, and Coding Amplitude Modulated Pulses," M. S. Thesis, Electrical Eng. Dept., Mass. Inst. Tech., Cambridge, Mass., 1949.

10.4 McMillan, B., "Two Inequalities Implied by Unique Decipherability," *IRE Trans. Inf. Theory*, IT-2, p. 115, 1956.

10.5 Karush, J., "A Simple Proof of an Inequality of McMillan," *IRE Trans. Inf. Theory*, IT-7, p. 118, 1961.

10.6 Gilbert, E. N., and E. F. Moore, "Variable-Length Binary Encoding," *Bell Systems Tech. J.* 38, p. 933, 1959.

10.7 Huffman, D. A., "A Method for the Construction of Minimum Redundancy Codes," *Proc. IRE*, 40, p. 1098, 1952.

10.8 Karp, R. M., "Minimum-redundancy Coding for the Discrete Noiseless Channel," *IRE Trans. Inf. Theory*, IT-7, p. 27, 1961.

10.9 Schützenberger, M. P., and R. S. Marcus, "Full Decodable Code-word Sets," *IRE Trans. Inf. Theory*, IT-5, p. 12, 1959.

10.10 Bandyopadhyay, G., "A Simple Proof of the Decipherability Criterion of Sardinas and Patterson," *Inform. and Control*, 6, p. 331, 1963.

10.11 Ash, R., *Information Theory*, Interscience, N.Y., N. Y., 1965.

10.12 Riley, J. A., "The Sardinas-Patterson and Levenshtein Theorem," *Inform. and Control*, 10, p. 120, 1967.

10.13 Levenshtein, V. I., "Certain Properties of Code Systems," *Soviet Physics Doklady*, 6, p. 858, 1962.

10.14 Abramson, N., *Information Theory and Coding*, McGraw-Hill Book Co., N. Y., N.Y., 1963.

10.15 Mason, S. J., "Feedback Theory—Further Properties of Signal Flow Graphs," *Proc. IRE*, 44, p. 920, 1956.

Supplementary Bibliography

Even, S., "Tests for Unique Decipherability," *IEEE Trans. Inform. Theory*, IT-10, 1964, p. 185.

Fano, R. M., *Transmission of Information*, MIT Press and John Wiley & Sons, Inc., N.Y., N.Y., 1961.

Jaynes, E. T., "A Note on Unique Decipherability," *IRE Trans. Inf. Theory*, IT-5, 1959, p. 98.

Jelinek, F., *Probabilistic Information Theory*, McGraw-Hill Book Co., N.Y., N.Y., 1968.

Levenshtein, V. I., "Some Properties of Coding and Self-Adjusting Automata for Decoding Messages," *Prob. Kibernetiki*, 11, 1964, p. 63.

Markov, Al. A., "Alphabet Coding," *Soviet Math.*, 1, 1960, p. 596.

Markov, Al. A., "On Alphabet Coding," *Soviet Phys. Doklady*, 6, 1962, p. 553.

Schützenberger, M. P., "On an Application of Semi-group Methods to Some Problems in Coding," *IRE Trans. Inf. Theory*, IT-2, 1956, p. 47.

chapter eleven

Synchronizability

11.1 Introduction

The decodability condition on a code dictionary provides that it be possible to separate any sequence of words from the dictionary uniquely into the individual words comprising the sequence *given the initial symbol of that sequence*. In many applications, however, the initial symbol may not be known; the message may already be in progress before the receiver begins receiving it, for example, or a previous error may have caused an earlier portion of the sequence to be improperly decoded. When this situation is possible, it may be necessary to require the dictionary to be *synchronizable* as well as decodable. A sequence of words beginning perhaps with a word suffix is said to be *synchronizable with delay* d_s if no more than d_s symbols need be observed before the initial symbol of at least one of its words can be unambiguously determined. A dictionary is said to be synchronizable with delay d_s if it is decodable and if every sequence of symbols resulting from some concatenation of words from the dictionary is synchronizable with delay d_s or less. When used without qualification, the term synchronizable will imply a finite delay.

This chapter begins with some general conditions which a dictionary must satisfy if it is to be synchronizable. These conditions are useful for testing whether or not a given dictionary has this property. In the next section, the concept of *self-synchronization* is introduced, and some of the consequences of this constraint are also explored. The following section then demonstrates an upper bound on the number of words of various lengths which can exist in any synchronizable dictionary. Finally, a construction is presented in Section 11.5 whereby this bound can always be attained.

11.2 Testing a Dictionary for Synchronizability

The requirement that a dictionary be synchronizable imposes one of two conditions on any received sequence of code words: Every symbol sequence obtained by deleting at least one initial symbol, but less than a complete word, from a sequence of words is either (1) not decodable, or (2) decodable but with the property that correct synchronization is automatically obtained after the first few words are decoded. An example of a dictionary which can exhibit both phenomena is one consisting of the words $w_1 = 01$, $w_2 = 100$, $w_3 = 101$, $w_4 = 1101$. If, for instance, the first two symbols of the sequence beginning $w_2 w_4 w_i \ldots$ were unavailable, the output of the decoder would be $w_1 w_3 w_i \ldots$ and synchronism is achieved by the time w_i is received, regardless of the index i. In contrast, if only the first symbol were omitted from this same sequence, the decoder would encounter the initial symbols 00. Since no code word begins with this prefix, the decoder would recognize that a synchronization error had been made, and select some other symbol as the initial symbol.

Theorem 11.1. A decodable dictionary is synchronizable with bounded delay only if it and its reverse are decodable with bounded delay.

    Proof. Suppose the decoding delay of the dictionary D is not bounded. Then there is some sequence of unlimited length which cannot be unambiguously decoded. In the notation of Section 10.4, with c_i representing a word and s_i a word suffix in D, a typical undecodable sequence might have the form

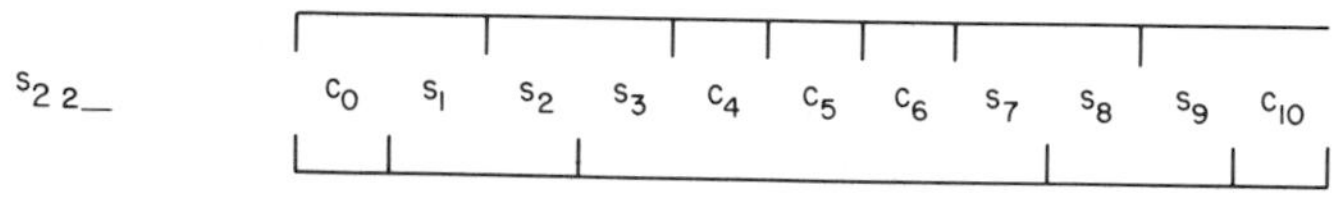

$$s_2 \, {}_2 \underline{\quad} \qquad c_0 \quad s_1 \quad s_2 \quad s_3 \quad c_4 \quad c_5 \quad c_6 \quad s_7 \quad s_8 \quad s_9 \quad c_{10} \qquad (11.2.1)$$

where the scoring indicates two possible decodings. Suppose the lower scoring corresponds to the transmitted code words. Then deleting the symbols comprising $c_1 = c_0 s_1$ from the prefix of this sequence leaves a sequence decodable for an unbounded duration, even though the synchronization is in error.

    Similarly, suppose the reverse dictionary D^* is not decodable with a bounded delay. Then, again in the notation of Section 10.4 there is some non-decodable sequence

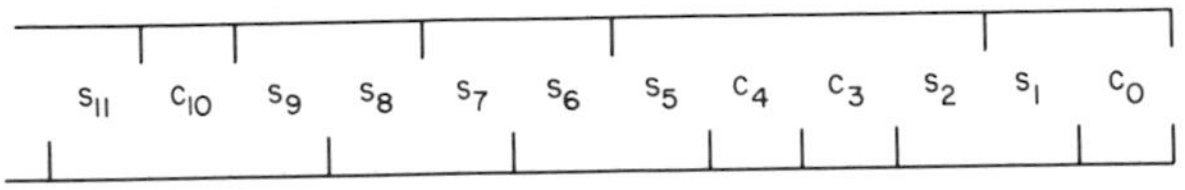

(11.2.2)

where the sequence is now read from right to left. (The c_i and $s_i \ldots s_k$ terms are words of the original code D. When read from right to left, they are words of the reverse code. Similarly, the terms s_i, read from right to left, are all suffixes and, since the decoding delay is unbounded, prefixes in D^*. These terms, read from left to right, are therefore both suffixes and prefixes in D.) Since s_{11}, for example, is the suffix of a code word in D, the sequence (11.2.2) read from left to right could occur as a sequence of words from D with the initial symbols deleted. The scoring under the sequence would correspond to the decoded words, that above to the transmitted words. Synchronization will not be obtained until c_1 is received. But if the decoding delay of the reverse code D^* is not bounded, the sequence (11.2.2) can be arbitrarily long. Thus, in this case too, the synchronization delay of the dictionary D is unbounded.

In Section 10.4, we defined the suffix decomposition of the dictionary D. Let S be the set of suffixes of D and consider the *suffix decomposition of the set S* obtained by using S rather than D as the first set S_0 of the decomposition. Again using the notation of Section 10.4, we observe that any decodable sequence of words beginning with a suffix can be written in the form

$$s_0 x_1 x_2 x_3 x_4 \ldots \qquad (11.2.3)$$

where, as before, x_i is either a code word c_i or a word suffix s_i. Again, any subsequence of this sequence of the form $s_i c_{i+1} c_{i+2} \cdots c_{i+j-1} s_{i+j}$ is a code word. Moreover, if none of the first m sets of the suffix decomposition is empty, a sequence of the form (11.2.3) exists containing m terms.

Theorem 11.2. A decodable dictionary D is synchronizable if, and only if, some set S_{n_0} in the suffix decomposition of the set S of suffixes of D is empty for some integer $n_0 \leq \min(N_s, N_s^*) + 1$. The terms N_s and N_s^* denote the number of distinct suffixes which are not also code words in the dictionaries D and D^*, respectively.

Proof. Consider a chain† of segments from the suffix decomposition of the set of suffixes S. Clearly, all sequences of code words which can be decoded after some initial symbols have been deleted are characterized by some such chain. If the dictionary is synchronizable one of the two situations described above must always occur for all of these sequences: either a point must be reached beyond which it is not possible to decode, or the decoding must come into synchronism with the true code words. The first of these events corresponds to a condition in which some segment in the chain gives rise to no entry in the next set. If the second situation

†Cf. Section 10.4.

occurs, a word must eventually be decoded which is either the suffix of the transmitted code word composing that part of the sequence or which has as its suffix that transmitted code word. In either case, the corresponding segment in the suffix decomposition will be a code word. But by Theorem 11.1, if a dictionary is synchronizable with finite delay it must also be decodable with finite delay. Thus, when a segment is a code word, the chain following that segment must, by Theorem 10.14, be of finite length. Consequently, all chains must be of finite length and some set must be empty if the dictionary is to be synchronizable. Conversely, if some set is empty, all chains are of finite length and it is either not possible to decode the corresponding sequence further with the given starting point, or the decoding must come into synchronism with the transmitted sequence of words.

To bound the number of segments which can occur in a chain, we again note that if all chains are to be of bounded length none can contain repeated segments. Since the set S_0 contains a suffix there are at most $N_s - 1$ remaining suffixes which can be included in any chain. Because of the second mode of attaining synchronism, one of these segments may be a code word rather than a suffix, but because the dictionary is decodable, only one segment in the chain can contain a word. Accordingly, the longest chain can contain at most $N_s + 1$ segments, and set S_{n_0} must be empty for some integer $n_0 \leq N_s + 1$.

Now suppose there were some chain containing more than $N_s^* + 1$ segments. This chain when read from right to left corresponds to some chain in the suffix decomposition of the suffix set of D^*. Because D^* is decodable, this chain can also contain at most one code word as a segment. If it contains more than $N_s^* + 1$ segments, some suffix of D^* must be repeated and an infinite chain can be constructed. Consequently $n_0 \leq N_s^* + 1$ also, and the theorem is proved.

As an example, let D_1 be the dictionary {01, 101, 110, 001}. Since D_1 has the prefix property it is clearly decodable. But the suffix decomposition of the suffix set of D_1 is

D_1	$S_0 = S$	S_1	S_2
01	0	1	1
101	1	01	01
110	01	10	10
001	10		

and since $S_2 = S_1$ all succeeding sets will be equal to S_1. The dictionary is not synchronizable. In particular, the sequence

1 0 1 1 0 1 1 0 1 1 0 1

can be continued indefinitely without ever resolving the correct synchronization if the starting position is not known.

In contrast, the dictionary $D_2 = \{101, 00110, 10111, 11001\}$ which was asserted in Section 10.1 to be decodable, is also synchronizable since

D_2	$S_0 = S$	S_1	S_2	S_3	S_4
101	0	1	1	1	01
00110	1	01	01	01	0111
10111	01	10	10	111	1001
11001	10	001	111	0111	
	11	0110	0111	1001	
	001	0111	1001		
	110	1001			
	111	111			
	0110				
	0111				
	1001				

and S_5 is empty.

Theorem 11.3. If a dictionary D is synchronizable, the synchronization delay d_s is bounded by

$$\max\left\{d_c - l_{\max}, d_c^* - l_{\max}, \left[\frac{n_0 - 1}{2}\right] l_{\min}\right\} + 1 \leq d_s \leq \left[\frac{n_0}{2} + 1\right] l_{\max} - 1 \quad (11.2.4)$$

where d_c and d_c^* are the decoding delays of dictionaries D and D^* respectively, $l_{\max}$ and $l_{\min}$ are the number of symbols in the maximum and minimum length code words in D, and S_{n_0} is the first empty set in the suffix decomposition of the set of suffixes of D.

Proof. The proof of the upper bound parallels that of Theorem 10.15, but is slightly complicated by the fact that the first m symbols $(0 \leq m \leq l_{\max} - 2)$ of a maximal length non-synchronizable sequence need not be involved in the corresponding suffix decomposition of the set of suffixes of D. [The first complete code word received might begin with the $(m + 1)$st symbol, the second with the kth for some k in the range $m < k < l_{\max}$. Clearly, if $k \geq l_{\max}$, the $(m + 1)$st symbol is uniquely established as the first symbol of the first true code word.] As before, when n_0 is even there are an even number of non-empty sets in the suffix decomposition (S_0 through $S_{n_0 - 1}$) and the longest sequence corresponding to these n_0 sets does not exceed $(n_0/2)l_{\max}$ symbols. Since the suffix in S_0 might actually be preceded by as many as $l_{\max} - 2$ additional symbols not actually involved in the suffix decomposition, $d_s \leq (n_0/2)l_{\max} + l_{\max} - 2$. When n_0 is odd, the length of the maximal-length sequence excluding the suffix in the set S_0 is bounded by $[(n_0 - 1)/2]$ $l_{\max}$. This sequence is preceded by a word suffix (at least part of which is in S_0) which cannot exceed $l_{\max} - 1$ symbols. When n_0 is odd, therefore, $d_s \leq ((n_0 + 1)/2)$ $l_{\max} - 1$ and the upper bound is proved.

If the dictionary has a decoding delay d_c there is a sequence of length $d_c - 1$ of the generic form (11.2.1) where neither of the two possible decodings is eliminated as a possibility until the d_cth symbol is observed. Deleting c_0 from the prefix of this sequence yields a sequence of length $d_c - 1 - (l_{\max} - 1)$ or greater which is not synchronizable. (c_0 can contain at most $l_{\max} - 1$ symbols since $c_0 s_1$ is a code word. If the whole of c_0 were not deleted, it might be possible to conclude on the basis of the remaining suffix that c_0 was actually received. And as both decodings of this sequence begin with the first symbol of c_0, there would, in this event, be no synchronization ambiguity.) Similarly, if the reverse dictionary has a synchronization delay d_c^*, then there is a sequence of length $d_c^* - 1$ of the generic form (11.2.2) where the decoding ambiguity has not yet been eliminated. This same sequence, read from left to right, is a sequence from D beginning with a suffix of a word in D and of length at least $d_c^* - 1 - (l_{\max} - 1)$, which is not synchronizable. As before, the term $l_{\max} - 1$ must be subtracted since it is conceivably possible to complete the sequence as soon as the first symbol of c_0 has been received. Finally, the argument establishing the lower bound of Theorem 10.15 can be used here to show that $d_s \geq [(n_0 - 1)/2]l_{\min}$. This concludes the proof of the lower bound.

11.3 Self-Synchronizing Code Dictionaries

We have observed two mechanisms whereby synchronization can be obtained. One of these mechanisms is operative when the received sequence is not decodable beyond a certain point; mis-synchronization is indicated (assuming none of the symbols is in error) and the decoder must begin decoding the sequence again commencing with a different initial symbol. This procedure has the advantage of putting into evidence the occurrence of a synchronization error. The words decoded when this error was in effect can be cancelled, and the received sequence decoded again when the correct synchronization has been determined.

It was also noted that in some cases synchronism is automatically obtained. The decoder begins with erroneous synchronization but in the process of decoding eventually comes into correct synchronism without ever reaching a point at which it cannot decode. No synchronization error is ever indicated so no attempt can be made to correct the words erroneously decoded before synchronization was obtained. Nevertheless, this method of synchronization does have the decided advantage of presenting no additional mechanization complexity. The decoder achieves synchronism without any modification whatsoever.

Sequences which have the property that the decoder automatically comes into correct synchronism even though the initial symbols comprise a word suffix rather than a code word will be referred to as *self-synchronizing* sequences. If any sequence of words from a decodable dictionary D beginning with any suffix of any word in D can be made into a self-synchronizing sequence by following it with some finite-length word sequence, D is called a *completely-self-synchronizing dictionary*. If some or none of these sequences can be made self-synchronizing, the dictionary is denoted, respectively, a *partially-* or *never-self-synchronizing dictionary.* No restriction has been placed on the amount of time necessary before the self-synchronization occurs; such codes may or may not be synchronizable with finite delay. Even if the delay is potentially infinite, however, the probability of an infinite sequence of words selected at random from a self-synchronizing dictionary not eventually coming into synchronism is zero. This follows because the probability of the necessary synchronizing sequence not being chosen at any particular stage is less than one, and this can happen an infinite number of times only with zero probability.

It is often desirable for a dictionary to be either completely- or never-self-synchronizing depending upon whether a simple decoding procedure is more or less important than the ability to detect when a synchronization error has occurred. A partially-self-synchronizing dictionary, since it must have the capability of synchronizing in the absence of a self-synchronizing sequence, offers no reduction in decoder complexity, nor does it assure the detection of all synchronization errors.

Theorem 11.4. A dictionary D is never-self-synchronizing if, and only if, none of the suffixes of D are also code words in D.

Proof. If there is a self-synchronizing sequence as, for example (again in the notation of Section 10.4), the sequence

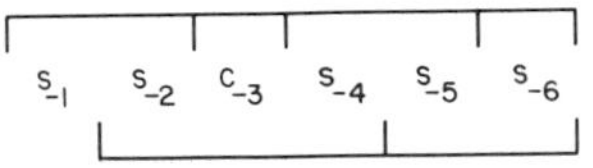

$$\overline{s}_1 \quad \overline{s}_2 \quad \overline{c}_3 \quad \overline{s}_4 \quad \overline{s}_5 \quad \overline{s}_6 \tag{11.3.1}$$

where the over-scoring (representing the decoded words in the presence of a synchronization error) and the under-scoring (indicating the transmitted code words) come into phase after the suffix s_6 is received, then s_6 must also be a code word. Conversely, whenever a suffix s_0 in D is also a code word c_0, the sequence s_0 obtained by deleting the prefix from a code word with this suffix is self-synchronizing.

The dictionary D_1 in Section 11.2 generates some self-synchronizing sequences as evidenced by the fact that all segments of the suffix decomposition of the suffixes of D_1 contain the word 01. Clearly, the one word sequence 101 is self-synchronizing when the first symbol is deleted since 01 is also a code word. In contrast, the dictionary D_2 in the same section is, by Theorem 11.4, never-self-synchronizing.

A completely-self-synchronizing dictionary D must exhibit two properties: (1) Any semi-infinite sequence of words from D must be decodable as another (perhaps different) sequence of words from D regardless of the number of initial symbols which have been deleted from that sequence. A dictionary having this property will be called *partitionable.* (2) Finite *synchronizing sequences* comprised of code words must exist which are capable of causing any of these partitionable sequences to achieve synchronism. That is, for every sequence of code words y preceded by any code word suffix x, there must be a synchronizing sequence z of code words such that the sequence xyz can be completely decoded. (Since z is comprised of code words only, and since the sequence xyz is required to end with a word and not a prefix, the decoder must achieve synchronism at least by the time the complete sequence xyz is decoded.)

The following theorem is an immediate consequence of the second of these properties.

Theorem 11.5. If a decodable dictionary is completely-self-synchronizing, it must have the prefix property.

Proof. Suppose the dictionary in question does not have the prefix property, that p is both a code word and the prefix of the word ps. Since the dictionary is completely-self-synchronizing there must be a decodable sequence sz where z is a sequence of code words. But psz must then be decodable in at least two ways: one decoding begins with the code word p, the second with the word ps. Consequently, the dictionary cannot be both self-synchronizing and decodable, unless it also has the prefix property.

The partitionability required of completely-self-synchronizing dictionaries is obviously satisfied when the dictionary is exhaustive, but unfortunately there are other points to be considered.

Theorem 11.6. The synchronization delay of decodable exhaustive dictionaries is infinite unless all words contain only one symbol.

Proof. If some word in the dictionary contains more than one symbol, there is at least one symbol (call it a) which is not a code word. This follows because the dictionary must have the prefix property (Theorem 10.4). Then the sequence $aaa \dots$ must exist as a concatenation of code words of length two or greater since the dictionary is exhaustive and the symbol a is not a code word. But this sequence is clearly not synchronizable. The same sequence would result were the first symbol omitted; there is no way of identifying the first symbol of any word.

Because of this, it may be useful to consider dictionaries which have the prefix property and which are partitionable, but not exhaustive. In order to characterize such dictionaries, it is convenient to define the *completion* D' of the dictionary D in such a way that the union of D and D' is an exhaustive, decodable dictionary. If D has the prefix property, but is not exhaustive, it is always possible to add words to it to yield a dictionary which is exhaustive and still retains the prefix property. A systematic approach for accomplishing this is to list all L-tuples (where L is the length of the longest word in D)

and to identify those which do not have words in D as prefixes. These L-tuples can be used as the set D'. While this construction for D' is not unique, it is the one which will be assumed in the ensuing discussion. Next, let S be the set of suffixes $\{s_i\}$ in D and let S_1 be the set of elements $\{x_i\}$ with x_i in S_1 if and only if $s_j x_i = c'$ for some L-tuple c' in D' and some suffix s_j in S. Then:

Theorem 11.7. A dictionary D having the prefix property is partitionable if, and only if, no element in the set S_1 (as defined in the preceding paragraph) is either a prefix in D, or can be represented as a sequence of words in D followed, perhaps, by a prefix in D.

Proof. If the stated condition is violated, there is clearly some suffix of some word in D which, when followed by a series of words in D, yields a word sequence containing a word in D'. Since the dictionary $D + D'$ has the prefix property, this same sequence cannot be decoded as words from D only and D would not be partitionable.

Conversely, suppose there is some sequence of words from D (beginning with a suffix in S) which is not partitionable. Then the first non-decodable portion of this sequence must begin with a word from D'. This can happen only if there is a suffix in S which can be followed by a series of zero or more words from D, concluding possibly with a prefix in D, to yield a word in D'. Such a sequence would be an element in S_1 of the form precluded by the statement of the theorem. (Note that no element in S can itself be a word in D' since the latter are all of length L, while the former are all of length less than L.) This proves the theorem.

To demonstrate that the set of partitionable but non-exhaustive dictionaries is not vacuous consider the dictionary $D_1 = \{1, 01, 001, 0001\}$. The only word needed to make the set exhaustive is $w = 0000$. But since $S = \{1, 01, 001\}$, S_1 is empty and D_1 is partitionable. As a second, perhaps less obvious example, let $D_2 = \{1, 001, 0001, 011, 0101, 01001, 010001\}$. Then $D' = \{000000, 000001, 000010, 000011, 010000\}$, $S = \{1, 01, 001, 11, 101, 1001, 0001, 10001\}$ and $S_1 = \{0000\}$. Since no sequence of words from D_2 begins with 0000, D_2 is partitionable.

Thus far we have been concerned with the first of the two conditions necessary to insure that a dictionary be self-synchronizing. The second condition, it will be recalled, demands the existence of a synchronizing sequence for every possible sequence of words beginning with a word suffix.

In order to facilitate the discussion of a necessary and sufficient condition for the existence of synchronizing sequences, let the *suffix set* Q be defined as the set of suffixes of a particular dictionary D which are neither words in D nor have words in D as prefixes. For example, if $D = \{0, 100, 101, 110, 1110, 1111\}$ then $Q = \{1, 10, 11, 111\}$.

Theorem 11.8. A partitionable, decodable dictionary D is completely-self-synchronizing if, and only if, each of the elements of the reverse Q^* of the suffix set Q is contained in at least one of the sets of the suffix decomposition of the reverse dictionary D^*.

Proof. Given the decomposition of D^* we can construct a sequence of the form (cf. Section 10.4)

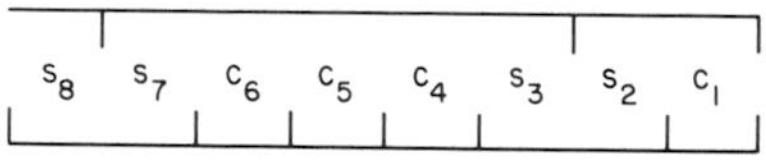

where the sequence is read from right to left. The scoring isolates subsequences which are words in D^* and, when read from left to right, words in D. In this specific example, s_8 is a suffix in D, and the over-scoring indicates a decoding, beginning with s_8, which comes into synchronism, as indicated by the underscoring, when the word c_1 is received. If such a sequence exists, there is clearly a synchronizing sequence for any series of code words which when decoded asynchronously, leaves as a suffix the element labeled, in this case, s_8. Since Q contains all possible suffixes of D which cannot be partially decoded themselves, the dictionary is completely-self-synchronizing if each element in Q^* occurs in some segment of the decomposition of D^*

Conversely, if there is a synchronizing sequence for some suffix in Q then it must be possible to decompose it into segments as was done in the specific example above. These segments, when read from right to left, define elements in the successive sets of the suffix decomposition of the reverse dictionary D^*. The suffix in question must be in some set of this decomposition, so the condition stated in the theorem is both necessary and sufficient.

Consider again the dictionary

$$D_1 = \{1, 01, 001, 0001\}$$

Here Q is empty so the condition in Theorem 11.8 is trivially satisfied. Clearly, since every suffix is a code word, the dictionary is completely self-synchronizing, and in fact is synchronizable with minimum delay.

The second partitionable dictionary considered was the dictionary

$$D_2 = \{1, 001, 0001, 011, 0101, 01001, 010001\}$$

so

$$Q = \{01\}, \qquad Q^* = \{10\}$$

and the first two sets in the decomposition of D_2^* are

$S_0 = D_2^*$	S_1
1	0
100	00
1000	000
110	10
1010	010
10010	0010
100010	00010

Since S_1 contains the only element in Q^* this dictionary is also completely-self-synchronizing.

In contrast, the dictionary $D_3 = \{11, 00, 01, 1000, 1001, 1010, 1011\}$ is exhaustive and decodable. But

$$Q = \{0, 1, 10\}$$
$$Q^* = \{0, 1, 01\}$$

and the decomposition of D_3^* is

$S_0 = D_3^*$	S_1	S_2	S_3
11	01	01	01
00			
10			
0001			
1001			
0101			
1101			

All subsequent sets contain only 01, so D_3 is not completely-self-synchronizing. In particular no word sequence can be adjoined to any sequence beginning with an odd-length suffix to make it self-synchronizing since all word sequences have even length and any sequence beginning with an odd-length suffix and followed by code words from D_3 will have an odd length. This argument can obviously be generalized to prove the following theorem.

Theorem 11.9. Any dictionary having words with lengths which are all multiples of some integer b greater than one cannot be completely-self-synchronizing. (Note, in particular, that this applies to all dictionaries having uniform word length $b > 1$.)

Since the set P of all prefixes of the words of a dictionary is perhaps more conveniently tabulated than the set Q the following theorem may be useful.

Theorem 11.10. If the set Q in the statement of Theorem 11.8 is replaced by P, the set of prefixes in the dictionary D, the theorem remains valid.

Proof. Clearly, all elements in Q must be in P if the dictionary is to be partitionable. The condition of Theorem 11.8 with Q replaced by P is certainly sufficient. To demonstrate the necessity of this condition with Q replaced by P, we show that each element in P^* must appear in some set of the suffix decomposition of D^* if this statement is true for each element in Q^*. This follows immediately because any element p^* in P^* can be used to construct a word either of the form

$w_1^* = w_2^* w_3^* \ldots w_i^* p^*$ or of the form $w_1^* = q^* w_2^* w_3^* \ldots w_i^* p^*$ where the w_i^* terms are words in D^* and q^* is in Q^*. In either case, since all the words in D^* appear in the set S_0 and since all the elements in Q^* appear in some set, so too must all the elements in P^*.

We now estimate the number of sets of the suffix decomposition of D^* which must be investigated before a dictionary D can be declared completely-self-synchronizing.

Theorem 11.11. If a partitionable, decodable dictionary D is completely-self-synchronizing, no more than n_1 sets of the suffix decomposition of D^* need be investigated before the condition of Theorem 11.8 can be established, where

$$n_1 \leq N_p$$

and where N_p is the number of prefixes in D.

Proof. First of all, note that $N_p = N_s^*$, the number of suffixes in D^*. The proof then parallels that of Theorem 11.2. Each entry in a set is dependent only upon the elements in the previous set. Therefore, no new elements will appear in the decomposition following any set which contains only elements which have already occurred in previous sets. It is necessary to investigate only so long as new segments appear in each set. Since there are only $N_s^* = N_p$ possible segments, the theorem is proved.

The most easily characterized partitionable dictionaries are the exhaustive dictionaries. The primary disadvantage of this class of dictionaries is in their unbounded synchronization delay (Theorem 11.6). This liability can be overcome, however, through the use of a *universal synchronizing sequence*. A universal synchronizing sequence is a synchronizing sequence which is effective regardless of the suffix preceding it. By inserting such a sequence periodically in the stream of code words, one is assured that the decoder will be in synchronism after decoding the sequence even if it was not before. This procedure, of course, adds redundancy to the system, but then so do all other synchronizing schemes discussed in this chapter. The other methods considered involved restricting the number of words in the dictionary beyond the limitation already imposed by the decodability condition. The universal synchronizing sequence method allows the complete decodable dictionary to be used, but necessitates the periodic inclusion of non-information-conveying code words. We now demonstrate that such sequences do in fact exist.

Theorem 11.12. There exists a universal synchronizing sequence for any completely-self-synchronizing dictionary.

Proof. Since the dictionary is completely-self-synchronizing, there is for each suffix s a sequence of code words σ such that any sequence of code words following the reception of the sequence $s\sigma$ will be synchronously decoded. We construct a universal synchronizing sequence as follows: Let w_L be one of the code words of maximum length L. It will be the first word in the synchronizing sequence. Any asynchronously decoded sequence when followed by w_L will leave one of $L - 1$ possible suffixes. If the last word decoded is w_L or a suffix of w_L, synchronism is obtained. Suppose the remainder after the last word is decoded is the suffix s_1 of w_L. Then there must be a synchronizing sequence σ_1 for this suffix s_1. Suppose, instead, the suffix s_2 remains. Then denote by s_2' the suffix remaining when $s_2\sigma_1$ is decoded and adjoin its synchronizing sequence σ_2 to the universal synchronizing sequence under construction. In general, the suffix s_i results in a suffix s_i' when $s_i\sigma_1\sigma_2 \ldots \sigma_{i-1}$ is decoded which is, in turn, brought into synchronism by the word sequence σ_i. Proceeding in this manner, we are able to construct a universal synchronizing sequence of the form $\Sigma = w_L\sigma_1\sigma_2\sigma_3 \ldots \sigma_{L-1}$ such that, if the suffix s_i of w_L remains when the decoder first encounters the sequence, it will be in synchronism by the time the subsequence σ_i has been received, and, of course, will remain in synchronism through the remainder of the synchronizing sequence Σ.

As an example, consider the exhaustive dictionary $D = \{00, 111, 1100, 1101, 1000, 1001, 1010, 1011, 0100, 0101, 0110, 0111\}$.

D	$P = Q$	$S_0 = D^*$	S_1	S_2
00	0	00	11	1
111	1	111	01	01
1100	11	0011	10	10
1101	100	1011	0	11
1000	01	0001		0
1001	101	1001		011
1010	10	0101		001
1011	010	1101		101
0100	011	0010		010
0101	110	1010		110
0110		0110		
0111		1110		

Since the set P^* is contained in S_2, the dictionary is completely self-synchronizing. A universal synchronizing sequence might begin with the 4-symbol word $w_L = 0111$. Then if the suffix remaining after the last word preceding this sequence is decoded is $s_1 = 111$, synchronism is already obtained. If the suffix is $s_2 = 11$, it will be synchronized by the single word 00:

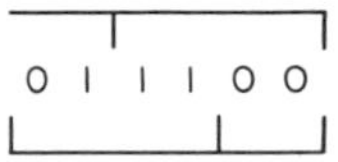

so that the universal synchronizing sequence now has the form 011100. Finally, if the suffix $s_3 = 1$ remains, the suffix s_3' is 100, and σ_3 can be equated to the word 0111. The complete synchronizing sequence is then $\Sigma =$ 0111000111. A different sequence could be constructed by beginning with some other 4-symbol word. A shorter universal synchronizing sequence, 1000111, results, for example, when $w_L = 1000$ is chosen.

Theorem 11.13. Every self-synchronizing code dictionary has a universal synchronizing sequence of length no greater than λ where

$$\lambda \leq \left[\frac{n_1 + 3}{2}\right] L(L - 1)$$

and where n_1 is as defined in Theorem 11.11, L is the maximum code word length, and the square brackets indicate the integer part of the enclosed fraction.

Proof. Given the decomposition of D^* we can, as in the proof of Theorem 11.8, construct a synchronizing sequence for any suffix in Q. The longest such sequence will involve n_1 segments plus the final code word. The sequence constructed from these segments (see Section 10.4) is such that no two adjacent segments contribute more than L symbols to its length. Hence, the length of any such sequence is no greater than $[(n_1/2) + 1]L$ symbols if n_1 is even and $[(n_1 + 3)/2]L$ symbols if n_1 is odd. From the construction of Theorem 11.12, the universal synchronizing sequence will involve no more than $L - 1$ such sequences. Thus, the theorem follows.

We have argued that dictionaries should generally be either completely-self-synchronizing or never-self-synchronizing. The former have the advantage of not requiring any additional equipment to obtain synchronization. Often, however, they are plagued with the possibility of an unlimited delay before synchronization is obtained. As was demonstrated in Theorem 11.12, this can be overcome by transmitting periodically a universal synchronizing sequence, such sequences existing for all completely-self-synchronizing codes. In addition, while all exhaustive codes have this deficiency (of unbounded delay), not all self-synchronizing codes do. Any dictionary such as $D_n = 1, 01, 001, 0001, 00001, \ldots$ (containing all words of the form $00 \ldots 01$ up to length n) in which all suffixes are also code words is clearly self-synchronizing with finite delay. In fact, these codes might be called *instantaneously synchronizable* since synchronism is automatically obtained as soon as any complete suffix has been received. The word lengths in this particular dictionary D_n incidentally are such that

$$\sum_{i=1}^{n} 2^{-l_i} = 1 - \frac{1}{2^n}$$

Consequently, D_n nearly attains the upper limit of the Kraft inequality for decodability. The synchronizability constraint does not necessarily increase the code redundancy significantly beyond that required for unique decodability alone. This question as to the maximum number of words possible in a synchronizable dictionary is explored further in the next section.

11.4 An Upper Bound on the Number of Words in a Synchronizable Dictionary

Let w represent an n-tuple and w^i the n-tuple obtained by cyclically permuting the symbols of w i positions to the right. That is, if

$$w = (\omega_0\, \omega_1\, \ldots\, \omega_{n-1})$$

$$w^i = (\omega_{n-i}\, \omega_{n-i+1}\, \ldots\, \omega_{n-1}\, \omega_0\, \ldots\, \omega_{n-i-1})$$

If $w_1 = w_2^i$ for any integer i, then w_1 and w_2 are said to be in the same *equivalence class*. The maximum number of distinct words of length n in an equivalence class is clearly n since $w^n = w$ for any word w. Some equivalence classes, called *degenerate*, contain fewer than n words. [The word $w = (0\ 0 \ldots 0)$, for example, is the only word in its equivalence class.] If w belongs to a degenerate equivalence class, then $w = w^i$ for some integer i in the range $1 \leq i \leq n - 1$.

If a dictionary is synchronizable (with finite delay) it evidently cannot contain two words from the same equivalence class. Except for the prefix, the word sequences $www \ldots$ and $w^iw^iw^i \ldots$ are identical, and there is no way to determine the initial symbol of such a sequence if both w and w^i are in the dictionary. Similarly, no synchronizable dictionary can contain *any* word from a degenerate equivalence class. For if $w = w^i$, the infinite sequence $www \ldots$ is unchanged when the first i symbols are deleted and there is no way to determine which symbol is the initial symbol of a code word.

We have therefore demonstrated the following lemma.

Lemma 11.1. The number of n-symbol words in any synchronizable code defined on an r-symbol alphabet is bounded by $N(n, r)$ where

$$N(n, r) = \frac{W(n, r)}{n} \tag{11.4.1}$$

and where $W(n, r)$ is the number of n-tuples belonging to non-degenerate equivalence classes.

As an example, consider the set of binary 4-tuples. The equivalence classes containing the 4-tuples 0000 and 1111 both have only one member, while that containing the 4-tuples 0101, 1010, has just those two members. The remaining $2^4 - 2 - 2 = 12$ 4-tuples belong to non-degenerate equivalence classes. There are therefore exactly $\frac{12}{4} = 3$ such classes, each contributing at most one word to any dictionary which is to be synchronizable. That a three-word synchronizable dictionary does exist is evidenced by the set {0001, 0011, 0111}.

If an n-tuple has the property that $w = w^{\nu}$ for some integer ν, it is said to be periodic with period ν. A word of length n belongs to a degenerate equivalence class if, and only if, it is periodic with some period $0 < \nu < n$. The following two lemmas will be useful in determining the number of words $W(n, r)$ in non-degenerate equivalence classes.

Lemma 11.2. If a sequence w of length n is periodic with periods $\nu_1, \nu_2, \ldots$, and if ν_1 is the smallest of these periods, then

(1) $\nu_1 \mid \nu_i$ all i
(i.e., $\nu_i = l_i \nu_1$ for some integer l_i) and
(2) $\nu_1 \mid n$.

Proof. Let $\nu_i = q_i \nu_1 + r_i$ with q_i a non-negative integer, and r_i an integer in the range $0 \leq r_i < \nu_1$. If w is periodic with both periods ν_i and ν_1, $w = w^{\nu_i} = w^{q_i \nu_1 + r_i} = w^{r_i}$. Since ν_1 is the smallest integer i for which $w^i = w$, it follows that $r_i \geq \nu_1$ or $r_i = 0$. But since r_i is in the range $0 \leq r_i < \nu_1$, $r_i = 0$ and $\nu_i = q_i \nu_1$ for some integer q_i. This holds for all periods ν_i of w, including the period $\nu_i = n$ and the lemma is proved.

Lemma 11.3. Let A be the subset of elements of the set S having the attribute a, and denote by $N(A)$ the number of elements in A. Then the number of elements in S having any of the attributes $a_1, a_2, \ldots, a_m$ (i.e., the number of elements in the set $A_1 + A_2 + \cdots + A_m$) is

$$N(A_1 + A_2 + \cdots + A_m) = \sum_i N(A_i) - \sum_i \sum_{j<i} N(A_i A_j) + \sum_i \sum_{j<i} \sum_{k<j} N(A_i A_j A_k) - \ldots - (-1)^m N(A_1 A_2 \ldots A_m) \tag{11.4.2}$$

where $N(A_i A_j \ldots A_\nu)$ is the number of elements having *all* the attributes $a_i, a_j, \ldots, a_\nu$ (i.e., the number of elements in the set $A_i A_j \cdots A_\nu$).

Proof. The proof follows by induction on m. Clearly, when $m = 2$,

$$N(A_1 + A_2) = N(A_1) + N(A_2) - N(A_1 A_2)$$

as the number of elements in $A_1 A_2$ is included in both $N(A_1)$ and $N(A_2)$ and hence is counted twice in the sum $N(A_1) + N(A_2)$. Now suppose the lemma is true for some $m \geq 2$. Then, letting $B = A_1 + A_2 + \ldots + A_m$, we have

$$N(A_{m+1} + B) = N(A_{m+1}) + N(B) - N(A_{m+1} B) \tag{11.4.3}$$

But since B and $A_{m+1}B$ both represent the sum (union) of m subsets, both $N(B)$ and $N(A_{m+1}B) = N(A_1 A_{m+1} + A_2 A_{m+1} + \ldots + A_m A_{m+1})$ satisfy Equation (11.4.2) by hypothesis. Combining Equations (11.4.2) and (11.4.3) yields an expression identical to that of Equation (11.4.2) but now involving $m + 1$ subsets, and the lemma follows by induction.

We are now in a position to prove the following theorem.

Theorem 11.14. The number of r-ary n-symbol words in a synchronizable code cannot exceed

$$N(n, r) = \frac{1}{n} \sum_{d \mid n} \mu(d) r^{n/d} \qquad (11.4.4)$$

where the summation is over the integers d which divide n, and where†

$$\mu(d) = \begin{cases} 1 & d = 1 \\ (-1)^k & d = p_1 p_2 \ldots p_k, \text{ with } p_1, p_2, \ldots, p_k \text{ all distinct primes} \\ 0 & \text{otherwise} \end{cases}$$

Proof. Suppose w is an n-tuple degenerate with minimum period ν. From Lemma (11.2), ν must evidently equal n/d with d some integral divisor of n. Obviously w is also degenerate with period $l\nu$ for any integer l. Since any integer d can be written as a product of prime powers, we conclude that $\nu = n/d < n$ is a period of w only if $l\nu = n/p$ is also, with p some prime divisor of n. To count the number $W_d(n, r)$ of degenerate n-tuples therefore, it is only necessary to count those n-tuples having any of the periods n/p_j, with $(p_1, p_2, \ldots, p_l)$ the set of prime divisors of n. But if w is an n-tuple with period n/d, it must be of the form $w = ss \ldots s$, with s some (n/d)-tuple. Since any n-tuple of this form is periodic with period n/d, there are exactly $r^{n/d}$ n-tuples having this period. Thus, using Lemma (11.3) and noting from Lemma (11.2) that w is degenerate with the periods n/p_j for all $j = j_1, j_2, \ldots, j_\mu$, if, and only if, it is also degenerate with the period $n/p_{j_1} p_{j_2} \ldots p_{j_\mu}$, we obtain

$$W_d(n, r) = \sum_i r^{n/p_i} - \sum_i \sum_{j<i} r^{n/p_i p_j} + \sum_i \sum_{j<i} \sum_{k<j} r^{n/p_i p_j p_k} \ldots -(-1)^l r^{n/p_1 p_2 \ldots p_l}$$

$$(11.4.5)$$

The number of non-degenerate n-tuples is therefore $W(n, r) = r^n - W_d(n, r)$, the maximum number of n-tuples in a synchronizable code is $1/n$th this value (Lemma 11.1), and the theorem is proved.

Theorem 11.14 is based on the fact that no two words of the same length in a synchronizable code can belong to the same equivalence class and that none can belong to a degenerate equivalence class. When the dictionary includes words of more than one length, this same reasoning can be used to improve the bound on the number of words possible if the code is to be synchronizable. For suppose two different word sequences were in the same equivalence class (e.g., suppose some cyclic permutation of an n-symbol

---

†The function $\mu(d)$ is called the Möbius μ function. It is frequently encountered in combinatorial contexts.

word could also be produced by following a k-symbol word by an $(n - k)$-symbol word, all words belonging to the same dictionary). If the two sequences S_1 and S_2 composed of different code words were in fact identical, the code would not even be uniquely decodable. If they were related by a cyclic permutation, then the infinite sequences $\ldots S_1S_1S_1 \ldots$ and $\ldots S_2S_2S_2 \ldots$ could not be distinguished without some extrinsic method for identifying an initial symbol; the code would not be synchronizable.

In order to state this condition more precisely, we define a *word* equivalence class, as opposed to a *symbol* equivalence class, as the set of all word sequences which differ only by a cyclic permutation on the *words* of the sequence. As an example, if $w_1 = 0011$, $w_2 = 01$, and $w_3 = 10$, the symbol sequences $w_1 = 0011$ and $w_2w_3 = 0110$ are in the same symbol equivalence class. The word sequences w_1 and w_2w_3 are not in the same word equivalence class, however, since one obviously cannot be produced by cyclically permuting the other. In contrast, the word sequences $w_1w_2w_3$, $w_3w_1w_2$ and $w_2w_3w_1$ are all in the same word equivalence class. Notice that the symbols which comprise the words are irrelevant in the definition of a word equivalence class.

The preceding argument concerning the requirement that different word sequences be in different symbol equivalence classes leads to the following conclusion.

Lemma 11.4. If a dictionary D is synchronizable, then the number of non-degenerate equivalence classes of words from D having a total length n cannot exceed the number of non-degenerate equivalence classes of symbol sequences of length n, for any integer n.

Proof. Suppose S_1 and S_2 are two different symbol sequences corresponding to the word sequences W_1 and W_2, respectively. Suppose, further, that S_1 and S_2 are in the same (symbol) equivalence class, but that W_1 and W_2 belong to two different word equivalence classes. The infinite sequence $\ldots S_1S_1 \ldots$ can be interpreted either as $\ldots W_1W_1 \ldots$ or as $\ldots W_2W_2 \ldots$ depending upon which symbol is taken as the initial symbol. But W_1 and W_2 are not related by a cyclic permutation. Two different decodings are therefore possible in the absence of prior synchronization, and the dictionary is not synchronizable. Similarly, if S_1 is degenerate, but W_1 is not, then either S_1 is not uniquely decodable, or else the sequence $\ldots S_1S_1 \ldots$ can be decoded in two different ways by beginning with different symbols. Synchronizability thus requires that any two word sequences in different non-degenerate word equivalence classes give rise to two symbol sequences which are in different non-degenerate symbol equivalence classes. This proves the lemma.

Note that unless constraints are placed on the allowable word *sequences* as well as on the words themselves, all cyclic permutations of any word sequence are obviously possible. Higher-order synchronizability might prohibit all but one word sequence in each word equivalence class from occurring.

That is, a second level of coding could be imposed on the code words themselves in order, for example, to enable frame as well as word synchronization to be determined directly from the symbol sequence.

In order to count the number of equivalence classes of words totaling n symbols in length, first consider the number of ways of writing n as a sum of positive integers (i.e., the number of *ordered partitions* of the integer n). As an example, the integer 4 can be written as

$$4 = \begin{array}{l} 1+1+1+1 \\ 1+1+2 \\ 1+2+1 \\ 2+1+1 \\ 2+2 \\ 1+3 \\ 3+1 \\ 4 \end{array}$$

We call two partitions of n equivalent if one is a cyclic permutation of the other. The set of partitions consisting of one representative of each equivalence class is denoted the *semiordered partitions* S of n. The semiordered partitions of 4, in particular, can be represented by the sequences

1111
112
22
13
4

A partition S of a number n is a sequence of integers. If ν is the minimum period of this sequence S, we denote by P the first ν integers in S and by $s(P)$ the unique integer such that $S = PP \ldots P = P^{s(P)}$, the sequence P repeated $s(P)$ times. Thus, when $n = 4$ we have

S	P	$s(P)$
1111	1	4
112	112	1
22	2	2
13	13	1
4	4	1

Now, suppose a synchronizable dictionary D contains σ_i words of length i, and let $S = i\,j \ldots k\,i\,j \ldots k \ldots k = P^{s(P)}$ be a partition of n. Then:

Lemma 11.5. The number of different non-degenerate equivalence classes represented by the words of the synchronizable dictionary D having respective lengths represented by the partition $S = P^{s(P)}$ of n is

$$N[s(P), \sigma_i\sigma_j \ldots \sigma_k] \tag{11.4.6}$$

where $N(n, r)$ is as defined in Equation (11.4.4).

Proof. There are $\rho = \sigma_i\sigma_j \ldots \sigma_k$ sets of words of D corresponding to the sequence $ij \ldots k$. Treating each of these sets as one symbol of a new alphabet, we can consider the original word sequence as a sequence of $s(P)$ of these new "symbols." Evidently, there are $N[s(P), \rho]$ non-degenerate equivalence classes represented by "words" of this form, which was to be proved.

With this result we can easily prove the following theorem.

Theorem 11.15. If D is a synchronizable dictionary having σ_i words of length i, then

$$\sum_S N[s(P), \sigma_i\sigma_j \ldots \sigma_k] \leq N(n, r) \qquad n = 1, 2, \ldots \tag{11.4.7}$$

where the summation is over all semiordered partitions $S = P^{s(P)}$ of n and where $P = ij \ldots k$.

Proof. From Lemma 11.4, the number of equivalence classes of symbol sequences of any length n must be at least as great as the number of equivalence classes of word sequences of that length. But for every partition S, there are $N[s(P), \sigma_i\sigma_j \ldots \sigma_k]$ word equivalence classes represented by words from D (Lemma 11.5). Clearly word sequences corresponding to different partitions S must be in different equivalence classes. The total number of word equivalence classes for any given n is therefore given by the summation on the left of the inequality (11.4.7). The number of symbol equivalence classes of this length is just $N(n, r)$, establishing the inequality (11.4.7) which must be satisfied for all integers n.

As an example, when $n = 4$, $r = 2$, we have

$$\sum_S N[s(P), \sigma_i\sigma_j \ldots \sigma_k] = N(4, \sigma_1) + N(1, \sigma_1^2\sigma_2) + N(2, \sigma_2) + N(1, \sigma_1\sigma_3) + N(1, \sigma_4)$$

and, from Theorem 11.15, this must be bounded by

$$N(4, 2) = \frac{2^4 - 2^2}{4} = 3$$

Thus, since $N(1, \sigma_4) = \sigma_4$, σ_4 cannot exceed 3, and if $\sigma_4 = 3$, σ_1 and σ_2 must be zero, because of the terms $N(4, \sigma_1)$ and $N(2, \sigma_2)$, respectively.

Similar arguments can be used to establish other constraints on the various word lengths. As is apparent, however, this bound is inconvenient to apply exhaustively (i.e., it is clearly tedious to enumerate all of the relevant sets $\{\sigma_i\}$ for which a synchronizable code is not precluded by this bound). (Remember that this bound must be satisfied for all sequence lengths, not just for $n = 4$.) Nevertheless, if the set $\{\sigma_i\}$ is such that the bound (11.4.7) is satisfied for all n, a synchronizable code having these word lengths σ_i can be constructed. This is demonstrated in the next section.

11.5 Maximal Synchronizable Dictionaries

It will now be shown that the bound in Theorem 11.15 is a sufficient as well as a necessary condition for the existence of synchronizable codes, that it is possible to construct a synchronizable code having σ_i words of length i if the inequality (11.4.7) is satisfied for all n. The construction, incidentally, will yield codes having the prefix property as defined in Chapter Ten. The tighter constraint of this chapter, of course, will result in generally smaller dictionaries than are attainable when only decodability is required.

The construction is as follows. Begin with the r-word dictionary D_0 in which each of the r symbols of the alphabet is itself a word. If $\sigma_1 = r$, D_0 is the desired dictionary. If $\sigma_1 < r$ (clearly σ_1 cannot be greater than r) let L be the maximum value of i for which $\sigma_i \neq 0$. Let $(D_0 - w_0)$ indicate the set of words remaining after the one-symbol word w_0 is removed from D_0 and $p(D_0 - w_0)$ the set obtained by prefixing by the symbol sequence p each of the words in $(D_0 - w_0)$. That is, if

$$D_0 = \begin{array}{l} a \\ b \\ c \\ d \\ \vdots \end{array}$$

then

$$p(D_0 - a) = \begin{array}{l} pb \\ pc \\ pd \\ \vdots \end{array}$$

The dictionary D_1 is defined as the set union

$$D_1 = (D_0 - w_0) + w_0(D_0 - w_0) + w_0w_0(D_0 - w_0) \ldots + w_0w_0 \ldots w_0(D_0 - w_0)$$

the maximum length prefix $w_0w_0 \ldots w_0$ having $L - 1$ symbols w_0. The choice of the word w_0 is arbitrary. As an example, if $r = 3$, $L = 4$,

$$D_0 = \begin{matrix} 0 \\ 1 \\ 2 \end{matrix} \qquad 0(D_0 - 0) = \begin{matrix} 01 \\ 02 \end{matrix} \qquad 00(D_0 - 0) = \begin{matrix} 001 \\ 002 \end{matrix} \qquad 000(D_0 - 0) = \begin{matrix} 0001 \\ 0002 \end{matrix}$$

and

$$D_1 = \begin{matrix} 1 \\ 2 \\ 01 \\ 02 \\ 001 \\ 002 \\ 0001 \\ 0002 \end{matrix}$$

In general, the ith iteration of this process involves finding the smallest value of j such that the number of words in D_{i-1} of that length exceeds σ_j, taking a word w_{i-1} of length j from D_{i-1} and using it as a prefix to construct D_i:

$$D_i = (D_{i-1} - w_{i-1}) + w_{i-1}(D_{i-1} - w_{i-1}) + w_{i-1}w_{i-1}(D_{i-1} - w_{i-1}) \ldots$$
$$+ w_{i-1}w_{i-1} \ldots w_{i-1}(D_{i-1} - w_{i-1})$$

The longest prefix $w_{i-1}w_{i-1} \ldots w_{i-1}$ which need be considered is that for which the shortest word in the set $w_{i-1}w_{i-1} \ldots w_{i-1}(D_{i-1} - w_{i-1})$ has length at least $L - j + 1$ since the next longer prefix would yield only words exceeding the maximum length of interest. This process, which we shall refer to as the *prefix construction procedure*, is terminated when the number of words of every length j in the last iteration equals or is exceeded by σ_j.

Continuing with the preceding example, suppose $r = 3$, $L = 4$, $\sigma_1 = 1$, $\sigma_2 = 2$, $\sigma_3 = 6$, $\sigma_4 = 9$, and $\sigma_j = 0$ for $j > 4$. The inequalities (11.4.7) are indeed satisfied by this set $\{\sigma_i\}$ as is easily verified. The dictionary D_1 is as defined above. Since it contains two words of length one, and since $\sigma_1 = 1$, we choose the word $w_1 = 1$ as a prefix and form D_2:

$$D_2 = \begin{matrix} 2 \\ 01 \\ 02 \\ 001 \\ 002 \\ 0001 \\ 0002 \\ 12 \\ 101 \\ 102 \\ 1001 \\ 1002 \\ 112 \\ 1101 \\ 1102 \\ 1112 \end{matrix}$$

Since D_2 contains three words of length 2, and since $\sigma_2 = 2$, we choose the word $w_2 = 01$ as a prefix and form D_3:

$$D_3 = \begin{array}{r} 2 \\ 02 \\ 001 \\ 002 \\ 0001 \\ 0002 \\ 12 \\ 101 \\ 102 \\ 1001 \\ 1002 \\ 112 \\ 1101 \\ 1102 \\ 1112 \\ 012 \\ 0102 \\ 0112 \end{array}$$

The dictionary D_3 contains one word of length one, two of length two, six of length three, and nine of length four. The prescribed conditions are met and the construction terminates.

As was mentioned earlier, this construction yields codes having the prefix property. This statement is easily proved by induction.

Theorem 11.16. The code dictionaries defined by the prefix construction procedure outlined in the preceding paragraphs all have the prefix property.

Proof. The dictionary D_0 clearly has the prefix property since all its words have length one. Assume D_i has the prefix property, and let w_i be used as the prefix in the construction of D_{i+1}. Then all words in D_{i+1} are of the form $w_i w_i \ldots w_i w_\nu$ where w_ν is a word in D_i and the number of repetitions of the prefix w_i is arbitrary. If the word $w_i w_i \ldots w_i w_\nu$ in D_{i+1} is a prefix of some other word $w_i w_i \ldots w_i w_\mu$, then either $w_i w_i \ldots w_i w_\nu$ must be a prefix of w_μ or w_ν must be a prefix of $w_i w_i \ldots w_i w_\mu$ (e.g., if $w_i w_i w_\nu$ is a prefix of $w_i w_\mu$, $w_i w_\nu$ is a prefix of w_μ). In either event, one of the words w_i, w_μ, or w_ν would have to be a prefix of one of the others, contrary to the hypothesis that D_i has the prefix property. Thus, by induction, all such dictionaries must have the prefix property, as was to be shown.

We must now prove that (1) this construction procedure does indeed produce synchronizable codes, and (2) it is possible to obtain at least σ_i code words of length i using this method, provided the inequalities (11.4.7) are all satisfied for the set of lengths $\{\sigma_i\}$.

To prove the first of these two statements, we first describe an algorithm for establishing synchronization with this class of codes. Let D_i be as defined above, and let D_ν be the dictionary actually being used. Further let D'_i, the *extension* of D_i, be defined as the dictionary that would have resulted at the ith step of the prefix construction procedure had no restriction been placed on the maximum word length. That is, in the example above, $D'_0 = D_0$, but

$$D'_1 = D_1 + \{00001, 00002, 000001, 000002, \ldots\}$$

etc., and each of the dictionaries D'_i, $i > 0$, contains an infinite number of words.

Now, consider an arbitrary sequence of words from D_ν possibly beginning with a suffix from D_ν. This sequence can obviously be separated into words from D'_0 by placing commas after each received symbol. Then, to separate the sequence into words from D'_1, it is only necessary to remove the commas following each occurrence of the word w_0 used in constructing D_1 from D_0. For any word except w_0 is in D'_1 and any word of the form $w_0 w_0 \ldots w_0 w_\mu$ with $\mu \neq 0$ is also in D'_1. By continuing this procedure, at the ith step deleting those commas following each appearance of the word w_{i-1}, we can successively divide the received sequence into words from $D'_0, D'_1, \ldots, D'_{\nu-1}$, and, finally, into words from D'_ν.

If the received sequence actually begins with a complete word from D_ν, it is apparent that this procedure would correctly separate this sequence into words from D_ν. For were this not the case, it would be possible to decode the sequence into two different sequences of words from D'_ν, a possibility prohibited by the prefix property of D'_ν. (Note that Theorem 11.16 places no constraint on the length of the words retained in any of the dictionaries D_i.) If, however, the received sequence begins with a suffix from D_ν, the comma following this suffix might be removed. This, in turn, alters the form of the first apparent word received and, as a consequence, can result in the removal of the comma following the first actual received word. Further, this erroneous removal of commas can propagate; if the comma following the jth received word is removed at some step, the comma following the $(j + 1)$st received word can be removed at some later step. The prefix property of D'_ν guarantees that no word is the prefix of another word. It does not guarantee, however, that one word cannot be contained in another word, or that the suffix of one word cannot be the prefix of another. Nevertheless, the prefix property does guarantee that the comma following any word in D_ν will not be removed until *after* the removal of the comma preceding it. The removal of the trailing comma before, or at the same time as, the leading comma would be possible only if a word in D'_ν were also a prefix in D'_ν. Thus, since only ν steps are required to divide the received sequence into words from D'_ν, the comma following the νth received word will not be removed. From this point on, the sequence must be correctly separated into words from D_ν, again because of the prefix property of this dictionary. The maximum synchronization delay, therefore, is bounded by $\nu L + L - 1$, with L denoting the maximum word length in D_ν and $L - 1$, of course, representing the longest possible suffix in D_ν.

So far then we have proved the following.

Theorem 11.17. The dictionaries D_ν defined by the prefix construction procedure are synchronizable with a synchronization delay not exceeding $(\nu + 1)L - 1$ symbols.

It remains to prove the second of the above statements concerning these dictionaries, that they are in fact maximal.

Theorem 11.18. Dictionaries of the form D_i can be constructed having σ_j words of length j for any set of integers $\{\sigma_i\}$ satisfying the bounds (11.4.7).

The proof of this theorem follows quite directly with the aid of the following lemma:

Lemma 11.6. If the dictionary D_i is constructed from the dictionary D_{i-1} by using a word w_{i-1} of length k as the prefix, then the bound 11.4.7 is satisfied with equality for all $n = k + 1, k + 2, \ldots, L$ (L again denoting the length of the longest code words retained in any of the dictionaries D_i).

Proof. The only word sequences which can be constructed from words in D_{i-1} but not from words in D_i are those ending with the word w_{i-1}. All such sequences having length greater than k are of the form

$$w_\alpha w_\beta \ldots w_{i-1} \tag{11.5.1}$$

where the w_l's are words in D_{i-1}. If $w_l \neq w_{i-1}$ for some w_l in this sequence, and the total sequence length does not exceed L, then some cyclic permutation of (11.5.1) is a word sequence ending in w_l, and one which can be constructed from words in D_i. If the sequence (11.5.1) is comprised of the word w_{i-1} only, then it belongs to a degenerate equivalence class. It is concluded, therefore, that all non-degenerate equivalence classes represented by word sequences from D_{i-1} of total length n where $k + 1 \leq n \leq L$ are also represented by sequences of words from D_i. All of the bounds (11.4.7) satisfied by the dictionary D_{i-1} for $k + 1 \leq n \leq L$ are also satisfied by D_i. Finally, since the bounds (11.4.7) are obviously satisfied with equality for all n by the dictionary D_0, and since the construction procedure is such that the length of the prefixes used in going from D_{i-1} to D_i is a non-decreasing function of i, the statement of the lemma follows by induction.

We can now prove Theorem 11.18.

Proof. Let s_j^i be the number of words of length j in D_i. If $\sigma_1 = r$, D_0 is the desired dictionary and the theorem is satisfied. If $\sigma_1 < r$, D_1 is constructed by using a word (symbol) from D_0 as a prefix as discussed. This can be continued until $\sigma_1 = s_1^i$. In general, words of length j are used as prefixes to construct the dictionary D_i until $s_j^i = \sigma_j$. In order for the process to continue until all of the necessary words are constructed, it must be shown that if $s_j^i = \sigma_j$ for all $j = 1, 2, \ldots, k$, then $s_{k+1}^i \geq \sigma_{k+1}$. If this were not true, none of the dictionaries D_l, $l > i$, would contain enough words of length $k+1$. If this condition does hold, then, by induction, some dictionary D_l will exist whereby $s_j^l \geq \sigma_j$ for all j and the theorem is proved. But from Lemma 11.6, we know that the dictionary D_i (constructed from D_{i-1} by using a prefix of length k) is such that

$$\sum_S N[s(P), s_\mu^i s_\nu^i \ldots s_\eta^i] = N(k + 1, r) \tag{11.5.2}$$

By hypothesis,

$$\sum_S N[s(P), \sigma_\mu \sigma_\nu \ldots \sigma_\eta] \leq N(k + 1, r) \tag{11.5.3}$$

Only words of length $k + 1$ or less can be involved in a sequence of total symbol length $k + 1$. Since $\sigma_j = s_j^i$ for all j in the range $1 \leq j \leq k$, the only terms in the two summations (11.5.2) and (11.5.3) which are not identical are those involving σ_{k+1} and s_{k+1}^i. Since $N(1, \sigma_{k+1})$ and $N(1, s_{k+1}^i)$ are the only such terms, it follows by eliminating identical terms from both sides of the inequality

$$\sum_S N[s(P), \sigma_\mu \sigma_\nu \ldots \sigma_\eta] \leq N(k + 1, r) = \sum_S N[s(P), s_\mu^i s_\nu^i \ldots s_\eta^i]$$

that

$$N(1, \sigma_{k+1}) \leq N(1, s_{k+1}^i)$$

and, hence, that

$$\sigma_{k+1} \leq s_{k+1}^i$$

as was to be proved.

Theorem 11.19. The number of words σ_i of length i, $1 \leq i \leq L$, generated by the prefix construction procedure is the coefficient of the term x^i in the polynomial

$$P(x) = 1 - \frac{1 - rx}{(1 - x)^{\alpha_1}(1 - x^2)^{\alpha_2}(1 - x^3)^{\alpha_3} \ldots (1 - x^l)^{\alpha_l} \ldots}$$
$$= \sum_{j=1}^{\infty} \sigma_j x^j \tag{11.5.4}$$

where α_i denotes the number of words of length i used as prefixes in the construction of the final dictionary.

Proof. We remark, first of all, that this expression is valid for all values of L, the length of the maximum length words retained in the construction procedure. Consequently, the terms σ_j, $j > L$, represent the number of words of length j which would be in the dictionary had those words been retained (i.e., the number of words of length j in the extension of the dictionary in question). With this in mind, we define $P(x) = \sum_{j=1}^{\infty} \sigma_j x^j$, and show that $P(x)$ is as given in Equation (11.5.4).

Let $P_\nu(x)$ represent the polynomial $\sum_{j=1}^{\infty} \sigma_j(\nu) x^j$ where $\sigma_j(\nu)$ is the number of words of length j in the dictionary D_ν (the dictionary obtained when some word in $D_{\nu-1}$ is used as a prefix to construct a new dictionary, as discussed above). Then, the number of words of length j in $(D_{\nu-1} - w_{\nu-1})$ is represented by the coefficient of x^j in the polynomial $P_{\nu-1}(x) - x^l$, with l the length of $w_{\nu-1}$. Similarly, the number of words of various lengths in the set $w_{\nu-1}w_{\nu-1} \ldots w_{\nu-1}(D_{\nu-1} - w_{\nu-1})$ is given by the polynomial $x^{\mu l}[P_{\nu-1}(x) - x^l]$ where μ is the number of repetitions of the prefix $w_{\nu-1}$. Thus,

$$P_\nu(x) = [P_{\nu-1}(x) - x^l] \sum_{\mu=0}^{\infty} x^{\mu l} = \frac{P_{\nu-1}(x) - x^l}{1 - x^l} = 1 - \frac{1 - P_{\nu-1}(x)}{1 - x^l}$$

so that

$$1 - P_\nu(x) = \frac{1 - P_{\nu-1}(x)}{1 - x^l}$$

and by induction

$$1 - P(x) = \frac{1 - P_0(x)}{(1 - x)^{\alpha_1}(1 - x^2)^{\alpha_2} \ldots (1 - x^l)^{\alpha_l} \ldots}$$

Since $P_0(x) = rx$, the theorem is proved.

As an example, when $r = 3$, $\alpha_1 = 2$ and $\alpha_2 = 1$ (this was the situation considered earlier),

$$P(x) = 1 - \frac{1 - 3x}{(1 - x)^2(1 - x^2)} = x + 2x^2 + 6x^3 + 9x^4 + 15x^5 + \ldots$$

It is interesting to note that this argument makes no assumptions concerning the order in which the prefixes are used; they do not need to be used in order of increasing length. Nor is Theorem 11.17 dependent upon any particular ordering of the prefixes. As a consequence, the preceding algorithm yields maximal synchronizable dictionaries regardless of the order of prefix removal. We repeat the construction of a dictionary with $r = 3$, $\sigma_1 = 1$, $\sigma_2 = 2$, $\sigma_3 = 6$, and $\sigma_4 = 9$ by removing the prefixes in an order different from before.

$D_0 =$	$D_1 =$	$D_2 =$	$D_3 =$
0	1	1	2
1	2	2	02
2	01	02	001
	02	001	002
	001	002	0001
	002	0001	0002
	0001	0002	011
	0002	011	012
		012	0102
		0102	12
			102
			1001
			1002
			1011
			1012
			112
			1102
			1112

Although this dictionary is synchronizable, and contains the same number of words of each length as the corresponding dictionary in the earlier example, the two dictionaries are not identical.

If the goal is to construct a synchronizable dictionary having the maximum possible number of words of length not exceeding L, it can clearly be achieved by using the shortest word as the prefix at every step of the construction. (That is, each word of length l in the range $L - (\nu + 1)j < l \leq L - \nu j$ in the set $(D_{i-1} - w_{i-1})$ yields ν words of length not exceeding L in D_i when w_{i-1} has length j. Clearly, the added number of words is a non-increasing function of j.) The total number of words of length L or less in D_i will not be less than the number of such words in D_{i-1} if, and only if, (1) the length j of the prefix used does not exceed $[L/2]$ and (2) when L is even there are at least two words in D_{i-1} of length $L/2$. If L is odd, all words of length $(L - 1)/2$ can be used as prefixes of all words of length $(L + 1)/2$ so the number of words in D_i is never less than that in D_{i-1} when $j \leq (L - 1)/2$. If L is even, then so long as there are at least two words of length $L/2$, one of them can be eliminated and used as the prefix to the other without decreasing the total number of words of length L or less. Accordingly, when L is odd, no words of length less than $(L + 1)/2$ will remain in the final dictionary D_i, while, when L is even, only one word of length $L/2$ and none of length less than this will be included. All of the word sequences of total length n in the range $[L/2] + 1 \leq n \leq L$ must be composed of single words. But, from Lemma 11.6, the number of such sequences is equal to the number of non-degenerate symbol equivalence classes for all n in this range. Thus, the total number of words in the final dictionary is just $\sum_{n=[L/2]+1}^{L} N(n, r)$, plus, when L is even, the one remaining word of length $L/2$. This result is summarized in the following theorem.

Theorem 11.20. The maximum number of words N_L of length L or less in a synchronizable dictionary is

$$N_L = \frac{1 + (-1)^L}{2} + \sum_{n=[L/2]+1}^{L} N(n, r)$$

In contrast, if only words of length L were used, the number of words would be simply $N(L, r)$.

This construction for maximal synchronizable dictionaries presented in this section is clearly not unique. A suffix construction, for example, in which deleted words are used as suffixes rather than prefixes would be equally effective. The construction considered here does have the prefix property with its inherent advantages, as discussed in Chapter Ten.

In concluding, we prove still another interesting property of the extended dictionaries D'_ν defined earlier.

Theorem 11.21. The extended prefix construction dictionaries D'_ν are exhaustive.

Proof. This theorem is easily proved by induction. Clearly, D'_0 is exhaustive. Since D'_ν has the prefix property, it is exhaustive, if, and only if, $\sum_{i=1}^{\infty} r^{-l_i(\nu)} = 1$ where $l_i(\nu)$ is the length of the ith word in D'_ν (cf. Theorem 10.4). Assume D'_μ is exhaustive. Then, since each word of length $l_i(\mu)$ in D'_μ (excluding w_μ) yields one word of length $l_i(\mu) + jl_\mu(\mu)$ in $D'_{\mu+1}$ for all $j = 0, 1, \ldots$ (with $l_\mu(\mu)$ the length of w_μ),

$$\sum_{i=1}^{\infty} r^{-l_i(\mu+1)} = \sum_{j=0}^{\infty} \sum_{i \neq \mu} r^{-[l_i(\mu)+jl_\mu(\mu)]} = \frac{\sum_{i \neq \mu} r^{-l_i(\mu)}}{1 - r^{-l_\mu(\mu)}} = 1$$

and the theorem is proved.

Thus, if no constraint is placed on the maximum word length, synchronizable dictionaries can meet with equality the same bound proved earlier for the less constrained decodable dictionaries. It should be noted, however, that whereas decodable dictionaries can be constructed having any set of word lengths meeting this bound, the lengths of the words in a synchronizable dictionary are also constrained by the inequalities (11.4.7). Moreover, the synchronization delay associated with the extended dictionaries D'_ν is unbounded for all $\nu > 0$ since the word length itself is unbounded (cf. Theorems 11.6 and 11.17).

Notes to Chapter Eleven

The techniques introduced in Section 11.2 for testing a dictionary for synchronizability are due to Levenshtein (Reference 11.1). Much of the material of Section 11.3, including the concept of a universal synchronizing sequence, is found in Reference 11.2. The bound on the number of words in a synchronizable dictionary was first presented in Reference 11.3 for uniform-length codes and in Reference 11.4 for variable-length codes. The construction for variable-length synchronizable dictionaries meeting these bounds (Section 11.5) is due to Scholtz (Reference 11.5). Other constructions for synchronizable variable-length codes can be found in References 11.6 through 11.10. As the titles suggest, the primary concern in these articles is the error control capabilities of such codes; i.e., their immunity to the unbounded propagation of decoding errors which could otherwise result from the insertion, deletion, or substitution of a symbol in the received symbol sequence. Other properties of variable-length synchronizable codes are discussed in Reference 11.12.

Problems

11.1 Show that, when $\mu(i)$ is as defined in Equation (11.4.4),

$$\sum_{i \mid n} \mu(i) = \begin{cases} 1, & n = 1 \\ 0, & n > 1 \end{cases}$$

Show further that if $W(d, r)$ denotes the number of non-degenerate r-ary d-tuples then

$$\sum_{d|n} W(d, r) = r^n = \sum_{d|n} \sum_{i|d} \mu(i) r^{d/i}$$

Use these two results to prove Theorem 11.14 by induction on n.

11.2 Identify those dictionaries (and their reverses) listed in Problem 10.1 which are synchronizable and determine whether each dictionary is never-, partially- or completely-self-synchronizing. Exhibit a sequence having maximum synchronization delay in every case. Construct a universal synchronizing sequence for those completely-self-synchronizing dictionaries which are not synchronizable with bounded delay.

11.3 (Reference 11.11). Let G be a code graph as defined in Problem 10.6, comprised of the $\nu + 2$ nodes $(S_0, S_1, \ldots, S_\nu, S_0^*)$, and let $T(G)$ be a graph associated with G and defined as follows. Each of the $\nu(\nu + 1)/2$ nodes of $T(G)$ is uniquely identified with an unordered pair (S_i, S_j) of distinct nodes of G (with S_0 and S_0^* for this purpose identified as the same node). A branch leads from the node (S_i, S_j) to the node (S_k, S_l) if, and only if, there are two branches in G bearing the same label, one of which leads from S_i to either S_k or S_l and the second from S_j to the other of the pair. The branch is labeled with the same label attached to the corresponding two branches in G. Thus, the graph $T(G)$ associated with the code graph of Problem 10.5 is as shown in Figure P11.3.

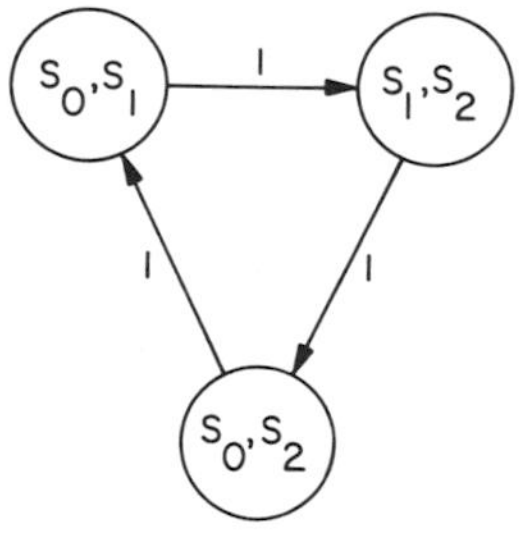

Figure P11.3

Prove that the (finite) dictionary D represented by G is synchronizable only if $T(G)$ contains no loops. Show further that if the longest path in $T(G)$ involves n branches, D is synchronizable with a delay not exceeding $n + d_c - 1$, with d_c its decoding delay. Apply this synchronizability test to some of the codes of Problem 10.1 and compare it in terms of the computational effort required to the test outlined in Section 11.5.

References

11.1 Levenshtein, V. I., "Certain Properties of Code Systems," *Soviet Physics Doklady*, 6, p. 858, 1962.

11.2 Gilbert, E. N., and E. F. Moore, "Variable-Length Binary Encodings," *Bell Systems. Tech. J.*, 38, p. 933, 1959.

11.3 Golomb, S. W., B. Gordon, and L. R. Welch, "Comma-free Codes," *Canadian J. Math.*, 10, p. 202, 1958.

11.4 Golomb, S. W., and B. Gordon, "Codes with Bounded Synchronization Delay," *Inform. and Control*, 8, p. 355, 1965.

11.5 Scholtz, R. A., "Codes with Synchronization Capability," *IEEE Trans. Inf. Theory*, IT-12, p. 135, 1966.

11.6 Neumann, P. G., "Efficient Error-Limiting Variable-Length Codes," *IRE Trans. Inf. Theory*, IT-8, p. 292, 1962.

11.7 Neumann, P. G., "On *a* Class of Efficient Error-Limiting Variable-Length Codes," *IRE Trans. Inf. Theory*, IT-8, p. S260, 1962.

11.8 Neumann, P. G., "Error Limiting Coding Using Information-Lossless Sequential Machines," *IEEE Trans. Inf. Theory*, IT-10, p. 108, 1964.

11.9 Mandelbrot, B., "Recurrent Noise-Limiting Codes," Symp. on Inform. Networks, Polytech. Inst. Brooklyn, Brooklyn, N. Y., p. 205, 1954.

11.10 Hatcher, T. R., "On a Family of Error-Correcting and Synchronizable Codes," *IEEE Trans. Inf. Theory*, IT-15, p. 620, 1969.

11.11 Even, S., "Test for Synchronizability of Finite Automata and Variable Length Codes," *IEEE Trans. Inf. Theory*, IT-10, p. 185, 1964.

11.12 Schützenberger, M. P., "On Synchronizing Prefix Codes," *Inform. and Control*, 11, p. 396, 1967.

chapter twelve

Synchronizable Block Codes

12.1 Introduction

When the code words in a dictionary are all constrained to have the same length the code is generally called a *block code* (cf. Section 10.1). A block code is obviously decodable. Moreover, the results of Section 11.5 assure a construction for maximal synchronizable dictionaries in this special case as well. Nevertheless, it is often desirable to impose further constraints on synchronizable dictionaries. It may be useful to place a tighter restriction on the allowable synchronization delay, for example, or to limit further the class of dictionaries in order to simplify the synchronization as well as the decoding procedure necessitated by the code. Several methods for accomplishing these various objectives are investigated in this chapter. We begin with a construction for synchronizable block code dictionaries for which the synchronization delay is considerably less than that guaranteed by the general construction of Section 11.5.

12.2 Comma-Free Codes

Let D be a block code dictionary consisting of the words w_i, $i = 1, 2, \ldots, N$, where

$$w_i = \omega_1^i \omega_2^i \ldots \omega_n^i \tag{12.2.1}$$

An *overlap* of the words w_i and w_j is defined as any n-tuple of the form

$$\omega_{k+1}^i \omega_{k+2}^i \ldots \omega_n^i \omega_1^j \omega_2^j \ldots \omega_k^j \tag{12.2.2}$$

for any k in the range $1 \leq k \leq n - 1$. A dictionary D is said to be *comma-free* if, and only if, no overlap of any two (not necessarily different) words in D is itself a word in D. Let $p_k^i = \omega_1^i \omega_2^i \ldots \omega_k^i$ denote a prefix and $s_{n-k}^i = \omega_{k+1}^i \omega_{k+2}^i \ldots \omega_n^i$ a suffix of the word w_i for some k is the range $1 \leq k \leq n - 1$. Then an obviously equivalent definition of a comma-free dictionary is the following: A dictionary D is comma-free if, and only if, for every prefix p_k^i and every suffix s_{n-k}^i such that

$$w_i = p_k^i s_{n-k}^i$$

is a word in D, either p_k^i is not a suffix in D or s_{n-k}^i is not a prefix in D.

It is evident that a comma-free dictionary D is synchronizable with a synchronization delay of at most $2n - 1$ symbols (with n the word length). For in any sequence of words from D, no word overlap can be mistaken for a code word. As soon as a code word is observed, synchronization is established. But any sequence of length $2n - 1$ must contain one complete word, so the maximum delay is indeed $2n - 1$. (Actually, this delay can be shortened by one symbol by noting that if $2n - 2$ consecutive symbols do not contain a word, a word must begin with the nth received symbol.)

The maximum number of n-symbol r-ary words in a synchronizable dictionary was found, in Section 11.4, to be bounded by

$$N(n, r) = \frac{1}{n} \sum_{d|n} \mu(d) r^{n/d} \tag{12.2.3}$$

This bound is, a fortiori, also applicable to comma-free codes. Since comma freedom is a considerably more severe restriction than synchronizability alone, it might be supposed that this bound could not generally be attained by comma-free codes. But, as we shall presently demonstrate, $N(n, r)$-word comma-free codes can in fact be constructed for all odd values of n. While some even-length $N(n, r)$-word comma-free codes have also been found, there exists no general construction for such codes. Indeed, no general construction is possible for all even n since sets of parameters n and r are known for which this bound cannot be realized. The maximum number of words in a two-symbol, comma-free dictionary defined over a four-symbol alphabet, for example, is known to be five, whereas $N(2, 4) = 6$. The maximal $n = 4$, $r = 4$ dictionary contains 57, not $N(4, 4) = 60$, words; other counterexamples, including an infinite range of values n, r for which the bound (12.2.3) cannot be attained, have also been demonstrated (Reference 12.1).

The construction for maximal, odd-length, comma-free codes to be presented here is simply a specialization of the prefix construction procedure of Section 10.5. The added restriction is that, at each step of the construction, the prefix word w_i must be the shortest *odd-length* word remaining in D_i. (If there are several words of this length, any one of them may be chosen.) As has already been shown (Lemma 11.6) the bound (11.4.7) is satisfied with equality by the words in D_i for all $n = l + 1, l + 2, \ldots, L$, with l the length of the prefix word w_{i-1}. Thus, by continuing the prefix construction procedure until the shortest word remaining in D_v is of length l for any l in the range $(n + 1)/2 \leq l \leq n \leq L$, we are assured that

$$\sum_{S} N[s(P), \sigma_i \sigma_j \ldots \sigma_k] = N(1, \sigma_n) = \sigma_n = N(n, r) \qquad (12.2.4)$$

[cf. Equation (11.4.7)] and the number σ_n of words of length n in D is indeed $N(n, r)$. The desired dictionary $D(n, r)$ is obtained by deleting from D_v all words except those of length n. That $D(n, r)$ is synchronizable with finite delay follows immediately from Section 11.5. As is demonstrated in the following theorem, however, it is actually comma-free.

Theorem 12.1. The r-ary n-symbol word dictionary $D = D(n, r)$ defined in the preceding paragraph is comma-free for any odd integer n and any integer r.

Proof. Since the words in D are of odd length, an overlap word can be identical to a code word only if some even-length prefix in D is also a suffix in D. To prove that D is comma-free, therefore, we need only show that this event cannot take place.

Let p be an even-length prefix in D and w an n-tuple having p as a suffix. As in Section 11.5, we place a comma after each symbol in both p and w, and then successively remove them, at the jth step removing those commas following each occurrence of the word w_j used in constructing D_{j+1} from D_j. After each such step, w and p can be expressed in the generic form

$$\begin{array}{lllllll} w = \ldots, & \ldots\ldots\ldots\ldots\ldots, & \ldots, & \ldots, & \ldots, & \ldots, & \ldots, \\ & c_{l+1} & c_l & c_{l-1} & c_{l-2} & & c_0 \\ & p = , & \ldots\ldots\ldots, & \ldots, & \ldots, & \ldots, & \ldots, \\ & & c_l^* & c_{l-1}^* & c_{l-2}^* & & c_0^* \end{array} \qquad (12.2.5)$$

where c_i and c_i^* denote the number of symbols between the ith comma (from the right) and the right-most comma in w and p, respectively, and the dots indicate the symbols themselves. (It will be convenient to refer to the different commas in w and p by their locations; whether c_i (or c_i^*) is used as a label for the ith comma in w (or p) or as a number corresponding to its location will be clear from the context.) After each comma is removed, the remaining commas are relabeled to conform to these rules, and l is redefined so that c_l^* always corresponds to the comma preceding the prefix as shown in (12.2.5)

The proof consists of a demonstration that if commas are simultaneously removed from w and p in accordance with the procedure mentioned above, c_0 will be removed at the same time as c_0^*. Since p is a prefix, c_0^* must eventually be removed; since c_0 is therefore also removed w cannot be a code word. To begin, we observe that if $c_v = c_v^*$ at any stage in the comma removal procedure, $c_i = c_i^*$ for all

$i < \nu$. Further, since p is a prefix, and since only odd-length words were used as prefixes in the construction of $D(n, r)$, $c_l^* - c_{l-1}^*$ must be odd. Were this not so, c_{l-1}^* could never be removed and p would have a word in D_ν as a prefix, an impossibility since D_ν has the prefix property (see Section 11.5).

We next show that after each step in the comma removal procedure, the remaining commas will be in one of three configurations:

(1) $c_l = c_l^*$

(2) $c_l > c_l^*$, $c_{l-1} = c_{l-1}^*$

(3) $c_l < c_l^* < c_{l+\nu}$, $c_{l-1} = c_{l-1}^*$, and $c_{l+i} - c_{l+i-1}$ is even for all $i = 0, 1, \ldots, \nu - 1$

In each case, c_l^* denotes the location of the comma preceding p, as shown in (12.2.5).

Initially the commas are clearly in configuration (1). If the commas are in configuration (1) at the jth step and any comma but c_l is removed, they remain in that configuration. (Note that c_ν and c_ν^* will be removed simultaneously for any $\nu < l$.) If c_l is removed, on the other hand, they must then be in configuration (2). Similarly, if they are in configuration (2) they will remain there unless, at some step, c_{l-1}^* or c_{l-1} is removed. But c_{l-1} cannot be removed at any step. This follows because the words used as prefixes in constructing D_{j+1} from D_j were selected in order of increasing (odd) length. Since $c_l - c_{l-1} > c_l^* - c_{l-1}^*$, and since $c_l^* - c_{l-1}^*$ is odd, c_{l-1}^* would therefore be removed before c_{l-1}, and the commas would then be in configuration (3). Moreover, c_{l-1}^* will not be removed in any configuration (assuming c_0^* has not already been removed) unless $c_{l-1}^* - c_{l-2}^*$ (and hence $c_{l-1} - c_{l-2}$) is even. If $c_{l-1}^* - c_{l-2}^*$ were odd, $c_l^* - c_{l-2}^*$ would be even, c_{l-2}^* could not subsequently be removed, and p could not be a prefix in D. As a result, if c_{l-1}^* is removed and the configuration changes from (2) to (3), then when the commas are relabeled, $c_l - c_{l-1}$ will be even. Further removals of the commas labeled c_{l-1}^* lead to the general condition stipulated for configuration (3). Thus, once in configuration (3), c_{l-1} cannot be removed before c_l. The commas will therefore remain in this configuration until c_l is removed, in which case they will return to configuration (2).

The point of all this is that after every step in the comma removal procedure, c_{l-1} and c_{l-1}^* will be identical at least until c_0^* is removed. (Remember that after each comma is removed, l is redefined so that c_l^* always corresponds to the comma preceding p.) Accordingly, since w and p must retain at least one comma at identical points to the left of c_0 and c_0^*, c_0 will be removed simultaneously with c_0^*. An exception to this last statement would be possible were c_{l-1}^* and c_0^* identical. But since $c_l^* - c_{l-1}^*$ is always odd and $c_l^* - c_0^*$ always even, this eventuality is not possible and the conclusion remains. The comma removal procedure will eventually result in c_0 being removed from w. Consequently, w cannot be a word in D, no even length prefix in D can also be a suffix in D, and D is comma-free, as was to be proved.

Thus, maximal comma-free dictionaries do exist for every odd integer n. Unfortunately, the proof of this fact, while constructive, does not offer much hope for an easily implemented procedure for encoding and decoding these codes.† For this reason, it may be useful to constrain further the

†Some progress has recently been made is this area; see Reference 12.2.

dictionary with the goal of simplifying the associated implementation, accepting the fact that these additional constraints will undoubtedly decrease the number of words in the dictionary. Several such constraints will be investigated in the next several sections.

The constraint to be considered in the next section will be made not specifically to simplify the encoding and decoding of the resulting codes, but rather to decrease even further the number of symbols needed to determine synchronization. The dictionaries obtained under this restriction will not only be somewhat easier to synchronize than ordinary comma-free dictionaries, but, in many cases, will have the unexpected advantage of easier encodability and decodability, even though this feature is in no way implied by the constraint.

Before proceeding, however, let us re-examine the comma-free constraint when the word length n is even. Unfortunately, Theorem 12.1 is limited to odd n. As mentioned earlier, the bound (12.2.1) has been shown to be unattainable for some values of n and r, so a general construction for even-length comma-free codes meeting this bound is obviously impossible. Nevertheless, a relatively simple construction exists for even word-length comma-free dictionaries which do come to within a small factor of reaching this bound.

We first construct a comma-free, 2-symbol word dictionary $D(2, r)$ as follows. The r-symbol alphabet is divided into three disjoint sets S_1, S_2, and S_3. The first symbol of any word $w = \omega_1\omega_2$ in $D(2, r)$ must be from either S_1 or S_2, but if ω_1 is in S_1, ω_2 must be in either S_2 or S_3, and if ω_1 is in S_2, ω_2 must be in S_3. The comma-free property of such a dictionary is easily established. Any overlap must be of the form $\omega_2\omega_1$ where ω_2 is in either S_2 or S_3 and ω_1 in either S_1 or S_2. No such overlap can be a code word. The number of words in $D(2, r)$ is clearly maximized by making the number of symbols in each set as nearly equal as possible. Thus, denoting by $N(S_i)$ the number of symbols in the set S_i, we let $N(S_1) = [r/3], N(S_2) = [(r + 1)/3]$, and $N(S_3) = [(r + 2)/3]$, the square brackets indicating the integer part of the enclosed fraction. The number of words in $D(2, r)$ is then

$$N(S_1)\{N(S_2) + N(S_3)\} + N(S_2)N(S_3) = \left[\frac{r^2}{3}\right] \tag{12.2.6}$$

To construct a comma-free dictionary of n-symbol words with $n = 2m$, $m > 1$, we let S_4 denote a subset of the words in $D(2, r)$ and S_5 the set of all ordered pairs of symbols not in S_4 (i.e., $S_4 + S_5$ contains all r^2 distinct pairs of symbols ab). The dictionary $D(2m, r)$ contains all words of the form $a_1a_2 \ldots a_m\, b_1b_2 \ldots b_m$ where the pair a_1b_1 is in S_4 and the pairs a_ib_i, $i \neq 1$, are in S_5. The dictionary $D(2m, r)$ is also obviously comma-free. The only overlap which could conceivably be a word from $D(2m, r)$ is one of the form $b_1b_2 \ldots b_m\, a_1a_2 \ldots a_m$, but the fact that the set of pairs a_1b_1 themselves constitute a comma-free dictionary prevents this from happening.

Using the same notation as before, we find the number of words in $D(2m, r)$ to be

$$N(S_4)\{N(S_5)\}^{(n/2)-1}$$

When $n \geq 6$, we can let $N(S_4)$ contain $[r^2/(n/2)]$ elements, thereby obtaining

$$\left[\frac{2r^2}{n}\right]\left\{r^2 - \left[\frac{2r^2}{n}\right]\right\}^{(n/2)-1}$$

words in $D(2m, r)$. When n is large, this expression becomes approximately

$$\frac{2r^n}{n}\left\{1 - \frac{2}{n}\right\}^{n/2} > \frac{2r^n}{en} \tag{12.2.7}$$

But, from Equation (12.2.3),

$$N(n, r) < \frac{r^n}{n} \tag{12.2.8}$$

for all n, so this construction yields dictionaries containing at least $100(2/e)\%$ $\approx 73.5\%$ as many words as are theoretically possible.

12.3 Path-Invariant Comma-Free Codes

The restriction on the code words to be imposed in this section is the following. If the block dictionary D contains n-symbol words, then no more than n consecutive symbols need be received before synchronization can be unequivocally determined. Clearly, such dictionaries will be comma-free. They are more restricted than comma-free codes, however, since the latter may require as many as $2n - 1$ symbols before synchronization is possible.

In order to delineate a construction for dictionaries with this property, we introduce the concepts of "paths" and "path-invariance" as applied to code dictionaries. To illustrate these concepts, we consider a comma-free code with parameters $r = 3$, $n = 3$:

$$\begin{array}{c} 100 \\ 101 \\ 102 \\ 200 \\ 201 \\ 202 \\ 211 \\ 212 \end{array} \tag{12.3.1}$$

Since all words in this dictionary are of the form cba with $c > b \leq a$ it is easily seen to be comma-free, and, since $N(3, 3) = 8$, the dictionary is maximal. We now define an $r \times n$-*coincidence matrix* of 0's and 1's, the ijth element of which is 1 if the ith symbol of the alphabet (under any ordering) appears in the jth symbol position of any word and 0 if this does not happen. Letting the first symbol of the alphabet of the dictionary (12.3.1) be 0, the second 1, and the third 2, we have associated with that dictionary the coincidence matrix

Symbol \ Position	1	2	3
0	0	1	1
1	1	1	1
2	1	0	1

(12.3.2)

A *path* through the coincidence matrix is defined as a line passing through one 1 in each column, as, for example,

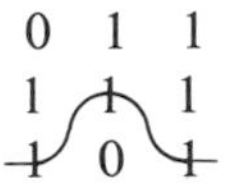

(12.3.3a)

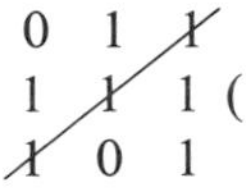

(12.3.3b)

There is a one-to-one correspondence between an r-ary n-tuple and a path through the coincidence matrix: if a path traverses the element $\alpha_{ij} = 1$, the jth symbol of the n-tuple is the ith symbol of the alphabet. In (12.3.3a) the path corresponds to the word 212, a word in the comma-free dictionary (12.3.1). In (12.3.3b), the word represented is 210 which is not in the dictionary.

This construction provides a means of testing a dictionary for comma freedom. Consider an overlap formed from a word suffix of length l and a word prefix of length $n - l$. All such overlaps correspond to paths through a coincidence matrix obtained by following the last l columns of the dictionary coincidence matrix by its first $n - l$ columns (i.e., by permuting the columns cyclically l positions to the right). All the original paths from the jth column to the $(j + 1)$st column remain unaltered (the jth column in the original matrix is now the $(j + l)$th column, modulo n) except, of course, when $j + l = n$. The paths may be completed by connecting each one in the lth column of the permuted matrix to each one in the $l + 1$st column. This represents the fact that any prefix can follow any suffix. These "overlap" matrices are illustrated below for the dictionary (12.3.1) together with the dictionary coincidence matrix itself, each path corresponding to a word or a word overlap:

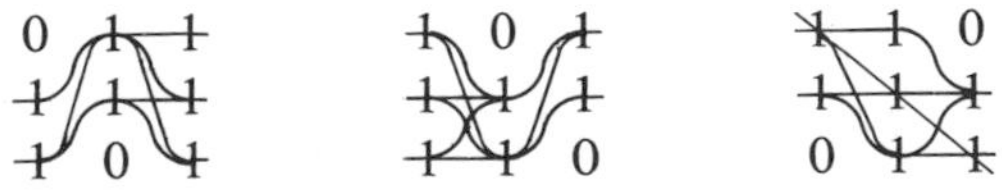

$l = 0$ $l = 1$ $l = 2$

(12.3.4)

Let us designate by M the original coincidence matrix and by P_M the set of paths through M which correspond to code words. Further, we denote by $M(j)$, the matrix M permuted cyclically j positions to the right and by $P_{M(j)}$ the set of paths through $M(j)$ as defined above.

Theorem 12.2. A code is comma-free if, and only if, none of the paths in the set P_M is in any of the sets $P_{M(j)}$.

Proof. This follows immediately because there is a one-to-one correspondence between each path in P_M and a code word and between each path in $P_{M(j)}$ and an overlap formed from the last j symbols of one word followed by the first $n - j$ symbols of another word.

Theorem 12.3. A dictionary of n-symbol words is synchronizable with a delay of no more than n symbols if, and only if, no path in any set $P_{M(j_1)}$ occurs in any other set $P_{M(j_2)}$, $j_1, j_2 = 0, 1, 2, \ldots, n - 1$, $j_1 \neq j_2$.

Proof. If the sets $P_{M(j_1)}$ and $P_{M(j_2)}$, $j_1 \neq j_2$, contain a path in common, and if the corresponding n-tuple is received, it is impossible to resolve whether the $(j_1 + 1)$st or $(j_2 + 1)$st symbol is the beginning of a true word. Conversely, if no path in one set occurs in any other, any n-tuple which can be formed from the last j_1 symbols of one word followed by the first $n - j_1$ symbols of a second word cannot be formed for any other value of j. Thus, any received n-tuple uniquely establishes the value of j and hence the correct synchronization.

While Theorem 12.2 suggests a method for determining whether a particular dictionary is comma-free or not, Theorem 12.3 demonstrates its greatest utility in its use as the basis for an algorithm for constructing the more restricted class of comma-free codes as discussed above. To this end, consider an $r \times n$ matrix M of ones and zeros with the property that *all* paths connecting ones in successive columns are in the set P_M and hence correspond to code words. The dictionary now is entirely determined by its coincidence matrix M. Since all possible paths correspond to code words, the set P_M need not be specified explicitly. For this reason, such codes are called *path-invariant codes.* We are interested in path-invariant codes which are also comma-free. Such codes, if they exist, will of course satisfy the condition of Theorem 12.2, but, in addition:

Theorem 12.4. Path-invariant, comma-free codes of word length n have a synchronization delay of, at most, n symbols. Moreover, every cyclic permutation of a path-invariant, comma-free code is another path-invariant comma-free code.

Proof. Both these statements follow from the observation that all possible paths in all matrices $M(j)$, $j = 0, 1, \ldots, n - 1$, correspond to code words or overlap words. Suppose a path through $M(j_1)$ were also a path through $M(j_2)$. Then a path through $M(j_1 - j_2)$ would be a path through $M(0) = M$ and the code would not be comma-free. Thus all paths through all the matrices $M(j)$ are unique, and by Theorem 12.3 the synchronization delay is, at most, n symbols. The same argument assures that every matrix $M(j)$ must also correspond to a comma-free code, and since all possible paths are connected, every such code is path-invariant.

It should be emphasized that comma-free dictionaries, in general, are not path-invariant. The path in (12.3.3b), for example, does not correspond to any code word in the dictionary (12.3.1).

One construction which generates path-invariant comma-free codes is the following. Let c and d be two binary column vectors and let $\bar{c}$ be the complement of c. Then consider the matrix

$$M = \{c \overbrace{\bar{c}\,\bar{c} \ldots \bar{c}}^{[n/2] \text{ colums}} \quad \overbrace{d\,d \ldots d}^{[(n-1)/2] \text{ columns}}\} \tag{12.3.5}$$

Any matrix $M(j)$ obtained by cyclically permuting the columns of M j positions to the right must be such that either its first column is $\bar{c}$ (if $[(n-1)/2] + 1 \leq j \leq n-1$) or the $(j+1)$st column is c (if $1 \leq j \leq [n/2]$). (If n is even, both situations occur when $j = n/2$.) In either case, it is apparent that no path through $M(j)$ can also be a path through M since at least one of the columns of M is replaced by its complement in $M(j)$. Hence we have the following theorem.

Theorem 12.5. The coincidence matrix (12.3.5) corresponds to a path-invariant comma-free code for all choices of c and d.

If c contains l ones and d contains m ones then there are

$$l(r-l)^{[n/2]} m^{[(n-1)/2]} \tag{12.3.6}$$

different paths through M and, consequently, that many different words in the dictionary. To maximize the number of code words, m should evidently equal r (d should be a column of 1's only) and l should be chosen so that

$$(l-1)(r-l+1)^{[n/2]} \leq l(r-l)^{[n/2]} \tag{12.3.7a}$$

and

$$(l+1)(r-l-1)^{[n/2]} < l(r-l)^{[n/2]} \tag{12.3.7b}$$

If l were a continuous variable, the expression $l(r-l)^{[n/2]}$ would be maximized when $l = r/(1 + [n/2])$. Since $l > 0$, if the dictionary is to have any words, the optimum value of l is unity whenever $[n/2] + 1 \geq r$. Thus, whenever n is large enough to satisfy this inequality, the number of code words N_w is

$$N_w = (r-1)^{[n/2]} r^{[(n-1)/2]} = \frac{n}{r}\left(1 - \frac{1}{r}\right)^{[n/2]} \frac{r^n}{n} \tag{12.3.8}$$

The ratio of the number of code words generated by this construction to the number of code words in a maximal comma-free dictionary goes to zero asymptotically with n. The major advantage of path-invariant comma-free dictionaries rests in the decreased synchronization delay and in the resulting possibilities for simplified implementation. For example, when n is large enough for l to equal unity in Equation (12.3.7), synchronization can be determined by searching for an occurrence of the symbol which must occur in the first symbol position and then observing whether or not it occurs in any of the next $[n/2]$ positions. If it does not, this symbol must represent the beginning of a code word.

An unexpected advantage of these dictionaries over ordinary comma-free dictionaries is in the relative simplicity of encoding and decoding using these codes. To illustrate, consider again the situation when l is equal to unity. The code words defined by Equation (12.3.5) all consist of a specific symbol, followed by $[n/2]$ symbols each of which can be any alphabet symbol but this initial symbol, followed by $[(n-1)/2]$ symbols which are unconstrained. The information can therefore be mapped onto a sequence of interleaved $[(n-1)/2]$-tuples defined over an r-symbol alphabet, and $[n/2]$-tuples defined over an $(r-1)$-symbol alphabet. The encoding and decoding operations are correspondingly easy to implement.

12.4 Prefix Codes

The next constraint to be investigated is also designed to simplify the synchronization operation by limiting the number of symbols needed to determine the correct synchronization. This is accomplished by using the same m-tuple to prefix every n-symbol code word. Synchronization is indicated by the occurrence of this prefix at some point in the received sequence. To insure that this prefix always indicates the beginning of a word, it is necessary to prohibit it from occurring as part of a code word or code word-prefix overlap. That is, if the prefix consists of the m symbols $\gamma_1\gamma_2\gamma_3 \ldots \gamma_m$ and a code word is represented by $\gamma_1\gamma_2 \ldots \gamma_m\alpha_1\alpha_2\alpha_3 \ldots \alpha_n$, none of the m-tuples

$$\begin{aligned}
&\gamma_{i+1}\gamma_{i+2} \ldots \gamma_m\alpha_1 \ldots \alpha_i && 1 \leq i \leq m-1 \\
&\alpha_{i+1}\alpha_{i+2} \ldots\ldots\ldots\ldots \alpha_{i+m} && 0 \leq i \leq n-m \\
&\alpha_{i+1}\alpha_{i+2} \ldots \alpha_n\gamma_1 \ldots \gamma_{i+m-n} && n-m+1 \leq i \leq n-1 \qquad (12.4.1)
\end{aligned}$$

is to equal $\gamma_1\gamma_2 \ldots \gamma_m$. Dictionaries satisfying this constraint will be referred to as *prefix code* dictionaries.†

This condition could be slightly relaxed without impairing the synchronizability of the dictionary. If, in particular, none of the l m-tuples

$$\begin{aligned}
&\alpha_{i+1}\alpha_{i+2} \ldots\ldots \alpha_{i+m} && n-l \leq i \leq n-m \\
&\alpha_{i+1} \ldots \alpha_n\gamma_1 \ldots \gamma_{i+m-n} && n-m+1 \leq i \leq n-1 \qquad (12.4.2)
\end{aligned}$$

are restricted, the first occurrence of the m-tuple $\gamma_1\gamma_2 \ldots \gamma_m$ would precede a word prefix by no more than l symbols, whereas the first such m-tuple occurring after a true prefix would not be observed until at least $n-l+m$

†The distinction between prefix code dictionaries and dictionaries having the prefix property (Chapter Ten) should be noted. Since prefix codes are a class of (strict-sense) block codes, they clearly have the prefix property.

additional symbols have been received following the prefix. The actual prefix could therefore be identified if $l < n - l + m$ or $l < (n + m)/2$. We shall subsequently comment on the *modified prefix code* dictionaries obtainable with this relaxed condition.

Note that both prefix and modified prefix code dictionaries are comma-free dictionaries of $(n + m)$-symbol words. This follows because every code word begins with the same m-tuple and no overlap of code words can have this property. (Or in the case of the modified prefix codes, every code word must begin with the same m-tuple which cannot occur again in a code word until at least $(m + n)/2$ symbols have been received. Clearly, no overlap of code words can have this property.) Since both prefix codes and modified prefix codes are comma-free with additional constraints (that to synchronize it is necessary to look at only m symbols at a time) the dictionaries should be expected to contain somewhat fewer words than the $N(n + m, r)$ of Equation (12.2.3). This is indeed the case, although asymptotically as the word length increases, the number of prefix code words is of the same order of magnitude as $N(n + m, r)$. This is in contrast to the path-invariant comma-free code dictionaries of the previous section. Unfortunately, prefix codes, while easier to synchronize than ordinary comma-free codes, do not share the encoding and decoding advantages that were demonstrated by path-invariant codes.

It is possible to derive a generating function for the number of words in a maximal prefix code dictionary. This will be facilitated by the following definitions: (1) An m-symbol prefix will be called *repetitive with period* ν $(0 < \nu \leq m)$, if $\gamma_1\gamma_2 \ldots \gamma_\nu\gamma_1\gamma_2 \ldots \gamma_{m-\nu} = \gamma_1\gamma_2 \ldots \gamma_m$. (Notice that all prefixes of length m are repetitive with period m. The distinction between repetitive sequences and periodic sequences should be emphasized. All periodic sequences are repetitive, but not conversely. The non-periodic sequence 101, for example, is repetitive with period two.) (2) An i-tuple $\alpha_1\alpha_2 \ldots \alpha_i$ will be said to be *properly terminated* with respect to some prefix $\gamma_1\gamma_2 \ldots \gamma_m$, if the sequence $\gamma_1\gamma_2\gamma_3 \ldots \gamma_m\alpha_1\alpha_2 \ldots \alpha_i$ does not have as a suffix the sequence $\gamma_1\gamma_2 \ldots \gamma_\nu$ with ν some repetitive period of that prefix.

Theorem 12.6. Let $\gamma_1\gamma_2 \ldots \gamma_m$, the prefix of the $(n + m)$-symbol r-ary words of a maximal prefix dictionary $D_m(n)$, have repetitive periods $\nu_1, \nu_2, \ldots, \nu_l, \nu_{l+1} \triangleq m$. Define

$$\Delta_i \triangleq \begin{cases} 1 & i = 0 \text{ or } i = \nu_j \quad \text{for some } j,\ 1 \leq j \leq l + 1 \\ 0 & \text{all other } i \leq m \end{cases}$$

Let $d_i = r\Delta_{i-1} - \Delta_i$ and let V_i denote the number of i-tuples properly terminated with respect to $\gamma_1\gamma_2 \ldots \gamma_m$. Then the number $N_m(n)$ of words in the dictionary $D_m(n)$ for all $n \geq 1$ is just the coefficient of x^n in the generating function

$$G(x) = \frac{(1 - rx)\sum_{i=1}^{m-1} V_i x^i + V_m x^m}{1 - \sum_{i=1}^{m} d_i x^i} \tag{12.4.3}$$

Proof. Consider the $(2m + n - 1)$-tuple

$$\gamma_1 \gamma_2 \ldots \gamma_m \alpha_1 \alpha_2 \ldots \alpha_n \gamma_1 \ldots \gamma_{m-1} \tag{12.4.4}$$

If $\alpha_1 \alpha_2 \ldots \alpha_n$ is properly terminated, there must be some greatest integer k, $0 \leq k < n$ such that either

$$\gamma_{k+1} \gamma_{k+2} \ldots \gamma_m \alpha_1 \alpha_2 \ldots \alpha_k = \gamma_1 \gamma_2 \ldots \gamma_m \qquad k < m$$

or

$$\alpha_{k-m+1} \alpha_{k-m+2} \ldots \alpha_k = \gamma_1 \gamma_2 \ldots \gamma_m \qquad k \geq m$$

Note that if $k < m$, k must be a repetitive period of the prefix $\gamma_1 \gamma_2 \cdots \gamma_m$, i.e., $\Delta_k = 1$. The $(n - k + m)$-tuple beginning with γ_{k+1} (if $k < m$) or with α_{k-m+1} (if $k \geq m$) and ending with α_n must then be a word in $D_m(n - k)$. Conversely, for each word in $D_m(n - k)$, $k < m$, there is a unique $(2m + n - 1)$-tuple of the form (12.4.4), with $\alpha_1 \alpha_2 \cdots \alpha_n$ a properly terminated n-tuple if, and only if, $\Delta_k = 1$. If $k \geq m$, there are exactly r^{k-m} such $(2m + n - 1)$-tuples for each word in $D_m(n - k)$ since the symbols $\alpha_1 \alpha_2 \cdots \alpha_{k-m}$ are then arbitrary. Thus

$$\sum_{k=0}^{n-1} \Delta_k N_m(n - k) = V_n \tag{12.4.5}$$

where

$$\Delta_k \triangleq r^{k-m} \qquad k \geq m$$

Further, if $n > m$ and if $\alpha_{n-m+1} \alpha_{m-n+2} \ldots \alpha_n$ is properly terminated, so also is $\alpha_1 \alpha_2 \ldots \alpha_n$, and conversely. Thus $V_n = r^{n-m} V_m$ for $n > m$.

Multiplying both sides of Equation (12.4.5) by x^n and summing over n, we obtain

$$\begin{aligned} \sum_{n=1}^{\infty} \sum_{k=0}^{n-1} \Delta_k N_m(n - k) x^n &= \sum_{k=0}^{\infty} \Delta_k x^k \sum_{n=k+1}^{\infty} N_m(n - k) x^{n-k} \\ &= \sum_{k=0}^{\infty} \Delta_k x^k \sum_{l=1}^{\infty} N_m(l) x^l = \sum_{n=1}^{\infty} V_n x^n \end{aligned} \tag{12.4.6}$$

Thus,

$$G(x) \triangleq \sum_{n=1}^{\infty} N_m(n) x^n = \frac{\sum_{n=1}^{\infty} V_n x^n}{\sum_{n=0}^{\infty} \Delta_n x^n} = \frac{\sum_{n=1}^{m-1} V_n x^n + V_m x^m/(1 - rx)}{\sum_{n=0}^{m-1} \Delta_n x^n + x^m/(1 - rx)}$$

and multiplying both the numerator and denominator of the expression on the right by $1 - rx$ yields the desired result.

Corollary. The number $N_m(n)$ of words in the maximal prefix dictionary $D_m(n)$ is given by the recursion

$$N_m(n) = \sum_{i=1}^{m} d_i N_m(n - i) \tag{12.4.7}$$

for all $n > m$.

Proof. This result follows immediately upon multiplying both sides of Equation (12.4.3) by $1 - \sum_{i=1}^{m} d_i x^i$ and equating the coefficients of x^n.

As an example, suppose the prefix is 1010 and $r = 2$. Since $\gamma_1\gamma_2 = \gamma_3\gamma_4$, the prefix is repetitive with period two. Since this is the only repetitive period, $\Delta_0 = 1$, $\Delta_1 = 0$, $\Delta_2 = 1$, $\Delta_3 = 0$, $\Delta_4 = 1$ and

$$N_4(n) = 2N_4(n-1) - N_4(n-2) + 2N_4(n-3) - N_4(n-4)$$

Further, as is easily verified, $V_1 = 2$, $V_2 = 3$, $V_3 = 6$, $V_4 = 12$, and, hence, from Equation (12.4.5) or from Equation (12.4.3), $N_4(1) = 2$, $N_4(2) = 3$, $N_4(3) = 4$, and $N_4(4) = 9$.

Although Equation (12.4.7) is useful for determining the number $N_m(n)$ of words in maximal prefix dictionaries for most word lengths $n + m$ of practical interest, the generating function demonstrates its greatest utility in indicating the asymptotic behavior of $N_m(n)$. To derive an asymptotic expression for $N_m(n)$ in fact, it is only necessary to find the root (or roots) of minimum modulus of the denominator $D(x)$ of $G(x)$. This statement is easily verified by making a partial fraction expansion of $G(x)$ in terms of the roots x_i of its denominator. A typical term in such an expansion is then

$$\frac{A_i}{x - x_i} = -\frac{A_i}{x_i\left(1 - \dfrac{x}{x_i}\right)} = -\frac{A_i}{x_i}\left(1 + \frac{x}{x_i} + \frac{x^2}{x_i^2} + \cdots\right) \tag{12.4.8}$$

and the coefficient of x^n in the polynomial $G(x)$ is of the form $-\sum_i (A_i/x_i^{n+1})$. (The same conclusion holds even if some of the roots are multiple.) Asymptotically, therefore, as $n \to \infty$, this coefficient will be completely dominated by those terms corresponding to the root or roots of minimum modulus.

As an example of the use of the generating function for determining the asymptotic behavior of $N_m(n)$, consider $G(x)$ when the prefix has no repetitive periods (i.e., when $\Delta_i = 0$, $i \neq 0, m$; the prefixes $aa \ldots ab$, $aa \ldots abb$, $aa \ldots abbb, \ldots, abb \ldots b$ all have this property). Clearly, all n-tuples are properly terminated with regard to a non-repetitive prefix for $1 \leq n < m$, and all but the prefix m-tuple itself when $n = m$. Thus, $V_n = r^n$, $1 \leq n < m$, $V_m = r^m - 1$, and, from Equation (12.4.3),

$$G(x) = \frac{1}{1 - rx + x^m} - 1 \tag{12.4.9}$$

The locations of all the roots of the denominator $D(x)$ of $G(x)$ can readily be established by using the root-locus method.† Consider the equation $K/x(x^{m-1} - r) = -1$. As $K \to 0$, the points $x = x_i$ for which this equation has a solution approach the roots of the equation $x(x^{m-1} - r) = 0$, i.e., the point $x = 0$ and the $m - 1$ equally spaced points $x = r^{1/(m-1)}e^{2\pi i[k+(1/2)]/m}$, $k = 0, 1, \ldots, m - 1$. The locus of roots shown in Figure 12.1 for $m = 10$ is typical of those for all m. As K is increased, the root located at $x = 0$ and

† Those unfamiliar with the root-locus method for locating the roots of a polynomial can consult Reference 12.3 for an alternative argument leading to the same conclusion.

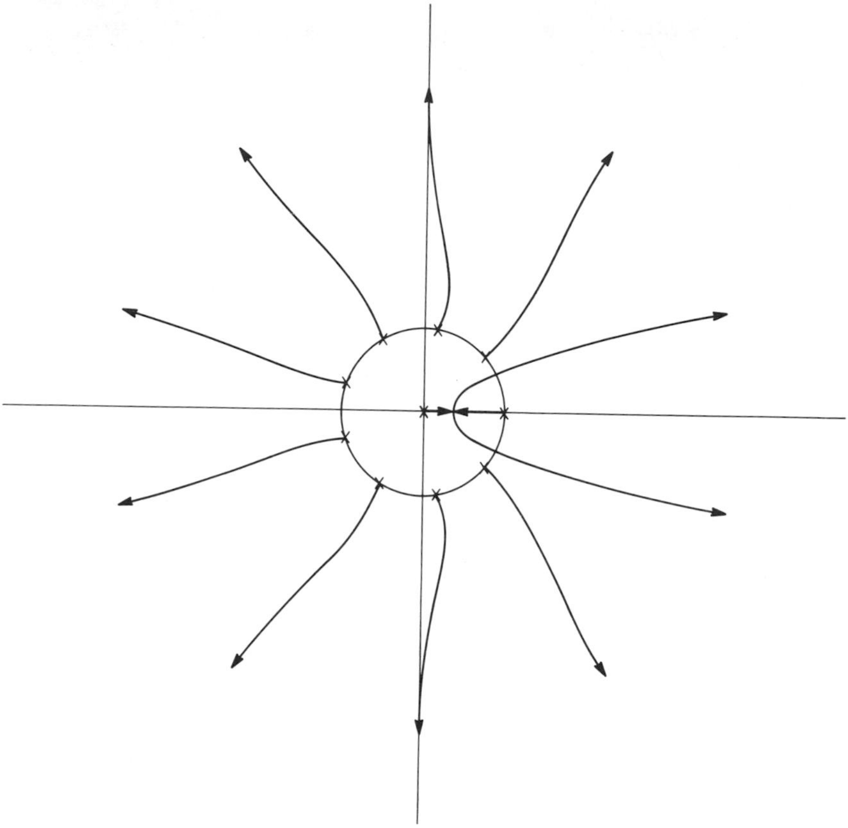

Figure 12.1

that at $x = r^{1/(m-1)}$ approach each other along the real axis, eventually meeting and splitting off into complex conjugate pairs. The moduli of the other $m - 2$ roots monotonically increase with K, and hence are at least $r^{1/(m-1)}$ for all K.

The situation of interest here occurs when $K = 1$. On the basis of the preceding paragraph, we conclude that $D(x)$ has either two, or no, positive real roots. But since $D(1/r) > 0$ and $D\{m/(m-1)r\} = -[1/(m-1)] \times[1 - m^m/(m-1)^{m-1}r^m] < 0$ for $m \geq 3$, $D(x)$ has a real root x_0 where $(1/r) < x_0 < [m/(m-1)](1/r)$. The other positive real root must occur at some point $x_1 \geq 1$. This is evidenced by expressing $D(x)$ in the form $(x - x_0)(x - x_1)f(x)$, m even, or $(x - x_0)(x - x_1)(x - x_2)f(x)$, m odd, where $f(x)$ is a polynomial having all the non-real roots of $D(x)$. [If m is odd, there is, in addition to the two positive real roots, a negative real root x_2 of $D(x)$.] Since the roots of $f(x)$ must occur in complex conjugate pairs, $f(1) > 0$, and since $D(1) \leq 0$ and $x_0 < 1$, x_1 must be greater than, or equal to, one.

Thus we have established that $D(x)$ has only one root (x_0) with modulus less than one. As a consequence, not only does $N_m(n)$ approach A_0/x_0^{n+1} as $n \to \infty$ [see Equation (12.4.8)] but the neglected terms contributing to $N_m(n)$ are asymptotically zero. (The contribution due to the possible root at $x = 1$ might be an exception to the statement. In any event, this contribution is negligible.) Moreover, since x_0 is a simple root, $A_0 = \lim_{x \to x_0} (x - x_0)G(x) = N(x_0)/D'(x_0)$ where $N(x)$ is the numerator of $G(x)$ and $D'(x)$ the derivative of its denominator.

In order to complete the derivation of an asymptotic expression for $N_m(n)$, however, we must refine our estimate of the location of the root x_0. To accomplish this, we define $y_0 = 1 - (1/rx_0)$. Then $D(x_0) = 0$ if, and only if,

$$y_0[1 - (m-1)y_0] < y_0(1 - y_0)^{m-1} = r^{-m} < y_0$$

From this it quickly follows that $y_0 = r^{-m} + O(mr^{-2m})$ and

$$x_0^{-1} = r[1 - r^{-m} + O(mr^{-2m})] \tag{12.4.10}$$

Thus, when the prefix has no repetitive periods less than m,

$$N_m(n) \sim -\frac{N(x_0)}{D'(x_0)}\frac{1}{x_0^{n+1}} = \frac{1}{r - mx_0^{m-1}}\frac{1}{x_0^{n+1}} \tag{12.4.11}$$

where x_0 is as defined in Equation (12.4.10).

Theorem 12.7. Asymptotically, for large word lengths N, a maximal prefix dictionary will contain at least $[(r-1)r^{1/(r-1)} \log_r e]^{-1}(r^N/N)$ words.

Proof. Let $n + m = N$ and $m = \log_r (N/c)$ where c is chosen so as to make m an integer. Then, using Equations (12.4.10) and (12.4.11), and treating c as a constant, we find

$$N_m(N - m) \sim \left(\frac{1}{x_0}\right)^n \sim \left(1 - \frac{c}{N}\right)^N r^{N-m} \sim ce^{-c}\left(\frac{r^N}{N}\right) \tag{12.4.12}$$

To maximize this expression subject to the condition that $m = \log_r (N/c)$ be an integer, observe that ce^{-c} monotonically increases with c for $c < 1$ and monotonically decreases with c for $c > 1$. Further, some value of c in the range $c_0 \le c \le rc_0$ must yield an integer value of m for any $c_0 > 0$. By setting $c_0 e^{-c_0} = rc_0 e^{-rc_0}$ or $c_0 = (\log_e r)/(r-1)$ we can satisfy the inequality

$$ce^{-c} \geq [(r-1)r^{1/(r-1)} \log_r e]^{-1}$$

for some c such that m is an integer. Then, asymptotically

$$\frac{1}{e}\frac{r^N}{N} \geq N_m(N - m) \geq [(r-1)r^{1/(r-1)} \log_r e]^{-1}\frac{r^N}{N} \tag{12.4.13}$$

the upper bound following because $ce^{-c} \le e^{-1}$.

Now consider the prefix $aa \ldots a$ (i.e., the prefix having all repetitive periods 1, 2, . . . , m). An n-tuple $\alpha_1 \alpha_2 \dots \alpha_n$, $n \leq m$, is properly terminated with respect to this prefix if, and only if, $\alpha_n \neq a$. Accordingly $V_n = (r-1)r^{n-1}$, $1 \leq n \leq m$, and, from Equation (12.4.3),

$$G(x) = \frac{(r-1)x(1-x)}{1 - rx + (r-1)x^{m+1}} \tag{12.4.14}$$

Since the denominator $D(x)$ of this generating function is of the same form as the one previously encountered, the same arguments concerning the location of its roots can be used virtually intact. In particular, $D(x)$ contains only one root

$$x_0' = \frac{1}{r}\left(1 - \frac{r-1}{r}\frac{1}{r^m} + O\left(\frac{m}{r^{2m}}\right)\right)^{-1} \tag{12.4.15}$$

having a modulus less than unity. Since this root is smaller than the root x_0 found in the previous example, the prefix *aa . . . a* promises a more rapid rate of growth of $N_m(n)$ than the non-repetitive prefix *aa . . . ab*.

Indeed, these two prefixes represent the extremes in the rate of growth of $N_m(n)$. That is, the root of minimum modulus x_0'' of the denominator $D(x)$ of the generating polynomial $G(x)$ falls in the interval $x_0' \leq x_0'' \leq x_0$, with x_0 and x_0' as given in Equations (12.4.10) and (12.4.15), respectively, for all prefixes of fixed length m. To verify this statement, we first note that the denominator

$$D(x) = 1 - \sum_{i=1}^{m} d_i x^i = (1 - rx)\sum_{i=0}^{m-1}{}' \Delta_i x^i + x^m \tag{12.4.16}$$

is positive for $x = 1/r$ and negative for $x = [m/(m-1)][1/r]$ (since $\Delta_0 = 1$). Thus the polynomial $f(x) \triangleq D(x)/(1 - rx)$ must have a root x_0'' somewhere in the interval $1/r < x_0'' < [m/(m-1)][1/r]$. But $f(x)$ is monotonically increasing in both x and the coefficients Δ_i for all x in this interval. Increasing any Δ_i therefore requires x_0'' to be decreased. Accordingly, x_0'' attains its minimum value (x_0') when $\Delta_1 = \Delta_2 = \cdots = \Delta_{m-1} = 1$ and its maximum value (x_0) when $\Delta_1 = \Delta_2 = \dots = \Delta_{m-1} = 0$.

Ironically, even though the prefix *aa . . . a* insures the most rapid growth of the dictionary size for a given value of m, it is always possible, at least when $r = 2$, to find a larger dictionary having the same word length ($n + m$) using the prefix *aa . . . ab* (or any other non-repetitive prefix), the one which promised the slowest rate of growth. The answer to this paradox is that the value of m in the two cases will not be the same. In fact, the optimum length of the prefix in the latter case will always be one greater than that in the former. The reason for this is apparent. The m-symbol prefix 111 . . . 1 must always be followed by 0 in any code word to prevent the prefix from also occurring as the m-tuple beginning with the second word symbol. Every word, in fact, must begin with the $(m + 1)$-symbol prefix 11 . . . 10. The same dictionary is synchronizable by searching for this prefix rather than the original. But if this prefix is to be used, some additional words, in particular those which end with a string of one or more *ones*, can now be included in the dictionary. Thus, for the same N, the dictionary size can be increased by going from the m-symbol prefix 111 . . . 1 to the $(m + 1)$-symbol prefix 111 . . . 10.

Before concluding this section, we shall comment briefly on the effect on the dictionary size when the prefix code constraint is weakened slightly to yield the modified prefix codes mentioned earlier. Modified prefix codes, it will be recalled, are codes whose N-symbol words all have the same m-symbol prefix, and are so constrained to prevent this prefix from occurring as any of the m-tuples starting with the ith symbol for any i in the range $2 \leq i \leq [(N/2) + 1]$. Once the true prefix is observed, no other prefix can be seen until at least $[(N/2) + 1]$ symbols have been received. If the prefix occurs in any other position, a second prefix will always be received before $[(N/2) + 1]$ symbols have passed. Thus, the determination of synchronization is but slightly more complicated than it was before, while the number of code words can be increased.

One method for constructing such a dictionary is to adjoin as suffixes to each word in an ordinary, length $m + [(n + m)/2]$, prefix-code dictionary all possible $[(n - m + 1)/2]$-tuples. Clearly, none of the m-tuples beginning with the ith symbol of any word in the resulting dictionary will be the restricted prefix for any i in the range $2 \leq i \leq m + [(n + m)/2] - (m - 1) = [(N/2) + 1]$, as required. This construction is over-constrained for some prefixes since it may place unnecessary restrictions on those m-tuples which begin with the $[(N/2) + 2]$nd through the $[(N/2) + m]$th symbols. Nevertheless such a construction yields, from Equation (12.4.12),

$$N_m\left(\left[\frac{N}{2}\right]\right) r^{\{[(N+1)/2]-m\}} \sim 2ce^{-c}\left(\frac{r^N}{N}\right) \tag{12.4.17}$$

code words. The effect of the modification is, asymptotically, to double the size of the dictionary. Interestingly, the upper bound here, obtained by equating c to 1, is just the lower bound [Equation (12.2.7)] attained by a non-prefix construction in Section 12.2.

It has been shown in this section that asymptotically there are nearly as many words in a prefix code dictionary as there are in an ordinary comma-free dictionary having the same word length [cf. Equations (12.2.3) and (12.4.13)]. As mentioned earlier, the primary advantage of prefix codes lies in the relative ease with which they can be synchronized. If the dictionary is simply comma-free, it is necessary to consider N-symbol sequences to determine if any of the code words have been received and to do this for all N possible initial positions. With prefix codes, in contrast, it is only necessary to observe successive m-tuples ($m \sim \log_r N$) and to identify the prefix when it occurs to obtain synchronization. Unfortunately, the encoding and decoding procedures may be as difficult using prefix codes as when comma-free codes are used. In the next section we shall investigate still another class of synchronizable block codes, a class which to a large extent avoids this problem.

12.5 Comma Codes

Another method for establishing word synchronization is to transmit periodically a *comma* or special m-symbol sequence to identify the correct synchronization position. A comma will be called *singular* with respect to the dictionary D if it can occur in a sequence of words from D only where it is explicitly inserted; under the same conditions, D will be referred to as a *comma code* dictionary. Thus, if the comma is to be singular, all m-tuples from a code word-code word, code word-comma, or comma-code word overlap must be other than the comma. (A modified comma code might allow code word-comma overlaps to be a comma but not comma-code word overlaps, or vice versa.)

A singular comma can be transmitted whenever it is felt synchronization information may be necessary. Once the comma is identified at the receiver, synchronization is established. There is no need to transmit the comma again until something causes a loss of synchronism, since the word length here, as in the other sections of this chapter, is constrained to be uniform. However, the comma would typically be transmitted periodically at moderately short intervals, after every k data words, say, to insure that if synchronization is lost not too much time elapses before it can be re-established.

If $k = 1$, the comma codes are just the prefix codes of the previous section. If $k > 1$, a new situation arises. It is now necessary to restrict the code words further to prevent an m-tuple formed from an overlap of any two of them from being the comma; the comma itself does not separate every pair of n-tuples as in the case of prefix codes.

These codes will be less efficient (more redundant) than the prefix codes. For if $N = nk + m$, the comma code can be compared to the prefix code with the same m-symbol prefix and with word length N. Obviously, no N-tuple not in the prefix code can occur as a sequence of words of the comma code. Some N-tuples of the prefix code cannot occur as any sequence of comma code words, however. None of the latter sequences, for example, can end with an i-tuple which is the beginning of the comma if any word also exists which has as its prefix the last $m - i$ symbols of the comma.

On the other hand, these codes may be easier to implement. Moreover, the code constraints are independent of k (for $k \geq (m + n - 1)/n$), so the frequency with which the comma is transmitted can be changed to suit the demands of the system without the necessity of significantly altering the encoding and decoding equipment.

The structure of the code words will of course depend upon the values of m and n. Although it is possible to obtain results for all relationships between m and n we shall limit the discussion to three of the more interesting cases.

Theorem 12.8. The maximal r-ary comma code dictionary, when the comma is of length m and the code words are of length n, contains $M_m(n)$ words where:

$$(1)\ M_1(n) = (r-1)^n$$

$$(2)\ M_n(n) = r^n - r^{[n/2]} - \frac{r^{[(n+1)/2]}-1}{r-1} \qquad n > 1$$

$$(3)\ M_{2n}(n) = r^n - 1$$

In the latter two cases, the optimum comma is an m-tuple of the form $aaa \ldots ab$ (or $ba \ldots a$), consisting of $m - 1$ repetitions of one r-ary symbol a followed by some other symbol $b \neq a$ (or its reverse).

Proof. *Case (1)* $m = 1$ (one symbol only is used as a comma). Then the $r - 1$ remaining symbols only can be used in the construction of the code words; there are $(r-1)^n$ such words.

Case (2) $m = n > 1$. Let $\gamma_1\gamma_2 \ldots \gamma_n$ be the comma. Then, for all i, $1 \leq i \leq n-1$, either no word in the dictionary can end with the i-tuple $s_i \triangleq \gamma_1\gamma_2 \ldots \gamma_i$, or else no word can begin with the $(n-i)$-tuple $p_{n-i} \triangleq \gamma_{i+1}\gamma_{i+2} \ldots \gamma_n$. Prohibiting s_i as a suffix eliminates r^{n-i} words from the dictionary while prohibiting p_i as a prefix eliminates r^i words. Suppose s_l is the shortest prohibited suffix. Then the $(n - l + 1)$-tuple p_{n-l+1} must be eliminated as a code word prefix and at least $r^{n-l} + r^{l-1} - 1$ n-tuples plus the comma are excluded from the dictionary. (The "-1" term is needed because one n-tuple could have both the suffix s_l and prefix p_{n-l+1}.) But

$$r^{n-l} + r^{l-1} > r^{[n/2]} + \frac{r^{[(n+1)/2]}-1}{r-1}$$

for all $l < [n/2]$. Thus, the shortest prohibited suffix, and by analogy the shortest prohibited prefix must contain at least $[n/2]$ symbols if the bound $M_n(n)$ is to be attainable.

If n is even, either a suffix or a prefix of length $n/2$ must be prohibited. Suppose $s_{n/2}$ is the shortest suffix and $p_{(n/2)+1}$ the shortest prefix prohibited (the argument is identical if the prefix $p_{n/2}$ and suffix $s_{(n/2)+1}$ are prohibited). Then no code word can have either of the prefixes $p_{(n/2)+1}$ or $p_{(n/2)+2}$, or either of the suffixes $s_{n/2}$ or $s_{(n/2)+1}$. Unless $s_{n/2}$ is the suffix of $s_{(n/2)+1}$, or $p_{(n/2)+1}$, the prefix of $p_{(n/2)+2}$, all but at most four pairs of the eliminated words will be distinct. But, as is easily verified, when $n \geq 8$ the $r^{n/2} + 2r^{(n/2)-1} + r^{(n/2)-2} - 3$ n-tuples thereby eliminated (including the comma itself) by themselves leave fewer than $M_n(n)$ words available for the dictionary. Thus, either $s_{n/2}$ is the suffix of $s_{(n/2)+1}$, $p_{(n/2)+1}$ is the prefix of $p_{(n/2)+2}$, or both, if the stated bound is to be attained. But both of these situations can occur simultaneously only if the comma consists of a single repeated symbol: $\gamma_1\gamma_2 \ldots \gamma_n = aa \ldots a$. The code word-comma overlap constraint, for example, would then eliminate from the dictionary all words having the one-symbol suffix a. This alone would preclude the attainment of the bound $M_n(n)$. Accordingly, either $s_{n/2}$ is the suffix of $s_{(n/2)+1}$ and the comma has the form $aa \ldots a \gamma_{(n/2)+2} \ldots \gamma_n$, or $p_{(n/2)+1}$ is the prefix of $p_{(n/2)+2}$ and the comma is of the form $\gamma_1 \gamma_2 \ldots \gamma_{(n/2)-2}$ $a \ldots a$. In the first situation, all the words having either the suffix $s_{n/2}$ or any of the prefixes p_{n-i}, $i = 0, 1, \ldots, (n/2) - 1$, will be distinct. In the second case, all words having either the prefix $p_{(n/2)+1}$ or any of the suffixes s_i, $i = n/2, (n/2) + 1, \ldots, n$, will be distinct. In either event, the number of remaining candidates for the dictionary is easily seen to be bounded by $M_n(n)$. The special cases $n = 2, 4$, and 6 are easily treated separately and shown to be subject to the same bound.

When n is odd essentially the same argument can be used to demonstrate that any comma giving rise to a maximal dictionary either begins or ends with $(n + 1)/2$ repetitions of the same symbol. (The details are left to the reader.) The stated bound then immediately follows.

Now consider the dictionary $D_n(n)$ containing all n-tuples except those beginning with any of the prefixes $p_{[n/2]+1}, p_{[n/2]+2}, \ldots, p_n$ where p_i is of the form $aa \ldots ab$ for all i, and except those ending with the suffix $s_{[(n+1)/2]} = a \ldots a$. Clearly, such a dictionary does contain $M_n(n)$ words. Letting the comma be $aa \ldots ab$, we observe that no code word-comma, comma-code word, or code word-code word overlap can produce the comma; accordingly, $D_n(n)$ is a comma code dictionary containing $M_n(n)$ words. (The same argument clearly holds for the reverse comma $baa \ldots a$.)

Case (3) $m = 2n$. If the comma is the $2n$-tuple $aaa \ldots ab$, only one n-tuple (the n-tuple $aa \ldots a$) need be eliminated from the list of code words. No comma-word or word-comma overlap can produce the comma, and if any overlap involving three words were to contain the comma, the middle word would have to be the eliminated n-tuple $aa \ldots a$. Thus, $r^n - 1$ n-tuples are permitted in the dictionary. Regardless of the value of m relative to n, at least one n-tuple must obviously be eliminated from the dictionary if the comma is to be unique. Consequently, the comma under discussion is optimum. (Again, the reverse of this comma is equally effective.)

The comma code with $m = 2n$ is particularly attractive from the point of view of implementation. It is only necessary to insure that the data never generate the code word $aa \ldots a$. This is generally quite easily accomplished in practice. Then the data can be transmitted directly with the comma inserted periodically. Synchronization is obtained whenever the n-tuple $aa \ldots a$ followed by the n-tuple $aa \ldots ab$ is observed.

Other relationships between n and m may also be of theoretical, if not practical, interest. When $[m/2] < n < m$, the argument used in the $m = n$ situation can be extended virtually intact to show that the number of words in the maximal comma dictionary is

$$M_m(n) = r^n - r^{n-[(m+1)/2]} - \frac{r^{n-[m/2]} - 1}{r - 1} \qquad \left[\frac{m}{2}\right] < n \leq m \qquad (12.5.1)$$

Again, the maximal dictionary is obtained when the comma $aa \ldots b$ (or $ab \ldots bb$) is used, and when all prefixes of the form $aa \ldots ab$ and of length $[m/2] + 1$ or greater and the suffix $aa \ldots a$ of length $[(m + 1)/2]$ are prohibited. Similarly, when $m \geq 2n$, the comma $aa \ldots ab$ can be used, and, as in the $m = 2n$ case, only the word $aa \ldots a$ need be eliminated from the dictionary. Thus,

$$M_m(n) = r^n - 1 \qquad n \leq \left[\frac{m}{2}\right] \tag{12.5.2}$$

Things become more complicated, however, when $n > m$, for now the comma must be prohibited from occurring within a code word as well as in the overlap of two or more code words. The situation is analogous to that encountered in the discussion of prefix codes. The maximum number of words in a dictionary is most readily determined by recursion. The interested reader is referred to Reference 12.4 for details.

We conclude this section with an investigation of the redundancy inherent in comma codes when the constraint lengths become large. To facilitate comparison with other synchronizable block codes, we denote by $N = m + kn$ the overall constraint length when an m-symbol comma is to be transmitted after every k n-symbol code words. The set of N-tuples obtained by forming every possible concatenation of k words from the comma-code dictionary and preceding them by the comma is obviously a comma-free dictionary of N-symbol "words." We have already argued that this set will be more redundant than ordinary comma-free or prefix codes of the same length due to the added constraint imposed on the comma codes. We now determine qualitatively just how much redundancy has been added vis-à-vis the other codes when N is large. To do so, we fix N and choose n so as to maximize the number $L_m(N)$ of N-tuples which can be formed using the words of an n-symbol word comma-code dictionary under the three constraints of greatest interest: $m = 1$, $m = n$, and $m = 2n$.

When $m = 1$, $L_1(N) = (r - 1)^{N-1}$ regardless of the value of n. For purposes of comparison with the comma-free bound this can be written

$$L_1(N) = \frac{N\left(1 - \dfrac{1}{r}\right)^N}{r - 1} \frac{r^N}{N} \tag{12.5.3}$$

When $m = n$ and n is even, we find

$$\begin{aligned} L_n(N) = [M_n(n)]^k &= \frac{r^N}{r^n}\left[1 - \frac{r}{r-1}\frac{an}{N}\left(1 - \frac{an}{rN}\right)\right]^{(N/n)-1} \\ &\sim \frac{r^N}{r^n}\exp\left(-\frac{r}{r-1}a\right) \end{aligned} \tag{12.5.4}$$

where $a = a(n) = N/nr^{n/2}$, and N/n is assumed large. It is easily verified that this last expression is maximized as a function of n (with N constant) when

$$1 = \frac{r}{r-1}a\left(\frac{\log_r e}{n} + \frac{1}{2}\right) \sim \frac{r}{r-1}\frac{a}{2}$$

Thus, when N (and hence n) is large, $L_n(N)$ is maximized for

$$N = anr^{n/2} \approx 2(r - 1)nr^{[(n/2)-1]} \tag{12.5.5}$$

Since for any $\epsilon > 0$, $r^{n/2} < nr^{n/2} < r^{(n/2)(1+\epsilon)}$ for sufficiently large n, it follows that, asymptotically, when N satisfies Equation (12.5.5)

$$r^n \sim \left[\frac{rN}{2(r-1)}\right]^2$$

and from Equation (12.5.4)

$$L_n(N) \sim \left(\frac{r-1}{r}\right)^2 \frac{4}{e^2 N} \frac{r^N}{N} \tag{12.5.6}$$

When n is odd, a similar argument establishes that

$$L_n(N) \sim r\left(\frac{r-1}{2r-1}\right)^2 \frac{4}{e^2 N} \frac{r^N}{N} \tag{12.5.7}$$

with $N \approx [2(r-1)/(2r-1)]nr^{(n+1)/2}$. Finally, when $m = 2n$,

$$L_{2n}(N) = (r^n - 1)^{(N/n)-2} = \frac{r^N}{r^{2n}}\left(1 - \frac{1}{r^n}\right)^{(N/n)-2} \sim \frac{r^N}{r^{2n}} e^{-a} \tag{12.5.8}$$

with $a = N/nr^n$. This last expression is maximum with respect to n when $a \approx 2$. Proceeding as before, therefore, we conclude

$$L_{2n} \sim \frac{4}{e^2 N} \frac{r^N}{N} \tag{12.5.9}$$

These asymptotic expressions, of course, are valid only for those N having the specified relationships to n, not for arbitrary N. The method used in Section 12.4 to underbound $N_m(n)$ [cf. Equations (12.4.12) and (12.4.13)] could be applied here as well, and the asymptotic value of $L_m(N)$ thereby bounded for all N. The only effect would be to decrease, slightly, the constant multiplying the term r^N/N^2 in these expressions. In any event, since here N was rather artificially introduced solely to establish a basis for comparison with other comma-free codes, a bound for arbitrary N is of dubious interest.

12.6 PSK-Synchronizable Block Codes

When an r-ary PSK signal is demodulated using a carrier reference derived directly from the modulated signal (e.g., using the methods of Chapter Eight, or using a differentially coherent approach) the detector output will exhibit an r-fold ambiguity. The output symbols can be expressed in the form $\ldots \alpha_{i-1} + \beta,\ \alpha_i + \beta,\ \alpha_{i+1} + \beta \ldots$ where the α's are known, but where, due to the lack of an absolute reference, β can be any one of the r symbols of the alphabet.

One technique for resolving this ambiguity is to encode the information differentially; that is, to let the difference, modulo-r, between successive channel symbols, rather than the symbols themselves, represent the information. This approach was presumed, for example, in the discussion of DPSK in Section 4.6. Interestingly, only $N + 1$ differentially encoded symbols are needed to convey N symbols of information. In effect, therefore, the minimum possible number of r-ary symbols (one) is used to resolve the r-fold ambiguity.

In any event, this lack of an absolute reference presents an additional problem so far as word synchronization is concerned. By way of illustration, consider the binary dictionary $D = \{00001, 01001, 01101, 10001, 11001, 11101\}$. This is a member of the class of dictionaries shown to be comma-free in the conventional sense in Section 12.2. Now, however, because of the reference uncertainty, the received words could be either from D or from $\bar{D}$, its complement dictionary. In the absence of prior synchronization, therefore, the sequence . . . 00001000010000100 . . . , for example, could correspond to either of the two word sequences . . . $w_1 w_1 w_1$ . . . or . . . $\bar{w}_6 \bar{w}_6 \bar{w}_6$ . . . (with $w_1 = 00001$ and $w_6 = 11101$). The potential synchronization delay is consequently unbounded. Evidently dictionaries which are synchronizable under ordinary circumstances are not necessarily *PSK-synchronizable* (i.e., synchronizable when used in conjunction with a PSK communication system lacking an absolute reference).

The preceding example demonstrates that, at the very least, we shall have to impose the following constraint on code dictionaries if they are to be PSK-synchronizable. If the dictionary D_n of n-symbol r-ary words contains the code word

$$w = \omega_1 \omega_2 \dots \omega_n \tag{12.6.1a}$$

then it cannot also contain any word of the form

$$w' = (\omega_{i+1} + \beta)(\omega_{i+2} + \beta) \dots (\omega_{i+n} + \beta) \tag{12.6.1b}$$

for any $i = 1, 2, \dots, n - 1$, and for any code symbol β ($\beta = 0, 1, 2, \dots, r - 1$). (The addition $\omega_{i+j} + \beta$ is taken modulo r, while the subscript addition is modulo n.) This, of course, precludes the possibility that w itself is of the form w' for some $i \neq 0$, modulo n.

The number of words in a PSK-synchronizable dictionary can be bounded, as in Section 11.4, by counting the number of non-degenerate equivalence classes of n-tuples. Only now an equivalence class must be redefined to include all words bearing the relationship of w to w' [Equation (12.6.1)]. That is, two words are now in the same equivalence class (call it a *PSK-equivalence class*) if one can be obtained from the other by permuting it cyclically by some amount and adding to it the constant n-tuple $\beta\beta \dots \beta$ for any β ($\beta = 0, 1, \dots, r - 1$). Any word belonging to an equivalence class containing fewer than $n \cdot r$ distinct n-tuples will be referred to as PSK-degenerate. Since any PSK-degenerate word can be converted to another word in its class by an operation of the sort considered here, we have the following lemma.

Lemma 12.1. A PSK-synchronizable dictionary can have at most one word from each non-degenerate PSK equivalence class.

We now prove a second lemma needed in bounding the number of words in a maximal PSK-synchronizable dictionary.

Lemma 12.2. The number of solutions k to the equation $k \cdot q = 0 \pmod r)\, 0 \leq k < r$, is (r, q) where (r, q) denotes the greatest common divisor of the integers q and r.

Proof. If $(r, q) = \nu$, then $r = \beta\nu$ and $q = \gamma\nu$ with β and γ integers having no common factors, i.e., $(\beta, \gamma) = 1$. The equation $k \cdot q = 0 \pmod r$ is then satisfied if, and only if, $k = l(\beta/\gamma)$, where l is an integer multiple of γ, and hence if, and only if, $k = \alpha\beta$ for some integer α. Since $\nu\beta = r$, the integers $k_i = i\beta < r$ are unique solutions to $kq = 0 \pmod r$ for all $i = 0, 1, 2, \ldots, \nu - 1$. There are therefore exactly $(r, q) = \nu$ such solutions.

Theorem 12.9. The number of words in a PSK-synchronizable dictionary of n-symbol r-ary words cannot exceed

$$N'(n, r) = \frac{1}{rn} \sum_{d \mid n} (r, d)\mu(d) r^{n/d}$$

where the summation is over all integer divisors of n, (r, d) denotes the greatest common divisor of r and d, and $\mu(d)$ is the Möbius μ function:

$$\mu(d) = \begin{cases} 1 & d = 1 \\ (-1)^\nu & d = p_1 p_2 \ldots p_\nu, \qquad p_1, p_2, \ldots, p_\nu \text{ all distinct primes} \\ 0 & \text{otherwise} \end{cases}$$

Proof. If w is in a PSK-degenerate equivalence class, there must be some smallest integer ν and some n-tuple $k_n \triangleq (kk \ldots k)$, $0 \leq k < r$, such that $w = w^\nu + k_n$, with w^ν indicating the n-tuple obtained by cyclically permuting the symbols of w ν positions to the left. Clearly, ν must be an integer divisor of n (cf. Lemma 11.2). Then also $w = (w^\nu + k_n)^\nu + k_n = w^{2\nu} + (2k)_n$, and, in general, $w = w^{i\nu} + (ik)_n$ for any integer i. Equating the first $d = n/\nu$ symbols of w and $w^{i\nu} + (ik)_n$, and letting s denote an arbitrary ν-tuple, we conclude that w must be of the form

$$w = (s, s + k_{n/d}, s + (2k)_{n/d}, \ldots, s + [(d-1)k]_{n/d}) \tag{12.6.2}$$

Further, since $w^{d\nu} + (dk)_n = w + (dk)_n = w$, the product dk must equal zero, modulo r. Thus, there are $r^{n/d}$ n/d-tuples s and (r, d) integers k (Lemma 12.2) representing distinct PSK-degenerate n-tuples of the form (12.6.2). These same n-tuples are degenerate with period n/d in the sense of Section 11.4 if, and only if, $k = 0$ (modulo r). Consequently, there are (r, d) n-tuples PSK-degenerate with minimum period n/d for every one n-tuple degenerate with the same period in the ordinary sense. The remainder of the proof entirely parallels the proof of Theorem 11.14; it is only necessary to substitute $(r, d)r^{n/d}$ here for $r^{n/d}$ in the proof of Theorem 11.14 to conclude that the number of PSK-non-degenerate n-tuples is

$$W'(n, r) = \sum_{d \mid n} (r, d)\mu(d) r^{n/d} \tag{12.6.3}$$

Since each PSK-non-degenerate equivalence class contains exactly $n \cdot r$ n-tuples and since a PSK-synchronizable dictionary can have at most one word from each equivalence class (Lemma 12.1) the theorem is proved.

We now demonstrate a construction for maximal PSK-synchronizable dictionaries D'_n for all word lengths n. Let D_n be a maximal synchronizable dictionary of n-symbol words. (Such a dictionary can be constructed by the techniques of Section 11.5. Moreover, if n is odd, D_n can actually be comma-free as shown in Section 12.2.) Let S_n be the subset of words of D_n with symbols summing to zero, modulo r. For example, if $r = 2$, and D_5 is the comma-free code {00001, 01001, 01101, 10001, 11001, 11101},

$$S_5 = \begin{matrix} 01001 \\ 10001 \\ 11101 \end{matrix}$$

The dictionary D'_n is composed of all the words of the form

$$\begin{aligned} w &= 0, \sigma_1, \sigma_1 + \sigma_2, \sigma_1 + \sigma_2 + \sigma_3, \ldots, \sigma_1 + \sigma_2 + \sigma_3 + \cdots + \sigma_{n-1} \\ &= \omega_1 \omega_2 \ldots \omega_n \end{aligned} \tag{12.6.4}$$

where $\sigma_1 \sigma_2 \ldots \sigma_n$ is a word in S_n.

Theorem 12.10. The dictionary D'_n is PSK-synchronizable.

Proof. Consider an arbitrary sequence of words from D'_n

$$\ldots \omega_i \omega_{i+1} \ldots \omega_n \omega'_1 \omega'_2 \omega'_3 \ldots \omega'_{i-1} \omega'_i \ldots$$

Taking the difference, modulo r, between successive symbols yields the sequence

$$\ldots \sigma_{i+1} \sigma_{i+2} \ldots \sigma_{n-1} \sigma_n \sigma'_1 \sigma'_2 \ldots$$

[*Note:* the first symbol of an output word is $0 - (\sigma_1 + \sigma_2 + \cdots + \sigma_{n-1}) = \sigma_n$, since by construction $\sigma_1 + \sigma_2 + \cdots + \sigma_n = 0 \pmod r$.] Since this is a sequence of words from the synchronizable dictionary D_n, it is synchronizable. The synchronization delay with D'_n is just one symbol greater than that associated with D_n in the absence of the r-fold symbol ambiguity.

Theorem 12.11. The dictionary D'_n contains $N'(n, r)$ words (see Theorem 12.9) and hence is maximal.

Proof. There are exactly r^{n-1} n-tuples $\sigma_1 \sigma_2 \ldots \sigma_n$ such that $\sigma_1 + \sigma_2 + \cdots + \sigma_n = 0 \pmod r$. The first $n - 1$ symbols may be chosen arbitrarily, but the last is then uniquely defined:

$$\sigma_n = -\sum_{i=1}^{n-1} \sigma_i$$

Any degenerate n-tuple (in the sense of Section 11.4) must have the property that

$$w = ss \ldots s$$

for some $s = \sigma_1 \sigma_2 \ldots \sigma_{n/q}$. The number of such n-tuples having integers summing to zero, modulo r, is $(r, q)r^{(n/q)-1}$ since the first $(n/q) - 1$ symbols of s can be chosen arbitrarily, while the last must be selected to satisfy the equation $q\left(\sum_{i=1}^{n/q} \sigma_i\right) = 0 \pmod r$, and from Lemma 12.2, this equation has (r, q) solutions. As in the proof of Theorem 11.14, it is necessary to count only those degenerate n-tuples having a period n/q where q is a product of primes in order to count every such n-tuple once and only once. Accordingly, there are

$$r^{n-1} - \sum_i (r, p_i)r^{(n/p_i)-1} + \sum_i \sum_{j>i} (r, p_i p_j)r^{(n/p_i p_j)-1} + \cdots + (-1)^m (r, p_1 p_2 \ldots p_m) r^{(n/p_1 p_2 \ldots p_m)-1} \tag{12.6.5}$$

non-degenerate n-tuples whose integers sum to zero, modulo r. Dividing Equation (12.6.5) by n, we find the number of distinct non-degenerate (again, in the Section 11.4 sense) equivalence classes having words with this property is $N'(n, r)$. Since each non-degenerate equivalence class produces one word in D_n (i.e., since D_n is maximal), each non-degenerate equivalence class containing n-tuples whose integers sum to zero produces one word in D'_n. Thus D'_n is maximal and the theorem is proved.

12.7 Concluding Remarks

A number of constraints have been investigated in this chapter for insuring the synchronizability of block code dictionaries. It is interesting to compare these different constraints in terms of the redundancy they introduce. Let D be a dictionary containing W r-ary N-symbol words. Then each symbol represents, on the average, $(\log_2 W)/N$ bits of information. If D were unconstrained, W could equal r^N and each symbol could represent $\log_2 r$ bits. The redundancy in D can therefore be measured in terms of the *redundancy ratio*

$$\lambda = \frac{\log_2 r - (\log_2 W)/N}{\log_2 r} = 1 - \frac{\log_r W}{N} \tag{12.7.1}$$

Synchronizability alone imposes the condition that $W \leq r^N/N$, and hence that $\lambda \geq (\log_r N)/N$. This lower bound is, in fact, virtually attainable not only by synchronizable codes, but by the considerably more severely constrained comma-free codes, as was shown in Section 12.2. Prefix codes, at least asymptotically, also achieve this same lower bound (see Section 12.4). The path-invariant comma-free codes of Section 12.3, while easier to implement, have a redundancy ratio $\lambda \approx \frac{1}{2}[1 - \log_r (r - 1)]$ [cf. Equation (12.3.8)] which, in contrast to that of ordinary comma-free codes, does not approach zero asymptotically with N.

Undoubtedly the most practical codes, from both the implementation and synchronization points of view, are the comma codes. If a comma is inserted after each k words, and k is properly chosen (see Section 12.5), the resulting dictionary of $m + kn = N$-symbol "words" is comma-free with a redundancy ratio of

$$\lambda \approx \begin{cases} \dfrac{2 \log_r N}{N} & m = n \text{ or } m = 2n \\ 1 - \log_r (r - 1) & m = 1 \end{cases} \tag{12.7.2}$$

Thus, when $m = n$ or $m = 2n$, the latter case leading to a particularly easy implementation, the redundancy ratio is only twice that achievable with any synchronizable dictionary. When $m = 1$, the resulting comma codes are twice as redundant as the path-invariant comma-free codes. The latter, in fact, when N is large, are just modified comma codes with $m = 1$, the modification allowing the single-symbol comma to be used in the latter half of each code word.

Finally, when the code is required to be PSK-synchronizable (Section 12.6) the redundancy ratio is bounded by $\lambda \geq (\log_r N + 1)/N$, a ratio which is asymptotically identical to that imposed by ordinary synchronizability.

Notes to Chapter Twelve

The comma-free code constraint on block codes was first imposed and investigated by Golomb, Gordon, and Welch (Reference 12.5; see also References 12.6 and 12.7). The first general construction for maximal, odd-length, comma-free codes is due to Eastman (Reference 12.8) although the construction presented in Section 12.2 is that Scholtz (Reference 12.9). The construction for even-length comma-free codes in Section 12.2 is discussed in Reference 12.1. The results of Section 12.3 on path-invariant comma-free codes are found in the paper by the same name (Reference 12.10). The prefix code results of Section 12.4 are due to Gilbert (Reference 12.3). The construction for the PSK-synchronizable codes of Section 12.6 can be found in Reference 12.11. Finally, comma codes were originally investigated by Kendall (Reference 12.12) and subsequently by the author (Reference 12.4).

An additional requirement which has been imposed on block codes with some success, but which was not discussed here, is the capability of correcting errors caused by the insertion or deletion of symbols in the received symbol sequence. That is, with such codes it must be possible not only to detect a synchronization error but also to reconstruct the entire received sequence, including the word in which the error actually took place. The reader is referred to References 12.13 through 12.17 for details.

Problems

12.1 (Reference 12.7). Let each r-ary n-tuple $\alpha^i = \{\alpha^i_1 \alpha^i_2 \cdots \alpha^i_n\}$ have associated with it a numerical value $n(\alpha^i) = \sum_{j=1}^{n} \alpha^i_j r^{n-j}$. A dictionary D is constructed by selecting from each non-degenerate equivalence class the unique n-tuple α^i having the smallest numerical value $n(\alpha^i)$. Show that D is synchronizable with a synchronization delay not exceeding

$$d_s \leq \min \left\{nr^{[n/2]}, \left[\frac{N(n, r)}{2} + 1\right]n\right\}$$

with $N(n, r)$ as defined in Equation (12.2.3).

12.2 (Reference 12.18). Symbol synchronization techniques of the sort discussed in Chapter Seven may be of limited effectiveness if symbol transitions in the data stream are too rare. For this reason it is sometimes useful to encode the message into a block code subject to the constraint that each code word have at least some minimum number of symbol transitions. Let $D_1(m, n)$ be a binary dictionary consisting of all n-symbol words which contain no strings of either *ones* or *zeros* exceeding m symbols in length, and let $N_1(m, n)$ be the number of such words.

(a) Consider the dictionary $D_2(m - 1, n - 1)$ consisting of all words of the form $w' = (\beta_1\beta_2 \cdots \beta_{n-1})$ where $\bar{\beta}_i = \alpha_i + \alpha_{i+1}$, with $w = (\alpha_1\alpha_2 \cdots \alpha_n)$ a word in $D_1(m, n)$. (The plus sign here denotes modulo 2 addition and $\bar{\beta}_i$ the complement of β_i.) Show that no word in $D_2(m - 1, n - 1)$ contains any strings of *ones* exceeding $m - 1$ symbols in length. Show further that $D_2(m - 1, n - 1)$ contains $N_2(m - 1, n - 1) = (\frac{1}{2})N_1(m, n)$ words, the total number of binary $(n - 1)$-tuples having this property.

(b) Now consider the following encoding scheme. The source symbols are each assigned a numerical value. The source symbol corresponding to the number l is then represented by the binary $(n - 1)$-tuple $(\gamma_1\gamma_2 \cdots \gamma_{n-1})$ where

$$l = \sum_{j=1}^{n-1} \gamma_j w_j^{(m)}$$

with

$$w_j^{(m)} = \begin{cases} 2^{j-1}, & 1 \leq j \leq m \\ w_{j-1} + w_{j-2} + \cdots + w_{j-m}, & m < j \end{cases}$$

and with

$$\gamma_j = \begin{cases} 1 & \text{if } l - \sum_{i=j+1}^{n-1} \gamma_i w_i^{(m)} \geq w_j^{(m)} \\ 0 & \text{otherwise} \end{cases}$$

Note that every number l in the range $0 \leq l < w_n^{(m)}$ is uniquely represented in this manner. Show that no more than $m - 1$ consecutive γ_i's can equal 1 and hence that the set of such $(n - 1)$-tuples is contained in the dictionary $D_2(n - 1, m - 1)$.

(c) It can be shown (Reference 12.19) that the number of *ordered partitions* of the integer n (cf. Section 11.4) involving no integer greater than m is just the number $w_n^{(m)}$ defined above. Use this fact to show that the encoding algorithm defined in part (b) is optimum; i.e., that no dictionary subject to the constraints imposed on $D_2(m - 1, n - 1)$ can contain more than $w_n^{(m)}$ words. On the basis of this, describe an algorithm for generating the dictionary $D_1(m, n)$.

(d) It is also shown in Reference 12.19 that the coefficient of x^j in the power series expansion of the generating function

$$W_m(x) = \frac{x(1 - x^m)}{1 - x(2 - x^m)}$$

is $w_j^{(m)}$. Using this result, derive an approximate expression for $N_1(2, n)$ (cf. Section 12.4) and compare it with the number of words in a maximal comma-free code of the same word length.

12.3 A heuristic estimate of the number $N_m(n)$ of words in a maximal prefix code dictionary can be made by noting that a random r-ary m-tuple will be other than the prefix with probability $1 - r^{-m}$. Thus, if one ignores the dependence between overlapping n-tuples, the probability that a random sequence of length n does not contain the prefix is just $(1 - r^{-m})^n$. Compare the resulting estimate of $N_m(n)$ with the exact value derived in Section 12.4.

12.4 The flow graph approach (cf. Problem 10.6) can also be used in conjunction with prefix codes. It is convenient in this case to identify the flow graph as a *state diagram*, the successive nodes or *states* defined by the last m symbols of the sequence in question, with m the prefix length. The next state is then uniquely determined by the present state and the next symbol. All paths of length n through such a state diagram which do not pass through the prefix state and which end in a state representing a properly terminated n-tuple clearly correspond to legitimate words in a $(n + m)$-symbol prefix code. Conversely, all words in a prefix code can be represented by paths through the corresponding state diagram. The binary state diagram associated with the three-symbol prefix 111, for example, is shown in Figure P12.4a where those forbidden paths leading back to state 111 have been omitted. Any path of length n beginning in the state 111 and ending in any of the properly terminated states (those having double borders) corresponds to a code word in an n-symbol prefix code.

(a) Use the technique of Problem 10.6 to derive a generating function for the number of code words of length $N = n + m$ in a maximal prefix code having the prefix 111. Compare the result with that of Section 12.4.

(b) (Reference 12.3). The state diagram described above contains 2^m states, with m the prefix length. Show that an equivalent state diagram can be defined

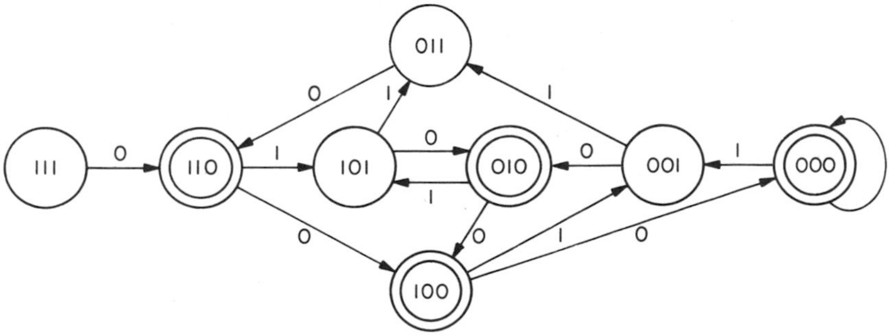

Figure P12.4a

involving only $m + 1$ states as follows: The input sequence leads to state S_k if (1) its last $m - k$ symbols are the first $m - k$ symbols of the prefix, and (2) this same statement cannot be made with k replaced by any other integer $k' < k$. (Thus, for example, the above state diagram can be replaced by the state diagram below. The two diagrams represent identical dictionaries.)

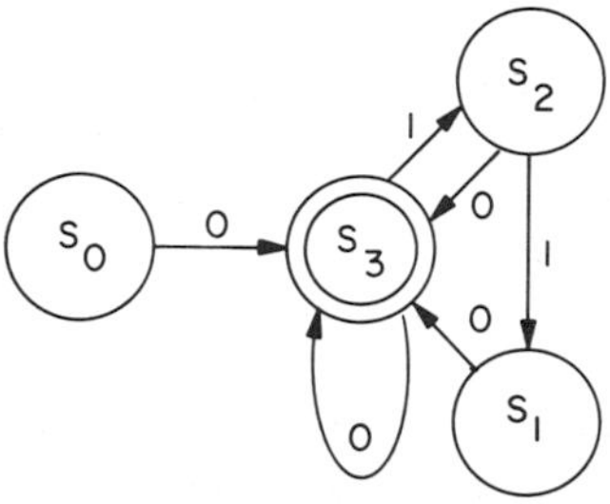

Figure P12.4b

(c) (Reference 12.20) Consider a *reinforced* prefix code in which not only is the m-symbol prefix prohibited from occurring (except as a prefix) but all m-tuples differing in fewer than d positions from the prefix are similarly precluded (e.g., if $d = 2$, and the (binary) prefix is 111, the code is to be so restricted that none of the m-tuples 111, 011, 101, and 110, occur within any code word or code word-prefix overlap). Draw a state diagram corresponding to the maximal $d = 2$ reinforced prefix code having the prefix 1101. Show that the generating function for the number $N_m(n, 2)$ of code words of length $n + m$ in such a code is $G(x) = x^3(1 + x)/(1 - x - x^4) = \sum_{n=1}^{\infty} N_m(n, 2)x^n$.

(d) (Reference 12.20). Describe how the state diagram approach could be used to determine the number of words in a maximal prefix dictionary in which the prefix is prohibited from occurring only over portions of the code word (e.g., the first l symbols following the true prefix and the last l symbols preceding the next prefix, as well as the overlaps involving these l-tuples and the prefix, might be restricted not to contain the prefix for some $l < N/2$ while the remaining symbols are unconstrained). Show, for example, that the maximal modified prefix binary code (cf. Section 12.4) having the prefix 11 and word length $N = 12$ ($n = 10$) contains 160 words, 32 words more than are obtained using the construction mentioned in Section 12.4.

12.5 (Reference 12.3) Show that the $(n + 1)$-symbol comma $c = (00 \cdots 01)$ is singular with respect to the dictionary containing all r-ary n-tuples ending with a non-zero symbol. Compare this construction in terms of redundancy with the comma code constructions of Section 12.5.

References

12.1 Jiggs, B. H., "Recent Results in Comma-Free Codes," *Can. J. Math.*, 15, p. 178, 1963.

12.2 Scholtz, R. A., and L. R. Welch, "The Mechanization of Codes with Bounded Synchronization Delay," *IEEE Trans. Inf. Theory*, IT-lb, p. 241, 1970.

12.3 Gilbert, E. N., "Synchronization of Binary Messages," *IRE Trans. Inf. Theory*, IT-6, p. 470, 1960.

12.4 Stiffler, J. J., "Instantaneously Synchronizable Block Code Dictionaries." Space Program Summary 37–32, IV, Jet Propulsion Lab., Calif. Inst. Tech., Pasadena, Calif., p. 268, 1965.

12.5 Golomb, S. W., B. Gordon, and L. R. Welch, "Comma-Free Codes," *Can. J. Math.*, 10, p. 202, 1958.

12.6 Golomb, S. W., L. R. Welch, and M. Delbruck, "Construction and Properties of Comma-free Codes," *Biol. Medd. Dan. Vid. Selsk*, 23, 1958.

12.7 Golomb, S. W., and B. Gordon, "Codes with Bounded Synchronization Delay," *Inform. and Control*, 8, p. 355, 1965.

12.8 Eastman, W. L., "On the Construction of Comma-Free Codes," *IEEE Trans. Inf. Theory*, IT-11, p. 263, 1965.

12.9 Scholtz, R. A., "Maximal and Variable Word-Length Comma-Free Codes," *IEEE Trans. Inf. Theory*, IT-15, p. 300, 1969.

12.10 Kendall, W. B., and I. S. Reed, "Path-invariant Comma-Free Codes," *IRE Trans. Inf. Theory*, IT-8, p. 350, 1962.

12.11 Eastman W. L., and S. Even, "On Synchronizable and PSK-Synchronizable Block Codes," *IEEE Trans. Inf. Theory*, IT-10, p. 351, 1964.

12.12 Kendall, W. B., "Optimum Synchronizing Words for Fixed Word-Length Code Dictionaries," Space Program Summary, IV, Jet Propulsion Lab., Calif. Inst. Tech., Pasadena, Calif., p. 37, 1964.

12.13 Sellers, F. F., "Bit Loss and Gain Correction Code," *IEEE Trans. Inf. Theory*, IT-8, p. 35, 1962.

12.14 Ullman, J. D., "Near-Optimal, Single Synchronization-Error-Correcting Codes," *IEEE Trans. Inf. Theory*, IT-12, p. 418, 1966.

12.15 Ullman, J. D., "On the Capabilities of Codes to Correct Synchronization Errors," *IEEE Trans. Inf. Theory*, IT-13, p. 95, 1967.

12.16 Levenshtein, V. I., "Binary Codes Capable of Correcting Deletions, Insertions, and Reversals," *Sov. Phys. Doklady*, 10, p. 707, 1966.

12.17 Levenshtein, V. I., "Binary Codes Capable of Correcting Spurious Insertions and Deletions of Ones," *Prob. Per. Inform.*, 1, p. 12, 1965.

12.18 Kautz, W. H., "Fibonacci Codes for Synchronization Control," *IEEE Trans. Inf. Theory*, IT-11, p. 284, 1965.

12.19 Riordan, J., *An Introduction to Combinatorial Analysis*, John Wiley & Sons, Inc., N.Y., N.Y., 1961.

12.20 Ramamoorthy, C. V., and D. W. Tufts, "Reinforced Prefixed Comma-Free Codes," *IEEE Trans. Inf. Theory*, IT-13, p. 366, 1967.

chapter thirteen

Coding to Prevent Errors

13.1 Introduction

Suppose a communication channel is subject to noise of an intensity sufficient to make it difficult to identify correctly the noise-corrupted symbols at the receiver. Errors will presumably occur quite frequently. To improve the reliability of the system, it may be worthwhile to encode the information to be transmitted, perhaps by transmitting in addition to the information symbols certain symbols which provide a check on their validity. The most elementary method is just to send each information symbol twice. If the corresponding two received symbols are identical, more confidence can be attached to that portion of the message than if the symbol had been transmitted only once. If, on the other hand, the two received symbols are different, at least one of them must be in error and the decision as to what was transmitted could be discounted accordingly.

The decoder for this single repetition code might operate in an *error-detecting* mode. If either one of the two symbols representing a single information symbol were in error, the decoder could detect this event and indicate to the user its inability to make a decision. If a double error occurs (i.e., if both repeated symbols are erroneously identified at the receiver), an undetected error could still be made, but this is hopefully a much less likely event.

Now consider the situation in which two repetitions rather than one are transmitted. The decoder may then accept as the correct symbol that which it identifies most often. That is, if in a series of three detector outputs it identifies the symbol s_1 twice (or three times) and the symbol s_2 once (or not at all) it assumes s_1 was actually transmitted (and repeated twice). Since two or three consecutive incorrect detections are presumably unlikely, the probability of an erroneous decision should thereby be decreased. Here, if only one error occurs in the three transmissions, the correct decision is still made; the decoder is said to be operating in an *error-correcting* mode.

To put the above discussion on a more quantitative level, assume there are only two channel symbols and let the probability of the incorrect detection of either of them be p. (The two error probabilities of course do not need to be equal, but most generally they will be.) Further, assume all symbol errors are statistically independent events. Then, when the code and the decoding algorithm just discussed are used, the probability of a correct decision is $P'_c = 1 - 3p^2(1 - p) - p^3$. This is greater than the probability $P_c = 1 - p$ of a correct decision when the symbol is transmitted only once so long as $0 < p < \frac{1}{2}$; that is, so long as the incorrect detection of a symbol is less likely than the correct detection.

In the case of the single-repetition code and the error-detecting decoder, the probability of an incorrect decision is just p^2 which is clearly less than the corresponding probability without repetition for all $0 < p < 1$. On the other hand, the probability P''_c of a correct decision using this code is $1 - 2p(1 - p) - p^2$ which is less than $P_c = 1 - p$ for all $0 < p < 1$. The advantage of using this code obviously lies in its tendency to exhibit errors when they occur, not in its ability to correct them.

This device of repetition can be extended indefinitely. If a symbol is repeated four times, single errors can be corrected and double errors detected or single, double, and triple errors can be detected. Similarly, five repetitions allow even two misidentifications in the five transmissions to be corrected. The cost of the repetitions is apparent. If every symbol is repeated n times, each symbol conveys only $1/n$th as much information. Stated in another way, $n - 1$ of the n-repeated symbols are *redundant*, inserted only to add protection against the noise in the channel.

It is obviously possible to insert useful redundancy in the transmitted signal in ways other than by mere repetition. For example, suppose each block of k *information* symbols is followed by a *parity-check* symbol defined as the sum of the previous k symbols. (This is meaningful only if the term "sum" is appropriately defined. The r symbols could be identified with the integers $0, 1, \ldots, r - 1$, for instance, and the sum interpreted modulo r. Or, if r is a prime power, the symbols could be identified with the elements of a finite field and the sum defined over this field.) Here only one in $k + 1$ symbols is redundant. Yet a single error can still be detected. This follows because if any one of these $k + 1$ symbols is erroneously detected, the $(k + 1)$st symbol will no longer be the sum of the previous k symbols; the occurrence of an error will be flagged. Two or more errors could escape undetected. If, however, the symbol error probability p is sufficiently small so that two or more errors in $k + 1$ transmissions are quite unlikely, this ability to detect one error may be satisfactory.

The parity-check concept can also be extended. The obvious generalization is to divide the set of k information symbols into a number of subsets and to transmit in addition to these k symbols a parity-check symbol for each of these subsets. For example, a sequence of three information symbols could be followed by a parity check on the first and second symbols and by a second parity check on the second and third symbols. (The repetition scheme discussed above, incidentally, can be regarded as a sequence of identical parity checks on a single information symbol.)

Each of these coding procedures is an example of a more general coding technique; viz., the mapping of blocks of k information symbols onto blocks of $n > k$ code symbols, called *code words*. (In the sequel, the ensemble of code words will be referred to as the *code dictionary*, or, more briefly, as the *code*. The ratio k/n, the fraction of information symbols per code word, is called the *information rate*.) There are many possible ways of defining such a mapping. One class of mappings, for example, would involve adjoining $n - k$ parity-check symbols to each k information symbols. The problem is to determine the best mapping for a given set of conditions, the best presumably that which either minimizes the probability of an erroneous reception of a code word or maximizes the probability of its correct reception. (That the two criteria are not necessarily the same can be inferred from the above discussion of the error-detecting mode of operation.) Unfortunately, these are rather awkward criteria to investigate. An easier measure of the merit of a particular encoding scheme follows upon observing the relationship between the probability of mistaking one code word for another and the number of corresponding symbols in which the two words disagree. Suppose d_{ij} of the corresponding symbols in the two code words w_i and w_j are unequal (e.g., if $w_i = 00111$ and $w_j = 10100$, $d_{ij} = 3$). Obviously, if one of these words is transmitted, at least $d_{ij}/2$ transmission errors must occur before the received word is "closer to" (differs in fewer symbols from) the second word than it is to the original. In general, if d is the number of symbols in which the two closest words disagree, up to at least $e = (d - 1)/2$ errors could be corrected by identifying the received n-tuple with the dictionary word closest to it. Similarly, any set of $d - 1$ or fewer errors could always be detected.

Parenthetically, we remark that one may elect to correct some errors and detect others. For example, all received words which are within a distance e of a code word might be assumed to have arisen from that word, whereas those received words differing from all code words by more than the distance

e might be regarded as not decodable. Then if d is the distance between the closest code words, all received words which have sustained at least $e + 1$ errors but not more than $d - e - 1$ errors will be recognized as possessing errors but not decoded. If a code is to correct up to e errors and also to identify all received words containing more than e but not more than $t = d - e - 1$ errors, d must evidently be at least $t + e + 1$.

In all of these cases, the probability of a decoding error is related to the minimum distance d between any two code words. This measure of distance, the number of corresponding symbols in which two code words differ, is defined as the *Hamming distance.*

In the ensuing search for error-correcting and error-detecting codes, dictionaries will be evaluated in terms of the associated minimum distance d separating any two words, the best dictionary for any given parameters n and k being one for which this minimum distance d attains its maximum. It must be acknowledged that the best dictionary in this sense is not necessarily the best in terms of maximizing the probability of a correct reception or minimizing the probability of an erroneous reception. These measures are related to the minimum Hamming distance, but they are not necessarily monotone functions of it. We shall return to this aspect of the problem subsequently.

13.2 Linear Codes

In order to describe more general constructions for error-correcting and error-detecting codes it is necessary to impose a structure on the code dictionary. The structure most successfully used involves identifying the code symbols with the elements of some finite field $GF(q)$ and requiring the code words to constitute a vector subspace† of n-tuples over $GF(q)$. Such a code is called a *linear* (n, k) *code*, n denoting the number of symbols in each code word, and k the dimension of the subspace.

The words of a linear code can be formed by taking all the linear combinations of a set $\{\mathbf{g}_j\}$ of k n-component basis vectors (or *generators*):

$$\mathbf{w}_i = \sum_{j=1}^{k} a_{ij}\mathbf{g}_j \tag{13.2.1}$$

where a_{ij} is an element in the field $GF(q)$ over which the code words are defined. The dictionary generated by k linearly independent vectors contains q^k code words. The $k \times n$ matrix $\mathbf{G}$, the rows of which are the code words

---

†For a brief discussion of some of the algebraic concepts used in coding theory and for a definition of those terms which are unfamiliar, see Appendix B.

$\mathbf{g}_j, j = 1, 2, \ldots, k$, will be referred to as the *generator matrix* of the linear (n, k) code.

Consider for example the linear dictionary $D = \{000, 202, 122, 021, 220, 110, 012, 101, 211\}$ defined over the ternary field. Since it contains nine code words, its dimension k is two ($3^2 = 9$). All the code words, for example, are linear combinations of the first two non-zero words. (Note that every linear dictionary, since it defines a vector space, must contain the identity, the vector all of whose components are zero.)

One of the chief advantages of restricting a code to be linear is the relative convenience of determining the minimum Hamming distance between two code words. Let the *weight* of a vector be defined as the number of its components which are not zero. Then the minimum Hamming distance separating any two code words in a linear code is equal to the weight of the minimum-weight, non-zero code word in its dictionary. This follows immediately from the observation that the Hamming distance between any two code words $\mathbf{w}_r$ and $\mathbf{w}_s$ is equal to the weight of the difference vector $\mathbf{w}_r - \mathbf{w}_s$; i.e., a component in $\mathbf{w}_r - \mathbf{w}_s$ is non-zero if, and only if, the corresponding components in $\mathbf{w}_r$ and $\mathbf{w}_s$ are different. But, since the code is linear, $\mathbf{w}_t = \mathbf{w}_r - \mathbf{w}_s$ is itself a word in the code dictionary and its weight equals the distance separating $\mathbf{w}_r$ and $\mathbf{w}_s$.

A second reason for giving special attention to the class of linear codes is that it includes the important class of parity-check codes discussed earlier. As we shall see presently, the two classes of codes are in fact identical if the parity-check concept is generalized as follows. Let $(\omega_1, \omega_2, \ldots, \omega_k)$ be a set of information symbols defined over some field $GF(q)$ and let the parity-check symbols ω_i, $k < i \leq n$, be determined by the equations

$$\sum_{j=1}^{k} \eta_{ij}\omega_j + \omega_i = 0$$

the summation taken over $GF(q)$. This can be rewritten as

$$\sum_{j=1}^{n} \eta_{ij}\omega_j = 0 \tag{13.2.2}$$

where $\eta_{ij} = 0$ for all $j > k$, except for $j = i$. The earlier definition of a parity-check symbol restricted η_{ij} to be -1 (or 0) for all $j \leq k$. The generalization introduced here is to let $\eta_{ij}, j \leq k$, be any element in the field $GF(q)$.

The set of Equations (13.2.2) can be expressed more conveniently using matrix notation. Let $\mathbf{w} = (\omega_1\ \omega_2 \ldots \omega_n)$ represent a code word and $\mathbf{h}_{i-k} = (\eta_{i1}\ \eta_{i2} \ldots \eta_{in})$, $i = k+1, k+2, \ldots, n$, $n-k$ *parity-check vectors.* Then there is a *parity-check matrix* $\mathbf{H}$ of the form

$$\mathbf{H} = [\mathbf{A}_{n-k,k}\mathbf{I}_{n-k}] \tag{13.2.3}$$

with $\mathbf{A}_{n-k,k}$ an $(n-k) \times k$ matrix and $\mathbf{I}_{n-k}$ the $(n-k) \times (n-k)$ identity matrix, such that any row $\mathbf{h}_i$ in $\mathbf{H}$ satisfies the equation $\mathbf{w} \cdot \mathbf{h}_i = 0$ for any $\mathbf{w}$

in the code. Conversely, any n-tuple $\mathbf{w}$ satisfying the equation $\mathbf{w}\cdot\mathbf{h}_i = 0$ for all $\mathbf{h}_i$ in $\mathbf{H}$ is a code word. Thus, a *parity-check code* dictionary D can be defined as the set of all $\mathbf{w}$ satisfying the equation†

$$\mathbf{w}\mathbf{H}^T = \mathbf{0} \triangleq (0, 0, \ldots, 0) \tag{13.2.4}$$

The proof of the equivalence of parity-check codes and linear codes is facilitated upon noting that all linear code generator matrices are combinatorially equivalent (equivalent except for a possible column permutation) to a generator matrix of the form

$$\mathbf{G} = [\mathbf{I}_k \mathbf{B}_{k,n-k}] \tag{13.2.5}$$

with $\mathbf{I}_k$ the $k \times k$ identity matrix and $\mathbf{B}_{k,n-k}$ some $k \times (n-k)$ matrix. This follows because, by definition, a linear code is generated by a $k \times n$ matrix $\mathbf{W}$ having the row rank (and hence the column rank) k. Expressing $\mathbf{W}$ in reduced echelon form yields a matrix of the form $\mathbf{G}$. The matrix $\mathbf{G}$ generates the same code as does $\mathbf{W}$, only possibly with a reordering of its symbols (i.e., a permutation of its columns). This clearly does not alter the distance separating any two of its words, so the resulting codes have identical distance properties. Codes which can be generated by a matrix of the form (13.2.5) are referred to as *systematic codes*.‡ The first k symbols of each code word can be identified with the information symbols. That is, if the k information symbols $(\alpha_1, \alpha_2, \ldots, \alpha_k) \triangleq \mathbf{a}$ are encoded into the word

$$\mathbf{w} = \mathbf{a}\mathbf{G} \tag{13.2.6}$$

the ith symbol of the code word is α_i itself for all i in the range $1 \leq i \leq k$, and the information symbols are all exhibited explicitly.

With these preliminaries, we can prove the following theorem.

Theorem 13.1. Every systematic code is a parity-check code and conversely.

Proof. First, if $\mathbf{G}$ is the generator of a systematic code dictionary D of dimension k, there must be a null space of dimension $n - k$ spanned by an $(n - k) \times n$ matrix $\mathbf{H}$ such that

$$\mathbf{G}\mathbf{H}^T = \mathbf{O} \tag{13.2.7}$$

with $\mathbf{O}$ a $k \times (n - k)$ matrix of zeros (cf. Appendix B, Theorem B.15). Since $\mathbf{G} = [\mathbf{I}_k\mathbf{B}_{k,n-k}]$ the null space is obviously spanned by a matrix $\mathbf{H}$ of the form $[-\mathbf{B}^T_{k,n-k}\mathbf{I}_{n-k}]$. Therefore, any word generated by $\mathbf{G}$ (i.e., any word in D) must satisfy an equation of the form (13.2.4). Moreover, since $\mathbf{G}$ spans the null space of $\mathbf{H}$, all vectors satisfying such an equation are in the set generated by $\mathbf{G}$, so every

†The transpose of a matrix is indicated here by the superscript T.

‡Some authors use instead the term *canonical* code, and let the term systematic code designate any code which is equivalent to a canonical code under a column permutation.

systematic code is a parity-check code. Conversely, if **D** is a parity-check code dictionary, it can be generated by any matrix **G** spanning the null space of its parity-check matrix **H**, and every parity-check code is a systematic code.

13.3 Bounds on the Minimum Distance between Code Words

The number of code dictionaries which could be constructed for almost any set of parameters is immense. Suppose it were desired to select from the q^n n-tuples over $GF(q)$ a subset of q^k of them as a dictionary. There are $N_1 = \binom{q^n}{q^k}$ ways of doing this, a number which is extremely large even for moderate values of n, k, and q. If only linear codes are of interest, there are still $N_2 = L(n, k, q)$ such codes to consider, where

$$L(r, s, q) \triangleq (q^r - 1)(q^r - q)(q^r - q^2) \ldots (q^r - q^{s-1})$$

is the number of ways of selecting a set of s linearly independent vectors from an r-dimensional vector space over $GF(q)$. [This expression for $L(r, s, q)$ follows because the first vector can be any one of the q^r vectors in the subspace except the identity, the second any of the $q^r - q$ vectors not linearly related to the first, the third any one of the $q^r - q^2$ vectors which are not linear combinations of the first two, etc.]. The number of unique linear codes is less than N_2, however, since all of them can be generated in more than one way, but even if a method is used to insure that the same linear code does not need to be considered twice, there still remain $N_3 = L(n, k, q)/L(k, k, q)$ unique codes. This is because, given a linear dictionary, there are $L(k, k, q)$ distinct ways of selecting its generator set. This represents the multiplicity with which a particular code will occur when all N_2 unique sets of generators are selected. Finally, the number of contending dictionaries can be further reduced by restricting consideration to systematic codes only. Since every linear code is combinatorially equivalent to a systematic code, this restriction does not actually eliminate any linear codes, but it does reduce somewhat the number of different dictionaries to be investigated. Since there are $q^{(n-k)k}$ different $k \times (n - k)$ matrices $\mathbf{B}_{k,n-k}$ which can be used to define a code generator of the form (13.2.5), there are $N_4 = q^{(n-k)k}$ distinct (but not necessarily combinatorially inequivalent) systematic (n, k) codes. But even N_4 is much too large to allow a search to be carried out with any degree of completeness. There are, for example, over 32 million different systematic binary (10, 5) codes and this number grows very rapidly with n, k, and q.

The point of all this is to emphasize the impossibility of an exhaustive investigation and consequently the importance of being able to evaluate a given code by some means other than by comparison with all other codes having the same parameters. Several upper bounds on the maximum minimum distance obtainable with code dictionaries are presented in this section. In addition, a method of constructing dictionaries having a specified minimum distance is described, establishing thereby a *lower* bound for the best possible construction.

13.3.1 Some Upper Bounds

In the following discussion let $D_q(n, N, d)$ denote a dictionary consisting of N n-symbol words defined over a q-symbol alphabet, any two words of which are separated by a Hamming distance of at least d. Further, let $V_n(q)$ be the set of all q^n n-tuples defined over the same q-symbol alphabet. If $D_q(n, N, d)$ is linear it will be referred to as an (n, k, d) dictionary over $GF(q)$, with $k = \log_q N$. In this case, $V_n(q)$ will be an n-dimensional vector space over $GF(q)$. The first bound to be derived is a consequence of the following two lemmas.

Lemma 13.1. Let D be an N-word dictionary defined over a q-symbol alphabet, and let $S_\rho(\mathbf{x})$ denote a *sphere* of Hamming radius ρ in $V_n(q)$ centered at $\mathbf{x}$ (i.e., $S_\rho(\mathbf{x})$ consists of all n-tuples differing from $\mathbf{x}$ in ρ or fewer components). Then there exists an $\mathbf{x}$ such that $S_\rho(\mathbf{x})$ contains at least N_ρ words from D, where

$$N_\rho = \left\{\frac{N}{q^n}\sum_{i=0}^{\rho}\binom{n}{i}(q-1)^i\right\}$$

with $\{l\}$ the smallest integer greater than or equal to l.

Proof. Let $N_\rho(\mathbf{x})$ be the number of words from D in $S_\rho(\mathbf{x})$. Then, since each code word $\mathbf{w}$ is contained in exactly $\sum_{i=0}^{\rho}\binom{n}{i}(q-1)^i$ different spheres of radius ρ,

$$\max_{\mathbf{x}\in V_n(q)} N_\rho(\mathbf{x}) \geq \operatorname*{ave}_{\mathbf{x}\in V_n(q)} N_\rho(\mathbf{x}) = \frac{1}{q^n}\sum_{\mathbf{x}\in V_n(q)} N_\rho(\mathbf{x}) = \frac{N}{q^n}\sum_{i=0}^{\rho}\binom{n}{i}(q-1)^i \qquad (13.3.1)$$

and the lemma is proved.

Lemma 13.2. If $S_\rho(\mathbf{x})$ contains $M \geq 2$ words from D, at least two of these words are within Hamming distance d of each other, where

$$d \leq \begin{cases} \min\left[\dfrac{M\rho}{M-1}\left(2-\dfrac{\rho}{n}\dfrac{q}{q-1}\right), 2\rho\right] & \rho \leq \dfrac{q-1}{q}n \\[2ex] \dfrac{M(q-1)}{(M-1)q}n & \rho \geq \dfrac{q-1}{q}n \end{cases}$$

Proof. Let $\mathbf{A}$ be an $M \times n$ rectangular array, the rows of which are the code words in $S_\rho(\mathbf{x})$. Let $s_j(i)$ be the number of occurrences of the symbol $\alpha_j(i)$ in the ith column of $\mathbf{A}$ with $\alpha_0(i) \triangleq x_i$, the ith component of $\mathbf{x}$, and with the other $\alpha_j(i)$ defined arbitrarily, subject, of course, to the condition that $\alpha_\mu(i) \neq \alpha_\nu(i)$, $\mu \neq \nu$.

Finally, let d_{ij} be the distance separating the code words $\mathbf{w}_i$ and $\mathbf{w}_j$, and I the set of indices i such that $\mathbf{w}_i$ is in $S_\rho(\mathbf{x})$. Then

$$\min_{\substack{i,j \in I \\ i \neq j}} M(M-1)d_{ij} \leq \sum_{\substack{i,j \in I \\ i \neq j}} d_{ij} = \sum_{i=1}^{n} \sum_{j=0}^{q-1} s_j(i)(M - s_j(i))$$

$$= M \sum_{i=1}^{n} \sum_{j=0}^{q-1} s_j(i) - \sum_{i=1}^{n} \sum_{j=1}^{q-1} s_j^2(i) - \sum_{i=1}^{n} s_0^2(i) \tag{13.3.2}$$

By Schwarz's inequality, $n(q-1) \sum_{i=1}^{n} \sum_{j=1}^{q-1} s_j^2(i) \geq \left(\sum_{i=1}^{n} \sum_{j=1}^{q-1} s_j(i)\right)^2$ and $n \sum_{i=1}^{n} s_0^2(i) \geq \left(\sum_{i=1}^{n} s_0(i)\right)^2$. But $\sum_{i=1}^{n} \sum_{j=1}^{q-1} s_j(i)$ is just the sum of the distances between $\mathbf{x}$ and the code words in $S_\rho(\mathbf{x})$, while $\sum_{i=1}^{n} \left(\sum_{j=1}^{q-1} s_j(i) + s_0(i)\right) = Mn$. Defining $\bar{\rho} = \frac{1}{M} \sum_{i=1}^{n} \sum_{j=1}^{q-1} s_j(i)$, and observing that $\bar{\rho} \leq \rho$, we have

$$\min_{\substack{i,j \in I \\ i \neq j}} M(M-1)d_{ij} \leq M^2 \bar{\rho}\left\{2 - \frac{\bar{\rho}}{n}\frac{q}{q-1}\right\} \leq \begin{cases} M^2 \rho\left\{2 - \dfrac{\rho}{n}\dfrac{q}{q-1}\right\} & \rho \leq n\dfrac{q-1}{q} \\ M^2 n \dfrac{q-1}{q} & \rho \geq n\dfrac{q-1}{q} \end{cases} \tag{13.3.3}$$

Since d is obviously over bounded by 2ρ, this proves the lemma.

Combining Lemmas 13.1 and 13.2 immediately leads to the following result.

Theorem 13.2 (The Elias bound). The dictionary $D_q(n, N, d)$ cannot exist unless

$$d \leq \min_{\rho_0 \leq \rho \leq [(q-1)/q]n} \left\{\frac{N_\rho}{N_\rho - 1}\rho\left(2 - \frac{\rho}{n}\frac{q}{q-1}\right), 2\rho\right\}$$

with N_ρ the smallest integer equal to or greater than $\frac{N}{q^n}\sum_{i=0}^{\rho}\binom{n}{i}(q-1)^i$ and with ρ_0 the smallest value of ρ for which $N_\rho \geq 2$.

Corollary (The Hamming bound). A necessary condition for the existence of the dictionary $D = D_q(n, N, d)$ is

$$N \leq \begin{cases} q^n \Big/ \displaystyle\sum_{i=0}^{(d-1)/2} \binom{n}{i}(q-1)^i & d \text{ odd} \\ q^{n-1} \Big/ \displaystyle\sum_{i=0}^{(d/2)-1} \binom{n-1}{i}(q-1)^i & d \text{ even} \end{cases} \tag{13.3.4}$$

Proof. From Theorem 13.2, $d \leq 2\rho_0$, with ρ_0 the smallest integer for which $N_{\rho_0} > 1$. The stated inequality must certainly be satisfied for all odd d, since were this not so, $N_{(d-1)/2}$ would exceed one, and hence $\rho_0 \leq (d-1)/2$, leading to a contradiction. If D is an N-word, minimum distance d, dictionary in $V_n(q)$, then D', the dictionary resulting when the last symbol is deleted from every word in D, is an N-word dictionary in $V_{n-1}(q)$ having minimum distance at least $d - 1$. Thus, when d

is even, the same inequality holds with d replaced by $d - 1$ and n by $n - 1$, and the corollary is proved.

Corollary (The Plotkin bound). The dictionary $D = D_q(n, N, d)$ cannot exist unless

$$d \leq \frac{N(q-1)}{(N-1)q} n \tag{13.3.5}$$

Proof. This bound follows immediately from Theorem 13.2 upon noting that, for any p, $N_p/(N_p - 1) \geq N/(N - 1)$.

Theorem 13.3. The existence of a dictionary $D = D_q(n, N, d)$ implies the existence of a dictionary $D_q(n - 1, \{N/q\}, d)$.†

Proof. Let D_i be the subset of words in D beginning with the symbol α_i. At least one such set must contain $\{N/q\}$ or more words. Deleting the first symbol from each word in this set yields a dictionary having the required properties.

Theorem 13.4 (The Griesmer bound). If D is a *linear* (n, k, d) dictionary over $GF(q)$, there also exists an $(n - d, k - 1, \{d/q\})$ linear dictionary over $GF(q)$.†

Proof. Let $\mathbf{G} = \{\mathbf{g}_i;\ i = 1, 2, \ldots, k\}$ be the set of generators of D and, with no loss of generality, assume $\mathbf{G}$ contains a vector $\mathbf{g}_k$ of weight d. For convenience, assume that the first $n - d$ components of $\mathbf{g}_k$ are zero. (This may require a relabeling of the component positions.) Suppose $\mathbf{w}$ is a non-zero word in D having d' of its first $n - d$ components non-zero. Then, since $\mathbf{w} + \alpha\mathbf{g}_k$ is also in D for any α in $GF(q)$, D must contain some word having exactly d' of its first $n - d$, and at most $d - \{d/q\}$ of its last d, components other than zero. Accordingly, if $\mathbf{w} \neq \alpha\mathbf{g}_k$ for any α in $GF(q)$, $d' + d - \{d/q\} \geq d$, and $d' \geq \{d/q\}$. Let D' be a linear $(n - d, k - 1)$ dictionary generated by $\mathbf{G}' = \{\mathbf{g}_i', i = 1, \ldots, k - 1\}$ with $\mathbf{g}_i'$ consisting of the first $n - d$ components of $\mathbf{g}_i$. Clearly, any two words in D' are separated by distance at least d', as was to be shown.

Corollary. An (n, k, d) linear code over $GF(q)$ is possible only if

$$n \geq \sum_{i=0}^{k-1} d_i \tag{13.3.6}$$

with $d_0 = d$ and $d_i = \{d_{i-1}/q\}$.†

Proof. By repeated application of Theorem 13.4, we conclude that an (n, k, d) code implies the existence of an $(n - d_0 - d_1 - \cdots - d_{k-2}, 1, d_{k-1})$ code. But this is obviously not possible unless $n - d_0 - d_1 - \cdots - d_{k-2} \geq d_{k-1}$.

These last two theorems can be used either independently, or in conjunction with Theorem 13.2, to bound the maximum minimum distance d

†Here again $\{l\}$ denotes the smallest integer equal to or greater than l.

attainable with a dictionary D containing N n-symbol words. It follows immediately from Theorem 13.3, for example, that the existence of a dictionary $D_q(n, N, d)$ implies the existence of a dictionary $D_q(n - t, \{N/q^t\}, d)$ for any $t \leq \{\log_q N\} - 1$. Setting $t = \log_q \{N - 1\}$, we conclude further that a dictionary cannot have N n-symbol words separated by minimum distance d unless

$$d \leq n - \{\log_q N\} + 1 \tag{13.3.7}$$

In the case of linear codes at least, Theorem 13.4 can sometimes result in a bound tighter than that of Theorem 13.2. The latter bound, for example, does not preclude the existence of the dictionary $D_2(10, 16, 5)$, but from Theorem 13.4, no linear binary code can have these parameters. Similarly, a binary linear (15, 8, 5) code could exist only if there were a (10, 7, 3) code (cf. Theorem 13.4). The latter code is precluded, however, by the bound in Theorem 13.2, although the dictionary $D_2(15, 256, 5)$ is not. Thus, a non-linear code with these parameters is possible, but not a linear code.†

We conclude the discussion of upper bounds by proving two more corollaries to Theorem 13.4.

Corollary. A linear (n, k) code over $GF(q)$ with distance

$$d = \beta_0 q^{k-1} - \sum_{i=1}^{k-1} \beta_i q^{l_i - 1} \qquad (\beta_i = 0, 1, 2, \ldots, q-1)$$
$$(1 \leq l_{j+1} < l_j < k - 1) \tag{13.3.8a}$$

can exist only if

$$n \geq \frac{\beta_0(q^k - 1)}{q - 1} - \frac{\sum_{i=1}^{k-1} \beta_i(q^{l_i} - 1)}{q - 1} \tag{13.3.8b}$$

Proof. From Theorem 13.4

$$d_j = \left\{\frac{d_{j-1}}{q}\right\} = \beta_0 q^{k-j-1} - \sum_{i=1}^{k-1} \beta_i q^{l_i - j - 1}$$

where q^l is taken to be zero for any $l < 0$. Substituting this expression for d_j into Equation (13.3.6) establishes the desired result.

Corollary.‡ If a linear (n, k) q-ary code has minimum distance $d \leq q^{k-1}$ then

$$n \geq \frac{q}{q-1}(d - 1) + k - \log_q d \tag{13.3.9}$$

---

†Interestingly, a non-linear code *does* exist with these parameters (cf. Reference 13.1). The best (15,8) linear code has distance 4. This is one of the few known cases in which a non-linear code is known to exist having a greater minimum distance than is possible with a linear code of the same word length and dimension.

‡This bound is also sometimes called the Plotkin bound.

Proof. From Equation (13.3.6), since $d_i \geq \max(d/q^i, 1)$ for all $i, 0 \leq i < k$,

$$n \geq \max_{1 \leq l \leq k} \left[d \sum_{i=0}^{l-1} \frac{1}{q^i} + k - l \right] = \max_{1 \leq l \leq k} \frac{q}{q-1} d\left(1 - \frac{1}{q^l}\right) + k - l \quad (13.3.10)$$

As is easily verified, this last expression has a unique maximum with respect to l (for $l \geq 0$) at some point $l = l_0$, and is a monotonically decreasing function of l for $l > l_0$, and a monotonically increasing function for $l < l_0$. The right side of the inequality (13.3.10) is the same for both $l = \log_q d$ and $l = \log_q d + 1$. Since there must be some integer l in the range $0 \leq \log_q d \leq l \leq \log_q d + 1 \leq k$, and since the expression on the right in (13.3.10) must be at least as great for any l in this region as it is at the end points, the corollary follows.

13.3.2 The Gilbert Lower Bound

Before leaving the discussion of bounds it is reassuring to be able to prove the existence of linear codes with at least some error-correcting or error-detecting properties. The proof of the Gilbert bound assures the constructability of (n, k) q-ary linear codes with a specified minimum distance. The motivation for the construction stems from the following lemma.

Lemma 13.3. If there is a code word of weight d generated by **G**, there must be a linearly dependent relationship involving exactly d columns of its null-space generator **H** and conversely.

Proof. Since $\mathbf{w}\mathbf{H}^T = \mathbf{0}$ for all $\mathbf{w} = (\omega_1\, \omega_2 \ldots \omega_n)$ in the dictionary D (generated by **G**), then, denoting by $\mathbf{h}_i^T$ the ith row of $\mathbf{H}^T$ (i.e., $\mathbf{h}_i$ is the ith column of **H**) we have $\sum \omega_i \mathbf{h}_i^T = \mathbf{0}$. Therefore, if there is a code word containing exactly d non-zero elements, there is a linear relationship involving exactly d columns of **H**. Conversely, if a subset of d columns of **H** is linearly dependent, $\sum_{i=1}^{n} \gamma_i \mathbf{h}^T = \mathbf{0}$ with exactly d of the terms γ_i are non-zero. Then $\mathbf{w}\,\mathbf{H}^T = \mathbf{0}$ where $\mathbf{w} = (\gamma_1, \gamma_2, \ldots, \gamma_n)$ must be a vector of weight d in the subspace generated by **G**.

This provides the basis for the Gilbert construction. To exhibit an (n, k) linear code having minimum distance at least d, it is only necessary to select the columns of an $(n - k) \times k$ matrix **H** in such a way that no set of $d - 1$ or fewer of them are linearly dependent. Thus, the first column $\mathbf{h}_1$ of **H** can be selected from any of the $(n - k)$-tuples over $GF(q)$ except the all-zeros vector $\mathbf{0}^T$. The second column $\mathbf{h}_2$ can be any except $\mathbf{0}^T$ or (if $d > 2$) any of the $q - 1$ non-zero multiples of $\mathbf{h}_1$; the third column $\mathbf{h}_3$ any but $\mathbf{0}^T$, the $(q - 1)$ multiples of either $\mathbf{h}_1$ or $\mathbf{h}_2$ (if $d > 2$), or the $(q - 1)^2$ linear combinations involving both $\mathbf{h}_1$ and $\mathbf{h}_2$ (if $d > 3$). In general, the $(j + 1)$st column of **H** can be selected from the remainder of the $(n - k)$-tuples after all possible linear combinations involving $d - 2$ or fewer of the j previously selected columns of **H** have been eliminated. This assures that no $d - 1$ or fewer of the selected columns are linearly dependent. In the worst case, all of these eliminated $(n - k)$-tuples are distinct. Hence, we can assuredly select the nth column of $\mathbf{H}$ if after all the possibly distinct

$$\sum_{j=0}^{d-2} \binom{n-1}{j}(q-1)^j \triangleq N(n, q, d)$$

$(n - k)$-tuples have been eliminated, at least one still remains. Since there are $q^{(n-k)}$ $(n - k)$-tuples, this is possible if $N(n, q, d) < q^{n-k}$. Since the rank of $\mathbf{H}$ cannot exceed $n - k$, the dictionary must contain at least q^k words. Consequently, we have proved the following theorem.

Theorem 13.5. It is possible to construct an $(n, k)q$-ary linear dictionary with minimum distance $d \geq 2$ for all n, k, and d satisfying the inequality

$$q^k < q^n \Big/ \sum_{i=0}^{d-2} \binom{n-1}{i}(q-1)^i \tag{13.3.11}$$

13.4 Cyclic Codes

With a few important exceptions, the search for good linear error-correcting code constructions has been fruitful only after even more constraints were placed upon the code structure. As the mathematical constraints which the code must satisfy become more rigid, the subset of eligible codes of course becomes smaller, and may even exclude at the outset the best codes from consideration. On the other hand, those codes which remain are easier to analyze, and good codes in this limited class can be more readily identified.

The constraint most effectively applied is the requirement that the linear code be *cyclic*. Denoting by $\mathbf{w}^i$ the cyclic permutation of the symbols of the code word $\mathbf{w}$ i positions to the left [if $\mathbf{w} = (w_0 w_1 \cdots w_{n-1})$, then $\mathbf{w}^i = (w_i w_{i+1} \cdots w_{n-1} w_0 w_1 \cdots w_{i-1})$], we refer to a linear dictionary D as cyclic if, and only if, for every word $\mathbf{w}$ in D there is also a word $\mathbf{w}^i$ in D for every integer i. All of these permutations need not be distinct, however. Both of the dictionaries

0000	0000
1111	0011
	0110
	1100
	1001
	0101
	1010
	1111

for example, are cyclic.

It is useful to associate with each code word $\mathbf{w} = (w_0\ w_1 \ldots w_{n-1})$ a polynomial

$$w(x) = w_0 + w_1 x + w_2 x^2 + \cdots + w_{n-1} x^{n-1} \tag{13.4.1}$$

If the code is cyclic, the polynomial $x^l w(x)$ must also represent a code word when the exponents are interpreted modulo n; $x^l w(x)$ corresponds to the word obtained by cyclically permuting the symbols of $\mathbf{w}$ l positions to the right ($n - l$ positions to the left). Moreover, since the code is linear, $p(x)w(x)$ represents a code word for any polynomial $p(x)$ with coefficients which are elements in the field over which the code is defined. Clearly, reducing modulo n the exponents of a polynomial $p(x)$ is equivalent to reducing the polynomial modulo the polynomial $x^n - 1$. [See Appendix B.3; if $p(x) = q(x)(x^n - 1) + r(x)$, with $r(x)$ a polynomial of degree less than n, $p(x) = r(x)$, modulo $x^n - 1$.]

Theorem 13.6. Every code word in an (n, k) cyclic code can be represented by a polynomial of the form $\alpha(x)g(x)$ with $\alpha(x)$ a polynomial of degree $k - 1$ or less particular to the code word in question, and $g(x)$ a unique monic† divisor of $x^n - 1$ of degree $n - k$.

Proof. Let $g(x)$ be a polynomial corresponding to a code word in the cyclic dictionary such that no other code word polynomial, except that associated with the all-zeros word, has a smaller degree. Then any code word polynomial $w(x)$ can be expressed as $w(x) = q(x)g(x) + r(x)$ where the degree of $r(x)$ is less than the degree of $g(x)$. Since the code is cyclic, and since both $w(x)$ and $g(x)$ are code word polynomials, so must $q(x)g(x)$ and $w(x) - q(x)g(x) = r(x)$ represent code words. But the degree of $r(x)$ is less than the degree of $g(x)$ which was by hypothesis the code word polynomial of minimum degree. Consequently, $r(x)$ must be identically zero, and $w(x) = q(x)g(x)$ (cf. Appendix B, Theorem B.4′).

Since $x^n - 1$ corresponds to the all-zeros word which is in any linear dictionary, there must exist some polynomial $h(x)$ such that $h(x)g(x) = x^n - 1$. Thus, $g(x)$ divides $x^n - 1$.

Suppose $g(x)$ is of degree $n - l$. Then, because any polynomial of degree n or greater is equivalent, modulo $x^n - 1$, to some polynomial of degree less than n, there are exactly q^l distinct polynomials modulo $x^n - 1$ of the form $\alpha(x)g(x)$ over the code field $GF(q)$. Since all words in the (n, k) cyclic dictionary can be represented by a polynomial of this form, l must equal k, and $g(x)$ must have degree $n - k$.

Finally, since the code is linear, all words of the form $\alpha g(x)$ must be in the dictionary for any element α in $GF(q)$. With no loss of generality, therefore, we can stipulate that $g(x)$ is a monic polynomial, and, since all other code word polynomials of degree $n - k$ must be multiples of $g(x)$, $g(x)$ is then unique.

The polynomial $g(x)$ defined in Theorem 13.6 will be referred to as the *generator polynomial* of the associated (n, k) cyclic code.

Now let the polynomial $v^*(x)$ be defined as

$$v^*(x) = v_{n-1} + v_{n-2}x + v_{n-3}x^2 + \cdots + v_0 x^{n-1} = \sum_{i=0}^{n-1} v_{n-1-i}x^i \qquad (13.4.2)$$

†A monic polynomial is a polynomial whose leading coefficient is 1; cf. Appendix B.

where $(v_0\ v_1\ \ldots\ v_{n-1})$ is an n-tuple in the null space of the code in question, and consider the product $w(x)v^*(x)$, with the exponents taken modulo n:

$$w(x)v^*(x) = \sum_{i=0}^{n-1} (v_0 w_{i+1} + v_1 w_{i+2} + \cdots + v_{n-1} w_i)x^i$$

The coefficient of x^i in the product polynomial is the inner product representing one of the parity-check equations which must be satisfied by the vector $\mathbf{w}^{i+1}$. Therefore, if $w(x)$ is a code word polynomial and $v^*(x)$ a parity-check polynomial as defined above, the product $w(x)v^*(x)$ must equal zero when the exponents are interpreted modulo n. In other words, $w(x)v^*(x)$ must be a multiple of $x^n - 1$:

$$w(x)v^*(x) = p(x)(x^n - 1) \tag{13.4.3}$$

Theorem 13.7. The null space of a cyclic code is also a cyclic code with the generator $h(x) = (x^n - 1)/g(x)$.

Proof. Clearly, $\alpha(x)h(x)\beta(x)g(x) = \alpha(x)\beta(x)(x^n - 1) = p(x)(x^n - 1)$ and any n-tuple corresponding to any of the polynomials $\alpha(x)h(x)$ is in the null space of the code generated by the polynomial $g(x)$. Conversely, suppose $v^*(x)$ corresponds to some n-tuple in the null space. Then $g(x)v^*(x) = p(x)(x^n - 1) = p(x)g(x)h(x)$, so

$$v^*(x) = p(x)h(x) \tag{13.4.4}$$

All polynomials representing n-tuples in the null space are of the form (13.4.4) and conversely, where $h(x) = (x^n - 1)/g(x)$, as was to be shown.

Notice that the space of the cyclic code generated by the polynomial $g(x) = \sum_{i=0}^{n-k} g_i x^i$ is spanned by the vectors

$$\mathbf{G} = \begin{bmatrix} g_0 & g_1 & \cdots & g_{n-k} & 0 & 0 & \cdots & 0 \\ 0 & g_0 & \cdots & g_{n-k-1} & g_{n-k} & 0 & \cdots & 0 \\ \cdot & \cdot & & & & & & \cdot \\ \cdot & \cdot & & & & & & \cdot \\ \cdot & \cdot & & & & & & \cdot \\ 0 & 0 & \cdots & g_0 & g_1 & g_2 & \cdots & g_{n-k} \end{bmatrix} \tag{13.4.5}$$

and its null space is spanned by

$$\mathbf{H} = \begin{bmatrix} h_0 & h_1 & \cdots & h_{k-1} & h_k & 0 & \cdots & 0 \\ 0 & h_0 & \cdots & h_{k-2} & h_{k-1} & h_k & \cdots & 0 \\ \cdot & \cdot & & & & & & \cdot \\ \cdot & \cdot & & & & & & \cdot \\ \cdot & \cdot & & & & & & \cdot \\ 0 & 0 & \cdots & h_0 & h_1 & h_2 & \cdots & h_k \end{bmatrix} \tag{13.4.6}$$

where $h(x) = \sum_{i=0}^{k} h_{k-i} x^i$ is the null space generator polynomial.

Moreover, the matrix **G** of Equation (13.4.5) can obviously be written in the form (13.2.5) by a series of elementary row operations only. Since these row operations leave the resulting dictionary unaltered, all cyclic codes are systematic.

A code word $\mathbf{w} = (w_0\ w_1 \ldots w_{n-1})$ is in an (n, k) cyclic code dictionary if, and only if,

$$\sum_{j=l}^{k+l} w_j h_{j-l} = \sum_{i=0}^{k} w_{i+l}\, h_i = 0 \tag{13.4.7}$$

for all $l = 0, 1, \ldots, n-k-1$, where $\mathbf{h} = (h_0\ h_1 \ldots h_k\ 0 \ldots 0)$ is an n-tuple in its null space. This follows immediately from the expression (13.4.6) for the null-space generator matrix. Accordingly, the symbols of any code word must satisfy a recursion of the form

$$w_{l+k} = -\frac{1}{h_k}\sum_{i=0}^{k-1} w_{i+l} h_i \qquad l = 0, 1, \ldots, n-k-1 \tag{13.4.8}$$

The symbols w_j for all $k \leq j \leq n-1$ are uniquely defined in terms of the first k symbols $w_0, w_1, \ldots, w_{k-1}$ by this difference equation.

The standard procedure for solving linear difference equations is to assume a solution of the form $w_l = x^l$ and to determine the roots of the resulting polynomial. In the present instance, the solution is given by $w_l = \beta^l$ where β is any root of the polynomial

$$\sum_{i=0}^{k} h_i x^{i+l} = x^{l+k}\sum_{j=0}^{k} h_{k-j} x^{-j} = x^{l+k} h\left(\frac{1}{x}\right) \tag{13.4.9}$$

with $h(x)$ as previously defined. The roots of $h(1/x)$ exist in some extension field of the code field $GF(q)$. (See Appendix B, Section B.3.)

One root of the polynomial (13.4.9), of course, is $x = 0$. Since $x^k h(1/x)$ is a polynomial of degree k, there are up to k other solutions $w_l = \beta_i^l$, where $h(\beta_i^{-1}) = 0$, and since the difference equation is linear, any linear combination of these k solutions must also be a solution. The general solution, therefore, is of the form

$$w_l = \sum_{i=1}^{k} c_i \beta_i^l \tag{13.4.10}$$

where the c_i are constants which must be chosen so as to satisfy the initial conditions

$$w_0 = \sum_{i=1}^{k} c_i \qquad w_1 = \sum_{i=1}^{k} c_i \beta_i \qquad w_2 = \sum_{i=1}^{k} c_i \beta_i^2 \ldots w_{k-1} = \sum_{i=1}^{k} c_i \beta_i^{k-1}$$

That is, the vector $\mathbf{c} = (c_1, c_2, \ldots, c_k)$ must satisfy the equation

$$\mathbf{M}\mathbf{c}^T = \tilde{\mathbf{w}}^T \tag{13.4.11}$$

where

$$\mathbf{M} = \begin{bmatrix} 1 & 1 & \cdots & 1 \\ \beta_1 & \beta_2 & \cdots & \beta_k \\ \beta_1^2 & \beta_2^2 & \cdots & \beta_k^2 \\ \cdot & \cdot & & \cdot \\ \cdot & \cdot & & \cdot \\ \cdot & \cdot & & \cdot \\ \beta_1^{k-1} & \beta_2^{k-1} & \cdots & \beta_k^{k-1} \end{bmatrix}$$

and $\tilde{\mathbf{w}} = (w_0, w_1, \ldots, w_{k-1})$.

Equation (13.4.11) will have a solution **c** if, and only if, the matrix **M** has rank k, and hence, if the determinant of **M** is non-zero (see Appendix B.6). But, for all $k > 1$,

$$\det \mathbf{M} = \prod_{\substack{i,j=1 \\ i>j}}^{k} (\beta_i - \beta_j) \tag{13.4.12}$$

To see this, observe that both sides of Equation (13.4.12) are polynomials $p(\beta_i)$ of degree $k - 1$ in any one of the β_i (i.e., $p(\beta_i) = \sum_{j=0}^{k-1} d_j \beta_i^j$). Moreover, if $\beta_i = \beta_j$ for any $j \neq i$, $\det \mathbf{M} = 0$, and hence $p(\beta_i)$ has $\beta_i - \beta_j$ as a factor. Since the two sides of Equation (13.4.12) are polynomials of the same degree and having the same factors, they differ at most by a constant. But since the coefficient of the term $1\beta_2 \beta_3^2 \ldots \beta_k^{k-1}$ is the same for both polynomials, they must in fact be equal. (This determinant, incidentally, is called a van der Monde determinant.)

We conclude therefore that $\det \mathbf{M} \neq 0$ if the roots β_i are all distinct. [If they are not distinct a solution can still be obtained, although it is not of the form (13.4.10). In all situations of interest here, however, the β_i will be distinct.] Thus, it is possible to solve for the coefficients c_i so as to satisfy the initial conditions. Since the terms w_l of Equation (13.4.10) do satisfy both the recursion relationship and the initial conditions, they must indeed be the symbols of the code word having those particular k information symbols.

So far we have only demonstrated another way of determining the symbols of a cyclic code word in terms of the information symbols. Since we could already do this considerably more simply by other techniques, the result hardly seems worth the effort. The utility of Equation (13.4.10), however, is demonstrated by the following remarks. First of all, the roots β_i of $h(1/x)$ can always be expressed in the form $\beta_i = \beta_0 \beta^{e_i}$ where e_i's are integers, and β_0 and β are elements in some field $GF(q^m)$. (In particular, $\beta_i = \alpha^{e_i}$ where α is a primitive element in this field.) But if $\beta_i = \beta_0 \beta^{e_i}$, Equation (13.4.10) can be written as

$$w_l = \beta_0^l \sum_{i=1}^{k} c_i (\beta^{e_i})^l \tag{13.4.13}$$

Accordingly, we can associate with each word in the dictionary a polynomial

$$g_w(x) = \sum_{i=1}^{k} c_i x^{e_i} \tag{13.4.14}$$

such that

$$w_l = \beta_0^l g_w(\beta^l)$$

(cf. Reference 13.2). These polynomials are generally referred to in coding theory literature as *Mattson-Solomon polynomials*.

Now suppose $\max_{1 \le i \le k} e_i = s$. Then $g_w(x)$ is a polynomial in x of degree s. There are at most s solutions $x = \beta^l$ for which $g_w(\beta^l) = 0$. If β is an element of order e, at least $[n/e](e - s)$ of the w_i must therefore be non-zero. Since this is true for every word in the dictionary, the dictionary itself must have minimum weight $d \geq [n/e](e - s)$. This is frequently a convenient method for underbounding the minimum weight of a cyclic code dictionary. Note that neither β nor the integers e_i are uniquely determined by the relationship $\beta_i = \beta_0 \beta^{e_i}$, and s may be different for different elements β. Nevertheless, the lower bound $d \geq [n/e](e - s)$ is valid for all β satisfying this relationship. We summarize these results in the following theorem.

Theorem 13.8. Let D be an (n, k) cyclic code dictionary, the null space of which is generated by the polynomial $h(x)$, and let $h(x)$ have the (distinct) roots $\beta_1^{-1}, \beta_2^{-1}, \ldots, \beta_k^{-1}$. Then for every word $\mathbf{w} = (w_0\ w_1 \ldots w_{n-1})$ in D, there is an associated Mattson-Solomon polynomial

$$g_w(x) = \sum_{i=1}^{k} c_i x^{e_i}$$

such that

$$w_l = \beta_0^l g_w(\beta^l)$$

[The integers e_j are defined by the relationship $\beta_0 \beta^{e_i} = \beta_i$, $i = 1, 2, \ldots, k$, and the coefficients c_i by Equation (13.4.11).] The weight of the minimum weight word in D is bounded by

$$d \geq [n/e](e - s)$$

where $s = \max_{1 \le i \le k} e_i$ and e is the order of β.

13.5 Some Important Cyclic Codes

If $h(x)$ is to generate the null space of an (n, k) cyclic code, $h(x)$ must divide $x^n - 1$; i.e., $(x^n - 1)/h(x)$ must be a polynomial with coefficients in the code field. One method for exhibiting all of the cyclic codes which exist for a given value of n, therefore, is to determine all of the irreducible factors of the polynomial $x^n - 1$ over the field in question. For example, since $x^7 - 1 = (x + 1)(x^3 + x + 1)(x^3 + x^2 + 1)$ over the binary field, (7, 1)(7, 3)(7, 4),

(7, 6), and, of course, (7, 7) cyclic codes exist. But since $x + 1$, $x^3 + x + 1$, and $x^3 + x^2 + 1$ are themselves irreducible over this field, no other length seven, cyclic, binary codes exist. These codes can easily be exhibited. Since the (7, 1) code corresponds to $h(x) = 1 + x$, the code is

0000000
1111111

The (7, 3) code corresponding to $h(x) = x^3 + x + 1$ is,

0000000
1110100
1101001
1010011
0100111
1001110
0011101
0111010

The (7, 4) code corresponding to $h(x) = (x^3 + x + 1)(x + 1) = x^4 + x^3 + x^2 + 1$ contains all the words in the (7, 3) code above, since, if the polynomial $w(x)(x^3 + x + 1)$ is identically zero modulo $x^7 - 1$, so also is $w(x)(x^3 + x + 1)(x + 1)$. In addition, the polynomial $w(x)$ representing the non-zero word in the (7, 1) code satisfies the condition $w(x)(x + 1) = 0$, mod $x^7 - 1$, so $w(x)(x + 1)(x^3 + x + 1) = 0$, mod $x^7 - 1$, also, and the vector with only ones as coordinates is also a word in the (7,4) code. The (7, 4) code is therefore the union of the previous (7, 3) code with its coset containing the all-ones n-tuple. Note that another (7, 3) code is defined by $h(x) = (x^3 + x^2 + 1)$, another (7, 4) code by $h(x) = (x^3 + x^2 + 1)(x + 1)$. These latter codes are the null spaces, respectively, of the (7, 4) and (7, 3) codes described above.

This approach, while instructive, is clearly impractical for large values of n. It is therefore important to be able to define general classes of codes which exist for many parameters n and k, and to determine their associated distance properties. This is the task of the present section.

We begin with the highly restricted class of codes in which all but the all-zeros word in any dictionary are cyclic permutations of any one of these words and in which all such permutations are distinct. Such codes will be called *maximal-length cyclic codes*. The significance of this term will be made evident in the following theorem.

Theorem 13.9. In a maximal-length (n, k) cyclic code defined over the field $GF(q)$, the length n of the code words must be $q^k - 1$.

Proof. If the code is linear it must contain q^k words for some integer k. But since all cyclic permutations of any non-zero word are distinct, there are at least n such words, and since n words must account for all but one word in the dictionary $n = q^k - 1$.

Clearly, if all n cyclic permutations of a word in a k-dimensional cyclic code are to be distinct, n cannot exceed $q^k - 1$. Hence, their designation as maximal-length cyclic codes.

Any word in a maximal-length cyclic code must, when added to any scalar multiple of any of its cyclic permutations, yield another cyclic permutation of itself. This follows because the code is linear (so all linear combinations of code words are also code words) and because all non-zero code words are cyclic permutations of one word. (Note the distinction between this code and the ordinary cyclic codes. In the latter, all cyclic permutations of each code word are code words, but all code words are not necessarily cyclic permutations of all other words.) As an example consider the (7,3) binary code exhibited above. This code is a maximal-length cyclic code. The sum of the word 1110100 and any of its cyclic permutations, say 1101001, must be another of its cyclic permutations, in this case 0011101. This feature of these codes is sometimes called the *cycle-and-add* property.†

Theorem 13.10. The polynomial $h(x)$ generates the null space of an (n, k) maximal-length cyclic code over $GF(q)$ if, and only if, $h(x)$ is a primitive polynomial (i.e., if, and only if, the roots of $h(x)$ are of order $n = q^k - 1$; cf. Appendix B.3).

Proof. If $h(x)$ divides $x^m - 1$, the difference between $g(x)$ and $x^m g(x)$ is zero, modulo $x^n - 1$. The corresponding code words are therefore identical. Consequently, $h(x)$ cannot divide any polynomial of the form $x^m - 1$ for any $m < n = q^k - 1$, and the roots of $h(x)$ must be of order n.

Conversely, if $h(x)$ is primitive, it does not divide any polynomial of the form $x^m - 1$ for any $m < n$. Thus $(x^m - 1)g(x) = [(x^m - 1)(x^n - 1)]/h(x) \neq r(x)(x^n - 1)$ unless $m \geq n$, and $g(x)$ is not identical to any of its cyclic permutations. Clearly, if $g(x)$ is distinct from all of its cyclic permutations, all of these permutations must be distinct from each other, and the theorem is proved.

Corollary. There exists a maximal-length $(q^k - 1, k)$ cyclic code for every $q = p^m$, with p a prime and m an integer, and for every integer k.

Proof. Every finite field has a primitive element and hence there exists a primitive polynomial $h(x)$ for every q and k (cf. Appendix B.3).

The distance properties of maximal-length cyclic codes are readily determined with the aid of the following lemma.

---

†That the words in a binary maximal-length code are all the cyclic permutations of a linear feedback shift-register sequence (see Section 6.6) will, if not already apparent, become so shortly.

Lemma 13.4. Let **D** be a $q^k \times n$ matrix, the rows of **D** being the code words of a linear code of length n over $GF(q)$. Then any column of **D** having any non-zero component contains each field element exactly q^{k-1} times.

Proof. Since the code represented by **D** is linear, it is unaltered (except for a reordering of the rows) when some scalar multiple of any one row of **D** is added to each of its rows. Suppose the ith column of **D** contains the non-zero field element α m times ($m > 0$) and suppose the jth row $\mathbf{d}_j$ of **D** has α as its ith component. Then adding $(\beta - \alpha)\alpha^{-1}\,\mathbf{d}_j$ to every row of **D**, we conclude that the ith column of **D** also contains the field element β m times, and as this is true for any β (including $\beta = 0$), it must contain every field element exactly m times. Since there are q field elements and q^k rows, $m = q^k/q = q^{k-1}$. The only situation in which this argument does not hold is when all elements in a column are zero.

Theorem 13.11. Each non-zero maximal-length cyclic code word contains each non-zero field element q^{k-1} times and the zero element $q^{k-1} - 1$ times. Hence, these codes have minimum weight $d = (q - 1)q^{k-1}$. Since this is the maximum possible minimum weight for a $(q^k - 1, k)$ code as given by the bound (13.3.5), these codes are optimum.

Proof. The matrix **D** representing a maximal-length code can contain no column of zeros only since the code is cyclic and every component of any row must occur once in every column position. Thus, there are $nq^{k-1} = (q^k - 1)q^{k-1}$ occurrences of every field element in **D** (cf. Lemma 13.4). But all non-zero rows of **D** must contain the same elements since they are all cyclic permutations of the same row. The all-zeros row of **D** accounts for $q^k - 1$ of the zeros. There are $[(q^k - 1)q^{k-1}]/(q^k - 1) = q^{k-1}$ occurrences of each non-zero element in each row (and $[(q^k - 1)q^{k-1} - (q^k - 1)]/(q^k - 1) = q^{k-1} - 1$ zeros) so each non-zero row in **D** has weight $d = (q - 1)q^{k-1}$ as was to be proved.

The following code words, when combined with all of their cyclic permutations, and the all-zeros n-tuple, are examples of maximal-length cyclic codes:

$(n = 1,\ k = 1,\ q = 2)$: 1
$(n = 2,\ k = 1,\ q = 3)$: 21
$(n = 4,\ k = 1,\ q = 5)$: 2431
$(n = 3,\ k = 2,\ q = 2)$: 110
$(n = 7,\ k = 3,\ q = 2)$: 1110100
$(n = 15,\ k = 4,\ q = 2)$: 100110101111000
$(n = 8,\ k = 2,\ q = 3)$: 11202210
$(n = 26,\ k = 3,\ q = 3)$: 20212210222001012112011100

Any non-zero maximal-length cyclic code word exhibits a number of properties which are characteristic of purely random q-ary sequences (cf. Problem 13.5). For this reason these n-tuples are often referred to as *pseudo-random* or *pseudo-noise* sequences.

In view of the discussion in Section 13.4 it should not be surprising that cyclic codes are sometimes more conveniently expressed in terms of the roots of their generating polynomial. The largest and most important class of cyclic codes, commonly called the BCH codes, are defined in this way. [The letters BCH refer to the names Bose and Ray-Chaudhuri (Reference 13.3) and Hocquenghem (Reference 13.4) who were the first to describe the construction and properties of these codes.]

Theorem 13.12. Let β be an element of order n in the field $GF(q^m)$ and let $g(x)$ be the polynomial of least degree with coefficients in $GF(q)$ having the roots $\beta^{m_0}, \beta^{m_0+1}, \beta^{m_0+2}, \ldots, \beta^{m_0+t-1}$. Then $g(x)$ generates an (n, k) cyclic code over $GF(q)$ for some $k \geq 0$, with minimum distance at least $d = t + 1$.

Proof. The n elements $\beta^{m_0}, \beta^{m_0+1}, \ldots, \beta^{m_0+n-1}$ are all roots of $x^n - 1$. This follows because $(\beta^l)^n = (\beta^n)^l = 1$ for all integers l. Moreover, since $\beta^{m_0+i} \neq \beta^{m_0+j}$ unless $\beta^{i-j} = 1$ (i.e., unless $i = j$, modulo n), these n elements are distinct and $x^n - 1$ has no other roots. All of the specified roots of $g(x)$ are roots of $x^n - 1$ so a polynomial $g(x)$ satisfying the necessary conditions must exist having degree $n - k$ for some integer $k \geq 0$, and can be used to generate an (n, k) cyclic code. The roots of $h(x) = (x^n - 1)/g(x)$ are contained in the subset $(\beta^{m_0+t}, \beta^{m_0+t+1}, \ldots, \beta^{m_0+n-1})$ and the roots of $h(1/x)$ in the subset $(\beta_0, \beta_0 \beta, \ldots, \beta_0\beta^{n-t-1})$ where $\beta_0 = \beta^{1-m_0}$. Thus, from Theorem 13.8, the minimum weight of any non-zero word in the code generated by $g(x)$ is bounded by $d \geq n - s$, where

$$s = \max_{1 \leq i \leq k} e_i \leq n - t - 1$$

and the theorem is proved.

These are the BCH codes. Notice that the word length n must divide $q^m - 1$ for some integer m. This relationship determines the smallest field $GF(q^m)$ containing β: viz., m is the smallest integer for which $q^m = 1$, modulo n.

The most important classes of BCH codes result when $m_0 = 0$ and when $m_0 = 1$. The dimension k is easily underbounded for both of these situations. Let $m_i(x)$ be a polynomial of smallest degree over $GF(q)$ having β^i as a root. Then the elements β^{iq^j} are also roots of $m_i(x)$ for all $j = 1, 2, \ldots, m$ (see Appendix B.4). Since every root of the form β^r with r divisible by q is the root of some polynomial $m_i(x)$ with $i < r$, at most $l - [l/q] = [\{(q-1)/q\}(l+1)]$ polynomials $m_i(x)$ are needed to account for all of the roots $\beta^i, i = 1, 2, \ldots, l$ (the brackets denote the integer part of the enclosed fraction). Since β is an element of $GF(q^m)$ all of the polynomials $m_i(x)$ are of degree not exceeding m (again, see Appendix B.4). The element β^0, of course, is a root of the minimal polynomial $x - 1$. Thus

$$\tilde{g}(x) = \prod_{\substack{i=1 \\ i \neq 0,\, \text{modulo } q}}^{t-1} m_i(x) \times \begin{cases} (x-1) & m_0 = 0 \\ m_t(x) & m_0 = 1 \end{cases}$$

is a polynomial of degree

$$\nu = \begin{cases} \left[\frac{q-1}{q}(d-1)\right]m + 1 & m_0 = 0 \\ \left[\frac{q-1}{q}d\right]m & m_0 = 1 \end{cases} \tag{13.5.1}$$

The generator polynomial $g(x)$ obviously divides $\tilde{g}(x)$; its degree $n-k$ therefore does not exceed ν, and $k \geq n - \nu$ establishing the desired bound.

As an example, consider the class of binary BCH codes with $m = 6$. The congruence classes and the orders e of the corresponding roots (cf. Appendix B.4) are

0						$e = 1$	
1	2	4	8	16	32	63	
3	6	12	24	48	33	21	
5	10	20	40	17	34	63	
7	14	28	56	49	35	9	
9	18	36				7	
11	22	44	25	50	37	63	(13.5.2)
13	26	52	41	19	38	63	
15	30	60	57	51	39	21	
21	42					3	
23	46	29	58	53	43	63	
27	54	45				7	
31	62	61	59	55	47	63	

If $g(x)$ is the minimum polynomial having the primitive root α, $t = 2$ [since α^2 is also a root of $g(x)$] and the resulting (63, 57) code has minimum distance $d \geq 3$. By successively including higher powers of the primitive element α in the definition of $g(x)$ we obtain, referring to the above congruence classes, codes with the following parameters:

n	k	d
63	57	3
63	51	5
63	45	7
63	39	9
63	36	11
63	30	13
63	24	15
63	18	21
63	16	23
63	10	27
63	7	31
63	1	63

Note that the last eight of these codes all have dimensions greater than those assured by Equation (13.5.1). This is typical when k is small relative to n. If the root α^0 is also included in the set of roots defining $g(x)$ (i.e., if $m_0 = 0$) the same set of parameters results but with k decreased by one, and d increased by one.

Non-primitive elements of $GF(2^6)$ can also be used to define BCH codes. Again referring to the set of congruence classes (13.5.2) and selecting the root $\beta = \alpha^3$ of order 21, we conclude that there exist BCH codes having the parameters

n	k	d
21	15	3
21	12	5
21	6	7
21	4	9
21	1	21

Similarly, the elements $\beta = \alpha^7$, $\beta = \alpha^9$, and $\beta = \alpha^{21}$ define BCH codes with $n = 9$, $k = 3$, $d = 3$, and $n = 9$, $k = 1$, $d = 9$; with $n = 7$, $k = 4$, $d = 3$, and $n = 7$, $k = 1$, $d = 7$; and with $n = 3$, $k = 1$, $d = 3$, respectively.

Actual construction of any of these codes, of course, requires the identification of the irreducible polynomial corresponding to each of the congruence classes in (13.5.2). This task is considerably facilitated, when $q = 2$, by the availability of tables of irreducible polynomials (cf. References 13.5, 13.6).

In concluding, we mention one subclass of BCH codes of special interest, the Reed-Solomon codes (Reference 13.7), resulting when $n = q - 1$. These BCH codes have dimension $k \geq n - d + 1$ [cf. Equation (13.5.1); because d must be less than $q = n + 1$, $[\{(q - 1)/q\}d] = d - 1$ for all d]. Since d can never exceed $n - k + 1$ [Equation (13.3.7)], these codes are optimum.†

13.6 Punctured Cyclic, Shortened Cyclic, and Hamming Codes

In this section we shall discuss briefly several other code constructions different from, but related to, the cyclic codes of the previous section. The first of these constructions involves the shortening of an (n, k) cyclic code by requiring i of the k information symbols to be always zero and deleting these symbols from the code word. Because a cyclic code is a systematic code (Theorem 13.12) such a set can be obtained by constraining the first i symbols

†Every BCH code over $GF(q)$ can be shown to be a subspace of a (possibly shortened) Reed-Solomon code over some larger field $GF(q^a)$, its code words consisting of all those words in the Reed-Solomon code having symbols in $GF(q)$ only (Reference 13.8). In this sense, Reed-Solomon codes constitute the more general class of codes.

of each word to be zero. The resulting *shortened cyclic code*, having length $n - i$ and dimension $k - i$ is clearly still a linear code but is no longer cyclic. Since the remaining code words are all shortened versions of words in the original cyclic dictionary, and since the i deleted symbols were identical in each of these words, the minimum distance separating any two words is at least as great as that in the original dictionary. As an example, the (7, 3) binary cyclic code with distance 4 (cf. Section 13.5) can be shortened by taking only those words whose initial symbol is zero and then deleting this symbol yielding the (6, 2) shortened cyclic code also with distance 4:

000000
100111
011101
111010

The generator polynomial $g(x)$ for cyclic codes divides the polynomial $x^n - 1$, the dictionary consisting of words corresponding to all polynomials of the form $p(x)g(x)$, modulo $x^n - 1$. A generalization of this class of codes results when $g(x)$ is defined instead as the divisor of some other nth-degree polynomial $f(x) \neq x^n - 1$ and the code words represented by polynomials of the form $p(x)g(x)$, modulo $f(x)$. Such codes are called *pseudo-cyclic* codes. It can be shown (Problem 13.6), however, that every pseudo-cyclic code with minimum weight greater than 2 is a shortened cyclic code, and conversely, every shortened cyclic code is a pseudo-cyclic code. Except when $d = 2$ the two classes of codes are therefore equivalent.

(Optimum $d = 2$ codes, incidentally, exist for all values of n and q. From the Hamming bound, when $d = 2$

$$q^k \leq q^{n-1}$$

The code obtained by adjoining to each of the $q^{n-1}(n - 1)$-tuples over $GF(q)$ a parity-check component so as to make the sum of all n components equal to zero clearly has minimum distance two and is therefore optimum.)

Another class of codes, the *punctured cyclic codes* (Reference 13.9), is obtained by deleting from the generator matrix of the maximum-length q-ary cyclic code certain of its columns. The dimension of the code remains the same, but some of the parity-check symbols are omitted. We begin the discussion of these codes by proving the following theorem.

Theorem 13.13. Let $\mathbf{G}_k$ be a $k \times q^k - 1$ matrix of elements of $GF(q)$, the rows of which generate a maximal-length cyclic code. Then each of the $q^k - 1$ q-ary k-tuples, excluding the all-zeros k-tuple, occurs once and only once as a column of $\mathbf{G}_k$.

Proof. As has already been remarked, since the dictionary generated by $\mathbf{G}_k$ is cyclic it cannot have an all-zero column. Hence, $\mathbf{G}_k$ cannot have the all-zero

k-tuple as one of its columns. Suppose two of the columns, the ith and the jth, of $\mathbf{G}_k$ are identical. Then the ith and jth columns of the matrix $\mathbf{D}$ generated by $\mathbf{G}_k$ are also identical. Since $\mathbf{D}$ represents a cyclic code, its νth row, after a suitable row permutation, can be expressed in the form $\mathbf{w}_\nu = (\omega_\nu, \omega_{\nu+1}, \ldots, \omega_{\nu+n-1})$ for all $\nu = 1, 2, \ldots, n = q^k - 1$. But if the ith and jth columns of $\mathbf{D}$ are identical, $\omega_{i+\nu}$ and $\omega_{j+\nu}$ must be equal for all ν. The ith and jth cyclic permutations of the code word $\mathbf{w}_\nu$ would then be identical and $\mathbf{D}$ could not represent a maximal-length cyclic dictionary. All columns of $\mathbf{D}$ and hence of $\mathbf{G}_k$ must be distinct and the theorem is proved.

Corollary. Every non-zero q-ary k-tuple occurs once and only once as k consecutive symbols $(\omega_i\, \omega_{i+1} \ldots \omega_{i+k-1})$ in any maximal-length $(n = q^k - 1)$ cyclic code word $\mathbf{w} = (\omega_0\, \omega_1 \ldots \omega_{n-1})$. [The subscripts on the symbols ω_i are to be reduced modulo n. For example, if $i = n - 2$, the relevant k-tuple is $(\omega_{n-2}\, \omega_{n-1}\, \omega_0\, \omega_1 \ldots \omega_{k-3})$].

Proof. Let $g(x) = \sum_{i=0}^{n-k-1} g_i x^i$ be the generator polynomial of the maximal-length dictionary D. Then, from Theorem 13.13, each of the non-zero k-tuples occurs as a column of the generator matrix $\mathbf{G}_k$. But each column of this matrix consists of k consecutive symbols of the code word $\mathbf{g} = (g_0, g_1, \ldots, g_{n-1})$. Every non-zero k-tuple is thus represented by k consecutive symbols of $\mathbf{g}$, and, since every word in D is a cyclic permutation of $\mathbf{g}$, the corollary follows.

Now let the columns of $\mathbf{G}_k$ be divided into $(q - 1)$ *classes* such that, if the column vector $\mathbf{u}$ is in class one, $\alpha_2\mathbf{u}$ is in class two, $\alpha_3\mathbf{u}$ in class three, etc. where α_i is an element of the field over which $\mathbf{G}_k$ is defined, with $\alpha_i \neq \alpha_j$, $i \neq j$, and $\alpha_i \neq 0$, $\alpha_1 = 1$. By suitably permuting the columns of $\mathbf{G}_k$ (which, of course, does not alter the distance properties of the resulting dictionary) an equivalent generator matrix $\mathbf{G}_k^*$ will have the form

$$\mathbf{G}_k^* = \alpha_1\mathbf{H}_k, \alpha_2\mathbf{H}_k, \alpha_3\mathbf{H}_k, \ldots, \alpha_{q-1}\mathbf{H}_k$$

where $\mathbf{H}_k$ is a $k \times m$ generator matrix with $m = (q^k - 1)/(q - 1) = q^{k-1} + q^{k-2} + \cdots + 1$.

Theorem 13.14. Every non-zero code word in the dictionary generated by $\mathbf{H}_k$ has weight $d_0 = q^{k-1}$.

Proof. Let $\mathbf{w}$ be a non-zero code word generated by $\mathbf{H}_k$ and suppose $\mathbf{w}$ has weight d_0. Then,

$$\mathbf{w}^* = \mathbf{w}, \alpha_2\mathbf{w}, \alpha_3\mathbf{w}, \ldots, \alpha_{q-1}\mathbf{w}$$

is a code word generated by $\mathbf{G}_k^*$ and the weight of $\mathbf{w}^*$ must be $(q - 1)d_0$. But all words generated by $\mathbf{G}_k$ and hence by $\mathbf{G}_k^*$ have weight $d = (q - 1)q^{k-1}$, and the theorem is proved.

Corollary. The code generated by the matrix obtained by deleting (puncturing) any r of the classes of the k-dimensional generator matrix $\mathbf{G}_k$ of a maximal-length cyclic code has word length $n = (q - 1 - r)[(q^k - 1)/(q - 1)]$ and distance $(q - 1 - r)q^{k-1}$.

Proof. Since each class generates code words of length $m = (q^k - 1)/(q - 1)$ and, except for the all-zeros word, of uniform weight q^{k-1}, and since after deleting r classes $q - 1 - r$ remain, the corollary follows.

These are examples of punctured cyclic codes. To extend the puncturing algorithm, let $\mathbf{G}_l$ be a matrix containing all the columns of $\mathbf{G}_k$ having zeros in the same $k - l$ of their component positions. The matrix $\mathbf{G}_l$ will generate some permutation of the words of a $(q^l - 1, l)$ maximal-length code but with each word repeated q^{k-l} times. This follows because all $q^l - 1$ non-zero l-tuples are represented by the non-zero segments of the columns of $\mathbf{G}_l$.

Now divide the generator matrix $\mathbf{G}_k$ into $q - 1$ classes as before. This procedure simultaneously divides the subset $\mathbf{G}_l$ of $\mathbf{G}_k$ into $q - 1$ classes also. Each row of each class of $\mathbf{G}_l$ contains either $(q^{l-1} - 1)/(q - 1)$ zeros, or (in the case of the all-zeros row) $(q^l - 1)/(q - 1)$ zeros. Consequently, each of these rows contains at most q^{l-1} non-zero elements.

The general puncturing algorithm is as follows: First delete $q - 1 - t$ of the *classes* of $\mathbf{G}_k$. Further delete from r of the remaining t classes of $\mathbf{G}_k$ all of those columns belonging to $\mathbf{G}_l$. The code words remaining at this point have length $n = [t(q^k - 1)/(q - 1)] - [r(q^l - 1)/(q - 1)]$ and minimum weight $d = tq^{k-1} - rq^{l-1}$; the dimension of the code is still k. This procedure is then extended; we select another subset $\mathbf{G}_{l'}$ of the columns of $\mathbf{G}_k$ consisting of all those columns having zeros in all but a specific set of l' of their component positions. If the sets of non-zero components of the columns of $\mathbf{G}_l$ and $\mathbf{G}_{l'}$ are disjoint, the sets $\mathbf{G}_l$ and $\mathbf{G}_{l'}$ will certainly be disjoint. (Clearly, this is possible if, and only if, $l + l' \leq k$.) Thus, we can also puncture from r' of the t classes of $\mathbf{G}_k$ all of the columns which are in $\mathbf{G}_{l'}$ leaving a code having dimension k, word length $n' = [t(q^k - 1)/(q - 1)] - [r(q^l - 1)/(q - 1)] - [r'(q^{l'} - 1)/(q - 1)]$ and minimum weight $d' = tq^{k-1} - rq^{l-1} - r'q^{l'-1}$. By continuing this process, we demonstrate the following theorem.

Theorem 13.15. Punctured (n, k) cyclic codes with distance d exist for all

$$n = r_0 \frac{q^k - 1}{q - 1} - \sum_{i=1}^{m} r_i \frac{q^{l_i} - 1}{q - 1}$$

and

$$d = r_0 q^{k-1} - \sum_{i=1}^{m} r_i q^{l_i - 1}$$

provided $q - 1 \geq r_0 \geq \max r_i \geq 0$ and

$$\sum_{i=1}^{m} l_i \leq k$$

These codes meet the bound of Equation (13.3.8) and hence are optimum.

Since the null space of a subspace is itself a subspace, it too defines a linear code. Consider, therefore, the null space of the code spanned by any one of the classes of the generator matrix of a maximal-length cyclic code. These codes are called Hamming codes.†

Theorem 13.16. The Hamming codes are $n=(q^s-1)/(q-1)$, $k=[(q^s-1)/(q-1)]-s$ linear codes with minimum distance $d = 3$ and exist for all positive integers s. No single-error correcting code of word length $n = (q^s - 1)/(q-1)$ contains more words than the Hamming code of the same word length.

Proof. First of all, since the Hamming code null space consists of vectors of length $n = (q^s - 1)/(q - 1)$ and dimension s, the code space consists of vectors of length n and dimension $n - s$. Further, since the null spaces exist for all integers s, so do the codes themselves. Now from Lemma 13.3 two code words in a code are separated by distance d if and only if there are d linearly dependent columns in the matrix generating its null space. But, by definition, no class contains any column which is a multiple of any other column in the same class. Thus, any two columns of a particular class are linearly independent, and using any one class as the generator of the null space of a code assures that any two of its words are separated by a distance of at least three. Finally from Equation (13.3.4), the number N of words in any n-symbol q-ary code capable of correcting $e = (d - 1)/2 = 1$ error must satisfy the inequality $N \leq q^n/[1 + (q - 1)n]$. But for the Hamming codes, $n = (q^s - 1)/(q - 1)$, and this inequality becomes

$$\log_q N \leq n - s = k$$

The maximum value of N is therefore attained with Hamming codes for any $n = (q^s - 1)/(q - 1)$.

13.7 Kronecker Product, Kronecker Sum, and Concatenated Codes

One rather tempting method for constructing large error-correcting code dictionaries is to try to combine two or more smaller dictionaries in such a way as to preserve their desired properties in the larger code. The major

†Actually Hamming limited his investigation to binary codes. Here we denote by Hamming codes all those which have the parameters of Theorem 13.16. The Hamming codes defined here are cyclic only in the binary case. Other cyclic Hamming codes do exist, although not for all possible parameters (cf. Reference 13.5).

difficulty encountered in the use of large codes is in the increased encoding and especially decoding complexity necessitated by them. When the large code is obtained by combining smaller codes, however, the encoding and decoding procedures are often little more complex than those associated with the individual component codes. Two methods for efficiently combining codes are discussed in this section.

The first general construction involves the *Kronecker product* of the generators $\mathbf{G}_1$ and $\mathbf{G}_2$ of the component codes. The Kronecker product $\mathbf{A} \times \mathbf{B}$ of two matrices $\mathbf{A} = \{a_{ij}\}$ and $\mathbf{B} = \{b_{ij}\}$, both with elements in the same field F, is defined as follows:

$$\mathbf{A} \times \mathbf{B} = \begin{bmatrix} a_{11}\mathbf{B} & a_{12}\mathbf{B} & \cdots & a_{1n}\mathbf{B} \\ a_{21}\mathbf{B} & a_{22}\mathbf{B} & \cdots & a_{2n}\mathbf{B} \\ \cdot & \cdot & & \cdot \\ \cdot & \cdot & & \cdot \\ \cdot & \cdot & & \cdot \\ a_{v1}\mathbf{B} & a_{v2}\mathbf{B} & \cdots & a_{vn}\mathbf{B} \end{bmatrix} \tag{13.7.1}$$

the matrix $\mathbf{A}$ having the dimensions $v \times n$. If $\mathbf{B}$ has the dimension $\mu \times m$, the Kronecker product has the dimensions $\mu v \times mn$. (The product $a\mathbf{B}$ indicates the matrix obtained by multiplying each element of $\mathbf{B}$ by the field element a, the multiplication of course being over the field F. The products $\mathbf{A} \times \mathbf{B}$ and $\mathbf{B} \times \mathbf{A}$, incidentally, are not in general equal, although these matrices are easily seen to be combinatorially equivalent.)

If $\mathbf{G}_1$ and $\mathbf{G}_2$ are generator matrices of two linear codes, the *Kronecker product code* is the code generated by the matrix $\mathbf{G}_1 \times \mathbf{G}_2$.

Theorem 13.17. If $\mathbf{G}_i$ is a $k_i \times n_i$ matrix which generates an (n_i, k_i) linear code D_i having minimum distance d_i over $GF(q)$, then the Kronecker product code generated by $\mathbf{G}_1 \times \mathbf{G}_2$ is an (n_1n_2, k_1k_2) code having minimum distance d_1d_2.

Proof. Consider an arbitrary linear combination of rows of $\mathbf{G}_1 \times \mathbf{G}_2$. Because any word in D_1 contains at least d_1 non-zero elements, any word in the code generated by $\mathbf{G}_1 \times \mathbf{G}_2$ (except the all-zeros word) must contain d_1 or more occurrences of some non-zero multiple of some word in D_2. But since any word in D_2 must contain at least d_2 non-zero elements, any word in the Kronecker product code must contain at least d_1d_2 non-zero elements. Because no linear combination of the k_1k_2 rows of $\mathbf{G}_1 \times \mathbf{G}_2$ produces the all-zeros vector, the dimension of $\mathbf{G}_1 \times \mathbf{G}_2$ must be k_1k_2. Finally, the length of a word in $\mathbf{G}_1 \times \mathbf{G}_2$ is obviously n_1n_2 and the theorem is proved.

Another useful construction procedure is related to the Kronecker product method but yields codes with smaller dimensions and with larger minimum distances than does the latter. For lack of a better term, we denote these codes as *Kronecker sum*† codes. The Kronecker sum $\mathbf{A} \boxplus \mathbf{B}$ of two matrices $\mathbf{A} = \{\alpha_{ij}\}$ and $\mathbf{B} = \{\beta_{ij}\}$ is defined as follows:

$$\mathbf{A} \boxplus \mathbf{B} = \begin{bmatrix} \alpha_{11} + \mathbf{B} & \alpha_{12} + \mathbf{B} & \cdots & \alpha_{1m} + \mathbf{B} \\ \cdot & \cdot & & \cdot \\ \cdot & \cdot & & \cdot \\ \cdot & \cdot & & \cdot \\ \alpha_{\mu 1} + \mathbf{B} & \alpha_{\mu 2} + \mathbf{B} & \cdots & \alpha_{\mu m} + \mathbf{B} \end{bmatrix} \tag{13.7.2}$$

where $\mathbf{A}$ is a $\mu \times m$ matrix and $\mathbf{B}$ a $\nu \times n$ matrix, both defined over the same field F. The sum $a + \mathbf{B}$ indicates the matrix obtained by adding a to every element of $\mathbf{B}$ over F.

Theorem 13.18. Let $\mathbf{A}$ be a matrix representing an M m-symbol word dictionary, with the Hamming distance separating any two words at least d_A and at most $m - d_A$. Similarly, let $\mathbf{B}$ represent a dictionary consisting of N n-symbol words having minimum Hamming distance d_B. Then $\mathbf{C} = \mathbf{A} \boxplus \mathbf{B}$ represents a dictionary having NM mn-symbol words and with minimum Hamming distance d_c where

$$d_c \geq \min\,[md_B, nd_A]$$

Proof. The distance between any two rows in $\mathbf{C}$ is equal to the number of non-zero elements in their difference; i.e., the distance between the ijth row of $\mathbf{C}$ $(\alpha_{i1} + \mathbf{b}_j, \alpha_{i2} + \mathbf{b}_j, \ldots, \alpha_{im} + \mathbf{b}_j)$ ($\mathbf{b}_j$ denoting the jth row of $\mathbf{B}$) and its $\mu\nu$th row $(\alpha_{\mu 1} + \mathbf{b}_\nu, \alpha_{\mu 2} + \mathbf{b}_\nu, \ldots, \alpha_{\mu m} + \mathbf{b}_\nu)$ is the number of non-zero elements in the mn-tuple

$$(\alpha_{i1} - \alpha_{\mu 1}) + (\mathbf{b}_j - \mathbf{b}_\nu), (\alpha_{i2} - \alpha_{\mu 2}) + (\mathbf{b}_j - \mathbf{b}_\nu), \ldots, (\alpha_{im} - \alpha_{\mu m}) + (\mathbf{b}_j - \mathbf{b}_\nu) \tag{13.7.3}$$

Let $d(A)$ be the distance between the ith and μth rows of $\mathbf{A}$ and $d(B)$ the distance between the jth and νth rows of $\mathbf{B}$. Whenever $\alpha_{il} - \alpha_{\mu l}$ equals zero, $d(B)$ of the elements in the corresponding term of (13.7.3) will be non-zero. When $\alpha_{il} - \alpha_{\mu l}$ is different from zero, the corresponding term of (13.7.3) will contain at least $n - d(B)$ non-zero elements. Letting $d(C)$ be the distance between these two rows in $\mathbf{C}$, we have

$$d(C) \geq [m - d(A)]d(B) + d(A)[n - d(B)] \tag{13.7.4}$$

Since at least one of the conditions $i \neq \mu$ or $j \neq \nu$ must be true if the two rows in $\mathbf{C}$ are to be distinct, we need consider only the following three situations:

(1) $i \neq \mu$ and $d_A \leq d(A) \leq m/2$. Then $d(C) \geq [m - 2d(A)]d(B) + nd(A) \geq nd_A$.

(2) $i \neq \mu$ and $m/2 \leq d(A) \leq m - d_A$. Then $d(C) \geq nd(A) + n[m - 2d(A)] = n[m - d(A)] \geq nd_A$.

(3) $i = \mu$. Then $d(A) = 0$ and $d(C) \geq md(B) \geq md_B$.

This proves the theorem.

---

†This term should not be confused with the term *direct-sum* as defined, for example, in Reference 13.10.

Corollary. If **A** and **B** represent (n_1, k_1) and (n_2, k_2) *linear* codes, respectively, with the respective minimum distances d_1 and d_2, and if the maximum weight of any row in **A** does not exceed $n_1 - d_1$, the Kronecker sum code represented by **C** is also linear with the corresponding parameters n_1n_2, $k_1 + k_2$, and $d_c = \min(n_1d_2, n_2d_1)$.

Proof. Let $\mathbf{a}_i^{(n_2)}$ be the n_1n_2-tuple obtained by repeating n_2 times each component of $\mathbf{a}_i$, the ith generator of **A**: $\mathbf{a}_i = \alpha_{i1}, \alpha_{i2}, \ldots, \alpha_{in_1}$; $\mathbf{a}_i^{(n_2)} = \alpha_{i1}, \alpha_{i1}, \ldots, \alpha_{i1}, \alpha_{i2}, \ldots, \alpha_{i2}, \alpha_{i3}, \ldots, \alpha_{in_1}$. Let $\mathbf{b}_j^{(n_1)} = \mathbf{b}_j, \mathbf{b}_j, \ldots, \mathbf{b}_j$ be the n_1n_2-tuple resulting from repeating the jth generator of **B**, n_1 times. Then **C** can be generated by the k_1 generators $\mathbf{a}_i^{(n_2)}$ combined with the k_2 generators $\mathbf{b}_j^{(n_1)}$ and hence is linear. Moreover, since **A** must contain at least one n_1-tuple of weight d_1, we can, without loss of generality, let $\mathbf{a}_1$ be a vector of this weight. Similarly, $\mathbf{b}_1$ can be a vector of weight d_2. But **C** contains both $\mathbf{a}_1^{(n_2)}$ of weight n_2d_1 and $\mathbf{b}_1^{(n_1)}$ of weight n_1d_2. Thus, the inequality of Theorem 13.18 is satisfied with equality when the dictionaries represented by **A** and **B** are both linear.

Kronecker product codes are sometimes also called *iterative* codes; they can be regarded as an iteration of the coding process. Suppose the information symbols are first grouped into blocks of k_2 symbols and encoded into words of length n_2. The process can then be iterated by grouping these words into blocks of k_1 words and adding $n_1 - k_1$ parity-check *words* to each block. If this is done by taking the ith symbol from each of these k_1 words and encoding these k_1 symbols into a word of length n_1 for each $i = 1, 2, \ldots, n_2$, the resulting (n_1n_2, k_1k_2) code is easily seen to be a Kronecker product code. This iteration process, of course, can be repeated. This is equivalent to defining Kronecker product codes in terms of codes which are themselves Kronecker product codes.

This interpretation of Kronecker product codes leads to a rather interesting and useful generalization. Suppose instead of imposing a second code structure on the *symbols* of the words resulting from the first encoding, we encode the *words* themselves, treating these words as elements in $GF(q^{k_2})$. This could be done by associating with the k_2 information symbols $\alpha_1, \alpha_2, \ldots, \alpha_{k_2}$ of each word a polynomial $p(x) = \sum_{i=1}^{k_2} \alpha_i x^{i-1}$ and carrying out the arithmetic modulo an irreducible polynomial of degree k_2. The parity-check "symbols" determined by the second code structure could then be represented as words from the original (n_2, k_2) code. [In practice, the encoding operations would presumably be reversed; the information first encoded over $GF(q^{k_2})$ and the resulting "symbols" then encoded over $GF(q)$.] If the second code is an (n_1, k_1) linear code (over the field of q^{k_2} elements) the code resulting after the combined coding is an (n_1n_2, k_1k_2) linear code over the field of q elements. Moreover, the distance separating any two words of this code is easily seen to be $d_1 d_2$, the product of the distances of the two individual codes (i.e., any two of the q^{k_2}-ary code words must differ in at least d_1 "symbol" positions and any two different "symbols" must differ in at least d_2 places). The resulting code is called a *concatenated code* (Reference 13.8).

At first glance it would appear that concatenated codes offer no advantage over Kronecker product codes. In both cases, combining an (n_1, k_1) code with minimum distance d_1 and an (n_2, k_2) code with minimum distance d_2 produces an $(n_1 n_2, k_1 k_2)$ code with minimum distance $d_1 d_2$. However, both of the Kronecker product component codes were defined over the field of q elements, while one of the codes giving rise to the concatenated code was defined over the field of q^{k_2} elements. If D_1 is an (n_1, k_1) code over $GF(q^k)$, the maximum attainable minimum distance separating any two of its words will be at least as great when $k > 1$ as it is when $k = 1$. In particular, D_1 can be realized as the Kronecker product of an (n_1, k_1) code with the (k, k) code, both defined over $GF(q)$. The dictionary produced by interpreting the resulting code with $k = k_2$ as an (n_1, k_1) code over $GF(q^{k_2})$ and concatenating it with an (n_2, k_2) code is identical to the dictionary obtained by forming the Kronecker product of the original (n_1, k_1) and (n_2, k_2) dictionaries. Every Kronecker product code, therefore, can be realized as a concatenated code. The converse of this statement is not true.

The major advantage in combining smaller codes to produce larger ones is that the resulting codes can be decoded by cascading the decoders associated with each of the component codes. Concatenated codes, for example, can be decoded by first dividing the received symbol sequence into blocks of n_2 symbols and decoding each block relative to the dictionary D_2. The output of this decoder can then be interpreted as a sequence of n_1-symbol words over $GF(q^{k_2})$ and decoded relative to D_1. Similar approaches can be used with Kronecker product and Kronecker sum codes. The decoder complexity therefore need increase only arithmetically while the code dimensions increase geometrically. The cascaded decoders, however, may not be able to correct all error patterns which are actually correctable. Certain error patterns, consisting of only $(d_1 + 1)(d_2 + 1)/4$ errors, for example, can cause a decoding error when concatenated or Kronecker product codes are decoded, whereas up to $(d_1 d_2 - 1)/2$ errors are in principle correctable. On the other hand, many error patterns of weight greater than $(d_1 + 1)(d_2 + 1)/4$ will be corrected with cascaded decoders, and, if the component codes are properly chosen, the overall results can be quite satisfactory.

One of the most interesting classes of codes resulting from the Kronecker sum construction are those obtained by repeatedly taking the Kronecker sum of the binary matrix

$$\mathbf{D}_1 = \begin{bmatrix} 00 \\ 01 \end{bmatrix}$$

with itself. The resulting matrix, after the ith iteration, has the form

$$\mathbf{D}_i = \mathbf{D}_1 \boxplus \mathbf{D}_{i-1} = \begin{bmatrix} \mathbf{D}_{i-1} & \mathbf{D}_{i-1} \\ \mathbf{D}_{i-1} & \bar{\mathbf{D}}_{i-1} \end{bmatrix}$$

where $\bar{\mathbf{D}}_{i-1}$ is the complement of $\mathbf{D}_{i-1}$ (the matrix obtained by replacing each symbol in $\mathbf{D}_{i-1}$ by its complement). Letting d_i indicate the minimum distance separating any two rows in $\mathbf{D}_i$, we conclude from the corollary to Theorem 13.18 that, since $d_1 = 1 = n_1/2$, $d_i = n_i/2$ for all i. These matrices, incidentally, when combined with their complements $\bar{\mathbf{D}}_i$, represent the well-known first-order *Reed-Muller codes* (Reference 13.11). We shall encounter these same codes in another context shortly.

In concluding this section we should mention that the methods discussed here for combining smaller codes to produce larger ones can also be applied to code null spaces to define new null spaces, sometimes with useful results (cf. Reference 13.12).

13.8 Convolutional Codes

Another technique for increasing the effective code word length without increasing correspondingly the complexity of the associated encoder and decoder is offered by the convolutional encoding approach. *Convolutional codes* are perhaps easiest to define in terms of their generator structure. The generator $\mathbf{G}$ of a convolutional code has the form

$$\mathbf{G} = \begin{bmatrix} \mathbf{G}_0 & \mathbf{G}_1 & \mathbf{G}_2 & \cdots & \mathbf{G}_{m-1} & \mathbf{O} & \mathbf{O} & \mathbf{O} & \cdots \\ \mathbf{O} & \mathbf{G}_0 & \mathbf{G}_1 & \cdots & \mathbf{G}_{m-2} & \mathbf{G}_{m-1} & \mathbf{O} & \mathbf{O} & \cdots \\ \mathbf{O} & \mathbf{O} & \mathbf{G}_0 & \cdots & \mathbf{G}_{m-3} & \mathbf{G}_{m-2} & \mathbf{G}_{m-1} & \mathbf{O} & \cdots \\ \cdot & \cdot & \cdot & & \cdot & \cdot & \cdot & \cdot & \\ \cdot & \cdot & \cdot & & \cdot & \cdot & \cdot & \cdot & \\ \cdot & \cdot & \cdot & & \cdot & \cdot & \cdot & \cdot & \end{bmatrix} \tag{13.8.1}$$

where $\mathbf{G}_l$, $l = 0, 1, \ldots, m-1$, is a $\mu \times \nu$ matrix with elements in some field $GF(q)$, $\mathbf{G}_l = \{g_{ij}^{(l)}\}$, and $\mathbf{O}$ represents a $\mu \times \nu$ matrix of zeros. Thus, if $\mathbf{x} = (x_0, x_1, \ldots)$ denotes the sequence of information symbols (also defined over $GF(q)$) and $\mathbf{y} = (y_0, y_1, \ldots)$ the sequence of symbols to be transmitted,

$$\mathbf{y} = \mathbf{x}\mathbf{G} \tag{13.8.2}$$

The information rate is μ/ν.

In principle, the matrix $\mathbf{G}$ can be semi-infinite. Consequently, a convolutional code, in contrast to the other codes discussed in this chapter, is not a block code; there is no one-to-one mapping between (finite) blocks of information symbols and blocks of code symbols. Notice, however, that any one of the first μ information symbols affects only the first $m\nu$ code symbols. As we shall soon discover, this *constraint length* $n \triangleq m\nu$ is in many ways analogous to the block length (code word length) in ordinary error-correcting codes. In the discussion to follow, $\mathbf{G}$ will be assumed to be in systematic form; that is

$$g_{ij}^{(l)} = \begin{cases} 1 & i = j = 1, 2, \ldots, \mu \qquad l = 0 \\ 0 & i, j \leq \mu, i \neq j \qquad l = 0 \\ 0 & \text{all } i, j \leq \mu \qquad l \neq 0 \end{cases} \tag{13.8.3}$$

Thus, the information symbols will appear explicitly in the code sequence $\mathbf{y}$ with $x_{i\mu+j} = y_{i\nu+j}$ for all $i \geq 0, 0 \leq j \leq \mu - 1$.

As with block codes, an indication of the relative merits of different convolutional encodings can be found in the minimum Hamming distance separating any two code words. Since convolutional code words are (potentially) infinitely long, the appropriate definition of distance is less obvious in this case, however, and in fact depends on the particular decoding algorithm to be used. One useful definition is the minimum distance d separating any two distinct *initial words* $\mathbf{y}_0 = \{y_0, y_1, \ldots, y_{n-1}\}$, at least one of which has some non-zero information symbol y_i, $0 \leq i \leq \mu - 1$. Since convolutional codes (as defined here) are linear, d represents the minimum weight of any initial word having at least one of its first μ symbols other than zero.

The significance of this definition of d is as follows: Because the code constraint length is $n = m\nu$, the first μ information symbols $x_i = y_i$, $0 \leq i \leq \mu - 1$, affect only the initial word $\mathbf{y}_0$. As soon as the first n symbols are available at the receiver, a decision can be made concerning these μ information symbols. (A more reliable decision is possible if the decision is delayed, however; see Problem 13.8.) In general, the ith set of information symbols, $x_{i\mu+j} = y_{i\nu+j}$, $j = 0, 1, \ldots, \mu - 1$, affects only the ith *word* $\mathbf{y}_i \triangleq \{y_{i\nu}, y_{i\nu+1}, \ldots, y_{n+i\nu-1}\}$. And if the previous $m - 1$ sets of information symbols have been correctly identified, their influence on $\mathbf{y}_i$ is known and $\mathbf{y}_i$ is effectively also an initial word. Thus, if no prior decoding errors have been committed, d furnishes a measure of reliability with which the ith set of information symbols can be identified. (A decoder which utilizes previously decoded symbols in its subsequent decoding efforts, as suggested here, is called a *feedback* decoder.)

Once a decoding error is made, of course, the effect of the preceding information symbols will not be correctly accounted for, and subsequent decoding errors are considerably more likely. Because of this tendency for errors to propagate, convolutional code words are often truncated in practice. This can be accomplished by dividing the information symbols into blocks of length, say, $K = (L - m + 1)\mu$, and following each block with $(m - 1)\mu$ zeros. The resulting truncated convolutional code is in actuality an (N, K) block code with block length $N = L\nu$, although, when L is large, this fact is of minor consequence.

Since the minimum distance d separating initial words having non-identical initial information symbols is a convolutional code counterpart to the block code distance discussed in earlier sections of this chapter, it is of interest to find bounds on this quantity analogous to those in Section 13.3. The following two theorems exhibit such bounds on this distance d which we shall hereafter refer to as the *initial distance.*

Theorem 13.19. There exists an $(n = m\nu, k = m\mu)$ q-ary convolutional code having a minimum distance at least d for any d satisfying the inequality

$$\sum_{j=1}^{d-1}\left[\binom{n}{j} - \binom{n-\mu}{j}\right](q-1)^j < q^{n-k} \tag{13.8.4}$$

Proof. Let $\mathbf{y}_0 = \{y_0, y_1, \ldots, y_{n-1}\}$ be a q-ary n-tuple having at least one non-zero component y_i for some $i < \mu$, and let $N(\mathbf{y}_0)$ be the maximum number of distinct $(m\nu, m\mu)$ convolutional codes in which any such n-tuple is an initial word. Further, let $N(d)$ denote the total number of n-tuples of the form $\mathbf{y}_0$ having weight less than d. Finally, let $N(m\nu, m\mu)$ be the number of distinct $(m\nu, m\mu)$ convolutional codes. Then if

$$N(\mathbf{y}_0)N(d) < N(m\nu, m\mu) \tag{13.8.5}$$

there must exist at least one $(m\nu, m\mu)$ convolutional code having no initial word (other than those beginning with μ or more zeros) of weight less than d, and hence, having an initial distance at least d.

Clearly, $N(d)$ can be obtained by subtracting from the total number of q-ary n-tuples of weight less than d, the number of such n-tuples beginning with μ or more zeros; i.e.,

$$N(d) = \sum_{j=1}^{d-1}\left[\binom{n}{j} - \binom{n-\mu}{j}\right](q-1)^j \tag{13.8.6}$$

Moreover, because there are exactly $q^{(\nu-\mu)\mu}$ distinct matrices $\mathbf{G}_l$ [see Equations (13.8.1) and (13.8.3)] and because m such matrices define an $(m\nu, m\mu)$ convolutional code,

$$N(m\nu, m\mu) = q^{m(\nu-\mu)\mu} \tag{13.8.7}$$

It remains only to determine $N(\mathbf{y}_0)$. To this end, let $\boldsymbol{\eta}_j$ be the m-component vector consisting of the $(i\nu + j)$th components of $\mathbf{y}_0$, $i = 0, 1, \ldots, m-1$; i.e., $\boldsymbol{\eta}_j \triangleq \{y_j, y_{\nu+j}, \ldots, y_{(m-1)\nu+j}\}$. Let $\mathbf{M}_i$ denote the $m \times m$ triangular matrix

$$\mathbf{M}_i = \begin{bmatrix} x_i & 0 & 0 & \cdots & 0 \\ x_{\mu+i} & x_i & 0 & \cdots & 0 \\ x_{2\mu+i} & x_{\mu+i} & x_i & \cdots & 0 \\ \cdot & \cdot & \cdot & & \cdot \\ \cdot & \cdot & \cdot & & \cdot \\ \cdot & \cdot & \cdot & & \cdot \\ x_{(m-1)\mu+i} & \cdot & \cdot & \cdots & x_i \end{bmatrix}$$

where $(x_{i\mu}, x_{i\mu+1}, \ldots, x_{(i+1)\mu-1}) = (y_{i\nu}, y_{i\nu+1}, \ldots, y_{i\nu+\mu-1})$ represents the ith block of information symbols in $\mathbf{y}_0$. And let

$$\boldsymbol{\gamma}_{ij} = (g_{ij}^{(0)}, g_{ij}^{(1)}, \ldots, g_{ij}^{(m-1)})^T$$

with $g_{ij}^{(l)}$ the ijth element of the matrix $\mathbf{G}_l$. Then Equation (13.8.2) implies

$$\boldsymbol{\eta}_j = \sum_{i=0}^{\mu-1} \mathbf{M}_i \boldsymbol{\gamma}_{ij} \qquad j = \mu, \mu + 1, \ldots, \nu - 1 \tag{13.8.8}$$

At least one of the terms x_i, say x_{i_0}, must be non-zero. Since $\mathbf{M}_{i_0}$ is then non-singular, Equation (13.8.8) can be rewritten as

$$\boldsymbol{\gamma}_{i_0 j} = \mathbf{M}_{i_0}^{-1}\left\{\boldsymbol{\eta}_j - \sum_{\substack{i=0\\ i\neq i_0}}^{\mu-1} \mathbf{M}_i \boldsymbol{\gamma}_{ij}\right\} \qquad j = \mu, \mu + 1, \ldots, \nu - 1 \tag{13.8.9}$$

These equations must all be satisfied for any convolutional code capable of generating the initial word $\mathbf{y}_0$. Accordingly, for any specific $\mathbf{y}_0$, $m(\nu - \mu)$ of the arbitrary elements of the m matrices $\mathbf{G}_l$ are uniquely specified in terms of the other $m(\nu - \mu)(\mu - 1)$ such elements, and

$$N(\mathbf{y}_0) = q^{m(\nu-\mu)(\mu-1)} \tag{13.8.10}$$

Substituting Equations (13.8.6), (13.8.7), and (13.8.10) into the inequality (13.8.5) yields the desired result.

This bound is the convolutional code analog of the Gilbert bound for block codes. When the block length is identified with the constraint length, the block code dimension with $m\mu$ (the constraint length measured in terms of information symbols), and the block code distance with the convolutional code initial distance, the two bounds are in fact identical when $\mu = 1$ and $q = 2$. The Gilbert bound, however, was constructive while its convolutional code analog presented here is only an existence theorem.

The upper bounds on the distance attainable with (n, k) block codes (Section 13.3) are also clearly applicable to the initial distance attainable with convolutional codes when n is replaced by the constraint length $m\nu$ and k by μ. Such bounds, of course, ignore the effect of the possibly non-zero information symbols $(x_{i\mu}, x_{i\mu+1}, \ldots, x_{(i+1)\mu-1})$, $i > 0$, on the weight of an initial word $\mathbf{y}_0$. A frequently tighter bound which does take this effect into account is demonstrated in the following theorem.†

Theorem 13.20. The minimum initial distance d of a q-ary $(n = m\nu, k = m\mu)$ convolutional code is overbounded by

$$d \leq \frac{q-1}{q}\left\{n + \frac{\nu}{q^\mu - 1}\right\} \tag{13.8.11}$$

Proof. Clearly, d cannot exceed the average weight d_{ave} of all the initial words $\mathbf{y}_0$ having at least one non-zero initial information symbol y_i, $i \leq \mu - 1$. These words are all contained in the subset of words generated by the matrix $\mathbf{M}_1$, consisting of the first $m\mu$ rows and $m\nu$ columns of the matrix (13.8.1), but not by the matrix $\mathbf{M}_2$, consisting of the last $(m - 1)\mu$ rows and $(m - 1)\nu$ columns of

†Generally stronger upper bounds can be found in References 13.13 and 13.35.

$\mathbf{M}_1$. If $\mathbf{M}_1$ (and hence $\mathbf{M}_2$) has no all-zeros columns (such columns would contribute nothing to the distance of the resulting code), it generates a dictionary containing a total of $N(m\nu, m\mu, q)$ non-zero q-ary symbols where $N(n, k, q) \triangleq nq^{k-1}(q-1)$ (see Lemma 13.4). Thus,

$$(q^{\mu} - 1)q^{(m-1)\mu}d_{\text{ave}} = N(m\nu, m\mu, q) - N[(m-1)\nu, (m-1)\mu, q] \qquad (13.8.12)$$

and the stated inequality immediately follows.

Techniques for the construction of good convolutional codes are much less developed than those for block codes. Most convolutional codes are constructed by some sort of search procedure or random selection combined with certain heuristic restrictions designed to limit the set of contending codes to those showing the most promise (cf. References 13.15, 13.16). One sometimes useful scheme for constructing convolutional codes from block codes is as follows: Let $\mathbf{G}^{(l)}$, $l = 1, 2, \ldots, m$ represent m sets of generators with $\mathbf{G}^{(l)}$ the generator of a $(\nu, l\mu)$ linear block code having a minimum distance d_l, and with the set $\mathbf{G}^{(l)}$ contained in the set $\mathbf{G}^{(l+1)}$ for all $l = 1, 2, \ldots, m-1$. Then the code obtained by including in the $\mathbf{G}_i$ of Equation (13.8.1) those generators in $\mathbf{G}^{(i+1)}$ but not in $\mathbf{G}^{(i)}$ is an $(m\nu, m\mu)$ convolutional code with a minimum initial distance obviously underbounded by

$$d \geq \sum_{l=1}^{m} d_l \qquad (13.8.13)$$

Note that the resulting convolutional code can be reduced to systematic form by so reducing $\mathbf{G}^{(m)}$.

As an example, let $\mu = 1$, and let $\mathbf{G}^{(m)}$ be the Kronecker product of the generator of a (q^{m-1}, m) punctured cyclic code having distance $d = (q-1)q^{m-2}$ (Section 13.6) with an r-tuple having only non-zero components. Letting $\mathbf{G}_0$ be the maximum weight rq^{m-1}-tuple in $\mathbf{G}^{(m)}$, we have

$$d_l = \begin{cases} rq^{m-1} & l = 1 \\ r(q-1)q^{m-2} & 1 < l \leq m \end{cases}$$

Thus, the construction described in the previous paragraph yields an (rmq^{m-1}, m) convolutional code with, from Equation (13.8.13), an initial distance bounded by

$$d \geq d_0 \triangleq rq^{m-2}[(q-1)m + 1]$$

But, from Equation (13.8.12), $d_0 = d_{\text{ave}}$, so $d = d_0$ and the resulting code attains its maximum possible initial distance. Since $d = d_{\text{ave}}$, all initial words having at least one non-zero initial information symbol must have weight d; such codes are called *uniform* convolutional codes.

Another class of convolutional codes results when $\mathbf{G}^{(l)}$ generates a BCH code or code subspace. That is, $\mathbf{G}^{(l)}$ can be defined as a subset of the generators of some (n, k) BCH code, with $k \geq \mu l$ the dimension of the smallest BCH code containing all of the generators in the sets $\mathbf{G}^{(i)}$ for $i < l$. Thus, if $q = 2$, $\nu = 63$, and $\mu = 1$, $\mathbf{G}^{(1)}$ can be taken as the generator of the (63, 1) binary BCH code, $\mathbf{G}^{(i)}$ as a subset of the generators of the (63, 7) BCH code for all $2 \leq i \leq 7$, etc. Proceeding in this manner, we obtain a set of distances (cf. Section 13.5)

$$d_l \geq \begin{cases} 63 & l = 1 \\ 31 & 2 \leq l \leq 7 \\ 27 & 8 \leq l \leq 10 \\ 23 & 11 \leq l \leq 16 \\ 21 & 17 \leq l \leq 18 \end{cases}$$

If $m = 18$, for example, the resulting convolutional code has a constraint length $n = 1134$ and a distance $d \geq 510$. For comparison, the upper bound of Theorem 13.20 limits d to 598 or less. Similarly, if $\mu = 7$, we find

$$d_l = \begin{cases} 31 & l = 1 \\ 23 & l = 2 \\ 15 & l = 3 \\ 13 & l = 4 \\ 11 & l = 5 \end{cases}$$

etc.

and, if $m = 5$, $n = 315$, and $d \geq 93$, while Theorem 13.20 overbounds d by 157.

In summary, it has been shown that the minimum initial distance attainable with convolutional codes satisfies bounds remarkably similar to those derived earlier for block codes having a block length equal to the convolutional code constraint length. While good general constructions for convolutional codes seem to be more elusive than those for block codes, reasonably good, and in some cases optimum, convolutional codes can sometimes be defined in terms of block code generators. Since the distance properties of convolutional and block codes are apparently similar, the advantages, if any, of the convolutional approach are not immediately obvious. We shall return to this point presently.

13.9 Encoding and Decoding Linear Codes

The merits of the various coding schemes discussed above cannot be properly assessed without answers to the following two questions: (1) How difficult is it to implement the required encoder and decoder? (2) What is the probability of a decoding error? The purpose of this section and of the one following is to provide at least partial answers to these questions.

Clearly, the answers to these two questions will be interrelated, both depending on the specific decoding algorithm to be used. As we will discover in Section 13.10, the error probability P_e is typically an exponentially decreasing function of the code word length n. The complexity of the optimum decoder, however, can be an exponentially *increasing* function of n. This provides a strong impetus for a search for more easily implementable, perhaps suboptimum, decoding procedures which still retain this exponential relationship between P_e and n. Even if the resulting error probability is larger than that attainable with an optimum decoder for a given n, the simpler decoder could conceivably be used with much longer code words, and hence yield superior performance. This decoding problem has been the subject of much active research, and a number of effective algorithms are known. A detailed description of these algorithms is beyond the scope of this chapter. The major features of the most important of these algorithms, however, will be outlined; more details can be found in the references.

Before beginning the discussion of specific decoders, it is well to ask whether similar limitations are not also imposed by the encoder. Most generally, this is not the case, at least when the code in question is linear. An encoder for a linear code is shown in Figure 13.1. The long blocks in

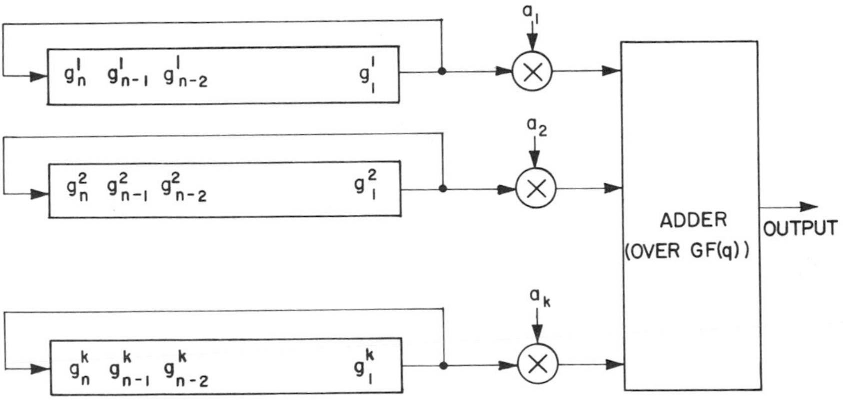

Figure 13.1 A linear code encoder.

Figure 13.1 represent circulating storage devices, and the circles product devices, the output of which is the product over $GF(q)$ of the two inputs. The components of the k-tuples $(a_1, a_2, \ldots, a_k)$ to be encoded are used as one of the inputs to each of the multipliers, the k generators $\mathbf{g}^i = (g_1^i, g_2^i, \ldots, g_n^i)$, $i = 1, 2, \ldots, k$, of the (n, k) code provide the other inputs. Thus, the output will be the n-tuple

$$\mathbf{g} = \sum_{i=1}^{k} a_i \mathbf{g}^i \tag{13.9.1}$$

as desired.

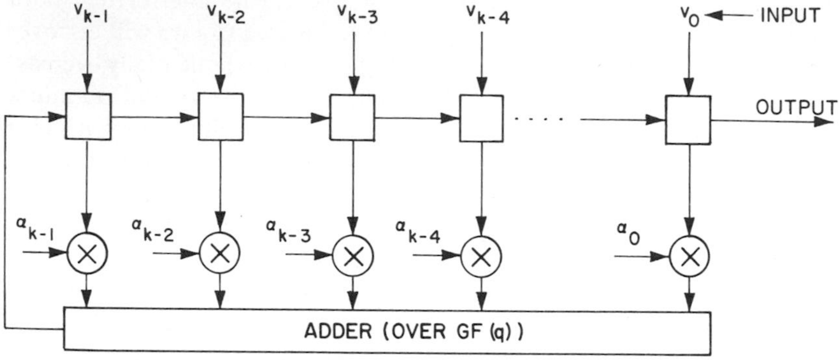

Figure 13.2 Cyclic code shift-register encoder.

Cyclic codes afford even greater encoder efficiencies. Recall from Section 13.4 that a cyclic code has associated with it a parity-check vector $\mathbf{h} = (h_0, h_1, \ldots, h_k)$ such that $\mathbf{v} = (v_0, v_1, \ldots, v_{n-1})$ is a code word if and only if

$$\sum_{i=r}^{k+r} h_{i-r}v_i = \sum_{i=0}^{k} h_i v_{i+r} = 0 \tag{13.9.2}$$

for all $r = 0, 1, \ldots, n-k-1$. Therefore, the symbols v_i, for $i \geq k$, are defined in terms of the information symbols $v_j, j = 0, 1, \ldots, k-1$, by the recursion relationship

$$v_{k+r} = -\frac{1}{h_k}\sum_{i=0}^{k-1} h_i v_{i+r} \triangleq \sum_{i=0}^{k-1} \alpha_i v_{i+r} \tag{13.9.3}$$

which gives rise to the encoder depicted in Figure 13.2. Such a device, called a *feedback shift-register*,† consists of k memory cells so connected that upon receipt of a clock pulse (not shown) the content of each cell shifts one cell to the right while their weighted sum [over $GF(q)$] shifts into the leftmost cell. When the initial input $v_0, v_1, \ldots, v_{k-1}$, representing the information symbols, is shifted into the k shift register cells, the first n output symbols from the cell on the extreme right of the diagram will comprise the desired code word. Thus, the entire encoder is extremely compact. At least in the case of binary codes, the encoder elements are readily constructed, the modulo-two multiplier being a closed connection if $\alpha_k = 1$ and an open circuit if $\alpha_k = 0$, and the modulo-two adder realizable with conventional logical elements.‡

†Again see Section 6.6. There the discussion began with the (binary) maximal-length shift-register and demonstrated the properties of the resulting sequences. Here we see the shift-register as the consequence of the cyclic nature of the code dictionary, and as such of considerably greater versatility than was indicated in Section 6.6. Note that the feedback connections are determined by the null-space generator polynomial $h(x)$.

‡It is also possible to encode cyclic codes with an $(n-k)$-stage shift-register (cf. Problem 13.2), an obvious advantage when $k > n/2$.

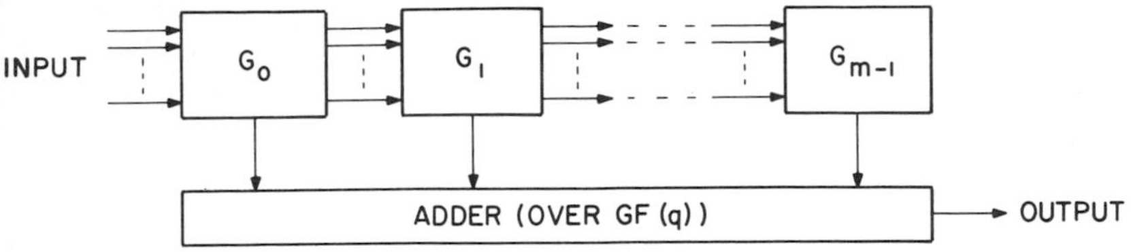

Figure 13.3 Convolutional code encoder.

Convolutional coders are equally amenable to efficient implementation, the device in Figure 13.3 representing one such approach. The boxes labeled $\mathbf{G}_i$, $i = 0, 1, \ldots, m - 1$, represent linear (ν, μ) block coders of the form shown in Figures 13.1 or 13.2. The information symbols are impressed at the input to the first of these encoders in blocks of μ symbols; after ν code symbols are generated, the same μ symbols are transferred, as a block, to the next encoder, and a new block of information symbols shifted into the first encoder. The code words obtained by summing [over $GF(q)$] the symbols produced at the outputs of each of these encoders are easily seen to be those generated by a matrix of the form (13.8.1).

Although any precise definition of encoder (or decoder) complexity must take into account the specific implementation used, an indication of this complexity can be obtained by counting the total number of operations (additions, multiplications, etc.) needed to encode (decode) each information symbol. Accordingly, the complexity of the encoder shown in Figure 13.1 is proportional to the word length n while that shown in Figure 13.2 is proportional to n/k. Similarly, the convolutional coder complexity is of the order of, at most, $m\nu = n$. Unless much tighter equipment constraints are applied to the encoder than to the decoder, the encoder complexity therefore is seldom a limiting factor.

Returning to the decoding problem, we remark that there is another important factor which must be taken into consideration in any discussion of the trade off between decoder complexity and performance; viz., the form of the detector output. At its simplest, the detector output is one of the q possible code symbols; at the other extreme it might consist of q voltage levels representing the outputs of filters matched to each of the q channel symbols. In the first instance, a hard decision as to each received symbol is made prior to the decoding operation. Presumably, fewer decoding errors would be made if the decoder were provided more complete information concerning the received symbols. The utilization of this information, on the other hand, would undoubtedly require a more complex decoder. To the extent that a choice as to the form of the detector output exists, this factor must therefore also be entered into the decoder performance versus complexity equation.

To accommodate this additional degree of freedom in a convenient way, we allow the detector output to be any one of the *detector symbols* r_ν, $\nu = 0, 1, \ldots, N - 1$. Not only does this model encompass the two extremes mentioned above (with $N = q$ and $N \to \infty$, respectively), it also includes many intermediate channels of practical interest. Except for these two extremes, however, we limit consideration here to a specialization of this general detector model. Under this more restricted model, each detector output is identified with one of the possible code symbols s_i, $i = 0, 1, \ldots, q-1$. In addition, it is assigned to a *reliability class* C_j, $j = 1, 2, \ldots, J$, as a measure of the confidence with which this identification can be made. Thus, the detector symbols r_ν assume the form $s_i^{(j)}$ the subscript denoting the code symbol with which r_ν is identified and the superscript its reliability class.

Let $\Pr\{s_i^{(j)} \mid s_\nu\}$ be the probability of receiving $s_i^{(j)}$ when s_ν is transmitted. The channel (which of course includes the detector) associated with this model will be assumed symmetric in the sense that the probabilities $\Pr\{s_i^{(j)} \mid s_\nu\} \triangleq \begin{cases} p_{cj} & \nu = i \\ p_{ej} & \nu \neq i \end{cases}$ are independent of both i and ν. For convenience, the reliability classes will be so ranked that $\Pr\{s_i \mid s_i^{(j)}\}$, the probability that the transmitted symbol was s_i when $s_i^{(j)}$ is received, is a decreasing function of j. In general, p_{cj} will also be assumed to be equal to or greater than p_{ej} for all j.

Except when $q = 2$, reliability class detectors usually destroy relevant information concerning the received symbol, even as the number J of classes approaches infinity. Nevertheless, this model does represent many situations of practical interest. If $J = 2$, for example, and the detector output is identified either with one of the q code symbols or else with a $(q + 1)$ st symbol which is equally likely to be any of these ($p_{c2} = p_{e2}$), the resulting channel is an *erasure* channel. More generally, this model includes the case in which the detector symbols are obtained by quantizing the output of the largest of q matched filters, again assuming the likelihood of mistaking one symbol for another is independent of the two symbols in question.

With these preliminaries, we can now outline some of the more important decoding algorithms.

13.9.1 Maximum-Likelihood Decoding

Let $\mathbf{w}_\nu$ denote the νth n-symbol word in the code dictionary and $\mathbf{v}$ an n-tuple of detector symbols corresponding to some received word. Then, assuming all code words are equally likely to be transmitted, the optimum decoder, in the sense of minimizing the word error probability, decodes $\mathbf{v}$ as $\mathbf{w}_l$ where

$$\max_\nu \Pr\{\mathbf{v} \mid \mathbf{w}_\nu\} = \Pr\{\mathbf{v} \mid \mathbf{w}_l\} \tag{13.9.4}$$

i.e., the optimum decoder is a *maximum-likelihood decoder*. If the channel statistics are independent from symbol to symbol

$$\Pr\{\mathbf{v} \mid \mathbf{w}_\nu\} = \prod_{\mu=1}^{n} \Pr\{\eta_\mu \mid \omega_\mu(\nu)\} \tag{13.9.5}$$

where $\omega_\mu(\nu)$ and η_μ represent, respectively, the μth symbols in $\mathbf{w}_\nu$ and $\mathbf{v}$. Generally, each of the code symbols s_i will occur as the μth symbol of some code word for all μ. The maximum-likelihood decoder must therefore determine, in some way or another, nq probabilities $\Pr\{\eta_\mu \mid \omega_\mu(\nu)\}$ and form q^k different products of these probabilities. The decoder complexity (per information symbol) is consequently proportional to $(n/k)q^k$ and, for a fixed rate k/n, is indeed an exponentially increasing function of the block length n.

The practicality with which the probabilities $\Pr\{\eta_\mu \mid \omega_\mu(\nu)\}$ can be calculated obviously depends on the channel statistics. As was demonstrated in Chapter Four, for example, they are relatively easily determined when the channel is a phase-coherent Gaussian channel, less so when the channel is phase-incoherent. In any event, it is frequently expedient to restrict the detector alphabet. A significant decoder simplification results when the detector is a reliability class detector of the type defined earlier. Then Equation (13.9.5) can be written as

$$\Pr\{\mathbf{v} \mid \mathbf{w}_\nu\} = \prod_{j=1}^{J} p_{ej}^{n_{ej}^\nu} p_{cj}^{n_{cj}^\nu} = \exp \sum_{j=1}^{J} \left\{ n_j \log_e p_{cj} - n_{ej}^\nu \log_e \left(\frac{p_{cj}}{p_{ej}} \right) \right\} \tag{13.9.6}$$

where n_{cj}^ν is the number of times a symbol of the form $s_i^{(j)}$ in $\mathbf{v}$ corresponds to a symbol s_i in $\mathbf{w}_\nu$, n_{ej}^ν is the number of times $s_i^{(j)}$ in $\mathbf{v}$ corresponds to s_μ in $\mathbf{w}_\nu$ for some $\mu \neq i$, and $n_j = n_{ej}^\nu + n_{cj}^\nu$ is independent of ν. The maximum-likelihood decoder, then, decodes $\mathbf{v}$ as $\mathbf{w}_l$ if

$$d_\nu \triangleq \sum_{j=1}^{J} n_{ej}^\nu \frac{\log_e \left(\frac{p_{cj}}{p_{ej}} \right)}{\log_e \left(\frac{p_{c1}}{p_{e1}} \right)} \triangleq \sum_{j=1}^{J} \alpha_j n_{ej}^\nu \tag{13.9.7}$$

attains its minimum value when $\nu = l$. (If this situation exists for more than one value of l, any one of the code words $\mathbf{w}_l$ can be selected.)

When $J = 1$, d_ν is simply the Hamming distance between the channel output $\mathbf{v}$ and the νth code word $\mathbf{w}_\nu$ and the maximum-likelihood decoder seeks the code word closest, in the Hamming sense, to the received n-tuple. This fact makes it possible to use the algebraic structure of the code to reduce somewhat the complexity of the maximum-likelihood decoder. Since $\mathbf{v}$ is now a q-ary n-tuple, it belongs to some coset of the linear code dictionary D (Appendix B). The maximum-likelihood decoder therefore needs only to determine the coset to which $\mathbf{v}$ belongs and to subtract from it the *coset leader* (here defined as any one of the minimum weight n-tuples in the coset). This follows because if $\mathbf{v} - \mathbf{c}$ is in D, $\mathbf{v}$ and $\mathbf{c}$ must be in the same coset; since $\mathbf{c}$ is an element of minimum weight in the coset containing $\mathbf{v}$, there can be no word in D closer to $\mathbf{v}$ than $\mathbf{v} - \mathbf{c}$.

A relatively simple method of identifying the coset of the received word is suggested by the following theorem. Let $\mathbf{H}$ be the generator matrix of the null space of an (n, k) code dictionary D. Then define the *syndrome* $\mathbf{s}$ of the vector $\mathbf{v}$ as $\mathbf{s} = \mathbf{v}\mathbf{H}^T$. The vector $\mathbf{s}$ is an $n - k$ dimensional vector over $GF(q)$.

Theorem 13.21. Two vectors $\mathbf{v}_1$ and $\mathbf{v}_2$ are in the same coset of a linear code dictionary D if, and only if, their syndromes, $\mathbf{s}_1$ and $\mathbf{s}_2$, are equal.

Proof. Suppose $\mathbf{s}_1 = \mathbf{s}_2$. Then

$$\mathbf{v}_1\mathbf{H}^T = \mathbf{v}_2\mathbf{H}^T$$

and $(\mathbf{v}_1 - \mathbf{v}_2)\mathbf{H}^T = \mathbf{0}$. Accordingly, $\mathbf{v}_1 - \mathbf{v}_2$ is in D, and $\mathbf{v}_1$ and $\mathbf{v}_2$ are in the same coset. Conversely, suppose $\mathbf{v}_1$ and $\mathbf{v}_2$ are in the same coset. Then $\mathbf{v}_1 = \mathbf{v}_2 + \mathbf{w}_i$ where $\mathbf{w}_i$ is in D, and

$$\mathbf{s}_1 = \mathbf{v}_1\mathbf{H}^T = \mathbf{v}_2\mathbf{H}^T + \mathbf{w}_i\mathbf{H}^T = \mathbf{v}_2\mathbf{H}^T = \mathbf{s}_2$$

Thus, decoding can be accomplished by storing the q^{n-k} coset leaders and identifying them by their associated syndromes. When $\mathbf{v}$ is received, the syndrome $\mathbf{s} = \mathbf{v}\mathbf{H}^T$ is determined. The corresponding coset leader is then subtracted from the original vector $\mathbf{v}$. In many cases, this can be considerably easier to implement than the more direct procedure of finding the distance between each dictionary word and the received word. This is particularly true in the case of cyclic codes, since the product $\mathbf{v}\mathbf{H}^T$ is realized by multiplying the corresponding polynomial $v(x)$ by the null space generator polynomial $h(x)$ and reducing the result modulo $x^n - 1$. This operation is not difficult to implement (Reference 13.5).

This is not to suggest that the code structure can be effectively utilized only when the code and detector alphabets are identical. Maximum-likelihood decoders have been designed for certain Kronecker sum codes requiring of the order of only $n \log_e n$ operations even for the coherent Gaussian channel (cf. Reference 13.14). (The rate of these codes approaches zero asymptotically with n, so this statement must be qualified accordingly.)

13.9.2 Algebraic Decoding

When the detector and code alphabets are identical, the received word $\mathbf{v}$ is of the form $\mathbf{v} = \mathbf{w} + \mathbf{e}$ where $\mathbf{w}$ is a code word and $\mathbf{e} = (e_1, e_2, \ldots, e_n)$ an error vector with components in the code field $GF(q)$. If the code in question is linear, and if $\mathbf{H}$ is the generator of its null space, then, with $\mathbf{v}$, $\mathbf{w}$, and $\mathbf{e}$ represented as row vectors,

$$\mathbf{v}\mathbf{H}^T = (\mathbf{w} + \mathbf{e})\mathbf{H}^T = \mathbf{e}\mathbf{H}^T \tag{13.9.8}$$

The code structure thus defines a set of $n - k$ equations in the n unknowns e_i. This set of equations will not have a unique solution; each such set in fact has q^k solutions, since if $\mathbf{e}$ is a solution, so also is $\mathbf{e} + \mathbf{w}$ for any code word $\mathbf{w}$. Nevertheless, the distance properties of the code guarantee at most one solution having $[(d - 1)/2]$ or fewer non-zero components. Accordingly, at least when the number of transmission errors does not exceed $[(d-1)/2]$, Equation (13.9.8) can be solved uniquely for $\mathbf{e}$ and the correct code word $\mathbf{w} = \mathbf{v} - \mathbf{e}$ identified.

The term *algebraic decoding* refers to a class of decoding algorithms designed to determine systematically the minimum weight solution $\mathbf{e}$ to Equation (13.9.8). Most generally, these algorithms are guaranteed to work only when there exists a solution $\mathbf{e}$ of weight $[(d-1)/2]$ or less. Highly efficient algebraic decoding procedures have been developed, in particular, for the BCH cyclic codes discussed in Section 13.6. The complexity of a BCH decoder can be on the order of only $n \log_e n$, a significant improvement over the exponential dependence on the code length typically exhibited by maximum-likelihood decoders. The required decoder becomes considerably more complex, however, if an attempt is made to correct more than $[(d-1)/2]$ errors. (For a thorough discussion of algebraic decoding techniques, see Reference 13.11.)

All algebraic decoding operations are performed in the field $GF(q)$ over which the code is defined. This implies that the detector and code alphabets must be identical. With only a slight increase in decoder complexity, however, algebraic decoders can also accommodate erasures. The arithmetical operations are still carried out over $GF(q)$, but the decoder also recognizes the possibility of "blank symbols" which it treats as completely unknown.

It is easily seen that any combination of e errors and t erasures can be corrected so long as $2e + t < d$ with d the minimum Hamming distance separating any two code words. For if two n-symbol words differ in at least d symbol positions, any subset of $n - t$ of their symbols must differ in at least $d - t$ positions. Thus at least $[(d - t + 1)/2]$ errors are needed to convert a word which has undergone t erasures into a vector which is at least as close to a second word as it is to the original.

Surprisingly, an algebraic decoder capable of decoding in the presence of both errors and erasures can also be used effectively, with only minor modification, in conjunction with a general reliability class detector. The key to this statement is in the following lemma.

Lemma 13.5. Let k_1 and k_2 and the sets $\{m_j\}$ and $\{n_j\}$, $j = 1, 2, \ldots, J$, represent arbitrary constants, and let $\{\alpha_j\}$ be a set of J coefficients, with $1 \geq \alpha_1 \geq \alpha_2 \geq \cdots \geq \alpha_J \geq 0$. Then the inequality

$$\sum_{j=1}^{J} \alpha_j m_j < k_1 - k_2 \sum_{j=1}^{J} (1 - \alpha_j) n_j \tag{13.9.9}$$

implies

$$\sum_{j=1}^{l} m_j < k_1 - k_2 \sum_{j=l+1}^{J} n_j \tag{13.9.10}$$

for some $0 \leq l \leq J$.†

†Sums of the form $\sum_{j=\nu}^{\nu-1} x_j$ are equated to zero.

Proof. Suppose for all l in the range $[0, J]$

$$\sum_{j=1}^{l} m_j \geq k_1 - k_2 \sum_{j=l+1}^{J} n_j$$

Then for any set of non-negative coefficients λ_l such that $\sum_{l=0}^{J} \lambda_l = 1$,

$$\sum_{l=0}^{J} \lambda_l \sum_{j=1}^{l} m_j \geq k_1 - k_2 \sum_{l=0}^{J} \lambda_l \sum_{j=l+1}^{J} n_j$$

or

$$\sum_{j=1}^{J} \beta_j m_j \geq k_1 - k_2 \sum_{j=1}^{J} (1 - \beta_j) n_j$$

with $\beta_j \triangleq \sum_{l=j}^{J} \lambda_l$. Thus, defining $\lambda_0 = 1 - \alpha_1, \lambda_j = \alpha_j - \alpha_{j+1}, j = 1, 2, \ldots, J - 1$, and $\lambda_J = \alpha_J$, so that $\beta_j = \alpha_j$ for all $j = 1, 2, \ldots, J$, we have shown that the inequality (13.9.9) cannot be satisfied unless at least one of the inequalities (13.9.10) also holds, and the lemma is proved.

Theorem 13.22. Let D be an (n, k) linear code dictionary having minimum Hamming weight d. Let $\mathbf{v}$ be the output of a reliability class detector corresponding to some transmitted code word $\mathbf{w}_\mu$, and let α_j be a weight attached to the jth reliability class C_j, with $1 \geq \alpha_1 \geq \alpha_2 \geq \cdots \geq \alpha_J \geq 0$. Further, let n_{ej}^{ν} be the number of times a jth reliability class symbol in the received n-tuple $\mathbf{v}$ differs from the corresponding symbol in the code word $\mathbf{w}_\nu$, and let n_j be the total number of symbols in $\mathbf{v}$ in this reliability class. Then if there exists a $\mathbf{w}_\nu$ such that

$$\sum_{j=1}^{J} \alpha_j n_{ej}^{\nu} < \frac{d - \sum_{j=1}^{J} (1 - \alpha_j) n_j}{2} \tag{13.9.11}$$

the following two statements hold:

(1) $\mathbf{w}_\nu$ is unique
(2) $\mathbf{w}_\nu$ will be included in the set of words obtained by algebraically decoding $\mathbf{v}$ with the symbols in the lowest $J - l$ reliability classes treated as erasures and the rest given weight one, for all l in the range $[1, J]$.

Proof. Statement (1). Suppose $\mathbf{w}_\nu$ were not unique, that there were two words $\mathbf{w}_\nu$ and $\mathbf{w}_\mu$ satisfying the inequality (13.9.11). Then

$$\sum_{j=1}^{J} \alpha_j (n_{ej}^{\nu} + n_{ej}^{\mu}) < d - \sum_{j=1}^{J} (1 - \alpha_j) n_j$$

and from Lemma 13.5,

$$\sum_{j=1}^{l} (n_{ej}^{\nu} + n_{ej}^{\mu}) < d - \sum_{j=l+1}^{J} n_j$$

for some $1 \leq l \leq J$. $\left(\text{Since } \sum_{j=1}^{J} n_j = n \geq d, \text{ the case } l = 0 \text{ can be ignored.}\right)$ But if the set of class C_j symbols in $\mathbf{v}$ differing from their counterparts in $\mathbf{w}_\nu$ were disjoint from the set differing from their counterparts in $\mathbf{w}_\mu$ for all $j \leq l$, and if $\mathbf{w}_\mu$ and $\mathbf{w}_\nu$ differed in all symbol positions corresponding to a class C_j symbol in $\mathbf{v}$ for all

$j > l$, the Hamming distance $d_{\nu\mu}$ separating $\mathbf{w}_\mu$ and $\mathbf{w}_\nu$ would be $\sum_{j=1}^{l} (n^{\nu}_{ej} + n^{\mu}_{ej}) + \sum_{j=l+1}^{J} n_j$. In any other event, $d_{\nu\mu}$ would be less than this value. Thus, if the inequality (13.9.11) were to hold

$$d_{\nu\mu} \leq \sum_{j=1}^{l} (n^{\nu}_{ej} + n^{\mu}_{ej}) + \sum_{j=l+1}^{J} n_j < d$$

for some $1 \leq l \leq J$, leading to a contradiction, and statement (1) is true.

Statement (2). If the $t_l \triangleq \sum_{j=l+1}^{J} n_j$ least reliable symbols in $\mathbf{v}$ are treated as erasures, and the rest are given weight one, an algebraic decoder can correct up to $[(d - t_l - 1)/2]$ errors, as discussed above. Since the Hamming distance separating the unerased positions of $\mathbf{v}$ and the corresponding positions of $\mathbf{w}_\nu$ is $e_l \triangleq \sum_{j=1}^{l} n^{\nu}_{ej}$, such a decoder will decode $\mathbf{v}$ as $\mathbf{w}_\nu$ if

$$2e_l + t_l < d \tag{13.9.12}$$

But, from Lemma 13.5, the inequality (13.9.11) holds only if

$$\sum_{j=1}^{l} n^{\nu}_{ej} < \frac{d - \sum_{j=l+1}^{J} n_j}{2}$$

for some $1 \leq l \leq J$. Thus, for some l in the range $[1, J]$, an algebraic decoder treating the t_l least reliable symbols as erasures will decode $\mathbf{v}$ as the unique $\mathbf{w}_\nu$ satisfying the inequality (13.9.11).

The maximum-likelihood decoder associated with a reliability-class detector, it will be recalled, decodes the received n-tuple $\mathbf{v}$ as $\mathbf{w}_\nu$ only if the *generalized minimum distance*

$$d_\mu = \sum_{j=1}^{J} \alpha_j n^{\mu}_{ej}$$

attains its minimum value for $\mu = \nu$. The significance of Theorem 13.22 is that if d_ν is less than $\{[d - \sum_{j=1}^{J} (1 - \alpha_j)n_j]/2\}$ for some ν, an algebraic decoder will be able to isolate $\mathbf{w}_\nu$ by treating all received symbols in the $J - l$ least reliable classes as erasures, and all the remaining symbols as maximally reliable. After each such decoding, the decoded word, if any, can be tested to see if it satisfies the inequality (13.9.11). If it does, the decoding operation stops; if it does not, the erasure criterion is changed and a new trial is made. Conceivably, such a decoder could find a word satisfying the inequality (13.9.12) for every l it tries. Nevertheless, at most one of these words will satisfy the inequality (13.9.11). Moreover, at most $[(d + 1)/2]$ decoding trials need be made. This follows because a decoder can correct as many errors with $t + 1$ erasures as it can with t when t and d are either both odd or both even. Thus, if $t + 1$ erasures result from erasing the symbols in the $(J - l)$ least reliable classes and t result from erasing those in the $(J - l - 1)$ least reliable classes, the second trial can be omitted; it cannot yield any word not already found in the previous trial. Further, if t exceeds $d - 1$, both trials can be omitted since in this event the condition (13.9.12) cannot possibly be satisfied.

The only conditions placed on the coefficients α_j throughout this discussion were that $0 \leq \alpha_j \leq 1$, and that they be non-increasing functions of j. Subject to these conditions, any set of coefficients can be used in the definition of the generalized distance d_ν. The optimum choice for α_j when maximum-likelihood decoding is attempted was found to be as given in Equation (13.9.7). This same choice, however, is not necessarily optimum for generalized minimum distance decoding. This statement will be further elaborated in Section 13.10.

13.9.3 Probabilistic Decoding

Algebraic decoding schemes attempt to reduce the amount of decoding effort vis-à-vis that required by the maximum-likelihood approach by exploiting the algebraic structure of the code. Probabilistic decoding strategies try to achieve the same goal by replacing one difficult decision (as to which code word was received) by a number of simpler ones. This is accomplished, in essence, by dividing the received word into segments and, perhaps tentatively, identifying a subset of code words likely to account for each successive segment. If these subsets are mutually consistent (i.e., if at least one code word belongs to every subset) the decoder continues until it has finally converged on a single word. If, at any time during this procedure, it encounters difficulty (if none of the code words still remaining in contention shows sufficient promise of being correct), it may backtrack, reversing a previous decision and selecting a different subset of code words for further consideration.

Thus a probabilistic decoder examines only a small portion of the received code word at any one time with a consequent reduction in the complexity of the associated implementation. The number of operations and hence the time needed to arrive at a decision, however, is now often a random variable (depending on the particular algorithm used) and may sometimes exceed the time available before a new word must be decoded.

The *sequential decoding algorithms*, designed for use with convolutional codes, are among the most successful probabilistic decoding schemes. The reason the probabilistic approach is particularly effective with convolutional codes is apparent. The first ν code symbols of a convolutional code word are determined entirely by the first μ information symbols (see Section 13.8). Thus, a preliminary decision as to these μ information symbols can be made after only ν code symbols have been received. This initial decision in effect reduces the number of contending code words by the factor $q^{-\mu}$, as does each succeeding decision, after each additional block of ν symbols has been received. If any of these blocks of ν symbols contains too many errors, an incorrect decision will be made at that point and the correct code word will be among those tentatively eliminated from consideration. Further decoding attempts, however, will presumably suggest an earlier error (none of the contending code words will satisfactorily account for the subsequent blocks of received symbols) and this erroneous decision with high probability will be re-examined at some later time.

The probabilistic approach can also be used effectively with some classes of linear block codes. In this case, the true identity of each successive symbol is estimated on the basis of all those received symbols involved with it through the parity-check relationships imposed by the code structure. If each estimate were correct, the word would be decoded after only n such operations. (Actually, k operations would be sufficient since only the information bits used must be determined.) Generally, however, some of these estimates will be wrong and several iterations are required, each estimate of the preceding iteration re-evaluated in terms of the other estimates made during that iteration. When all of the estimates are consistent (i.e., when all the parity-check constraints are satisfied by the current estimates) the received n-tuple has been decoded. In effect, the number of contending code words is tentatively reduced at each stage of the decoding operation by estimating one of the received symbols. If the estimate is correct, the received word must belong to the subset of code words having that symbol in that position; if the estimate is in error it will hopefully be corrected on some subsequent iteration.

Details concerning probabilistic decoding algorithms can be found in the references (cf. References 13.17 through 13.20). Let it suffice here to observe that such techniques can be quite effective in many applications. While the number of operations needed to decode a word algebraically is nearly constant, independent of the number of errors it contains, probabilistic decoding algorithms usually require a random number of operations, a number which is highly dependent on the specific channel perturbations. Nevertheless, the average number of decoding operations may be less for a given word or constraint length n with a probabilistic as opposed to an algebraic algorithm. In fact, the average number of operations per information symbol needed to decode a convolutional code sequentially seems to be a function only of the parameters μ and ν (cf. Section 13.9) and of the channel statistics; it is essentially independent of the constraint length n.

13.9.4 Threshold Decoding

Before concluding the discussion of decoding techniques, we mention one other approach to the decoding problem which doesn't quite fall into either the probabilistic or algebraic categories, but embodies some features of both. This decoding strategy, called *threshold decoding*, is particularly effective with certain classes of codes, including both block and convolutional codes.

In order to apply this algorithm, it is necessary to select from the set of parity-check equations generated by the parity-check matrix **H** a subset E_i of equations *orthogonal* on the ith code word symbol w_i. That is, each equation in the set E_i must involve w_i in a parity-check relationship with other symbols w_j, $j \neq i$, with no symbol other than w_i involved in more than one of these equations.

Suppose the sequence $(w_1 + e_1, w_2 + e_2, \ldots, w_n + e_n)$ is received, with w_j representing the jth symbol of some code word and e_j an error symbol. Each equation in the orthogonal set E_j then provides an independent estimate $\hat{w}_i$ of w_i; i.e., $\hat{w}_i(\nu)$ is the code symbol needed to satisfy the νth equation in the set E_i. If E_i contains $d'_i - 1$ equations then since the ith received symbol $w_i + e_i$ also provides an independent estimate of w_i, a decoder can readily determine d'_i such estimates. If fewer than $[(d'_i + 1)/2]$ of the error symbols e_j involved in these equations are non-zero, more than half of the resulting estimates will certainly be correct and a simple majority decision among these estimates will establish the correct value of w_i. Accordingly, if each set E_i, $i = 1, 2, \ldots, n$ contains at least $d' - 1$ equations and if the received code word contains $[(d' - 1)/2]$ or fewer errors, the word will be correctly decoded by a decoder operating on this principle. This is the basis of the threshold decoding algorithm.

If sets E_i can be found for each i containing at least $d - 1$ equations, with d the minimum Hamming distance separating any two words in the code dictionary, the code is said to be *completely orthogonalizable*. When a code is completely orthogonalizable, and some codes are, a threshold decoder can correct any error pattern which could be corrected by most algebraic decoders. Moreover, many error patterns containing more than $e = [(d - 1)/2]$ errors can be corrected with a threshold decoder. While an algebraic decoder can sometimes also be modified to correct more than e errors, this can be done only at the cost of greatly increased complexity.

Generally, the equations in the set E_i do not all involve the same number of received symbols. The reliability of an estimate $\hat{w}_i(\nu)$ will obviously depend, to some extent, on the number of received symbols used to determine this estimate. As a consequence, the performance of a threshold decoder can sometimes be improved by weighting each estimate $\hat{w}_i(\nu)$ in accordance with the probability that it is correct, rather than basing the decision on the majority of these estimates as suggested above. Greater detail concerning this refinement, as well as a much more thorough discussion of threshold decoding in general, can be found in Reference 13.15.

13.10 Decoding Error Probabilities

The probability of a decoding error depends not only on the channel statistics and the specific code used, but obviously on the decoding algorithm as well. Even with this information, exact expressions for the error probability are frequently difficult to determine. The error probability associated with the maximum-likelihood decoding of linear codes, for example, can be calculated only if the complete weight structure of the code is known. The probability of mistaking one word for another is a function of the distance separating those particular two words, regardless of the minimum distance of the code.

The situation is further complicated when the number of detector symbols is large relative to the number of code symbols. Useful bounds on the error probability, however, are generally much easier to determine. The bounds of Chapter Four, for example, are directly applicable here when the channel is a phase-coherent, Gaussian channel. The maximum-likelihood decoder for such a channel seeks the maximum over ν of a function of the form $\log_e \Pr\{\mathbf{v} \mid \mathbf{w}_\nu\} = \sum_\mu \log_e \Pr\{\eta_\mu \mid \omega_\mu(\nu)\}$ (see Section 13.9). Since $\log_e \Pr\{\eta_\mu \mid \omega_\mu(\nu)\}$ is a linear function of the correlation between the received signal and the signal representing $\omega_\mu(\nu)$, the decision as to the received word is the same whether the decoding proceeds on a symbol-by-symbol basis (so long as no information is discarded) or whether the received *word* is correlated against each of the possible code words. The only significance of the code structure on the received signal is in the potential reduction it affords in receiver complexity.

In this section, we derive an upper bound on the probability of a word error when using the generalized minimum distance decoding rule defined in Section 13.9.2. We concentrate on this particular case for several reasons. First, because of the generality of this decoding rule, the bound itself is quite versatile. In addition, it reveals in a convenient way the relationship between the error probability and the pertinent code parameters. And finally, it also serves as a bound for the error probabilities associated with many other decoding rules of interest, including not only maximum-likelihood decoding but all those in which the detector output includes (and the decoder utilizes) more information than is provided by a reliability class detector alone.

The bound is stated in the following theorem.

Theorem 13.23. If a q-ary n-symbol word from a dictionary D having minimum distance d is transmitted over a noisy channel and decoded in accordance with the generalized minimum distance decoding rule, the probability P_e of a decoding error is bounded by

$$P_e \leq \min_{s \geq 0} \exp\left\{-n\left[s\frac{d}{n} - \mu(s)\right]\right\} \tag{13.10.1}$$

where $\mu(s) = \log_e \sum_{j=1}^{J} \{p_{cj}e^{s(1-\alpha_j)} + (q-1)p_{ej}e^{s(1+\alpha_j)}\}$ with p_{ej}, p_{cj}, and α_j as previously defined (cf. page 424 and Theorem 13.22).

Proof. A decoding error is committed only if the inequality (13.9.11) is violated when $\mathbf{w}_v$ is the word actually transmitted. Thus

$$P_e = \Pr\left\{\sum_j \alpha_j n_{ej}^v \geq \frac{d - \sum_j (1-\alpha_j)n_j}{2} \middle| v\right\}$$
$$= \Pr\{\sum_j [(1+\alpha_j)n_{ej}^v + (1-\alpha_j)n_{cj}^v] \geq d \mid v\} \qquad (13.10.2)$$

Let y_i be a discrete random variable with

$$p(y_i) = \sum_j \{(q-1)p_{ej}\delta[y_i - (1+\alpha_j)] + p_{cj}\delta[y_i - (1-\alpha_j)]\}$$

Then y_i assumes the value $(1+\alpha_j)$ with the probability $(q-1)p_{ej}$, the probability that the detector symbol is in class C_j and is identified with *any* q-ary symbol other than the one actually transmitted, and the value $(1-\alpha_j)$ with the probability p_{cj}, the probability that the detector symbol is in class C_j and is identified with the transmitted symbol. Accordingly,

$$\Pr\{\sum_j [(1+\alpha_j)n_{ej}^v + (1-\alpha_j)n_{cj}^v] \geq d \mid v\} = \Pr\{z \geq d\} \qquad (13.10.3)$$

where $z = \sum_{i=1}^{n} y_i$. Let $\{z_j\}$ denote the set of values which can be assumed by the random variable z. Then, for any $s \geq 0$,

$$\Pr\{z \geq d\} = \sum_{\text{all } z_j \geq d} p(z_j) \leq e^{-sd} \sum_{\text{all } z_j} e^{sz_j} p(z_j) \qquad (13.10.4)$$

the inequality following because $e^{-s(d-z)}$ is non-negative for all z and is at least one for all $z \geq d$. Continuing, we have

$$P_e = \Pr\{z \geq d\} \leq e^{-sd} E(e^{sz}) = e^{-sd} E\left(\prod_{i=1}^{n} e^{sy_i}\right)$$
$$= e^{-sd} \prod_{i=1}^{n} E(e^{sy_i}) \qquad (13.10.5)$$

This last equation is a consequence of the assumed independence of the successive symbol errors, and hence, of the random variables y_i. Since

$$E(e^{sy_i}) = \sum_{j=1}^{J} [(q-1)p_{ej}e^{s(1+\alpha_j)} + p_{cj}e^{s(1-\alpha_j)}] \qquad (13.10.6)$$

the theorem is proved.

The tightest upper bound on P_e results when the exponent in Equation (13.10.1) is maximized over s; i.e., when the derivative $\mu'(s) = d/n$. As is easily verified, $\mu''(s)$ is positive for all s (except in the trivial case in which all the α_j's are identical and $\sum_j p_{cj} = 0$ or 1). Thus, since $\mu'(\infty) = \max_j (1+\alpha_j) > d/n$, the function $s(d/n) - \mu(s)$ has a unique maximum for some s, $0 < s < \infty$, only if $\mu'(0) < d/n$. But if $\mu'(0) < d/n$, the upper bound (13.10.1) is an exponentially decreasing function of the word length n.

The parameters α_j can also be chosen to minimize the bound on P_e. Differentiating the exponent of the bound (13.10.1) with respect to α_j and recalling that α_j is constrained to lie between 0 and 1, we find that the smallest bound is obtained when

$$\alpha_j = \begin{cases} 1 & \log_e \dfrac{p_{cj}}{(q-1)p_{ej}} \geq 2s \\ 0 & \log_e \dfrac{p_{cj}}{(q-1)p_{ej}} \leq 0 \\ \dfrac{1}{2s} \log_e \dfrac{p_{cj}}{(q-1)p_{ej}} & \text{otherwise} \end{cases} \tag{13.10.7}$$

The tightest upper bound on the error probability [Equation (13.10.1)] results when the α_j are as defined above and the parameter s then chosen to satisfy the equation $\mu'(s) = d/n$.

It is interesting to compare this definition of the optimum parameters α_i with the one found in Section 13.9 for maximum-likelihood decoding, viz.:

$$\alpha_j = \log_e \left(\frac{p_{cj}}{p_{ej}}\right) \Big/ \log_e \left(\frac{p_{c1}}{p_{e1}}\right) \tag{13.10.8}$$

Although the two definitions are identical when $J = 1$ (and $p_c > \frac{1}{2}$), they will usually be different reflecting the difference in the two decoding rules (and the fact that the α_j are chosen here to minimize a bound on P_e, not P_e itself).

When $J = 1$, and $p_c \geq 1 - (d/2n)$, the bound (13.10.1) becomes

$$P_e \leq \exp\left\{-n\left[\left(1 - \frac{d}{2n}\right)\log_e\left(\frac{1 - d/2n}{p_c}\right) + \frac{d}{2n}\log_e\left(\frac{d/2n}{1 - p_c}\right)\right]\right\} \tag{13.10.9}$$

and is an exponentially decreasing function of n for all $p_c > 1 - (d/2n)$. Since the expected number of errors in n received symbols is $n(1 - p_c)$, the condition for an exponentially decreasing error probability bound is just the condition that the expected number of symbol errors be less than $d/2$, an obviously necessary as well as sufficient condition for a small error probability. In general, the bound of Theorem 13.23 can be shown to be exponentially tight; no bound of the form $P_e \leq e^{-cn}$ with $c > \max_{s \geq 0} [s(d/n) - \mu(s)]$ is possible for arbitrarily large n. The bound [Equation (13.10.4)] used in establishing this result, incidentally, is generally called the Chernoff bound.

The relationship between the minimum distance d associated with a code dictionary and its error-correcting capabilities is particularly evident in the bound of Theorem 13.23. Clearly, the probability of a decoding error is a monotonically decreasing function of d when the decoding decisions are based on the generalized minimum distance rule. This need not be the case, however, when the decoding is maximum-likelihood, as is demonstrated by the following two dictionaries:

$$D_1 = \begin{matrix} 0000 \\ 0011 \\ 1100 \\ 1111 \end{matrix} \qquad\qquad D_2 = \begin{matrix} 0000 \\ 0001 \\ 1110 \\ 1111 \end{matrix}$$

When the detector symbols are restricted to be binary, any word from D_1 is correctly decoded by the maximum-likelihood rule with probability 1 if the error vector $\mathbf{e}$ has weight zero, with probability $\frac{1}{2}$ if it has weight one, with probability $\frac{1}{4}$ if it is any vector of weight two other than a code word, and is erroneously decoded otherwise. Similarly, any word from D_2 is decoded correctly if $\mathbf{e} = 0000, 0010, 0100,$ or 1000, and incorrectly otherwise. Thus, the decoding error probabilities with D_1 and D_2 are

$$P_1 = 1 - (1 - p)^4 - 2p(1 - p)^3 - p^2(1 - p)^2$$

and

$$P_2 = 1 - (1 - p)^4 - 3p(1 - p)^3$$

respectively, with p the symbol error probability, and $P_2 < P_1$ for all $p < \frac{1}{2}$. Yet D_1 has minimum distance 2 while D_2 has a minimum distance of only 1. The general relationship between the probability of a decoding error and the minimum distance associated with a code dictionary is nevertheless reflected in the bound of Theorem 13.23, a bound which holds for maximum-likelihood as well as minimum distance decoding.

Similar bounds can be derived relating the constraint length of convolutional codes to the probability of a decoding error. Indeed, if a convolutional code is to be decoded using a feedback decoding algorithm, the bound of Theorem 13.23 itself applies to the probability of the *first* decoding error when n is identified with the constraint length and d with the initial distance of the code in question. Once an error is committed, however, subsequent decoding errors will be much more probable. On the other hand, if the decoding rule allows earlier decisions to be reversed, the first decoding error may be corrected subsequently, and the probability of a decoding error can be even smaller than with a block code having the same parameters (cf. Reference 13.21). This negative exponential dependence of the error probability on the convolutional code constraint length n, coupled with the near independence of the complexity of a sequential decoder on n when the rate μ/ν is held constant, constitutes the major argument in favor of this approach. The constraint length can be increased to the point where the probability of a decoding error is entirely negligible. The parameter of primary concern then becomes the probability of a decoding *failure*. This possibility arises because the number of operations required to decode an information symbol is a random variable. Under adverse conditions the decoder may not be able to decode as fast as the symbols are being received, and hence must store these symbols until it can get to them. A decoding failure occurs when the storage capacity of the decoder is insufficient to contain the backlog of received symbols.

Before concluding this section, we call attention to two assumptions which have been made throughout this discussion but which are not necessarily always valid. The first is that the sole measure of a code's effectiveness is the probability of a decoding error. As observed in Section 13.1, for example, it may be more important for a code to *detect* the occurrence of certain errors in the received symbols than to correct these errors. This is particularly true when requests for repeat transmissions are possible. When such is the case, the decoding operation might proceed as before, but with the added restriction that if the decoded and received words differ by more than a specified amount (under whatever measure is being used) the decoded word is discarded. The error probability is obviously not a sufficient measure of performance in this event; it could in fact be made zero simply by discarding all received words. A more meaningful measure in this situation might be the probability of a correct decision.

The second assumption permeating the above discussion is that the noise contributions to the successive detector outputs are statistically independent. If this assumption is not valid, several alternatives are available. One method is to interleave the symbols of several words, thereby separating the symbols of any one word by an amount sufficient to assure the independence of the noise affecting them. Another method is to accept the fact that the noise can cause symbol errors to occur in bursts, and to select the code accordingly.

A number of constructions for *burst-error-correcting* codes are known (cf. Reference 13.5). In addition, many ordinary error-correcting codes, particularly cyclic codes, can be shown to be effective over channels subject to bursts of errors. This is demonstrated in the following theorem.

Theorem 13.24. Any (n, k) cyclic code is capable of detecting any burst of errors of length† not exceeding $n - k$, of detecting any two bursts of combined length not exceeding $n - 2k + 1$, or of correcting any single burst of length not exceeding $(n - 2k + 1)/2$.

Proof. The basis for this proof lies in the observation that no code word in an (n, k) cyclic dictionary, other than the all-zeros word, can contain k or more consecutive zeros. For if there were such a word, it would have a cyclic permutation beginning with k consecutive zeros, and hence could only be the all-zeros word. Note that so far as this statement is concerned, the first symbol in a code word immediately follows the last. Since an error pattern is undetectable only if it converts one code word into another, and is consequently itself a code word, no burst of errors affecting $n - k$ or fewer consecutive bits can cause an undetectable error. Similarly, any two bursts of combined length not exceeding $n - 2k + 1$ must leave at least k consecutive symbols unaffected and therefore also cannot duplicate a code word. From this it immediately follows that any error burst of length $(n - 2k + 1)/2$ or less can be corrected, if the decoder is designed to correct rather than detect bursts of errors.

†The *length* of a burst of errors refers to the number of symbols separating, and including, the first and last code symbols affected by the burst. The intervening symbols may or may not be in error.

Codes defined over larger alphabets with each code symbol represented by several channel symbols can also be effective in combatting bursts of errors. If a code having minimum distance d is defined over $GF(q^l)$, for example, and each code symbol is represented by l q-ary channel symbols, a noise burst would have to affect at least $l[(d-3)/2]+2$ channel symbols in order to alter $[(d+1)/2]$ code symbols and hence to cause a decoding error.

13.11 Coding for the White Gaussian Channel—Polyphase Codes

Previous sections of this chapter have been concerned with the problem of coding for channels which are basically *symmetric*. The likelihood of one symbol being mistaken for any other was assumed to be independent of the two symbols involved or at least sufficiently so that the merit of a particular code could be estimated in terms of the minimum Hamming distance separating any two of its words. This is a valid model, for example, when the code symbols are represented by orthogonal, equal energy, equal time duration waveforms.

In many cases, of course, this channel model is not applicable. Investigation of one such situation is the subject of this section. Specifically, it is desired to construct an N-symbol alphabet for use over a white Gaussian channel by forming certain concatenations of the r-ary PSK signal elements

$$\sqrt{2}\,\sin\left(\omega_c t + \frac{2\pi x_\nu}{r}\right) \qquad iT_s < t < (i+1)T_s \tag{13.11.1}$$

where $\omega_c = k\pi/T_s$ with k either an integer, or else large as compared to unity, and where x_ν is an integer, $0 \le x_\nu \le r-1$ (cf. Chapter Four). The symbols themselves then assume the form

$$x_l(t) = \sqrt{2}\,\sin\left(\omega_c t + \frac{2\pi x_\nu^l}{r}\right) \tag{13.11.2}$$

for $(\nu-1)T_s < t < \nu T_s$, $\nu = 1, 2, \ldots, n$, with $x_\nu^l \in \{0, 1, \ldots, r-1\}$. (In general, of course, N will be less than r^n so some concatenations will not be recognized as symbols. The advantage of this type of symbol construction will become apparent shortly.)

If the received signal is to be demodulated coherently (cf. Section 4.2), the error probability associated with such a signal set at a given signal-to-noise ratio is a function only of the normalized correlation coefficients

$$\begin{aligned}\tilde{\rho}_{lm} &= \frac{1}{nT_s}\int_0^{nT_s} x_l(t)x_m(t)\,dt = \frac{1}{n}\sum_{\nu=1}^{n}\cos\frac{2\pi}{r}(x_\nu^l - x_\nu^m) \\ &= \operatorname{Re}\left\{\frac{1}{n}\sum_{\nu=1}^{n} e^{(2\pi i/r)(x_\nu^l - x_\nu^m)}\right\}\end{aligned} \tag{13.11.3}$$

where $\operatorname{Re}(z)$ denotes the real part of z. In the case of incoherent demodulation (cf. Section 4.3) the parameters

$$\hat{\rho}_{lm} = \frac{1}{nT_s}\int_0^{nT_s} x_l(t)\hat{x}_m(t)\,dt = \frac{1}{n}\sum_{\nu=1}^{n} \sin\frac{2\pi}{r}(x_\nu^l - x_\nu^m)$$
$$= \operatorname{Im}\left\{\frac{1}{n}\sum_{\nu=1}^{n} e^{(2\pi i/r)(x_\nu^l - x_\nu^m)}\right\} \tag{13.11.4}$$

are also important. Here $\hat{x}_m(t) = \sqrt{2}\cos[\omega_c t + (2\pi x_\nu^m/r)]$, $(\nu - 1)T_s < t < \nu T_s$, and $\operatorname{Im}(z)$ is the imaginary part of the complex quantity z. In either case, the sum $\frac{1}{n}\sum_{\nu=1}^{n} e^{(2\pi i/r)(x_\nu^l - x_\nu^m)}$ is the measure of significance.†

The task just outlined can thus be regarded as a coding problem in which the code symbols are elements in the set $\{\zeta(\nu)\} \triangleq \{\exp(2\pi i\nu/r)\}$. The goal is to combine these symbols into N n-symbol code words, $\zeta^l = \{\zeta(x_1^l), \zeta(x_2^l), \ldots, \zeta(x_n^l)\}$, with the effectiveness of a particular encoding measured in terms of the inner products $\zeta^l \cdot \zeta^m$, $l \neq m$ (cf. Equations 13.11.3 and 13.11.4). Such codes will be denoted *polyphase*, or more simply, *phase* codes.

It is convenient to impose a structure on the set of n-tuples $\mathbf{x}^\alpha = (x_1^\alpha, x_2^\alpha, \ldots, x_n^\alpha)$ comprising the exponents of the phase code words. In particular, suppose the set $G = \{\mathbf{x}^\alpha\}$ forms an Abelian group under the operation of term-by-term addition, modulo r: $\mathbf{x}^\alpha + \mathbf{x}^\beta = (x_1^\alpha + x_1^\beta, x_2^\alpha + x_2^\beta, \ldots, x_n^\alpha + x_n^\beta)$. The resulting phase code will be referred to as a *group* code and G as the code *exponent group*.‡ (Note that if r is a prime, G is also a vector space.) Then all of the correlation coefficients (13.11.3) and (13.11.4) are contained in the set

$$\rho_\alpha = \frac{1}{n}\sum_{\nu=1}^{n} e^{(2\pi i/r)x_\nu^\alpha} \qquad \alpha = 1, 2, \ldots, N \tag{13.11.5}$$

As was the case with error-conducting codes, the imposed structure simplifies the task of identifying "good" codes. A phase code, of course, is characterized not by its weight structure, but rather in terms of the set of parameters ρ_α, as defined in Equation (13.11.5).§

Theorem 13.25. Let $\boldsymbol{\xi}^\alpha = (\xi_1^\alpha, \xi_2^\alpha, \ldots, \xi_n^\alpha)$, $\alpha = 1, 2, \ldots, N$ be N n-component vectors defined over the field of complex numbers, and let

---

†A related measure which has resulted in some interesting codes is called the *Lee metric;* cf. Reference 13.10.

‡The phase code dictionary itself is a group of n-tuples under term-by-term *multiplication.* It is generally more convenient, however, to characterize the code in terms of its exponents.

§Note that when $r = 2$, $\rho_\alpha = (n - 2d_\alpha)/n$ where d_α is the Hamming weight of the n-tuple $\mathbf{x}^\alpha$. Since a binary vector space is also a group, all the binary codes discussed in the preceding sections can be used to define phase codes. The same comment can be made in regard to the relationship between ternary error-correcting and ternary phase code dictionaries in the case of phase-coherent detection. For in this case, $\tilde{\rho}_\alpha = (n - d_\alpha)/n$ where d_α is as previously defined.

$$\rho_{\alpha\beta} = \frac{1}{n}\boldsymbol{\xi}^\alpha \cdot \boldsymbol{\xi}^\beta = \frac{1}{n}\sum_{\nu=1}^{n} \xi_\nu^\alpha (\xi_\nu^\beta)^*$$

with $\rho_{\alpha\alpha} = 1$ for all α, and with $(\xi_\nu^\alpha)^*$ the complex conjugate of ξ_ν^α. Further let $\tilde{\rho}_{\alpha\beta} = \mathrm{Re}(\rho_{\alpha\beta})$ and $\hat{\rho}_{\alpha\beta} = \mathrm{Im}(\rho_{\alpha\beta})$. Then:

(1) If $\rho_{\alpha\beta} = 0$ for all $\alpha \neq \beta$, then $N \leq n$.
(2) If $\rho_{\alpha\beta} = -1/n$ for all $\alpha \neq \beta$, then $N \leq n + 1$.
(3) If $\tilde{\rho}_{\alpha\beta} = 0$ for all $\alpha \neq \beta$, then $N \leq 2n$.
(4) Phase codes containing N n-symbol words exist for all n satisfying the constraints (1), (2), and (3), in each case with N attaining its upper bound.

Proof. Statement (1). The space V containing the n-component vectors $\boldsymbol{\xi}^\alpha$ is in the space of n-tuples over the complex field and consequently of maximum dimension n. Thus, the maximum number of linearly independent vectors in V is n (cf. Appendix B, Theorem B.10). But if the set of vectors $\{\boldsymbol{\xi}^\alpha\}$ satisfies the conditions $\rho_{\alpha\beta} = 0$ for all $\alpha \neq \beta$, they are obviously linearly independent. The maximum number of vectors satisfying the conditions of statement (1) is therefore n and the statement is proved.

Statement (2). Suppose the n-component vectors $\boldsymbol{\xi}^\alpha$ satisfy the conditions of statement (2). Then the correlation $\rho'_{\alpha\beta}$ between the vectors $\boldsymbol{\eta}^\alpha$ and $\boldsymbol{\eta}^\beta$, where $\boldsymbol{\eta}^l = (1, \xi_1^l, \xi_2^l, \ldots, \xi_n^l)$, is

$$\rho'_{\alpha\beta} = \frac{1}{n+1} + \frac{n}{n+1}\left(-\frac{1}{n}\right) = 0$$

for all $\alpha \neq \beta$. Thus, from statement (1), $N \leq n + 1$.

Statement (3). Suppose the N vectors $\boldsymbol{\xi}^\alpha$ satisfy the conditions of the third statement, and define the $2n$-component vectors

$$\boldsymbol{\eta}^\alpha = \{\boldsymbol{\xi}^\alpha, (\boldsymbol{\xi}^\alpha)^*\} = \{\xi_1^\alpha, \xi_2^\alpha, \ldots, \xi_n^\alpha, (\xi_1^\alpha)^*, (\xi_2^\alpha)^*, \ldots, (\xi_n^\alpha)^*\}$$

Then

$$\frac{1}{2n}\boldsymbol{\eta}^\alpha \cdot \boldsymbol{\eta}^\beta = \frac{1}{2n}\{\boldsymbol{\xi}^\alpha \cdot \boldsymbol{\xi}^\beta + (\boldsymbol{\xi}^\alpha \cdot \boldsymbol{\xi}^\beta)^*\} = \frac{1}{n}\mathrm{Re}\{\boldsymbol{\xi}^\alpha \cdot \boldsymbol{\xi}^\beta\} = \tilde{\rho}_{\alpha\beta} = 0$$

Accordingly, statement (1) assures a maximum of $2n$ such vectors $\boldsymbol{\eta}^\alpha$, and hence a maximum of $2n$ vectors $\boldsymbol{\xi}^\alpha$ as was to be proved.

Statement (4). Let D_0 be a dictionary defined by the exponent group G_0 consisting of all multiples, modulo r, of the r-tuple $(0, 1, 2, \ldots, r-1)$. Then the correlation between any two different words in D_0 is

$$\rho_{\alpha\beta} = \frac{1}{r}\sum_{\nu=0}^{r-1} e^{(2\pi i/r)(\alpha-\beta)\nu} = 0$$

for all $\alpha \neq \beta$. Since D_0 contains r words of r symbols each, it meets the bound of statement (1). Similarly, if G_1 is the group of $(r-1)$-tuples obtained by deleting the leading zero from every element in G_0, $\rho_{\alpha\beta} = -1/(r-1)$, $\alpha \neq \beta$, and the corresponding dictionary D_1 meets the bound (2). Finally, the dictionary $D_2 = D_1 \cup (iD_1)$ (with iD_1 the dictionary obtained by multiplying each symbol in D by i) contains $2n$ words with $\tilde{\rho}_{\alpha\beta} = 0$, $\alpha \neq \beta$, and hence meets bound (3). This completes the proof of the theorem.

Dictionaries meeting the bound of statement (1) will be called *orthogonal*, those meeting the bound of statement (2) will be designated *transorthogonal*, and those meeting the bound of statement (3) *phase-coherent orthogonal*.

Theorem 13.26. Let **A** and **B** be $M \times m$ and $N \times n$ matrices whose rows are words in the group phase codes A and B, respectively. Let $S_A = \{\rho_\alpha(A)\}$ be the set of correlation coefficients between any two words in A and $S_B = \{\rho_\beta(B)\}$ the set for B. If $\mathbf{C} = \mathbf{A} \times \mathbf{B}$ is the Kronecker product of **A** and **B**, the rows of **C** form a group phase code C of MN mn-symbol words, and if $S_C = \{\rho_\gamma(C)\}$ is the set of correlation coefficients associated with C, then

$$\rho_\gamma(C) = \rho_\alpha(A)\rho_\beta(B)$$

where $\gamma = \alpha N + \beta$ refers to the γth row of the matrix **C**, α to the αth, and β to the βth row of **A** and **B**, respectively.

Proof. That the rows of $\mathbf{C} = \mathbf{A} \times \mathbf{B}$ form a group follows upon noting the equivalence between the Kronecker product of the phase code matrices and the Kronecker sum of the associated exponent matrices. The Kronecker sum of two groups produces another group as is easily verified.

Accordingly, the set of correlation coefficients of C is the set $\{\rho_\gamma(C)\}$ where

$$\rho_\gamma(C) = \frac{1}{mn} \sum_{\nu=1}^{mn} e^{(2\pi i/r)z_\nu^\gamma} \tag{13.11.6}$$

and where $(z_1^\gamma, z_2^\gamma, \ldots, z_{mn}^\gamma)$ is the element in the exponent group corresponding to the γth row in **C**. But, if $\gamma = \alpha N + \beta$

$$\begin{aligned} \frac{1}{mn} \sum_{\nu=1}^{mn} e^{(2\pi i/r)z_\nu^\gamma} &= \frac{1}{mn} \sum_{\mu=1}^{m} \sum_{\nu=1}^{n} e^{(2\pi i/r)(y_\nu^\beta + x_\mu^\alpha)} \\ &= \left(\frac{1}{m} \sum_{\mu=1}^{m} e^{(2\pi i/r)x_\mu^\alpha}\right)\left(\frac{1}{n} \sum_{\nu=1}^{n} e^{(2\pi i/r)y_\nu^\beta}\right) \end{aligned} \tag{13.11.7}$$

where $(x_1^\alpha, x_2^\alpha, \ldots, x_m^\alpha)$ and $(y_1^\beta, y_2^\beta, \ldots, y_n^\beta)$ are the exponent group elements corresponding to the αth and βth rows of **A** and **B**, respectively. Thus

$$\rho_\gamma(C) = \rho_\alpha(A)\rho_\beta(B)$$

and the theorem is proved.

Corollary. The (r^k, k) (i.e., $n = r^k, N = r^k$) dictionary found by taking the Kronecker product of the $(r, 1)$ orthogonal dictionary with itself k times is also an orthogonal dictionary.

Proof. This follows immediately from Theorem 13.26 since either $\rho_\alpha(A) = 0$ or $\rho_\beta(B) = 0$, or both for any $\gamma = \alpha N + \beta \neq 0$.

Corollary. Transorthogonal $(r^k - 1, k)$ phase codes exist for all r and k and can be obtained by deleting the first component from each of the words of the appropriate orthogonal dictionaries.

Proof. The first component of each of the words of the $(r, 1)$ orthogonal dictionary as defined in Theorem 13.25 is 1. This property clearly holds for the Kronecker product of any two of these dictionaries. Thus, deleting this first symbol from each word, we have $\rho'_\gamma = (N\rho - 1)/(N - 1) = -1/(N - 1)$ as was to be proved.

As mentioned in Chapter Four, transorthogonal codes are presumably optimum for the the phase-coherent (and orthogonal codes for the phase-incoherent) white Gaussian channels. Since N-word biphase (0° and 180°) orthogonal and transorthogonal, and $2N$-word quadraphase (0°, 90°, 180°, and 270°) phase-coherent orthogonal dictionaries exist for all $N = 2^k$, the advantage in considering other coding schemes may be questioned. But each bit of information is represented by $n/\log_2 N$ code symbols when the information is conveyed via a phase code consisting of N n-symbol words. If the information rate is fixed, the required bandwidth is increased, relative to that needed to transmit the same information in encoded binary form, by the same factor. Orthogonal (and transorthogonal) codes, therefore, can be used only if a bandwidth expansion factor of $n/\log_2 n$ is tolerable. This obviously imposes a limit on the acceptable word length n. Yet, as was shown in Chapter Four, the error probability is an exponentially decreasing function of n, even when $\rho_{\max}$ (or $|\rho_{\max}|$) is greater than zero. This suggests that it may be expedient, when the available bandwidth is the limiting constraint, to use other than orthogonal codes, accepting larger correlation coefficients in exchange for a smaller bandwidth expansion factor. This exchange, of course, is advantageous only if the word length n can be increased sufficiently to compensate for these larger correlation coefficients without violating the bandwidth limitation. Although little is known, in general, about how to effect the most favorable exchange, phase-coding techniques have been used in this endeavor with some success (cf. References 13.22 through 13.25).

One scheme for increasing the number of words in a phase code dictionary for a fixed word length is as follows: Let G be the group of exponents of an orthogonal phase code and adjoin to G its *cosets* $C_1, C_2, \ldots, C_{r-1}$, with C_ν denoting the coset of G containing the n-tuple $(\nu, \nu, \ldots, \nu)$ .The number of words in the corresponding phase code dictionary is thereby increased by a factor of r. The correlation between any word with exponents in C_ν and any word with exponents in C_μ is

$$\rho_{\mu\nu} = \frac{1}{n} \sum_{l=1}^{n} e^{(2\pi i/r)(x_l^\alpha - x_l^\beta)}\, e^{(2\pi i/r)(\nu-\mu)} \tag{13.11.8}$$

where $\mathbf{x}^\alpha$ and $\mathbf{x}^\beta$ are n-tuples in G. Thus

$$\rho_{\nu\mu} = \begin{cases} 0 & \alpha \neq \beta \\ e^{(2\pi i/r)(\nu-\mu)} & \alpha = \beta \end{cases}$$

and

$$\max_{\nu \neq \mu} \tilde{\rho}_{\nu\mu} = \cos(2\pi/r) \quad r > 3$$

When $r = 2$, this simply doubles the number of words in the dictionary by also including the negatives of these words. This, of course, yields a *biorthogonal* dictionary (see Section 4.2). By analogy, we shall call such dictionaries, for an arbitrary value of r, *r-orthogonal* dictionaries.

This same technique, incidentally, can be used to increase the size of a code even when the basic orthogonal code is not a phase code (e.g., when it is a set of orthogonal FSK symbols) or when the phase code exponents are defined modulo a different value of r from that used in defining the cosets. Rather impressive reductions in equipment complexity, however, are often possible when group phase codes, as opposed for example to FSK techniques, are used to realize the desired signal set. The information can be encoded as though it were to be transmitted over an r-ary symmetric channel, and the resulting sequence of r-ary symbols used to phase modulate a sinusoid. The encoder can be of the form shown in Figure 13.1.

The exponent group of a $(2^{k-1}, k)$ binary biorthogonal code, for example, is easily seen (cf. Theorem 13.26) to consist of the 2^k linear combinations, over the binary field, of the k 2^{k-1}-tuples

$$\begin{array}{l} 111111111 \cdots\cdots 11 \\ 010101010 \cdots\cdots 01 \\ 001100110 \cdots\cdots 11 \\ \vdots \\ 000000000 \cdots\cdots 11 \end{array} \tag{13.11.9†}$$

i.e., of the binary sequences of periods 1, 2, 4, 8, . . ., 2^{k-1}. The storage devices in Figure 13.1 can therefore be replaced, in this case, by a k-stage binary counter. The outputs of successive stages of such a counter are precisely the desired sequences. The corresponding biorthogonal code, therefore, is particularly easy to generate.

The receiver implementation can also be simplified significantly when phase codes are used. This is due to the fact that if $y(t)$ is a received noise-corrupted signal of the form (13.11.2), then the correlation between it and the lth code word is

$$\frac{1}{n}\sum_{\nu=1}^{n} \xi_\nu^l \eta_\nu \tag{13.11.10}$$

where $\xi_\nu^l = e^{(2\pi i/r)x_\nu^l}$, x_ν^l denoting the phase of the lth code word during the interval $(\nu - 1)T_s < t < \nu T_s$, and where

$$\eta_\nu = \frac{1}{T_s}\int_{(\nu-1)T_s}^{\nu T_s} y(t) \sin \omega_c t \, dt - \frac{i}{T_s}\int_{(\nu-1)T_s}^{\nu T_s} y(t) \cos \omega_c t \, dt$$

It is only necessary to determine the quantities η_ν and to carry out the matrix operation

$$\mathbf{My} = \mathbf{z} \tag{13.11.11}$$

where $\mathbf{M}$ is the $N \times n$ matrix $\{\xi_\nu^l\}$, $l = 1, 2, \ldots, N$; $\nu = 1, 2, \ldots, n$, and $\mathbf{y} = (\eta_1, \eta_2, \ldots, \eta_n)^T$. The component of $\mathbf{z}$ with the largest real part (or, in the case of phase-incoherent reception, with the largest absolute value) then indicates the word most likely received. The matrix multiplication can be

†Note that these vectors generate the $(2^{k-1}, k)$ Reed-Muller code defined in Section 13.8.

done digitally by either a special-purpose or a general-purpose digital computer. Particularly efficient algorithms are known for accomplishing this when the code dictionary is a Kronecker product dictionary (cf. Reference 13.14).

The error probabilities associated with orthogonal, biorthogonal, and transorthogonal signal sets have already been investigated in Chapter Four. The bounds developed in that chapter, of course, can be applied to any phase code when the noise is additive, white, and Gaussian. Tighter bounds for r-orthogonal codes can be found in References 13.26 and 13.27.

Notes to Chapter Thirteen

There is a vast literature on error-correcting codes including several excellent books; in particular, Peterson's book (Reference 13.5) and the recent book by Berlekamp (Reference 13.10). The exposition by Solomon, Reference 13.28, is also highly recommended. Since these books both include extensive bibliographies of coding theory literature, only those books and articles most relevant to the presentation of this chapter are listed below.

Much of the introductory material in this chapter concerning the structure of linear codes originates with Reed (Reference 13.11) and Slepian (Reference 13.29). The basic concepts leading to the bounds of Section 13.3 originate with the people with whose names they are associated, although, in most cases, the proofs presented there differ somewhat from the originals. In particular, the proof of the Elias bound is taken from Reference 13.30, that of the Griesmer bound from Reference 13.9, and the proof of the Gilbert bound is due to Sacks (Reference 13.31; see also Reference 13.32). (For other upper bounds, which in some cases are tighter than those of Section 13.3, see References 13.33 and 13.34.) The treatment of the distance properties of cyclic codes in Section 13.4 originates with the work of Mattson and Solomon (Reference 13.2). The punctured cyclic codes of Section 13.6 were first introduced in Reference 13.9. Much of the material in Section 13.8 on convolutional codes can be found in Massey's monograph (Reference 13.15). The generalized minimum distance coding concept as well as the associated error probability bounds (Sections 13.9 and 13.10) are due to Forney (Reference 13.8). Additional material on polyphase codes can be found in References 13.22 through 13.25.

Problems

13.1 Let $f(x) = \Pi_i g_i^{\nu_i}(x)$ with the ν_i positive integers and the polynomials $g_i(x)$ irreducible over some field $GF(q)$ of characteristic p. Show that if the roots of $g_i(x)$ are in $GF(q^{m_i})$ and have multiplicative order n_i, then (cf. Appendix B):

(a) The roots of $f(x)$ are in the field field $GF(q^m)$ where m is the least common multiple of the m_i.

(b) $f(x)$ generates a cyclic code of word length n equal to p^a times the least common multiple of the n_i, with a the smallest integer such that $p^a \geq \max_i \nu_i$. It is not the generator of any cyclic code having word length less than n.

(c) If $f(x)$ generates an (n, k, d) p^m-ary cyclic dictionary D (p a prime, m a positive integer), then $f^p(x)$ generates a (pn, pk, d) cyclic dictionary which is equivalent to the dictionary generated by the Kronecker product of the generators of D with the generators of the $(p, p, 1)$ p-ary code.

13.2 (cf. Reference 13.5). Let $g(x) = \sum_{i=0}^{n-k} g_i x^i$ be the generator of an (n, k) cyclic code. Show that any one of the $(n - k)$-stage shift-registers shown in Figure P13.2 could be used as an encoder for such a code, but that only the third shift-register results in a systematic encoding. (The k information symbols

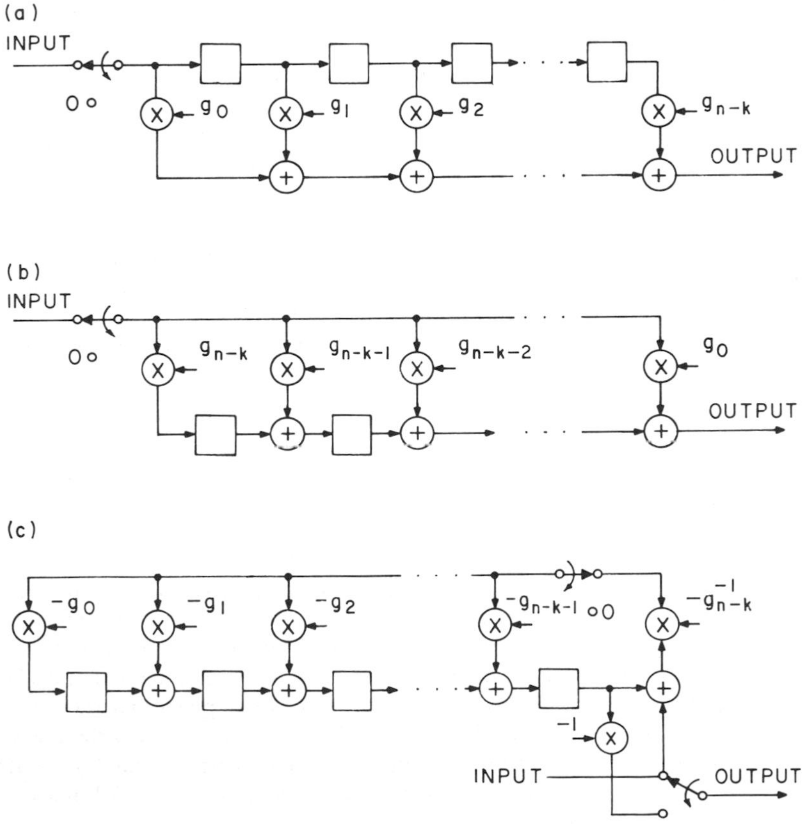

Figure P13.2

are first read serially into each of these devices; the switches are then all switched to the second position and left there for $n - k$ clock cycles. The n output symbols constitute the code word.) [*Hint:* Note that if the input to the third shift-register were added instead to the input to its first memory cell, the resulting device would effectively divide its input, considered as the successive coefficients of a polynomial over $GF(q)$, by the polynomial $g(x)$ and store the remainder.]

13.3 Let $w(x)$ represent a code word from the (n, k) cyclic dictionary generated by $g(x)$ and let $e(x) = \sum_{i=0}^{n-1} e_i x^i$ be an arbitrary polynomial with coefficients in the code field.

(a) Show that the syndrome of the m-tuple corresponding to $v(x) = w(x) + e(x)$ can be identified with the coefficients of the residue polynomial $v(x)$ modulo $g(x)$. Thus, show that the syndrome of an n-tuple with respect to any (n, k) cyclic code can be determined using an $(n - k)$-stage shift-register (see Problem 13.2).

(b) Show that if $g(x)$ has the roots β_i, $i = 1, 2, \ldots, r$ the code null space is spanned by the matrix

$$\mathbf{H} = \begin{bmatrix} 1 & \beta_1 & \beta_1^2 & \cdots & \beta_1^{n-1} \\ 1 & \beta_2 & \beta_2^2 & \cdots & \beta_2^{n-1} \\ \cdot & & & & \\ \cdot & & & & \\ \cdot & & & & \\ 1 & \beta_r & \beta_r^2 & \cdots & \beta_r^{n-1} \end{bmatrix}$$

(i.e., show that $\mathbf{v}$ is a code word if and only if $\mathbf{vH}^T = 0$).

13.4 (a) Show that there exists a (q^{m-1}, m) punctured cyclic code for every prime power q and every integer $m \geq 2$ having one word of weight zero, one word of weight q^{m-1}, and $q^m - 2$ words of weight $(q - 1)q^{m-2}$.

(b) What can be said in general concerning the weights of the words in a punctured cyclic code?

13.5 Consider an ensemble of random q-ary n-symbol sequences (i.e., sequences of n symbols in which each successive symbol is equally likely to be any one of the q possible symbols regardless of its predecessors or successors in the sequence). It is evident that such sequences exhibit the following characteristics: (1) Each symbol occurs n/q times on the average. (2) The expected number of "runs of length i" (i.e., occurrences of i successive repetitions) of a given symbol is n/q^i and is independent of the symbol in question. (3) If the ith and $(i + \nu)$th symbols (modulo n) of such a sequence are compared for all $i = 1, 2, \ldots, n$ and for any $\nu \neq 0$, modulo n, they will be equal, on the average, n/q times. Show that any non-zero maximal-length cyclic code word has, as nearly as possible, all three of these properties (i.e., show that (1) each symbol occurs either $[n/q]$ or $[(n + 1)/q]$ times; (2) there are either $[n/q^i]$ or $[(n + 1)/q^i]$ runs of length i of each symbol; and (3) the ith and $(i + \nu)$th symbols of such a sequence are equal exactly $[n/q]$ times for any $\nu \neq 0$, modulo n).

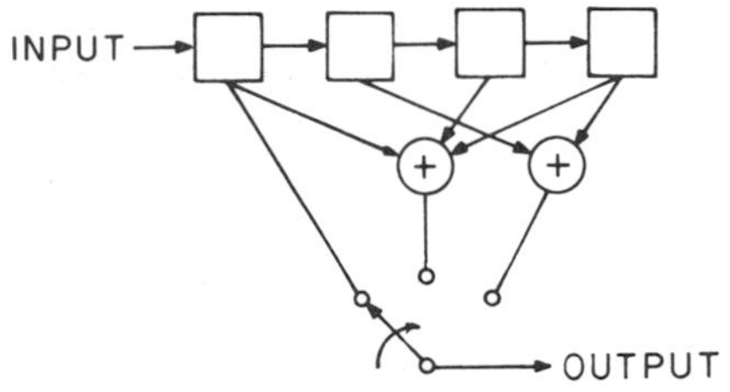

Figure P13.7

13.6 (cf. Reference 13.5). Prove that every pseudo-cyclic code with minimum weight $d > 2$ is a shortened cyclic code and that every shortened cyclic code is a pseudo-cyclic code. [*Hint:* Show that if $g(x)$ generates an (n, k) pseudo-cyclic code with $d > 2$, it also generates an (n', k) cyclic code with $n' > n$ which, if shortened by $n' - n$ symbols, is identical to the first code. Conversely, show that if $g(x)$ generates an (n', k) cyclic code which is shortened by $n' - n$ symbols, the resulting code can be represented by the set of polynomials $p(x)g(x)$, modulo $x^n - r(x)$, with $r(x)$ a polynomial of degree less than that of $g(x)$ defined by the equation $x^n = g(x)q(x) + r(x)$.]

13.7 Consider the binary convolutional code generated by the matrix (13.8.1) with $m = 4$ and with $\mathbf{G}_0 = 110$, $\mathbf{G}_1 = 001$, $\mathbf{G}_2 = 010$, $\mathbf{G}_3 = 011$. (a) Find the minimum initial distance of the code and compare with the bounds (13.8.4) and (13.8.11). (b) Show that an encoder for such a code can be realized as shown in Figure P13.7. (The output switch is switched to each of the three positions in sequence following each successive input.) (c) Describe a general convolutional encoder of the form shown in Figure P13.7 and compare its complexity, as a function of μ, ν, and m, with that of the encoder suggested by Figure 13.3.

13.8 (Reference 13.21). Consider the following algorithm for decoding an $(m\nu, m\mu)$ convolutional code: (1) Upon receiving the first $qm\nu$ transmitted symbols, the decoder determines the value of the associated likelihood function for each of the $q^{m\mu}$ possible transmitted sequences. This set of possible sequences is separated into $q^{(m-1)\mu}$ subsets, with two sequences in the same subset if and only if the last $(m - 1)\mu$ symbols in the corresponding information sequences are identical, and only one sequence, the sequence giving rise to the largest value of the likelihood function, is selected from each set. (2) The decoder then accepts the next ν received symbols and again evaluates the likelihood of the entire received sequence under $q^{m\mu}$ different hypotheses, one corresponding to each of the q^{μ} possible ν-symbol extensions of the $q^{(m-1)\mu}$ sequences not previously discarded. (3) These sequences are divided into $q^{(m-1)\mu}$ subsets as in step (1) and all but one sequence in each subset are eliminated from further consideration. The decoder then returns to step (2), the process continuing in this manner until the transmission is truncated by transmitting a sequence of $(m - 1)\nu$ *zeros*. Show that this decoding algorithm is a maximum-likelihood algorithm, i.e., that the decoder would select the same information sequence were the decision made by determining the likelihood of all possible transmitted sequences and selecting the largest. [*Hint:* Observe the basic structure indicated by Equation (13.8.1).]

The appropriate distance measure for this decoding algorithm (and the sequential decoding algorithm as well) is the minimum distance, called the *free distance*, separating any two distinct, untruncated, encoded sequences. Show that the free distance in any linear convolutional code is overbounded by the minimum over all positive integers l of the quantity $d[(l+m-l)v, l_\mu]$, with $d(n, k)$ the maximum attainable distance in a linear (n, k) block code over $GF(q)$. Using the bound (13.3.5) on $d(n, k)$, compare this bound on the free distance with that of Equation (13.8.11) on the initial distance.

13.9 (Reference 13.8). A code word from a binary dictionary with minimum Hamming distance d is transmitted over a white Gaussian noise channel. The two code symbols are represented by antipodal equal-energy waveforms and detected using a matched filter. The ratio of the symbol energy to the noise spectral density is denoted R_s. Prove the following statements using the appropriate results from Section 13.10. (a) If a "hard" decision is made concerning each received symbol (i.e., if only one reliability class is utilized) the word error probability is overbounded, approximately, for large R_s, by $P_e \le \exp[-(d/2)R_s]$. [*Hint:* $1 - \text{erf}(x) \le (1/\sqrt{\pi}\, x) \exp(x^2)$, $x \ge 0$.]
(b) If two reliability classes are used with $\alpha_1 = 1$, $\alpha_2 = 0$ (i.e., if erasures are also recognized) the tightest bound of the form 13.10.1 results when all symbols, and only those symbols, yielding a detector output z such that $|\log_e[p(z|1)/p(z|0)]| < s$ are identified as erasures. In this case, the word error probability is bounded for large R_s by $P_e \le \exp(-.686dR_s)$.
(c) If the number J of reliability classes is unbounded and the weights α_j are chosen optimally (cf. Equation 13.10.7) the word error probability is bounded for large R_s by $P_e \le \exp(-dR_s)$.

Determine a lower bound on the word error probability using maximum likelihood decoding (cf. Equation 4.2.19 and Problem 13.13) and compare with these results.

13.10 Show that maximal-length binary cyclic codes are completely orthogonalizable (cf. Section 13.9) and describe a decoding algorithm exploiting this fact. [*Hint:* Remember the cycle-and-add property.]

13.11 Prove that any binary cyclic code with minimum distance d can correct any burst of errors of length not exceeding $b = (3d - 5)/4$. [*Hint:* (cf. Section 14.12). Let $f(x)$ represent an error pattern resulting from two arbitrary bursts each of length b or less. If some burst of length b or less is not correctable, some such polynomial must also represent a code word, and, since the code is cyclic, so must $(x + 1)f(x)$.]

13.12 (cf. Reference 13.10). The (shortened) *generalized rth-order Reed-Muller code* of length $n = q^m - 1$ over $GF(q)$ is defined by the generator polynomial

$$g(x) = \prod_j (x - \alpha^j)$$
$$0 \le w(j) < (q-1)m - r$$
$$0 \le j < q^m - 1$$

with α a primitive element in $GF(q^m)$ and $w(j)$ the (ordinary) sum of the coefficients in the q-ary expansion of j (e.g. if $q = 3$ and $j = 46 = 1\cdot 3^3 + 2\cdot 3^2 + 0\cdot 3^1 + 1\cdot 3^0$, $w(j) = 4$.) (The non-shortened code has length $n' = n + 1$ and is obtained by adding an all-zeros column to the generator matrix of the shortened code and then adjoining an all-ones row.) Show that: (a) The coefficients of $g(x)$ are indeed elements in $GF(q)$. (b) The code generated by $g(x)$ has minimum weight at least $d = (q - R)q^{(m-Q-1)}$ where Q and R are defined by the relationship $r = Q(q - 1) + R$, $R < q - 1$. [*Hint:* Show that the elements α^j are roots of $g(x)$ for all integers j in the interval $0 \leq j \leq (q - R)q^{(m-Q-1)} - 2$ and use Theorem 13.8.] (c) The code generated by $g(x)$ has $k = \sum_{i=0}^{r} P(i, m, q - 1)$ information symbols with $P(i, m, q - 1)$ denoting the number of ordered partitions of i into exactly m non-negative integers none of which exceed $q - 1$. $\left[\text{Note that } P(i, m, 1) = \binom{m}{i}.\right]$ (d) When $r = 1$, $g(x)$ generates a maximal-length cyclic code. Moreover, when $q = 2$,the non-shortened version of the first-order Reed-Muller codes defined here are combinatorially equivalent to the Kronecker sum codes introduced at the conclusion of Section 13.7.

13.13 Let G be a group of r-ary n-tuples, as defined in Section 13.11, and let G' be the union of G with its cosets $C_1, C_2, \ldots, C_{r-1}$, with C the coset of G having the coset leader $(v, v, \ldots, v)$. (a) Show that if the minimum Hamming distance separating any two n-tuples in G is d then G is the exponent group of a polyphase code with

$$\max_{\mu \neq \nu} \operatorname{Re}(\rho_{\mu\nu}) \leq \frac{n-d}{n} + \frac{d}{n}\cos\frac{2\pi}{r}$$

(Note that when $r = 2$ this inequality is actually an equality.) (b) Show that if the minimum Hamming distance separating any two n-tuples in G' is $d\,(d < n)$, then G is the exponent group of a polyphase code with

$$\max_{\mu \neq \nu} \operatorname{Re}(\rho_{\mu\nu}) \leq \operatorname{Re}(\rho_0)$$

and

$$\max_{\mu \neq \nu} |\rho_{\mu\nu}| \leq |\rho_0|$$

where

$$\rho_0 = \frac{n-d}{n}\left\{1 + 2\sum_{\nu=1}^{l} \cos\frac{2\pi\nu}{r} + ([\delta](1-\delta) + \delta)\exp\left((l+1)\frac{2\pi i}{r}\right) + [\delta](\delta - 1)\exp\left(-(l+1)\frac{2\pi i}{r}\right)\right\}$$

with $l = [d/2(n-d)]$, $\delta = (d/(n-d)) - 2l$, and with the square brackets denoting the integer part of the enclosed quantity.

13.14 Let $\tilde{\rho}_{\alpha\beta}$, $\hat{\rho}_{\alpha\beta}$, N, and n be as defined in Theorem 13.25 and prove the following statements:

(i) If $\hat{\rho}_{\alpha\beta} = 0$, all α, β, and $\tilde{\rho}_{\alpha\beta} \leq \rho$, all $\alpha \neq \beta$, then

$$N \leq \begin{cases} -\dfrac{1-\rho}{\rho}, & \rho \leq -\dfrac{1}{n} \\ n+1, & -\dfrac{1}{n} \leq \rho < 0 \\ 2n, & \rho = 0 \end{cases}$$

(ii) If $\tilde{p}_{\alpha\beta} \leq p$, all $\alpha \neq \beta$, with $\hat{p}_{\alpha\beta}$ arbitrary, N is still subject to the same bounds but with n replaced everywhere by $2n$. [*Hint:* See Equation (4.2.14) for the first bound. Prove the remaining two bounds by induction on n.]

References

13.1 A. W. Nordstrom and J. P. Robinson, "An Optimum Nonlinear Code," *Inform. and Control*, **11**, p. 613, 1967.

13.2 H. F. Mattson and G. Solomon, "A New Treatment of Bose Chaudhuri Codes," *Soc. Ind. Appl. Math.*, **9**, p. 654, 1961.

13.3 R. C. Bose and D. K. Ray-Chaudhuri, "On a Class of Error-Correcting Binary Group Codes," *Inform. and Control*, **3**, p. 68, 1960; "Further Results on Error-Correcting Binary Group Codes," *Inform. and Control*, **3**, p. 279, 1960.

13.4 A. Hocquenghem, "Codes Correcteurs d'Erreurs," *Chiffres*, **2**, p. 147, 1959.

13.5 W. Peterson, *Error-Correcting Codes*, Mass. Inst. Tech. Press, Cambridge, Mass., 1961.

13.6 R. W. Marsh, "Table of Irreducible Polynomials over *GF*(2) through Degree 19," National Security Agency,Washington, D. C., 1957.

13.7 I. S. Reed and G. Solomon, "Polynomial Codes over Certain Finite Fields," *J. Soc. Ind. Appl. Math.*, **8**, p. 300, 1960.

13.8 G. D. Forney, Jr., *Concatenated Codes*, Mass. Inst. Tech. Press, Cambridge, Mass., 1966.

13.9 G. Solomon and J. J. Stiffler, "Algebraically Punctured Cyclic Codes," *Inform. and Control*, **8**, p. 170, 1965.

13.10 E. R. Berlekamp, *Algebraic Coding Theory*, McGraw-Hill Book Co., N. Y., N. Y., 1968.

13.11 I. S. Reed, "A Class of Multiple-Error-Correcting Codes and the Decoding Scheme," *IEEE Trans. Inf. Theory*, IT-4, p. 38, 1954.

13.12 J. K. Wolf, "On Codes Derivable from the Tensor Product of Check Matrices," *IEEE Trans. Inf. Theory*, **11**, p. 281, 1965.

13.13 J. P. Robinson, "An Upper Bound on the Minimum Distance of a Convolutional Code," *IEEE Trans. Inf. Theory*, IT-11, p. 567, 1965.

13.14 R. R. Green, "A Serial Orthogonal Decoder," Jet Prop. Lab. Space Programs Summary, 37–39, Vol. IV, p. 247, 1966.

13.15 J. L. Massey, *Threshold Decoding*, Mass. Inst. Tech. Press, Cambridge, Mass., 1963.

13.16 J. J. Bussgang, "Some Properties of Binary Convolutional Code Generators," *IEEE Trans. Inf. Theory*, IT-11, p. 90, 1965.

13.17 R. M. Fano, "A Heuristic Discussion of Probabilistic Decoding," *IEEE Trans. Inf. Theory*, IT-9, p. 64, 1963.

13.18 R. G. Gallager, *Low Density Parity Check Codes*, Mass. Inst. Tech. Press, Cambridge, Mass., 1963.

13.19 F. Jelinek, "Fast Sequential Decoding Algorithm Using a Stack," *IBM J. Research and Devel.*, **13**, p. 675, 1969.

13.20 J. Wozencraft and B. Reiffen, *Sequential Decoding*, Mass. Inst. Tech. Press, Cambridge, Mass., 1961.

13.21 A. J. Viterbi, "Error Bounds for Convolutional Codes and an Asymptotically Optimum Decoding Algorithm," *IEEE Trans. Inf. Theory*, IT-13, p. 260, 1967.

13.22 L. H. Zetterberg, "A Class of Codes for Polyphase Signals on a Bandlimited Gaussian Channel," *IEEE Trans. Inf. Theory*, IT-11, p. 385, 1965.

13.23 I. Ingemarsson, "Signal Sets Generated by Orthogonal Transformations of the Signal Space," Royal Inst. of Technology, Stockholm 70, Sweden, Tech. Report 21, 1968.

13.24 A. J. Viterbi, "On a Class of Polyphase Codes for the Coherent Gaussian Channel," *IEEE Int. Conv. Record*, Part 7, p. 209, 1965.

13.25 D. Slepian, "Group Codes for the Gaussian Channel," *Bell Syst. Tech. J.*, **47**, p. 575, 1968.

13.26 I. S. Reed and R. A. Scholtz, "*N*-orthogonal Phase-Modulated Codes," *IEEE Trans. Inf. Theory*, IT-12, p. 388, 1966.

13.27 A. J. Viterbi and J. J. Stiffler, "Performance of *N*-orthogonal Codes," *IEEE Trans. Inf. Theory*, IT-13, p. 521, 1967.

13.28 G. Solomon, "Algebraic Coding Theory," in *Communication Theory*, A. V. Balakrishnan, Ed., McGraw-Hill Book Co., N. Y., N. Y., 1968.

13.29 D. Slepian, "A Class of Binary Signaling Alphabets," *Bell System Tech. J.*, **35**, p. 203, 1956.

13.30 R.J. McEliece and H. Rumsey, Jr., "Sphere-Packing in the Hamming Metric," *Bull. of the Amer. Math. Soc.*, **75**, p. 32, 1969.

13.31 G. E. Sacks, "Multiple Error Correction by Means of Parity Checks," *IRE Trans. Inf. Theory*, IT-4, p. 145, 1958.

13.32 R. R. Varsharmov, "Estimate of the Number of Signals in Error-Correcting Codes," *Doklady A. N.*, S.S.S.*P.*, 117, p. 739, 1957.

13.33 S. M. Johnson, "A New Upper Bound for Error-Correcting Codes," *IRE Trans. Inf. Theory*, IT-8, p. 203, 1962.

13.34 N. Wax, "On Upper Bounds for Error Detecting and Error Correcting Codes of Finite Length," *IRE Trans. Inf. Theory*, IT-5, p. 168, 1959.

13.35 J. L. Massey, "Some Algebraic and Distance Properties of Convolutional Codes," in *Error Correcting Codes*, H. B. Mann, Ed., John Wiley & Sons, Inc., 1969.

chapter fourteen

Synchronizable Error-Correcting Codes

14.1 Introduction

Error-correcting codes, like the less redundant codes considered in Chapter Ten, are of considerably greater practical value if they are also synchronizable. Since the words in error-correcting codes generally all have the same length, they are necessarily uniquely decodable. But they are rarely synchronizable without alteration; all linear codes, for example, contain the all-zeros code word. Altering error-correcting codes in order to insure their synchronizability is the subject of this chapter.

As originally defined in Chapter Eleven, the term synchronizable was applied to dictionaries so constrained that the decoder could determine word synchronization directly from the received symbol stream in the absence of any prior information concerning this synchronization, and after only a limited number of symbols had been received. This definition, of course, must be appended with the phrase "in the absence of symbol errors" (or, at least, "in the absence of certain error patterns") to be of use in connection with noisy channels, since critically located symbol errors could obviously delay word synchronization indefinitely, regardless of the code being used. Indeed, because of the possibility of symbol errors, ordinary synchronizability may not be sufficient for codes which are to be used over noisy channels. Nevertheless, the synchronizability constraint does restrict the received symbol sequence so that, presumably, correct synchronization can be deduced after a sufficiently long period of observation, even in the presence of errors. This point will be further elaborated later (cf. Section 14.5). The first few sections of this chapter will be concerned with techniques for altering error-correcting codes to render them synchronizable in the original sense. This investigation has the added function of laying the groundwork for later sections in which this concept will be extended and modified.

Although we will not explicitly explore the possible synchronization properties of convolutional codes in this chapter, we mention in passing that at least some of the techniques to be discussed here can be readily extended to include these codes as well. This is obviously true when, as is frequently done, the encoded symbol sequence is periodically truncated (see Section 13.8) and hence converted into a block code of some (generally large) block length N. Regardless of whether or not the code is truncated, however, the decoder must at least be able to separate the received sequence into ν-symbol segments corresponding to the μ-symbol blocks of information symbols impressed at the input to the encoder. (See Section 13.8 for a definition of terminology.) This task is sometimes referred to as *node* or *branch* synchronization. One method for facilitating this is to treat the encoded symbol stream as though it were a sequence of words from a $(\nu, m\mu)$ linear block code. (This, of course, ignores the dependence between successive ν-symbol segments imposed by the convolutional code structure.) The methods to be developed in this chapter for enabling a decoder to distinguish true (block) code words from overlap words are then directly applicable and can be used to provide convolutional codes with the additional structure needed to establish node synchronization more reliably. As we shall discover, the efficacy of these techniques is a generally decreasing function of the information rate. Consequently, it may actually be advantageous to regard the received symbol sequence as a sequence of words from an $(i\nu, (m + i - 1)\mu)$ block code for some integer $i > 1$, thereby exploiting the code structure more effectively. Whether the resulting decrease in synchronization time makes the concomitant increase in synchronizer complexity worthwhile, however, will depend both on the code and on the channel parameters. In any event, the applicability to the node synchronization problem of the techniques to be discussed below should be apparent.

14.2 Prefix and Comma Error-Correcting Codes

One method investigated in Chapter Twelve for rendering block codes synchronizable was to prefix each word with some l-tuple and insist that that l-tuple not occur elsewhere in any code word sequence. A second scheme involved the periodic transmission of a comma in the sequence of code words, the dictionary chosen so as to assure the non-occurrence of this comma except when specifically transmitted. These two methods are here applied to error-correcting code dictionaries.

Each of the q^k different k-tuples occurs as the initial k symbols of some word in any *systematic* q-ary (n, k) code dictionary. This fact applies equally to any coset of a systematic code. Consequently, if a q^k-word code is systematic (and frequently even if it is not), any prefix, if it is to be unique, must contain at least $k + 1$ symbols. Adjoining such a prefix to each word therefore increases its redundancy by at least $k + 1$ symbols without at all improving the distance properties of the code. When k is small as compared to n this may not be objectionable, but as k increases this added redundancy becomes more serious. Nevertheless, the following theorem does assure the existence of a simple technique whereby any *cyclic* code can be made synchronizable with an easily mechanized synchronization procedure.

Theorem 14.1. Any (n, k) cyclic code can be converted to a prefix code by preceding each word with the prefix $000 \ldots 0a$ consisting of k zeros followed by any non-zero symbol in the alphabet. No shorter prefix will suffice.

Proof. Since the prefix $00 \ldots 0a$ has no repetitive period other than $k + 1$ (cf. Section 12.4), no overlap word formed from two prefixed code words can contain this prefix. Because the original code is systematic, the $k + 1$st and succeeding symbols are determined by parity checks on the first k symbols. If the first k symbols are zeros so must be the $k + 1$st symbol. Since the code is cyclic, all sequences of k consecutive zeros must be followed by a zero. Therefore, the pattern $000 \ldots 0a$ can occur only at the prefix, never elsewhere, and the extended code (i.e., the code resulting when each word is preceded by the prefix under discussion) is a prefix code. That no shorter prefix can be used to achieve this property has already been established. (Note that this result holds even when $k = n$.)

The extended code, of course, is no longer linear. This is of little significance since the prefix would be used only to synchronize the received word sequence; it could be ignored so far as the encoding and decoding operations are concerned, and the advantages of the linear property can still be exploited.

A weaker result can be obtained for linear codes in general by observing that if the code has a minimum Hamming distance d, all code words (except the all-zeros word) have at least d non-zero elements. The length of the longest string of zeros in any but the all-zeros code word therefore cannot exceed $n - d$. This establishes the following theorem.

Theorem 14.2. Any linear code dictionary having n-symbol words with minimum distance d can be made into a prefix code by preceding every code word with the prefix $000 \ldots 0a$ of length $n - d + 2$ where a is any non-zero symbol of the alphabet.

Since for any linear code $n - d + 2 \geq k + 1$ (cf. Section 13.4), this method is always less efficient than that described for cyclic codes.

Another method considered in Chapter Twelve for synchronizing block codes is to transmit periodically an m-symbol *comma.* Any comma which is to be singular (cf. Section 12.5) with respect to an (n, k) cyclic code must have length at least $2k + 1$. This follows because every k-tuple can occur both at the beginning of some code word and at the end of some word. The overlap words, therefore, must contain every possible $2k$-tuple.

Theorem 14.3. There exists a comma of length $2k + 1$ which is singular with respect to any (n, k) cyclic code for all n and k, $k \leq n - 1$. No shorter comma has this property.

Proof. The last statement has already been established. (This statement, incidentally, is equally applicable to any coset of a cyclic code.) The existence of a singular comma of length $2k + 1$ is evidenced by considering the comma $0000 \ldots 0a0 \ldots 0$ consisting of k zeros, followed by any non-zero symbol of the alphabet, followed by k more zeros. To prove this comma is singular, we observe that (1) no permutation of the n-tuple $x = 000 \ldots 0a$ is a code word, and (2) neither of the $(k + 1)$-tuples $s_1 = 000 \ldots 0a$ nor $s_2 = a00 \ldots 0$ occur in any code word. Both statements follow immediately from the fact that in a cyclic code, no word except the all-zeros word contains a sequence of more than $k - 1$ consecutive zeros. But if an overlap of two or more code words is to produce the comma, either one of the words must contain s_1, or one must contain s_2, or one must equal some cyclic permutation of x. None of these situations is possible. Similarly, if some overlap involving both code words and the comma is to produce the comma, at least $k + 1$ of these symbols must be contributed by one code word; again, this code word would have to contain either s_1 or s_2, or some permutation of x and consequently does not exist. This proves the theorem.

An alternative technique for designing a comma-code also having error-correcting properties is as follows.

Theorem 14.4. Let D be any cyclic (n, k) error-correcting code dictionary and D' the dictionary obtained by deleting from D the all-zeros code word. Then the comma $000 \ldots 0a$, consisting of $2k - 1$ zeros followed by any non-zero element, is singular with respect to D'.

Proof. No word in D' can contain the all-zeros k-tuple. Yet, if any overlap of code words is to yield the comma, one of the words would have to contain the all-zeros k-tuple. Clearly, no overlap involving the comma can produce the comma, and the theorem is proved. (Here it is not necessary to require $k < n$; cf. Section 12.5.)

As in the case of linear prefix codes, it is possible to prove a somewhat weaker result for general linear comma codes.

Theorem 14.5. The comma $000\ldots0a$ having length $2(n-d)+2$ is singular with respect to any linear dictionary containing n-symbol words separated by a minimum distance d, and from which the all-zeros word has been deleted.

Proof. Since the longest sequence of zeros formed from an overlap of two words would have length $2(n-d)$ (if the last $n-d$ symbols of one word and the first $n-d$ symbols of a second word were all zeros), $2(n-d)+1$ zeros indicate the presence of the comma, the end of which is signaled by the symbol a.

14.3 Comma-Free Error-Correcting Codes

One condition which we found useful to impose upon block codes was that of comma freedom. By using only about $1/n$th of the total number of available n-tuples, it was possible to keep all possible overlaps of code words in the set of excluded n-tuples. Since error-correcting codes also depend upon the use of only a subset of all n-tuples as code words, it might be hoped that few additional n-tuples would have to be excluded from the set in order to render it comma-free. In Section 14.4, we shall actually prove, at least when the code is cyclic and sufficiently redundant, that some of its cosets will be comma-free. Thus, it can be very simply altered to make it comma-free without excluding *any* additional n-tuples from the dictionary and without losing any of the algebraic properties which are useful in encoding and decoding. In this section, we present a rather simple test for determining whether or not there exists a coset of a linear code which is comma-free.

A code will be referred to as *invulnerable to synchronization at position m* if the sequence

$$b_{m+1}b_{m+2}\ldots b_n a_1 a_2 \ldots a_m \tag{14.3.1}$$

is not a code word when $a_1a_2\ldots a_n$ and $b_1b_2\ldots b_n$ are any two (not necessarily different) words from the code. If a code is invulnerable at all positions m ($m = 1, 2, \ldots, n-1$) then, of course, it is comma-free. If the code is linear, it clearly is not invulnerable at any position since it must contain the word $00\ldots0$. However, it may be possible to add a vector $\mathbf{c}$ to each code word, thereby forming a coset which is invulnerable at least at some positions. The coset has the same error-correcting properties as the original code, since it has the same distance properties.

Let $\mathbf{G}$ be a $k \times n$ matrix, the rows of which are the generators of a subspace over $GF(q)$. Then any vector in the subspace (represented as a row vector) can be expressed as $\mathbf{xG}$ where $\mathbf{x}$ is a q-ary k-component row vector. If the code is a coset of the subspace generated by $\mathbf{G}$, any code word $\mathbf{w}$ is represented by

$$\mathbf{w} = \mathbf{xG} + \mathbf{c} \tag{14.3.2}$$

for some vector $\mathbf{x}$. Now, let $\mathbf{G}$ be partitioned into $\mathbf{G}_1$ and $\mathbf{G}_2$ of dimensions $k \times m$ and $k \times (n-m)$, respectively:

$$\mathbf{G} = (\mathbf{G}_1, \mathbf{G}_2) \tag{14.3.3}$$

and similarly, let $\mathbf{c}$ be partitioned into the vectors $\mathbf{c}_1$ and $\mathbf{c}_2$ of m and $n - m$ components each:

$$\mathbf{c} = (\mathbf{c}_1, \mathbf{c}_2) \tag{14.3.4}$$

The number of coordinate agreements between any code word and any word overlap of the form (14.3.1) can then be determined by evaluating the number of zeros in a vector of the form

$$\begin{aligned} &\mathbf{xG} + \mathbf{c} - \mathbf{y}(\mathbf{G}_2, \mathbf{0}_m) - \mathbf{z}(\mathbf{0}_{n-m}, \mathbf{G}_1) - (\mathbf{c}_2, \mathbf{c}_1) \\ &\qquad = [\mathbf{x}, -\mathbf{y}, -\mathbf{z}] \begin{bmatrix} \mathbf{G} \\ \mathbf{G}_2, \mathbf{0}_{n-m} \\ \mathbf{0}_m, \mathbf{G}_1 \end{bmatrix} + \mathbf{c} - [\mathbf{c}_2, \mathbf{c}_1] \end{aligned} \tag{14.3.5}$$

The matrices $\mathbf{0}_l$ are $k \times l$ matrices all of whose elements are zero and $\mathbf{x}$, $\mathbf{y}$, and $\mathbf{z}$ are k-component vectors. If the coset which is to be used as the dictionary is vulnerable to synchronization at position m there will exist some $3k$ dimensional q-ary column vector $\mathbf{v}$ such that

$$\mathbf{Mv} = \mathbf{b} \tag{14.3.6}$$

where

$$\mathbf{M} = \begin{bmatrix} \mathbf{G} \\ \mathbf{G}_2, \mathbf{0}_{n-m} \\ \mathbf{0}_m, \mathbf{G}_1 \end{bmatrix}^T \qquad \mathbf{b} = [(\mathbf{c}_2, \mathbf{c}_1) - \mathbf{c}]^T$$

In order to prove that a code is invulnerable at position m, it is sufficient to show that there is no vector $\mathbf{v}$ for which the above equation has a solution. But this equation has a solution (Appendix B) if, and only if,

$$\rho(\mathbf{M}) = \rho(\mathbf{M}, \mathbf{b}) \tag{14.3.7}$$

where $\rho(\mathbf{A})$ denotes the rank of $\mathbf{A}$ and $(\mathbf{M}, \mathbf{b})$ is the $n \times (3k + 1)$ matrix obtained by adjoining the column vector $\mathbf{b}$ to the matrix $\mathbf{M}$. The following lemma will prove useful in investigating the invulnerability properties of linear codes.

Lemma 14.1. Let $\mathbf{M}$ be an $n \times l$ matrix, and $\mathbf{b} = (b_1, b_2, \ldots, b_n)^T$ a column vector, with elements in $GF(q)$ and let $\mathbf{m}_i = (m_{i1}, m_{i2}, \ldots, m_{il})$, $i = 1, 2, \ldots, n$, denote the ith row of $\mathbf{M}$. Then,

$$\rho(\mathbf{M}) \neq \rho(\mathbf{M}, \mathbf{b})$$

if and only if there exists a set of elements $\{\alpha_i\}$ in $GF(q)$ such that

$$\sum_{i=1}^{n} \alpha_i \mathbf{m}_i = \mathbf{0}$$

(with $\mathbf{0}$ the all-zeros vector) while

$$\sum_{i=1}^{n} \alpha_i b_i \neq 0$$

Proof. If $\rho(\mathbf{M}) = \rho$ there must exist ρ linearly independent vectors $\mathbf{m}_j$ such that

$$\mathbf{m}_i = \sum_{j=1}^{\rho} \beta_j^{(i)} \mathbf{m}_j \qquad \text{all } i \tag{14.3.8a}$$

Clearly, the corresponding set of vectors $\{\mathbf{m}'_j\}$ in $(\mathbf{M}, \mathbf{b})$ is also a linearly independent set. If $\rho(\mathbf{M}, \mathbf{b}) = \rho$ also, this set $\{\mathbf{m}'_j\}$ must span the space of $(\mathbf{M}, \mathbf{b})$, and

$$\mathbf{m}'_i = \sum_{j=1}^{\rho} \beta_j^{(i)} \mathbf{m}'_j \qquad \text{all } i \tag{14.3.8b}$$

For any given i, the coefficients $\beta_j^{(i)}$ in Equation (14.3.8a) are uniquely defined. Were this not so, it would be possible to exhibit a linearly dependent relationship among the vectors $\{\mathbf{m}_j\}$. Consequently, since the first l components of $\mathbf{m}_i$ and $\mathbf{m}'_i$ are identical, so too are the coefficients $\beta_j^{(i)}$ in Equations (14.3.8a) and (14.3.8b). Now

$$\sum_{i=1}^{n} \alpha_i \mathbf{m}_i = \sum_{j=1}^{\rho} \gamma_j \mathbf{m}_j \tag{14.3.9a}$$

with $\gamma_j = \sum_{i=1}^{n} \alpha_i \beta_j^{(i)}$, and

$$\sum_{i=1}^{n} \alpha_i \mathbf{m}'_i = \sum_{j=1}^{\rho} \gamma_j \mathbf{m}'_j \tag{14.3.9b}$$

By hypothesis, the summation in Equation (14.3.9a) yields the all-zero vector $\mathbf{0}$. This implies that $\gamma_j = 0$, all j, since the vectors $\{\mathbf{m}_j\}$ are linearly independent. This, in turn, implies that the summation in Equation (14.3.9b) also equals $\mathbf{0}$, contrary to the hypothesis that $\sum_i \alpha_i b_i \neq 0$. Thus, if $\sum_i \alpha_i \mathbf{m}_i = \mathbf{0}$ and $\sum_i \alpha_i \mathbf{b}_i \neq 0$, for any set of coefficients $\{\alpha_i\}$ $\rho(\mathbf{M}, \mathbf{b}) > \rho(\mathbf{M})$. Conversely, if for every set $\{\alpha_i\}$ for which $\sum_i \alpha_i \mathbf{m}_i = \mathbf{0}$, $\sum_i \alpha_i b_i = 0$ also, then obviously $\rho(\mathbf{M}, \mathbf{b}) = \rho(\mathbf{M})$ and the lemma is proved.

We therefore conclude that a dictionary is invulnerable to synchronization at position m if, and only if, the conditions of the preceding lemma are satisfied for some set of elements $\{\alpha_i\}$ with $\mathbf{M}$ and $\mathbf{b}$ as defined in Equation (14.3.6). Let $\{\alpha_i\}$ be a set of elements of $GF(q)$ such that

$$\sum_i \alpha_i \mathbf{m}_i = \mathbf{0} \tag{14.3.10}$$

Then, recalling the definition of the matrix $\mathbf{M}$ [Equation (14.3.6)], we find that the following three equations must hold,

$$\begin{aligned} &\sum_{i=1}^{n} \alpha_i \mathbf{g}_i = \mathbf{0} \\ &\sum_{i=n-m+1}^{n} \alpha_i \mathbf{g}_{i-(n-m)} = \sum_{i=1}^{m} \alpha_{i+n-m} \mathbf{g}_i = \mathbf{0} \\ &\sum_{i=1}^{n-m} \alpha_i \mathbf{g}_{i+m} = \sum_{i=m+1}^{n} \alpha_{i-m} \mathbf{g}_i = \mathbf{0} \end{aligned} \tag{14.3.11}$$

where $\mathbf{g}_i$ represents the ith column of $\mathbf{G}$. But, by definition, a vector $\mathbf{h} = (h_1, h_2, \ldots, h_n)$ is in the null space of the code generated by $\mathbf{G}$, if and only if

$$\sum_{i=1}^{n} h_i \mathbf{g}_i = 0 \tag{14.3.12}$$

Consequently, Equation (14.3.10) is satisfied by the set $\{\alpha_i\}$ if, and only if, the null space of the code generated by $\mathbf{G}$ contains the vectors $\mathbf{h} = \{\alpha_1, \alpha_2, \ldots, \alpha_n\}$, $\mathbf{h}^{(1)} = \{\alpha_{n-m+1}, \alpha_{n-m+2}, \ldots, \alpha_n, 0, 0, \ldots, 0\}$ and $\mathbf{h}^{(2)} = \{0, 0, \ldots, 0, \alpha_1, \alpha_2, \ldots, \alpha_{n-m}\}$. This fact, combined with the above lemma, proves the following theorem.

Theorem 14.6. If an (n, k) linear code has a coset which is invulnerable to synchronization at position m, its null space must contain three vectors $\mathbf{h} = \{h_1, h_2, \ldots, h_n\}$, $\mathbf{h}^{(1)} = \{h_{n-m+1}, h_{n-m+2}, \ldots, h_n, 0, 0, \ldots, 0\}$ and $\mathbf{h}^{(2)} = \{0, 0, \ldots, 0, h_1, h_2, \ldots, h_{n-m}\}$ such that $\mathbf{h}\cdot\mathbf{c} \neq \mathbf{h}\cdot\mathbf{c}^m$, where $\mathbf{c} = \{c_1, c_2, \ldots, c_n\}$ is some coset leader, and $\mathbf{c}^m \triangleq \{c_{m+1}, c_{m+2}, \ldots, c_n, c_1, \ldots, c_m\}$.

It is thus possible to establish whether a linear code can be made invulnerable to synchronization at any position m by observing if its null space generates three vectors satisfying the above conditions. This vastly reduces the amount of effort needed to determine the possible comma-free properties of such a code. When the code is cyclic it is possible to demonstrate even simpler necessary and sufficient conditions for the existence of comma-free cosets. This is the subject of the next section.

14.4 Comma-Free Cyclic Codes

If $\mathbf{G}$ generates an (n, k) cyclic code, then the code null space is also cyclic. Further (Section 13.4) the code null space is generated by an $(n - k) \times n$ matrix of the form $\mathbf{H} = \{h_{ij}\} = \{h_{j-i+1}\}$ with $h_l = 0$ unless $1 \leq l \leq k + 1$, but with neither h_1 nor h_{k+1} equal to zero. Combining this fact with the results of the previous section enables us to prove the following theorem.

Theorem 14.7. Any (n, k) cyclic code can be made invulnerable to synchronization at all positions m for $1 \leq |m| \leq n - k - 1$, $[|m| \triangleq \min(m, n - m)]$ by simply adding some non-zero element of the code field to the first symbol of each code word.

Proof. If $1 \leq m \leq n - k - 1$, the code null space obviously contains the three vectors $\mathbf{h} = (h_1, h_2, \ldots, h_{k+1}, 0, 0, \ldots, 0)$, $\mathbf{h}^{(1)} = (0, 0, \ldots, 0)$ and $\mathbf{h}^{(2)} = (0, 0, \ldots, h_1, h_2, \ldots, h_{k+1}, 0, \ldots, 0)$. Since $\mathbf{c} = (\alpha, 0, \ldots, 0)$ for some $\alpha \neq 0$,

$$\mathbf{h}\cdot\mathbf{c} = h_1\alpha \neq \mathbf{h}\cdot\mathbf{c}^m = 0 \tag{14.4.1}$$

if $1 \leq m \leq n - k - 1$, and the conditions of Theorem 14.6 are satisfied. An identical argument leads to the same conclusion for $1 \leq n - m \leq n - k - 1$. Consequently, Equation (14.3.6) will have no solution, when **c** is as defined, for any m satisfying $1 \leq |m| \leq n - k - 1$, and the theorem is proved.

Cosets other than those obtained by adding an element to the first column may also work. For example, adding a non-zero element to each term in the last column is equally effective.

It is also possible to prove a converse to Theorem 14.7.

Theorem 14.8. All cosets of any (n, k) systematic code are vulnerable to synchronization at all positions m where $n - k \leq m \leq k$.

Proof. The first k columns of the generator **G** of any systematic code must be linearly independent. If $n - k \leq m \leq k$ no subset of the first $n - m \leq k$ columns of **G** and hence of the first $n - m$ rows of **M** can be linearly dependent. But, in addition, none of the last $m \leq k$ rows of **M** can be included in a linearly dependent combination since the columns of $\mathbf{G}_1$ are linearly independent. Hence, $\rho(\mathbf{M}) = n$ and the conditions of Theorem 14.6 do not hold.

Corollary. If $k \leq (n - 1)/2$, any (n, k) cyclic code can be made comma-free. If $k > (n - 1)/2$ no coset of any (n, k) cyclic code is comma-free.

Proof. This follows because all erroneous synchronization positions can be made invulnerable when $n - k - 1 \geq k$; i.e., when $k \leq (n - 1)/2$. Since all cyclic codes are systematic codes, Theorem 14.8 is applicable, and if $k > (n - 1)/2$, all those values of m in the range $(n - k, k)$ are vulnerable to synchronization regardless of the coset under consideration.

The question arises as to whether non-cyclic, error-correcting linear codes have comma-free cosets, and, in particular, whether such cosets exist for $k > (n - 1)/2$. Intuitively, it might be supposed that the conditions of Theorem 14.6 make cyclic codes superior to non-cyclic codes so far as their comma freedom is concerned. It seems difficult to make a general statement concerning the comma-free properties of cosets of non-cyclic codes, however, although specific codes or classes of codes are sometimes easily investigated using the techniques outlined here.

Consider as an example the class of shortened $(n - \nu, k - \nu)$ cyclic codes, obtained by deleting ν information bits from an (n, k) cyclic code (see Section 13.6). The null space of such a code is generated by the matrix $\mathbf{H}_\nu$ resulting when the first ν columns of the null space generator of the original cyclic code are deleted. With this in mind, we need only duplicate the argument of Theorem 14.7 (letting **h** be the $(\nu + 1)$st row of $\mathbf{H}_\nu$) to prove the following.

Theorem 14.9. Any $(n - \nu, k - \nu)$ shortened cyclic code can be made invulnerable to synchronization at all positions m for $1 \leq |m| \leq n - k - \nu - 1$ by simply adding some non-zero element to the first symbol of each code word.

While Theorem 14.8 is also applicable to shortened cyclic codes, a stronger converse to Theorem 14.9 is apparently not generally possible. The binary (7, 3) cyclic code, for example, can be made invulnerable to synchronization at position m for all $1 \leq |m| \leq 3$, and hence is comma-free. Theorem 14.9 establishes the existence of a (6, 2) shortened cyclic code which is invulnerable at positions $1 \leq |m| \leq 2$ and a (5, 1) code invulnerable only at positions $|m| = 1$. It is easy to verify that the rank of the matrix **M** (see Section 14.3) corresponding to the (6, 2) shortened cyclic code is 6 when $m = 3$, so all of its cosets are indeed vulnerable at position $m = 3$, even though $k < (n - 1)/2$. In contrast, the (5, 1) shortened cyclic code has a comma-free coset.

As a second example of a class of non-cyclic codes let **G** be an $n \times k$ matrix, the rows of which generate the words of an (n, k) cyclic binary Hamming code ($2^{n-k} = n + 1$; cf. Section 13.6) and let $\mathbf{G}'$ be obtained from **G** by some column permutation. Then $\mathbf{G}'$ also generates an (n, k) binary code having the same distance properties, and hence the same error-correcting capabilities, as the code generated by **G**. All such codes will be referred to here as Hamming codes although, in general, they will not be cyclic. (It can be verified, incidentally, that all binary, linear, single-error-correcting codes having $2^k = 2^n/(n+1)$ words can be generated by a matrix of the form $\mathbf{G}'$.) Since $k > (n - 1)/2$ for all such codes except the (3, 1), no coset of any cyclic Hamming code will be comma-free for any $k > 1$. Conceivably, however, some coset of a non-cyclic Hamming code could be comma-free even for $k > 1$.

To show that this is not possible, it is again sufficient to prove $\rho(\mathbf{M}') = n$ for some position of vulnerability m where $\mathbf{M}'$ is defined in terms of $\mathbf{G}'$ by Equation (14.3.6). Every row of the null space matrix of a binary Hamming code contains exactly $(n + 1)/2$ *ones* (cf. Section 13.6). Consider the matrix $\mathbf{M}'$ when $m = (n - 1)/2$. Since exactly $(n + 1)/2$ columns of $\mathbf{G}'$ must be involved in any parity-check relationship and since $\mathbf{G}'_1$, in this case, contains only $(n - 1)/2$ columns, no subset of the bottom $(n - 1)/2$ rows of $\mathbf{M}'$ is linearly dependent. The only remaining possibility is that the top $(n + 1)/2$ rows of $\mathbf{M}'$ are so related. This is readily seen to imply:

$$\sum_{i=1}^{(n+1)/2} \mathbf{g}'_i = \sum_{i=(n+1)/2}^{n} \mathbf{g}'_i = 0$$

where $\mathbf{g}_i'$ represents the ith column of the generator matrix $\mathbf{G}'$. This in turn requires

$$\sum_{\substack{i=1 \\ i \neq (n+1)/2}}^{n} \mathbf{g}_i' = 0$$

i.e., a linear combination of $n - 1$ columns of $\mathbf{G}'$ must be identically zero. This is contrary to the condition that exactly $(n + 1)/2$ columns must be included in any such relationship unless $n - 1 = (n + 1)/2$; i.e., unless $n = 3$. Consequently, the (3, 1) is the only binary Hamming code which has a comma-free coset.

A frequently useful alternative expression for the vulnerability equation [Equation (14.3.6)] is afforded by representing the vectors involved as polynomials as in Section 13.4. Equation (14.3.7) becomes, in this notation,

$$w(x) + q_m(x) + x^{n-m}p_m(x) = c(x)(x^{n-m} - 1) \quad (\text{modulo } x^n - 1) \tag{14.4.2}$$

where $w(x) = \sum_{i=0}^{n-1} w_i x^i$ represents some word $\mathbf{w} = (w_0\ w_1 \ldots w_{n-1})$ in the dictionary D generated by $\mathbf{G}$, $q_m(x)$ is a polynomial of degree at most $n - m - 1$ corresponding to a suffix in D, and $p_m(x)$ a polynomial of degree at most $m - 1$ representing a prefix in D. The polynomial $c(x)$ represents the coset leader. (Note that here the term "coset leader" denotes *an* element in the coset, not necessarily the minimum weight element.)

If the code is cyclic, $p_m(x)$ is also a suffix, and $q_m(x)$ a prefix in D. When this is the case, the invulnerability equation can be expressed more simply in either of the forms

$$w(x) = c(x)(x^{n-m} - 1) - x^{n-m}p_m(x)$$

or

$$w(x) = c(x)(x^{n-m} - 1) - q_m(x) \tag{14.4.3}$$

The dictionary then is vulnerable to synchronization at position m if, and only if, the right side of Equation (14.4.3) is a code word. Recalling from Section 13.4 that a polynomial $p(x)$ represents a word from the cyclic code dictionary D generated by the polynomial $g(x)$ if, and only if, the syndrome $\{p(x)\} \triangleq r(x)$ is identically zero [where $p(x) = q(x)g(x) + r(x)$], we have, as the condition for invulnerability at position m,

$$\{c(x)(x^{n-m} - 1) - x^{n-m}p_m(x)\} \neq 0$$

or

$$\{c(x)(x^{n-m} - 1) - q_m(x)\} \neq 0 \tag{14.4.4}$$

Using this approach we can prove Theorems 14.7 and 14.8 (for cyclic codes) quite simply. First, if $c(x) = 1$, and if $m \leq n - k - 1$, $\{x^{n-m} - 1 - x^{n-m}p_m(x)\} = \{x^{n-m}[1 - x^m - p_m(x)]\} \neq 0$ since $p_m(x)$ has degree at most $m - 1$ and $g(x)$ is of degree $n - k$. Similarly, if $1 \leq n - m \leq n - k - 1$,

$\{x^{n-m} - 1 - q_m(x)\} \neq 0$. Second, to prove Theorem 14.8 for cyclic codes, we need only observe that $\{c(x)(x^{n-m} - 1) - x^{n-m}p_m(x)\}$ is a polynomial of degree at most $n - k - 1$. Since $p_m(x)$ represents a prefix in D, its first k coefficients correspond to information symbols and hence are arbitrary. If $n - k \leq m \leq k$, there must exist a polynomial $p_m(x)$ such that $\{x^{n-m}[c(x)(1 - x^m) - p_m(x)]\} = 0$ regardless of the polynomial $c(x)$.

14.5 Synchronization Over Noisy Channels

When noise is not a factor, word synchronization is usually determined, if it can be determined at all, with perfect reliability. Once a synchronizing pattern is observed or once all but one of the possible decodings of a sequence are eliminated because they result at some point in a non-word, synchronization has been obtained. Noise can modify the synchronization problem in several ways. If synchronization is to be established by recognizing a particular m-tuple, for example, and the code dictionary has been carefully selected to prohibit the occurrence of this m-tuple at any unwanted position, noise can cause false synchronization by altering an m-tuple occurring in the body of the word sequence so that it appears at the receiver as the synchronization pattern. Moreover, the noise can increase the synchronization delay by changing the synchronization pattern causing it to pass unrecognized. Similarly, if word synchronization relies upon the non-existence of erroneously decodable sequences exceeding some finite length, the noise may nullify this property. A non-decodable sequence may be changed to a decodable sequence by the noise, while formerly decodable sequences might become non-decodable.

The possibility of erroneous synchronization in the presence of noise even with synchronizable code dictionaries necessarily alters the synchronization procedure. A statistical decision must now be made. The "most nearly" decodable sequence must be selected, or the pattern most closely resembling the synchronization pattern and recurring at the correct intervals must be determined. If comma-free codes of word length n are used, for example, synchronization is established by observing all received n-tuples to determine if they are words or not. Since, in the absence of noise, no overlap word is ever a code word, it is more likely even in the presence of noise that true words will be recognized as code words while overlaps will not. And since true code words should be observed after every n symbols are received, it should be possible to increase the reliability of the synchronization estimate by successive recognitions of code words separated by exactly n symbols. As an example, suppose words from the comma-free dictionary $D = \{010, 011, 020, 021, 022, 120, 121, 122\}$ are transmitted over a noisy channel and the sequence

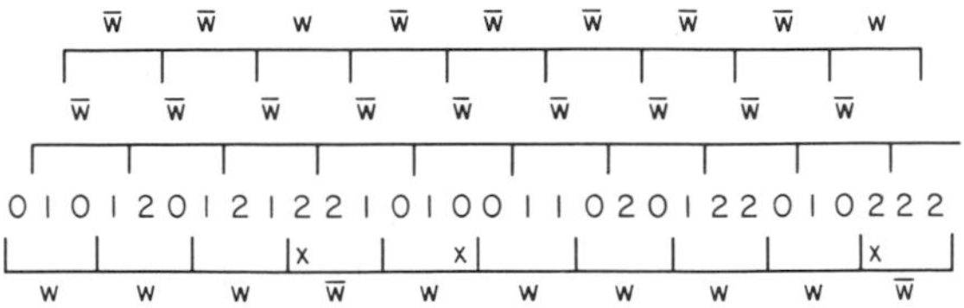

is received with the symbols indicated by the x's in error. Since the decoding represented by the underscoring encounters only two non-words ($\bar{w}$'s) while the other two possibilities result almost entirely in non-words, the underscoring would seem to correspond to the correct synchronization. The same observations are applicable to the other synchronization methods discussed in the previous chapters.

This change in the synchronization search procedure necessitated by the presence of noise may change the relative attractiveness of the various constraints imposed on block codes in order to make them synchronizable. Two variations on the synchronization constraints of Chapter Twelve, relevant when the communication channel is subject to noise, are examined in the next several sections.

14.6 Barker Sequences

The first variation to be considered actually involves a weakening of one of the constraints of Chapter Twelve. A comma code was constrained so as to prevent the occurrence of some particular m-tuple in any arbitrary sequence of words from its dictionary. Whenever this m-tuple, or comma, was observed at the receiver, it must have been inserted deliberately and hence provided the desired synchronization. As already observed, the situation is not so simple when noise is present. Multiple observations of the comma are presumably necessary to increase the probability that it was actually received and not just an effect of noise. An obviously simpler method of word synchronization, and one often as acceptable when several observations are needed anyway, is to transmit a comma periodically as before, but to place no restriction on the remainder of the transmitted sequence.

Suppose an m-tuple is used as a comma and is transmitted after every $N - m$ information symbols. The comma can now occur in the information sequence even if there were no noise. But if the information consists of a random sequence of r-ary symbols selected independently and with equal probability, the probability of any particular information m-tuple being the comma is r^{-m}. Further, since the noise forces multiple observations anyway, the probability of the comma occurring l times, separated by exactly N symbols (as it must if it is to be mistaken for the true comma) is r^{-lm}. Thus, a long enough observation should discern the true pattern, which does repeat periodically, although sometimes altered by noise, even when the information symbols are unconstrained. Indeed, the possible decrease in the time necessary for a reliable decision attainable by placing constraints on the information sequence is often insignificant.

At first glance, one might suppose any one comma would be as good as any other since all sequences of a given length will occur randomly with the same probability. This would be essentially true† were it not also necessary to distinguish between the comma and those m-tuples beginning with any of its $m-1$ suffixes or ending with any of its $m-1$ prefixes. The comma consisting of m repetitions of the same symbol, for example, would clearly be a poor choice. If such a comma were either preceded or followed by this same symbol, at least two successive m-tuples would be identified as commas.

Let d_i be the Hamming distance separating the i-symbol suffix of the comma from its i-symbol prefix. The average distance between the comma and an m-tuple beginning with its i-symbol suffix and ending with $m-i$ independent r-ary data symbols is, in the absence of errors,

$$d_i + \frac{r-1}{r}(m-i) = \frac{r-1}{r}\left[m - \left(i - \frac{r}{r-1}d_i\right)\right] \tag{14.6.1}$$

(This is also the distance between the comma and an m-tuple beginning with $m-i$ data symbols and ending with the i-symbol comma prefix.) In view of the preceding remarks, then, an appropriate choice for a comma would be some m-tuple for which the distances (14.6.1) are maximized for all i. This implies selecting a sequence for which the quantity

$$c_i \triangleq i - \frac{r}{r-1}d_i \tag{14.6.2}$$

attains its minimum possible value for all i, $0 < i < m$.

The bound established in the following theorem will be useful in the sequel.

Theorem 14.10. Let c_i be as defined in Equation (14.6.2). Then, for any r-ary m-symbol comma,

$$\max_{1 \le i \le m-1} c_i \ge \operatorname*{ave}_{1 \le i \le m-1} c_i \ge \frac{m}{2} - \frac{r}{2(r-1)(m-1)}\left[m^2\left(\frac{r-1}{r}\right)\right]$$

the square brackets indicating the integer part of the enclosed fraction.

Proof. Let D be a dictionary comprised of the m cyclic permutations of the comma. The average Hamming distance d_{ave} between the comma and its $m-1$ distinct cyclic permutations is clearly equal to the average distance separating any

†The reason for the qualifier is made clear in Problem 14.3.

two words in D. This distance is limited by the Plotkin bound [Equation (13.3.5)] to $d_{\text{ave}} \leq m^2(r-1)/(m-1)r$. Now, from Equation (14.6.2)

$$\sum_{i=1}^{m-1} c_i = \frac{m(m-1)}{2} - \frac{r(m-1)}{2(r-1)} d_{\text{ave}} \tag{14.6.3}$$

Since $(m-1)d_{\text{ave}}$ must be an integer, the theorem immediately follows.

The first investigations in this area were due to Barker (Reference 14.1) and were limited to sequences over the binary alphabet. Under the original constraint, sequences were acceptable only if $\max c_i \leq 0$. We first examine the consequences of this restriction.

Theorem 14.11. If a binary sequence of length m exists having $c_i \leq 0$ for all i, $i = 1, 2, \ldots, m-1$, then

$$c_i = \begin{cases} 0 & i \text{ even} \\ -1 & i \text{ odd} \end{cases}$$

Moreover, such sequences can exist having length $m > 2$ only if $m = 4t - 1$ for some integer t.

Proof. When $r = 2$, the quantity $c_i = i - 2d_i$ must be an even integer if i is even and an odd integer if i is odd. Suppose $\gamma_1 \gamma_2 \ldots \gamma_m$ is a sequence for which

$$c_i = \begin{cases} -1 & i \text{ odd} \\ 0 & i \text{ even} \end{cases}$$

Then

$$\sum_{i=1}^{m-1} c_i = -\left[\frac{m}{2}\right]$$

and meets the bound of Theorem 14.10. Any sequence $\gamma'_1 \gamma'_2 \ldots \gamma'_m$ having some parameters $c'_i < c_i$ would require at least one of the terms c'_j to be greater than the corresponding term c_j. This implies $c'_j \geq c_j + 2 \geq 1$ and hence violates the condition that $\max c_i \leq 0$.

To prove the second statement of the theorem, we observe that, for any binary sequence $\gamma_1 \gamma_2 \ldots \gamma_m$,

$$\begin{aligned} a_i \triangleq c_i + c_{m-i} &= \sum_{\nu=1}^{m} (1 - 2\gamma_\nu)(1 - 2\gamma_{\nu+i}) \\ &= m - 4\sum_{\nu=1}^{m} \gamma_\nu + 4\sum_{\nu=1}^{m} \gamma_\nu \gamma_{\nu+i} = m - 4l \end{aligned} \tag{14.6.4}$$

for some integer l. (The subscripts of the terms γ_i are interpreted modulo m.) But, as we have just shown, when m is even, and $\max c_i \leq 0$, $a_i = c_i + c_{m-i} = \begin{cases} 0 & i \text{ even} \\ -2 & i \text{ odd} \end{cases}$. This violates the condition of Equation (14.6.4), unless $m = 2$, so m cannot be an even integer greater than 2. If m is odd, $a_i = -1$ for all $i \neq 0$, modulo m, and, from Equation (14.6.4), $m = 4l - 1$ as was to be shown.

Barker was able to construct sequences with the desired property (with $m > 2$) for lengths $m = 3, 7$, and 11, and conjectured that no other such sequences exist. (When $m = 1$ the trivial sequence 1 clearly satisfies the required conditions, and when $m = 2$, the sequence 10 does likewise.) This conjecture was later proved to be true. The sequences which Barker found are

$m = 3$	110
$m = 7$	1110010
$m = 11$	11100010010

(Equivalent sequences result upon reversing these sequences and upon complementing either their odd- or their even-numbered symbols. Barker observed as an aid in constructing these sequences, incidentally, that the relationships $\gamma_i \neq \gamma_{m+1-i}$ for odd i and $\gamma_i = \gamma_{m+1-i}$ for even i must always be satisfied. This can easily be proved by induction on i.)

Since only three inequivalent sequences exist under Barker's original constraint, efforts have been made to find additional sequences under somewhat more relaxed restrictions. The first obvious weakening of the constraint on the coefficients c_i is to let $c_i = 1$ for some values of i. Moreover, since it does not seem to be advantageous to let c_i have large negative values (cf. Theorem 14.11) and since a large negative c_i can sometimes be detrimental, the constraint $|c_i| \leq 1$, $0 < i < m$, immediately suggests itself. Binary sequences satisfying this constraint, generally called *Barker sequences*, have been found for lengths $m = 1, 2, 3, 4, 5, 7, 11$, and 13. It has been shown (Reference 14.2) that no such sequences exist for odd length $m > 13$. (This, of course, proves, a fortiori, the non-existence of any sequences satisfying Barker's original constraint for any lengths $m > 11$.)

Theorem 14.12. Let $\gamma_1 \gamma_2 \ldots \gamma_m$ be a binary sequence having coefficients c_i bounded in absolute value by unity for all $i = 1, 2, \ldots, m - 1$. Then, if m is an even integer greater than 2, $m = 4v^2$ with v an integer. If m is odd, $c_i = (-1)^{(m-1)/2}$, i odd, and $c_i = 0$, i even, $0 < i < m$.

Proof. If $|c_i| \leq 1$, $|c_i| = 1$ for all odd i and $c_i = 0$ for all even i. Thus, when m is even,

$$a_i = c_i + c_{m-i} = \begin{cases} 0 & \text{i even, } i \neq 0 \\ 0, +2, \text{ or } -2 & i \text{ odd} \end{cases}$$

But from Equation (14.6.4), $a_i - a_j = 4l$ for some integer l and for all integers i, j. Accordingly, if m is even and greater than 2, $a_i = 0$ for all $i = 1, 2, \ldots, m - 1$. Defining $\alpha_i = 1 - 2\gamma_i$, we find

$$\sum_{i=1}^{m} a_i = m = \sum_{i=1}^{m} \sum_{j=1}^{m} \alpha_j \alpha_{j+i} = \left(\sum_{j=1}^{m} \alpha_j \right)^2 = t^2 \qquad (14.6.5)$$

for some integer t. Since m is even, t must be even and $m = (2v)^2$ for some integer v. If m is odd, and $|c_i| \leq 1$, $a_i = (-1)^{(m-1)/2}$ [cf. Equation (14.6.4)]. Thus, since c_i is even when i is even, and odd otherwise, $c_i = (-1)^{(m-1)/2}$, i odd, and $c_i = 0$, i even.

It has been shown (cf. Reference 14.3) that no even length Barker sequence exists for any v in the range 2 to 54 ($16 \leq m \leq 11{,}664$).† The existence of any Barker sequences with lengths other than those mentioned above is highly improbable.

A further generalization is motivated by observing that for larger values of m it might be quite satisfactory to relax further the constraint on $|c_i|$. Theoretical results seem to be quite difficult to obtain, however, when $|c_i|$ is not bounded by 1. Sequences for which $|c_i| \leq 2$ have been found for all lengths $m \leq 21$ and for $m = 25$ and 28; when the constraint $|c_i| \leq 3$ is imposed, sequences for all lengths $m \leq 34$ are known (Reference 14.3).

Obviously, the coefficient c_i cannot be uniformly small if the term $a_i = c_i + c_{m-i}$ is not. The converse of this statement, that c_i will be uniformly small if a_i is small, of course does not follow. Nevertheless, the search for Barker-type sequences might well begin with sequences which are known to have small values of a_i; viz., the maximal-length shift-register sequences (see Theorem 13.11). Since the coefficients c_i are different for different cyclic permutations of a sequence, each maximal-length sequence of length m may generate up to m inequivalent synchronizing sequences. As an example, consider the sequence $S_0 = 1110100$ and define S_i to be the sequence S_0 permuted cyclically i positions to the right. The coefficients c_i for these seven sequences are tabulated in Table 14.1. While all these sequences are reasonably

Table 14.1 Coefficients c_i [Equation (14.6.2)] of the cyclic permutations of the maximal-length sequence 1110100

	S_0	S_1	S_2	S_3	S_4	S_5	S_6
c_6	0	−2	0	0	0	−2	−2
c_5	1	1	−1	−3	−1	−1	−1
c_4	0	−2	−2	−2	0	0	2
c_3	−1	1	1	1	−1	−1	−3
c_2	−2	−2	0	2	0	0	0
c_1	−1	1	−1	−1	−1	1	1

well-behaved so far as the maximum coefficient c_i is concerned, only $S_4 = 0100111$ is a Barker sequence.

Very little is known about Barker-type sequences in the non-binary case. Again, however, the set of maximal-length shift-register sequences (when r is a prime or a prime power) seems to be a reasonable set to investigate in the search for Barker-type sequences. The average value of the coefficients c_i associated with these sequences at least attains the lower bound given in Theorem 14.10, as is easily verified.

---

†The sequence 1110 and all of its permutations are Barker sequences of length four.

14.7 The Index of Comma Freedom

Another method of obtaining word synchronization in the presence of noise is again to encode the information, but under a constraint more severe than ordinary comma freedom. The comma-free constraint insures that every overlap word differs from every code word in at least one symbol. Noise could very possibly eradicate this difference and prolong the synchronization search. However, if every overlap word differed from every code word in at least $s > 1$ symbols, the probability of the noise nullifying this difference should be correspondingly decreased. Consequently, the code words could be more readily distinguished from overlap words with a resulting increase in the rapidity and reliability of word synchronization. Codes which satisfy this requirement that every overlap word differs from every code word in at least s positions will be called *comma-free with index s*.

Some of the techniques of Chapter Twelve for constructing codes with the comma-free property can be extended to yield codes having larger indices of comma freedom. We will not pursue this particular approach further, however, because of a rather serious limitation as to the practicality of such codes; viz., the lack of protection against errors after synchronization is obtained. Two code words in a large index comma-free code may differ by only one symbol. The noise is presumably such that one symbol is relatively easily mistaken for another or the larger index of comma freedom would not be necessary. Under such conditions, then, many decoding errors would be made. While missynchronization could be corrected by further observation, the erroneously decoded code word will not usually be repeated for a second chance. Consequently, it is reasonable to impose the condition that the minimum distance between code words be at least as great as the minimum distance between code and overlap words. Generally, in fact, we will be interested in dictionaries for which the former distance is considerably greater than the latter.

A rather simple bound on the number of code words such dictionaries can contain follows.

Theorem 14.13. If the r-ary dictionary D, containing N n-symbol words, has an index of comma freedom s and minimum distance at least s, then

$$N \leq N_s(n, r) = \frac{W_s(n, r)}{n} \tag{14.7.1}$$

where $W_s(n, r)$ is the maximum number of n-symbol r-ary words in a (not necessarily linear) dictionary having minimum Hamming distance s.

Proof. Every cyclic permutation of each code word must disagree with every cyclic permutation of every other word, and with every other cyclic permutation of itself, in at least s symbols if the code is both comma-free with index s and has minimum distance s. The dictionary containing all n cyclic permutations of every code word in D then has nN code words which must be separated by a minimum Hamming distance s. There can be no more than $W_s(n, r)$ such words.

Note that when $s = 1$, this yields the bound

$$N \leq \frac{r^n}{n}$$

which is, of course, a valid, but not a minimal bound on the number of words in ordinary comma-free code dictionaries.

Corollary. A dictionary containing N n-symbol, r-ary words cannot have an index of comma freedom exceeding

$$s \leq \min_{\rho_0 \leq \rho \leq [(r-1)/r]n} \left\{ \frac{N_\rho}{N_\rho - 1} \rho \left(2 - \frac{\rho}{n} \frac{r}{r-1}\right), 2\rho \right\} \tag{14.7.2}$$

where N_ρ is the smallest integer equal to or exceeding $(nN/r^n) \sum_{i=0}^{\rho} \binom{n}{i} (r-1)$, and ρ_0 is the smallest integer such that $N_{\rho_0} \geq 2$.

Proof. This follows immediately upon using the bound of Theorem 13.2 on $W_s(n, r)$ in Theorem 14.13.

In some cases, it may be more important to be able to detect the loss of synchronism than to have the capability of determining it initially directly from the code word sequence. This would be the situation for example when a synchronization mode precedes the communication mode, thereby establishing the necessary initial synchronization, but when the possibility of a subsequent loss of synchronism is not negligible. When synchronism is lost, it is often because the local word timing reference has slipped by a few symbol positions from the correct position.

It is therefore useful to define an *index of invulnerability s_m at position m* indicating that any overlap beginning with the $(m + 1)$st symbol of one word and ending with the mth symbol of a second word must differ in at least s_m symbols from any true code word. The preceding paragraph suggests some advantage in using a dictionary having a larger index of invulnerability at all positions m in some range $-\mu \leq m \leq \mu$ at the expense of perhaps no invulnerability at all for other values of m. We shall characterize by an *index of invulnerability s of degree μ* any dictionary having an index of invulnerability of at least s at all positions m, $1 \leq |m| \leq \mu$. Clearly, if $\mu = [n/2]$ where n is the code word length, the dictionary is comma-free with index s. The bound of Theorem 14.13 can easily be generalized to encompass codes having indices of invulnerability of arbitrary degree.

Theorem 14.13a. If an r-ary dictionary D, containing N n-symbol words, has an index of invulnerability s of degree $\mu \leq n - 1$ and minimum distance at least s, then

$$N \leq \frac{W_s(n, r)}{\mu + 1}$$

where $W_s(n, r)$ is as previously defined [Equation (14.7.1)].

Proof. Let $\mathbf{w}$ be a word in D and $\mathbf{w}^m$ the n-tuple obtained by cyclically permuting the symbols of $\mathbf{w}$ m positions to the right. Let S be the set of all n-tuples $\mathbf{w}^m$ with $\mathbf{w}$ any word in D and m any integer in the range $-[\mu/2] \leq m \leq [(\mu + 1)/2]$. If any two of these n-tuples, say $\mathbf{w}_1^{m_1}$ and $\mathbf{w}_2^{m_2}$, are within distance σ of each other, then the code word $\mathbf{w}_1$ is within distance σ of the permuted code word $\mathbf{w}_2^{m_2 - m_1}$. Since, by construction, $|m_2 - m_1| \leq \mu$, the $(\mu + 1)N$ n-tuples in S must be mutually separated by a distance of at least s, and the theorem is proved.

14.8 On Testing the Index of Invulnerability of Cosets of Linear Codes

Using the notation of Section 14.3, we recall that the coset of a linear code generated by the columns of the matrix $\mathbf{G}$ is vulnerable to synchronization at position m if, and only if, the equation

$$\mathbf{M}\mathbf{v} = \mathbf{b} \tag{14.8.1}$$

has a solution; i.e., if, and only if,

$$\rho(\mathbf{M}) = \rho(\mathbf{M}, \mathbf{b}) \tag{14.8.2}$$

with $\mathbf{M}$ and $\mathbf{b}$ as defined in Equation 14.3.6. Now suppose for some value of m and some vector $\mathbf{v}$

$$\mathbf{M}\mathbf{v} - \mathbf{b} = \mathbf{e}(\mathbf{v}, m) \tag{14.8.3}$$

Suppose, further, $\mathbf{G}$ is such that

$$\min_{\substack{\mathbf{v}, m \\ 1 \leq |m| \leq \mu}} |\mathbf{e}(\mathbf{v}, m)| = |\mathbf{e}(\mathbf{v}_0, m_0)| = |\mathbf{e}| = s \tag{14.8.4}$$

where $|\mathbf{e}(\mathbf{v}, m)|$ denotes the Hamming weight of $\mathbf{e}(\mathbf{v}, m)$. The coset obtained by adding $\mathbf{c}$ to the dictionary generated by $\mathbf{G}$ then has an index of invulnerability s of degree μ. In this case,

$$\rho(\mathbf{M}) = \rho(\mathbf{M}, \mathbf{b} + \mathbf{e}) \tag{14.8.5}$$

for $m = m_0$, but this equation does not hold for any $\mathbf{e}$ of lesser weight for any m in the range $1 \leq |m| \leq \mu$.

Unfortunately, no very convenient method exists for determining the minimum value of $|\mathbf{e}|$ when it is not zero. The following approach is useful for establishing an upper bound on $|\mathbf{e}|$ for a given $\mathbf{G}$, for evaluating $|\mathbf{e}|$ for a particular coset, and in some cases as a method of constructing an optimum coset leader for a given generator **G**. The calculation involved, however, becomes increasingly tedious as the code length increases, and the first of these situations is probably the only one for which the technique is useful for any but relatively small code dictionaries.

The procedure is as follows. An operator matrix **A** is determined which reduces **M** to echelon canonical form:

$$\mathbf{AM} = \begin{pmatrix} \mathbf{B} \\ \mathbf{O} \end{pmatrix} \tag{14.8.6}$$

where **B** is a $\rho_0 \times 3k$ matrix with ρ_0 the rank of **M** and **O** is a matrix of all zeros. Multiplying both sides of Equation (14.8.3) by **A**, one obtains

$$\mathbf{AMv} = \mathbf{Ab} + \mathbf{Ae} \tag{14.8.7}$$

Since $\rho(\mathbf{M}) = \rho_0$, the first ρ_0 of the above equations can be solved. The left side of each of the remaining $n - \rho_0$ equations is always zero, so **e** must be chosen to satisfy the equation

$$\widetilde{\mathbf{Ab}} = -\widetilde{\mathbf{Ae}} \tag{14.8.8}$$

where the "tilde" indicates the truncated vectors consisting of the last $n - \rho_0$ components only. There are q^n ways of selecting **e** over $GF(q)$, but only $q^{n-\rho_0}$ vectors $\widetilde{\mathbf{Ae}}$. We can systematically generate all of these vectors by beginning with the **e** of lowest weight and proceeding to higher weight vectors until the complete set of $q^{n-\rho_0}$ vectors $\widetilde{\mathbf{Ae}}$ has been represented at least once. The weight s of the maximum weight vector needed to complete the set provides a bound on the index of invulnerability of the dictionary at position m. This index can be realized if a vector **c** exists satisfying Equation (14.8.8). If, for every other value of m in the range $1 \leq |m| \leq \mu$, **c** also necessitates a vector **e** of weight at least s, the dictionary has an index of invulnerability s of degree μ.

Equation (14.8.7), incidentally, suggests another rather useful bound on the index of invulnerability attainable with a linear code.

Theorem 14.14. The index of invulnerability s_m at position m of any coset of the linear code generated by **G** cannot exceed $n - \rho_0$, where $\rho_0 = \rho(\mathbf{M})$ is the rank of the matrix **M** of Equation (14.8.1).

Proof. In order to reduce **M** to echelon canonical form, it is only necessary to select ρ_0 linearly independent rows of **M**, to add suitable multiples of these rows to each other and to the remaining rows of **M**, and finally to permute the rows of the resulting matrix. If **A** is a matrix which accomplishes this, and **e** is a vector having zeros in the ρ_0 positions corresponding to these linearly independent rows of **M** and arbitrary components elsewhere, then **Ae** is simply a permutation of **e**, with the non-zero components of **e** permuted to the last $n - \rho_0$ component positions. Since these components are arbitrary, Equation (14.8.8) can therefore always be satisfied with a vector **e** of weight not exceeding $n - \rho_0$.

Corollary. The index of invulnerability of an (n, k) systematic code at any position m cannot exceed

$$s_m \leq \begin{cases} n - |m| - k & |m| \leq k, n - k \\ n - 2k & |m| > k \end{cases} \tag{14.8.9}$$

with $|m| \triangleq \min(m, n - m)$.

Proof. If $\mathbf{G}$ generates a systematic code, its first k columns are linearly independent. Assume $m \leq n/2$. Then, if $m \leq k$ and $m \leq n - k$, the first m columns, and, if $k < m < n - k$, the first k columns, of $\mathbf{G}_1$ in Equation (14.3.6) are also linearly independent. The matrix $\mathbf{M}$ must therefore contain at least $k + m$ and $2k$ linearly independent rows, respectively, in the two cases, and the rank ρ_0 of $\mathbf{M}$ is underbounded accordingly. If m is replaced by $|m|$ the same bound clearly holds for $m > n/2$. Since $s_m \leq n - \rho_0$ (Theorem 14.14) the corollary follows.

14.9 The Index of Invulnerability in Cyclic Codes

It was argued in Section 14.4 that a coset of the cyclic code dictionary D is invulnerable to synchronization at position m unless either

$$c(x)(x^{n-m} - 1) - x^{n-m}p_m(x) \tag{14.9.1}$$

or, equivalently,

$$c(x)(x^{n-m} - 1) - q_m(x)$$

represents a code word in D, modulo the polynomial $x^n - 1$. The polynomial $c(x)$ corresponds to the coset leader while $p_m(x)$ and $q_m(x)$ are polynomials of degrees not exceeding $m - 1$ and $n - m - 1$, respectively. Extending the same argument, we conclude that the minimum number of coefficients of any of the polynomials in Equations (14.9.1) which must be changed to produce a code word polynomial is equal to the index of invulnerability s_m at position m of the particular coset in question. Note, incidentally, that for any coset of any cyclic code, $s_{n-m} = s_m$. This follows because if s_m of the corresponding coefficients of the polynomials

$$c(x)(x^{n-m} - 1) - x^{n-m}p_m(x) \quad \text{and} \quad w(x)$$

differ, so also do s_m of the coefficients of

$$c(x)(1 - x^m) - p_m(x) \quad \text{and} \quad x^m w(x) = w'(x)$$

But if $w(x)$ represents a word in the cyclic code (not in the coset), $w'(x)$ does also, and if $p_m(x)$ is a prefix in this code, it must also be a suffix.

Theorem 14.15. Let D_0 and D be (n, k_0) and (n, k) cyclic dictionaries generated by $g(x)$ and $m(x)g(x)$, respectively, with $m(x)$ a polynomial of degree $k_0 - k$ and having at least one root of order ϵ. If the minimum distance separating any two words in D_0 is d_0, the index of invulnerability s_m of the coset of D having the n-tuple represented by $g(x)$ as a coset leader is bounded by

$$s_m \geq \begin{cases} d_0 - |m| & |m| \leq \min(n - d_0, \epsilon - 1) \\ 2d_0 - n & n - d_0 \leq |m| \leq \epsilon - 1 \end{cases}$$

Proof. If $0 < m < \epsilon$, all the roots of $x^{n-m} - 1$ are of order less than ϵ, and $x^{n-m} - 1$ is not a multiple of $m(x)$. Consequently, the n-tuple represented by $c(x)(x^{n-m} - 1)$ [with $c(x) = g(x)$], and hence the word represented by $w'(x) = c(x)(x^{n-m} - 1) - w(x)$ [with $w(x)$ corresponding to a word in D], are both in D_0 but not in D. Let $p_m(x)$ represent a prefix of length m in D. Then since any word in D containing this prefix must differ from any other word in D_0 in at least d_0 places, at most $\min(m, n - d_0)$ of the corresponding coefficients of the polynomials $x^{n-m}p_m(x)$ and $w'(x)$ can be equal. Thus, $c(x)(x^{n-m} - 1) - w(x) - x^{n-m}p_m(x)$ is a polynomial having at least $d_0 - \min\{m, n - d_0\}$ non-zero coefficients. Any overlap word must therefore differ from any code word by at least this much, and the theorem is proved.†

As an example, consider the $n = 15$ binary BCH codes (cf. Section 13.5). The congruence classes of exponents and the corresponding polynomials are shown in Table 14.2.

Table 14.2 The factors of $x^{15} - 1$ over $GF(2)$

Congruence classes	*Corresponding polynomials*	*Order of roots of these polynomials*
1 2 4 8	$p_1(x) = x^4 + x + 1$	15
3 6 12 9	$p_2(x) = x^4 + x^3 + x^2 + x + 1$	5
5 10	$p_3(x) = x^2 + x + 1$	3
7 14 13 11	$p_4(x) = x^4 + x^3 + 1$	15
0	$p_5(x) = x + 1$	1

Thus, if we let $g(x) = p_1(x)p_2(x)$, and $m(x) = p_3(x)$ we obtain a coset of the (15,5) BCH code having minimum distance 7 and an index of invulnerability $s_m \geq 5 - |m|$ for all $|m| \leq 2 = \epsilon - 1$. Similarly, by defining $g(x) = p_1(x)p_2(x)$ and $m(x) = p_4(x)$ we produce a (15, 3) code having minimum distance at least 5 and index of invulnerability $s_m \geq 5 - |m|$ for all $m \leq 4$.

Notice that the minimum distance d separating any two words in D may or may not be greater than d_0, depending on the polynomial $m(x)$. In some cases, it may be more important for ϵ to be large than for d to exceed d_0 and $m(x)$ must be chosen accordingly.

Parameters of some of the other codes obtained by the construction of Theorem 14.15 are summarized in Table 14.3.

This approach is particularly propitious in the case of the Reed-Solomon

†For a slightly stronger statement of this theorem, see Problem 14.4.

Table 14.3 Examples of BCH codes having cosets with invulnerability properties. [$\hat{s}_m$ = upper bound on s_m; Equation (14.8.9).]

n	k	d	d_0	ϵ	s_m	$\|m\|_{max}$	$\hat{s}_m$	
15	11	3	1	15	$1-\|m\|$	0	$4-\|m\|$	
15	9	3	3	3	$3-\|m\|$	2	$6-\|m\|$	
15	7	5	3	5	$3-\|m\|$	2	$8-\|m\|$	
15	5	7	5	3	$5-\|m\|$	2	$10-\|m\|$	$\|m\|\leq 5$
							5	$\|m\|\geq 5$
15	3	5	5	15	$5-\|m\|$	4	$12-\|m\|$	$\|m\|\leq 3$
							9	$\|m\|\geq 3$
15	2	10	6	15	$6-\|m\|$	5	$13-\|m\|$	$\|m\|\leq 2$
							11	$\|m\|\geq 2$
15	1	15	7	15	$7-\|m\|$	6	13	

codes (Section 13.5). Letting $g(x) = \prod_{i=1}^{n-k-1} (x - \alpha^i)$, with $n = q - 1$ and α a primitive element in $GF(q)$, and letting $m(x) = (x - \alpha^{n-k})$, for example, we obtain an (n, k) code with minimum distance $n - k + 1$, and with an index of invulnerability

$$s_m \geq \begin{cases} n - k - |m| & |m| \leq \min(k, \epsilon - 1) \\ n - 2k & k \leq |m| \leq \epsilon - 1 \end{cases} \tag{14.9.2}$$

where ϵ is the order of α^{n-k}. If $\epsilon \geq n - k + 1$, the resulting code has an index of invulnerability meeting, at all positions m, the upper bound of Equation (14.8.9). The (4, 1) Reed-Solomon code over $GF(5)$ generated by the polynomial $g(x) = (x - 3)(x - 4)(x - 2)$, for example, has a coset D [with $c(x) = (x - 3)(x - 4)$] having an index of comma freedom $s = 2$:

$$D = \begin{matrix} 2310 \\ 3421 \\ 4032 \\ 0143 \\ 1204 \end{matrix}$$

In restricting our attention to vectors belonging to some (n, k_0) cyclic code, we are obviously limiting the set of possible coset leaders which could conceivably be used. The advantages of doing so are essentially those associated with cyclic codes in general: (1) The mathematical structure thereby imposed makes it possible to obtain results and (2) the coset leaders are relatively easy to generate. Nevertheless, other approaches are also sometimes useful. One such approach is investigated in the remainder of this section.

Lemma 14.2. If D is an (n, k) cyclic code dictionary with minimum distance d, and if δ_m of the first (lowest order) $n - m$ coefficients of the polynomial $c(x)(x^{n-m} -$

1) [i.e., of the polynomial $\sum_{i=0}^{n-m-1} (c_{i+m} - c_i)x^i$] are non-zero, the coset of D having the coset leader $\mathbf{c} = \{c_0\ c_1\ \ldots\ c_{n-1}\}$ has an index of invulnerability s_m underbounded by

$$s_m \geq \min\{\delta_m, d - |m| - \delta_m\} \tag{14.9.3}$$

Proof. Regardless of the polynomial $p_m(x)$, the polynomial (14.9.1) must have at least δ_m and at most $m + \delta_m$ non-zero coefficients. Since any non-zero code word polynomial must have at least d non-zero coefficients, and since for cyclic codes $s_m = s_{n-m}$, the inequality (14.9.3) is valid for all m, $1 \leq m \leq n - 1$.

The following lemma (Reference 14.4) is therefore of interest.

Lemma 14.3. Let

$$c(x) = \sum_{i=\delta_0}^{[(\delta-1)/2]} x^{i(\mu+1)-\delta_0} + x^{n-1} \tag{14.9.4}$$

where $\delta_0 = \begin{cases} 0 & \delta \text{ even} \\ 1 & \delta \text{ odd} \end{cases}$ and where $\mu < \left(n - \left[\frac{\delta}{2}\right]\right) / \left[\frac{\delta+1}{2}\right]$ (with $[\alpha]$ denoting the integer part of α). Then exactly δ of the first $n - m$ coefficients of the polynomial $(x^{n-m} - 1)c(x)$ are non-zero for all m in the interval $1 \leq m \leq \mu$.

Proof. Consider the polynomial

$$(x^{n-m} - 1)c(x) = \sum_{i=\delta_0}^{[(\delta-1)/2]} (x^{i(\mu+1)-\delta_0-m} - x^{i(\mu+1)-\delta_0}) + x^{n-m-1} - x^{n-1}$$

Since $m \leq \mu < \left(n - \left[\frac{\delta}{2}\right]\right) / \left[\frac{\delta+1}{2}\right]$, the largest exponent occurring under the summation is less than $n - m - 1$. Moreover, no two terms under the summation have the same exponent so long as $1 \leq m \leq \mu$. When δ is odd, therefore, the summation contributes $2[(\delta - 1)/2] = \delta - 1$ non-zero coefficients corresponding to exponents in the range $(0, n - m - 1)$. When δ is even, $\delta_0 = 0$ and, of the δ terms under the summation sign, only the first has an exponent not in the required range. Since the coefficient of x^{n-m-1} is not zero, exactly δ of the first $n - m$ coefficients of $(x^{n-m} - 1)c(x)$ are non-zero in either case, as was to be shown.

Theorem 14.16. Any cyclic (n, k) code with minimum distance d can be made invulnerable to synchronization with index $s_m = s$ at all positions $|m| \leq \mu$ where

$$\mu \leq \min\left\{d - 2s, \left(n - 1 - \left[\frac{s}{2}\right]\right) / \left[\frac{s+1}{2}\right]\right\} \tag{14.9.5}$$

Proof. Using the coset leader represented by the polynomial $c(x)$ of Lemma 14.3, we find, from Lemma 14.2, that

$$s \geq \min\{\delta, d - \mu - \delta\}$$

Thus, if $\mu \leq d - 2\delta$, we are assured of an index of invulnerability $s \geq \delta$. The second upper bound on μ is a condition imposed by Lemma 14.3.

Note that when $s = 1$, $c(x) = x^{n-1}$. The corresponding coset leader, of course, makes any (n, k) cyclic code invulnerable to synchronization at all positions m, $|m| \leq n - k - 1$, although Theorem 14.16 asserts this to be the case only for $|m| \leq d - 2$. When $s = 2$, $c(x) = 1 + x^{n-1}$ and the resulting code will be invulnerable with index 2 at all positions m for $|m| \leq d - 4$. If $d \geq [n/2] + 4$, such a code is therefore comma-free with index 2. This construction, however, can never guarantee indices of comma freedom greater than 2, since μ must be less than $(n - [s/2])/[(s + 1)/2]$.

The coset leader represented by the polynomial $c(x)$ of Equation (14.9.4) is not the only one having the desired properties. The basic principle underlying its choice, however, is apparent; the non-zero coefficients of $c(x)$ and those of $x^{n-m}c(x)$ are disjoint for all $|m| \leq \mu$. As a result, δ_m (Lemma 14.2) is constant for all m in the specified interval; regardless of the field over which $c(x)$ is defined. Although the constant δ_m condition is easily achieved, it is not necessarily the most efficacious. Note, for example, that δ_m can increase as m decreases without adversely affecting the index of invulnerability [see Equation (14.9.3)]. This fact can often be used to advantage, particularly if $c(x)$ is tailored to a specific code or class of codes. Regardless of the polynomial $c(x)$, however, the maximum index of invulnerability guaranteed by this construction cannot exceed $(d - |m|)/2$. [Again, see Equation (14.9.3).] Thus, a construction of the sort described in Theorem 14.15 is generally preferable whenever d_0 exceeds $(d + |m|)/2$.

For purposes of comparison, the results obtained upon applying the Theorem 14.16 construction to the BCH (15, k) codes are summarized in

Table 14.4 The attainable invulnerability properties of some (15, k) cyclic codes using the coset leader of Equation (14.9.4). (The asterisk indicates comma-free codes.)

n	k	d	s	$\lvert m \rvert_{max}$
15	11	3	1	3
15	9	3	1	5
15	7	5	1	7*
15	7	5	2	1
15	5	7	1	7*
15	5	7	2	3
15	5	7	3	1
15	1	15	2	7*
15	1	15	4	6
15	1	15	5	4
15	1	15	6	3
15	1	15	7	1

Table 14.4 (cf. Table 14.3). (The values of $|m|_{\max}$ for the $s = 1$ case are those demonstrated possible by Theorem 14.7.)

14.10 Synchronization-Error-Correcting Codes

In introducing the concept of synchronization invulnerability, we remarked that when a once synchronized system loses synchronization it is often by only a few symbol positions in either direction. For this reason, it was argued, it is sometimes useful to select a coset leader so as to insure a large index of invulnerability at all positions $|m| \leq \mu$ for some relatively small μ rather than attempting to make the code comma-free. For the same reason, it is also advantageous to be able to determine not only that a synchronization error has occurred, but also which particular position m is actually being observed. This would allow the error to be corrected in one step, obviating the need for a sometimes time-consuming search procedure. Codes which enable the decoder to correct synchronization errors in this manner will be referred to as *synchronization-error-correcting* codes.† In this section we consider several methods for constructing such codes.

The approach taken in the previous sections to facilitate the detection of synchronization errors was to require that all code word overlaps of the form

$$a_{m+1}a_{m+2} \ldots a_n b_1 b_2 \ldots b_m$$

differ in as many symbol positions as possible from all the code words for all $|m|$ of interest. The obvious generalization of this constraint needed to enable synchronization errors to be corrected as well as detected is to require all such overlaps with $m = m_1$ to differ as much as possible not only from all true code words but from all other overlaps with $m \neq m_1$, again for all $|m|$ of interest. If the distance between any two overlaps corresponding to different values of m, $|m| \leq \mu$, is t or greater, the magnitude and direction of any synchronization error of magnitude $|m| \leq \mu$ can always be discerned correctly so long as fewer than $[(t + 1)/2]$ of the corresponding symbols are erroneously detected. This is the basis for the constructions to be considered here.‡

†These codes should not be confused with the insertion- and deletion-error-correcting codes mentioned in Chapter Twelve. If the cause of the synchronization error is the insertion or deletion of one or more symbols in some code word, the techniques to be investigated here apply to the first received n-tuple following such an error. No attempt will be made to correct the word containing the inserted or deleted symbols.

‡Another term, synchronization *recovery* codes, is sometimes used in the literature to denote codes which indicate, if not the magnitude of the synchronization error, at least the sign of this error (i.e., such codes are able to distinguish between a synchronization error of $+m_1$ symbols and one of $-m_2$ symbols, for all $0 < m_1, m_2 \leq \mu$.) Since there seems to be little difference in either the redundancy or the implementation complexity imposed by synchronization-error-correcting codes and synchronization recovery codes, only the former will be considered in any detail here.

The first construction is just that of Theorem 14.15. If the maximum value of $|m|$ is further restricted, these codes can be used to correct synchronization errors, as the following theorem demonstrates.

Theorem 14.17. Let D_0, D, $g(x)$, $m(x)$, ϵ, and d_0 be defined as in Theorem 14.15, and again consider the coset of D having the vector represented by $g(x)$ as a coset leader. The distance between any overlap word occurring at any position m_1 and any overlap word at any other position m_2, for such a dictionary, is at least

$$t \geq \begin{cases} d_0 - 2|m| & |m| \leq n - d_0 \\ 3d_0 - 2n & |m| \geq n - d_0 \end{cases}$$

where $|m| = \max\{|m_1|, |m_2|\} < \epsilon/2$.

Proof. Any overlap word occurring at any position m can be represented by the polynomial:

$$\begin{aligned} Q_m(x) &= x^{n-m}c(x) + x^{n-m}p_m(x) + q_m(x) = x^{n-m}c(x) + w(x) + x^{n-m}p'_m(x) \\ &= x^{n-m}c(x) + w'(x) + q'_m(x) \end{aligned} \tag{14.10.1}$$

[The notation is as defined in Equation (14.4.3); i.e., $c(x) = g(x)$ is the coset leader polynomial, $w(x)$ and $w'(x)$ represent words in D; $p_m(x)$ and $p'_m(x)$ are polynomials of degree $m - 1$ or less representing prefixes in D, and $q_m(x)$ and $q'_m(x)$ are polynomials of degree $n - m - 1$ or less representing suffixes in D.] The distance between an overlap word occurring at position m_1 and an overlap word at position m_2 is equal to the number of non-zero coefficients in the polynomial $Q_{m_1}(x) - Q_{m_2}(x)$. To simplify the argument, we assume $m_1 \leq n/2$ and $m_2 \geq n/2$. The minimum distance separating the two overlap words under this assumption provides a lower bound on the distance in general, as is easily verified. Then

$$\begin{aligned} Q_{m_1}(x) - Q_{m_2}(x) &= c(x)x^{n-m_1}(1 - x^{|m_1|+|m_2|}) \\ &\quad + w''(x) + x^{n-m_1}p'_{m_1}(x) - q'_{m_2}(x) \end{aligned} \tag{14.10.2}$$

with $w''(x)$ representing a word in D and $|m| = \min(m, n - m)$. If $|m_1| + |m_2| < \epsilon$, the first term, and hence the sum of the first two terms, of Equation (14.10.2) represent words in D_0 but not in D (see proof of Theorem 14.15) and thus have at least d_0 non-zero coefficients. Then, paralleling the proof of Theorem 14.15, we conclude that the distance t separating the overlap words in question is bounded by

$$t \geq \begin{cases} d_0 - (|m_1| + |m_2|) & |m_1|, |m_2| \leq n - d_0 \\ d_0 - 2(n - d_0) & |m_1|, |m_2| \geq n - d_0 \end{cases} \tag{14.10.3}$$

and the theorem is proved.

In the case of (n, k) Reed-Solomon codes, $d_0 = n - k$, and if $\epsilon > 2\mu$,

$$t \geq \begin{cases} n - k - 2\mu & \mu \leq k \\ n - 3k & \mu \geq k \end{cases}$$

This result cannot be significantly improved upon as the following theorem demonstrates.

Theorem 14.18. Let D' be a coset of an (n, k) cyclic code dictionary. Then the minimum distance t_μ separating any overlap of its words at position μ from any overlap at position $-\mu$ is bounded by

$$t_\mu \leq \begin{cases} n - k - 2\mu & \mu \leq k, \frac{n-k}{2} \\ n - 3k & k \leq \mu \leq \frac{n-k}{2} \\ 2(\mu - k) & \frac{n-k}{2}, k \leq \mu \\ 0 & \frac{n-k}{2} \leq \mu \leq k \end{cases}$$

Proof. Consider the two vectors

$$\mathbf{v}_1 = a_{n-\mu+1}, a_{n-\mu+2}, \ldots, a_n, b_1, b_2, \ldots, b_{n-\mu}$$
$$\mathbf{v}_2 = c_{\mu+1}, c_{\mu+2}, \ldots, c_n, d_1, d_2, \ldots, d_\mu$$

with $\mathbf{a} = (a_{n-\mu+1}, \ldots, a_n)$ and $\mathbf{c} = (c_{\mu+1}, \ldots, c_n)$ suffixes, and $\mathbf{b} = (b_1, \ldots, b_{n-\mu})$ and $\mathbf{d} = (d_1, \ldots, d_\mu)$ prefixes in D'. Since D' is the coset of a cyclic code, there is some μ-tuple $\mathbf{a}$ such that the first $\delta = \min(\mu, k)$ components of $\mathbf{v}_1$ are identical to the first δ components $\mathbf{v}_2$. Similarly, and for the same reason, the last δ components of $\mathbf{v}_2$ can be identical to the last δ components of $\mathbf{v}_1$. Finally, since the $(n - \mu)$-tuple $\mathbf{b}$ is as yet unconstrained we can choose it so that its first $\gamma = \min(k, n - 2\mu)$ components equal the corresponding components in $\mathbf{v}_2$. Thus, $t_\mu \leq n - 2\delta - \gamma$ and the statement of the theorem follows.

Corollary. No (n, k) cyclic code is capable of correcting all possible synchronization errors unless $k \leq [(n-1)/3]$.

Proof. If $k \geq n/3$, $(n - k)/2 \leq k$. The corollary follows immediately upon setting $\mu = (n - k)/2$, since then, from Theorem 14.18, $t_\mu = 0$.

This condition, in contrast to the one found earlier for comma freedom (Section 14.4) is necessary but *not* sufficient, as an exhaustive examination of the (4, 1) code $\begin{Bmatrix} 0000 \\ 1111 \end{Bmatrix}$ will verify. Nevertheless, any (n, k) cyclic code has a coset which is synchronization-error-correcting at all positions $|m| \leq (n - k - 2)/2$ (cf. Theorem 14.18), as the following theorem demonstrates.

Theorem 14.19. Let D' be the coset of an (n, k) cyclic code dictionary D having the coset leader represented by $c(x) = 1 + x^{n-1}$. Then all overlaps of words from D' at position m_1 are distinct from all overlaps at position m_2 for all $m_1 \neq m_2$, $\mu \triangleq \max\{|m_1|, |m_2|\} \leq (n - k - 2)/2$.

Proof. As has already been noted, the minimum distance separating two overlap words, one at position m_1, $0 < m_1 \leq n/2$ and one at position m_2, $n/2 <$

$m_2 \leq n$, is equal to the minimum number of non-zero coefficients in the polynomial (14.10.2). The two overlap words can be equal, then, only if

$$xc(x)(1 - x^{m_1+|m_2|}) + xp'_{m_1}(x) - x^{m_1+1}q'_{m_2}(x)$$

represents a code word in D. But if $c(x) = 1 + x^{n-1}$, this polynomial is of degree $m_1 + |m_2| + 1$ or less. If $n - m_1 - |m_2| - 2 \geq k$, therefore, it cannot represent a word in D unless it is the all-zeros word, an impossibility here because the coefficient of x^0 is clearly not zero. Virtually identical arguments hold when $0 \leq m_2 < m_1 < n/2$ and when $n/2 \leq m_1 < m_2 \leq n$ proving the theorem.†

The coset leader used in Theorem 14.16 also results in a dictionary with some synchronization-error-correction capability.

Theorem 14.20. Let D be a coset of an (n, k) cyclic code with minimum distance d having a coset leader of the sort represented by the polynomial (14.9.4), but with μ replaced by 2μ. Then any overlap of its words at any position m_1 can be separated by a distance at least t from any overlap word at any other position $m_2 \neq m_1$, for all $|m_1|, |m_2| \leq \mu$, where

$$\mu \leq \min \left\{ \frac{d - 2t - 1}{2}, \frac{n - \left\lceil \frac{t+3}{2} \right\rceil}{2\left\lceil \frac{t+2}{2} \right\rceil} \right\}$$

Proof. By paralleling in all essentials the proofs of Lemmas 14.2 and 14.3, one can easily verify that: (1) The distance separating a word overlap at position m_1 from an overlap at position m_2 is underbounded by

$$t_{m_1, m_2} \geq \min \{\delta_{m_1, m_2}, d - \mu' - \delta_{m_1, m_2}\}$$

where

$$\mu' \triangleq \begin{cases} |m_1| & 0 \leq m_2 < m_1 \leq \frac{n}{2} \text{ or } \frac{n}{2} \leq m_1 < m_2 \leq n \\ |m_1| + |m_2| & 0 < m_1 \leq \frac{n}{2} < m_2 \leq n \end{cases}$$

and δ_{m_1, m_2} is the number of non-zero terms in the first $n - \mu'$ coefficients of the polynomial

$$Q(x) \triangleq \begin{cases} (x^{n-m_1} - x^{n-m_2})c(x) & 0 \leq m_2 < m_1 \leq \frac{n}{2} \\ (1 - x^{n-|m_1|+|m_2|})c(x) & \frac{n}{2} \leq m_1 < m_2 \leq n \\ (x^{n-m_1-|m_2|} - 1)c(x) & 0 < m_1 \leq \frac{n}{2} < m_2 \leq n \end{cases}$$

---

†Note that, by the same argument, D' would have synchronization *recovery* capability for all $\mu \leq (n - k - 1)/2$ were $c(x)$ defined as $c(x) = 1$ (or x^{n-1}).

$c(x)$, of course, representing the coset leader. (2) If $c(x)$ is as defined in Equation (14.9.4), but with μ replaced everywhere by 2μ, then

$$\delta - 1 \leq \delta_{m_1, m_2} \leq \delta \qquad |m_1|, |m_2| \leq \mu < \frac{n - [\delta/2]}{2[(\delta + 1)/2]}$$

Combining these two statements, we find

$$t \triangleq \min_{m_1, m_2} t_{m_1, m_2} \geq \min \{\delta - 1, d - 2\mu - \delta\}$$

where $\mu \triangleq \max \{|m_1|, |m_2|\} \leq \dfrac{n - 1 - [\delta/2]}{2[(\delta + 1)/2]}$. The stated result is then obtained by requiring $d - 2\mu - \delta \geq \delta - 1$.

Thus far little has been said concerning methods for exploiting the synchronization-error correction and detection potential of linear codes demonstrated in this and the preceding several sections of this chapter. Actually, the relative desirability of the various possible approaches is strongly dependent on the operative physical constraints. The most satisfactory solution when the synchronization problem is primarily one of initial acquisition with a loss of synchronization occurring only infrequently will not necessarily be so when synchronization loss occurs quite regularly.

In the first of these situations, it may be expedient for the decoder to operate always as though the synchronization were correct. The validity of this assumption could then be checked by monitoring the distance between each received n-tuple and the corresponding decoder output as suggested in Section 14.5. If the code being used has a sufficiently large index of invulnerability at the position in question, this fact should be reflected in a discernible statistical bias in this observed distance. If, on the basis of these observations, a synchronization error is detected, a synchronization search could be initiated (or continued). This technique is explored further in Sections 14.16 and 14.17. If the code being used has synchronization correction as well as detection capabilities, further observation could presumably determine the correct word synchronization without the need for a search, although the implementation of this approach might be difficult.

When loss of synchronization is a relatively frequent event, the statistical approach just outlined may be ineffective; the average time between synchronization losses might well be insufficient for a reliable decision. In this event, a more reasonable scheme would be to divide the set of received n-tuples into two or more subsets. If a received n-tuple falls into a subset of n-tuples which are sufficiently similar to actual code words, correct synchronization could be assumed and the decoding operation could proceed. If not, a search could again be initiated, or, depending upon the subset to which the received n-tuple belongs, a synchronization correction could be immediately effected. This approach, in contradistinction to the previous one, almost inevitably decreases the synchronous error-correction capability of the code. Or, stated another way, more redundant codes must be used to achieve the same synchronous performance. This is to be expected since more than error correction (or detection) is being demanded of the code.

In general, this latter synchronization scheme can pose sizable implementation problems. As in ordinary error correction, however, these problems can often be considerably simplified by exploiting the algebraic structure of the code being used at the expense of a possibly reduced synchronization-error correction or detection capability. This approach is explored in the next section.

14.11 Algebraically Synchronizable Error-Correcting Codes

To illustrate the application of algebraic techniques to the simultaneous correction of substitution and synchronization errors, let us consider once again the code defined in Theorem 14.15. The decoding algorithm is as follows. The decoder first subtracts the coset leader from each received n-tuple and decodes the result as though it were a word in the dictionary D_0 (cf. Theorem 14.15 for a definition of terms). The syndrome of the decoded word is then evaluated relative to the dictionary D and used to determine both the magnitude and direction of the synchronization error.

To see how, and under what conditions, this algorithm works, assume the decoder actually begins decoding at position m. The received word, after the coset leader has been subtracted, is then represented by the polynomial

$$c(x)(x^{n-m} - 1) + w(x) - x^{n-m}p_m(x) + e(x)$$

where $e(x)$ is a polynomial representing the substitution errors and $c(x) = g(x)$, the generator of D_0. Let the number of non-zero coefficients in $e(x)$ (i.e., the number of errors) be e. Then if $e + m \leq (d_0 - 1)/2$, the decoder output will be

$$c(x)(x^{n-m} - 1) + w(x)$$

(Remember that the decoding is done as though the received n-tuple were a word in D_0.) The syndrome $s_m(x)$ is obtained by dividing this polynomial by $g(x)m(x)$. Thus

$$s_m(x) = \{c(x)(x^{n-m} - 1) + w(x)\}_{m(x)g(x)} = \{x^{n-m} - 1\}_{m(x)}$$

the last expression indicating the remainder after dividing $(x^{n-m} - 1)$ by $m(x)$. Suppose β is a root of $m(x)$ of order ϵ. Then $s_m(\beta) = \beta^{n-m} - 1$. Clearly, no two syndromes $s_{m_1}(x)$ and $s_{m_2}(x)$, $m_1 \neq m_2$, can be equal so long as $|m_1| + |m_2| < \epsilon$; each synchronization error of magnitude not exceeding $\mu \leq (\epsilon - 1)/2$, therefore, gives rise to a unique syndrome. To complete the decoding operation it is only necessary to alter the decoder synchronization in accordance with this syndrome. If the syndrome is zero, the original decoder output was the correct word (assuming the actual number of errors did not exceed $(d_0 - 1)/2 - |m|$) and the decoding process can continue with the next received word.

One penalty in using this decoding algorithm is the reduction of the effective Hamming distance of the dictionary D from d to d_0 (or, equivalently, the decrease of the dimension of D_0 from k_0 to k). It is, of course, possible to decode the received word as a word from D rather than D_0 and to avoid this loss in synchronous performance. This considerably complicates the decision as to whether or not synchronization has been lost, however, and in effect forces the decoder to make a statistical decision in this regard, as discussed previously. The reason for this is that if the received symbol stream were decoded with respect to D all correctable error patterns would indeed be corrected. Only the average distance between the original n-tuples and their decoded versions, and perhaps the number of uncorrectable error patterns, would indicate the possibility of a synchronization error.

This inability to realize the full synchronous potential of an algebraically synchronizable error-correcting code, however, can sometimes be overcome, as the following construction demonstrates (cf. Reference 14.4). Let D_0 be an (n, k) cyclic code generated by $g(x)$ and having minimum distance $d_0 = d$ and let D be an $[n - (2\mu + 1), k - (2\mu + 1)]$ shortened cyclic code obtained by retaining only those words in D_0 which both begin with $\mu + 1$ zeros and end with μ zeros, and then deleting those zeros. Further, let D' be a coset of D having a coset leader represented by the polynomial $c(x) = w_0(x)$, with $w_0(x)$ any word in D_0 beginning with the $(\mu + 1)$-tuple 00 . . . 01 and ending with the μ-tuple 00 . . . 0. (Note that such a word does exist since by tacit assumption $2(\mu + 1) \leq k$.) Then the following is true.

Theorem 14.21. It is possible to decode algebraically words from the dictionary D' defined in the preceding paragraph in such a way that: (1) All error patterns containing $(d - 1)/2$ or fewer errors will be corrected when the decoder is synchronous with the received word sequence; and (2) If the decoder actually begins decoding at position m, with $|m| \leq \mu$ both the magnitude and direction of this synchronization error will be determined provided the number of substitution errors does not exceed $[(d - 1)/2] - 2|m| - 1$.

Proof. The decoding algorithm is as follows: (1) The received symbol sequence is separated into blocks of length $n - 2(\mu + 1)$. (2) Each block is prefixed with the $(\mu + 1)$-tuple 00 . . . 01 and suffixed with the μ-tuple 00 . . . 0, and the resulting n-tuple is decoded with respect to D_0. (3) If its first non-zero symbol appears in the ith position, the decoded word is cyclically permuted $\mu + 1 - i$ positions to the right, $(i \leq \mu + 1)$, or $i - \mu - 1$ positions to the left $(i > \mu + 1)$, and the coset leader is subtracted from the result. (4) If the first $(\mu + 1)$ and last μ symbols of the resulting n-tuple are all zeros, these symbols are deleted and the remainder accepted as the decoded word. In addition, the decoder is assumed to have begun decoding at position $m = \mu + 1 - i$ and the appropriate adjustment is made before the next $(n - 2\mu - 1)$-tuple is decoded. (5) If the conditions in step (4) do not hold, the result is recorded as a decoding failure and the decoder proceeds to the next $(n - 2\mu - 1)$-tuple without any synchronization adjustment.

To verify that the performance realized by this decoder is as stated in the theorem, consider the situation when the decoder begins decoding at position $m \leq \mu$. The n-tuple formed by the decoder can be represented by the polynomial

$$Q(x) = x^{\mu} + x^{\mu+1}q_{m+2\mu+1}(x) + x^{n-\mu-m}p_m(x) + e(x)$$

where $q_l(x)$ and $p_l(x)$ represent an $(n - l)$-symbol suffix in D and an l-symbol prefix in D', respectively, and $e(x)$ the substitution error pattern. Since

$$x^{\mu} + x^{\mu+1}p'_m(x) + x^{\mu+1+m}q_{m+2\mu+1}(x) = w(x) + w_0(x)$$

with $w(x)$ representing a word in D, and $p'_m(x)$ an m-symbol prefix in D', and with $w_0(x)$ as previously defined, we have

$$Q(x) = x^{\mu} + x^{n-m}(w(x) + w_0(x)) - x^{\mu-m} - x^{\mu-m+1}p'_m(x) + x^{n-\mu-m}p_m(x) + e(x)$$

The decoder output will therefore be the word represented by $w'(x) = x^{n-m}[w(x) + w_0(x)]$ provided the number of non-zero coefficients in the polynomial $Q(x) - w'(x)$ does not exceed $(d - 1)/2$. And this will certainly be so if $e + 2m + 1 \leq (d - 1)/2$, $m \neq 0$, or if $e \leq (d - 1)/2$, $m = 0$, with e the number of substitution errors. Under these conditions, then, the first non-zero symbol in the decoded word will appear at position $i = \mu + 1 - m$, and both the synchronization error and the received word will be correctly identified. An identical argument clearly applies when the decoder begins decoding at position $m \geq n - \mu$ and the theorem is proven.

Thus, shortened cyclic codes do have cosets which enable synchronization errors to be corrected without a diminished synchronous substitution-error-correction capability. The trick is to choose the coset leader in such a way that synchronization errors are reflected as substitution errors in symbols which were never actually transmitted. This statement should be qualified on two counts, however. First, the shortened code may actually have minimum distance greater than that of the original cyclic code; if so, the decoding algorithm used here does not exploit this fact. Second, the vulnerability of the synchronization-error-correction procedure to substitution errors increases twice as rapidly with $|m|$ as it did with the previous construction.

The third construction (Reference 14.5) to be presented here for algebraically decodable synchronization-correction codes represents an interesting contrast to the preceding scheme; rather than shortening a cyclic code, we actually lengthen it. This construction, while increasing the code redundancy without increasing its substitution-error-correction capability, results in a code which is no more vulnerable to substitution errors in false synchronism than it is when the decoder is in true synchronism with the received sequence.

The motivation for this construction stems from the observation that the vulnerability to substitution errors of the codes described in Theorems 14.15 and 14.17 increases with m only when the overlap word is composed of two different code words [i.e., when $p_m(x) \neq 0$]. Prefixing and suffixing each transmitted word with suitable "buffer sequences" can eliminate this defect.

Specifically, let D, D_0, $g(x)$, $m(x)$, d_0, and ϵ be as defined in Theorems 14.15 and 14.17. A new $(n + m_r + m_l, k)$ dictionary D_s is constructed consisting of words of the form

$$\mathbf{w}_s = w_{n-m_l}w_{n-m_l+1} \ldots w_{n-1}w_0w_1 \ldots w_{n-1}w_0w_1 \ldots w_{m_r-1} \tag{14.11.1}$$

where $\mathbf{w} = (w_0\ w_1 \ldots w_{n-1})$ is a word in the coset of D having the coset leader represented by the polynomial $g(x)$. The dictionary D_s differs from this coset only in the repetition of its last m_l symbols and its first m_r symbols.

Now let the decoding algorithm be as follows: First, the initial m_l and the final m_r symbols of the received $(n + m_l + m_r)$-symbol code word are discarded. The coset leader is then subtracted, and the remaining n-symbol word decoded as though it were a word from D_0. If the synchronization is correct, and if $(d_0 - 1)/2$ or fewer substitution errors occurred during transmission, the transmitted word will be correctly decoded. But, in addition, if the synchronization is in error by not more than m_r symbols to the right or m_l symbols to the left, the decoder input, except for the $(d_0 - 1)/2$ or fewer substitution errors which are corrected, will be some cyclic permutation of some word in the coset of D generated by $g(x)$. The argument in the proof of Theorem 14.17 is directly applicable to the present situation; now, however, the prefix polynomials $p_m(x)$ and the suffix polynomials $q_m(x)$ are all identically zero. Repeating the proof of Theorem 14.17, therefore, with this alteration, yields the following result.

Theorem 14.22. The $(n + m_r + m_l, k)$ dictionary D_s defined in the preceding paragraphs can correct up to $(d_0 - 1)/2$ substitution errors, and simultaneously correct all synchronization errors in the range $-m_l \leq m \leq m_r$, $m_l + m_r < \epsilon$.

14.12 Synchronizable Burst-Error-Correcting Codes

It should not be surprising that codes designed to combat bursts of errors should also be effective in detecting and correcting synchronization errors. Indeed, when the code is cyclic, a synchronization error can itself be represented as a burst of errors. By using the coset of this code having the coset leader represented by $c(x) = 1 + x^{n-1}$ (with n the word length), we can insure that any error burst due to erroneous synchronization presents itself to the decoder as an "end-around" burst; i.e., as a burst beginning in the suffix and ending in the prefix of the same code word. In synchronous operation, this same error pattern would normally be observed only if two successive bursts appeared separated by less than a word length, a presumably unlikely event. Since cyclic burst-error-correcting codes are nevertheless capable of correcting end-around errors as well as ordinary burst errors, this unused capacity can be used to correct or detect synchronization errors. Specifically we prove the following.

Theorem 14.23. Let D be an (n, k) cyclic dictionary capable of detecting all pairs of error bursts of total length not exceeding b_d,† and let D' be the coset of D having the coset leader represented by the polynomial $c(x) = 1 + x^{n-1}$. Then, if the possibility of end-around error bursts is excluded, the decoder for D' can be designed either (1) to detect the simultaneous occurrence of a burst of length b or less and a synchronization error of magnitude μ or less for any $\mu + b \leq b_d - 2$; or (2) to correct a synchronization error of magnitude not exceeding $\mu \leq \max\{(b_d - 4)/2, (n - k - 2)/2\}$ and a burst of errors of length $b \leq b_d/2$, with $\mu + b \leq b_d - 2$, provided they do not occur simultaneously.

Proof. The proof of this result follows almost immediately from the comments in the preceding paragraph. The decoder can decode the received symbol sequence by subtracting from each successive n-tuple the coset leader and decoding the result with respect to D. If it begins decoding at position m, the n-tuple at the decoder input is, in the absence of an error burst, representable by the polynomial

$$c(x)(x^{n-m} - 1) + w(x) - x^{n-m}p_m(x) \qquad 0 \leq m \leq \frac{n}{2}$$

or

$$c(x)(x^{n-m} - 1) + w(x) - q_m(x) \qquad \frac{n}{2} \leq m \leq n$$

[cf. Equation (14.10.1)]. In either event, the effect of the synchronization error is to introduce an end-around burst of length $|m| + 2$, with disjoint sets of bursts corresponding to each m, $|m| < (n - 1)/2$. Thus, it immediately follows that the simultaneous occurrence of both types of errors can be detected if $|m| + 2 + b \leq b_d$, and that either error can be corrected if it occurs alone and both $|m| + 2$ and b are overbounded by $b_d/2$ (cf. Section 13.10). The two types of errors are distinguishable because a synchronization error always produces an end-around burst while by hypothesis a true burst error never does. That synchronization errors with magnitudes as large as $(n - k - 2)/2$ can actually be corrected in the absence of other errors has already been shown in Theorem 14.19. If synchronization errors larger than $(b_d/2) - 2$ are to be corrected, however, the burst error correction capability must decrease accordingly, so that burst and synchronization errors can still be distinguished. This requirement is clearly satisfied so long as $\mu + 2 + b \leq b_d$ with $\mu = \max |m|$.

†Recall from Section 13.10 that $b_d \geq n - 2k + 1$ for any (n, k) cyclic code.

Obviously, the results of the preceding sections on the synchronization-error-correcting and detecting capabilities of linear codes in the presence of arbitrary substitution errors apply as well when the errors are constrained to occur in bursts. In some cases, these results can be strengthened when the errors occur only in bursts, as the next theorem demonstrates.

Theorem 14.24. The coset dictionary defined in Theorem 14.15 is capable of correcting all error bursts of length b or less and of simultaneously correcting all synchronization errors of magnitude not exceeding $\mu < \epsilon/2$ so long as

$$\mu + b \leq (n - 3k_0 + 2)/2$$

Proof. Paralleling the proof of Theorem 14.17, we conclude that any two overlap words, one beginning at position m_1 and the other at position m_2 and containing error bursts of lengths b_1 and b_2, respectively, will be distinct if

$$c(x)x^{n-m_1}(1 - x^{|m_1|+|m_2|}) + w''(x) + x^{n-m_1}p'_{m_1}(x) - q'_{m_2}(x) + b_1(x) + b_2(x) \not\equiv 0 \tag{14.12.1}$$

for any word $w''(x)$ in D. The notation is identical to that of Equation (14.10.2) except here the two polynomials $b_1(x)$ and $b_2(x)$ represent error bursts of lengths b_1 and b_2. As in Theorem 14.17 we asume $0 \leq m_1 \leq n/2 < m_2 \leq n$. The same argument holds with only minor modifications, and actually under more relaxed conditions, for the other cases as well. Now for any $|m_1| + |m_2| \leq \epsilon$ (with ϵ as defined in Theorem 14.15), the first term, and hence the sum of the first two terms, in Equation (14.12.1) represent words in D_0 but not in D. Thus, the condition in Equation (14.12.1) will be violated only if

$$P(x) \triangleq x^{n-m_1}p'_{m_1}(x) - q'_{m_2}(x) + b_1(x) + b_2(x) = w_0(x) \tag{14.12.2}$$

with $w_0(x)$ some non-zero word in D_0 (but not in D). Since $x^{n-m_1}p'_{m_1}(x) - q'_{m_2}(x)$ can be interpreted as a burst of length $m_1 + |m_2|$, Equation (14.12.2) cannot be satisfied unless some non-zero word in D_0 can be represented as the sum of three error bursts of lengths $|m_1| + |m_2|$, b_1, and b_2. But if $|m_1| + |m_2| + b_1 + b_2 \leq n - 3k_0 + 2$, at least one of the gaps separating two of these error bursts represents k_0 or more symbols. Because D_0 is an (n, k_0) cyclic dictionary, only the all-zeros word can contain a sequence of k_0 or more consecutive zeros. If $|m_1|, |m_2| \leq \mu$ and if $b_1, b_2 \leq b$, Equation (14.12.2) cannot be satisfied under the stipulated conditions, and the theorem is proved.

If the dictionary in the preceding theorem is defined over the binary field, the bound on the maximum allowable sum $b + \mu$ can in some cases be increased. Suppose the polynomial $P(x)$ as defined in Equation (14.12.2) contains a total of γ "ones." Obviously $P(x)$ cannot represent a non-zero word in D_0 unless $\gamma \geq d_0$, the weight of the minimum weight word in D_0. But if $P(x)$ is a word in D_0 so also is $(1 + x)\, P(x)$, and if $P(x)$ contains a total of γ ones, $(1 + x)\, P(x)$ can have at most $2(|m_1| + |m_2| + b_1 + b_2 - \gamma) + 6$ ones (i.e., a symbol in $(1 + x)\, P(x)$ is a one only if a zero appears as one of the corresponding symbols in either $P(x)$ or $xP(x)$ and a *one* as the other). If $(1+x)P(x)$ is to represent a word in D_0 either $\gamma = n$ or $2(|m_1| + |m_2| + b_1 + b_2 - \gamma) \geq d_0 - 6$. If both $P(x)$ and $(1+x)P(x)$ are to represent words in D_0, then,

$$|m_1| + |m_2| + b_1 + b_2 \geq \min\{3(d_0 - 2)/2, n\} = 3(d_0 - 2)/2$$

This last equality follows directly from the bound 13.3.5, i.e., since here $k_0 \geq 2$, $n \geq 3d_0/2$. This establishes the following result.

Theorem 14.25. If the dictionary of Theorem 14.24 is defined over $GF(2)$, then it can correct all error bursts of length b or less and all synchronization errors of magnitude μ or less where

$$\mu + b \leq (3d_0 - 7)/4$$

14.13 The Index of Comma Freedom in Kronecker Product, Kronecker Sum, and Concatenated Codes

One useful method for constructing codes having larger indices of comma freedom is to form the Kronecker product or Kronecker sum (see Section 13.9) of two smaller codes both of which are comma-free with an index of at least one. The derivation of a bound on the index of comma freedom of the resulting code in terms of the indices of the component codes is the first task of this section. The results are summarized in the following theorem.

Theorem 14.26. Let the $M \times m$ and $N \times n$ matrices $\mathbf{A}_m$ and $\mathbf{B}_n$ represent two linear code dictionaries, both containing the all-ones code word $1\,1 \ldots 1$. Let $\mathbf{c}_m$ be an m-tuple and $\mathbf{d}_n$ an n-tuple such that the respective codes $\mathbf{A}_m + \mathbf{c}_m$ and $\mathbf{B}_n + \mathbf{d}_n$ have indices of comma freedom s_m and s_n. (The matrix $\mathbf{A}_m + \mathbf{c}_m$ represents the dictionary obtained by adding, term-by-term over the field $GF(q)$ over which both matrices $\mathbf{A}_m$ and $\mathbf{B}_n$ are defined, the n-tuple $\mathbf{c}_m$ to each row in $\mathbf{A}_m$, and similarly for the matrix $\mathbf{B}_n + \mathbf{d}_n$.) Further, let $\mathbf{C}_{mn}$ represent the dictionary generated by the Kronecker product of the generators of $\mathbf{A}_m$ and $\mathbf{B}_n$, and let $\mathbf{D}_{mn}$ be the Kronecker sum of $\mathbf{A}_m$ and $\mathbf{B}_n$. Then, the index of comma freedom s_{mn} in the dictionaries represented by both $\mathbf{C}_{mn} + (\mathbf{c}_m \boxplus \mathbf{d}_n)$ and $\mathbf{D}_{mn} + (\mathbf{c}_m \boxplus \mathbf{d}_n)$ is bounded by

$$s_{mn} \geq \min(ns_m, ms_n)$$

The mn-tuple $\mathbf{c}_m \boxplus \mathbf{d}_n$ is the Kronecker sum of $\mathbf{c}_m = (\gamma_1\,\gamma_2 \ldots \gamma_m)$ and $\mathbf{d}_n = (\delta_1\,\delta_2 \ldots \delta_n)$; i.e., $\mathbf{c}_m \boxplus \mathbf{d}_n = (\gamma_1 + \mathbf{d}_n, \gamma_2 + \mathbf{d}_n, \ldots, \gamma_m + \mathbf{d}_n)$, where $\gamma_i + \mathbf{d}_n = (\gamma_i + \delta_1, \gamma_i + \delta_2, \ldots, \gamma_i + \delta_n)$.

Proof. The theorem can be proved quite easily upon observing the following two properties of any word $\mathbf{w}$ in either $\mathbf{C}_{mn}$ or $\mathbf{D}_{mn}$: (1) $\mathbf{w}$ is of the form

$$\mathbf{w} = \mathbf{b}_{i_1}\,\mathbf{b}_{i_2} \ldots \mathbf{b}_{i_m}$$

where the $\mathbf{b}_i$ are words in $\mathbf{B}_n$. (2) The symbols ω_{in+j}, $i = 0, 1, \ldots, m-1$ of the word

$\mathbf{w} = (\omega_1\, \omega_2 \ldots \omega_{mn})$ constitute a code word in $\mathbf{A}_m$ for every $\mathbf{w}$ and every $j = 1, 2, \ldots, n$. [Both of these statements simply reflect the fact that if $\mathbf{w}$ is a word in a linear dictionary D defined over the field $GF(q)$ and if D contains the all-ones word $\mathbf{w}_0 = (1\,1 \ldots 1)$ then both $\alpha\mathbf{w}$ and $\alpha\mathbf{w}_0 + \mathbf{w}$ are words in D for any α in $GF(q)$.] As an immediate consequence of these two statements we have for any word $\mathbf{w}'$ in either $\mathbf{C}_{mn} + (\mathbf{c}_m \boxplus \mathbf{d}_n)$ or $\mathbf{D}_{mn} + (\mathbf{c}_m \boxplus \mathbf{d}_n)$ the statements: (1′) $\mathbf{w}'$ is of the form

$$\mathbf{w}' = \mathbf{b}'_{i_1}\, \mathbf{b}'_{i_2} \ldots \mathbf{b}'_{i_m}$$

where the $\mathbf{b}'_i$ are words in $\mathbf{B}_n + \mathbf{d}_n$. (2′) The symbols ω'_{in+j}, $i = 0, 1, \ldots, m-1$ of the word $\mathbf{w}' = (\omega'_1\, \omega'_2 \ldots \omega'_{mn})$ constitute a code word in $\mathbf{A}_m + \mathbf{c}_m$ for every $\mathbf{w}'$ and every $j = 1, 2, \ldots, n$.

Now let $\mathbf{w}$ and $\mathbf{w}'$ both be words in either $\mathbf{W} = \mathbf{C}_{mn} + (\mathbf{c}_m \boxplus \mathbf{d}_n)$ or $\mathbf{W} = \mathbf{D}_{mn} + (\mathbf{c}_m \boxplus \mathbf{d}_n)$ and consider any overlap of $\mathbf{w}\mathbf{w}'$ beginning with the $(i+1)$st symbol of $\mathbf{w}$ and ending with the ith symbol of $\mathbf{w}'$. If $i \neq 0$, modulo n, this overlap is a sequence of m n-tuples all of which could occur as overlaps of words of $\mathbf{B}_n + \mathbf{d}_n$. Accordingly, all of these overlap words must differ in at least s_n places from any word in $\mathbf{B}_n + \mathbf{d}_n$, and at least ms_n symbols of any such overlap must differ from the corresponding symbols in any word in $\mathbf{W}$. Similarly, if $i = 0$, modulo n ($i = \nu n$, $\nu \neq 0$), the symbols $\omega_{\nu n+j}\omega_{(\nu+1)n+j}, \ldots, \omega_{(m-1)n+j}\omega'_j\omega'_{n+j}, \ldots, \omega'_{(\nu-1)n+j}$ constitute, for each $j = 1, 2, \ldots, n$, an m-tuple which could occur as an overlap of two words from $\mathbf{A}_m + \mathbf{c}_m$. Thus, each of these n m-tuples must differ in at least s_m places from any word in $\mathbf{A}_m + \mathbf{c}_m$, and the corresponding overlap of $\mathbf{w}\mathbf{w}'$ must differ from any word in $\mathbf{W}$ in at least ns_m places. Since all non-trivial overlaps of any two words in either $\mathbf{C}_{mn} + (\mathbf{c}_m \boxplus \mathbf{d}_n)$ or $\mathbf{D}_{mn} + (\mathbf{c}_m \boxplus \mathbf{d}_n)$ have been accounted for, we have shown that

$$s_{mn} \geq \min\,[ns_m, ms_n] \tag{14.13.1}$$

and the theorem is proved.

The statement of this theorem requires both component codes to contain the all-ones word. This is to assure that all code word overlaps occurring in the Kronecker sum or Kronecker product dictionary could also have occurred in the component dictionaries. Clearly, if $\tilde{D}$ is a dictionary contained in D and $\tilde{D}$ is used rather than D in defining the Kronecker sum or product dictionary, the resulting dictionary has an index of comma freedom at least as great as the dictionary defined in terms of D. Thus, there is no conflict between the hypotheses of Theorems 14.26 and 13.18. (The latter, it will be recalled, demanded that the words of one of the two dictionaries involved in the Kronecker sum have maximum weight less than n.) Note, too, that if at least one of the two component dictionaries in a Kronecker sum and if both of the dictionaries in a Kronecker product contain the all-ones code word, so also does the resulting dictionary. Accordingly, the process can be iterated; comma-free Kronecker product and Kronecker sum dictionaries can be used to generate still larger comma-free Kronecker product and sum dictionaries.

A similar result holds for concatenated codes. Let D_1 be an (n_1, k_1) code defined over $GF(q^{k_2})$ and having an index of comma freedom s_1. Similarly, let D_2 be an (n_2, k_2) code over $GF(q)$, having an index of comma freedom s_2. A concatenated code results upon uniquely identifying with each element of $GF(q^{k_2})$ a word from D_2. Nothing is changed, so far as the synchronous properties of this code are concerned, by requiring D_1 and D_2 to be comma-free. The concatenated code, however, is now also comma-free with (in terms of the q-ary symbols) an index of comma freedom

$$s_{n_1 n_2} \geq \min\{n_1 s_2, d_2 s_1\} \tag{14.13.2}$$

where d_2 is the minimum distance separating any two words in D_2. For again consider an overlap word beginning with the $(i+1)$st symbol of one word and ending with the ith symbol of another. If $i = 0$, modulo n_2 (but not modulo $n_1 n_2$), the comma freedom of D_1 guarantees that this overlap word will differ from any code word in at least s_1 q^{k_2}-ary "symbols," and hence, in at least $s_1 d_2$ q-ary symbols. If $i \neq 0$, modulo n_2, the distance between the overlap word and any true word is obviously at least $n_1 s_2$.

Actually, this feature of concatenated codes is rather irrelevant, since the purpose of concatenation is not simply to produce larger q-ary codes, but in addition to realize greater decoding efficiencies (cf. Section 13.9). Suppose the comma-free dictionaries D_1 and D_2 are combined by concatenation, or for that matter by forming their Kronecker product or Kronecker sum. Then any sequence of words from this combined dictionary is also a sequence of words from D_2 (or D_1, depending upon the order in which the two dictionaries are combined). Word synchronization can thus be accomplished in two stages, the first stage designed to identify the initial symbols of the words from D_2, the second to resolve the remaining ambiguity. The first stage of the search can proceed as though D_2 were used alone. The second stage then involves searching through the n_1 remaining positions and relying on the comma-free properties of D_1. As a result, the total number of positions which must be observed is $n_1 + n_2$ rather than the $n_1 n_2$ needed when this substructure is ignored.

In concluding this section, we summarize some of the known results concerning the comma-free properties of the first-order Reed-Muller codes mentioned in Section 13.7. These codes, it will be recalled, can be generated by taking repeatedly the Kronecker sum of the matrix

$$\mathbf{D}_1 = \begin{bmatrix} 00 \\ 01 \end{bmatrix}$$

with itself and then adjoining to the resulting dictionary its complement. Thus, the dictionary D_i is represented by the matrix

$$\mathbf{D}_i = \begin{bmatrix} \mathbf{D}_{i-1} & \mathbf{D}_{i-1} \\ \mathbf{D}_{i-1} & \bar{\mathbf{D}}_{i-1} \end{bmatrix}$$

and the $(2^{k-1}, k)$ Reed-Muller dictionary is simply $D_{k-1} \cup \bar{D}_{k-1}$. The same dictionary D_{k-1} can equally well be defined by taking the Kronecker sum of any of the matrix pairs $\mathbf{D}_j$ and $\mathbf{D}_{k-1-j}$, $1 \leq j \leq k-1$, as is readily verified.

Consequently, in order to demonstrate the existence of comma-free Reed-Muller codes of all dimensions, it is only necessary to exhibit them for the smaller dimensions. In fact, if the dictionary $D_1 \cup \bar{D}_1$ could be made comma-free, Theorem 14.26 would assure that so, too, could the Reed-Muller codes for all higher dimensions.

Unfortunately, the dictionary $D_1 \cup \bar{D}_1$ contains all binary 2-tuples and clearly cannot be made comma-free. Moreover, the bound of Theorem 14.13 demonstrates that none of the dictionaries $D_{k-1} \cup \bar{D}_{k-1}$ can be comma-free for any $k < 4$. The $k = 4$ Reed-Muller code cannot be made comma-free either, as can easily be verified using the techniques of Section 14.3. (The rank of the matrix $\mathbf{M}$ defined in Equation 14.3.6 is 8, for example, when $m = 5$.) To demonstrate the existence of comma-free Reed-Muller codes for all dimensions $k \geq 5$, then, we must exhibit them for dimensions $k = 5, 6, 7$, and 8 (e.g., the Kronecker sum of the dictionary D_4 with itself produces the dictionary D_8 and $D_8 \cup \bar{D}_8$ is the $k = 9$ Reed-Muller dictionary). Such codes have been found and their index of comma freedom tested using a computer. The results, along with upper bounds on these indices, are summarized in Table 14.5.

Table 14.5 The best known index of comma freedom s and the best known upper bound s' on this index attainable with $(2^{k-1}, k)$ first-order Reed-Muller dictionaries.

k	s	s'
5	2	2
6	7	7
7	16	22
8	34	56

The first two upper bounds ($k = 5$ and 6) were obtained using the techniques described in Section 14.8, the latter two by various other considerations (References 14.6 and 14.7). All of these bounds are tighter than those established in Equation (14.7.1). Coset leaders producing optimal $k = 6$ dictionaries were found by a screening process (via computer), again in connection with the approach discussed in Section 14.8 (Reference 14.6). Examples of coset leaders achieving the index of comma freedom s given in Table 14.5 for $k = 5$ and 8 are given in Reference 14.8, for $k = 6$ in Reference 14.6, and for $k = 7$ in Reference 14.7.

14.14 Comma-Free Phase Codes

The synchronizability of phase codes (Section 13.11) can also be improved by altering the dictionaries to render them comma-free. The measure of significance in this case is the correlation between a word and a word overlap rather than the Hamming distance between them. That is, if $\mathbf{y} = (y_1\, y_2 \ldots y_n)$ denotes the exponents of a received n-tuple, and $\mathbf{x}^\mu = (x_1^\mu x_2^\mu \ldots x_n^\mu)$ the exponents of a code word, consider the *asynchronous correlation coefficient*

$$\rho_\mu(\mathbf{y}) = \frac{1}{n} \sum_{\nu=1}^{n} e^{(2\pi i/r)(x_\nu^\mu - y_\nu)} \tag{14.14.1}$$

Since the decoder is presumably a correlation detector, these (noise-perturbed) coefficients are the basis for its decisions as to the word received. Word synchronization will clearly be more readily obtainable from the data sequence itself if the set of correlation coefficients $\rho_\mu(\mathbf{y})$ is significantly different depending upon whether the n-tuple $\mathbf{y}$ represents a true code word, or whether it represents an overlap of two code words.

The following theorem demonstrates a relationship between these correlation coefficients and the index of comma freedom associated with the code exponent group (cf. Section 13.11 for a definition of terminology).

Theorem 14.27. Let C be some coset of a group G consisting of n-tuples of integers modulo r, and let each of the n-tuples in C determine the exponents of a word in the phase code dictionary D. Suppose C, considered as a dictionary of n-symbol r-ary words, has an index of comma freedom s. Then

$$\max_{\mathbf{y},\mu} \operatorname{Re}\{\rho_\mu(\mathbf{y})\} \leq 1 - \frac{s}{n}\left(1 - \cos\frac{2\pi}{r}\right)$$

where $\rho_\mu(\mathbf{y})$ is as defined in Equation (14.14.1). Moreover, if G contains the all-ones n-tuple $\mathbf{g}_0 = (1\,1 \ldots 1)$,

$$\max_{\mathbf{y},\mu} |\rho_\mu(\mathbf{y})| \leq \left[1 - 2\frac{s}{n}\left(1 - \frac{s}{n}\right)\left(1 - \cos\frac{2\pi}{r}\right)\right]^{1/2}$$

Proof. The first of these statements is immediately apparent, as

$$\max_{\mathbf{y},\mu} \operatorname{Re}\{\rho_\mu(\mathbf{y})\} \leq \operatorname{Re}\left\{\frac{n-s}{n} + \frac{s}{n}e^{2\pi i/r}\right\} = 1 - \frac{s}{n}\left(1 - \cos\frac{2\pi}{r}\right) < 1 \tag{14.14.2}$$

The proof of the second statement is nearly as direct. First, note that $|\rho_\mu(\mathbf{y})| = |\rho_\mu(\mathbf{y} - \gamma\mathbf{g}_0)|$ for any integer γ. Let γ be the value (or one of the values) assumed by the greatest number, say l, of the components of $\mathbf{y} - \mathbf{x}^\mu$. Then

$$|\rho_\mu(\mathbf{y} - \gamma\mathbf{g}_0)| \leq \begin{cases} \left|\dfrac{l}{n} + \dfrac{n-l}{n}e^{2\pi i/r}\right| & l \geq \dfrac{n}{2} \\ \dfrac{1}{2}\left|1 + e^{2\pi i/r}\right| & l \leq \dfrac{n}{2} \end{cases} \tag{14.14.3}$$

since $|\rho_\mu(\mathbf{y} - \gamma\mathbf{g}_0)|$ clearly attains its maximum value when the arguments $y_\nu - x_\nu^\mu - \gamma$, $\nu = 1, 2, \ldots, n$, are as nearly equal as possible. Since G contains $\mathbf{g}_0$ and is comma-free with index s, $l \leq n - s$, and the stated upper bound on $|\rho_\mu(\mathbf{y})|$ follows. [This bound can be tightened if s is greater than $n/2$; see Problem 13.5.]

As a consequence, the results concerning comma-free codes obtained in the previous sections are still meaningful here. It should be emphasized, however, that except in the binary case, and in the ternary phase-coherent case, the two measures are different and that the best comma-free codes do not necessarily generate the best phase codes.

The phase codes of primary interest are undoubtedly the orthogonal and r-orthogonal phase codes discussed in Section 13.11. They correspond to highly redundant error-correcting codes and, one would expect, it should be possible to find such codes with asynchronous correlation coefficients having magnitudes considerably less than unity. As was shown in Section 13.11, r-orthogonal phase codes can be constructed for arbitrary dimensions by the Kronecker product method. The following theorem concerning the maximum asynchronous correlation coefficient of Kronecker product dictionaries is therefore of interest.

Theorem 14.28. Let A_m and B_n be M and N word phase code group (coset) dictionaries of m- and n-symbol words, respectively, both defined over the set of symbols $\{\exp(2\pi i\gamma/r)\}$. Further, let the exponent cosets of both of these dictionaries be cosets of groups which contain the all-ones m- (n-) tuple $(1\ 1\ \ldots\ 1)$. Finally, let ρ_m and ρ_n be asynchronous correlation coefficients having the maximum absolute value associated with these two dictionaries. Then

$$|\rho_{mn}| \leq \max\{|\rho_m|, |\rho_n|\}$$

where ρ_{mn} is the asynchronous correlation coefficient of the Kronecker product of the two dictionaries having the maximum absolute value.

Proof. The proof follows immediately from the proof of Theorem 14.26. The Kronecker product of phase codes corresponds to the Kronecker sum of their exponent group cosets (i.e., to the coset obtained by adding to each of the elements in the Kronecker sum of their groups the Kronecker sum of their respective coset leaders). Thus, when $i \neq 0$, modulo n, $|\rho_{mn}| \leq |\rho_n|$ and when $i = 0$, modulo n, $|\rho_{mn}| \leq |\rho_m|$, and the theorem is proved.

We next prove a useful lower bound on the asynchronous correlation coefficient attainable with orthogonal dictionaries (and, a fortiori, with r-orthogonal dictionaries).

Theorem 14.29. Let D be an n word orthogonal phase-code dictionary of n-symbol words with an associated maximum (in absolute value) asynchronous correlation coefficient ρ_n. Then

$$|\rho_n| \geq \frac{1}{\sqrt{n}}$$

Proof. Denoting by $\{\boldsymbol{\xi}^\mu\}$ the set of words in D and by $\boldsymbol{\eta}$ any overlap of any two words of D, we have

$$\boldsymbol{\eta} = \sum_{\mu=1}^{n} \rho_\mu \boldsymbol{\xi}^\mu \tag{14.14.4}$$

This follows because the set of orthogonal vectors $\boldsymbol{\xi}^\mu$ spans the space of n-tuples defined over the complex field. Furthermore, since $\boldsymbol{\xi}^\nu \cdot \boldsymbol{\eta} = n\rho_\nu$,

$$\frac{1}{n}(\boldsymbol{\eta} \cdot \boldsymbol{\eta}) = 1 = \sum_{\mu=1}^{n} |\rho_\mu|^2 \tag{14.14.5}$$

Now suppose $\max_\mu |\rho_\mu| < 1/\sqrt{n}$. Then $\sum_{\mu=1}^{n} |\rho_\mu|^2 < 1$ leading to a contradiction and proving the theorem.†

In the binary case, $|\rho_n| = |1 - (2s/n)|$ where s is the index of comma freedom; the two measures, in this case, are equivalent. The first-order Reed-Muller codes are biorthogonal under the measure of this section, so the discussion of the preceding section concerning comma-free $(2^{k-1}, k)$ Reed-Muller codes is directly applicable here. The magnitude of the asynchronous correlation coefficients attainable with these codes can be obtained directly from Table 14.5.

When $r = p$ is a prime greater than 3, p-ary, p-orthogonal (p^{k-1}, k) codes having asynchronous correlation coefficients with maximum absolute values less than unity are easily shown to exist for all $k \geq 2$. If such codes exist for $k = 2$, then the existence of all the others is assured by the Kronecker product theorem. To verify that it is possible to render the $k = 2$ codes comma-free, it is only necessary to observe that the code generated by taking all linear combinations over $GF(p)$ of the two p-tuples

$$\begin{bmatrix} 0123\ldots(p-1) \\ 1111\ldots1 \end{bmatrix} \tag{14.14.6}$$

is cyclic. Consequently, the results of Section 14.4 demonstrate that the coset obtained by adding any non-zero integer to the last (or first) column of this code is comma-free. This is true so long as $k \leq (n-1)/2$, i.e., so long as $p \geq 5$. Actually, when $p > 7$, the coset leader of Theorem 14.16 can be used to render these codes comma-free with index 2. Higher indices of comma freedom are undoubtedly possible for larger values of p.

14.15 Barker Phase Sequences

Suppose information is to be transmitted over a Gaussian channel as a sequence of T_s-second PSK symbols $s_\nu(t) = \sqrt{2}\sin(\omega_c t + (2\pi/r)\xi_\nu)$, $\xi_\nu \in (0, 1, \ldots, r-1)$, and that synchronization is to be obtained by

†This bound, combined with some additional refinements, is the basis for the last two upper bounds given in Table 14.5.

identifying at the receiver a periodically inserted comma sequence $\{z_\nu(t) = \sqrt{2}\sin(\omega_c t + (2\pi/r)x_\nu), (\nu - 1)T_s < t < \nu T_s\}$, $\nu = 1, 2, \ldots, n$. (We shall refer to the sequence $\{x_\nu\}$ as the *comma phase sequence*.) If the information can be assumed to be composed of mutually independent symbols, uniformly distributed over the signal alphabet, the argument of Section 14.6 can immediately be extended to suggest that the measure of concern is the *aperiodic correlation coefficient*

$$\lambda_k = \sum_{\nu=1}^{n-k} \zeta_\nu \zeta_{\nu+k}^* \tag{14.15.1}$$

where $\zeta_\nu = e^{(2\pi i/r)x_\nu}$. A "good" synchronizing sequence in this case might be defined as one for which $|\lambda_k| \leq 1$ for all $k = 1, 2, \ldots, n - 1$. This is the obvious analog of the condition imposed in Section 14.6.

The parameter λ_k is identical to the parameter c_k of Section 14.6 when $r = 2$ and in the phase-coherent case, when $r = 3$. Thus, the results of Section 14.6 are directly applicable here in those two situations. Few Barker-type sequences (sequences for which $|\lambda_k| \leq 1$, all k) of any length are known for values of r greater than 2. When $r = r_1 r_2$, sequences can obviously be constructed over the r-ary alphabet which are at least as good as those obtainable over either the r_1-ary or the r_2-ary alphabets. This can be accomplished simply by using the phase sequence $\{r_2 x_\nu\}$ where $\{x_\nu\}$ is a comma phase sequence over the r_1 symbol alphabet. Thus, when $r = 4$, sequences satisfying the constraint $|\lambda_\nu| \leq 1$, all k, exist with lengths $n = 2, 3, 4, 5, 7, 11$, and 13, since binary sequences exist for those lengths. In addition, there is a quaternary sequence with $|\lambda_k| \leq 1$, $1 \leq k \leq n - 1$, for $n = 15$. Other non-binary PSK sequences satisfying this constraint are known when $r = 6$ for all n in the range $2 \leq n \leq 13$ (Reference 14.9).

Again, generally acceptable Barker-type phase sequences can be obtained by selecting some cyclic permutation of a pseudo-noise sequence. For any phase sequence, we have the inequalities

$$\begin{aligned}\max_{0<k<n} |\lambda_k + \lambda_{n-k}| &\geq \operatorname*{ave}_{0<k<n} |\lambda_k + \lambda_{n-k}| \\ &= \frac{1}{n-1}\left|\sum_{k=1}^{n}\sum_{j=1}^{n} z_k z_{k+j}^* - n\right| \\ &= \frac{1}{n-1}\left\{\left|\sum_{k=1}^{n} z_k\right|^2 - n\right\} \geq -\left[\frac{n}{n-1}\right]\end{aligned} \tag{14.15.2}$$

The pseudo-noise phase sequences, in meeting this bound, would seem to be a logical source of Barker-type sequences.

Another method for constructing longer Barker-type phase sequences is to take the Kronecker sum of two shorter phase sequences, as the following theorem suggests.

Theorem 14.30. The aperiodic correlation coefficients λ_k associated with the Kronecker sum phase sequence

$$x_1 + y_1, x_1 + y_2, \ldots, x_1 + y_m, x_2 + y_1, x_2 + y_2, \ldots, x_l + y_m$$

are defined in terms of the aperiodic correlation coefficients $\lambda_k(x)$ and $\lambda_k(y)$ of the two component sequences $\{x_\mu\}$, $\mu = 1, 2, \ldots, l$, and $\{y_\nu\}$, $\nu = 1, 2, \ldots, m$, by

$$\lambda_k = \lambda_j(x)\lambda_h(y) + \lambda_{j+1}(x)\lambda^*_{m-h}(y) \qquad \begin{array}{l} k = jm + h \\ 0 \leq h < m \end{array}$$

Proof. Let $\alpha_\nu = e^{(2\pi i/r)x_\nu}$ and $\beta_\nu = e^{(2\pi i/r)y_\nu}$. Then with $k = jm + h$, $0 \leq h < m$, we have

$$\begin{aligned} \lambda_k &= \sum_{\mu=1}^{l-j} \sum_{\nu=1}^{m-h} \alpha_\mu \beta_\nu (\alpha_{\mu+j}\beta_{\nu+h})^* + \sum_{\mu=1}^{l-(j+1)} \sum_{\nu=m-h+1}^{m} \alpha_\mu \beta_\nu (\alpha_{\mu+j+1}\beta_{\nu-(m-h)})^* \\ &= \sum_{\mu=1}^{l-j} \alpha_\mu \alpha^*_{\mu+j} \sum_{\nu=1}^{m-h} \beta_\nu \beta^*_{\nu+h} + \sum_{\mu=1}^{l-(j+1)} \alpha_\mu \alpha^*_{\mu+j+1} \sum_{\nu'=1}^{h} \beta^*_{\nu'} \beta_{\nu'+m-h} \\ &= \lambda_j(x)\lambda_h(y) + \lambda_{j+1}(x)\lambda^*_{m-h}(y) \end{aligned} \tag{14.15.3}$$

as was to be shown.

Note that the results are not necessarily the same if the roles of the two sequences in the Kronecker sum are reversed. Note, too, that the maximum of the normalized aperiodic correlation coefficients λ_k/n cannot be decreased using this construction.

14.16 Word and Frame Synchronization—Choice of Statistics

In the previous sections of this chapter, and in Chapters Eleven and Twelve, various means of establishing word synchronization were investigated. The approaches considered can be separated into two categories:† (1) A special synchronization sequence (or prefix or comma) is periodically inserted into the data stream. Its identification at the receiver establishes the desired synchronization. For convenience, we shall refer to this category as the *prefix method.* (2) The difference between true words and overlap words is exploited in order to distinguish between the two situations. Although this approach is not limited to comma-free dictionaries (cf. Chapter Seven) it is more effective when the dictionaries are comma-free and will be referred to subsequently as the *comma-free method.*

The generic equivalence of word and frame synchronization should be emphasized. By substituting "word" for "symbol" and "frame" for "word" all of the word synchronization techniques become equally valid for frame synchronization. If k q-ary information symbols are represented by each word,

†A third method equally applicable to word, frame, or symbol synchronization is to use an auxiliary synchronization channel (or, equivalently, to use the communication channel in two different modes, a synchronization mode and a data transmission mode). This approach was investigated in some detail in Chapter Six and will not be considered here.

a frame can be regarded as a word composed of q^k-ary symbols. The relative attractiveness of the various approaches, of course, may well be altered when frame, as opposed to word, synchronization is being considered, but the essential equivalence of the two problems remains.

In the next section we estimate the synchronization delay inherent in the use of both the prefix and comma-free methods. By synchronization delay we mean here the number of words or symbols needed to acquire synchronization initially with a preassigned reliability. (The distinction between this definition of synchronization delay and that used in Chapter Eleven, in the noise-free situation, should be noted.) The related problem of reacquiring synchronization after it has been lost differs from the initial acquisition problem only in the altered a priori probabilities of the different contending synchronization positions, and will not be considered here explicitly.

Before proceeding, however, we must select a statistic upon which to base the decision as to whether or not synchronization has been obtained. To this end we represent the signal by a discrete set of observables $Y = \{y_i\}$. (In the case of the Gaussian channel, these terms y_i can be samples of the received signal at uniformly separated instances of time; if the channel is already discrete, as is the q-ary symmetric channel, these terms are the symbols produced at the receiver output.) As in symbol synchronization discussions, it is also convenient to divide the total period of ambiguity T into a finite number N of epochs ν, $\nu = 0, 1, 2, \ldots, N-1$, with $N = T/\Delta t$. Generally prior, lower-order synchronization will already have been established, so this quantization of time will be quite natural.

First, consider the situation in which a synchronizing sequence or prefix is inserted periodically into the information stream. We divide the set of samples y_i into two subsets Y_1 and Y_2, $Y_1 = Y_1(\nu)$ containing those samples y_i taken from the prefix portion of the received signal assuming the νth epoch to be the correct one, and $Y_2 = Y_2(\nu)$ the set remaining after Y_1 is removed from Y. Letting $p(A\,|\,B)$ denote the probability (or probability density) of the event A conditioned on the event B and assuming the noise affects the successive samples independently, we then have

$$p(Y\,|\,\nu) = \sum_D p(Y_2\,|\,D, \nu)p(D)p(Y_1\,|\,S, \nu) \tag{14.16.1}$$

Each of the sets D is a set of information samples which would be observed in the absence of noise with the probability $P(D)$. Similarly, S is the set of samples characterizing the noise-free prefix. The optimum (maximum-likelihood) decision is to select that epoch ν for which the probability $p(Y\,|\,\nu)$ is maximized. (This is the situation when the same set of observables Y is used to test all epochs, i.e., in the parallel search situation. As in Chapter Six, however, we shall use this same decision rule for both parallel and serial searches.)

The summation over D in Equation (14.16.1) is most inconvenient from a practical point of view, and we would like to be able to equate it to a constant independent of ν. Unfortunately, this is rarely strictly true. If the information sequence were comprised of independent symbols, with any one such sequence as likely as any other, and if the samples y_i were taken after hard decisions were made concerning each received symbol, the summation would, indeed, be independent of ν. More generally, however, this is not the case. Except when hard decisions have already been made the noise will inevitably cause some sets of samples $Y_2(\nu)$ to resemble an information sequence more closely than others. The same statement holds when the information sequence is in any way constrained, as, for example, when it is encoded for error protection, or to prevent the occurrence of the synchronizing sequence.

Nevertheless, because the resulting synchronizer is considerably more practical so far as mechanization is concerned, and because the analysis of the simplified synchronizer is much more tractable, we shall ignore the summation over the information sequence in Equation (14.16.1). It must be acknowledged, however, that the resulting synchronizer, now relying solely on the decision variable

$$z_\nu = p(Y_1 \,|\, S, \nu) \tag{14.16.2}$$

is generally suboptimum.

When the channel is a white Gaussian noise channel, the decision variable z_ν becomes, after discarding those terms not dependent upon ν, and upon taking the limit as the sampling density becomes infinite,

$$z_\nu = \sum_j \int_{jT_r+\nu\Delta t}^{jT_r+T_s+\nu\Delta t} y(t)s(t-\nu\Delta t)\,dt \tag{14.16.3}$$

Here T_r is the time interval separating successive occurrences of the prefix, T_s the period of the prefix, $y(t)$ the received signal, and $s(t)$ the prefix signal (with $s(t + iT_r) = s(t)$ for all i).

The q-ary symmetric channel decision variable is also easily determined. If the Hamming distance $n_j(\nu)$ separates the jth repetition of the presumed prefix in the set of samples $Y_1(\nu)$ from the actual prefix as represented by the set S, and if $\sum_{j=1}^{M} n_j(\nu) = d_\nu$,

$$z_\nu = p^{d_\nu}(1-p)^{Mn-d_\nu}$$

where p is the probability of a symbol error, n the number of symbols in the synchronizing sequence, and M the number of repetitions of the prefix used in the decision. If $p < \frac{1}{2}$, therefore, the epoch selected by the maximum-likelihood criterion is the one minimizing the summation

$$d_\nu = \sum_j n_j(\nu) \tag{14.16.4}$$

With the second synchronization scheme mentioned above, the comma-free method, the probability $p(Y\,|\,\nu)$ is conveniently expressed as

$$p(Y\,|\,\nu) = \sum_D p(Y|D, \nu)P(D) \tag{14.16.5}$$

where the notation is as in Equation (14.16.1). Since there is no synchronizing sequence, all the information is now contained in the summation over D. Because we are interested in the application of this approach when the information is encoded using a comma-free dictionary, we shall assume the information is represented by a sequence of words, each word selected at random from a dictionary of N words w_i, $i = 1, 2, \ldots, N$. Then, dividing Y into disjoint subsets $Y_j = Y_j(\nu)$ corresponding to the interval over which the jth word would be received were the νth epoch the correct one, we have

$$p(Y \mid \nu) = \frac{1}{N} \prod_j \sum_i p(Y_j \mid w_i, \nu) \tag{14.16.6}$$

This, of course, is identical to the situation encountered in the investigation of symbol synchronization techniques (cf. Chapter Seven). It is especially true in the case of word synchronization that one of the terms $p(Y_j \mid w_i, \nu)$ can be expected to be significantly larger than the rest when ν is indeed the correct epoch. (Were this not the situation, the synchronous word error probability would generally be unacceptably high.) Consequently, if the νth epoch is the correct one, we have

$$p(Y \mid \nu) \approx \frac{1}{N} \prod_j \max_i p(Y_j \mid w_i, \nu) \tag{14.16.7}$$

The maximum-likelihood decision is to select that epoch for which the probability $p(Y \mid \nu)$ attains its maximum. Since $p(Y \mid \mu)$ can be closely approximated by the expression of Equation (14.16.7) when μ is the correct epoch, and since

$$p(Y \mid \nu) \geq \frac{1}{N} \prod_j \max_i p(Y_j \mid w_i, \nu)$$

for any epoch ν (the neglected terms are all non-negative), the decision should not be significantly less reliable if it is based on the approximation (14.16.7) rather than on the exact expression of Equation (14.16.6). The advantages in using the approximation are apparent. (The same decision variable would result, incidentally, if a joint maximum-likelihood decision were made as to both the received word sequence $\{w_i\}$ and the epoch ν.)

If the communication channel is the white Gaussian noise channel, the decision variable resulting after taking the logarithm of Equation (14.16.7) and neglecting those terms independent of ν is

$$z_\nu = \sum_i \max_j \int_{iT_w + \nu\Delta t}^{(i+1)T_w + \nu\Delta t} [y(t)w_j(t - \nu\Delta t) - \tfrac{1}{2}w_j^2(t - \nu\Delta t)]\, dt \tag{14.16.8}$$

where T_w denotes the word period.† The synchronizer selects the μth epoch if $z_\mu > z_\nu$ for all $\nu \neq \mu$.

†This is the phase-coherent synchronizer decision variable. The extension to the phase-incoherent case parallels in all essentials the corresponding extension in Chapter Seven.

The decision variable for the q-ary symmetric channel resulting from the approximation (14.16.7) is

$$d_\nu = \sum_i \min_j n_i(j; \nu) \tag{14.16.9}$$

where $n_i(j; \nu)$ is the Hamming distance between the sequence received over the ith interval and the jth word of the code dictionary. The synchronizer then chooses the μth epoch if $d_\mu < d_\nu$ for all $\nu \neq \mu$.

14.17 Word and Frame Synchronization—Synchronization Delay

Although in some special cases exact answers can be obtained relatively easily, we shall limit consideration here to a method for approximating the delay needed to attain word or frame synchronization. The justification for this approach lies in the generality and tractability of the results and in the fact that more exact estimates can be derived, if at all, only at the expense of considerably more detailed analysis.

The approximation involves first using the bound

$$P_0 \leq P_e \leq (N-1)P_0 \tag{14.17.1}$$

on the probability of an erroneous synchronization decision, where

$$P_0 = \begin{cases} \max\limits_{\nu \neq \mu} \Pr\{z_\mu < z_\nu\} & \text{Gaussian channel} \\ \max\limits_{\nu \neq \mu} \Pr\{d_\mu > d_\nu\} & q\text{-ary symmetric channel} \end{cases}$$

with μ indicating the correct epoch. For convenience, and without loss of generality, we let $\mu = 0$ in the subsequent discussion. We next observe that the statistics z_0, z_ν, d_0, and d_ν are generally sums of a large number of independent random variables and apply the central limit theorem to the difference

$$y_\nu = \begin{cases} z_0 - z_\nu & \text{Gaussian channel} \\ d_\nu - d_0 & q\text{-ary symmetric channel} \end{cases} \tag{14.17.2}$$

The synchronization delay can then be estimated, as in Chapters Six and Seven, by satisfying the equation

$$r^2 \triangleq \min_{\nu \neq 0} \frac{E^2(y_\nu)}{\operatorname{Var}(y_\nu)} = 2\kappa_N \log_e \frac{1}{P_e} \tag{14.17.3}$$

The constant κ_N, it will be recalled, is bounded by $1 \leq \kappa_N \leq 1 + [\log_e(N-1)/\log_e(1/P_e)]$, so in most cases, $\kappa_N \approx 1$.

This method of estimating the synchronization delay is valid either when observations pertaining to the different epochs are made simultaneously, or when they are made serially in conjunction with a fixed-sample-size search (cf. Chapter Six). Although it will not be investigated here, the sequential search algorithm could presumably also be used to advantage in determining the word or frame epoch. Since the observables generally satisfy rather unwieldy distributions, however, approximations to the optimum test must usually be made.

We now consider some specific examples.

14.17.1 The Prefix Method—q-ary Symmetric Channel

If the information symbols can be assumed mutually independent and uniformly distributed, then when $\nu \geq m$ (m denoting the number of symbols in a prefix) and when M observations are made prior to a decision

$$\begin{aligned} E(d_0) &= Mm\,p \\ E(d_\nu) &= Mm\frac{q-1}{q} \\ \operatorname{Var}(d_\nu - d_0) &= \operatorname{Var}(d_\nu) + \operatorname{Var}(d_0) = Mm\left[p(1-p) + \frac{q-1}{q^2}\right] \end{aligned} \tag{14.17.4}$$

with p the probability of a symbol error. Presumably, the prefix will be so chosen that the epochs $\nu < m$ will be more easily distinguishable from the true epoch than when $\nu \geq m$. Indeed, Barker-type sequences are constrained to have precisely this property. From Equation (14.17.3), then, we have

$$M = \frac{p(1-p) + \dfrac{q-1}{q^2}}{m\left(1 - \dfrac{1}{q} - p\right)^2} 2\kappa_N \log_e \frac{1}{P_e} \tag{14.17.5}$$

This is the approximate number of times the prefix must be received to attain the error probability P_e when all epochs are observed simultaneously. Serial observation, of course, takes N times as long.

The question arises as to the advantage of prohibiting the prefix from occurring in the random data stream as could be accomplished by using either prefix or comma-codes. The advantage is rather obvious when the error probability is very small and only a few observations are needed for a reliable decision. In contrast, when the symbol error probability is of sufficient magnitude to require a comparatively large number of observations, and the approximations leading to Equation (14.17.5) are thereby valid, these advantages tend to disappear. The only effect of these restrictions on the data sequence is to increase slightly the term $E(d_\nu)$ and to decrease somewhat the term $\operatorname{Var}(d_\nu)$. Neither of these changes will significantly alter the result in Equation (14.17.5). (This same conclusion would not follow, however, were the information-bearing portion of the received signal not ignored in the definition of the decision variable; cf. Equation (14.16.1) and the ensuing discussion.

In this event, the synchronization procedure would belong to the comma-free class rather than to the prefix class as these terms have been defined here.)

14.17.2 The Prefix Method—White Gaussian Channel

If the information-bearing signal $x(t)$ is statistically orthogonal to the prefix $s(t)$, and if $s(t)$ represents unity average power, then

$$E(z_0) = M\,A\,T_s$$
$$E(z_\nu \mid \nu\Delta T > T_s) = 0$$
$$\text{Var}\,(z_0) = MN_0T_s/2$$
$$\text{Var}\,(z_\nu \mid \nu\Delta T > T_s) = MN_0T_s/2 + MA^2T_s^2\sigma_0^2$$

where

$$\sigma_0^2 = E\left\{\left[\frac{1}{T_s}\int_0^{T_s} x(t + \nu\Delta T)s(t)\,dt\right]^2 \middle| \nu\Delta T > T_s\right\} \tag{14.17.6}$$

and N_0 represents the single-sided noise spectral density. Then, letting $R_s = A^2T_s/N_0$, we find the required number of prefix repetitions to be approximately

$$M = \left(\frac{1 + R_s\sigma_0^2}{R_s}\right)2\kappa_N \log_e \frac{1}{P_e} \tag{14.17.7}$$

For a more specific example, suppose both the prefix and the information are represented by sequences of binary symbols of the form $\pm\sqrt{2}\sin\omega t$, each symbol having a duration of T_s/m seconds. Then, if the successive information symbols are assumed mutually independent and uniformly distributed,

$$\sigma_0^2 = \frac{1}{2^m}\sum_{l=0}^{m}\binom{m}{l}\left(1 - \frac{2l}{m}\right)^2 = \frac{1}{m} \tag{14.17.8}$$

14.17.3 The Comma-Free Method—q-ary Symmetric Channel—Error-correcting Codes

The comma-free synchronization delay is generally difficult to estimate without a detailed examination of the particular dictionary in question. The difficulty arises because now the decision is based not on the distances between the received n-tuples at a particular epoch and a known prefix, but rather on the distance between these n-tuples and the code words closest to them. There are in effect q^k prefixes instead of just one.

Suppose the distance between a received word overlap and the nearest code word, say w_i, is, in the absence of errors, δ. The expected distance between this n-tuple and w_i when the probability of a symbol error is p is then

$$(n - \delta)p + \delta\left(1 - \frac{p}{q - 1}\right) = np + \delta\left(1 - \frac{q}{q - 1}p\right) \tag{14.17.9}$$

This, unfortunately, is only an upper bound on the expected distance between the received n-tuple and the *nearest* code word, since this nearest word, after the errors have been made, may no longer be w_i.

Nevertheless, if the dictionary in question has minimum distance d, and index of comma freedom s, if s is small relative to $d/2$, and if the (synchronous) word error probability is small, Equation (14.17.9) (with $\delta = s$) does provide a reasonable estimate of the expected minimum distance between a word overlap and a word. If the overlap word actually does differ in only s symbols from any code word, Equation (14.17.9) overbounds the expected minimum distance, as we have already noted. But if fewer than $(d/2) - s$ errors are made, the original nearest word will remain the nearest word, and since the probability of $(d/2) - s$ or more errors must be small under the stated assumptions, this effect should likewise be relatively small. Moreover, while the pre-error minimum distance is guaranteed to be s, it most generally will exceed s, and since the expected distance from Equation (14.17.9) is an increasing function of δ, the estimate obtained by equating δ to s is in this sense conservative.

For an estimate of the synchronization delay, we let

$$\begin{aligned} E(d_0) &= M\,n\,p \\ E(d_\nu) &= M\,n\,p + Ms\left(1 - \frac{q}{q-1}p\right) \\ \operatorname{Var}(d_0) &= \operatorname{Var}(d_\nu) = M\,n\,p\,(1-p) \end{aligned} \tag{14.17.10}$$

The variances given here are the variances of the distance between the n-tuple of interest and the pre-error closest word; consequently, they are upper bounds on the actual variances of d_0 and d_ν. Note that d_0 and d_ν are not independent if the two epochs are observed simultaneously. Nevertheless, since for any two random variables x and y,

$$\begin{aligned} E(x-y)^2 &= E(x^2) + E(y^2) - 2E(xy) \le E(x^2) + E(y^2) \\ &\quad + 2E^{1/2}(x^2)E^{1/2}(y^2) = [E^{1/2}(x^2) + E^{1/2}(y^2)]^2 \end{aligned}$$

we have

$$\operatorname{Var}(y_\nu) \le \begin{cases} 2M\,n\,p\,(1-p) & \text{serial observation} \\ 4M\,n\,p\,(1-p) & \text{parallel observation} \end{cases} \tag{14.17.11}$$

Accordingly, we find as an approximate estimate of the total number of code words which must be received to insure a reliable decision,

$$M = \begin{cases} \dfrac{2p(1-p)}{\left[\dfrac{s}{n}\left(1 - \dfrac{q}{q-1}p\right)\right]^2} 2\kappa_n \log_e \dfrac{1}{P_e} & \text{serial observation} \\[2ex] \dfrac{4p(1-p)}{n\left[\dfrac{s}{n}\left(1 - \dfrac{q}{q-1}p\right)\right]^2} 2\kappa_n \log_e \dfrac{1}{P_e} & \text{parallel observation} \end{cases} \tag{14.17.12}$$

It must be emphasized that these results provide only an estimate of the synchronization delay; they are neither upper nor lower bounds on this delay.

14.17.4 The Comma-Free Method—White Gaussian Channel—Orthogonal Codes

If D is an $(n = r^k, k)$ orthogonal dictionary having a maximum asynchronous correlation coefficient ρ_m, we can immediately conclude, using the results of Appendix A.3, that

$$E(z_0) \approx M\,A\,T_w \tag{14.17.13}$$

$$E(z_v) \leq M\,A\,T_w\rho_m + M(N_0T_w/2)^{1/2}\sqrt{2\log_e n} + O(1/\sqrt{\log_e n})$$

Since each word represents equal power, $P_w = A^2$, the second term in the integral defining z_v [Equation (14.16.8)] has been omitted here. The term T_w, of course, represents the word period and N_0 the single-sided noise spectral density. Similarly,

$$\begin{aligned} \mathrm{Var}\,(z_0) &= M\,N_0\,T_w/2 \\ \mathrm{Var}\,(z_v) &\approx M\pi^2 N_0\,T_w/24\log_e n \end{aligned} \tag{14.17.14}$$

Accordingly, from Equation (14.17.3),

$$M \approx \begin{cases} \dfrac{n(1 + \pi^2/12\log_e n)2\kappa_n \log_e(1/P_e)}{2R_b(\log_2 n)[(1 - \rho_m) - (\log_e 2/R_b)^{1/2}]^2} & \text{serial observation} \\[2ex] \dfrac{(1 + \pi/\sqrt{12\log_e n})^2\, 2\kappa_n \log_e(1/P_e)}{2R_b(\log_2 n)[(1 - \rho_m) - (\log_e 2/R_b)^{1/2}]^2} & \text{parallel observation} \end{cases} \tag{14.17.15}$$

with $R_b = A^2T_w/N_0\log_2 n$ the ratio of the signal energy per bit to the noise spectral density.

An evidently sufficient condition for the synchronizability of orthogonal codes is

$$R_b > \log_e 2/(1 - \rho_m)^2 \tag{14.17.16}$$

The probability of a word error using an orthogonal signal set was found, in Section 4.2, to approach zero asymptotically with the number of words n only if $R_b > \log_e 2$. Since, presumably, ρ_m should approach zero as $n \longrightarrow \infty$ (although this has not been proved), the condition for the synchronizability of orthogonal codes is, at least asymptotically, identical to the condition for a small word error probability.

Equation (14.17.16) nevertheless provides an extremely conservative estimate of the signal-to-noise ratio required for synchronization. One of the main reasons for this is the assumption, implicit in Equation (14.17.13), that every overlap word has the maximum possible correlation with every code word. Except perhaps for very large orthogonal dictionaries, such an assumption necessarily violates Theorem 14.29. Most overlap words, in fact, will not correlate that closely with any dictionary word, and certainly not with all of them. Considerably tighter estimates of the synchronization delay associated with orthogonal comma-free codes have been derived using the constraint of Theorem 14.29 (cf. References 14.8 and 14.10). Even these bounds tend to be quite conservative, however. Experimental evidence (Reference 14.11) indicates that typically only a few (five to ten) received words are required to make a decision concerning any particular epoch with an acceptably small error probability.

Notes to Chapter Fourteen

Most of the material on Barker sequences in Section 14.6 can be found in either the original paper by Barker (Reference 14.1) or in the article by Turyn (Reference 14.3). The first construction for synchronizable, cyclic error-correcting codes presented in Sections 14.9 and 14.10 is an extension of an approach taken by Solomon (Reference 14.12). The second construction as well as that for shortened cyclic codes in Section 14.11 is due to Tong (Reference 14.4). See also the paper by Levy (Reference 14.13) in this respect. The synchronization-error-correcting codes using the buffering technique, as discussed in Section 14.11, were first investigated by Bose and Caldwell (Reference 14.5). The burst-error-correcting code constructions are either taken from or motivated by the work of Tong (Reference 14.4) and Tavares (Reference 14.14). Other Section 14.10 and 14.11 results, including Theorem 14.19, can also be found in this latter reference and in References 14.15 through 14.17. Barker phase sequences (Section 14.15) are discussed in Reference 14.9. Much of the remainder of the material in this chapter is explored, in some cases in greater detail, in References 14.18 through 14.20 and Reference 14.8.

Problems

14.1 The concept "invulnerability to synchronization at position m" was introduced in Section 14.3. A weaker property of a code dictionary, which might be called "νth-order invulnerability at position m," obtains if a decoder, beginning at position m in an arbitrary sequence of code words, encounters at most $\nu - 1$ consecutive code words before it finds a non-word. If a dictionary is νth-order invulnerable at all positions $m \neq 0$, then it is obviously synchronizable (in the sense of Chapter Eleven) with a delay not exceeding $(\nu + 1)n - 1$ symbols. Show that the results of Section 14.3 still apply if the phrase "invulnerable to synchronization at position m" is replaced with "νth-order invulnerable at position m," and if $\mathbf{M}$, $\mathbf{x}$, and $\mathbf{b}$ are properly redefined as a $\nu n \times (2\nu + 1)k$

matrix and as $(2\nu + 1)k$- and νn-component vectors, respectively. Using this fact, show that no coset of any binary Hamming code dictionary other than the (3, 1) is synchronizable with finite delay. Note that this statement holds even if each successive code word is chosen from a different coset.

14.2 Binary information is to be transmitted over a white Gaussian channel using antipodal signals [i.e., a *one* is to be represented by the signal $s(t)$, $0 \leq t \leq T_b$, and a *zero* by $-s(t)$]. A clock signal, coherent with the bit stream but with a period $T_0 = T_b/n$, is transmitted on a separate channel and recovered with negligible phase error. Show that, if loss of bit synchronization is to be detected by observing the magnitude of the bit decision variables, the optimum signal $s(t)$ is a Barker (or Barker-type) sequence of length n. Determine the bit sync acquisition time when $s(t)$ is a Barker sequence and the techniques of Chapter Seven are used. Compare with the sync acquisition time when $s(t)$ is rectangular.

14.3 (a) Use the results of Section 12.4 to determine the relationship between the number m of symbols in a comma and the likelihood that it occurs in a random symbol sequence of a given length. Show that when the number n of information symbols separating two successive commas in a random r-ary sequence is large the probability P that the comma inadvertently appears in the sequence is, with $N = n + m$,

$$P \approx \begin{cases} 1 & m \leq \log_r N \\ N/r^m & m \gg \log_r N \end{cases}$$

regardless of the structure (repetitive periods) of the comma (cf. Problem 12.3).
(b) Show that a sequence is repetitive with period ν if and only if the quantity $c_{m-\nu}$ defined in Equation (14.6.2) equals $m - \nu$.

14.4 Show that Theorems 14.15 and 14.17 remain valid when ϵ is redefined as the least common multiple of the orders of the roots of $m(x)$. Under what conditions might it be advantageous for $m(x)$ to be reducible?

14.5 (cf. Reference 14.21). Let D_0 be an (n, k_0) cyclic code dictionary over $GF(q)$ and $D_\nu(l)$ a q-ary prefix code of l-symbol words all having the same ν-symbol prefix (cf. Section 12.4). Define D as the dictionary of all words in D_0 whose first l symbols ($l \leq k_0$) are also words in $D_\nu(l)$, and D' as the same dictionary but with the symbols of each word cyclically permuted $[(l + 1)/2]$ positions to the left. Show that D is invulnerable to synchronization at all positions $|m| \leq l - 1$ and determine the index of invulnerability s_m as a function of ν. Similarly, show that with D', it is possible to correct synchronization errors not exceeding $[(l - 1)/2]$ symbols in magnitude. Show that this construction in general introduces more redundancy than that of Theorem 14.15, but that both reduce the dimension of D_0 by at least $\log_q \mu$ in order to provide synchronization invulnerability at all positions $|m| \leq \mu$. Describe a similar construction in terms of a comma code rather than a prefix code.

14.6 (Reference 14.16). Show that the term $d - 2s$ in the statement of Theorem 14.16 can be replaced by $\max\{d - 2s, (3d - 7s - 1)/2\}$ when the code in question is binary. [*Hint:* Show that at least one, and hence both, of the polynomials $v(x) = (x^{n-m} - 1)c(x) + x^{n-m}p_m(x) + e(x)$ and $(1 + x^{\mu-m+1})v(x)$ can not represent a non-zero code word, under the imposed conditions, when $e(x)$ is an error vector of weight not exceeding $s - 1$.]

14.7 Derive the index of invulnerability s_m of the dictionary D_s of Theorem 14.22. Note that m need not necessarily be limited to the region $(-m_l, m_r)$.

14.8 It was generally tacitly assumed in Sections 14.9, 14.10, and 14.11 that a synchronization error of $-m$ symbols was as likely to occur as one of $+m$ symbols. How could the results of those sections be improved if this were not the case?

14.9 Let D_0, D, n, d_0, s_m, and $g(x)$ be as defined in Theorem 14.15. An ln-symbol comma is defined as a sequence of l repetitions of the word in D_0 represented by $g(x)$. Show that any ln-tuple in an arbitrary sequence of words from D must be Hamming distance d_c or greater from this comma where

$$d_c \geq (l-1)d_0 + \min_m s_m$$

14.10 (Reference 14.22). Let a_ν and b_ν be defined by $\nu = a_\nu n + b_\nu$, $0 \leq b_\nu < n$, $\nu = 0, 1, \ldots, n^2 - 1$, and consider the phase sequence $\{x_\nu = a_\nu b_\nu\}$. Show in particular that the aperiodic correlation coefficients

$$\lambda_k = \sum_{\nu=0}^{n^2-1-k} \zeta_\nu \zeta_{\nu+k}^* \qquad \begin{cases} \zeta_\nu \triangleq e^{(2\pi i/n)x_\nu} \\ k = 1, 2, \ldots, n^2 - 1 \end{cases}$$

are bounded by

$$|\lambda_k| \leq |\lambda_{[(n+1)/2]}| = \left| \sum_{\nu=0}^{[(n-1)/2]} e^{(2\pi i/n)\nu} \right| \triangleq b_n$$

[*Hint:* Let $k = qn + r$, $0 \leq q, r \leq n - 1$ and verify that

(1) $$\lambda_k = \frac{|\zeta^{r(q+1)} - 1|^2}{(\zeta^{-r} - 1)(\zeta^{-(q+1)} - 1} - \frac{|\zeta^{rq} - 1|^2}{(\zeta^{-r} - 1)(\zeta^{-q} - 1)}$$

with $\zeta = \zeta_{n+1} = e^{(2\pi i/n)}$;

(2) $$\left| \sum_{\nu=0}^{Q-1} \zeta^{-r\nu} \right| = \left| \frac{\zeta^{-rQ} - 1}{\zeta^{-r} - 1} \right| \leq \left| \frac{\zeta^{Q} - 1}{\zeta - 1} \right| = \left| \sum_{\nu=0}^{Q-1} \zeta^{\nu} \right|, \; Q \leq (n+1)/2;$$

(3) $\Lambda_Q \triangleq \sum_{q=0}^{Q-1} \lambda_{qn+r} \leq b_n$;

(4) $\arg(\Lambda_Q/\Lambda_{Q-1}) = \pi/n$;

(5) hence, for $n \geq 3$, $|\lambda_k| \leq b_n$. (The case $n = 2$ can be treated separately.).]

References

14.1 R. H. Barker, "Group Synchronization of Binary Digital Systems," in *Communication Theory*, W. Jackson, Ed., Butterworth, London, England, p. 273, 1953.

14.2 J. Storer, and R. Turyn, "On Binary Sequences," *Proc. Amer. Math. Soc.*, **12**, p. 394, 1961.

14.3 R. Turyn, "Sequences with Small Correlation," in *Error Correcting Codes*, H. B. Mann, Ed., John Wiley & Sons, Inc., 1969.

14.4 S. Y. Tong, "Synchronization Recovery Techniques for Binary Cyclic Codes," *Bell System Tech. J.*, vol. 45, p. 561, 1966.

14.5 R. C. Bose, and J. G. Caldwell, *Synchronizable Error-Correcting Codes*, Inst. Statistics, Mimeo Series No. 498, Univ. North Carolina, Chapel Hill, N. C., Nov., 1966.

14.6 L. Baumert, and H. C. Rumsey, "The Index of Comma Freedom for the Mariner Mars 1969 High Rate Telemetry Code," *Space Programs Summary*, 37–46, Vol. IV, Jet Propulsion Lab., Calif. Inst. Tech., Pasadena, Calif., p. 221, 1967.

14.7 L. Baumert, and H. C. Rumsey, "The Maximum Indices of Comma Freedom for the High Data Rate Telemetry Codes," *Space Programs Summary*, 37–51, Vol. III, Jet Propulsion Lab., Calif. Inst. Tech., Pasadena, Calif., p. 260, 1968.

14.8 S. W. Golomb *et al.*, *Digital Space Communications*, Prentice-Hall, Inc., Englewood Cliffs, N. J., 1964.

14.9 S. W. Golomb, and R. A. Scholtz, "Generalized Barker Sequences," *IEEE Trans. on Inf. Theory*, IT-11, **4**, p. 533, 1965.

14.10 J. J. Stiffler, "Word Synchronization Time Using Orthogonal Comma-Free Codes," *Space Programs Summary*, 37–38, Vol. IV, Jet Propulsion Lab., Calif. Inst. Tech., Pasadena, Calif., p. 265, 1966.

14.11 M. Easterling *et al.*, "High Rate Telemetry Project," *Space Programs Summary*, 37–48, Vol. II, Jet Propulsion Lab., Calif. Inst. Tech., Pasadena, Calif., p. 83, 1967.

14.12 G. Solomon, "Self-synchronizing Reed-Solomon Codes," *IEEE Trans. on Inf. Theory*, IT-14, p. 608, 1968.

14.13 J. E. Levy, "Self-Synchronizing Codes Derived from Binary Cyclic Codes," *IEEE Trans. Inf. Theory*, IT-12, p. 286, 1966.

14.14 S. Tavares, "A Study of Synchronization Techniques for Binary Cyclic Codes," Ph. D. Thesis, Dept. of Elec. Eng., McGill Univ., Montreal, Canada, July, 1968.

14.15 S. E. Tavares, and M. Fukada, "Matrix Approach to Synchronization Recovery for Binary Cyclic Codes," *IEEE Trans. Inf. Theory*, IT-15, p. 93, 1969.

14.16 S. E. Tavares, and M. Fukada, "Further Results on the Synchronization of Binary Cyclic Codes," *IEEE Trans. Inf. Theory*, IT-16, p. 238, 1970.

14.17 S. E. Tavares, and M. Fukada, "Synchronization of Cyclic Codes in the Presence of Burst Errors," *Inform. and Control*, **14**, p. 423, 1969.

14.18 J. J. Stiffler, "Synchronization of Telemetry Codes," *IEEE Trans. Space Elec. and Telemetry*, SET-8, p. 112, 1962.

14.19 J. J. Stiffler, "Comma-free Error-Correcting Codes," *IEEE Trans. Inf. Theory*, IT-11, p. 107, 1965.

14.20 J. J. Stiffler, "Synchronization Methods for Block Codes," *IRE Trans. on Inf. Theory*, IT-8, p. 25, 1962.

14.21 D. Mandelbaum, "Some Hybrid Methods for Synchronization of Cyclic Codes," to be published.

14.22 R. Turyn, "The Correlation Function of a Sequence of Roots of 1," *IEEE Trans. Inf. Theory*, IT-13, p. 524, 1967.

Supplementary Bibliography

Calabi, L., and W. E. Hartnett, "A Family of Codes for the Correction of Substitution and Synchronization Errors," *IEEE Trans. Inf. Theory*, IT-15, p. 102, 1969.

Chase, D., "Coding Theorems for the Nonsynchronized Channel," *IEEE Trans. Inf. Theory*, IT-16, p. 241, 1970.

Frey, A. H., "Message Framing and Error Control," *IEEE Trans. on Military Electronics*, MIL-9, p. 143, 1965.

Hachett, C. M., Jr., "Synchronization of Cyclic Codes," *First Ann. Conf. Inform. Sci. Systems*, Princeton, N. J., 1967.

Jaffe, R. M., "Digilock Telemetry System for the Air Force Special Weapons Center's Blue Scout," *IRE Trans. on Space Elect. and Telemetry*, SET-8, p. 44, 1962.

Kasahara, Y., and M. Kasahara, "Notes on Synchronization for Burst Error Correction Codes," *J. Inst. Elec. Commun. Eng.* (Japan), **47**, p. 85, 1964.

Kasahara, Y., and M. Kasahara, "Synchronization," *J. Inst. Elec. Commun. Eng.* (Japan), **49**, p. 33, 1966.

Mandelbaum, D., "On Synchronization of Convolutional Codes," *IEEE Trans. Comm. Tech.* (to appear).

Mandelbaum, D., "A Note on Synchronizable Error-Correcting Codes," *Inform. and Control*, **13**, p. 429, 1968.

Natali, F. D., "Synchronization Properties of a Near-Comma-Free (8, 4) Code," *IEEE Trans. on Comm. Tech.*, Com.-17, p. 500, 1969.

Sekimoto, T., and H. Kaneko, "Group Synchronization for Digital Transmission Systems," *IRE Trans. an Commun. Systems*, CS-10, p. 381, 1962.

Shiva, S. G. S., and G. Seguin, "Synchronizable Error-Correcting Binary Codes," *IEEE Trans. Inf. Theory*, IT-16, p. 241, 1970.

Stiffler, J. J., "On the Use of Comma-Free Codes for Symbol Insertion and Deletion Correction," *Space Programs Summary*, 37-34, Vol. IV, Jet Propulsion Lab., Calif. Inst. of Tech., Pasadena, Calif., p. 221, 1965.

Tong, S. Y., "Correction of Synchronization Errors with Burst-Error-Correcting Cyclic Codes," *IEEE Trans. Inf. Theory*, IT-15, p. 106, 1969.

Van Horn, J. H., "A Theoretical Synchronization System for Use with Noisy Digital Signals," *IEEE Trans. on Comm. Tech.*, Com.-12, p. 82, 1964.

Weldon, E. J., Jr., "A Note on Synchronization Recovery with Extended Cyclic Codes," *Inform. and Control*, **13**, p. 354, 1968.

appendix a

Some Useful Results in Probability Theory and Statistics

A.1 On the Number of Observations Needed for a Sequential Decision

The discussion in this section has to do with the following situation. A sequential probability ratio test is to be conducted based on the statistics $z_i = \log_e [p_0(y_i)/p_1(y_i)]$ and the decision thresholds A and B. If for some M, the sum $Z_M = \sum_{i=1}^{M} z_i \geq \log_e A$, while $\log_e B < Z_m < \log_e A$ for all $m < M$, the null hypothesis H_0 is accepted; if $Z_M \leq \log_e B$ under the same condition, H_0 is rejected (the alternative hypothesis H_1 is accepted). The observables y_i are assumed independent and identically distributed in accordance with the density function $p_\theta(y)$. The quantity of primary concern here is the random variable $M = M(\theta)$.

Lemma A.1. Let $\Pr\{M > n\}$ be the probability that more than n observations are needed for a decision in a sequential probability ratio test. Then if $E(z_i^2) \neq 0$ and if $C \triangleq |\log_e A| + |\log_e B| < \infty$,

$$\lim_{n\to\infty} n^r \Pr\{M > n\} = 0 \tag{A.1.1}$$

for any finite r.

Proof. Since the variables z_i are independent and identically distributed with either a non-zero variance or a non-zero mean (or both), the probability that more than k observations are needed to complete the test is less than one for sufficiently large k:

$$\Pr\{M > k\} \le \Pr\left\{\left|\sum_{i=1}^{k} z_i\right| < C\right\} = p < 1$$

for some finite k. Then $\Pr\{M > km\} \le p^m$ and $\Pr\{M > n\} \le p^{[n/k]} \le p^{(n/k)-1}$. Consequently,

$$\lim_{n\to\infty} n^r \Pr\{M > n\} \le \lim_{n\to\infty} \frac{n^r (p^{1/k})^n}{p} = 0$$

for any finite r.

Lemma A.2. Let $p_\theta(y)$ be the probability density function of the random variable y, let $z = \log_e [p_0(y)/p_1(y)]$, and assume that the following three conditions hold:

(1)
$$\phi(h) \triangleq E(e^{hz}) \tag{A.1.2}$$
exists for all real h.

(2) The events

$$e^z > 1 + \delta \quad \text{and} \quad e^z < 1 - \delta \tag{A.1.3}$$

both occur with some non-zero probability for some positive δ.

(3)
$$\phi'(0) = E(z) \neq 0$$
Then the equation

$$\phi(h) = 1 \tag{A.1.4}$$

has one and only one non-zero solution $h = h(\theta)$.

Proof. If $h > 0$,

$$\phi(h) > \Pr\{e^z > 1 + \delta\}(1 + \delta)^h$$

and, from condition (2),

$$\lim_{h\to\infty} \phi(h) = \infty$$

Similarly, if $h < 0$

$$\phi(h) > \Pr\{e^z < 1 - \delta\}(1 - \delta)^h$$

and

$$\lim_{h\to-\infty} \phi(h) = \infty$$

And, since

$$\phi''(h) = E(z^2 e^{hz}) > 0$$

for all real h, $\phi(h)$ has a unique minimum in the interval $-\infty < h < \infty$. Since $\phi'(0) \neq 0$, by condition (3), this minimum does not occur at the point $h = 0$. There must therefore be one and only one value of $h = h(\theta)$, other than $h = 0$, for which $\phi(h) = 1$ and the lemma is proved.

The *complementary density function*

$$p_{\bar\theta}(y) \triangleq \left[\frac{p_0(y)}{p_1(y)}\right]^{h(\theta)} p_\theta(y) \tag{A.1.5}$$

will prove useful in the ensuing discussion. Note that $p_{\bar{\theta}}(y)$ is indeed a density function, i.e., $p_{\bar{\theta}}(y) \geq 0$ and $\int p_{\bar{\theta}}(y)\, dy = \phi[h(\theta)] = 1$. Further, the equation

$$1 = \phi[h(\bar{\theta})] = \int_{-\infty}^{\infty} \left[\frac{p_0(y)}{p_1(y)}\right]^{h(\bar{\theta})} p_{\bar{\theta}}(y)\, dy = \int_{-\infty}^{\infty} \left[\frac{p_0(y)}{p_1(y)}\right]^{h(\bar{\theta})+h(\theta)} p_{\theta}(y)\, dy$$

has the unique non-zero solution $h(\bar{\theta}) = -h(\theta)$.

Theorem A.1. Let $Z_n = \sum_{i=1}^{n} z_i$, with the z_i as previously defined, and let the random variable M denote the value of n when Z_n first reaches either of the (finite) thresholds $\log_e A$ or $\log_e B$. Then

$$E\{\phi^{-M}(h)e^{hZ_M}\} = 1 \tag{A.1.6}$$

for all h for which $|\phi(h)| \geq 1$. (This equation is known as *Wald's identity*.)

Proof. For any fixed n,

$$E\{e^{hZ_n}\} = \phi^n(h) = E\{e^{hZ_n} \mid M \leq n\}\Pr\{M \leq n\} + E\{e^{hZ_n} \mid M > n\}\Pr\{M > n\} \tag{A.1.7}$$

If $n \geq m$, $Z_n - Z_m$ is independent of Z_m, and

$$\begin{aligned} E\{e^{hZ_n} \mid M \leq n\} &= \sum_{m=1}^{n} E\{e^{hZ_m}e^{h(Z_n - Z_m)} \mid M = m, M \leq n\}\Pr\{M = m \mid M \leq n\} \\ &= \sum_{m=1}^{n} E\{e^{hZ_m} \mid M = m, M \leq n\}\phi^{n-m}(h)\Pr\{M = m \mid M \leq n\} \\ &= \phi^n(h)E\{e^{hZ_M}\phi^{-M}(h) \mid M \leq n\} \end{aligned} \tag{A.1.8}$$

Note that the last expectation is taken with respect to both Z_M and M. Combining Equations (A.1.7) and (A.1.8) and dividing both sides by $\phi^n(h)$, we obtain

$$1 = E\{e^{hZ_M}\phi^{-M}(h) \mid M \leq n\}\Pr\{M \leq n\} + \phi^{-n}(h)E\{e^{hZ_n} \mid M > n\}\Pr\{M > n\} \tag{A.1.9}$$

Now, $E\{e^{hZ_n} \mid M > n\} < \max(A^h, B^h)$, where A and B are the two decision thresholds, and from Lemma A.1, $\lim_{n\to\infty} \Pr\{M > n\} = 0$. Thus, in the limit as $n \to \infty$, Equation (A.1.9) yields the desired result for any h such that $|\phi(h)| \geq 1$.

Corollary. Let $P(\theta)$ denote the probability of accepting the hypothesis H_0 when the observables are characterized by the density function $p_\theta(y)$. Then, to the extent that the amount by which the likelihood ratio exceeds threshold A or falls below threshold B at the time of a decision can be neglected,

$$P(\theta) = \frac{1 - B^{h(\theta)}}{A^{h(\theta)} - B^{h(\theta)}} \tag{A.1.10}$$

where $h(\theta)$ is as defined in Lemma A.2.

Proof. Since $\phi[h(\theta)] = 1$, Equation (A.1.6) is applicable, and

$$\begin{aligned} 1 &= E\{\phi^{-M}[h(\theta)]e^{h(\theta)Z_M}\} = E\{e^{h(\theta)Z_M}\} \\ &= P(\theta)e^{h(\theta)\log_e A} + [1 - P(\theta)]e^{h(\theta)\log_e B} \end{aligned}$$

and the stated result follows immediately.

Corollary.

$$E\left\{\frac{\partial^r}{\partial h^r}[e^{hZ_M}\phi^{-M}(h)]\right\} = 0 \tag{A.1.11}$$

for any integer $r > 0$ and for any $|\phi(h)| \geq 1$.

Proof. This statement follows upon differentiation of both sides of Equation (A.1.9) r times with respect to h. Since, from Lemma A.1

$$\lim_{n\to\infty} n^r \Pr\{M > n\} = 0$$

for all real r, the rth derivative of the right-hand term in Equation (A.1.9) vanishes as $n \to \infty$. Interchanging the order of expectation and differentiation of the remaining term establishes the stated result.

Equation (A.1.11) can be used to obtain an expression for the rth moment of M for any r. The following theorem, however, is useful in reducing these expressions to a more convenient form.

Theorem A.2. Let $P(M|\theta, R)$ and $P(M|\theta, A)$ be the probabilities that exactly M observations are needed for a decision in a sequential probability ratio test, given that the underlying density function is $p_\theta(y)$ and that the null hypothesis is, respectively, rejected and accepted. Then

$$P(M|\theta, R) = P(M|\bar{\theta}, R)$$

and

$$P(M|\theta, A) = P(M|\bar{\theta}, A)$$

where $\bar{\theta}$ is as defined in Equation (A.1.5).

Proof. Since, from Equation (A.1.5), $p_{\bar{\theta}}(y)/p_\theta(y) = [p_0(y)/p_1(y)]^{h(\theta)}$, it follows that a sequential probability ratio test, identical to the one hypothesized at the beginning of this section, results when the statistics $Z'_m = \sum_{i=1}^{m} z'_i = \sum_{i=1}^{m} \log_e [p_{\bar{\theta}}(y)/p_\theta(y)]$ are compared to the thresholds $\log_e (A^{h(\theta)})$ and $\log_e (B^{h(\theta)})$. Now, the sequential test is so designed that hypothesis H_0 is rejected after M observations if, and only if, the set of observables $Y(M) = \{y_1, y_2, \ldots, y_M\}$ is in the region S_M defined by the equation:

$$\frac{P[Y(M) \in S_M | H_0]}{P[Y(M) \in S_M | H_1]} = \frac{P(M, R|H_0)}{P(M, R|H_1)} = B$$

where, as before, the possibility that this ratio can be less than B has been ignored. Thus, by analogy,

$$\frac{P(M, R|\bar{\theta})}{P(M, R|\theta)} = \frac{P(M|\bar{\theta}, R)P(R|\bar{\theta})}{P(M|\theta, R)P(R|\theta)} = B^{h(\theta)} \tag{A.1.12}$$

But, $P(R|\theta) = 1 - P(\theta)$ and $P(R|\bar{\theta}) = 1 - P(\bar{\theta})$ where $P(\theta)$ is as defined in Equation (A.1.10), and $P(\bar{\theta})$ is as defined by the same equation with $h(\theta)$ replaced by $h(\bar{\theta}) = -h(\theta)$. Accordingly,

$$\frac{P(M \mid \bar{\theta}, R)}{P(M \mid \theta, R)} = \frac{1 - P(\theta)}{1 - P(\bar{\theta})} B^{h(\theta)} = 1 \tag{A.1.13}$$

An identical argument holds for the ratio of the distributions $P(M \mid \bar{\theta}, A)$ and $P(M \mid \theta, A)$, and the theorem is proved.

Corollary.

$$P(M \mid R, \theta) = P(M \mid R, \bar{\theta}) = \frac{P(\theta)P(M \mid \bar{\theta}) - P(\bar{\theta})P(M \mid \theta)}{P(\theta) - P(\bar{\theta})} \tag{A.1.14}$$

and

$$P(M \mid A, \theta) = P(M \mid A, \bar{\theta}) = \frac{[1 - P(\bar{\theta})]P(M \mid \theta) - [1 - P(\theta)]P(M \mid \bar{\theta})}{P(\theta) - P(\bar{\theta})}$$

with $P(M \mid \theta)$ the probability that the test terminates after exactly M observations given that the observables are characterized by $p_\theta(y)$, and similarly for $P(M \mid \bar{\theta})$.

Proof. The result follows immediately from Theorem A.2 and the relationship

$$P(M \mid \theta) = P(M \mid \theta, A)P(\theta) + P(M \mid \theta, R)[1 - P(\theta)]$$

The preceding results will now be used to obtain expressions for the first and second moments of the number of observables needed for a sequential decision. Setting $r = 1$ and $r = 2$, respectively, in Equation (A.1.11) yields

$$0 = E\{Z_M e^{hZ_M} \phi^{-M}(h)\} - E\{M e^{hZ_M} \phi^{-(M+1)}(h)\phi'(h)\}$$

and

$$0 = E\{Z_M^2 e^{hZ_M} \phi^{-M}(h)\} - 2E\{M Z_M e^{hZ_M} \phi^{-(M+1)}(h)\phi'(h)\}$$
$$+ E\{M(M+1) e^{hZ_M} \phi^{-(M+2)}(h)[\phi'(h)]^2\} - E\{M e^{hZ_M} \phi^{-(M+1)}(h)\phi''(h)\}$$

Thus, setting $h = 0$ and assuming $E(z \mid \theta) \triangleq E(z_i \mid \theta) \neq 0$, we have

$$E(M \mid \theta) = \frac{E(Z_M \mid \theta)}{E(z \mid \theta)} = \frac{P(\theta) \log_e A + [1 - P(\theta)] \log_e B}{E(z \mid \theta)} \tag{A.1.15}$$

and

$$\operatorname{Var}(M \mid \theta) = \frac{E(M \mid \theta) \operatorname{Var}(z \mid \theta) - \operatorname{Var}(Z_M \mid \theta) + 2[E(MZ_M \mid \theta)E(z \mid \theta) - E^2(Z_M \mid \theta)]}{E^2(z \mid \theta)} \tag{A.1.16}$$

But

$$\operatorname{Var}(Z_M \mid \theta) = P(\theta)[1 - P(\theta)](\log_e A - \log_e B)^2$$

and

$$E(MZ_M \mid \theta) = E(MZ_M \mid \theta, A)P(\theta) + E(MZ_M \mid \theta, R)[1 - P(\theta)]$$
$$= E(M \mid A, \theta)P(\theta) \log_e A + E(M \mid R, \theta)[1 - P(\theta)] \log_e B$$

Thus, from Equations (A.1.14), (A.1.15), and (A.1.16)

$$\operatorname{Var}(M \mid \theta) = \frac{E(M \mid \theta)\operatorname{Var}(z \mid \theta) - P(\theta)[1 - P(\theta)](\log_e A - \log_e B)^2}{E^2(z \mid \theta)} + \frac{2P(\theta)[1 - P(\theta)]}{P(\theta) - P(\bar{\theta})} \frac{(\log_e A - \log_e B)}{E(z \mid \theta)} [E(M \mid \theta) - E(M \mid \bar{\theta})] \tag{A.1.17}$$

Finally, from Equation (A.1.14)

$$E(M \mid R, \theta) = E(M \mid R, \bar{\theta}) = \frac{P(\theta)E(M \mid \bar{\theta}) - P(\bar{\theta})E(M \mid \theta)}{P(\theta) - P(\bar{\theta})}$$

$$E(M \mid A, \theta) = E(M \mid A, \bar{\theta}) = \frac{[1 - P(\bar{\theta})]E(M \mid \theta) - [1 - P(\theta)]E(M \mid \bar{\theta})}{P(\theta) - P(\bar{\theta})} \tag{A.1.18}$$

and

$$\operatorname{Var}(M \mid R, \theta) = \operatorname{Var}(M \mid R, \bar{\theta}) = \frac{\operatorname{Var}(M \mid \bar{\theta})P(\theta) - \operatorname{Var}(M \mid \theta)P(\bar{\theta})}{P(\theta) - P(\bar{\theta})} - \frac{P(\theta)P(\bar{\theta})}{[P(\theta) - P(\bar{\theta})]^2} [E(M \mid \theta) - E(M \mid \bar{\theta})]^2$$

$$\operatorname{Var}(M \mid A, \theta) = \operatorname{Var}(M \mid A, \bar{\theta}) = \frac{\operatorname{Var}(M \mid \theta)[1 - P(\bar{\theta})] - \operatorname{Var}(M \mid \bar{\theta})[1 - P(\theta)]}{P(\theta) - P(\bar{\theta})} - \frac{[1 - P(\theta)][1 - P(\bar{\theta})]}{[P(\theta) - P(\bar{\theta})]^2} [E(M \mid \theta) - E(M \mid \bar{\theta})]^2 \tag{A.1.19}$$

A.2 Efficient and Asymptotically Efficient Estimates

Let $\hat{\alpha} = \hat{\alpha}(x_1, x_2, \ldots, x_n)$ be an estimate of the parameter α based on the identically distributed observables x_i, $i = 1, 2, \ldots, n$. Further, let $p(X \mid \alpha) = p(x_1, x_2, \ldots, x_n \mid \alpha) = p(x_1 \mid \alpha)p(x_2 \mid \alpha) \ldots p(x_n \mid \alpha)$ be the conditional probability density function of the observables x_i given α. In the following manipulations it will frequently be necessary to differentiate both sides of the equation

$$\int_{-\infty}^{\infty} p(X \mid \alpha)\, dX = 1$$

with respect to α. Formally, we obtain

$$\frac{d}{d\alpha} \int_{-\infty}^{\infty} p(X \mid \alpha)\, dX = \int_{-\infty}^{\infty} \left[\frac{\partial \log_e p(X \mid \alpha)}{\partial \alpha}\right] p(X \mid \alpha)\, dX = 0 \tag{A.2.1}$$

If this equation is satisfied for all α in some open interval A, $p(X \mid \alpha)$ is said to be *regular with respect to its first α-derivative in A*. Similarly, if the equation

$$\frac{d^2}{d\alpha^2}\int_{-\infty}^{\infty} p(X \mid \alpha)\, dX = \int_{-\infty}^{\infty}\left[\frac{\partial^2 \log_e p(X \mid \alpha)}{\partial \alpha^2}\right] p(X \mid \alpha)\, dX + \int_{-\infty}^{\infty}\left[\frac{\partial \log_e p(X \mid \alpha)}{\partial \alpha}\right]^2 p(X \mid \alpha)\, dX = 0 \qquad \text{(A.2.2)}$$

is valid for all α in A (with the two integrals on the right both finite), $p(X \mid \alpha)$ is said to be *regular with respect to its second α-derivative in A*. Sufficient conditions for the regularity of $p(X \mid \alpha)$ can be found, for example, in References A.1 and A.5.

Theorem A.3. If $p(X \mid \alpha)$ is regular with respect to its first α-derivative in some open interval A which includes the point α, and if $E(\hat{\alpha} \mid \alpha) = \alpha + b(\alpha)$, then

$$E\{[\hat{\alpha} - \alpha - b(\alpha)]^2 \mid \alpha\} \geq \frac{\left(1 + \dfrac{db(\alpha)}{d\alpha}\right)^2}{nE\left[\left(\dfrac{\partial \log_e p(x \mid \alpha)}{\partial \alpha}\right)^2 \middle| \alpha\right]} \qquad \text{(A.2.3)}$$

with equality if, and only if,

$$\frac{\partial \log_e p(X \mid \alpha)}{\partial \alpha} = k[\hat{\alpha} - \alpha - b(\alpha)] \qquad \text{(A.2.4)}$$

where k is independent of $\hat{\alpha}$ (but not necessarily independent of α).

Proof. Since $E(\hat{\alpha} \mid \alpha) = \alpha + b(\alpha)$

$$\int_{-\infty}^{\infty} [\hat{\alpha} - \alpha - b(\alpha)] p(X \mid \alpha)\, dX = E[\hat{\alpha} - \alpha - b(\alpha) \mid \alpha] = 0 \qquad \text{(A.2.5)}$$

Since $p(X \mid \alpha)$ is regular, we can differentiate under the integral in Equation (A.2.5) obtaining

$$\int_{-\infty}^{\infty} [\hat{\alpha} - \alpha - b(\alpha)]\left[\frac{\partial \log_e p(X \mid \alpha)}{\partial \alpha}\right] p(X \mid \alpha)\, dX = \int_{-\infty}^{\infty}\left(1 + \frac{db(\alpha)}{d\alpha}\right) p(X \mid \alpha)\, dX = 1 + \frac{db(\alpha)}{d\alpha} \qquad \text{(A.2.6)}$$

Squaring both sides of Equation (A.2.6), and using Schwarz's inequality† yields‡

$$\left(1 + \frac{db(\alpha)}{d\alpha}\right)^2 \leq \int_{-\infty}^{\infty} [\hat{\alpha} - \alpha - b(\alpha)]^2 p(X \mid \alpha)\, dX \int_{-\infty}^{\infty}\left[\frac{\partial \log_e p(X \mid \alpha)}{\partial \alpha}\right]^2 p(X \mid \alpha)\, dX$$

$$= E\{[\hat{\alpha} - \alpha - b(\alpha)]^2 \mid \alpha\} \int_{-\infty}^{\infty}\left[\frac{\partial \log_e p(X \mid \alpha)}{\partial \alpha}\right]^2 p(X \mid \alpha)\, dX \qquad \text{(A.2.7)}$$

---

†$[\int f(x)g(x)\, dx]^2 \leq \int f^2(x)\, dx \int g^2(x)\, dx$ with equality if, and only if, $f(x) = kg(x)$ for some k independent of x.

‡This relationship is of interest in itself as it establishes an inequality of the form (A.2.3) but valid even when the variables x_i are not statistically independent.

But, since $p(X \mid \alpha)$, and hence $p(x_i \mid \alpha)$, are regular

$$0 = \frac{\partial}{\partial \alpha} \int_{-\infty}^{\infty} p(x_i \mid \alpha)\, dx_i = \int_{-\infty}^{\infty} \frac{\partial \log_e p(x_i \mid \alpha)}{\partial \alpha} p(x_i \mid \alpha)\, dx_i \tag{A.2.8}$$

and

$$\int_{-\infty}^{\infty} \left[\frac{\partial \log_e p(X \mid \alpha)}{\partial \alpha}\right]^2 p(X \mid \alpha)\, dX = \int_{-\infty}^{\infty} \left[\sum_{i=1}^{n} \frac{\partial \log_e p(x_i \mid \alpha)}{\partial \alpha}\right]^2 p(X \mid \alpha)\, dX$$

$$= \sum_{i=1}^{n} \int_{-\infty}^{\infty} \left[\frac{\partial \log_e p(x_i \mid \alpha)}{\partial \alpha}\right]^2 p(x_i \mid \alpha)\, dx_i \triangleq nE\left[\left(\frac{\partial \log_e p(x \mid \alpha)}{\partial \alpha}\right)^2 \middle| \alpha\right] \tag{A.2.9}$$

the cross-terms all disappearing due to Equation (A.2.8). Combining Equations (A.2.7) and (A.2.9) establishes the desired inequality. Since Schwarz's inequality holds with equality if, and only if, one of the integrands is a constant multiple of the other, the condition for equality here is that

$$\frac{\partial \log_e p(X \mid \alpha)}{\partial \alpha} = k[\hat{\alpha} - \alpha - b(\alpha)] \tag{A.2.10}$$

and the theorem is proved.

Now let $\hat{\alpha}$ be a maximum-likelihood estimate of α; i.e., $\hat{\alpha}$ is a solution to the equation

$$\frac{\partial \log_e p(X \mid \alpha)}{\partial \alpha} = 0 \tag{A.2.11}$$

(Note that $\hat{\alpha}$ is not necessarily the optimum estimate according to the maximum-likelihood principle; e.g., if $\hat{\alpha}$ is restricted to some finite interval, $p(X \mid \alpha)$ may attain its maximum value at one of the end points of that interval.)

The following corollary is an immediate consequence of Theorem A.3.

Corollary. The maximum-likelihood estimate will be efficient if an efficient estimate exists.

Proof. By definition,† the variance of an efficient estimate $\hat{\alpha}$ satisfies Equation (A.2.3) with equality, and with $b(\alpha) = 0$. Thus, if an efficient estimate exists, the equation

$$0 = \frac{\partial \log_e p(X \mid \alpha)}{\partial \alpha} = k[\hat{\alpha} - \alpha] \tag{A.2.12}$$

has the unique solution $\alpha = \hat{\alpha}$.

Theorem A.4. If $p(X \mid \alpha)$ is regular with respect to its second α-derivative over some region A containing the true value α_0 of α, then the maximum-likelihood estimator $\hat{\alpha}$ [Equation (A.2.11)] converges in probability to α_0 as $n \to \infty$. Moreover, $\hat{\alpha}$ is

†See Section 2.9.

an asymptotically efficient estimate of α, and is asymptotically Gaussianly distributed.

Proof. Consider the truncated Taylor series expansion of $\partial \log_e p(X \mid \alpha)/\partial\alpha$ about the point $\alpha = \alpha_0$:

$$\frac{1}{n}\frac{\partial \log_e p(X \mid \alpha)}{\partial \alpha} = B_0 + B_1(\alpha - \alpha_0) \tag{A.2.13}$$

where

$$B_0 = \frac{1}{n}\sum_{i=1}^{n}\left(\frac{\partial \log_e p(x_i \mid \alpha)}{\partial \alpha}\right)_{\alpha=\alpha_0} \quad \text{and} \quad B_1 = \frac{1}{n}\sum_{i=1}^{n}\left(\frac{\partial^2 \log_e p(x_i \mid \alpha)}{\partial \alpha^2}\right)_{\alpha=\alpha^*}$$

for some α^* such that $|\alpha - \alpha^*| \leq |\alpha - \alpha_0|$. The maximum-likelihood estimator $\hat{\alpha}$ is therefore a solution to the equation

$$B_0 + B_1(\alpha - \alpha_0) = 0 \tag{A.2.14}$$

Denoting by η_i and $(1/n)\sigma_i^2$ the mean and variance of the random variables B_i, we have, by the Chebycheff inequality,

$$\Pr\{|B_i - \eta_i| \geq \epsilon\} \leq \frac{1}{n}\frac{\sigma_i^2}{\epsilon^2} \tag{A.2.15}$$

(It will be assumed here that σ_1^2 is finite, although this is not essential to the proof; cf. Reference A.5. The regularity of $p(x|\alpha)$ assures the existence of the other moments of interest, as will be seen presently.) Thus

$$\begin{aligned}
\Pr\{|[B_0 + (\alpha - \alpha_0)B_1] - [\eta_0 + (\alpha - \alpha_0)\eta_1]| \geq \epsilon\} \\
\leq \Pr\{|B_0 - \eta_0| + |(B_1 - \eta_1)(\alpha - \alpha_0)| \geq \epsilon\} \\
\leq \Pr\left\{|B_0 - \eta_0| \geq \frac{\epsilon}{2}\right\} + \Pr\left\{|(B_1 - \eta_1)(\alpha - \alpha_0)| \geq \frac{\epsilon}{2}\right\} \\
\leq \frac{1}{n}\frac{4}{\epsilon^2}\{\sigma_0^2 + (\alpha - \alpha_0)^2\sigma_1^2\}
\end{aligned} \tag{A.2.16}$$

and Equation (A.2.14) converges in probability to the equation

$$\eta_0 + (\alpha - \alpha_0)\eta_1 = 0 \tag{A.2.17}$$

Since $p(x|\alpha)$ is regular with respect to its second α-derivative, it follows from Equations (A.2.1) and (A.2.2) that $\eta_0 = 0$ and, at the point $\alpha^* = \alpha_0$, $\eta_1 = -\sigma_0^2$. Thus there exists some subinterval of A for which Equation (A.2.17) has the unique solution $\alpha = \alpha_0$. The solution to Equation (A.2.14) therefore converges in probability to α_0.

Moreover, letting $\alpha = \hat{\alpha}$ be the actual solution to Equation (A.2.14), and noting that B_1 converges in probability to $-\sigma_0^2$, we have, asymptotically,

$$\hat{\alpha} = \alpha_0 + \frac{B_0}{\sigma_0^2} \tag{A.2.18}$$

But B_0 is the average of n identically distributed random variables each having the mean 0 and variance σ_0^2. Thus, by the central limit theorem, $\hat{\alpha}$ is asymptotically Gaussianly distributed with mean α_0 and variance $1/n\sigma_0^2 = \{nE[(\partial \log_e p(x \mid \alpha)/\partial\alpha)^2_{\alpha=\alpha_0} \mid \alpha_0]\}^{-1}$. By Theorem A.3 $\hat{\alpha}$ is therefore asymptotically efficient.

A.3 Extreme Value Statistics†

Let $x_1, x_2, \ldots, x_n$ be sample values of n independent random variables, all identically distributed according to the density function $p(x)$, and let x_{i_ν} be the νth largest of these samples. The problem is to determine the distribution of x_{i_ν}. Denoting by $P(x)$ the distribution of x,

$$P(x) = \int_{-\infty}^{x} p(\xi)\, d\xi$$

we have

$$\begin{aligned} P_\nu(x) &\triangleq \Pr(x_{i_\nu} \le x) = \sum_{l=n-\nu+1}^{n} \Pr \begin{Bmatrix} \text{exactly } l \text{ variables } x_i \text{ are} \\ \text{less than or equal to } x \end{Bmatrix} \\ &= \sum_{l=n-\nu+1}^{n} \binom{n}{l} [P(x)]^l [1 - P(x)]^{n-l} \end{aligned} \tag{A.3.1}$$

Moreover,

$$p_\nu(x) \triangleq \frac{dP_\nu(x)}{dx} = n\binom{n-1}{\nu-1}[P(x)]^{n-\nu}[1 - P(x)]^{\nu-1} p(x) \tag{A.3.2}$$

The goal here is to obtain expressions for the mean and variance of $x = x_{i_\nu}$, the νth largest of the variables x_i, when n is large. To this end let

$$\xi = n[1 - P(x)] \tag{A.3.3}$$

Then ξ is a random variable with the density function

$$q_\nu(\xi) = \begin{cases} \binom{n-1}{\nu-1}\left(\frac{\xi}{n}\right)^{\nu-1}\left(1 - \frac{\xi}{n}\right)^{n-\nu} & 0 \le \xi \le n \\ 0 & \text{otherwise} \end{cases} \tag{A.3.4}$$

As $n \to \infty$, with ν constant,

$$\left(1 - \frac{\xi}{n}\right)^{n-\nu} \to e^{-\xi}$$

and

$$\frac{1}{n^{\nu-1}}\binom{n-1}{\nu-1} = \frac{\left(1 - \frac{1}{n}\right)\left(1 - \frac{2}{n}\right)\cdots\left(1 - \frac{\nu-1}{n}\right)}{(\nu-1)!}$$

so that

$$\lim_{n\to\infty} q_\nu(\xi) = \frac{\xi^{\nu-1} e^{-\xi}}{(\nu-1)!} \qquad \xi \ge 0 \tag{A.3.5}$$

To find the first two moments of x, we first use Equation (A.3.3) to obtain an expression for x in terms of ξ. We will discover, for the functions

†cf. Reference A.1.

$P(x)$ of interest here, that x will be of the form $x = \alpha + \beta \log_e \xi$. Thus, to determine the moments of x we need the corresponding moments of the random variable $\log_e \xi$. But, when n is large,

$$q_\nu(\xi) \approx \frac{\xi^{\nu-1} e^{-\xi}}{(\nu - 1)!} \qquad \xi \geq 0$$

so that

$$E(\log_e \xi) \approx \frac{1}{(\nu - 1)!} \int_0^\infty \xi^{\nu-1} \log_e \xi \, e^{-\xi} \, d\xi \tag{A.3.6}$$

Recalling the definition of the gamma function

$$\Gamma(\nu) = \int_0^\infty x^{\nu-1} e^{-x} \, dx$$

and observing that $\Gamma(\nu) = (\nu - 1)!$ when ν is a positive integer, we find

$$\Gamma(\nu) E(\log_e \xi) \approx \frac{d\Gamma(\nu)}{d\nu} = \Gamma'(\nu) \tag{A.3.7}$$

Since $\Gamma(\nu) = (\nu - 1)\Gamma(\nu - 1)$, and hence $\Gamma'(\nu) = \Gamma(\nu - 1) + (\nu - 1)\Gamma'(\nu - 1)$, then

$$\begin{aligned} E(\log_e \xi) &\approx \frac{\Gamma'(\nu)}{\Gamma(\nu)} = \frac{1}{\nu - 1} + \frac{\Gamma'(\nu - 1)}{\Gamma(\nu - 1)} \\ &= \frac{1}{\nu - 1} + \frac{1}{\nu - 2} + \cdots + 1 + \frac{\Gamma'(1)}{\Gamma(1)} \end{aligned} \tag{A.3.8}$$

But

$$\frac{\Gamma'(1)}{\Gamma(1)} = \int_0^\infty \log_e x e^{-x} \, dx = -C$$

where C is Euler's constant, $C = 0.5772 \ldots$. Consequently,

$$E(\log_e \xi) \approx \sum_{i=1}^{\nu-1} \frac{1}{i} - C \tag{A.3.9}$$

Similarly,

$$E[(\log_e \xi)^2] \approx \frac{\Gamma''(\nu)}{\Gamma(\nu)} = \frac{1}{(\nu - 1)!} \int_0^\infty \xi^{\nu-1} \log_e^2 \xi \, e^{-\xi} \, d\xi$$

and

$$\begin{aligned} \operatorname{Var}(\log_e \xi) &\approx \frac{\Gamma''(\nu)}{\Gamma(\nu)} - \left(\frac{\Gamma'(\nu)}{\Gamma(\nu)}\right)^2 = \frac{\Gamma''(\nu - 1)}{\Gamma(\nu - 1)} - \left(\frac{\Gamma'(\nu - 1)}{\Gamma(\nu - 1)}\right)^2 - \frac{1}{(\nu - 1)^2} \\ &= \cdots = \frac{\Gamma''(1)}{\Gamma(1)} - \left(\frac{\Gamma'(1)}{\Gamma(1)}\right)^2 - \sum_{i=1}^{\nu-1} \frac{1}{i^2} \\ &= \frac{\pi^2}{6} - \sum_{i=1}^{\nu-1} \frac{1}{i^2} = \sum_{i=\nu}^{\infty} \frac{1}{i^2} \end{aligned} \tag{A.3.10}$$

We now investigate the asymptotic distribution of the νth largest of a sample of size n for two cases of special interest.

(1) The Gaussian distribution

Suppose the samples $x_1, x_2, \ldots, x_n$ are Gaussianly distributed with zero mean and unity variance. Then

$$\xi = \frac{n}{\sqrt{2\pi}} \int_x^\infty e^{-t^2/2}\, dt = \frac{n}{\sqrt{2\pi}} \frac{1}{x} e^{-x^2/2}\left(1 + O\left(\frac{1}{x^2}\right)\right)$$

and it can be verified that

$$x = \sqrt{2 \log_e n} - \frac{\log_e \log_e n + \log_e 4\pi + 2 \log_e \xi}{2\sqrt{2 \log_e n}} + O\left(\frac{1}{\log_e n}\right) \quad \text{(A.3.11)}$$

If y is the νth largest of the Gaussian variables $y_1, y_2, \ldots, y_n$, all having the mean m and variance σ^2, then $x = (y - m)/\sigma$ satisfies Equation (A.3.11). Consequently, using Equations (A.3.9), (A.3.10), and (A.3.11), we find for the mean and variance of the νth largest of n samples (with n large) from a Gaussian distribution (with mean m and variance σ^2) the values

$$E(y) = m + \sigma\left\{\sqrt{2 \log_e n} - \frac{\log_e \log_e n + \log_e 4\pi + 2(A_\nu - C)}{2\sqrt{2 \log_e n}} + O\left(\frac{1}{\log_e n}\right)\right\}$$

$$\text{Var}(y) = \frac{\sigma^2}{2 \log_e n}\left(\frac{\pi^2}{6} - B_\nu\right) + O\left(\frac{1}{\log_e n}\right) \quad \text{(A.3.12)}$$

where

$$A_\nu = \sum_{i=1}^{\nu-1} \frac{1}{i}, \qquad B_\nu = \sum_{i=1}^{\nu-1} \frac{1}{i^2} \qquad A_1 = B_1 = 0 \quad \text{and} \quad C = 0.5772\ldots$$

(2) The chi-squared distribution (two degrees of freedom)

If t is a chi-squared distributed random variable,

$$p(t) = \begin{cases} \dfrac{1}{2\sigma^2} e^{-t/2\sigma^2} & t \geq 0 \\ 0 & \text{otherwise} \end{cases} \quad \text{(A.3.13)}$$

we have

$$\xi = \frac{n}{2\sigma^2} \int_z^\infty e^{-t/2\sigma^2}\, dt = n e^{-z/2\sigma^2}$$

and, hence,

$$z = 2\sigma^2(\log_e n - \log_e \xi) \quad \text{(A.3.14)}$$

Thus, if z is the νth largest of the chi-squared distributed random variables $z_1, z_2, \ldots, z_n$,

$$E(z) = 2\sigma^2(\log_e n + C - A_\nu) \quad \text{(A.3.15a)}$$

and

$$\text{Var}(z) = 4\sigma^4\left(\frac{\pi^2}{6} - B_\nu\right) \quad \text{(A.3.15b)}$$

where A_ν, B_ν, and C are as previously defined.

References and Supplementary Bibliography

A.1 Cramér, H., *Mathematical Methods of Statistics*, Princeton Univ. Press, 1946.

A.2 Deutsch, R., *Estimation Theory*, Prentice-Hall, Inc., Englewood Cliffs, N. J., 1965.

A.3 Lehman, E. L., *Testing Statistical Hypotheses*, John Wiley, N. Y., N. Y., 1959.

A.4 Wald, A., *Sequential Analysis*, John Wiley, N. Y., N. Y., 1947.

A.5 Wilks, S. S., *Mathematical Statistics*, John Wiley, N. Y., N. Y., 1962.

appendix b

Some Mathematical Concepts Related to Coding Theory

B.1 Introduction

The emphasis in this appendix is on those algebraic concepts which have direct bearing on coding theory. Consequently, these concepts will be developed only so far as required for this particular application. To facilitate the discussion, we begin by defining some terms.

Let S be a *set* of objects $s_1, s_2, \ldots, s_n$ where n may or may not be finite. Define an *operator* f on pairs of elements in S. The set S is said to be *closed* under f if the element produced by the operation of f on any two elements in the set is also an element in the set: $f(s_i, s_j) = s_k$ where s_k is in S if s_i and s_j are in S. The set S contains an *identity* ϵ under f if $f(s_i, \epsilon) = f(\epsilon, s_i) = s_i$ for all s_i in S. The *inverse* $s_{i'}$ of an element s_i in S has the property that $f(s_i, s_{i'}) = f(s_{i'}, s_i) = \epsilon$. The elements of S satisfy the *associative law* under f if $f[f(s_i, s_j), (s_k)] = f[s_i, f(s_j, s_k)]$. The elements of S are *commutative* under f if $f(s_i, s_j) = f(s_j, s_i)$.

If a set contains an identity element ϵ under some operation f then ϵ must be unique. For suppose the contrary, that S contains two identities ϵ_1 and ϵ_2. Then $f(\epsilon_1, \epsilon_2) = \epsilon_1 = \epsilon_2$. Similarly, if s is in a set S which satisfies the associative law and s has an inverse in S it must also be unique. Again assume the contrary, that both s_1 and s_2 are inverses of s. Then $s_1 = f(s_1, \epsilon) = f[s_1, f(s, s_2)] = f[f(s_1, s), s_2] = f(\epsilon, s_2) = s_2$.

For an example illustrating these concepts let S_1 be the infinite set of all real numbers and S_2 the infinite set of all integers (positive *and* negative), and let f_1 be the operation of conventional addition, $f_1(s_1, s_2) = s_1 + s_2$, and f_2 the operation of conventional multiplication, $f_2(s_1, s_2) = s_1 s_2$. Then both S_1 and S_2 are closed under both f_1 and f_2. The identity in both S_1 and S_2 is 0 under f_1 and 1 under f_2. Every element in both S_1 and S_2 has an inverse in their respective sets under f_1 but only those elements in S_1 (with the exception of the additive identity 0) contain their own inverses under f_2. Finally, elements of both S_1 and S_2 satisfy the associative and commutative laws under both operators f_1 and f_2.

B.2 Groups, Subgroups, and Cosets

A group G is a set of elements which, under the operation f, is closed, contains an identity element, and in which each element has an inverse and all elements obey the associative law. The group is *Abelian* if, in addition, the elements of G are commutative under f. The *order* of a group simply denotes the number of elements it contains. A subset H of a group G is called a *subgroup* if it satisfies all the axioms of the group.

Let $G = \{g_1, g_2, \ldots, g_n\}$ be a group and $H = \{h_1, h_2, \ldots, h_m\}$ a subgroup of G, and, for convenience, let $f(a, b) = ab$ denote the group operation. Then if g is an element in G but not in H, the set $gH = \{gh_1, gh_2, \ldots, gh_m\}$ is referred to as a (left) *coset* of H and g as the *coset leader*. (Similarly Hg defines a right coset of H; if G is Abelian, $Hg = gH$.)

Theorem B.1. Two elements g_1 and g_2 in G belong to the same left coset of H if, and only if, $g_1^{-1}g_2$ is in H, g_1^{-1} denoting the inverse of g_1.

Proof. Let g_1 and g_2 be in the coset $H_1 = \{gh_1, gh_2, \ldots, gh_m\}$ of H. Then $g_1 = gh_i$ and $g_2 = gh_j$ for some i and j. Further, $g_1^{-1} = h_i^{-1}g^{-1}$, since $h_i^{-1}g^{-1}gh_i = h_i^{-1}h_i$ is the identity. Consequently, $g_1^{-1}g_2 = h_i^{-1}g^{-1}gh_j = h_i^{-1}h_j$ which is in H. Conversely, if $g_1^{-1}g_2 = h_k$ for some h_k in H, then $g_2 = g_1h_k$ and since H must contain the identity ϵ, $g_2 = g_1h_k$ and $g_1 = g_1\epsilon$ are both in the coset having g_1 as a coset leader.

Theorem B.2. If two cosets of H contain any elements in common, they are identical.

Proof. Let $H_i = \{g_ih_1, g_ih_2, \ldots, g_ih_m\}$, $i = 1, 2$, represent two cosets of H and assume $g_1h_l = g_2h_j$. Then $g_1 = g_2h_jh_l^{-1} = g_2h_k$, with $h_k = h_jh_l^{-1}$ an element in H, and every element g_1h_μ in H_1 is in H_2 since $g_1h_\mu = g_2h_kh_\mu = g_2h_\nu$.

As an example, the set of all integers clearly forms a group under addition, with 0 the identity element. Now consider a subgroup G of this group.

If the subgroup contains some integer a, it must also contain all multiples of a, for if a is in G, so also is $a + a = 2a$, $a + 2a = 3a$, etc. All elements, in fact, in any subgroup of integers must be multiples of its smallest positive integer. To demonstrate this, we first observe that if s and r are two positive integers, s can be written in the form $s = ar + b$ with a the largest integer such that $ar \leq s$, and with b some integer in the range $0 \leq b < r$. Furthermore a and b are both uniquely defined by this relationship. For if $s = ar + b = a'r + b'$, then $(a - a')r = b' - b$, and since $-r < b' - b < r$, $a = a'$ and hence $b = b'$. Now let r be the smallest positive integer in some subgoup G of integers and let s be an arbitrary integer in G. (With no loss of generality s can be assumed to be positive, since any subgroup containing s must also contain its inverse, $-s$.) Then G must also contain $s - ar = b$ with $0 \leq b < r$. But since r was by hypothesis the smallest positive integer in G, b must equal zero and s is a multiple of r. Since these observations will prove useful in the sequel, we restate them as theorems.

Theorem B.3. (Euclidean Division Algorithm). Let s and r be two positive integers Then $s = ar + b$ with a a unique non-negative integer and b a unique integer in the range $0 \leq b < r$.

Theorem B.4. If G is a subgroup of integers, all elements in G are multiples of its smallest positive integer.

If t is an integer and s an integer multiple of t, t is said to *divide* s. The largest integer t which divides both r and s is called the *greatest common divisor* of r and s and is usually written as $t = (r, s)$. If $(r, s) = 1$, r and s are *relatively prime*. If no integer other than 1 and s divides s, s is a *prime*. The following is an immediate corollary of Theorem B.4.

Corollary. If r and s are relatively prime, there exist integers a and b such that $ar + bs = 1$.

Proof. Consider the subgroup containing all integers of the form $ar + bs$, with a and b arbitrary integers. From Theorem B.4, the smallest positive integer in this subgroup must divide both r and s. Since r and s are relatively prime, this integer must be 1, and the corollary follows.

This leads to a second important corollary.

Corollary. (Unique factorization). Any integer $n > 1$ can be factored uniquely as a product of primes.

Proof. If a prime p divides a product of two integers, say rs, it must divide at least one of these integers. For suppose p divides rs, but not r. Then for some pair of integers a and b, $ar + bp = 1$, and $ars + bps = s$. Since p divides the left side of this equation, it must divide the right. By extension, if p divides an arbitrary product of integers, it must divide at least one of the factors in this product. Now, suppose some integer n can be written as a product of primes in two different ways: $n = p_1 p_2 \ldots p_k = p'_1 p'_2 \ldots p'_{k'}$. Since every prime on the left side of this last equation must divide some prime on the right, the two sides must be identical except for ordering.

Before concluding this section, we prove one more statement concerning groups which will be of subsequent use. Let g be an element in some group G, with the group operation again represented in multiplicative notation. Then $gg \triangleq g^2, gg^2 \triangleq g^3$, etc., are also elements in G. If the order of G is finite, therefore, there must be some integer ν such that $g^\nu = \epsilon$. The smallest such integer is called the *order* of the element g.

Theorem B.5. Let ν be the maximum of the orders of the elements in an Abelian group G. Then if μ is the order of some element in G, μ must divide ν. Further, ν divides n, the order of G.

Proof. Since the set of integers m for which $g^m = \epsilon$ clearly forms a group under addition, all such integers must, from Theorem B.4, be multiples of the order of g. Now suppose G contains two elements, g_1 and g_2, of relatively prime orders ν_1 and ν_2, respectively. Then the element $g = g_1 g_2$ is of order $\nu = \nu_1 \nu_2$. This follows because: (1) $g^{\nu_1 \nu_2} = \epsilon$, so ν must divide $\nu_1 \nu_2$; (2) $\epsilon = (g^\nu)^{\nu_1} = (g^{\nu_1})^\nu = g_2{}^{\nu_1 \nu}$ and ν_2 must divide $\nu_1 \nu$; (3) by symmetry, ν_1 must divide $\nu_2 \nu$. But since $(\nu_1, \nu_2) = 1$, statements (2) and (3) are both possible only if ν is a multiple of $\nu_1 \nu_2$, and hence, from statement (1), $\nu = \nu_1 \nu_2$.

Now let ν be the maximum order of any element in G and assume there exists some element in G of order μ not dividing ν. Then there must be some prime p such that $\nu = ap^i$ and $\mu = bp^j$ with $j > i \geq 0$ and with p not a factor of a. Were this not the case, ν would obviously be a multiple of μ. Thus, if $g_1^\nu = \epsilon = (g_1^{p^i})^a$, with g_1 in G, G must also contain an element of order a. Similarly, if $\epsilon = g_2^\mu = (g_2^b)^{p^j}$, with g_2 in G, G contains an element of order p^j. But since p^j and a are relatively prime, G must also contain an element $g_3 \triangleq g_1^{p^i} g_2^b$ of order $ap^j > \nu$ which was by hypothesis the largest order of any element in G. Thus, the order of any element in G must divide ν.

To show that ν divides n, the order of G, we observe that, if ν is the order of the element g in G, the set $G_1 = \{\epsilon, g, g^2, g^3, \ldots, g^{\nu-1}\}$ is a subgroup of G. Let g' be any element in G but not in G_1. Then $g'G_1 = \{g', g'g, g'g^2, \ldots, g'g^{\nu-1}\}$ is a coset of G_1 containing no elements in common with G_1 (Theorem B.2). Further, if g'' is any element in G but in neither G_1 nor $g'G_1$, we can form a second coset of G_1 having no elements in common with either G_1 or $g'G_1$. We can, in fact, continue this procedure, forming new cosets of G_1 until all elements in G are contained uniquely in some coset. Since each coset contains ν elements and since each of the n elements of G is in one and only one coset, ν must divide n, as was to be shown.

B.3 Galois Fields

A *field F* is a set of elements for which *two* operators are defined, addition and multiplication (although not necessarily conventional addition and multiplication): $f_1(s_1, s_2) = s_1 + s_2$ and $f_2(s_1, s_2) = s_1 s_2$. The set of elements in a field F forms an Abelian group under addition and the set of elements in F excluding the additive identity (denoted 0) form a non-empty Abelian group under multiplication. They also satisfy the *distributive* law under f_1 and f_2; $s_1(s_2 + s_3) = s_1 s_2 + s_1 s_3$.

Note that even though 0 does not have a multiplicative inverse, the product $0s = s0$ is defined and is in F. Specifically, $0s = s0 = 0$ for all s in F. This follows because for any s_1, s_2 in F, $s_1 s_2 = s_1(s_2 + 0) = s_1 s_2 + s_1 0$, so $s_1 0$ must equal 0.

As an example we have essentially already verified that the real numbers form a field under conventional multiplication and addition. Similarly, the rationals and the complex numbers also constitute fields. Since a field must possess an additive identity (0) and at least one other element, it contains a minimum of two elements. That such a field exists is demonstrated in Tables B.1 (a) and (b).

Table B.1 The binary field

	0	1
0	0	1
1	1	0

(a) Addition

	0	1
0	0	0
1	0	1

(b) Multiplication

This, of course, is the familiar binary field.

Consider now the arithmetic of integers modulo n. That is, only n integers $0, 1, \ldots, n - 1$ are recognized, the sum or product of any two of these integers identified with its remainder when divided by n. When $n = 2$ we have the binary field; when $n = 3$, Tables B.2 (a) and (b) apply

Table B.2 The ternary field

	0	1	2
0	0	1	2
1	1	2	0
2	2	0	1

(a) Addition

	0	1	2
0	0	0	0
1	0	1	2
2	0	2	1

(b) Multiplication

and clearly satisfy the conditions defining a field, in this case, the ternary field. The integers modulo four, however, yield Tables B.3 (a) and (b)

Table B.3 The integers modulo four

	0	1	2	3
0	0	1	2	3
1	1	2	3	0
2	2	3	0	1
3	3	0	1	2

(a) Addition

	0	1	2	3
0	0	0	0	0
1	0	1	2	3
2	0	2	0	2
3	0	3	2	1

(b) Multiplication

Since 2 does not have a multiplicative inverse, the integers modulo four do not form a field. It is easy to show, in fact, that the integers modulo any non-prime n do not form a field. For let $n = r \cdot s$ where r and s are both integers in the range $(2, n - 1)$. If the integers modulo n are to form a field, r in particular must have a multiplicative inverse r^{-1}. Thus, $r^{-1} \cdot 0 = 0 = r^{-1} \cdot r \cdot s = s$ which is a contradiction. As we shall show presently, q element fields do exist for finite values of q if, and only if, q is a prime power; i.e., if, and only if, $q = p^m$ for some prime p and integer m. Such fields are called *Galois fields* and are denoted $GF(q)$. The integers modulo p^m do not form such a field, however, unless $m = 1$.

To verify that the integers modulo a prime p do form a field, we need only to demonstrate the existence of a multiplicative inverse for each such integer. The other conditions are trivially satisfied. But the existence of a multiplicative inverse is an immediate consequence of Theorem B.4. For consider any integer r, $1 \leq r \leq p - 1$. The first corollary of Theorem B.4 asserts the existence of two integers a and b such that $ar + bp = 1$, so $ar = 1$, modulo p, and a is the multiplicative inverse of r, modulo p. Since r is an arbitrary integer in the range $(1, p - 1)$, all non-zero integers modulo p have multiplicative inverses and the set of such integers is a field.

Thus, p element fields do exist for all primes p. The key needed to demonstrate the existence of p^m element fields, with $m > 1$, can be found by examining polynomials of the form

$$f(x) = \sum_{i=0}^{k} a_i x^i \tag{B.3.1}$$

where the coefficients a_i are elements in some finite field. As we shall verify, polynomials with coefficients in a field of q elements (called the *ground* field) can be used to construct a field containing q^k elements (an *extension* field) for any positive integer k. Since the integers modulo a prime p define a field of p elements, this construction demonstrates the existence of fields containing p^m elements for all integers m and all primes p.

In the following discussion, the terms polynomial addition and polynomial multiplication will retain their conventional meanings, but with the proviso that the arithmetic operations on the coefficients are carried out over some finite field F. Moreover, if the words "polynomial" and "degree"† are substituted for "integer" and "magnitude," respectively, Theorem B.3 is easily seen to be valid for polynomials as well as integers. Similarly, Theorem B.4 also applies to polynomials when the term "subgroup" is replaced by "ideal," an *ideal* of polynomials defined as a subgroup closed with respect to the operation $a(x)r(x) + b(x)s(x)$ where $r(x)$ and $s(x)$ are any two polynomials in the ideal and $a(x)$ and $b(x)$ are arbitrary polynomials over F. We restate these theorems here as applied to polynomials.

Theorem B.3′. If $s(x)$ and $r(x)$ are polynomials with respective degrees m and k, $m, k > 0$, there exist unique polynomials $a(x)$ and $b(x)$ such that $s(x) = a(x)r(x) + b(x)$, with $b(x)$ of degree less than k.

Theorem B.4′. If I is an ideal, all polynomials in I are multiples of any one of its minimum degree polynomials.

As an illustration of polynomial operations over a finite field, the following equations are all identities when the coefficient field F is the binary field: $(x + 1)^2 = x^2 + 1$; $(x + 1) + x = 1$; $x^3 + x + 1 = (x^2 + x)(x + 1) + 1$. When F is the ternary field the corresponding identities become: $(x + 1)^2 = x^2 + 2x + 1$; $(x + 1) + x = 2x + 1$, $(x^3 + x + 1) = (x^2 + 2x + 2)(x + 1) + 2$.

Now consider the polynomial

$$x^7 + 1 = (x + 1)(x^3 + x + 1)(x^3 + x^2 + 1) \qquad \text{(B.3.2)}$$

where the coefficients are elements in the binary field. Evidently, $x^7 + 1$ can be factored as the product of three polynomials over the binary field; however, as is readily verified, none of these three factors can be further factored over this field. The polynomial $x^7 + 1$ is called *reducible* since it can be factored while the three polynomials on the right side of Equation (B.3.2) are called *irreducible*. (The reducibility concept is applicable to polynomials over any field, finite or infinite. The polynomial $(x^2 - 1) = (x + 1)(x - 1)$, for example, is reducible over the field of real numbers while the polynomial $x^2 + 1$ is irreducible over the same field.)

The leading coefficient (the coefficient of the highest power of x) of any binary polynomial $f(x)$ is obviously 1. But regardless of the coefficient field, it is always possible to find some field element α such that $f(x) = \alpha f'(x)$ where $f'(x)$ has leading coefficient 1. Such polynomials are called *monic*.

†If the highest power of x in the polynomial $f(x)$ is x^m, $f(x)$ is said to be of *degree m*.

Irreducible monic polynomials play a role in the set of polynomials over F analogous to that of primes in the set of integers. In particular, the following two corollaries to Theorems B.3′ and B.4′ are entirely analogous to those of their integer counterparts. (The coefficients of all polynomials in the statements of these corollaries are defined over the same finite field F.)

Corollary. If $r(x)$ and $s(x)$ are relatively prime [i.e., if no polynomial of positive degree divides both $r(x)$ and $s(x)$], there exist two polynomials $a(x)$ and $b(x)$ such that $a(x)r(x) + b(x)s(x) = 1$.

Corollary. Any monic polynomial is uniquely factorable as a product of irreducible monic polynomials.

Suppose $f(x)$ is an irreducible polynomial of degree m with coefficients in a field of q elements and consider the set S of polynomials defined over this same field modulo the polynomial $f(x)$. [The polynomial $s(x)$ modulo $f(x)$ is equal to its remainder when divided by $f(x)$; if $s(x) = p(x)f(x) + b(x)$, then $s(x) = b(x)$, modulo $f(x)$. As already noted, $b(x)$ is uniquely defined by this relationship.] There are q^m distinct polynomials modulo the polynomial $f(x)$ of degree m. For any two polynomials of the form $s_1(x) = \sum_{i=0}^{m-1} a_i x^i$ and $s_2(x) = \sum_{i=0}^{m-1} b_i x^i$ are distinct modulo $f(x)$ unless $a_i = b_i$ for all i, while any polynomial of degree $n \geq m$ is equal modulo $f(x)$ to some polynomial of degree less than m. Obviously, if $s_1(x)$ and $s_2(x)$ are in S, then both $s_1(x)s_2(x)$ and $s_1(x) + s_2(x)$ are in S, so S is closed under both addition and multiplication.

Associativity and commutativity of the elements of S are also obviously satisfied under both addition and multiplication and the distributive law clearly holds. The identity under addition is 0 while the multiplicative identity is 1 (the additive and multiplicative identitites, respectively, in the coefficient field). The additive inverse of $s(x)$ is $-s(x)$ and must also be in S since -1 is the additive inverse of 1 and hence is in the coefficient field F. The only other condition to be imposed on S if it is to be a field is that it contain multiplicative inverses for all of its elements except 0. This condition is also satisfied by S as the reader can verify by an argument analogous to that used to demonstrate the existence of multiplicative inverses in the set of integers modulo p. [Since $f(x)$ is irreducible, there exist two polynomials $a(x)$ and $b(x)$ such that $1 = a(x)r(x) + b(x)f(x)$ for any polynomial $r(x)$. Consequently, $r(x)$ must have an inverse in S; $r(x)a(x) = 1$, modulo $f(x)$.] Thus, S is a field containing q^m elements. Note that we have not yet demonstrated the existence of p^m element fields for all integers m since we have only assumed the existence of an irreducible polynomial of degree m over the ground field.

Now consider any field K containing both the q-element ground field of $f(x)$ and one of its *roots* α (i.e., an element α such that $f(\alpha) = 0$). Certainly K must contain all elements of the form $\sum_{i=0}^{m-1} a_i\alpha^i$ with the a_i arbitrary elements in the ground field. All q^m of these elements must be distinct, since, were this not true, α would also be a root of a polynomial of degree less than m, an impossibility if $f(x)$ is irreducible (i.e., the set of polynomials for which α is a root clearly forms an ideal, so all such polynomials must be multiples of any one of the polynomials of minimum degree in this ideal. Since $f(x)$ is irreducible, this minimum degree cannot be less than m). Moreover, if $s(\alpha)$ denotes a polynomial in α, if $\beta_1 = s_1(\alpha)$, and $\beta_2 = s_2(\alpha)$, $\beta_3 \triangleq \beta_1\beta_2 = s_1(\alpha)s_2(\alpha) = a(\alpha)f(\alpha) + b(\alpha) = b(\alpha)$. Thus, K contains as a subfield a set of elements which are in one-to-one correspondence with the elements of the field of polynomials modulo $f(x)$, and which obey the same rules of addition and multiplication. Such fields are said to be *isomorphic*.

In order to determine the values of r for which an r-element field can exist, we first itemize, in the following three lemmas, some of the more important consequences of the field axioms as they relate to finite fields.

Lemma B.1. Any r element field K must contain as a subfield elements isomorphic to the integers modulo some prime divisor of r.

Proof. Both 0 and 1 are elements in any field, and hence so are $1 + 1 = 2$, $2 + 1 = 3$, etc. Since the field is finite, there must be some integer i such that $i + 1 = 0$. The set of integers $(0, 1, 2, \ldots, i)$ therefore forms an additive subgroup of the field, which is itself an additive group of r elements. From Theorem B.5, $i + 1$ divides r. Further, since this set of integers is closed under both addition and multiplication (multiplication of integers being equivalent to repeated addition), it is also a subfield of the field K. Consequently, $i + 1$ must be a prime p, and the field operations are those of ordinary arithmetic with the result reduced modulo p.

A field containing a subfield isomorphic to the field of integers modulo p is said to be of *characteristic p*.

Lemma B.2. In a field K of characteristic p, $(\alpha + \beta)^q = \alpha^q + \beta^q$ where α and β are any two elements in K, and q is any power of p.

Proof. This follows upon expanding $(\alpha + \beta)^p$ via the binomial expansion, $(\alpha + \beta)^p = \sum_{i=0}^{p} \binom{p}{i}\alpha^i\beta^{p-i}$, and noting that the coefficient of every term in this expansion except the first and the last is a multiple of p. But $p = 0$ in a field of characteristic p, so $(\alpha + \beta)^p = \alpha^p + \beta^p$, $(\alpha + \beta)^{p^2} = (\alpha^p + \beta^p)^p = \alpha^{p^2} + \beta^{p^2}$, and, in general, $(\alpha + \beta)^q = \alpha^q + \beta^q$.

Lemma B.3. Any element in an r-element field K is a root of the polynomial $x^r - x$.

Proof. The non-zero elements in K form a multiplicative Abelian group of order $r - 1$. Let e be the maximum of the multiplicative orders of any non-zero element in K. Then, from Theorem B.5, e divides $r - 1$ and is a multiple of the multiplicative order of any other non-zero element in K. Thus, if α is a non-zero element in K, $\alpha^e = \alpha^{r-1} = 1$, and $\alpha^r - \alpha = 0$ for any element α in K, including $\alpha = 0$.

While e must divide $r - 1$, this does not preclude the possibility that $e = r - 1$. Any element in $GF(r)$ having multiplicative order $r - 1$ is called *primitive*.

With these preliminaries, we can now prove the following theorem.

Theorem B.6. An r element finite field exists if, and only if, $r = p^m$ with m a positive integer and p a prime. Moreover, all finite fields contain a primitive element, and any two such fields having the same number of elements are isomorphic.

Proof. If an r-element field K exists, all its elements are roots of the polynomial $x^r - x$ (Lemma B.3). Let p be the characteristic of K and factor $x^r - x$ over $GF(p)$. If one of these factors, say $f_1(x)$, has degree $m_1 > 1$, it can be used to construct a larger field K_1 containing p^{m_1} elements including some root α_1 of $f_1(x)$. Clearly, $x^r - x$ can be factored over K_1 as $(x - \alpha_1)g_1(x)$, with $g_1(x)$ a polynomial of degree $r - 1$. By extending this argument, using an irreducible factor of $g_1(x)$ over K_1 to construct a still larger field, etc., we eventually find a $p^{m'}$-element field K' over which all the irreducible factors of $x^r - x$ are linear: $x^r - x = \prod_{i=1}^{r} (x - \alpha_i)$. Moreover, the r roots α_i will be distinct. For suppose two (or more) of them were identical. Then $x^r - x$ would be of the form $(x - \alpha)^2q(x)$ with α the duplicated root, and the derivative, $rx^{r-1} - 1$, of $x^r - x$ would also have α as a root. But since r must be a multiple of the field characteristic p, $r = 0$ in K', so $rx^{r-1} - 1 = -1$ and hence has no roots.

We next observe that the roots of $x^{r-1} - 1$ form a multiplicative Abelian group. [If $\alpha^{r-1} = 1$ and $\beta^{r-1} = 1$, then $(\alpha\beta)^{r-1} = 1$. The identity element 1 is clearly also a root of $x^{r-1} - 1$.] Hence, all roots have some order dividing the maximum order e of any root. Moreover, $e = r - 1$ since otherwise all roots of $x^{r-1} - 1$ would also be roots of $x^e - 1$, with $e < r - 1$, contradicting the earlier observation that $x^r - x$ has r distinct roots. Any r-element field therefore contains a primitive element α whose first $r - 1$ powers account for all non-zero roots of $x^r - x$.

Since $x^r - x$ has a root α of order $r - 1$, one of its irreducible factors over $GF(p)$ must also have this root. This polynomial can therefore be used to construct a p^m element field containing α, and, as a consequence, all of the roots of $x^r - x$. Furthermore this field is isomorphic to a subfield of any field containing both α and $GF(p)$. Thus, the r-element field K, if it exists, must be a subfield of a p^m-element field. Since the elements of a field form an additive Abelian group, the elements of K form a subgroup of a p^m-element group, and r must divide p^m (Theorem B.5). An r-element field, therefore, can exist only if $r = p^k$ for some integer k.

Finally, if $r = p^m$ with m any positive integer, the roots of $x^r - x$ do form a field. The only field axiom not obviously satisfied by these roots is that they form an additive group. But if α and β are both roots of $x^{p^m} - x$, $(\alpha + \beta)^{p^m} = \alpha^{p^m} + \beta^{p^m} = \alpha + \beta$ (Lemma B.2) so the set of roots is indeed closed under addition. That any two q-element fields are isomorphic follows from the fact that any *primitive* factor of x^q-x (i.e., any irreducible factor having a primitive root α) generates a field F isomorphic to a subfield of any field containing both α and the ground field. Since every non-zero element in any q-element field K is some power of α, K is contained in F, and hence must be isomorphic to it. (Note that this also assures the existence of an irreducible polynomial of every degree over $GF(p)$.)

B.4 Polynomials over Finite Fields

The previous section was concerned with demonstrating the existence of a Galois field $GF(q)$ for every prime power q. In the course of the discussion, use was made of the fact that the polynomial $x^{q^m} - x$ could be factored into a product of irreducible polynomials over $GF(q)$. Since this fact is of major importance in the study of error-correcting codes, we now examine more thoroughly the details of this factorization.

Theorem B.7. Let $p(x)$ be a polynomial of degree m irreducible over the ground field $GF(q)$ and let α be a root of $p(x)$. Then $\alpha, \alpha^q, \alpha^{q^2}, \ldots, \alpha^{q^{m-1}}$ are all roots, and are the only roots, of $p(x)$.

Proof. If $p(x) = \sum_{i=0}^{m} a_i x^i$, then, from Lemma B.2,

$$[p(x)]^q = \left(\sum_{i=0}^{m} a_i x^i\right)^q = \sum_{i=0}^{m} a_i^q (x^q)^i = \sum_{i=0}^{m} a_i (x^q)^i = p(x^q)$$

The next to the last step follows because the coefficients a_i are in $GF(q)$ and hence, $a_i^q = a_i$. If $p(\alpha) = 0$, $[p(\alpha)]^q = 0 = p(\alpha^q)$ and α^q is also a root of $p(x)$. The elements $\alpha, \alpha^q, \alpha^{q^2}, \ldots, \alpha^{q^{m-1}}$ are consequently all roots of $p(x)$, and since $p(x)$ has at most m roots, it is only necessary to verify that these elements are distinct to complete the proof of the theorem. Suppose, to the contrary, that $\alpha^{q^i} = \alpha^{q^j}$ for some $0 \leq i < j \leq m - 1$. Then, since α is an element in $GF(q^m)$,

$$\alpha = \alpha^{q^m} = (\alpha^{q^j})^{q^{m-j}} = (\alpha^{q^i})^{q^{m-j}} = \alpha^{q^{m-j+i}}$$

and α is also an element in $GF(q^{m-j+i})$. This field is isomorphic to the field of polynomials modulo an irreducible polynomial of degree $m - j + i < m$. If this were true, α would be a root of such a polynomial and $f(x)$ would not be irreducible.

Corollary. All roots of an irreducible polynomial over $GF(q^m)$ have the same order.

Proof. Let α be one of the roots of maximum order e of the irreducible polynomial $f(x)$. All other roots of $f(x)$ are of the form α^{q^j}. Suppose one of these roots has order $e' < e$. Then $\alpha^{q^j e'} = 1$ and, from Theorem B.4, e must divide $q^j e'$. Since $e' < e$, e has a factor in common with q^j; i.e., since $q = p^k$, e has p as a factor. But e must also divide $q^m - 1$, and since p and $q^m - 1$ are relatively prime, this results in a contradiction.

Theorem B.8. The polynomial $x^{q^m} - x$ factors over $GF(q)$ uniquely as the product of all irreducible monic polynomials in $GF(q)$ whose degree divides m.

Proof. An irreducible polynomial $p(x)$ is a factor of $x^{q^m} - x$ if, and only if, all its roots are also roots of $x^{q^m} - x$. Consider any such polynomial $p(x)$ of degree k over $GF(q)$. This polynomial generates a q^k-element field isomorphic to the field containing $GF(q)$ and any root of $p(x)$. All roots β of $p(x)$ must be in this field and satisfy the relationship $\beta^{q^k} = \beta$. Now the set of integers i for which $\beta^{q^i} = \beta$ clearly form an additive Abelian group so all such integers must be multiples of the least positive integer for which this relationship holds. If $p(x)$ is a factor of $x^{q^m} - x$, any root β of $p(x)$ must also satisfy the condition $\beta^{q^m} = \beta$. The smallest integer i for which $\beta^{q^i} = \beta$ must therefore divide both k and m. Accordingly, if m is not a multiple of k, $\beta^{q^i} = \beta$ for some $i < k$, and from Theorem B.7, $p(x)$ would not be irreducible. If k does divide m, it is apparent that any root satisfying $x^{q^k} = x$ also satisfies $x^{q^m} = x$, so, in this case, all roots of $p(x)$ are indeed also roots of $x^{q^m} - x$. Thus, an irreducible monic polynomial of degree k is a factor of $x^{q^m} - x$ if, and only if, k divides m. Since no two such factors can have any roots in common (if they did at least one of the factors would be reducible), and since every root of $x^{q^m} - x$ must also be a root of one of its monic factors, $x^{q^m} - x$ is identically equal to the product of all irreducible polynomials whose degree divides m, as was to be shown.

In summary, all elements of the field $GF(q^m)$ are roots of the polynomial $x^{q^m} - x$. This polynomial is factorable over $GF(q)$ into the product of all monic irreducible polynomials whose degree divides m. If α is a root of one of these polynomials $p(x)$ of degree k, then so too are $\alpha^q, \alpha^{q^2}, \ldots, \alpha^{q^{(k-1)}}$, and these roots are all distinct. Now suppose α is a primitive element of $GF(q^m)$. Then the m roots α^{q^i}, $i = 0, 1, 2, \ldots, m - 1$ are all roots of the same irreducible m-degree polynomial. Suppose, further, that α^ν is not included in this set. Then all field elements of the form $\alpha^{\nu q^i}$, $i = 0, 1, \ldots, k - 1$ are roots of a second irreducible polynomial of degree $k \leq m$, with k the largest integer such that $\alpha^{\nu q^{k-1}} \neq \alpha^\nu$. Continuing in this manner, we can exhaust the set of non-zero field elements α^i, $i = 1, 2, \ldots, q^m - 1$, and thereby determine the degrees of all of the irreducible factors of $x^{q^m} - x$.

As an example, consider the factors of $x^9 - x$ over $GF(3)$. Let α be a primitive element of $GF(3^2)$. Then,

α and α^3 are the roots of an irreducible polynomial $p_1(x)$
α^2 and α^6 are the roots of an irreducible polynomial $p_2(x)$
α^4 is the root of an irreducible polynomial $p_3(x)$
α^5 and α^7 are the roots of an irreducible polynomial $p_4(x)$

and, of course, 0 and $\alpha^0 = 1$ are the roots of the irreducible polynomials x and $p_0(x) = x - 1$, respectively.

The exponents of α are thereby divided into classes in such a way that i and j are in the same class if, and only if, $i = q^r j$, modulo $q^m - 1$, for some integer r. These are called *congruence* classes. In the preceding example the integers, modulo 8, are divided into the five congruence classes

$$\begin{array}{c} 0 \\ 1,3 \\ 2,6 \\ 4 \\ 5,7 \end{array}$$

The polynomial $x^9 - x$ then must factor into six irreducible polynomials over $GF(3)$. As the reader can easily verify, the polynomials x, $p_0(x) = x + 2$, $p_1(x) = x^2 + x + 2$, $p_2(x) = x^2 + 1$, $p_3(x) = x + 1$, and $p_4(x) = x^2 + 2x + 2$ represent such a factorization. The polynomials $p_1(x)$, $p_2(x)$, and $p_4(x)$ are all irreducible 2nd degree polynomials and, as the reader can also verify, each of these polynomials generates the field $GF(3^2)$. The roots of $p_1(x)$ and $p_4(x)$ are of order 8 and hence are primitive; the roots of $p_2(x)$ are of order 4, and the root of $p_3(x)$ is of order 2.

B.5 Vector Spaces

Let V be a set of elements $\{\mathbf{v}_i\}$ called *vectors* and let the set $\{a_i\}$ represent elements (or *scalars*) from the field F. Two operations on V are defined: vector addition denoted $\mathbf{v}_i + \mathbf{v}_j$, and multiplication by a scalar indicated by $a_i\mathbf{v}_j$ (where $1\,\mathbf{v}_j = \mathbf{v}_j$). If the set V is closed under both vector addition and multiplication by a scalar (the product $a_i\mathbf{v}_j$ also being a vector), and if the vectors are associative and commutative under addition and satisfy the conditions

$$\begin{aligned} &(1)\ a_r(\mathbf{v}_i + \mathbf{v}_j) = a_r\mathbf{v}_i + a_r\mathbf{v}_j \\ &(2)\ (a_r + a_s)\mathbf{v}_i = a_r\mathbf{v}_i + a_s\mathbf{v}_i \\ &(3)\ (a_r a_s)\mathbf{v}_i = a_r(a_s\mathbf{v}_i) \end{aligned} \tag{B.5.1}$$

the set is said to be a *vector space*.

Closure under addition and multiplication by a scalar assures both that V contains an identity and that each vector in V has an inverse in V. This follows because the field must contain the elements 0 and -1. But $a\mathbf{v} = (a + 0)\mathbf{v} = a\mathbf{v} + 0\mathbf{v}$ and hence $0\mathbf{v} \triangleq \mathbf{0}$ is the identity vector. Moreover $1\mathbf{v} + (-1)\mathbf{v} = (1 - 1)\mathbf{v} = 0\mathbf{v} = \mathbf{0}$ and $(-1)\mathbf{v}$ is the inverse of $\mathbf{v}$.

As an example of a vector space, let V be a set of n-tuples of elements of the field F; i.e., let the elements of V be of the form $\mathbf{v} = (a_1, a_2, \ldots, a_n)$ with the a_i in F. Addition of n-tuples is defined as term-by-term addition, $(a_1, a_2, \ldots, a_n) + (b_1, b_2, \ldots, b_n) = (a_1 + b_1, a_2 + b_2, \ldots, a_n + b_n)$, and multiplying an n-tuple by a scalar is equivalent to multiplying each of its terms by that scalar, $b(a_1, a_2, \ldots, a_n) = (ba_1, ba_2, \ldots, ba_n)$. The set of all such n-tuples over a field F does indeed satisfy the conditions for a vector space as is easily verified. The identity of the vector space is the all-zero n-tuple $(0, 0, \ldots, 0) = \mathbf{0}$

A *linear combination* of the vectors $\mathbf{v}_1, \mathbf{v}_2, \ldots, \mathbf{v}_m$ is defined as the sum

$$a_1\mathbf{v}_1 + a_2\mathbf{v}_2 + \cdots + a_m\mathbf{v}_m$$

where the a_i are scalars. A set of m vectors is called *linearly dependent* if for some set of scalars $\{a_i\}$ at least one of which is not zero

$$\sum_{i=1}^{m} a_i\mathbf{v}_i = \mathbf{0} \tag{B.5.2}$$

A set of vectors is *linearly independent*, on the other hand, if the above equation is not true for any set of scalars $\{a_i\}$ not all equal to zero.

A subset of a vector space is called a *subspace* if it satisfies the definition of a vector space. To ascertain whether or not a subset of a vector space is indeed a subspace, one need only verify closure under addition and scalar multiplication. The remaining conditions clearly hold because all vectors in the subset are also in the vector space.

Theorem B.9. The vector space formed by taking all linear combinations of a set of vectors $\mathbf{v}_1, \mathbf{v}_2, \ldots, \mathbf{v}_m$ in V is a subspace of V.

Proof. Any vector in the subset can be written $\mathbf{v}_j = \sum_{i=1}^{m} a_i\mathbf{v}_i$ for some set of scalars a_i. Clearly,

$$\mathbf{v}_j + \mathbf{v}_k = \sum_{i=1}^{m} a_i\mathbf{v}_i + \sum_{i=1}^{m} b_i\mathbf{v}_i = \sum_{i=1}^{m} (a_i + b_i)\mathbf{v}_i = \sum_{i=1}^{m} c_i\mathbf{v}_i$$

and

$$d_j\mathbf{v}_j = \sum_{i=1}^{m} d_j a_i\mathbf{v}_i = \sum_{i=1}^{m} e_i\mathbf{v}_i$$

where a_i, b_i, c_i, d_i, e_i are all elements of the field F. Hence, the subset is closed under both addition and scalar multiplication and consequently is a subspace.

The number of linearly independent vectors used to define or *span* a subspace is the *dimension* of that subspace. Any set of linearly independent vectors which span a subspace is called a *basis* for the subspace.

Theorem B.10. If U is an m-dimensional subspace, then: (1) no set of m' vectors spans U for any $m' < m$; (2) no set of m'' vectors in U is linearly independent for any $m'' > m$; and (3) *any* set of m linearly independent vectors in U spans U.

Proof. Suppose U is spanned by the linearly independent vectors $\mathbf{u}_1$, $\mathbf{u}_2, \ldots, \mathbf{u}_m$ and also by the linearly independent vectors $\mathbf{v}_1, \mathbf{v}_2, \ldots, \mathbf{v}_r$ for some $r < m$. For every vector $\mathbf{u}_i$ there is a set of scalars $a_j(i)$ such that

$$\mathbf{u}_i = \sum_{j=1}^{r} a_j(i)\mathbf{v}_j$$

Since $\mathbf{u}_i \neq \mathbf{0}$, we can order the vectors $\mathbf{v}_j$ in such a way that $a_1(1) \neq 0$. Then

$$\mathbf{v}_1 = \frac{\left(\mathbf{u}_1 - \sum_{j=2}^{r} a_j(1)\mathbf{v}_j\right)}{a_1(1)}$$

Continuing this procedure, we similarly express $\mathbf{v}_2$ in terms of $\mathbf{u}_1, \mathbf{u}_2, \mathbf{v}_3, \ldots, \mathbf{v}_r$ (or some subset of these vectors), $\mathbf{v}_3$ in terms of $\mathbf{u}_1, \mathbf{u}_2, \mathbf{u}_3, \mathbf{v}_4, \ldots, \mathbf{v}_r$, etc., until we have an expression for $\mathbf{u}_{r+1}$ in terms of $\mathbf{u}_1, \mathbf{u}_2, \ldots, \mathbf{u}_r$. But this is a contradiction to the assumed linear independence of the vectors $\mathbf{u}_1, \mathbf{u}_2, \ldots, \mathbf{u}_m$. This proves the first statement of the theorem.

Similarly, if the vectors $\mathbf{v}_1, \mathbf{v}_2, \ldots, \mathbf{v}_r$ are in U, each of them can be expressed in terms of the basis vectors $\mathbf{u}_1, \mathbf{u}_2, \ldots, \mathbf{u}_m$. If $r > m$, the manipulations of the preceding paragraph can be repeated to obtain an expression for $\mathbf{v}_{m+1}$ as a linear combination of the vectors $\mathbf{v}_1, \mathbf{v}_2, \ldots, \mathbf{v}_m$. The set of vectors $\mathbf{v}_1, \mathbf{v}_2, \ldots, \mathbf{v}_r$ cannot be linearly independent, therefore, for any $r > m$ and statement (2) is proved.

Finally let $\mathbf{u}_1, \mathbf{u}_2, \ldots, \mathbf{u}_m$ be any set of linearly independent vectors in U and assume they do not span U. Then there is some vector $\mathbf{u}_{m+1}$ in U such that

$$\mathbf{u}_{m+1} \neq \sum_{i=1}^{m} a_i\mathbf{u}_i$$

for any set of a_i and hence

$$\mathbf{0} \neq \sum_{i=1}^{m+1} b_i\mathbf{u}_i$$

for any set of b_i. Thus, there are $m + 1$ linearly independent vectors in U contrary to statement (2), and the third statement of the theorem is proved.

The *order* of a subspace simply denotes the number of distinct vectors in it.

Theorem B.11. The order of a subspace defined over a field F of q elements is q^k where k is its dimension.

Proof. Since a subspace of dimension k can be generated by taking all linear combinations of its k basis vectors

$$\mathbf{v}_r = \sum_{i=1}^{k} a_i\mathbf{v}_i$$

and since $\mathbf{v}_r$ cannot equal $\mathbf{v}_s = \sum_{i=1}^{k} a'_i\mathbf{v}_i$ unless $a_i = a'_i$ for all i (otherwise there would be a linearly dependent relation among the basis vectors, contrary to the definition of a basis), there is a unique vector in the space for every k-tuple $(a_1, a_2, \ldots, a_k)$ where a_i is in F. Further, every vector in the subspace is defined by some k-tuple $(a_1, a_2, \ldots, a_k)$. Since there are q^k possible k-tuples over the field of q elements, the order of the subspace is q^k.

It will be observed that any vector space is an Abelian group closed under addition. Conversely, any group of n-tuples defined over a field of p elements (p a prime) and closed under term-by-term addition, is a vector space, with scalar multiplication effected by repeated addition of a vector to itself. In much of the coding theory literature the term *group* is used rather than *vector space*.

B.6 Matrices

An $(m \times n)$-dimensional *matrix* $\mathbf{M}$ is defined as an ordered set of mn elements belonging to some field F. These elements are conventionally arranged in a rectangular array of m rows and n columns:

$$\mathbf{M} = \begin{bmatrix} a_{11} & a_{12} & \cdots & a_{1n} \\ a_{21} & a_{22} & \cdots & a_{2n} \\ \cdot & & & \\ \cdot & & & \\ \cdot & & & \\ a_{m1} & a_{m2} & \cdots & a_{mn} \end{bmatrix} \tag{B.6.1}$$

The rows and columns of $\mathbf{M}$ can be considered to be n and m component vectors respectively over the field F.

The *row space* of the matrix $\mathbf{M}$ is the subspace spanned by its rows and its *row rank*, denoted $\rho_r(\mathbf{M})$, is the dimension of this subspace. Similarly the *column space* of $\mathbf{M}$ is the subspace spanned by its columns, and its *column rank*, $\rho_c(\mathbf{M})$, is the dimension of this subspace. It can be proved that $\rho_r(\mathbf{M}) = \rho_c(\mathbf{M})$ for all matrices $\mathbf{M}$ so the *rank* of $\mathbf{M}$ can be defined as $\rho(\mathbf{M}) = \rho_r(\mathbf{M}) = \rho_c(\mathbf{M})$ (cf. Reference B.3).

Theorem B.12. None of the following operations change the row (column) space of a matrix, nor hence its rank:

(1) Interchanging any two rows (columns).
(2) Multiplying any row (column) by a non-zero element of the field.
(3) Addition of any multiple of any row (column) to any other row (column).

Proof. That the first two operations do not change the row (column) space is obvious. To see that the third leaves the space invariant, compare the space V_1 generated by the vectors $\mathbf{v}_1, \mathbf{v}_2, \ldots, \mathbf{v}_n$ and the space V_2 generated by $\mathbf{v}_1 + a\mathbf{v}_2$, $\mathbf{v}_2, \ldots, \mathbf{v}_n$, where a is an element in F. Since both $\mathbf{v}_2$ and $\mathbf{v}_1 + a\mathbf{v}_2$ are in V_2 so is

$\mathbf{v}_1 + a\mathbf{v}_2 - a\mathbf{v}_2 = \mathbf{v}_1$. Thus V_2 contains all vectors which generate V_1. Similarly, V_1 contains $\mathbf{v}_1$ and $\mathbf{v}_2$, hence $\mathbf{v}_1 + a\mathbf{v}_2$ and therefore all the vectors which generate V_2. Consequently V_1 must be equal to V_2.

The operations of Theorem B.12 are called *elementary row* (*column*) *operations*. Any sequence of elementary row operations does not change the composition of the row space generated by the matrix in question (i.e., every vector in the row space of $\mathbf{M}$ is in the row space of the matrix $\mathbf{M}'$ obtained by elementary row operations on $\mathbf{M}$, and conversely). The same statement applies to elementary column operations and the column space of a matrix.

Moreover, a sequence of elementary row operations can be used to reduce any matrix $\mathbf{M}$ to a matrix $\mathbf{M}'$ in *echelon canonical form* (also sometimes called *Hermite canonical form*) in which the following conditions are satisfied:

(i) The first element in any non-zero row of $\mathbf{M}'$ is a *one* and is the only non-zero element in its column.

(ii) If the first non-zero element in the ith row of $\mathbf{M}'$ is in the jth column, the first non-zero element in the $(i + 1)$st row is in the $(j + l)$th column where $l \geq 1$.

To accomplish this reduction we first find some non-zero element (say a_{ij}) in the first non-zero row of $\mathbf{M}$ and, using elementary row operation (2) (Theorem B.12), change this element to *one*. We then use operation (3) to change to zero all other elements in the column containing this *one*. Continuing to the next non-zero row, we repeat the same two operations. Since the jth column now contains a zero in all but the ith row, it is unaffected when operations (2) and (3) are used to reduce other columns to the desired form. Thus, the procedure can be continued until the matrix $\mathbf{M}'$ is obtained, the second condition on $\mathbf{M}'$ being satisfied by a final row permutation [operation (1)]. Since only elementary row operations were used, the row spaces of $\mathbf{M}$ and $\mathbf{M}'$ are identical.

By subjecting $\mathbf{M}'$ to a sequence of column permutations only, $\mathbf{M}$ can be further reduced to the form

$$\mathbf{M}'' = \begin{bmatrix} \mathbf{I}_\rho & \mathbf{A}_{\rho, n-\rho} \\ \mathbf{0}_{m-\rho, \rho} & \mathbf{0}_{m-\rho, n-\rho} \end{bmatrix} \tag{B.6.2}$$

sometimes called the *reduced echelon form* of $\mathbf{M}$. Here $\mathbf{I}_\rho$ denotes the $(\rho \times \rho)$-dimensional *identity matrix* (the element in the ith row and jth column of $\mathbf{I}$ is *zero* unless $i = j$, in which case it is *one*). The matrix $\mathbf{A}_{\rho, n-\rho}$ is an arbitrary $[\rho \times (n - \rho)]$-dimensional matrix and $\mathbf{0}_{\nu,\mu}$ denotes a $(\nu \times \mu)$-dimensional matrix of zeros only. The row rank of $\mathbf{M}''$ (and hence of $\mathbf{M}$) is obviously the dimension of the identity matrix.

The matrices $\mathbf{M}$ and $\mathbf{M}''$ span the same row space except for a reordering of the components of the vectors in the space spanned by $\mathbf{M}''$ as compared to those in the space spanned by **M**. Such matrices are said to be *combinatorially equivalent.*

The *dot-product* (or *inner product*) of two n-component vectors $\mathbf{v}_i = (\gamma_1, \gamma_2, \ldots, \gamma_n)$ and $\mathbf{v}_j = (\delta_1, \delta_2, \ldots, \delta_n)$ is defined as the scalar†

$$\mathbf{v}_i \cdot \mathbf{v}_j = \sum_{\nu=1}^{n} \gamma_\nu \delta_\nu \tag{B.6.3}$$

If $\mathbf{v}_i \cdot \mathbf{v}_j = 0$, $\mathbf{v}_i$ and $\mathbf{v}_j$ are said to be *orthogonal.* The product of two matrices is defined in terms of the dot products of its vectors: if $\mathbf{M} = \mathbf{M}_1\mathbf{M}_2$ then the element in the ith row and jth column of **M** is the dot product of the ith row vector of $\mathbf{M}_1$ with the jth column vector of $\mathbf{M}_2$. The number of rows in $\mathbf{M}_2$ must be equal to the number of columns in $\mathbf{M}_1$ for such a product to be defined. If $\mathbf{M}_1$ is an $m_1 \times n$ matrix and $\mathbf{M}_2$ an $n \times m_2$ matrix, **M** is an $m_1 \times m_2$ matrix.

The *transpose* $\mathbf{M}^T$ of a matrix **M** is defined as follows. If the element in the ith row and jth column of **M** is a_{ij} then the element of the jth row and ith column of $\mathbf{M}^T$ is also a_{ij} (or, in the case of the complex field, a_{ij}^*).

Theorem B.13. If the equation $\mathbf{Mx} = \mathbf{b}$, when **M** (an $m \times n$ matrix) and **b** (an $m \times 1$ column vector) are both specified, is to have a solution **x** over the field F then

$$\rho(\mathbf{M}) = \rho(\mathbf{M}, \mathbf{b}) \tag{B.6.4}$$

where $(\mathbf{M}, \mathbf{b})$ is the $m \times (n + 1)$ matrix obtained by adjoining **b** to **M** as its $(n + 1)$st column.

Proof. Suppose there *is* a solution $\mathbf{x}^T = (a_1, a_2, \ldots, a_n)$ where the a_i are elements in F. Then

$$\mathbf{Mx} = \sum_{i=1}^{n} a_i \boldsymbol{\gamma}_i$$

where $\boldsymbol{\gamma}_i$ is the ith column of **M**. Thus, if $\mathbf{Mx} = \mathbf{b}$, **b** must be in the column space generated by the columns of **M**. Consequently, the dimension of the column space of $(\mathbf{M}, \mathbf{b})$ must be the same as that of **M** and the ranks of the respective matrices must be equal.

B.7 Null Spaces

An n-tuple **v** is said to be orthogonal to a subspace U if it is orthogonal to all vectors **u** in U.

Theorem B.14. All the vectors in the vector space V orthogonal to a subspace U form a subspace.

---

†If the vectors are defined over the complex field, the components of the second vector are conjugated ($\delta_\nu \to \delta_\nu^*$) in forming the dot product.

Proof. Let W be the set of all vectors orthogonal to the subspace U so that for any $\mathbf{w}$ in W and $\mathbf{u}$ in U, $\mathbf{w}\cdot\mathbf{u} = 0$. Let $\mathbf{w}_1$ and $\mathbf{w}_2$ be two vectors in W and a and b be any two elements in the field. Then

$$(a\mathbf{w}_1 + b\mathbf{w}_2)\cdot\mathbf{u} = a\mathbf{w}_1\cdot\mathbf{u} + b\mathbf{w}_2\cdot\mathbf{u} = 0$$

and $a\mathbf{w}_1 + b\mathbf{w}_2$ must also be in W for any a and b. Accordingly, W is closed under both vector addition and multiplication by a scalar and hence is a subspace.

Clearly, if $\mathbf{w}$ is orthogonal to all the vectors $\mathbf{u}_1, \mathbf{u}_2, \ldots, \mathbf{u}_l$ it is orthogonal to all the vectors in the space spanned by $\mathbf{u}_1, \mathbf{u}_2, \ldots, \mathbf{u}_l$. The set W of vectors orthogonal to all vectors in U is called the *null space* of U.

Theorem B.15. If the dimension of a subspace of n-tuples over $GF(q)$ is k, the dimension of its null space is $n - k$.

Proof. Let U be a k-dimensional space of n-tuples over $GF(q)$ and W its null space. Then U can be represented by a $(k \times n)$-dimensional matrix $\mathbf{G}$, the rows of which span U. Since the row rank of $\mathbf{G}$ is k, its column rank is also k. Suppose, for convenience in identifying them, the first k columns of $\mathbf{G}$ are linearly independent. (This may entail permuting the columns of $\mathbf{G}$: the equation $\mathbf{g}\cdot\mathbf{w} = 0$ is unaffected by permutation if the components of $\mathbf{g}$ and $\mathbf{w}$ are identically permuted, so there is no loss of generality in this assumption.) Then the last $n - k$ columns of $\mathbf{G}$, and any linear combinations of these columns, can be expressed as a linear combination of the first k columns.

If $\mathbf{G}\mathbf{w}^T = \sum_{i=1}^{n} \omega_i \mathbf{g}^i = \mathbf{0}$ where $\mathbf{g}^i$ is the ith column of $\mathbf{G}$ then $\mathbf{w} = (\omega_1, \omega_2, \ldots, \omega_n)$ is an n-tuple in W. This equation can be rewritten in the form

$$\sum_{i=1}^{k} \omega_i \mathbf{g}^i = -\sum_{i=k+1}^{n} \omega_i \mathbf{g}^i$$

But any linear combination of the last $n - k$ columns of $\mathbf{G}$ can be expressed as a linear combination of the first k. The last $n - k$ components of $\mathbf{w}$ can be chosen arbitrarily, and the first k components then selected to satisfy the equation. There must be at least q^{n-k} vectors $\mathbf{w}$ in W (if the n-tuples are defined over the field of q elements), and the dimension of W must be at least $n - k$.

Now suppose W has dimension m. Then there must be an $m \times n$ matrix $\mathbf{H}$ spanning W. The matrix $\mathbf{H}$, in turn, can be expressed in echelon canonical form, and, if the columns of $\mathbf{H}$ are permuted (inducing a corresponding reordering of the components of the n-tuples in U), in reduced echelon form. Since $\mathbf{G}\mathbf{w}^T = \mathbf{0}$ must be satisfied for every row $\mathbf{w}$ in $\mathbf{H}$, we can express each of the first m columns of $\mathbf{G}$ (or of some column permutation of $\mathbf{G}$) as a linear combination of its last $n - m$ columns. Thus, $n - m \geq k$ and $m \leq n - k$. But, as has already been shown, $m \geq n - k$. Consequently, $m = n - k$ and the theorem is proved.

Corollary. If W is the null space of U then U is the null space of W.

Proof. Clearly, all the vectors in U are orthogonal to all those in W. It remains to show that no vector orthogonal to the subspace W is not in U. But this follows immediately from Theorem B.13. If the dimension of W is $n - k$ the dimension of its null space is k. Since U is contained in the null space of W and has dimension k it must be the null space.

As an example of some of these concepts, let F be the field of three elements (0, 1, 2) and let V be the space of all 2-tuples defined over F. Then

$$V = (0, 0), (0, 1), (0, 2), (1, 0), (1, 1), (1, 2), (2, 0), (2, 1), (2, 2)$$

and the order of V is 9. Since $3^2 = 9$, the dimension of the space is two. The space is generated, for example, by the linearly independent vectors (0, 1) and (1, 0). (Any other linearly independent set of vectors in V, of course, also generates V; e.g., (0, 1) and (1, 1) or (0, 2) and (2, 1), etc.). Now let U be the subspace generated by the single vector (0, 1);

$$U = (0, 0), (0, 1), (0, 2)$$

the cosets of U are

$$U_1 = U + (1, 0) = (1, 0), (1, 1), (1, 2)$$
$$U_2 = U + (2, 0) = (2, 0), (2, 1), (2, 2)$$

and contain no elements in common. Since all nine 2-tuples are either in U, U_1, or U_2, there are no other cosets of U. Finally, let W be the subspace generated by vector (1, 0). Then $\mathbf{u} \cdot \mathbf{w} = 0$ for any vector $\mathbf{u}$ in U and any vector $\mathbf{w}$ in W. The sum of the dimensions of U and W is two, the dimension of the space V. Consequently, W is the null space of U and conversely.

References and Supplementary Bibliography

B.1 Albert, A.A., *Fundamental Concepts of Higher Algebra*, University of Chicago Press, Chicago, Ill., 1956.

B.2 Birkhoff, G., and S. MacLane, *A Survey of Modern Algebra*, Macmillan, New York, N. Y., 1941.

B.3 Perlis, S., *Theory of Matrices*, Addison-Wesley Publishing Co., Inc., Reading, Mass., 1952.

B.4 Peterson, W. W., *Error-Correcting Codes*, Massachusetts Institute of Technology Press and John Wiley & Sons, Inc., New York, N. Y., 1961.

Message Units (S_0)	*Probabilities* S_0	S_1	S_2	S_3	S_4	S_5
A_1	0.40	0.40	0.40	0.40	0.40	1.00
A_2	0.20	0.20	0.20	0.20	0.40	
A_3	0.10	0.10	0.12	0.18	0.20	
A_4	0.08	0.08	0.10	0.12		
A_5	0.05	0.05	0.08	0.10		
A_6	0.05	0.05	0.05			
A_7	0.05	0.05	0.05			
A_8	0.03	0.04				
A_9	0.02	0.03				
A_{10}	0.01					
A_{11}	0.01					

INDEX